HANDBOOK OF AQUEOUS ELECTROLYTE THERMODYNAMICS

HANDBOOK OF AQUEOUS ELECTROLYTE THERMODYNAMICS

Theory & Application

Joseph F. Zemaitis, Jr.
Chem Solve, Inc.

Diane M. Clark
OLI Systems, Inc.

Marshall Rafal
OLI Systems, Inc.

Noel C. Scrivner
E.I. duPont de Nemours & Co., Inc.

WILEY-INTERSCIENCE

A JOHN WILEY & SONS, INC., PUBLICATION

NSRDS | AIChE | DIPPR

A publication of the
Design Institute for Physical Property Data (DIPPR)
Sponsored by the
American Institute of Chemical Engineers
345 East 47th Street · New York, New York 10017

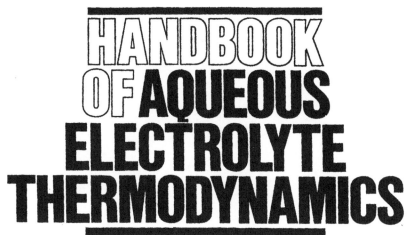

American Institute of Chemical Engineers, Inc.
345 East Forty-Seventh Street
New York, New York 10017

ISBN 978-0-8169-0350-4

Dedication: This book is dedicated to the memory of Dr. Joseph F. Zemaitis, Jr. Dr. Zemaitis, a colleague and friend for many years, was responsible for the outline of this book and the writing of the first three chapters. In a larger sense, he provided for the synthesis of a very diverse body of work into a coherent framework for problem solving. We dearly hope that our dedication to and respect for his memory is reflected in the content of this work.

Diane M. Clark
Marshall Rafal
Noel C. Scrivner

Acknowledgment: The authors wish to express their gratitude to Lisa Perkalis. This book could not have been completed without her word processing skills, patience and dedication in the face of never-ending "small" changes to the manuscript.

Sponsors: The DIPPR sponsors of this project, and the technical representatives who served on the steering committee are listed below. Stars indicate those companies which supported the project throughout its three year duration.

Sponsors of DIPPR Project 811 (1981–1983)

Participant	Technical Representative
*Air Products & Chemicals, Inc.	Dr. M.S. Benson
Allied Corporation	Mr. W.B. Fisher
*Amoco Chemicals Corporation	Dr. D.A. Palmer
Chevron Research Company	Dr. W.M. Bollen
*Chiyoda Chemical Engineering & Construction Co., Ltd.	Mr. T. Maejima
*E.I. du Pont de Nemours & Company, Inc.	Dr. N.C. Scrivner
*El Paso Products Company	Mr. K. Claiborne
Exxon Research and Engineering Co.	Dr. C. Tsonopoulos
*Halcon SD Group, Inc.	Dr. C.F. Chueh
Hoffmann-LaRoche, Inc.	Dr. R. Schefflan
Hooker Chemical Company	Dr. W.L. Sutor
Institut Français du Pétrole	Dr. J. Vidal
*Institution of Chemical Engineers	Dr. B. Edmonds
M.W. Kellogg Company	Mr. H. Ozkardesh
Kennecott Copper Corporation	Mr. D.S. Davies
*Kerr-McGee Chemical Corporation	Dr. D.S. Arnold
*Olin Chemicals Group	Mr. W.M. Clarke
*Phillips Petroleum Company	Mr. M.A. Albright
*Shell Development Company	Dr. R.R. Wood
*Simulation Sciences, Inc.	Dr. C. Black
*The Standard Oil Company (SOHIO)	Mr. T.B. Selover
*Texaco, Inc.	Dr. C. Wu
*Texasgulf, Inc.	Mr. S.H. DeYoung
Union Carbide Corporation	Mr. E. Buck
Federal Grantor	
*National Bureau of Standards	Dr. H.J. White, Jr.

*Supported Project for 3 year life

Foreword:

An international conference on the "Thermodynamics of Aqueous Systems" sponsored by the American Institute of Chemical Engineers (AIChE), the National Science Foundation (NSF), and the National Bureau of Standards (NBS), was held in Warrenton, Virginia, on October 22–25, 1979. The papers presented reflected a great deal of research on electrolyte solutions. However, it was apparent that there was no fundamental document to tie all of the different information together and so to form a framework for solving real problems.

Therefore, AIChE's Design Institute for Physical Property Data (DIPPR) decided to publish this book to meet such a need. Through a cooperative effort by participating corporations, different correlations have been compiled and objectively compared to experimental data, in regions of industrial interest. Effective methods of finding and using data are also described. The *Handbook* incorporates and extends previous work in a well-organized, easy to understand format, with a focus on applications to serious industrial problems. It will become a cornerstone in the study of aqueous electrolyte thermodynamics.

Electrolyte mixtures come in various forms and add another dimension to the normal complexities of nonelectrolyte solutions: entirely new species can form in water, some of which are not obvious; components can precipitate; soluble components can affect the vapor pressure of the solution very significantly. In industrial applications, the solutions are often highly concentrated and encounter high pressures and temperatures. Therefore expertise in electrolyte systems has become increasingly critical in oil and gas exploration and production, as well as in the more traditional chemical industry operations. A variety of correlations are available that can solve the problems that are encountered in industry, but which ones work best? This comprehensive handbook not only provides easy access to available data but also presents comparative studies of various correlations up to extreme conditions.

As the Chairman of the Technical Committee of AIChE's Design Institute for Physical Property Data, I conceived this cooperative research project and chose as its leader, Dr. Noel C. Scrivner of the E.I. duPont de Nemours Company, one of the leading practitioners of electrolyte thermodynamics. With his expertise and enthusiasm, Dr. Scrivner defined the work that needed to be accomplished, promoted the project until it was funded, and directed its completion. He was elected to head the steering committee of representatives from the supporting companies.

The initial work was carried out in 1981 by the Electrolyte Data Center of the National Bureau of Standards, under the direction of Dr. B.R. Staples. That work continued throughout the project, generating the bibliographies for electrolyte systems. One of these bibliographies, developed by R.N. Goldberg, is included in this volume.

A significant portion of the project funding came from the Office of Standard Reference Data of the NBS (Dr. David R. Lide, Jr., Director). The NBS liaison to DIPPR was Dr. Howard J. White, Jr. DIPPR provided additional funding along with administrative and technical assistance.

The task of creating the actual handbook was given to the late Dr. Joseph F. Zemaitis,

Jr., owner of Chem Solve, Inc. His intellectual contribution to this project will remain as his legacy. He wrote the first few chapters and carefully outlined the remainder of the book. His principal colleague in this work, Ms. Diane M. Clark, dedicated several years of creative effort to take the outline and complete the project. Another colleague of Dr. Zemaitis, Dr. Marshall Rafal, owner of OLI Systems, Inc., assumed the contractor's responsibility for execution of the handbook and contributed some of the writing. Dr. Scrivner gave technical direction and technical contributions to assure that the work would meet the standards set by the steering committee.

This book is dedicated to Dr. J.F. Zemaitis. Dr. David W.H. Roth, Jr., the administrative committee chairman of DIPPR, and I would like to express sincere appreciation to the other authors, the steering committee members, the corporate sponsors, and the National Bureau of Standards.

David A. Palmer, Chairman

DIPPR Technical Executive Committee and
DIPPR Technical Committee

TABLE OF CONTENTS

NOMENCLATURE

A - Debye-Hückel constant, log base 10, equation (4.31)

A_ϕ - Debye-Hückel constant for osmotic coefficients, log base e, equation (4.64)

A^γ - Debye-Hückel constant for activity coefficients, log base e, equation

a_i - activity of species i, equation (2.21)

a_t - parameter used by Criss and Cobble's correspondence principle, equation (3.32)

a_w - water activity

a - distance of closest approach or core size, equation (4.33)

B - parameter for Guggenheim's activity coefficient equation, log 10 basis, equation (4.39)

B - interaction parameter for Bromley's activity coefficient equation, equation (4.44)

B^γ - parameter in Pitzer's activity coefficient equation, equations (4.63), (4.66)

b_t - parameter used by Criss and Cobble's correspondence principle, equation (3.32)

C^ϕ - Pitzer parameter, equation (4.65)

Cp - heat capacity

$\overline{Cp^\circ}$ - partial molal heat capacity

c - molarity

C - salt concentration in the Setschénow equation (7.1)

D - dielectric constant, equations (4.1), (4.98)

d - solution density, equation (2.34), Chapter VIII

d_0 - solvent density

d_w - pure water density, equation (4.98)

E - intrinsic energy

e - electronic charge, equation (4.2)

F - NRTL parameter of Chen's activity coefficient equation, equations (4.70), (4.71)

f - rational activity coefficient (mole fraction based), equation (2.33c)

f^{γ} - Pitzer electrostatic function, equation (4.62)

f_V - fugacity of vapor species V, equation (3.20)

g_i - number of grams of i, i=0 for solvent

G - Gibbs free energy

$\overline{G_i}$ - partial molar Gibbs free energy, equation (2.14)

g_{ij} - radial distribution function, equation (4.51)

H - enthalpy

H - Henry's constant

$\overline{H_i}$ - partial molar enthalpy, equation (2.15)

I - ionic strength, equation (4.27)

K - equilibrium constant, dissociation constant

K_{aq} - equilibrium constant, equation (3.18)

K_{sp} - solubility product, equation (3.13)

K_T - thermodynamic equilibrium constant, equation (3.6)

$\overline{k^{\circ}}$ - partial molal compressibility, equation (3.40)

k - Boltzmann's constant, equation (4.2)

k - Setschénow salt coefficient, equation (7.1)

L - Ostwald coefficient

M - solute molecular weight, equation (2.34)

M_s - solvent molecular weight, equation (2.34)

m - molality

N_A - Avagadro's number, equation (4.2)

n_i - number of moles of i

P - pressure, total vapor pressure

p - partial pressure

pK = $-\log K$

q - Meissner's interaction parameter, equation (4.46)

R - gas law constant

r - radius of the field around an ion, equation (4.1a)

S - entropy

S° - gas solubility in pure water, Setschénow equation (7.1)

S - gas solubility in salt solution, Setschénow equation (7.1)

T - temperature, Kelvins

t - temperature, Celsius

V - volume

$\overline{V_i}$ - partial molar volume, equation (2.16)

x - liquid mole fraction

y - vapor mole fraction

Y - any extensive thermodynamic property, equation (2.8)

y - molar activity coefficient, equation (2.33b)

z_i - number of charges on ion i

α - Debye-Hückel constant, log base e

α - nonrandomness factor in Chen's equation for activity coefficients, equation (4.70)

α_t - parameter in correspondence principle equation, equation (3.36a)

β - parameter for Guggenheim's activity coefficient equation, equation (4.38)

β_t - parameter in correspondence principle equation, equation (3.36b)

β_0 - Pitzer interaction coefficient

β_1 - Pitzer interaction coefficient

β_2 - Pitzer interaction coefficient

Γ - reduced activity coefficient, equation (4.45)

γ - molal activity coefficient, equation (2.22)

δ - Bromley parameter for the additive quality of ion interaction, equation (4.45)

ε_{gas} - Lennard-Jones energy interaction, equation (7.18)

θ_{ij} - Pitzer's coefficient for like charged ion interactions, equations (5.32), (5.33)

μ_i - partial molar Gibbs free energy or chemical potential, equation (2.17)

$\mu_i°$ - reference state chemical potential, equation (2.21)

ν - sum of cation and anion stoichiometric numbers

$\underline{\nu}$ - harmonic mean of ν_+ and ν_-, equation (4.38)

ν_i - stoichiometric number of ion i

Π - osmotic pressure, equation (4.51)

ρ - charge density, equation (4.1)

τ - NRTL binary interaction parameter of Chen's activity coefficient equation, equations (4.70), (4.71)

ϕ - osmotic coefficient, equation (2.30)

ψ_a - ionic atmosphere potential, equation (4.14)

ψ_{ijk} - Pitzer triple ion interaction parameter, equation (5.32)

ψ_r - ionic potential

ψ_t - total electric field potential, equation (4.1)

$\emptyset$ - apparent molal volume

Subscripts:

A - anion

a - anion

C - cation

c - cation

g - gas

i - any species

j - any species

L - liquid

m - molecular species

s - salt

u - molecular (undissociated species

V - vapor

W - water

1,3,5,... - cations

2,4,6,... - anions

± - molecular species

Superscripts:

el - electrostatic

ex - excess

lc - local composition

P - Pitzer long range contribution of Chen's equation, equation (4.87)

tr - trace amount

° - indicates reference state of pure (single solute) solutions

∞ - infinite dilution

I:

Introduction

INTRODUCTION

In the past several years, interest in electrolyte phase equilibria has grown significantly. This growth in interest can be attributed to a number of evolving application areas and factors among which are:

- o Recognition of the necessity to reduce pollutant levels in process waste water streams. The removal of sulfur by formation of gypsum is an example of such an application.
- o Development of new flue gas scrubbing systems using regenerative processes. Scrubbing of Cl_2 from incinerator streams and SO_2 from flue gases are specific application examples.
- o Recent escalation of the prices of oil and gas leading to the study and development of synthetic fuel processes in which ammonia, carbon dioxide, and hydrogen sulfide are produced as by-products which usually condense to form aqueous solutions. Sour water strippers and amine scrubbers are specific processes developed in this area.

Most of the application areas mentioned above concern the vapor-liquid phase equilibria of weak electrolytes. However, in the past several years, considerable interest has also developed in the liquid-solid equilibria of both weak and strong electrolytes. Application areas and factors that have affected this growth in interest include:

- o Hydrometallurgical processes, which involve the treating of a raw ore or concentrate with an aqueous solution of a chemical reagent.
- o The need of corrosion engineers to predict the scale formation capabilities of various brines associated with oil production or geothermal energy production.
- o The need of petroleum engineers to predict the freezing or crystallization point of clear brines containing sodium, calcium, and zinc chlorides and bromides to high concentrations.
- o The need for waste water clean up customarily done by precipitation of heavy metals.
- o Sea water desalination.
- o Crystallization from solution in the manufacture of inorganic chemicals.

- o Specific ion electrolytes
- o Ion exchange

Specific processes which typify these application areas are:

- o Treatment of gypsum which is formed in waste water cleanup.
- o Several processes involving formation of $Cr(OH)_3$. These processes include:
  - cooling tower blowdown
  - plating processes
  - manufacture of chrome pigment

  Use of a simple solubility product (e.g. Lange's Handbook) for $Cr(OH)_3$ is invalid since precipitation involves intermediate complexes which form to a significant degree.

These are just a few of the application areas of electrolyte phase equilibria which have generated an interest in developing a better understanding of aqueous chemistry. In contrast to other systems, in particular hydrocarbon systems, design-oriented calculation methods are not generally available for electrolyte systems. In the undergraduate education of chemical engineers, little if any mention of electrolyte thermodynamics is made and most chemical engineering thermodynamic texts ignore the subject completely. If an engineer is exposed to electrolyte thermodynamics at all during undergraduate education, the subject is taught on a rudimentary level so that many misconceptions may arise. For example, in manufacturing chrome pigment, noted above, use of the solubility product in order to determine solubility leads to very large errors since the solubility product approach totally ignores formation of complexes and their attendant effect on system state. By contrast, for hydrocarbon systems, most engineers are presented with a basic groundwork that includes design-oriented guidelines for the calculation of vapor-liquid equilibria of simple as well as complex mixtures of hydrocarbons. In addition, during the last decade, the education of a chemical engineer has usually included the introduction to various computer techniques and software packages for the calculation of phase equilibria of hydrocarbon systems. This has not been the case for electrolyte systems and even now it is generally not possible for the engineer to predict phase equilibria of aqueous systems using available design tools.

For processes involving electrolytes, the techniques used in the past and even some still in use today at times rely heavily on correlations of limited data which are imbedded into design calculation methods oriented towards hydrocarbon systems. Worse yet, until recently, limited use of the limited data available has occasionally led to serious oversights. For example, in the Cl_2 scrubbing system noted earlier, the basic data published in Perry's Chemical Engineers' Handbook which has been used for years is actually in error. Hampering the improvement of the design tools or calculative techniques have been several restrictions including:

- o A lack of understanding of electrolyte thermodynamics and aqueous chemistry. Without this understanding, the basic equations which describe such systems cannot be written.

- o The lack of a suitable thermodynamic framework for electrolytes over a wide range of concentrations and conditions.

- o The lack of good data for simple mixtures of strong and/or weak electrolytes with which to test or develop new frameworks.

- o The diversity in data gathering because of the lack of a suitable thermodynamic framework. As a result of different thermodynamic approaches, experimental measurements to develop fundamental parameters have often led to the results being specific to the system studied and not generally useful for applications where the species studied are present with other species.

Fortunately, in the last decade, because of the renewed interest in electrolyte thermodynamics for the reasons described earlier, a considerable amount of work in the field of electrolyte thermodynamics has been undertaken. Techniques for the calculation of the vapor-liquid equilibria of aqueous solutions of weak electrolytes are being published. New, improved thermodynamic frameworks for strong electrolyte systems are being developed on a systematic basis. With these developments, DIPPR established, in 1980, Research Project No. 811. The objective of this project is to produce a "data book" containing recommended calculation procedures and serving as a source of thermodynamic data either through recommended tabulated values or through annotated bibliographies which point to suitable sources.

In order to meet the objectives of this project, several phases were established. The specific phases involved:

1) Definition of the project scope.

2) Gathering available data and literature references for the preparation of test data sets and data tables contained in the report.

3) Review of thermodynamics, techniques and recent developments in order to select those techniques to be evaluated in the final report.

4) Testing and comparison of the various techniques against selected test data sets.

5) Development of the handbook in order to present the results of the project in a useful and readable form.

This book is a result of DIPPR Research Project No. 811. In it, the reader and user will find a systematic presentation of electrolyte thermodynamics, from the basic definitions of equilibrium constants of ionic reactions, to the prediction of activity coefficients of various species in multicomponent aqueous solutions of strong and/or weak electrolytes and the resulting phase equilibria calculative techniques. For several systems, data are presented and calculative techniques are illustrated. The goal of this book is for the engineer, faced with the need for solving industrially important problems involving aqueous solutions of electrolytes, to be able to understand the possible alternatives available and apply them to the problem at hand, either with available data or through available data prediction and analysis techniques. Several examples will be used to illustrate the calculative techniques necessary for different types of problems. The examples chosen are of a size that can be solved with limited computer facilities. The techniques can be expanded for more complex problems.

In order to better understand the basis for the chapters which follow, let us consider the formulation of a predictive model for a particular aqueous based electrolyte system. The example chosen involves water-chlorine. The reactions to be considered are:

$H_2O(vap) = H_2O(aq)$

$Cl_2(vap) = Cl_2(aq)$

$Cl_2(aq) + H_2O = H(ion) + Cl(ion) + HClO(aq)$

$HClO(aq) = H(ion) + ClO(ion)$

$H_2O(aq) = H(ion) + OH(ion)$

The problem is to predict the resulting phase distribution and phase compositions. Or, in other words:

Given: Temperature (T), Pressure (P) and inflow quantities $H_2O(in)$ and $Cl_2(in)$

Determine:

1) Total vapor rate, V
2) Rate of $H_2O(aq)$
3) Vapor phase partial pressures, pH_2O and pCl_2
   Liquid phase concentrations, usually expressed in molality (gm moles solute per 1000 gms solvent-$H_2O(aq)$), $m_{HClO(aq)}$, $m_{Cl_2(aq)}$, $m_{H(ion)}$, $m_{OH(ion)}$, $m_{Cl(ion)}$, $m_{ClO(ion)}$

The problem, stated above for water-chlorine, is typical of all calculations involving electrolytes.

For the water-chlorine system above, a set of ten equations is required in order to solve for the ten unknowns just described. These equations are:

Equilibrium Equations: Equations 1-5

Equilibrium K equations are written, one for each reaction. As we shall see, these equations are of the form:

$$K = \frac{\prod_{iP} (\gamma_{iP})^{\nu_{iP}} (m_{iP})^{\nu_{iP}}}{\prod_{iR} (\gamma_{iR})^{\nu_{iR}} (m_{iR})^{\nu_{iR}}}$$

where,

K = The thermodynamic equilibrium constant; a function of T and P.

γ_{iP}, γ_{iR} = Activity coefficient or, for vapors, fugacity coefficient of the ith product and reactant respectively; a function of T, P and composition

ν_{iP}, ν_{iR} = Stochiometric coefficient of the ith product and reactant respectively

m_{iP}, m_{iR} = Molality or, for vapors, partial pressure of ith product and reactant respectively.

For our H_2O-Cl_2 system (using γ for activity coefficient, a for activity, f for fugacity coefficient and p for vapor partial pressure) we thus have five such equations:

$$K_{H_2O(vap)} = \frac{a_{H_2O(aq)}}{f_{H_2O}\, p_{H_2O(vap)}}$$

$$K_{Cl_2(vap)} \quad \frac{\gamma_{Cl_2(aq)}\, m_{Cl_2(aq)}}{f_{Cl_2}\, p_{Cl_2(vap)}}$$

$$K_{Cl_2(aq)} = \frac{\gamma_{H(ion)}\, m_{H(ion)}\, \gamma_{Cl(ion)}\, m_{Cl(ion)}\, \gamma_{HClO(aq)}\, m_{HClO(aq)}}{\gamma_{Cl_2(aq)}\, m_{Cl_2(aq)}\, a_{H_2O(aq)}}$$

$$K_{HClO(aq)} = \frac{\gamma_{H(ion)}\, m_{H(ion)}\, \gamma_{ClO(ion)}\, m_{ClO(ion)}}{\gamma_{HClO(aq)}\, m_{HClO(aq)}}$$

$$K_{H_2O(aq)} = \frac{\gamma_{H(ion)}\, m_{H(ion)}\, \gamma_{OH(ion)}\, m_{OH(ion)}}{a_{H_2O(aq)}}$$

Electroneutrality Equation: Equation 6

The electroneutrality equation states that the solution is, at equilibrium, electrically neutral. Generally stated, the equation is:

total molality of cations = total molality of anions

or, for our H_2O-Cl_2 system:

$$m_{H(ion)} = m_{OH(ion)} + m_{Cl(ion)} + m_{ClO(ion)}$$

Material Balances: Equations 7-10

The requisite material balances. In this case, four such balances are needed to make the number of equations equal to the number of unknowns. The equations are:

Overall Material

$$3H_2O(in) + 2Cl_2(in) = 3H_2O(aq) + H_2O/55.51\ (3m_{HClO(aq)} + m_{Cl(ion)} + 2m_{OH(ion)} + m_{H(ion)} + 2m_{ClO(ion)} + 2m_{Cl_2(aq)}) + V$$

Vapor Phase

$$P = p_{H_2O(vap)} + p_{Cl_2(vap)}$$

Chlorine

$$2Cl_2(in) = H_2O/55.51\ (m_{HClO(aq)} + m_{Cl(ion)} + m_{ClO(ion)}) + V\ p_{Cl_2(vap)}/P$$

Hydrogen

$$2H_2O(in) = 2H_2O(aq) + H_2O/55.51\ (m_{HClO(aq)} + m_{H(ion)} + m_{OH(ion)}) + V\ p_{H_2O(vap)}/P$$

These ten equations can, with a reasonable computer, be solved for the ten unknowns in question. Alternatively, by carefully organizing the calculations and making some simplifying assumptions, for a simple system such as Cl_2-H_2O trial and error using a calculator is also feasible. What has been understated thus far is that, embedded in equations 1-5, the K equations, is the essential complexity of the electrolyte calculations. The variables, K(T,P) and γ(T,P,m) are often highly nonlinear functions of the state variables shown. The purpose of this book is thus to describe:

1) The underlying physical chemistry theory which governs determination of K and γ.
2) Practical methods for calculation, estimation or extrapolation of these values.

II:

Thermodynamics of Solutions

THERMODYNAMICS OF SOLUTIONS

In order to calculate the equilibrium composition of a system consisting of one or more phases in equilibrium with an aqueous solution of electrolytes, a review of the basic thermodynamic functions and the conditions of equilibrium is important. This is particularly true inasmuch as the study of aqueous solutions requires consideration of chemical and/or ionic reactions in the aqueous phase as well as a thermodynamic framework which is, for the most part, quite different from those definitions associated with nonelectrolytes. Therefore, in this section we will review the definition of the basic thermodynamic functions, the partial molar quantities, chemical potentials, conditions of equilibrium, activities, activity coefficients, standard states, and composition scales encountered in describing aqueous solutions.

Basic Thermodynamic Functions

The thermodynamic properties of a system at equilibrium consists of two types of properties, intensive and extensive properties. The most common intensive properties encountered are the temperature, T, and pressure, P, which are independent of the size of a measurement sample and are constant throughout the system. In fact, our definition of true equilibrium, to be described later requires T and P to be uniform throughout the system and the constituent phases. The most common extensive properties are volume, V and mass. As one would suspect, these extensive properties are proportional to the size of a measurement sample.

The thermodynamic properties most often encountered in describing phase equilibria of a system are functions of the state of the system. This is important since the calculation of these thermodynamic properties depends only on the existing state of the system and not the route by which this state has been reached. The following energy and energy related properties are extensive properties if they refer to the system as a whole:

The intrinsic energy	E
The enthalpy	$H = E + PV$

The entropy S

The Gibbs free energy $G = H - TS$

The heat capacity at constant pressure $C_p = \left(\frac{\partial H}{\partial T}\right)_P$

The above thermodynamic properties are intensive properties when their values are expressed on a per mole basis. Other useful relationships which we will encounter are obtained through differentiation of these basic thermodynamic functions and include:

$$dE = TdS - PdV \tag{2.1}$$

$$dH = TdS + VdP \tag{2.2}$$

$$dG = -SdT + VdP \tag{2.3}$$

These relationships express the first and second laws of thermodynamics for a closed system. Furthermore, from the last expression above we can also obtain

$$\left(\frac{\partial G}{\partial P}\right)_T = V \tag{2.4}$$

$$\left(\frac{\partial G}{\partial T}\right)_P = -S \tag{2.5}$$

Rewriting the definition of the Gibbs free energy,

$$G = H - TS = H + T\left(\frac{\partial G}{\partial T}\right)_P \tag{2.6}$$

This equation can be arranged to give the Gibbs-Helmholtz equation, an expression which will be very useful in calculating the effect of temperature on equilibrium constants and is given by:

$$\left(\frac{\partial G/T}{\partial T}\right)_P = \frac{-H}{T^2} \tag{2.7}$$

Solutions - Basic Definitions and Concepts

The pure substances from which a solution can be made are called the components, or constituents of a solution. The extensive properties of a solution are determined by the pressure, temperature, and the amount of each constituent. The intensive properties of a solution are determined by the pressure, temperature and the relative amounts of each constituent, or in other words by the pressure, temperature and composition of the solution. For aqueous solutions, the most commonly used measurement of composition of the solution is the molality, m. Molality is defined as the number of moles of a solute in one kilogram of the solvent, and for aqueous solutions the solvent is water. One of the advantages of using the molality scale for concentration is that it is independent of temperature and thus, the density of the solution does not need to be known in order to determine the composition on a mole basis as would be required with the unit of concentration, molarity. The molality of a solute i in water is given by $1000g_i/(M\ g_o)$ or $1000\ n_i/g_o$ where g_i and g_o are the number of grams of solute and solvent, M is the solute molecule weight and n_i is the number of gm-moles of solute.

The thermodynamic analysis of solutions is facilitated by the introduction of quantities that measure how the extensive thermodynamic quantities (V, E, H, G, ...) of the system depend on the state variables T, P, and n_i. This leads to the definition of partial molar quantities where, if we let Y be any extensive thermodynamic property, we can define the partial molar value of Y for the ith component as:

$$\bar{Y}_i = \left(\frac{\partial Y}{\partial n_i}\right)_{T,P,n_{j\neq i}} \qquad (2.8)$$

where n_j stands for all the mole quantities except n_i. It is important to note that the partial molar (or partial molal, which has the same meaning) quantities pertain to the individual components of the system and are also properties of the system as a whole. Furthermore

$$dY = \left(\frac{\partial Y}{\partial n_1}\right)_{T,P,n_{j\neq 1}} dn_1 + \left(\frac{\partial Y}{\partial n_2}\right)_{T,P,n_{j\neq 2}} dn_2 + \ldots$$

$$dY = \bar{Y}_1\ dn_1 + \bar{Y}_2\ dn_2 + \bar{Y}_3\ dn_3 + \ldots$$

or

$$dY = \sum_i \bar{Y}_i\ dn_i \qquad (2.9)$$

Partial molar quantities are intensive properties of the solution since they depend only on the composition of the solution, not upon the total amount of each component. If we add the several components simultaneously, keeping their ratios constant, the partial molal quantities remain the same. We can thus integrate the above expression keeping n_1, n_2, ... in constant proportions and find, while holding temperature and pressure constant, that

$$Y = \bar{Y}_1 n_1 + \bar{Y}_2 n_2 + \ldots$$

or

$$Y = \sum_i \bar{Y}_i n_i \qquad (2.10)$$

Furthermore, we can differentiate this expression to obtain

$$dY = \bar{Y}_1\ dn_1 + n_1 d\bar{Y} + \bar{Y}_2\ dn_2 + n_2 d\bar{Y}_2 + \ldots$$

or $$dY = \sum_i \bar{Y}_i \, dn_i + \sum_i n_i \, d\bar{Y}_i \tag{2.11}$$

Since

$$dY = \sum_i \bar{Y}_i \, dn_i \tag{2.9}$$

as has previously been shown, we can then substitute this expression for dY into our total differential, equation (2.11), and obtain

$$0 = \sum_i n_i \, d\bar{Y}_i$$

Since Y represents any extensive property, if we substitute the Gibbs free energy, G, for Y we obtain the Gibbs-Duhem equation

$$0 = \sum_i n_i \, d\bar{G}_i \tag{2.13}$$

which is very useful in the thermodynamics of aqueous solutions.

The partial molar quantities of interest in aqueous solutions are the partial molar Gibbs free energy, enthalpy and volume which are defined respectively as:

$$\bar{G}_i = \left(\frac{\partial G}{\partial n_i}\right)_{T,P,n_{j \neq i}} \tag{2.14}$$

$$\bar{H}_i = \left(\frac{\partial H}{\partial n_i}\right)_{T,P,n_{j \neq i}} \tag{2.15}$$

$$\bar{V}_i = \left(\frac{\partial V}{\partial n_i}\right)_{T,P,n_{j \neq i}} \tag{2.16}$$

The partial molar Gibbs free energy is also known or defined as the chemical potential, μ_i

$$\mu_i = \left(\frac{\partial G}{\partial n_i}\right)_{T,P,n_{j \neq i}} \tag{2.17}$$

Equilibrium - Necessary Conditions

With aqueous solutions of electrolytes we have two types of equilibrium to consider: phase equilibrium and chemical or ionic reaction equilibrium. Phase equilibrium of interest are primarily vapor-liquid and liquid-solid, though vapor-liquid-solid is often of great importance as, for example, in carbonate systems. The necessary condition of phase equilibrium is that the chemical potential of any species i in phase a is equal to the chemical potential of that same species i in phase b or

$$\mu_{i,a} = \mu_{i,b} \tag{2.18}$$

For chemical or ionic equilibria in a particular phase, the condition of equilibrium is of the same form as the chemical equation. Thus if the reaction at equilibrium is represented by

$$aA + bB = cC + dD \tag{2.19a}$$

the condition of chemical equilibrium in a particular phase would be denoted by

$$a\mu_A + b\mu_B = c\mu_C + d\mu_D \tag{2.19b}$$

which can be represented in a generalized form as

$$\Sigma\, \nu_i\, \mu_i = 0 \tag{2.20}$$

where ν_i is the stoichiometric coefficient of species i in the reaction of interest; it is positive if the species is a product and negative if a reactant.

The two conditions of equilibrium for phase and chemical equilibrium can be combined to represent the heterogeneous liquid-solid equilibrium of an aqueous solution of a salt B in equilibrium with the solid of salt B

$$\mu_{B,s} = \mu_{B,aq}$$

The chemical potential of the solid crystal salt B is in phase equilibrium with the dissolved salt B in the liquid or aqueous phase. In aqueous systems we are primarily dealing with salts of strong electrolytes, which in water dissociate completely to the constituent cations and anions of the salt. The chemical potential of the dissolved salt is then given by

$$\mu_{B,aq} = \nu_c\, \mu_c + \nu_a\, \mu_a$$

where ν_c and ν_a represent the stoichiometric number of cations and anions while μ_c and μ_a are the chemical potential of the cation and anion respectively. Thus the condition of equilibrium for a strong electrolyte dissolved in water and in equilibrium with its crystalline phase becomes

$$\mu_{B,s} = \nu_c\, \mu_c + \nu_a\, \mu_a$$

Activities, Activity Coefficients and Standard States

It is generally more convenient in aqueous solution thermodynamics to describe the chemical potential of a species i, in terms of its activity, a_i. The basic relationship between activity and chemical potential was developed by G.N. Lewis who first established a relationship for the chemical potential for a pure ideal gas, and then generalized his results to all systems to define the chemical potential of species i in terms of its activity a_i as

$$\mu_i(T) = \mu_i^o(T) + RT \ln (a_i) \tag{2.21}$$

Here μ^o_i is a reference chemical potential or the standard chemical potential at an arbitrarily chosen standard state. The activity is a measure of the difference between the component's chemical potential at the state of interest and at its standard state. Thus as the chemical potential of component i approaches the chemical potential of component i at its arbitarily chosen standard state, the component's activity approaches unity.

For aqueous solutions in which the composition of the solution is expressed in terms of molality, the arbitrarily chosen standard state is the hypothetical ideal solution of unit molality at the system temperature and pressure. It is chosen so that as the molality approaches zero, the ratio of a_i/m_i tends to unity. This ratio a_i/m_i is called the molal activity coefficient, γ_i.

$$\gamma_i = a_i/m_i \qquad \gamma_i \longrightarrow 1 \text{ as } m_i \longrightarrow 0$$

and

$$\mu_i = \mu_i^o + RT \ln(\gamma_i m_i) \tag{2.22}$$

The standard state chosen must be such that these equations hold at all pressures and temperatures. In these equations there appears to be an inconsistency of units, since the activity coefficient, γ_i, is dimensionless, whereas molality has the units moles/kilogram. To avoid this inconsistency the activity coefficient should be defined as

$$\gamma_i = a_i m^o / m_i \tag{2.23}$$

where m^o is the unit molality. For convenience, particularly in writing expressions for equilibrium constants, the normal convention in aqueous chemistry is to omit the writing of m^o and use the form $a_i = \gamma_i m_i$ with the understanding that the activity and the activity coefficient are dimensionless.

We have defined the activity and activity coefficient of a species i. Unfortunately, in solutions of electrolytes we find that we cannot make a solution containing only cations or only anions and need to introduce a mean or average activity coefficient. For example, when one mole of NaCl is dissolved in a kilogram of water, we have created a one molal solution of NaCl which fully dissociates to form one mole of sodium ions and one mole of chloride ions. The chemical potential of the dissolved sodium chloride is given by

$$\mu_{NaCl_{aq}} = \mu_{Na^+_{aq}} + \mu_{Cl^-_{aq}}$$

or in the more general

$$\mu_{salt_{aq}} = \nu_c \mu_c + \nu_a \mu_a$$

$$\mu_{salt_{aq}} = (\nu_c \mu_c^o + \nu_a \mu_a^o) + \nu_c RT \ln(\gamma_c m_c) + \nu_a RT \ln(\gamma_a m_a) \quad (2.24)$$

Rearranging and designating $(\nu_c \mu_c^o + \nu_a \mu_a^o)$ as $\mu^o_{salt_{aq}}$, we obtain

$$\mu_{salt_{aq}} = \mu^o_{salt_{aq}} + RT \ln(\gamma_c^{\nu_c} \gamma_a^{\nu_a}) + RT \ln(m_c^{\nu_c} m_a^{\nu_a}) \quad (2.25)$$

From this expression comes the definition of the mean activity coefficient, $\gamma_\pm$, in terms of the ionic activity coefficients γ_c and γ_a. The mean activity coefficient is the property which is determined or calculated from experimental measurements. A similar expression results for the mean molality $m_\pm$, which is not generally used in reporting experimental measurements:

$$\gamma_\pm = (\gamma_c^{\nu_c} \gamma_a^{\nu_a})^{1/\nu} \quad (2.26)$$

$$m_\pm = (m_c^{\nu_c} m_a^{\nu_a})^{1/\nu} \quad (2.27)$$

where $\nu = \nu_c + \nu_a$, the stoichiometric number of moles of ions in one mole of salt. The expression for the chemical potential of the dissolved salt could be rewritten as

$$\mu_{salt_{aq}} = \mu^o_{salt_{aq}} + RT \ln(\gamma_\pm m_\pm)^\nu \quad (2.28)$$

Since we are concerned with solutions we need also to determine the activity of the solvent which, for our particular interest in aqueous solutions, is the activity of water, a_w. The activity of water is related to the chemical potential of water by

$$\mu_w = \mu_w^o + RT \ln a_w \quad (2.29)$$

where the standard state is, by convention, pure water at the system temperature and pressure. Thus for the pure solvent, water, the activity $a_w = 1$.

In real solutions, the activity of water is quite close to one as the dilution ratio of the dissolved salts increases. In order to accurately represent the activity of water for dilute solutions, several significant digits would be required. To avoid this problem, in many compilations of data it is common to tabulate data in terms of the osmotic coefficient, ϕ. The osmotic coefficient is defined for any aqueous solution as:

$$\phi = \frac{-1000 \ln a_w}{18.0153 \sum_i \nu_i m_i} \tag{2.30}$$

The osmotic coefficient tends to approach unity as the solution becomes infinitely dilute. The Gibbs-Duhem equation, equation (2.13), which was presented earlier at a constant temperature and pressure, can be expressed as:

$$\sum_i n_i \mu_i = 0 \tag{2.31}$$

This relation can be applied to an aqueous solution of a single solute salt in water and, after suitable manipulation, the following expression representing the interrelationship between the water activity, a_w, the mean activity coefficient of the dissolved salt, $\gamma_\pm$, and the osmotic coefficient, ϕ, can be obtained:

$$m\, d(\ln(m\, \gamma_\pm)) = -55.50837\, d(\ln a_w) = \nu d(m\, \phi) \tag{2.32}$$

This expression can be used in different ways. In the treatment of experimental data one of the most common methods of determining activity coefficient data involves the measurement of the water vapor pressure of the aqueous solution as a function of concentration through various experimental techniques. Through proper manipulation of this equation and the acceptance of a theory of ionic solution behavior, the above expression can be used to calculate the mean activity coefficient from the experimentally determined osmotic coefficient. This expression is also useful for calculating the water activity as a function of concentration if one has developed a model of the behavior of the activity coefficients with concentration and the parameters of the mean activity coefficient model have been experimentally determined.

The relationships expressed in this chapter, while discussed on a molality scale, hold true for calculations based on other scales. The basic definition of the activity coefficient is:

on a molal scale: $$\gamma_i = \frac{a_i(m)}{m_i} \tag{2.33a}$$

γ_i - molal activity coefficient

m_i - molality

$a_i(m)$ - activity on molal basis

on a molar scale: $$y_i = \frac{a_i(c)}{c_i} \qquad (2.33b)$$

y_i - molar activity coefficient

c_i - molarity

on a mole fraction scale: $$f_i = \frac{a_i(x)}{x_i} \qquad (2.33c)$$

f_i - rational activity coefficient

x_i - mole fraction

$a_i(x)$ - activity on a mole fraction basis

When necessary, activity coefficients may be converted from one scale to another via the following relationships presented by Robinson and Stokes (1):

$$f_\pm = (1. + .001\ M_s\ \nu\ m)\ \gamma_\pm \qquad (2.34a)$$

$$f_\pm = \left(\frac{d + .001\ c\ (\nu\ M_s - M)}{d_0}\right) y_\pm \qquad (2.34b)$$

$$\gamma_\pm = \left(\frac{d - .001\ c\ M}{d_0}\right) y_\pm = \left(\frac{c}{m d_0}\right) y_\pm \qquad (2.34c)$$

$$y_\pm = (1. + .001\ m\ M)\ \frac{d_0}{d}\ \gamma_\pm = \left(\frac{m d_0}{c}\right) \gamma_\pm \qquad (2.34d)$$

where: ν - stoichiometric number = $\nu_+ + \nu_-$

d - solution density

d_0 - solvent density

M - molecular weight of the solute

M_s - molecular weight of the solvent

Throughout this book the molecular weights used are from the NBS Tech Note 270 Series (11-16).

Occasionally, reference is made to the "reduced activity coefficient", as in Meissner's work discussed beginning in Chapter IV. It is defined as:

$$\Gamma \equiv \gamma^{1/z_+z_-}$$

where z_+ and z_- are the absolute value of the charges on the cation and anion.

REFERENCES/FURTHER READING

1. Robinson, R.A. and R.H. Stokes, Electrolyte Solutions, 2nd ed. revised, Butterworth and Co., London, 1970

2. Guggenheim, E.A., Thermodynamics, 5th ed. revised, North-Holland Publishing Co., Amsterdam, 1967

3. Lewis, G.N. and M. Randall, revised by Pitzer, K.S. and Brewer, L., Thermodynamics, 2nd edition, McGraw-Hill Book Co., Inc., New York, 1961

4. Smith, J.M. and H.C. Van Ness, Introduction to Chemical Engineering Thermodynamics, McGraw-Hill Book Co., Inc., New York, 1959

5. Bett, K.E., J.S. Rowlinson and G. Saville, Thermodynamics for Chemical Engineers, The MIT Press, Cambridge, Mass., 1975

6. Sandler, S.I., Chemical and Engineering Thermodynamics, J. Wiley and Sons, New York, 1977

7. Harned, H.S. and B.B. Owen, The Physical Chemistry of Electrolytic Solutions, third edition, Reinhold Publishing Corp., New York, 1958

8. Guggenheim, E.A. and R.H. Stokes, Equilibrium Properties of Aqueous Solutions of Single Strong Electrolytes, Pergamon Press, Oxford, 1969

9. Strumm, W. and J.J. Morgan, Aquatic Chemistry - An Introduction Emphasizing Chemical Equilibria in Natural Waters, 2nd ed., J. Wiley and Sons, New York, 1981

10. Denbigh, K., The Principles of Chemical Equilibrium, 2nd ed., Cambridge University Press, Cambridge, 1966

III:
Equilibrium Constants

EQUILIBRIUM CONSTANTS

In order to compute:

- o the solubility of a salt such as NaCl in an aqueous solution, or,
- o the vapor-liquid equilibria of a system containing several volatile species such as NH_3, H_2O and CO_2, or,
- o the concentrations of the various dissolved molecular and ionic species in a system such as H_2SO_4 and H_2O

we need to satisfy the necessary conditions for equilibrium.

In the previous chapter the necessary conditions for equilibrium were introduced in terms of the chemical potentials of the constituent species in the various phases. In this chapter we will relate these chemical potentials to a more convenient form, that of the equilibrium constant. Furthermore, we will discuss the application of equilibrium constants to the three types of equilibria which occur in our overall vapor-liquid-solid model:

- o ionic and/or chemical equilibria in aqueous solutions
- o solubility equilibria between crystals and saturated aqueous solutions
- o vapor-liquid equilibria between gas and aqueous solutions

In addition we will review techniques for computing these equilibrium constants from tabulations of selected values of thermodynamic properties and from experimental data. We will also examine the effects of temperature and pressure on these equilibrium constants.

Ionic and/or Reaction Equilibrium in Aqueous Solutions

For a chemical or ionic equilibrium occurring in an aqueous solution with reaction equilibrium represented by:

$$aA + bB = cC + dD \qquad (2.19a)$$

the condition of chemical or ionic equilibrium was shown to be:

$$a\,\mu_A + b\,\mu_B = c\,\mu_C + d\,\mu_D \qquad (2.19b)$$

Recalling that the chemical potential of an aqueous species can be represented by reference to a standard state or:

$$\mu_i(T) = \mu_i^o(T) + RT\ \ln(\gamma_i m_i) \tag{2.21}$$

we can expand our general expression for the equilibrium expressed above to be:

$$a(\mu_A^o + RT\ \ln(\gamma_A\ m_a)) + b(\mu_B^o + RT\ \ln\ (\gamma_B\ m_B)) \tag{3.1}$$

$$= c(\mu_C^o + RT\ \ln(\gamma_C\ m_C)) + d(\mu_D^o + RT\ \ln(\gamma_D\ m_D))$$

By combining terms:

$$a\ \mu_A^o + b\ \mu_B^o - c\ \mu_C^o - d\ {}^{o}_{D} = RT(c\ \ln(\gamma_C\ m_C) + d\ \ln(\gamma_D\ m_D) - a\ \ln(\gamma_A\ m_A) - b\ \ln(\gamma_B\ m_B)) \tag{3.2}$$

Further simplification results in:

$$a\mu_A^o + b\ \mu_B^o - c\ \mu_C^o - d\ \mu_D^o = RT\ \ln \frac{(\gamma_C\ m_C)^c(\gamma_D\ m_D)^d}{(\gamma_A\ m_A)^a(\gamma_B\ m_B)^b} \tag{3.3}$$

Recalling that μ_i^o is a reference chemical potential or the standard chemical potential at an arbitrarily chosen standard state and that the partial molar Gibbs free energy is also known or defined as the chemical potential, we can substitute for $\mu_i^o(T)$ the quantity $\bar{G}_i^o(T)$. Tabulations of partial molar Gibbs free energy are available. These are given in the form of tabulations of ΔG_f^o for a substance, which represents the free energy when one gram-formula weight of the substance is formed, isothermically at the indicated temperature, from the elements, each in its appropriate reference state. By reviewing our definitions, we can see that ΔGf_i^o is a valid form of $\bar{G}_i^o$. Thus the left hand side of the equation above can be represented by:

$$a\ \bar{G}_A^o + b\ \bar{G}_B^o - c\ \bar{G}_C^o - d\ \bar{G}_D^o \tag{3.4}$$

which is equivalent to:

$$a\ \Delta G_{f_A}^o + b\ \Delta G_{f_B}^o - c\ \Delta G_{f_C}^o - d\ \Delta G_{f_D}^o \tag{3.5}$$

The thermodynamic equilibrium constant for this reaction is defined as:

$$K_T = \exp\ \{[a\ \Delta G_{f_A}^o + b\ \Delta G_{f_B}^o - (c\ \Delta G_{f_C}^o + d\ \Delta G_{f_D}^o\ )]/RT\} \tag{3.6}$$

and thus the complete expression for the equilibrium is given by:

$$K_T = \frac{(\gamma_C\, m_C)^c (\gamma_D\, m_D)^d}{(\gamma_A\, m_A)^a (\gamma_B\, m_B)^b} \tag{3.7}$$

Frequently, the standard free energy change for reaction is defined as:

$$\Delta G^o_{RXN}(T) = c\,\Delta G^o_{f_C}(T) + d\,\Delta G^o_{f_D}(T) - (a\,\Delta G^o_{f_A}(T) + b\,\Delta G^o_{f_B}(T)) \tag{3.8}$$

or generalizing:

$$\Delta G^o_{RXN}(T) = \sum_i \nu_i \Delta G^o_{f_i}(T) - \sum_j \nu_j \Delta G^o_{f_j}(T) \tag{3.9}$$

where i represents the products and j represents the reactants; then the equilibrium constant K_T is given by:

$$K_T = \exp(-\Delta G^o_{RXN}(T) \;/\; RT) \tag{3.10}$$

Values of the standard free energy of formation of most inorganic species have been evaluated by thermodynamic experts and are given for a standard state of 25° Celsius and one atmosphere pressure, with the standard state of a solute in aqueous solution taken as the hypothetical ideal solution of unit molality. Two comprehensive series of thermochemical tables containing ΔG_f^o, the Gibbs free energy of formation of a substance, have been published. These are "Selected Values of Chemical Thermodynamic Properties" (D1-D7) issued by the National Bureau of Standards and "Thermal Constants of Compounds" (D8-D17) issued by the VINITI Academy of Sciences of the USSR. Both series use the same standard states, however values cannot be used interchangeably since the data analysis that went into the creation of these series reflects the expert opinion of the compilers and differences do occur for some species. The Russian compilation is more complete in that the primary and secondary references used for determining the thermodynamic properties have been published simultaneously with each volume in the series, whereas the NBS has yet to publish their primary and secondary references.

As an example, let us calculate the thermodynamic equilibrium constant for the ionic equilibrium representing the dissociation of the bisulfate ion in aqueous solution given by:

$$HSO_4^- = H^+ + SO_4^=$$

species	ΔG^o_f (kcal/mole) - NBS
H^+	0.0
HSO_4^-	-180.69
$SO_4^=$	-177.97

$$\Delta G^o_{RXN}(298.15) = \Delta G^o_{f_{H^+}}(298.15) + \Delta G^o_{f_{SO_4^=}}(298.15) - \Delta G^o_{f_{HSO_4^-}}(298.15)$$

$$= 0.0 + (-177,970.) - (-180,690.)$$

$$= 2720.\ \text{cal/mole}$$

$$K_T = \exp(-\Delta G^o_{RXN}(298.15) / RT)$$

$$= \exp(-2720./(1.987*298.15))$$

$$= \exp(-4.5913)$$

$$= .01014$$

This value, calculated with the standard free energies of formation compiled by the NBS, represents the dissociation constant of the bisulfate ion at 25° Celsius (298.15 Kelvins) which is the reference temperature for their compilation. The accepted value of this equilibrium constant as given in Robinson and Stokes (1) determined experimentally is .0104 ± .0003. The experimental result agrees quite well with that calculated from the standard free energies of formation and the difference between the results represents a difference in ΔG^o_{RXN} of approximately 15 cal/mole. Since the tabulated values are only given to the nearest 10 cal/mole in this case the difference is insignificant.

Solubility Equilibria Between Crystals and Saturated Solutions

We next consider the solubility of a pure salt, strong electrolyte, in an aqueous solution saturated with respect to that salt. The simple dissociation reaction can be represented as:

$$M = aA + bB \tag{3.11}$$

where M is the pure salt and A and B are the constituent ions.

From earlier developments we know that

$$K_{sp} = a_A^a\ a_B^b\ /\ a_M \tag{3.12}$$

Assuming the salt is pure and is of activity one,

$$K_{sp} = a_A^a \, a_B^b \tag{3.13}$$

or,

$$K_{sp} = (\gamma_A m_A)^a \, (\gamma_B m_B)^b \tag{3.14}$$

As in the previous section we eventually arrive at

$$K_{sp} = \exp(-\,\Delta G^o_{K_{sp}} \;/\; RT) \tag{3.15}$$

where,

$$\Delta G^o_{K_{sp}} = a\,\Delta G^o_A + b\,\Delta G^o_B - \Delta G^o_M \tag{3.16}$$

The thermodynamic K_{sp} can be evaluated from tabulated values of ΔG_f^o which are found in several sources. For example:

ΔG_f^o:	NaCl(s)	= -91,790 cal/gmol
	Na^+	= -62,589 cal/gmol
	Cl^-	= -31,372 cal/gmol

So for NaCl:

$$\Delta G^o_{K_{sp}} = -62589. - 31372. - (-91790.) = -2171.\ \text{cal/gmol}$$

$$K_{sp} = \exp\left[\frac{-(-2171.)}{1.987*298.15}\right] = 39.04$$

The thermodynamic solubility product of NaCl, assuming our values of the free energy of formation are correct, is for the thermodynamic temperature of 25° Celsius. If we make the simplifying assumption of a solution where γ's =1, then the salt solubility is 6.248 gmols/kg H_2O which is reasonably close to the value given in Linke/Seidell (2) of 6.146 gmoles/kg H_2O.

Vapor-Liquid Equilibria in Aqueous Solutions

We next consider the equilibrium or solubility of a vapor in aqueous solution. The equilibrium can be represented as:

$$V = L \tag{3.17}$$

where V is the vapor species and L is the aqueous molecular form.

As before, we get:

$$K_{aq} = a_L \;/\; a_V \tag{3.18}$$

or,

$$K_{aq} = \gamma_L \, m_L \;/\; a_V \tag{3.19}$$

The activity of a vapor species can be expressed, analogous to aqueous species, by the form

$$a_V = f_V p_V \tag{3.20}$$

where:

f_V = fugacity coefficient of species V

p_V = partial pressure of species V

Once again, we have a coefficient which captures the nonideality in terms of temperature, pressure and composition effects. Combining (3.19) and (3.20) we get

$$K_{aq} = \gamma_L m_L \; / \; f_V p_V \tag{3.21}$$

As before, we eventually arrive at:

$$K_{aq} = \exp(- \Delta G^o_{K_{aq}} \; / \; RT) \tag{3.22}$$

where,

$$\Delta G^o_{K_{aq}} = \Delta G^o_L - \Delta G^o_V \tag{3.23}$$

Temperature Effects on the Equilibrium Constant

We have seen earlier, that

$$\ln K = - \frac{\Delta G^o}{RT} \tag{3.10}$$

Differentiating, we get

$$R \frac{d \ln K}{dT} = \frac{d(\Delta G^o(T)/T)}{dT} \tag{3.24}$$

Since

$$dG = \frac{\partial G}{\partial T} dT + \frac{\partial G}{\partial P} dP + \frac{\partial G}{\partial n_i} dn_i$$

it can be seen that at constant pressure and composition

$$\frac{\partial}{\partial T}\left(\frac{\Delta G^o}{T}\right) = \frac{d}{dT}\left(\frac{\Delta G^o}{T}\right)$$

equation (3.24) can be restated using equation (2.7)

$$R \frac{d \ln K}{dT} = \frac{\Delta H}{T^2} \tag{3.25}$$

ΔH can be expressed as a function of temperature in terms of heat capacity, Cp:

$$\Delta H = \Delta H^o + \int \Delta Cp^o \, dT \tag{3.26}$$

and, assuming constant ΔCp,

$$\Delta H = \Delta H^o + \Delta Cp^o \; (T - T^o) \tag{3.27}$$

Combining (3.25) and (3.27) we get

$$R\frac{d\ \ln K}{dT} = \frac{\Delta H^\circ}{T^2} \quad \Delta Cp^\circ\left(\frac{1}{T} - \frac{T^\circ}{T^2}\right) \tag{3.28}$$

Finally, integrating between the limits of the reference temperature, T°, and T we get:

$$\ln K - \ln K^\circ = -\Delta H^\circ\left(\frac{1}{RT} - \frac{1}{RT^\circ}\right) - \frac{\Delta Cp^\circ}{R}\left(\ln T - \frac{T^\circ}{T} - \ln T^\circ + 1\right) \tag{3.29}$$

or,

$$\ln K = \ln K^\circ - \frac{\Delta H^\circ}{R}\left(\frac{1}{T} - \frac{1}{T^\circ}\right) - \frac{\Delta Cp^\circ}{R}\left(\ln\frac{T}{T^\circ} - \frac{T^\circ}{T} + 1\right) \tag{3.30}$$

But, we saw earlier that

$$\ln K^\circ = -\frac{\Delta G^\circ}{RT^\circ} \tag{3.10}$$

and thus,

$$\ln K = \left(-\frac{\Delta G^\circ}{RT^\circ}\right) - \frac{\Delta H^\circ}{R}\left(\frac{1}{T} - \frac{1}{T^\circ}\right) - \frac{\Delta Cp^\circ}{R}\left(\ln\frac{T}{T^\circ} - \frac{T^\circ}{T} + 1\right) \tag{3.31}$$

provides us with the desired definition of a temperature dependent K. Note, of course, that our assumption of constant heat capacity could have been modified by substituting a Cp(T) into (3.26) and an improved version of (3.31) could have been evolved.

The usefulness of (3.31) in predicting the temperature dependence of K is that ΔG°, ΔH° and ΔCp° are tabulated for a great many species in the same sources cited earlier in this chapter.

Estimating Temperature Effects on Heat Capacity and Other Thermodynamic Properties

In the previous section, the heat capacity was treated as being independent of temperature during equilibrium constant calculations. Although most thermodynamic property data is measured at 25°C, industrial applications demand accurate data for higher temperatures. The desirability of a method for accurately predicting elevated temperature thermodynamic properties in order to avoid the need for experimental measurements is obvious. In 1964, Criss and Cobble (9, 10) proposed the "correspondence principle" as such a method. They summarized it as follows (9):

"A standard state can be chosen at every temperature such that the partial molal entropies of one class of ions at that temperature are linearly releated to the corresponding entropies at some reference temperature."

Criss and Cobble noted that the entropies of ions can be expressed as functions of mass, charge and ionic size, and that the functional dependences appear to be the same at 25°C and higher temperatures. This suggested that it would not be necessary to know the complete functional dependences and the entropies at elevated temperatures could be determined from 25°C entropies. They suggested that the following equation could be used to determine the standard state entropy at t_2 having an experimental standard state entropy at temperature t_1:

$$\overline{S^o_{t_2}} = a_{t_2} + b_{t_2} \overline{S^o_{t_1}} \qquad (3.32)$$

In testing their theory they experienced the problem of the lack of reliable data at elevated temperatures. They noted that the best thermodynamic function for calculating the partial molal entropies are the partial molal heat capacities, $\overline{Cp^o}$, and that few systems had been studied at temperatures greater than 100°C. Acceptable estimates of $\overline{Cp}_{t_2}$ could be obtained given the partial molal heat capacities over an extended temperature range. Average heat capacity can also be estimated given a value for free energy at an elevated temperature t_2:

$$\Delta G^o_{t_2} = \Delta G^o_{t_1} + \Delta Cp^o \Big]_{t_1}^{t_2} \Delta T - \Delta S^o_{t_1} \Delta T - T_2 \Delta Cp^o \Big]_{t_1}^{t_2} \ln \frac{T_2}{T_1} \qquad (3.33)$$

$\Delta Cp^o \Big]_{t_1}^{t_2}$ is the average value of ΔCp^o between temperatures t_1 and t_2 .

Based on the experimental partial molal entropies, Criss and Cobble presented values for the a and b parameters of equation (3.32), which are tabulated in Appendix 3.1. They note a surprising accuracy of ± 0.5 cal. mole^{-1} deg^{-1} for simple ions up to 150°C.

The drawback to equation (3.33) is that, while $\overline{Cp^o}$ varies greatly with temperature and many values are available at 25°C, few studies of the temperature dependence of $\overline{Cp^o}$ have been made. Criss and Cobble (10) therefore extended their correspondence principle to enable the prediction of elevated temperature heat capacities.

Since the average value of the partial molal heat capacity between 25°C and t_2 can be described by:

$$\overline{Cp^\circ}\Big]_{25}^{t_2} = \frac{\overline{S^\circ_{t_2}} - \overline{S^\circ_{25}}}{\ln(T_2/298.15)} \qquad (3.34)$$

and the correspondence principle defines the entropy at elevated temperature t_2 as:

$$\overline{S^\circ_{t_2}} = a_{t_2} + b_{t_2}\,\overline{S^\circ_{25}} \qquad (3.32)$$

the average value of the partial molal heat capacity can be expressed in terms of the correspondence principle by substituting equation (3.32) for the $\overline{S_t^\circ}$ of equation (3.34):

$$\overline{Cp^\circ}\Big]_{25}^{t_2} = \frac{a_{t_2} - \overline{S^\circ_{25}}\,(1. - b_{t_2})}{\ln(T_2/298.15)} \qquad (3.35)$$

The appearance of this equation can be simplified by defining:

$$\alpha_{t_2} = \frac{a_{t_2}}{\ln(T_2/298.15)} \qquad (3.36a)$$

$$\beta_{t_2} = \frac{-(1. - b_{t_2})}{\ln(T_2/298.15)} \qquad (3.36b)$$

so that equation (3.35) becomes:

$$\overline{Cp^\circ}\Big]_{25}^{t_2} = \alpha_{t_2} + \beta_{t_2}\,\overline{S^\circ_{25}} \qquad (3.37)$$

Criss and Cobble define $\overline{S_{25}}^\circ$ as the "absolute" ionic entropy which is:

$$\overline{S^\circ_{25}} = \overline{S^\circ_{25}}\ (\text{conventional}) - 5.\ z \qquad (3.38)$$

where z is the ionic charge. Values for α and β have been calculated for various temperatures and are tabulated in Appendix 3.1.

As an example of the usage of equation (3.37), the average heat capacity of potassium between 25° and 60° is calculated:

$\alpha = 35.$

$\beta = -.41$

$\overline{S^\circ_{25}} = 24.5 - 5. * 1. = 19.5$

The value for $\overline{S_{25}}°$ (conventional) is from the NBS Tech Note 270-8 (D6).

$$Cp° \Big]_{25}^{60} = 35. + (-.41 \times 19.5) = 27.005 \text{ cal. mole}^{-1} \text{ deg.}^{-1}$$

Equilibrium Constants from Tabulated Data

In cases where ΔG°, ΔH° and ΔCp° are not tabulated for all species comprising a reaction, it is still sometimes possible to obtain K_{sp} values. The requirements are:

1) actual solubility data, and,
2) either,
   a) experimental activity coefficient data, or,
   b) estimated activity coefficient data.

It is then possible to use numerical regression to curve fit an equation of suitable form. The form

$$\ln K_{sp} = A + B/T + CT + DT^2 \quad (3.39)$$

is suggested as a reasonable one.

Pressure Effects on the Equilibrium Constant

In recent years, researchers have attempted to use molal volume and compressibility data to estimate the effect of pressure on ionic equilibria or nonideal solubility. The method of using V and k data to estimate the effect of pressure on ionic equilibria or the solubility of salts was first demonstrated by Owen and Brinkley (3):

$$K(P,T) = K(T) \exp((-\Delta V°(P-1) + .5\,\Delta k°(P-1)^2)/(RT)) \quad (3.40)$$

where:

K(P,T) - equilibrium constant at specified pressure and temperature

K(T) - equilibrium constant at specified temperature and standard state pressure of 1 atmosphere

ΔV° - the algebraic difference between the partial molal volumes of the products and reactants in their standard state taking into account the stoichiometry of the reaction (cm^3/mole)

P - the pressure of interest (atm)

Δk° - the algebraic difference between the partial molal compressibilities of the products and reactants in their standard state taking into account the stoichiometry of the reaction (cm^3/(mole atm))

R - the gas law constant and is 82.054 cm^3 atm/mole K

T - Kelvins

Over the years many things have been said about this equation because of the difficulty in getting data. The most common simplification is to avoid the molal compressibility term since k° for a mineral is small and remains relatively constant with temperature. Also the molal compressibilities of ions are hard to find published in the literature and are limited primarily to a range of 0 - 50° Celsius. At lower pressures ignoring this term is probably suitable, but for high pressure the term becomes more important.

In 1968 Lown, Thirsk and Wynne-Jones (4) have shown that the effect of pressure on ionic or nonideal equilibrium to pressures below 1000 bars can be estimated by:

$$K(P,T) = K(T) \exp((-\Delta V^\circ P + .5\,\Delta k^\circ P^2)/(RT)) \qquad (3.41)$$

To show the effect of pressure on the simple reaction:

$$NaCl(s) = Na^+ + Cl^-$$

we will evaluate ΔV°, Δk° and K(P,T)/K(T) at 25°C:

$$\Delta V^\circ = \bar{V}^\circ_{Na} + \bar{V}^\circ_{Cl} - \bar{V}^\circ_{NaCl}$$

$$= -1.21 + 17.82 - 27.013$$

$$= -10.405 \text{ cm}^3/\text{mole}$$

$$\Delta k^\circ = \bar{k}^\circ_{Na} + \bar{k}^\circ_{Cl} - \bar{k}^\circ_{NaCl}$$

$$= 3.94\text{x}10^{-3} + .74\text{x}10^{-3} - (1.1\text{x}10^{-4})$$

$$= 4.57\text{x}10^{-3} \text{ cm}^3/\text{mole bar}$$

or

$$= 4.630\text{x}10^{-3} \text{ cm/mole atm}$$

The resulting K(P,T)/K(T) at 25°C is K(P,25°C)/K(25°C) and is given, per equation (3.40), by the following table.

TABLE 3.1

Pressure (atm)	K(P,25°C)/K(25°C) $k^o = 0$	K(P,25°C)/K(25°C) $k^o = 4.6306 \times 10^{-3}$
100	1.043	1.042
200	1.088	1.0843
300	1.136	1.126
400	1.185	1.187
500	1.236	1.208
600	1.298	1.247
700	1.346	1.285
800	1.405	1.322
900	1.466	1.358
1000	1.52	1.392

APPENDIX 3.1

CRISS AND COBBLE PARAMETERS

Table 1: a and b Parameters for Using Criss and Cobble's Entropy Equation

From: C.M. Criss and J.W. Cobble, "The Thermodynamic Properties of High Temperature Aqueous Solutions. IV. Entropies of the Ions up to 200° and the Correspondence Principle", J.A.C.S., 86, 5385 (1964)

t, °C	simple cations		simple anions and OH^-		oxy anions	
	a_t	b_t	a_t	b_t	a_t	b_t
25	0	1.000	0	1.000	0	1.000
60	3.9	0.955	-5.1	0.969	-14.0	1.217
100	10.3	0.876	-13.0	1.000	-31.0	1.476
150	16.2	0.792	-21.3	0.989	-46.4	1.687
200	(23.3)	(0.711)	(-30.2)	(0.981)	(167.0)	(2.020)

t, °C	acid oxy anions		standard state (entropy of $H^+(aq)$ assigned)
	a_t	b_t	
25	0.	1.000	-5.0
60	-13.5	1.380	-2.5
100	-30.3	1.894	2.0
150	(-50.0)	(2.381)	6.5
200	(-70.0)	(2.960)	(11.1)

The constants in parentheses were estimated by extrapolation of corresponding values of a_t and b_t from lower temperatures and are subject to greater error.

From: C.M. Criss and J.W. Cobble, "The Thermodynamic Properties of High Temperature Aqueous Solutions. V. The Calculation of Ionic Heat Capacities up to 200°. Entropies and Heat Capacities above 200°", J.A.C.S., 86, 5390 (1964)

Table 2: Heat Capacity Parameters α and β

t, °C	cations		anions and OH^-		oxy anions		acid oxy anions	
	α_t	β_t	α_t	β_t	α_t	β_t	α_t	β_t
60	35.	-0.41	-46.	-0.28	-127	1.96	-122	3.44
100	46	-0.55	-58.	0.00	-138	2.24	-135	3.97
150	46	-0.59	-61.	-0.03	-133	2.27	(-143)	(3.95)
200	(50)	(-0.63)	(-65)	(-0.04)	(-145)	(2.53)	(-152)	(4.24)

Values in parentheses are from extrapolated values of the a and b entropy parameters and are subject to greater error.

Table 3: Estimated Entropy Parameters above 150°

T, °C	cations		anions and OH^-		oxy anions		acid oxy anions	
	a_t	b_t	a_t	b_t	a_t	b_t	a_t	b_t
200	23.3	0.711	-30.2	0.981	-67	2.020	-70.	2.960
250	29.9	0.630	-38.7	0.978	-86.5	2.320	-90.	3.530
300	36.6	0.548	-49.2	0.972	-106.	2.618	--	--

REFERENCES/FURTHER READING

1. Robinson, R.A. and R.H. Stokes, Electrolyte Solutions, 2nd ed., Butterworth and Co., London, (1970)

2. Linke, W.R., Seidell, A., Solubilities - Inorganic and Metal-Organic Compounds, Vol. I, American Chemical Society, Washington, D.C., (1958) Vol. II, (1965)

3. Owen, B.B. and S.R. Brinkley, Jr., "Calculation of the Effect of Pressure Upon Ionic Equilibria in Pure Water and In Salt Solutions", Chem. Rev., 29, 461 (1941)

4. Lown, D.A., H.R. Thirsk and L. Wynne-Jones, "Effect of Pressure on Ionization Equilibria in Water at 25°C", Trans. Faraday Soc., 64, 2073 (1968)

5. Owen, B.B. and S.R. Brinkley, Jr., "Extrapolation of Apparent Molal Properties of Strong Electrolytes", Ann. N.Y. Acad. Sci., 51, 753 (1949)

6. Marshall, W.L., "Correlations in Aqueous Electrolyte Behavior to High Temperatures and Pressures", Record of Chem. Prog., 30, 61 (1969)

7. Millero, F.J., "The Use of the Specific Interaction Model to Estimate the Partial Molal Volumes of Electrolytes in Sea Water", Geoch. Cosmo. Acta, 41, 215 (1977)

8. Millero, F.J., "The Effect of Pressure on the Solubility of Minerals in Water and Seawater", Geoch. Cosmo. Acta, 46, 11 (1982)

9. Criss, C.M. and J.W. Cobble, "The Thermodynamic Properties of High Temperature Aqueous Solutions. IV. Entropies of the Ions up to 200° and the Correspondence Principle", J. Am. Chem. Soc., 86, 5385 (1964)

10. Criss, C.M. and J.W. Cobble, "The Thermodynamic Properties of High Temperature Aqueous Solutions. V. The Calculation of Ionic Heat Capacities above 200°", J. Am. Chem. Soc., 86, 5390 (1964)

11. Šťěrbáček, Z., B. Biskup and P. Tausk, Calculation of Properties Using Corresonding-State Methods, Elsevier Scientific Pub. Co., New York (1979)

DATA REFERENCES

D1. Wagman, D.D.; W.H. Evans; V.B. Parker; I. Halow; S.M. Bailey; R.H. Schumm; Selected Values of Chemical Thermodynamic Properties - Tables for the First Thirty-Four Elements in the Standard Order of Arrangement, National Bureau of Standards Technical Note 270-3 (1968)

D2. Wagman, D.D.; W.H. Evans; V.B. Parker; I. Halow; S.M. Bailey; R.H. Schumm; Selected Values of Chemical Thermodynamic Properties - Tables for Elements 35 through 53 in the Standard Order of Arrangement, NBS Tech Note 270-4 (1969)

D3. Wagman, D.D.; W.H. Evans; V.B. Parker; I. Halow; S.M. Bailey; R.H. Schumm; K.L. Churney; Selected Values of Chemical Thermodynamic Properties - Tables for Elements 54 through 61 in the Standard Order of Arrangement, NBS Tech Note 270-5 (1971)

D4. Parker, V.B.; D.D. Wagman and W.H. Evans; Selected Values of Chemical Thermodynamic Properties - Tables for the Alkaline Earth Elements (Elements 92 through 97 in the Standard Order of Arrangement), NBS Tech Note 270-6, (1971)

D5. Schumm, R.H.; D.D. Wagman; S. Bailey; W.H. Evans; V.B. Parker, Selected Values of Chemical Thermodynamic Properties - Tables for the Lanthanide (Rare Earth) Elements (Elements 62 through 76 in the Standard Order of Arrangement), NBS Tech Note 270-7 (1973)

D6. Wagman, D.D.; W.H. Evans; V.B. Parker; R.H. Schumm; R.L. Nuttall; Selected Values of Chemical Thermodynamic Properties - Compounds of Uranium, Protactinium, Thorium, Actinium, and the Alkali Metals, NBS Tech Note 270-8 (1981)

D7. Wagman, D.D.; W.H. Evans; V.B. Parker; R.H. Schumm; I. Halow; S.M. Bailey; K.L. Churney; R.L. Nuttall; The NBS tables of chemical thermodynamic properties - Selected values for inorganic and C_1 and C_2 organic substances in SI units, J. Phys. Chem. Ref. Data, vol. 11, supp. 2 (1982)

D8. Glushko, V.P., dir.; Thermal Constants of Compounds - Handbook in Ten Issues, VINITI Academy of Sciences of the U.S.S.R., Moscow

Volume I (1965) - O, H, D, T, F, Cl, Br, I, At, ^{3}He, He, Ne, Ar, Kr, Xe, Rn

D9. Volume II (1966) - S, Se, Te, Po

D10. Volume III (1968) - N, P, As, Sb, Bi

D11. Volume IV-1 (1970) - C, Si, Ge, Sn, Pb
Volume IV-2 (1971)

D12. Volume V (1971) - B, Al, Ga, In, Tl

D13. Volume VI-1 (1972) - Zn, Cd, Hg, Cu, Ag, Au, Fe, Co, Ni, Ru, Rh, Pd, Os, Ir, Pt
Volume VI-2 (1973)

D14. Volume VII-1 (1974) - Mn, Tc, Re, Cr, Mo, W, V, Nb, Ta, Ti, Zr, Hf
Volume VII-2 (1974)

D15. Volume VIII-1 (1978) - Sc, Y, La, Ce, Pr, Nd, Pm, Sm, Eu, Gd, Tb, Dy, Ho, Er, Tm, Yb, Lu, Ac, Th, Pa, U, Np, Pu, Am, Cm, Bk, Cf, Es, Fm, Md, No
Volume VIII-2 (1978)

D16. Volume IX (1979) - Be, Mg, Ca, Sr, Ba, Ra

D17. Volume X-1 (1981) - Li, Na, K, Rb, Cs, Fv
Volume X-2 (1981)

D18. Vargaftik, N.B., Tables on the Thermophysical Properties of Liquids and Gases in Normal and Dissociated States, 2nd edition, Hemisphere Publishing Co., Washington (1975)

D19. Karapet'yants, M. Kh. and M.L. Karapet'yants, Thermodynamic Constants of Inorganic and Organic Compounds, Ann Arbor-Humphrey Science Publishers, Inc., Ann Arbor (1970)

D20. Barner, H.E. and R.V. Scheuerman, Handbook of Thermochemical Data for Compounds and Aqueous Species, John Wiley and Sons, Inc., New York (1978)

D21. Pankratz, L.B., Thermodynamic Properties of Elements and Oxides, U.S. Bureau of Mines Bulletin 672 (1982)

IV:

Activity Coefficients of Single Strong Electrolytes

ACTIVITY COEFFICIENTS OF SINGLE STRONG ELECTROLYTES

This chapter traces the history of activity coefficient models for aqueous solutions of single strong electrolytes. Of great importance is the Debye-Hückel equation, which has been the cornerstone of more recent models. In addition to Debye-Hückel, the following, more recent methods are outlined:

1. Guggenheim's equation - a one parameter equation with published parameter values for calculations at 0° and 25° C

2. Davies' equation - a suggested equation for the Guggenheim parameter

3. Bromley's equation - a suggested equation for the Guggenheim parameter

4. Meissner's equation and plot - a one parameter equation for reduced activity coefficients and a generalized family of curves from which it is proposed graphical predictions can be made

5. Pitzer's equation - a three to four parameter model for mean molal activity coefficients

6. Chen's equation - a model incorporating short and long range interactions between molecules and ions using two parameters

While these methods were based on activity coefficient calculations at 25°C, some of the authors suggested methods for adapting their models for any temperature. These methods are outlined.

All of the models presented have been used for a number of strong electrolytes. The resultant activity coefficients are plotted with experimental data for comparison.

Much of the material in this chapter is quite theoretical. For those interested in the application of these methods, more attention should be paid to the subsections beginning with the one entitled "Application".

HISTORY

Early in this century process calculations were done with experimentally determined phase diagrams. While such graphical techniques are successful for simple systems, there are some major drawbacks. These include the need for extensive experimentation and the difficulty in visualizing systems composed of more than four components.

Presented in 1887, Svanté Arrhenius'(1) theory of electrolytic dissociation, that partial dissociation of the solute into negatively and positively charged ions takes place, and his proposed method of calculating the degree of dissociation helped open the way for organized theoretical and experimental investigations of electrolyte solutions. This theory held that these ions in solution are in a state of chaotic motion similar to that in an ideal gas and that the interaction of ions in a solution does not affect their distribution and motion.

This was fine for dilute solutions of weak electrolytes but the contradiction with experimental results for strong electrolytes led to the realization that the electrostatic force between ions must be taken into account.

In 1923, Peter Debye and Erich Hückel presented their theory of interionic attraction which has been the foundation for the work done since then. Since strong electrolytes are highly dissociative in solution, the ion concentration is higher with the resulting distance between them smaller than for weak electrolytes. This increase in concentration results in the tendency towards an orderly distribution of ions as the electrostatic forces cause mutual attraction between oppositely charged ions. The potential energy of the ionic attraction must therefore be accounted for in considering electrolyte solutions.

The dependence of the total electric potential on the charge distribution is shown by Poisson's equation:

$$\nabla \psi_t = -\frac{4\pi\rho}{D} \qquad (4.1)$$

where: ψ_t - total electric field potential

ρ - charge density

D - dielectric constant

The Laplacian operator ∇ symbolizes differentiation, as in the form:

$$\frac{\partial^2\psi}{\partial x^2} + \frac{\partial^2\psi}{\partial y^2} + \frac{\partial^2\psi}{\partial z^2}$$

As the field being considered around an ion is spherical, the x, y and z coordinates can be replaced by a single coordinate, r, representing the spherical radius. Equation (4.1) can thus be expressed as:

$$\nabla\psi_t = \frac{1}{r^2}\frac{d}{dr}\left(r^2\frac{d\psi_t}{dr}\right) = -\frac{4\pi\rho}{D} \qquad (4.1a)$$

In accordance with Boltzmann's law, the numbers of positive and negative ions per unit volume of solution around the central ion at a distance r will be:

$$N_A n'_+ = N_A n_+ + \exp\left(\frac{-e\psi_t}{kT}\right) \quad \text{and} \quad N_A n'_- = N_A n_- \exp\left(\frac{+e\psi_t}{kT}\right) \qquad (4.2)$$

where: $n'_\pm$ - actual ion concentration in given volume

$n_\pm$ - mean ion concentration

N_A - Avagadro's number = 6.0232×10^{23} mole^{-1}

e - electronic charge = 4.8029×10^{-10} e.s.u. or

= 1.60206×10^{-19} coulomb

ψ_t - electrical potential

k - Boltzmann's constant = R/N_A = 1.38045×10^{-16} erg/deg

T - absolute temperature

The ionic charge is found by multiplying the number of ions by the charge, ze. For example, for cation of charge ze:

$$\text{ionic charge} = ze\, N_A n_+ \exp\left(\frac{-e\psi_t}{kT}\right); \qquad z = \text{the charge number}$$

The charge density ρ is the sum of all the ionic charges:

$$\rho = \sum_i z_i e\, n_i\, N_A \exp\left(-\frac{z_i e\psi_t}{kT}\right) \qquad (4.3)$$

Replacing the ρ in the equation (4.1) with equation (4.3) yields:

$$\nabla\psi_t = \frac{1}{r^2}\frac{d}{dr}\left(r^2\frac{d\psi_t}{dr}\right) = -\frac{4\pi e}{D}\sum_i z_i n_i N_A \exp\left(-\frac{z_i e\psi_t}{kT}\right) \tag{4.4}$$

In order to simplify, the exponential function can be expanded and approximated as:

$$\exp\left(-\frac{z_i e\psi_t}{kT}\right) = 1 - \frac{z_i e\psi_t}{kT} + \frac{z_i^2 e^2 \psi_t^2}{2k^2T^2} \tag{4.5}$$

For dilute solutions, especially at higher temperatures, where $z_i e\,\psi_t << kT$, the third term can be dropped. Substituting this back into (4.4) results in:

$$\nabla\psi_t = -\frac{4\pi}{D}\sum_i z_i n_i N_A e\left(1 - \frac{z_i e\psi_t}{kT}\right) \tag{4.6}$$

Since $\Sigma z_i n_i e = 0$ in an electrically neutral system, the equation is:

$$\nabla\psi_t = \frac{1}{r^2}\frac{d}{dr}\left(r^2\frac{d\psi_t}{dr}\right) = \frac{4\pi e^2}{DkT} N_A \Sigma z_i^2 n_i \psi_t$$

or

$$\nabla\psi_t = \frac{1}{r^2}\frac{d}{dr}\left(r^2\frac{d\psi_t}{dr}\right) = K^2 \psi_t \tag{4.7}$$

with

$$K^2 = \frac{4\pi e^2}{DkT} N_A \Sigma z_i^2 n_i \tag{4 7a}$$

Integration of equation 4.7 yields:

$$\psi_t = c_1 \frac{\exp(-Kr)}{r} + c_2 \frac{\exp(Kr)}{r} \tag{4.8}$$

As the value of ψ approaches zero as r increases, and the term exp(Kr)/r becomes indefinitely large, constant c_2 must equal 0. In order to determine c_1 the assumption is made that in very dilute solutions the distances between ions are so great in comparison with the ions' radii that the ions may be considered as points. Coulomb's law states that the energy needed to move a cation from infinity to point r, the potential ψ_r, is:

$$\psi_r = -\int_\infty^r E\,dr = -\int_\infty^r \frac{e}{Dr}\,dr = \frac{e}{Dr} \tag{4.9}$$

where:

$$E = -\frac{d\psi_a}{dr}$$

is the field strength or intensity. So in a dilute solution an ion of charge ez has the potential

$$\psi = \frac{ez}{Dr} \tag{4.10}$$

This means that as r approaches zero the total potential, ψ_t, becomes the ionic potential, ψ_r:

$$\psi_t = c_1 \frac{\exp(-Kr)}{r} = \frac{ez}{Dr} \tag{4.11}$$

Since the r's in the denominators cancel out, and as r approaches 0 the term exp(-Kr) approaches 1, then $c_1 = ez/D$.

$$\psi_t = \frac{ez}{D} \frac{\exp(-Kr)}{r} \tag{4.12}$$

When r>0 the total potential is the sum of the ion potential and the potential of the surrounding ionic atmosphere, ψ_a:

$$\psi_t = \psi + \psi_a \tag{4.13}$$

Substituting (4.10) and (4.12) into this expression gives an expression for the ionic atmosphere potential:

$$\psi_a = \psi_t - \psi = \frac{ez}{D} \frac{\exp(-Kr)}{r} - \frac{ez}{Dr} = -\frac{ez}{D}\left(\frac{1 - \exp(-Kr)}{r}\right) \tag{4.14}$$

As seen in Chapter II, the activity of a species is related to the chemical potential:

$$\mu_i = \mu_i^\circ + RT \ln (a_i) \tag{2.21}$$

The electrostatic chemical potential is part of the chemical potential or partial molar Gibbs free energy:

$$\mu_i = \left(\frac{\partial G}{\partial n_i}\right)_{T,P,n_{j \neq i}} \tag{2.17}$$

The expression for the ionic atmosphere potential (4.14) can be used to get the electrostatic chemical potential μ^{el}. Here again the exponential function is expanded and truncated, simplifying (4.14) to:

$$\psi_a = -\frac{ez}{D} K \tag{4.15}$$

To find the change in the electrostatic energy of interaction of an ionic atmosphere, the charge of each ion, ez, is changed from 0 to ez by a variable charge. Using a factor, α, which varies from 0 to 1, the quantity K is shown to change with the charge:

$$K' = \sqrt{\frac{4\pi e^2 N_A}{DKT}} \sqrt{\Sigma z_i^2 \alpha^2 n_i} = \alpha \sqrt{\frac{4\pi e^2 N_A}{DKT}} \sqrt{\Sigma z_i^2 n_i} = \alpha K \tag{4.16}$$

The ionic atmosphere potential at this charge is:

$$\psi'_a = -\frac{e\,\alpha\,z}{D}K' = -\frac{e\,\alpha^2\,z}{D}K \qquad (4.17)$$

So the change in the electrostatic energy for an ion as the charge changes is:

$$dE_i^{el} = \psi'_a\,de' = \psi'_a\,ez\,d\alpha = -\frac{e^2\,\alpha^2\,z^2}{D}K\,d\alpha \qquad (4.18)$$

By integrating (4.18) the ionic atmosphere energy of formation for one ion is found:

$$E_i^{el} = -\int_0^1 \frac{e^2\,z^2\,K\,\alpha^2 d\alpha}{D} = -\frac{e^2\,z^2\,K}{3D} \qquad (4.19)$$

Multiplying (4.19) by $n_i N_A$ will give E^{el} for all the ions in solution:

$$E^{el} = \Sigma\,n_i\,N_A\,E_i^{el} = -\frac{e^2\,N_A}{3\,D}\,\Sigma\,n_i\,z_i^2\,K \qquad (4.20)$$

As discussed in Chapter II, as the molality of a species i approaches zero, the activity coefficient approaches unity so that, at infinite dilutions, equation (2.22) becomes:

$$\mu_{i,id} = \mu_i^\infty + RT\ln m_i \qquad (4.21)$$

For a more concentrated solution (but still very dilute), the activity coefficient does not equal unity so:

$$\mu_i = \mu_i^\infty + RT\ln a_i = \mu_i^\infty + RT\ln m_i + RT\ln\gamma_i$$

$$\mu_i = \mu_{i,id} + RT\ln\gamma_i$$

$$\mu_i - \mu_{i,id} = RT\ln\gamma_i \qquad (4.22)$$

The assumption is made that the dilute solution, having the same molality m_i as the infinitely dilute solution, deviates from equation (4.21) due to the electrostatic interaction of the ions. The expression $\mu_i - \mu_{i,id}$ is then the electrostatic energy μ^{el} for one more of species i. As defined in Chapter II, μ is the derivative, with respect to mass n_i, of the Gibbs free energy G:

$$\mu_i = \left(\frac{\partial G}{\partial n_i}\right)_{T,P,n_{j\neq i}} \qquad (2.17)$$

so that μ^{el} can be obtained by differentiating equation (4.20) with respect to the number of moles of i:

$$\left(\frac{\partial E^{el}}{\partial n_i}\right) = -\frac{e^2\,N_A\,z_i^2}{3D}K - \frac{e^2\,N_A\,\Sigma\,n_i\,z_i^2}{3D}\,\frac{\partial K}{\partial n_i} \qquad (4.23)$$

Differentiating (4.7a):

$$2K \frac{\partial K}{\partial n_i} = \frac{4\pi e^2 z_i^2 N_A}{DkT} = \frac{K^2 z_i^2}{\Sigma z_i^2 n_i}$$

$$\frac{\partial K}{\partial n_i} = \frac{z_i^2}{2 \Sigma n_i z_i^2} K \tag{4.24}$$

By putting (4.24) into (4.23):

$$\frac{\partial E^{el}}{\partial n_i} = -\frac{e^2 N_A z_i^2}{3D} K - \frac{e^2 N_A \Sigma n_i z_i^2}{3D} \frac{z_i^2}{2 \Sigma n_i z_i^2} K = -\frac{e^2 N_A z_i^2}{2D} K \tag{4.25}$$

Up to this point the calculations have been done in terms of the moles of species i. In order to get the molal activity coefficient, γ, the expression for K^2, (4.7a) must be converted to a molality basis:

since: $n_i = \frac{c_i}{1000}$ — c_i being the molarity in moles per liter

and $c_i = \frac{m_i 1000 d}{1000 + m_i M_i}$ — d - density of solution, g/cm^3

M_i - molecular weight

m_i - molality

for dilute solutions were $m_i M_i << 1000$

$$c_i = m_i d$$

$$n_i = \frac{m_i d_0}{1000}$$ d_0 - solvent density

The expression (4.7a) for K^2 becomes:

$$K^2 = \frac{4\pi e^2 N_A d_0}{1000\, DkT} \Sigma z_i^2 m_i \tag{4.26}$$

This expression can now incorporate Lewis and Randall's (3) expression for ionic strength:

$$I = 1/2 \sum_i m_i z_i^2 \qquad (4.27)$$

$$K^2 = \frac{8 \pi e^2 N_A d_0}{1000 DkT} I \qquad (4.26a)$$

By using this expression for K in (4.25), the molal activity coefficient, as described by equation (4.22), is:

$$\ln \gamma_i = \frac{1}{RT}(\mu_i - \mu_{i,id}) = \frac{1}{RT}\left(\frac{\partial E^{el}}{\partial n_i}\right)$$

$$= -\frac{1}{RT} \frac{e^2 N_A z_i^2}{2D} \sqrt{\frac{8 \pi e^2 N_A d_0}{1000 DkT}} \sqrt{I}$$

$$\log \gamma_i = -\frac{1}{2.303}\left(\frac{e}{\sqrt{DkT}}\right)^3 \sqrt{\frac{2 \pi d_0 N_A}{1000}} z_i^2 \sqrt{I} \qquad (4.28)$$

In accordance with the definition of the mean activity coefficient given in expression (2.26):

$$\log \gamma_\pm = \frac{\nu_c \log \gamma_c + \nu_a \log \gamma_a}{\nu_c + \nu_a} \qquad (4.29)$$

When inserted into (4.28) this yields:

$$\log \gamma_\pm = -\frac{1}{2.303}\left(\frac{e}{\sqrt{DkT}}\right)^3 \sqrt{\frac{2 \pi d_0 N_A}{1000}} \frac{\nu_c z_c^2 + \nu_a z_a^2}{\nu_c + \nu_a} \sqrt{I}$$

$$= -\frac{1}{2.303}\left(\frac{e}{\sqrt{DkT}}\right)^3 \sqrt{\frac{2 \pi d_0 N_A}{1000}} |z_+ z_-| \sqrt{I} \qquad (4.30)$$

In order to simplify this expression, the coefficient A is created:

$$A = \frac{1}{2.303}\left(\frac{e}{\sqrt{DkT}}\right)^3 \sqrt{\frac{2 \pi d_0 N_A}{1000}} \qquad (4.31)$$

This coefficient will appear in all of the more recent activity coefficient calculation methods discussed. It is referred to as the Debye-Hückel constant and is temperature dependent.

By using this coefficient we are left with a simple expression for the mean molal activity coefficient:

$$\log \gamma_{\pm} = - A \left| z_{+} z_{-} \right| \sqrt{I} \tag{4.32}$$

which is called the Debye-Hückel limiting law.

LIMITATIONS AND IMPROVEMENTS TO THE DEBYE-HUCKEL LIMITING LAW

Due to the assumptions and simplifications made in deriving the ionic atmosphere potential equation, the Debye-Hückel limiting law is valid only for very dilute solutions of ionic strength .001 molal or less. Debye and Hückel, recognizing this, added a correction term to the limiting law. In assuming the ions to be point charges they ignored the fact that it is impossible for ions to infinitely approach one another. To compensate for this fact they introduced into the expression for the ionic atmosphere potential (4.15) a factor, *a*, to account for the distance of closest approach:

$$\psi_a = - \frac{ez}{D} K \frac{1}{1 + aK} \tag{4.33}$$

By using this equation for the ionic atmosphere potential in the derivation, the molal activity coefficient equation (4.32) becomes:

$$\log \gamma_{\pm} = - \frac{A \left| z_{+} z_{-} \right| \sqrt{I}}{1 + \beta a \sqrt{I}} \tag{4.34}$$

$$\text{where: } \beta = \sqrt{\frac{8 \pi e^2 N_A d_0}{1000 DkT}} \tag{4.35}$$

Calculations done with this equation can be satisfactory up to an ionic strength of .1 molal. It also works best for 1-1 electrolytes, decreasing in applicability for 1-2, 2-2 etc. electrolytes. The distance *a* is also assumed to be the same for all ions in the system, although this may not be the case. Problems also arise as this parameter is not a measurable quantity, but by choosing values close to the hydrated radius, often in the range of 3.5 to 6.2 Å (suggested by Robinson and Stokes (4)), reasonable results may be obtained.

The numerator of equation (4.34) deals with the long range interaction effects and the denominator makes a correction for short range effects. Nonetheless, the short range effects are not adequately compensated for due to the fact that possible ion-solvent molecule and possible ion-ion interactions are ignored. Hückel attempted to improve equation (4.34) by adding a term meant to account for the reduction of the dielectric constant D by increased concentration:

$$\log \gamma_{\pm} = - \frac{A\left|z_+ z_-\right| \sqrt{I}}{1 + \beta\, a \sqrt{I}} + CI \tag{4.36}$$

The general usage of this equation has usually had an empirically chosen C.

A comparison of the calculated activity coefficients using the limiting law (4.32), the extended Debye-Hückel (4.34) and the Hückel equation (4.36) for sodium chloride at 25° Celsius can be found in figures 4.1 and 4.2. Also plotted are experimental values of the molal activity coefficient as published by Robinson and Stokes (5). The value used for a is 4.0 Å and, as suggested by Robinson and Stokes, C = .055 $\mathrm{l.mole^{-1}}$.

FURTHER REFINEMENTS

The a factor makes usage of equations (4.34) and (4.36) difficult inasmuch as the 'distance of closest approach' is not truly measurable. It also assumes, incorrectly, that all of the ions in the system have equal radii. By setting a standard value for a = 3.04 Å so that the quantity βa in the denominator becomes unity, equation (4.34) can be simplified, as suggested by Güntelberg (6):

$$\log \gamma_{\pm} = - \frac{A\left|z_+ z_-\right| \sqrt{I}}{1 + \sqrt{I}} \tag{4.37}$$

This version of the Debye-Hückel equation holds quite well up to an ionic strength of .1 molal.

In 1935, Guggenheim (7) proposed another version of this equation based on the mole fractions scale (rational coefficients) and published values for the interaction coefficient β. This interaction parameter β is unrelated to the β of equation (4.36). With the advent of improved data, Guggenheim published revised

Figure 4.1

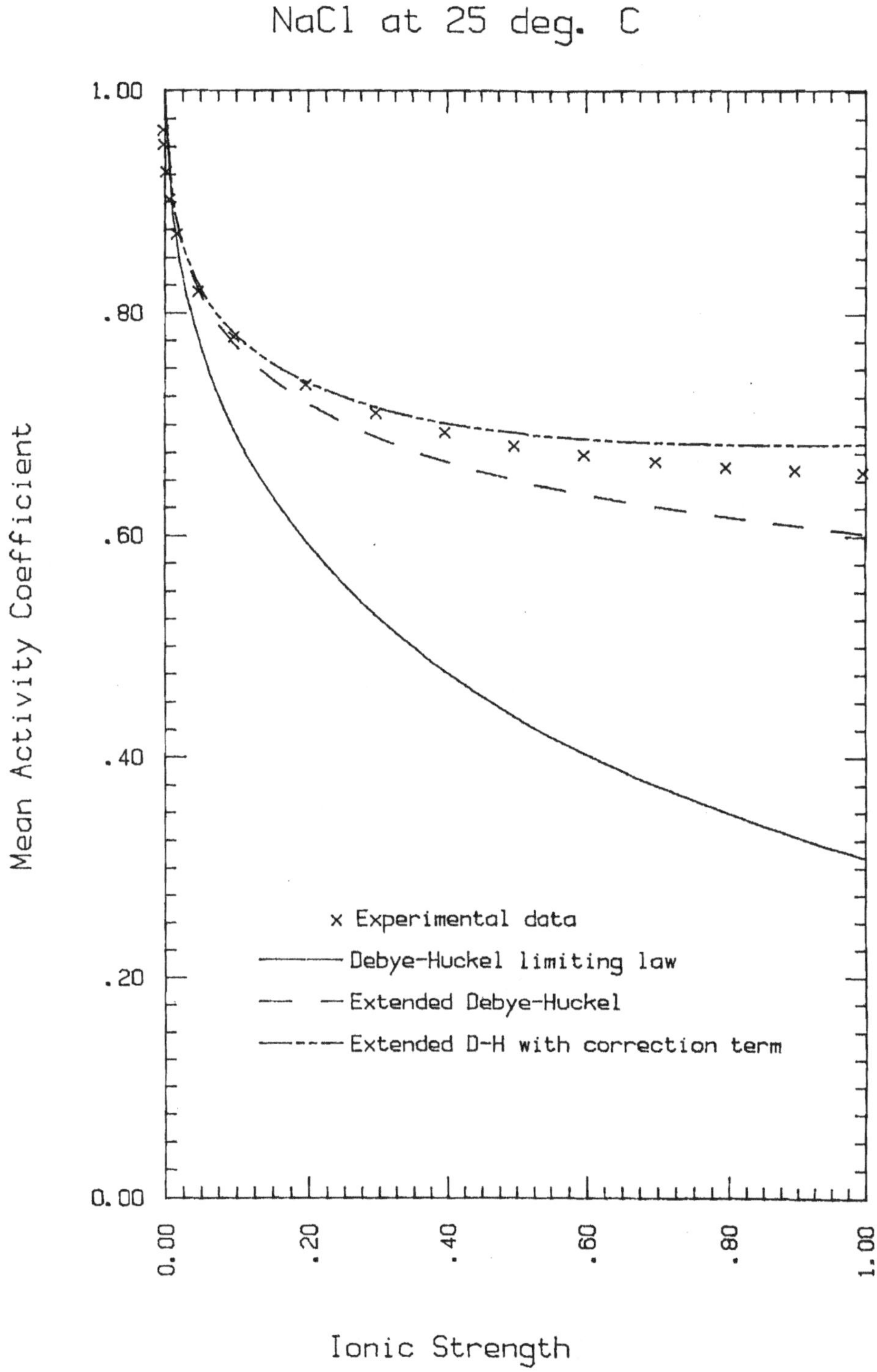
NaCl at 25 deg. C
1.00
.80
.60
.40
.20
0.00
Mean Activity Coefficient
0.00
.20
.40
.60
.80
1.00
Ionic Strength
× Experimental data
Debye-Huckel limiting law
Extended Debye-Huckel
Extended D-H with correction term

Figure 4.2

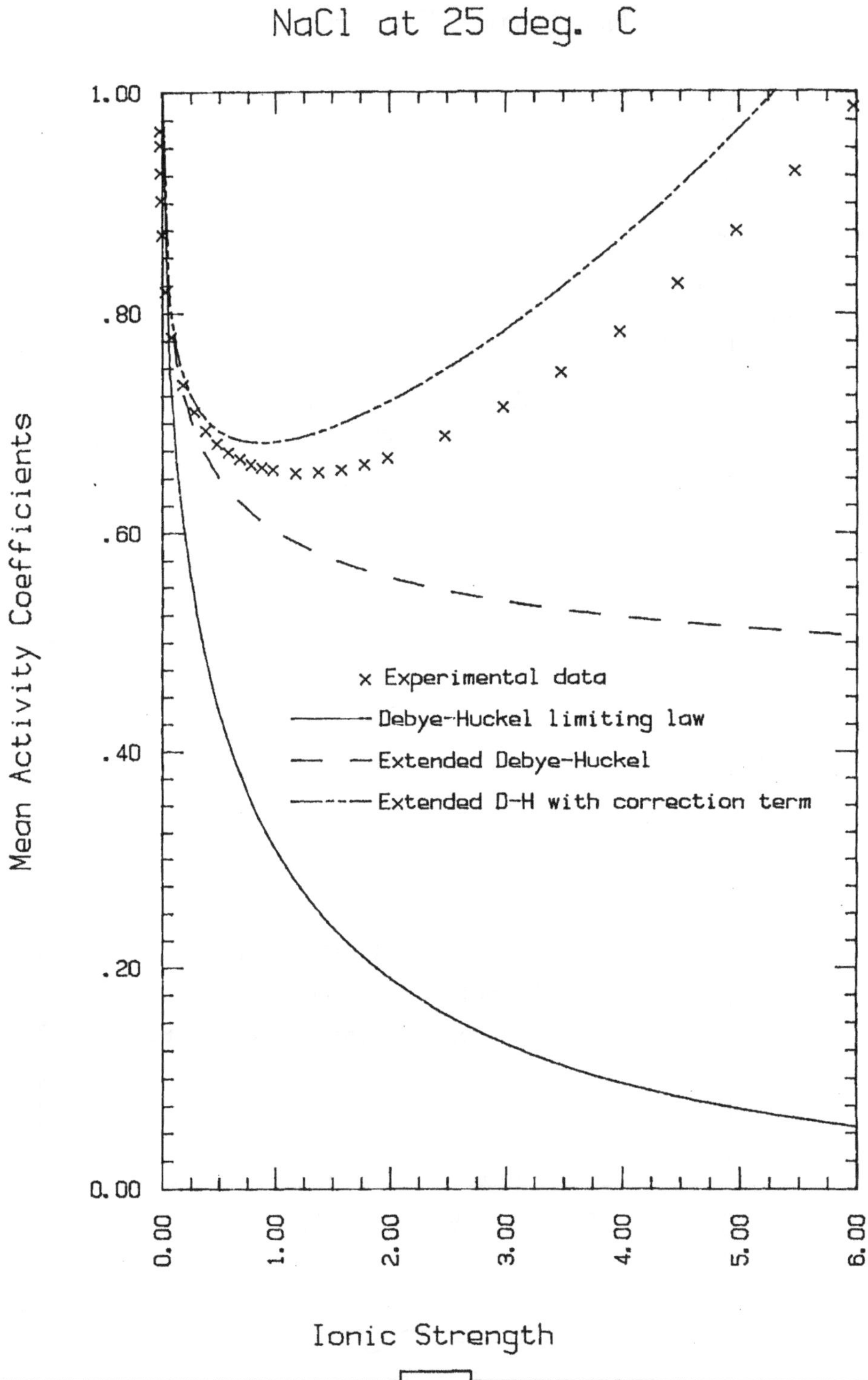
NaCl at 25 deg. C
Mean Activity Coefficients
1.00
.80
.60
.40
.20
0.00
0.00
1.00
2.00
3.00
4.00
5.00
6.00
Ionic Strength
× Experimental data
Debye-Huckel limiting law
Extended Debye-Huckel
Extended D-H with correction term

β's (8), based upon the more widely used ionic strength scale. The mean activity coefficient for an electrolyte with cation c and anion a is:

$$\ln \gamma_{c,a} = -\frac{\alpha |z_+ z_-| \sqrt{I}}{1 + \sqrt{I}} + \frac{2\nu_+}{\nu_+ + \nu_-} \sum_{a'} \beta_{c,a'} m_{a'} + \frac{2\nu_-}{\nu_+ + \nu_-} \sum_{c'} \beta_{c',a} m_{c'}$$

For a single electrolyte solution this reduced to:

$$\ln \gamma_\pm = -\frac{\alpha |z_+ z_-| \sqrt{I}}{1 + \sqrt{I}} + 2 \underline{\nu} \beta m \qquad (4.38)$$

where: α - log base e Debye-Hückel constant

ν_+ - number of cations per molecule of electrolyte

ν_- - number of anions per molecule of electrolyte

$\underline{\nu}$ - harmonic mean of ν_+ and ν_- $= \dfrac{2\nu_+\nu_-}{\nu_+ + \nu_-}$

I - ionic strength in molality scale

z_+ - number of charges on the cation

z_- - number of charges on the anion

β - interaction coefficient

m - electrolyte molality

Equation (4.38), converted to log base 10, is:

$$\log \gamma_\pm = \frac{A |z_+ z_-| \sqrt{I}}{1 + \sqrt{I}} + B m \qquad (4.39)$$

where: A - Debye-Hückel constant, log 10 basis, equation (4.31)

B - interaction parameter $= \dfrac{2\underline{\nu}\beta}{\ln 10}$

In 1938, C.W. Davies (9) suggested a value of $B = .1|z_1 z_2|$ for equation (4.39). He later (10) amended this to $B = .15|z_1 z_2|$. This method is relatively successful for solutions up to .1 M ionic strength but, since it ignores possible ionic association, is useful only for solutions in which the ions are non-associative.

In contrast, the β's published by Guggenheim and Turgeon (8) are meant to provide for ionic associations. They state that, for 1-1, 1-2 and 2-1 electrolytes, calculations to .1 M ionic strength should show good accuracy. In Appendix 4.1, their tables for β are reproduced, along with a brief description of how they were calculated.

Figures 4.3 and 4.4 show the mean activity coefficients of NaCl at 25°C calculated with the equations of Güntelberg, Guggenheim and Davies plotted against Robinson and Stokes (5) data. The results for a 1-2 electrolyte, K_2SO_4, are shown in Figure 4.5. The experimental data is smoothed data from Goldberg (13).

Figure 4.3

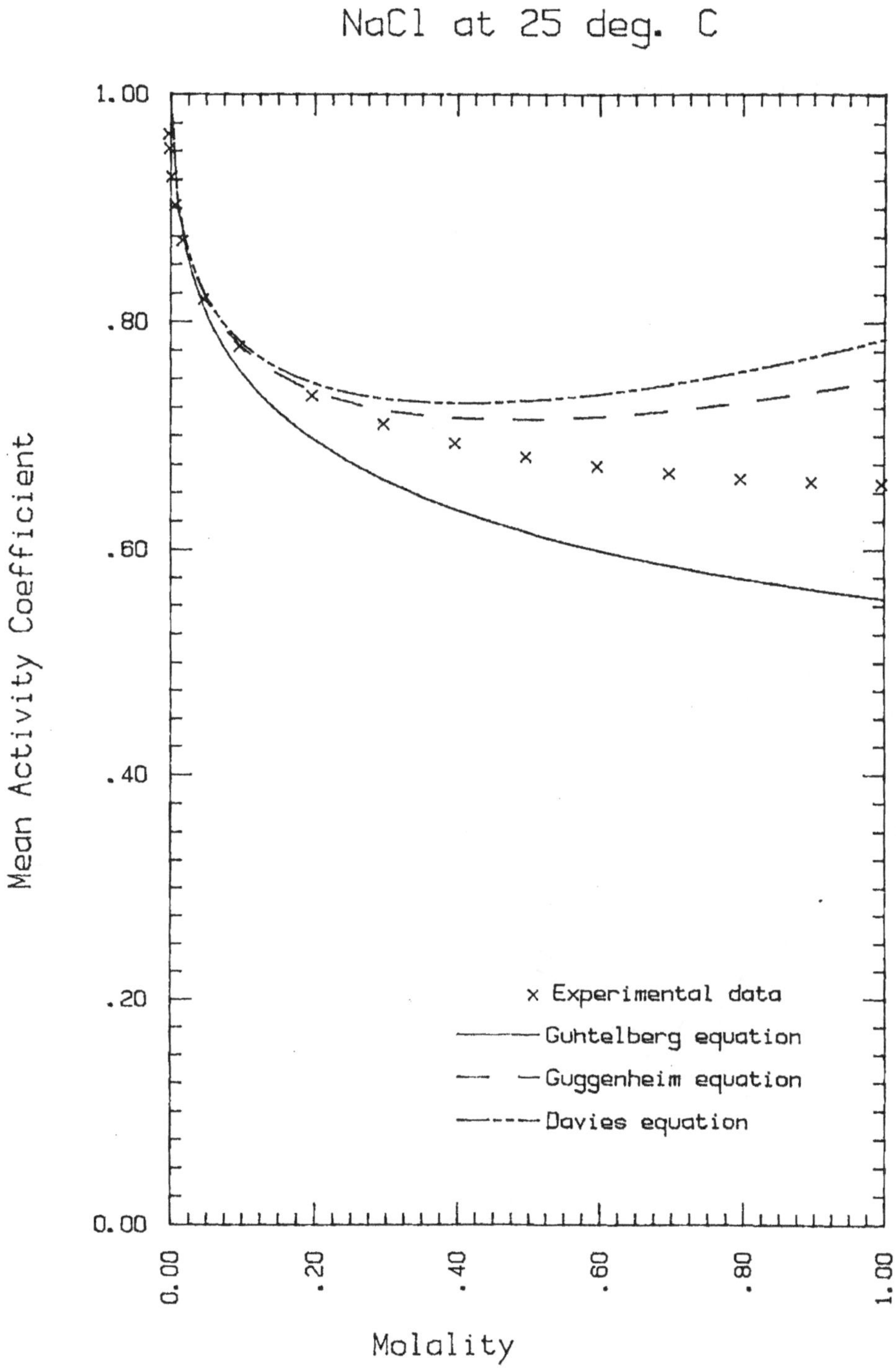
NaCl at 25 deg. C
Mean Activity Coefficient
1.00
.80
.60
.40
.20
0.00
0.00
.20
.40
.60
.80
1.00
Molality
× Experimental data
Guhtelberg equation
Guggenheim equation
Davies equation

Figure 4.4

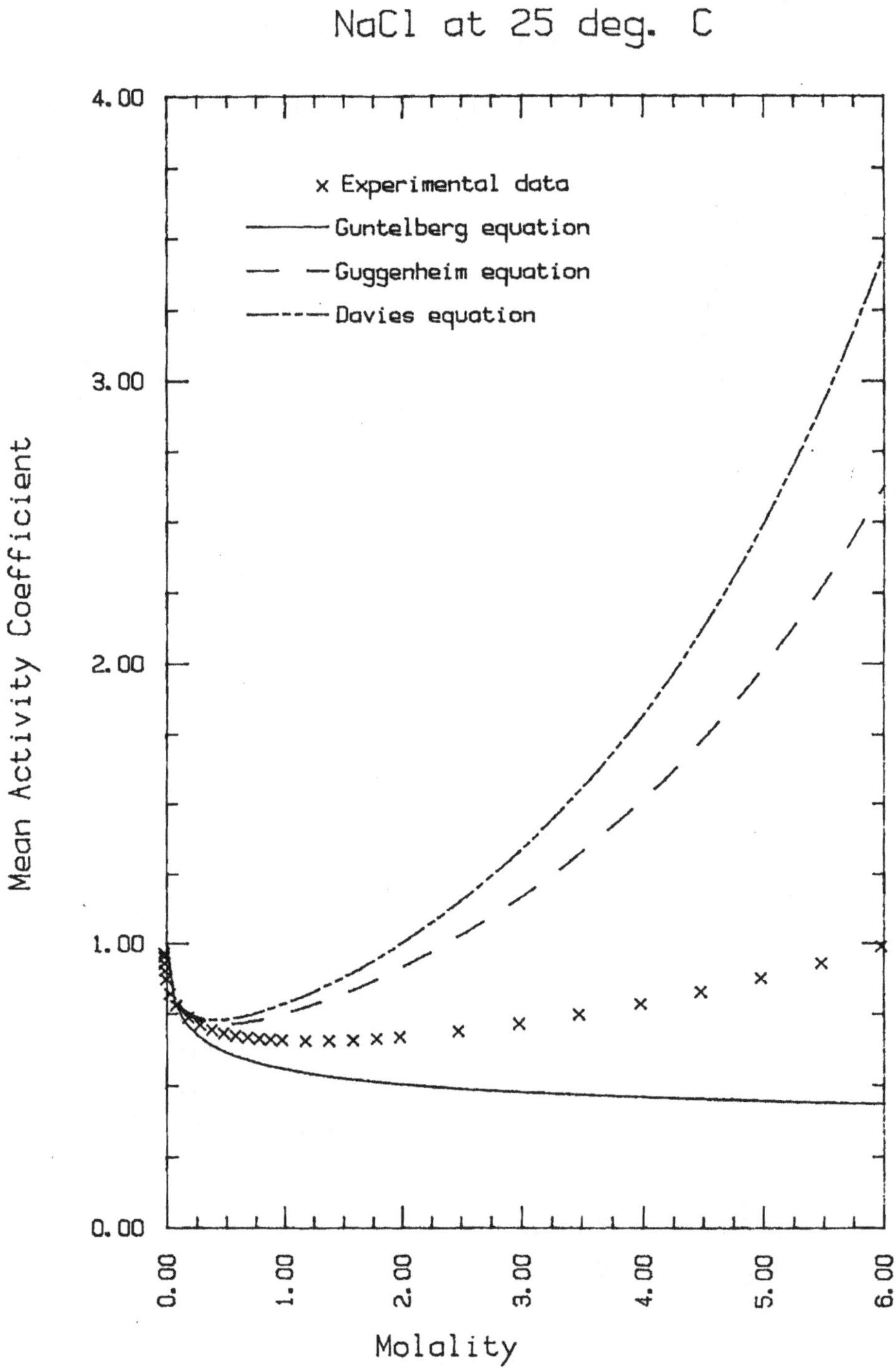
NaCl at 25 deg. C
Mean Activity Coefficient
Molality
4.00
3.00
2.00
1.00
0.00
0.00
1.00
2.00
3.00
4.00
5.00
6.00
× Experimental data
Guntelberg equation
Guggenheim equation
Davies equation

Figure 4.5

K2SO4 at 25 deg. C

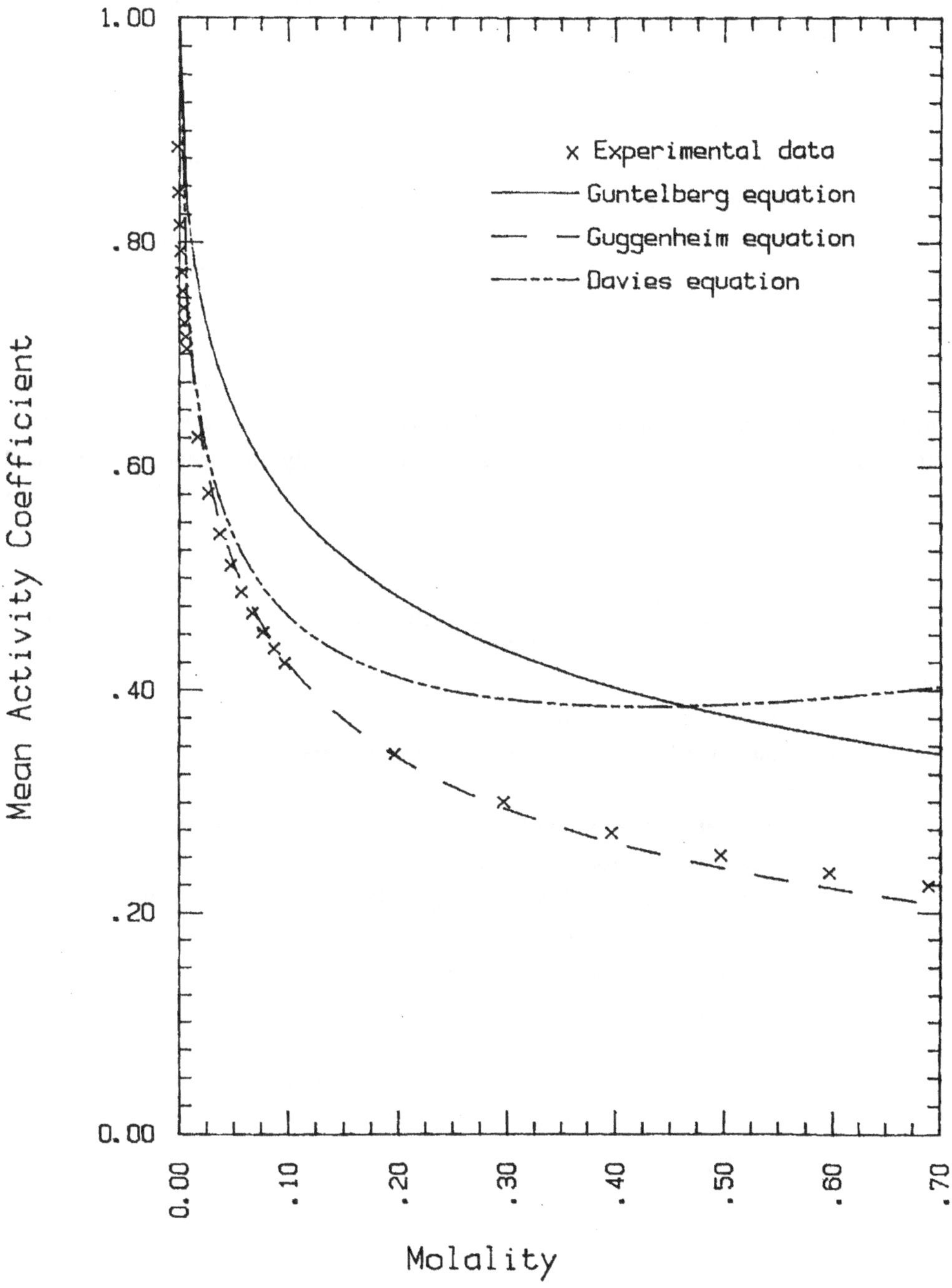

BROMLEY'S METHOD

In 1972, L.A. Bromley had published a paper in which he demonstrated that the β or B interaction parameter of Guggenheim's extended Debye-Hückel equation may be approximated by summing β values for the individual ions for uni-univalent solutions:

$$\beta_{\pm} \approx \beta_{+} + \beta_{-} \qquad (4.40)$$

While the average error he found was .05 kg/mole, the β's he found with equation (4.40), for a few species, greatly differed with reported values. He concluded that, while the ion-solvent interaction is of greatest importance, specific ion-ion interactions need to be taken into account as, particularly at high concentrations, they can be important (B1).

Bromley was led to believe that the interaction parameters are linear in ionic strength, particularly at low molalities, after viewing Figure 22-8 in the Pitzer and Brewer revision of Thermodynamics (B2). Consequently, he presented a method for calculating activity coefficients that takes this dependence into account (B3). He found that the best correlation to experimental data for strong electrolytes was:

$$\log \gamma_{\pm} = -\frac{A|z_{+}z_{-}|\sqrt{I}}{1 + \rho\sqrt{I}} + \frac{(B_0 - B)I}{(1 + aI^2)} + BI + CI^2 \qquad (4.41)$$

where:
- γ - activity coefficient
- A - Debye-Hückel constant, equation (4.31)
- z - number of charges on the cation or anion
- I - ionic strength
- ρ - constant in Debye-Hückel's equation (4.34) related to the distance of closest approach
- B_0 - the value of the ionic strength dependent interaction parameter at ionic strength = 0
- B - constant for ion interaction
- a - constant
- C - constant

The constant ρ was set to 1.0 as suggested by Guggenheim and others. The constant a is dependent upon the valence number:

$$a = 1.5/|z_{+}z_{-}| \qquad (4.42)$$

As the constant C seemed to bear no relation to the B interactions and Bromley found that the values of C formed a close to normal probability distribution around 0.0, the C is set equal to zero.

Bromley determined that, at 25°C, the best relationship between B and B_0 is:

$$\frac{B_0 - B}{|z_+z_-|} = 0.06 + 0.6\,B \qquad (4.43)$$

Using equations (4.42) and (4.43) and the determined values of the constants, equation (4.41) becomes:

$$\log \gamma_{\pm} = \frac{A|z_+z_-|\sqrt{I}}{1+\sqrt{I}} + \frac{(0.06 + 0.6B)|z_+z_-|\,I}{\left(1 + \frac{1.5}{|z_+z_-|}\,I\right)^2} + BI \qquad (4.44)$$

Bromley notes that equation (4.44) gives good results for strong electrolytes to ionic strengths of 6 molal, but due to the exponential quality of the expression, attempts to extrapolate to higher ionic strengths will show increasing error in the activity coefficient. Values for B can be found in Appendix 4.2.

Based on his work in 1972, Bromley introduced an improved version of equation (4.40) showing the additive quality of individual ion B's:

$$B = B_+ + B_- + \delta_+\,\delta_- \qquad (4.45)$$

The B and δ terms Bromley presented are also tabulated in Appendix 4.2. He cautions, however, that these values are meant to be used only for strong electrolytes which show complete dissociation since no allowance is made for the strong ion association seen in compounds such as sulfuric acid, bivalent metal sulfates, and zinc and cadmium halides. He also warns that, because of incomplete dissociation, equation (4.45) does not do as well with thalium and ammonium compounds, and displays some erratic behavior with nitrates.

In 1973, Bromley (B3) suggested a change to his equation (4.44) to compensate for some of these strong ion associations:

for 1-1 and 2-2 salts:

$$\log \gamma_{\pm} = -\frac{A|z_+z_-|\sqrt{I}}{1+\sqrt{I}} + \frac{(0.06+0.6B)|z_+z_-|I}{\left(1+\frac{1.5}{|z_+z_-|}I\right)^2} + BI - E\alpha\sqrt{I}\{1-\exp(-\alpha\sqrt{I})\} \qquad (4.44a)$$

for 1-2, 2-1 and other unsymmetrical salts:

$$\log \gamma_{\pm} = -\frac{A|z_+z_-|\sqrt{I}}{1+\sqrt{I}} + \frac{(0.06+0.6B)|z_+z_-|I}{\left(1+\frac{1.5}{|z_+z_-|}I\right)^2} + BI - E\ln(1+\alpha^2 I) \qquad (4.44b)$$

where: α - an arbitrary constant; approximately equal to one for 1-1 salts, set to 70 for 2-2 salts.

E - a new constant

Bromley determined values for the E constant and resulting new B parameters for some bivalent metal sulfates. These are tabulated in Appendix 4.2.

MEISSNER'S METHOD

Meissner et al. (M1) showed that, in plotting the reduced activity coefficient, Γ, versus the ionic strength, a family of curves forms. They define the reduced activity coefficient as:

$$\Gamma \equiv \gamma_{\pm}^{1/z_+z_-} \qquad (4.45)$$

where:

Γ - reduced activity coefficient

$\gamma_{\pm}$ - mean ionic activity coefficient

z_+ - absolute number of charges on the cation

z_- - absolute number of charges on the anion

They proposed that, having one value of $\gamma_{\pm}$ above the Debye-Hückel concentration range and using equation (4.45) to plot the Γ on the family of curves, the mean activity coefficient for any concentration, from low to saturated, could be graphically predicted. A method of dealing with temperature effects was also proposed in 1972 (M3).

After expansion and refinement of the original method (M4,M6,M8), including an expansion for handling multicomponent systems which will be discussed in the next chapter (M2,M5,M9), a useful method for predicting the activity coefficients of strong electrolytes over a range of temperatures and ionic strengths was presented (M7,M10).

The generalized isothermal curves of Meissner's plot, Figure 4.6, are said to represent the activity coefficient of a particular pure electrolyte in aqueous solution. In 1978, Meissner and Kusik (M7) presented a set of equations, based on a parameter q, for computer application of the method:

$$\Gamma^{o} = \{1. + B(1. + 0.1\ I)^{q} - B\}\ \Gamma^{*} \qquad (4.46)$$

with:

$$B = 0.75 - 0.065q \qquad (4.46a)$$

$$\log\ \Gamma^{*} = \frac{-.5107\ \sqrt{I}}{1 + C\ \sqrt{I}} \qquad (4.46b)$$

$$C = 1. + 0.055q\ \exp(-0.023I^{3}) \qquad (4.46c)$$

where:

$\Gamma°$ - reduced activity coefficient of pure solution at 25°C

I - ionic strength = $1/2 \sum_i m_i z_i^2$ or $1/2\ m_{\pm} \sum_i \nu_i z_i^2$

These equations describe the curves of Figure 4.6 for the q's as shown on the plot.

While some suggested values for q are given in Appendix 4.3, the q value for a species can be found if one experimental value for the activity coefficient at 25°C is available, preferably at an ionic strength greater than one. Given a value for the mean ionic activity coefficient $\gamma_{\pm}$, $\Gamma°$ is found using equation (4.45), and equations (4.46) are solved for q.

If no experimental value for the activity coefficient is known, Meissner suggests that a prediction can be made using Bromley's method, or if vapor pressure data is available, the $\Gamma°$ may be estimated due to the following:
The water activity, which is defined as:

$$a_w^o = \frac{\text{vapor pressure of } H_2O \text{ over the solution}}{\text{vapor pressure of pure } H_2O} \tag{4.47}$$

was expressed by Meissner (M10) as a variation of the Gibbs-Duhem equation (2.32):

$$-55.51 \log a_w^o = \frac{2I}{|z_+ z_-|} + 2 \int_0^{\Gamma°} I \, d(\ln \Gamma°) \tag{4.48}$$

The dotted lines on Figure 4.6 represent equation (4.48) solved for $a_w°$ for 1-1 electrolytes. Meissner claims that the water activities for higher electrolytes can be estimated using these lines. If a value for $\Gamma°$ at a given I is known, a value for $(a_w°)_{1-1}$ can be graphically obtained from Figure 4.6. The higher electrolyte water activity, $(a_w°)_{z_+z_-}$, is related as follows (M4):

$$\log (a_w^o)_{z_+z_-} = 0.0156\ I \left(1 - \frac{1}{|z_+ z_-|}\right) + \log (a_w^o)_{1-1} \tag{4.49}$$

Therefore, having the solution vapor pressure at ionic strength I, the $\Gamma°$ and q can be graphically determined using $(a_w°)_{1-1}$. The water activity can also be obtained from osmotic coefficient data:

$$\log (a_w^o)_{z_+z_-} = - \frac{\phi \sum_i \nu_i m_i}{55.51 \times 2.303} = - \frac{\phi\ I}{55.51 \times 2.303\ |z_+ z_-|} \tag{4.50}$$

When first setting up the family of curves, Meissner and Tester (M1) noted that not all the species they attempted to plot fit into the developing pattern. The following species' reduced activity coefficient plots crossed over other lines on the plot: sulfuric acid, sodium and potassium hydroxide, thorium nitrate, lithium chloride, magnesium iodide, uranyl fluoride, lithium nitrate, beryllium sulfate, and most of the cadmium and zinc salts. In 1973, Meissner and Peppas (M6) determined that this method could be applied to strong polybasic acids, such as H_3AsO_4 and H_3PO_4, if they were properly classified with respect to their ion charges. For example, a solution of H_3PO_4 contains mainly H^+ and $H_2PO_4^-$ ions. Very few HPO_4^{2-} and PO_4^{3-} ions are present. H_3PO_4 is therefore treated as a 1-1 electrolyte. Sulfuric acid, having large first and second dissociation constants, can be treated as a 1-2 electrolyte. Meissner and Peppas concluded that:

a) strong polybasic acids are best treated as 1-1 electrolytes when their second and third dissociation constants are small
b) dibasic acids with large first and second dissociation constants are best treated as 1-2 electrolytes.

They found similar behavior for the acid salts of most strong and weak polybasic acids. The primary alkali salts exhibited 1-1 electrolyte behavior, the secondary salts could be treated as 1-2 electrolytes and the ternary salts as 1-3. The potassium, sodium and ammonium salts of H_3PO_4 and H_3AsO_4 therefore fit into the Meissner plot. They also claimed that monosodium salts of succinic and malonic acids were treatable as 1-1 electrolytes; the disodium salts of fumaric and maleic acid and the bisulfates could be treated as 1-2 electrolytes.

While improvements in results were exhibited when applying the above generalizations, Meissner and Kusik (M7) suggested, in 1978, using the method for zinc and cadmium chlorides, bromides and iodides, sulfuric acid, and thorium nitrate only when experimental data is unavailable. The user should also be aware that estimating Γ° from vapor pressures and extrapolation over a wide ionic strength range can increase the calculation error.

Figure 4.6

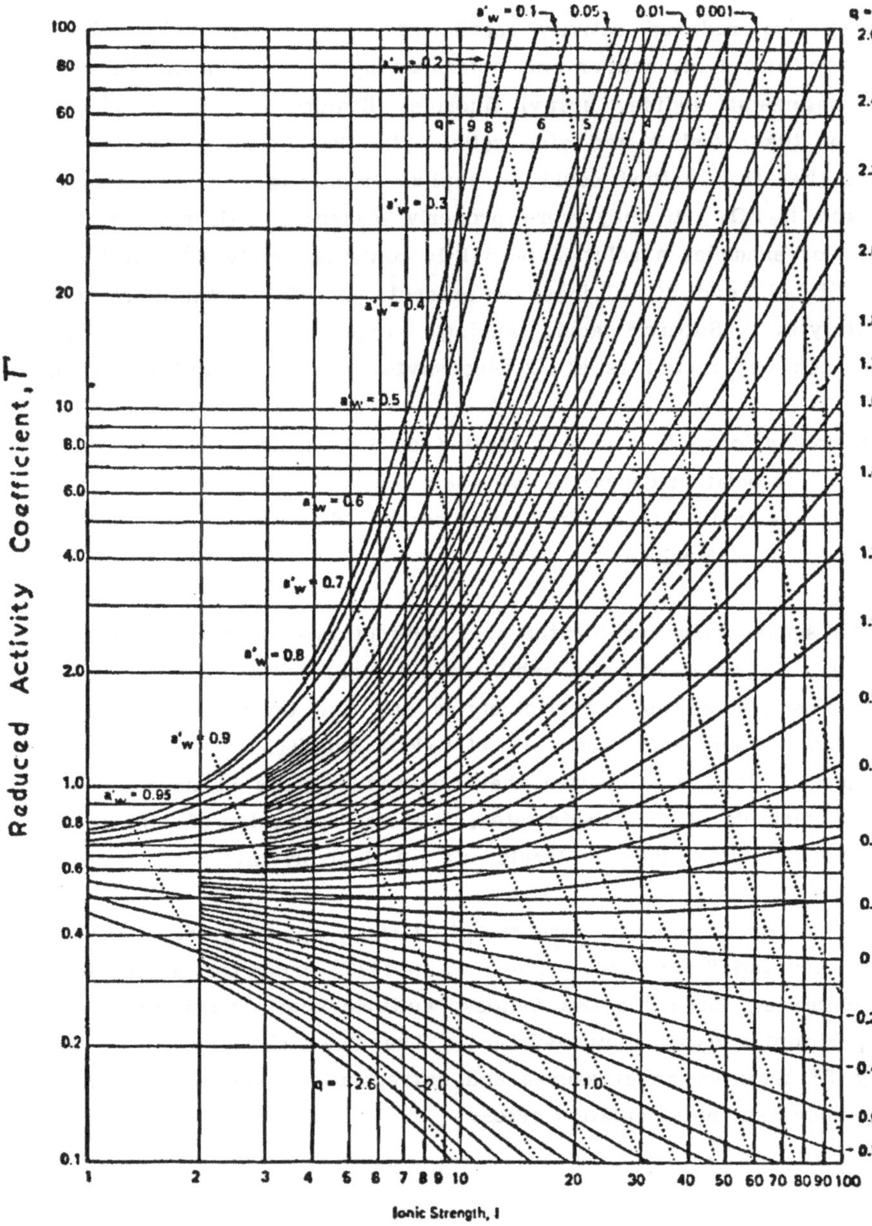

Reproduced by permission of the American Institute of Chemical Engineers "Electrolyte Activity Coefficients in Inorganic Processing," Kusik, C.L. and H.P. Meissner, AIChE Symp. Ser. 74(173), p. 17 (1978).

PITZER'S METHOD

In 1973, Pitzer el al. began presenting a series of papers (P1 - P12) reporting their development of a system of equations for electrolyte thermodynamic properties. In expanding the Debye-Hückel method, terms were added to account for the ionic strength dependence of the short-range forces effect in binary interactions.

While the Debye-Hückel derivation was based on a charging process which took into account the distance of closest approach, a, the kinetic effect of the hard core could not adequately be dealt with. In order to include these effects, Pitzer proposed using the osmotic pressure equation (P1):

$$\Pi - ckT = -\frac{1}{6} \sum_i \sum_j c_i c_j \int_0^\infty \frac{\partial \psi_{ij}}{\partial r} g_{ij}\, 4\pi r^3 dr \qquad (4.51)$$

where:

- c_i, c_j - concentrations of species i and j
- c - concentration of all solutes
- Π - osmotic pressure
- ψ_{ij} - intermolecular potential
- g_{ij} - radial distribution function
- r - distance

The potential energy between ions i and j is:

$$\psi_{ij} = \infty, \; r < a$$

$$\psi_{ij} = \frac{z_i z_j e^2}{Dr}, \; r \geq a \qquad (4.52)$$

where:

- z - charge number
- e - electronic charge
- D - dielectric constant
- a - distance of closest approach, or core size

Pitzer found equations (4.51) and (4.52) to yield:

$$\Pi - ckT = \frac{e^2}{6D} \sum_i \sum_j c_i c_j z_i z_j \int_a^\infty g_{ij}(r)\, 4\pi r\, dr + \frac{2}{3}\pi a^3 T \sum_i \sum_j c_i c_j g_{ij}(a) \qquad (4.53)$$

with the second term describing the kinetic effect of the hard core. The term $g(a)$ is the radial distribution function at contact. Using the Debye-Hückel K defined earlier:

$$K^2 = \frac{4\pi e^2}{DKT} \sum_i z_i^2 c_i \tag{4.7a}$$

and finding q_{ij} as the exponent in the Boltzmann function for the radial distribution function, $g_{ij}(r) = \exp\{-q_{ij}(r)\}$:

$$q_{ij}(r) = \frac{z_i z_j e^2}{DKT(1+Ka)} \frac{\exp(-K(r-a))}{r} \tag{4.54}$$

the expression for charge density ρ_i (equation (4.3) becomes:

$$\rho_i = \sum_i c_i z_i e \exp\{-q_{ij}(r)\} \tag{4.55}$$

The charge distribution described by Debye-Hückel is found by expanding the exponential term as before.

Pitzer then used the same kind of expansion for the $g_{ij}(a)$ term for the hard core effect of equation (4.53). Here the second term of the expansion dropped out, but the third term was kept. He then found the osmotic coefficient to be:

$$\phi - 1 = \frac{\Pi}{cKT} - 1 = \frac{-K^3}{24\pi c(1+Ka)} + c\left[\frac{2\pi a^3}{3} + \frac{1}{48\pi}\frac{K^4 a}{c^2(1+Ka)^2}\right] \tag{4.56}$$

The first term of equation (4.56) describes the electrostatic energy, and the second the hard core effects. This hard core effect term shows dependence on the ionic strength of the solution due to the K. This indicates the need for a series of virial coefficients that are functions of the ionic strength, and justifies Bromley and Guggenheim's additions to the Debye-Hückel equation.

As described in Chapter II and earlier in this chapter, many thermodynamic properties may be derived from the Gibb's excess free energy equation. For this reason, Pitzer began with it defined as:

$$\frac{G^{ex}}{RT} = n_w f(I) + \frac{1}{n_w} \sum_i \sum_j \lambda_{ij}(I)\, n_i n_j + \frac{1}{n_w^2} \sum_i \sum_j \sum_k \Lambda_{ijk}\, n_i n_j n_k \tag{4.57}$$

where:

n_w - kilograms of solvent

n - moles of solutes i, j and k

f(I) - function describing the long-range electrostatic effects as a function of temperature

$\lambda_{ij}(I)$ - term for describing the short-range inter-ionic effects as a function of ionic strength to display the type of behavior caused by the hard core effects

Λ_{ijk} - term for triple ion interactions which ignores any ionic strength dependence

The matrices of λ and Λ are assumed to be symmetrical.

Taking the derivative to get the mean ionic activity coefficient and converting to a molality basis:

$$\ln \gamma_i = \frac{1}{RT}\frac{\partial G^{ex}}{\partial n_i} = \frac{z_i^2}{2} f' + 2\sum_j \lambda_{ij} m_j + \frac{z_i^2}{2} \sum\sum_{jk} \lambda'_{jk} m_j m_k + 3\sum\sum_{jk} \Lambda_{ijk} m_j m_k \tag{4.58}$$

with $f' = \frac{df}{dI}$ and $\lambda'_{ij} = \frac{d\lambda_{ij}}{dI}$. For the electrolyte CA:

$$\ln \gamma_{CA} = \frac{|z_+ z_-|}{2} f' + \frac{2\nu_+}{\nu_+ + \nu_-}\sum_j \lambda_{Cj} m_j + \frac{2\nu_-}{\nu_+ + \nu_-}\sum_j \lambda_{Aj} m_j +$$

$$\frac{|z_+ z_-|}{2}\sum\sum_{jk} \lambda'_{jk} m_j m_k + \frac{3\nu_+}{\nu_+ + \nu_-}\sum\sum_{jk} \Lambda_{Cjk} m_j m_k + \frac{3\nu_-}{\nu_+ + \nu_-}\sum\sum_{jk} \Lambda_{Ajk} m_j m_k \tag{4.59}$$

For a single electrolyte this is reduced to:

$$\ln \gamma_\pm = \frac{|z_+ z_-|}{2} f' + \frac{m}{\nu}\{2\nu_+\nu_-(2\lambda_{CA} + I\lambda'_{CA}) + \nu_+^2(2\lambda_{CC} + I\lambda'_{CC}) +$$

$$\nu_-^2(2\lambda_{AA} + I\lambda'_{AA})\} + \frac{9\nu_+\nu_- m^2}{\nu}(\nu_+\Lambda_{CCA} + \nu_-\Lambda_{CAA}) \tag{4.60}$$

The term $\nu = \nu_+ + \nu_-$; the triple interaction terms between ions of the same sign were dropped.

The appearance of this equation can be simplified by factoring out the interactions to create:

$$\ln \gamma_\pm = |z_+ z_-| f^\gamma + m\left[\frac{2\nu_+\nu_-}{\nu}\right] B_\pm^\gamma + m^2\left[\frac{2(\nu_+\nu_-)^{1.5}}{\nu}\right] C_\pm^\gamma \tag{4.61}$$

where:

$$f^{\gamma} = 1/2\ f'$$

$$B_{\pm}^{\gamma} = 2\lambda_{CA} + I\lambda'_{CA} + \frac{\nu_+}{2\nu_-}(2\lambda_{CC} + I\lambda'_{CC}) + \frac{\nu_-}{2\nu_+}(2\lambda_{AA} + I\lambda'_{AA})$$

$$C_{\pm}^{\gamma} = \frac{9}{2\sqrt{\nu_+\nu_-}}(\nu_+\Lambda_{CCA} + \nu_-\Lambda_{CAA})$$

It is noted that this equation is of the same form as that presented by Guggenheim (equation (4.38)) with the addition of a third virial coefficient. The second virial coefficient, B, is now dependent upon ionic strength.

The following equations were chosen as having the greatest success in duplicating experimental results:

$$f^{\gamma} = -A_{\phi}\left(\frac{\sqrt{I}}{1 + b\sqrt{I}} + \frac{2}{b}\ln(1 + b\sqrt{I})\right) \quad \text{with } b = 1.2 \qquad (4.62)$$

$$B_{\pm}^{\gamma} = 2\beta_0 + \frac{2\beta_1}{\alpha^2 I}\{1. -(1. + \alpha\sqrt{I} - .5\alpha^2 I)\exp(-\alpha\sqrt{I})\} \qquad (4.63)$$

$$A_{\phi} = \frac{1}{3}\left(\frac{e}{\sqrt{DkT}}\right)^3\sqrt{\frac{2\pi d_0 N_A}{1000}} \qquad (4.64)$$

which is the Debye-Hückel constant for osmotic coefficients on a log e basis

$$C_{\pm}^{\gamma} = \frac{3}{2}C_{\pm}^{\phi} \qquad (4.65)$$

The parameter b, chosen as equal to 1.2 for best fit, is theoretically related to the distance at which strong ion-ion repulsive forces begin and is held constant for all solutes and temperatures as a matter of convenience. The α parameter was also chosen with regard to best fit for the classes of electrolytes. For 1-1, 2-1, 1-2, 3-1, 4-1 and 5-1 electrolytes $\alpha = 2.0$. Pitzer and Mayorga (P3) found that 2-2 electrolytes require the second virial coefficient to be expanded due to the electrostatic ion pairing tendencies of these species. For 2-2 electrolytes, they proposed the following equation for the best results:

$$B_{\pm}^{\gamma} = 2\beta_0 + \frac{2\beta_1}{\alpha_1^2 I}\{1 - (1 + \alpha_1\sqrt{I} - .5\alpha_1^2 I)\exp(-\alpha_1\sqrt{I})\} +$$

$$\frac{2\beta_2}{\alpha_2^2 I}\{1 - (1 + \alpha_2\sqrt{I} - .5\alpha_2^2 I)\exp(-\alpha_2\sqrt{I})\} \qquad (4.66)$$

here the $\alpha_1 = 1.4$; $\alpha_2 = 12$.

Values for the parameters of many species have been published by Pitzer and other authors. Some of these can be found in Appendix 4.4. It should be noted that they include the stoichiometric coefficients of the B and C terms; for 1-1 electrolytes these are unity, for 2-1 electrolytes

$$\frac{2\nu_+ \nu_-}{\nu} = \frac{4}{3} \quad \text{and} \quad \frac{2(\nu_+\nu_-)^{1.5}}{\nu} = \frac{2^{2.5}}{3}$$

etc. Hence, when using these published values, equation (4.61) becomes:

$$\ln \gamma_\pm = |z_+ z_-| f^\gamma + mB^\gamma_\pm + m^2 C^\gamma_\pm \tag{4.67}$$

The user should be aware of the maximum molality the parameters were fit to. In many cases experimental data to 6 M was available, but due to significant error, the range may have been reduced. Considerable error may occur when using the parameters at concentrations greater than this maximum.

CHEN'S METHOD

Although Pitzer's method of calculating activity coefficients has been applied with some success to solutions containing weak electrolytes (OP1, OP14), it's form is best suited to strong electrolyte solutions. This is because, in solutions of weak electrolytes, significant concentrations of molecular solutes are present. While Pitzer considered the ion-ion interactions, ion-molecule and molecule-molecule interactions, which are important in solutions of weak electrolytes, are ignored. In 1979, Chen et al. (C1) presented an extension to Pitzer's method to allow for both molecular and ionic solutes. More recently they proposed the addition of a local composition expression to account for the short range interactions between all species (C2, C3, C4).

The molecule-molecule interactions, such as electrostatic forces between permanent dipoles or the dispersion interaction between molecules, and the ion-molecule interactions, such as ion-dipole electrostatic forces, are only significant at close range and their effects drop rapidly as the separation distance increases. The interionic electrostatic forces retain importance over a much greater distance. These long range forces therefore have the dominant effect in dilute solutions, but as concentrations increase, so does the importance of the short range effects.

Chen et al. proposed that the excess Gibbs free energy and activity coefficients could then be expressed as the sum of the long range and short range contributions:

$$\frac{G^{ex*}}{RT} = \frac{G^{ex*,P}}{RT} + \frac{G^{ex*,lc}}{RT} \tag{4.68}$$

$$\ln f = \ln f^{P} + \ln f^{lc} \tag{4.69}$$

Short Range Interaction Model

Chen based this contribution on the Nonrandom, Two Liquid (NRTL) model proposed by Renon and Prausnitz in 1968 (C5). He felt it is valid since the heat of mixing for electrolyte systems is very large and, in comparison, the nonideal entropy of mixing is negligible. The algebraic simplicity and the fact that no specific area or volume data are needed were further recommendations. He described his derivation in the following manner.

Two assumptions are made to define the local composition:

1) The like-ion repulsion assumption - that is, due to large repulsive forces between ions of the same charge, the area immediately surrounding a cation will not contain other cations and the area immediately surrounding an anion will not contain other anions.

2) The local electroneutrality assumption - which states that, around a central solvent molecule, the arrangement of cations and anions will be such that the net ionic charge is zero.

These assumptions were displayed pictorially for a solution of one solvent with one completely dissociated electrolyte as:

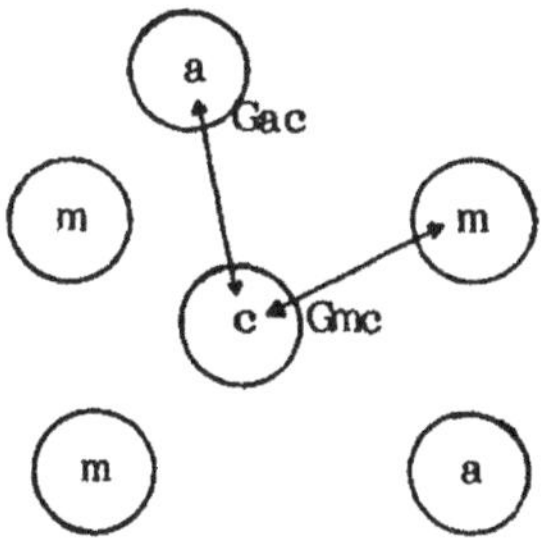

central cation cell

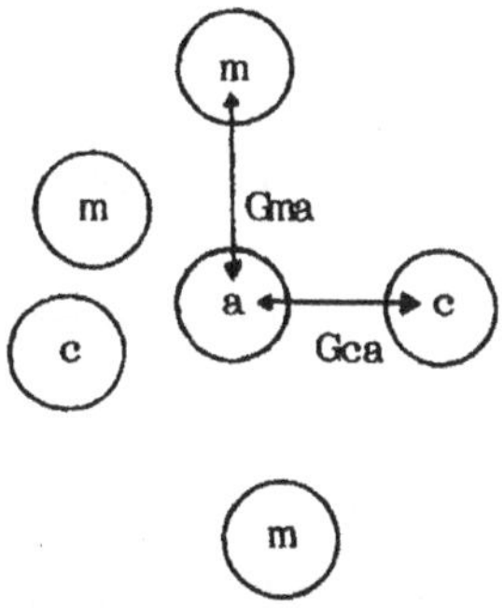

central anion cell

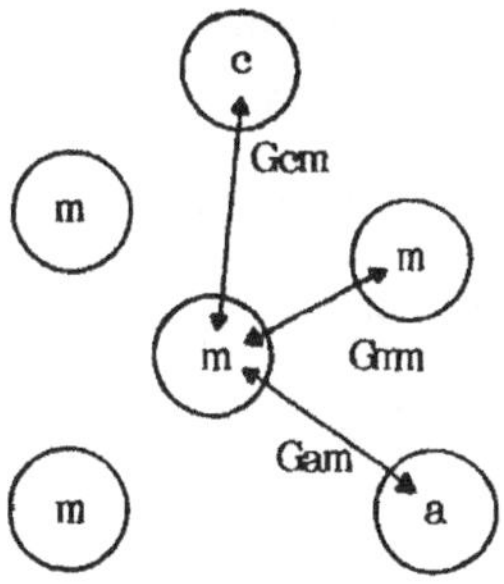

central molecule cell

In order to define the local mole fractions, the following expressions are given:

$$\frac{x_{ji}}{x_{ii}} = \frac{x_j}{x_i} F_{ji} \qquad (4.70a)$$

$$F_{ji} = \exp(-\alpha\, \tau_{ji})$$

$$\tau_{ji} = \frac{G_{ji} - G_{ii}}{RT} \qquad (4.70c)$$

where:

- i - central species
- j - species around central species i
- x_i, x_j - overall mole fractions
- x_{ji}, x_{ii} - local mole fractions
- F_{ji} - NRTL parameter
- α - nonrandomness factor, fixed equal to 0.2
- τ_{ij} - NRTL binary interaction energy parameter
- G_{ji}, G_{ii} - energies of interaction between j-i and i-i. These are symmetric in nature, $G_{ji} = G_{ij}$

The following equations were proposed for representing other local mole fractions:

$$\frac{x_{ji}}{x_{ki}} = \frac{x_j}{x_k} F_{ji,ki} \qquad (4.71a)$$

$$F_{ji,ki} = \exp(-\alpha\, \tau_{ji,ki}) \qquad (4.71b)$$

$$\tau_{ji,ki} = \frac{G_{ji} - G_{ki}}{RT} \qquad (4.71c)$$

Due to the like-ion repulsion assumption, the local mole fractions $x_{cc} = x_{aa} = 0$. The local compositions are thus:

around a central solvent molecule: $x_{cm} + x_{am} + x_{mm} = 1$ (4.72a)

around a central cation: $x_{mc} + x_{ac} = 1$ (4.72b)

around a central anion: $x_{ma} + x_{ca} = 1$ (4.72c)

Using equations (4.70a), (4.71a) and (4.72), Chen derived the following equations to define the local mole fractions in terms of the overall mole fractions:

around central solvent molecule m:

$$x_{im} = \frac{x_i F_{im}}{x_a F_{am} + x_c F_{cm} + x_m F_{mm}} \qquad i = c, a \text{ or } m \tag{4.73a}$$

around central cation c:

$$x_{ac} = \frac{x_a}{x_a + x_m F_{mc,ac}} \tag{4.73b}$$

around central anion a:

$$x_{ca} = \frac{x_c}{x_c + x_m F_{ma,ca}} \tag{4.73c}$$

The aim of this derivation is to find the local composition Gibbs free energy,

$$G^{ex*,lc}/RT,$$

as used in equation (4.68). Therefore, Chen defined three quantities, $G^{(a)}$, $G^{(c)}$ and $G^{(m)}$, for the residual Gibbs energies per mole of central anion, central cation and central solvent molecule cells as the sum of the local energies of interaction between pairs of species. Expressed in terms of the local mole fractions, they are:

$$G^{(a)} = z_-(x_{ma}G_{ma} + x_{ca}G_{ca}) \tag{4.74a}$$

$$G^{(c)} = z_+(x_{mc}G_{mc} + x_{ac}G_{ac}) \tag{4.74b}$$

$$G^{(m)} = x_{am}G_{am} + x_{cm}G_{cm} + x_{mm}G_{mm} \tag{4.74c}$$

Chen included the charge numbers z_+ and z_- in order to relate the number of ions and molecules, the coordination number, found around a central ion. The ratio of the charge numbers is equal to the ratio of the coordination numbers. The coordination number of a central univalent ion is assumed to be five; a multivalent ion's coordination number is then the charge number multiplied by the univalent coordination number. This allows for the use of the same nonrandomness factor, α, for all ions.

Since the short range contribution was developed as a symmetric model, the reference states were pure solvent and completely ionized pure electrolyte, which may be hypothetical. The reference state Gibbs free energies per mole are thus expressed as:

$$G^{(c)}_{ref} = z_+ G_{ac} \tag{4.75a}$$

$$G^{(a)}_{ref} = z_- G_{ca} \tag{4.75b}$$

$$G^{(m)}_{ref} = G_{mm} \tag{4.75c}$$

The molar excess Gibbs free energy is defined as:

$$G^{ex,lc} = x_m(G^{(m)} - G^{(m)}_{ref}) + x_c(G^{(c)} - G^{(c)}_{ref}) + x_a(G^{(a)} - G^{(a)}_{ref}) \tag{4.76}$$

This expresses the change in residual Gibbs energy that results when x_m, x_c and x_a moles of solvent, cation and anion are moved to their respective cells in solution from their reference states. After substituting equations (4.70c), (4.71c), (4.74) and (4.75) into (4.76), the excess Gibbs free energy is expressed

$$\frac{G^{ex,lc}}{RT} = x_m x_{cm} \tau_{cm} + x_m x_{am} \tau_{am} + x_c x_{mc} z_+ \tau_{mc,ac} + x_a x_{ma} z_- \tau_{ma,ca} \tag{4.77}$$

By applying the local electroneutrality assumption to a central solvent molecule cell:

$$x_{am} z_- = x_{cm} z_+ \tag{4.78}$$

Therefore, it can be seen that:

$$F_{am} = F_{cm} \tag{4.79a}$$

$$G_{am} = G_{cm} \tag{4.79b}$$

using equations (4.70) in (4.78). As stated earlier, the interaction energies are symmetrical. A new method of subscripting τ may be used:

$$\tau_{am} = \tau_{cm} = \tau_{ca,m} \quad \text{for central molecule cells} \tag{4.80a}$$

$$\tau_{mc,ac} = \tau_{ma,ca} = \tau_{m,ca} \quad \text{for central ion cells} \tag{4.80b}$$

This leaves the system with only two adjustable parameters for a single electrolyte in solution, $\tau_{ca,m}$ and $\tau_{m,ca}$, expression (4.77) is thus:

$$\frac{G^{ex,lc}}{RT} = x_m (x_{cm} + x_{am}) \tau_{ca,m} + (x_c x_{mc} z_+ + x_a x_{ma} z_-) \tau_{m,ca} \tag{4.81}$$

for symmetric local composition. This must be normalized to the infinite dilution states for the ions to obtain the unsymmetric local composition in order that it later may be combined with the unsymmetric Pitzer model. The unsymmetrical Gibbs expression is:

$$\frac{G^{ex*,lc}}{RT} = \frac{G^{ex,lc}}{RT} - x_c \ln f_c^{\infty} - x_a \ln f_a^{\infty} \tag{4.82}$$

Taking the partial derivatives of equation (4.81) to get $\ln f_c^{\infty}$ and $\ln f_a^{\infty}$, the mole fraction activities, equation (4.82) is:

$$\frac{G^{ex*,lc}}{RT} = x_m(x_{cm} + x_{am})\tau_{ca,m} + (x_c x_{mc} z_+ + x_a x_{ma} z_-)\tau_{m,ca}$$
$$- x_c(z_+\tau_{m,ca} + F_{cm}\tau_{ca,m}) - x_a(z_-\tau_{m,ca} + F_{am}\tau_{ca,m}) \tag{4.83}$$

From this equation the mole fraction based activity coefficients of the cation, anion and solute are derived:

$$\ln f_c^{lc} = \frac{x_m^2\,\tau_{ca,m}\,F_{cm}}{(x_c F_{cm} + x_a F_{am} + x_m)^2} - \frac{z_-\,x_a\,\tau_{m,ca}\,x_m\,F_{ma}}{(x_c + x_m F_{ma})^2}$$
$$+ \frac{z_+\,x_m\,\tau_{m,ca}\,F_{mc}}{x_a + x_m F_{mc}} - z_+\,\tau_{m,ca} - F_{cm}\,\tau_{ca,m} \tag{4.84}$$

$$\ln f_a^{lc} = \frac{x_m^2\,\tau_{ca,m}\,F_{am}}{(x_c F_{cm} + x_a F_{am} + x_m)^2} - \frac{z_+\,x_c\,\tau_{m,ca}\,x_m\,F_{mc}}{(x_a + x_m F_{mc})^2}$$
$$+ \frac{z_-\,x_m\,\tau_{m,ca}\,F_{ma}}{x_c + x_m F_{ma}} - z_-\,\tau_{m,ca} - F_{am}\,\tau_{ca,m} \tag{4.85}$$

$$\ln f_m^{lc} = x_{cm}\tau_{ca,m} + x_{am}\,\tau_{ca,m} + \frac{z_+\,x_c\,F_{mc}\,\tau_{m,ca}\,x_a}{(x_a + F_{mc}x_m)^2} + \frac{z_-\,x_a\,F_{am}\,\tau_{m,ca}\,x_c}{(x_c + F_{ma}x_m)^2} -$$
$$\frac{x_c\,x_m\,F_{cm}\,\tau_{ca,m}}{(x_c F_{cm} + x_a F_{am} + x_m)^2} - \frac{x_a\,x_m\,F_{am}\,\tau_{ca,m}}{(x_c F_{cm} + x_a F_{am} + x_m)^2} \tag{4.86}$$

Long Range Interaction Model

In his thesis, Chen suggested using the Debye-Hückel model:

$$\ln \gamma_i^{dh} = -\frac{\alpha z_i^2 \sqrt{I}}{(1 + \beta a\sqrt{I})}$$

$$\ln \gamma_{ca}^{dh} = -\frac{\alpha |z_+ z_-| \sqrt{I}}{(1 + \beta a\sqrt{I})}$$

where:

$$\alpha^2 = \frac{2\pi N_A d_0}{1000}\left(\frac{e^2}{DkT}\right)^3$$

$$\beta^2 = \frac{8\pi e N_A d_0}{1000\ DkT}$$

$$a = 4\ \text{Å}$$

$$I = 1/2 \sum_i m_i z_i^2$$

for the long range contribution, but he found that using the Pitzer Debye-Hückel equation improved the results. Chen et al. stated in 1982 that they chose the Pitzer form because it gives some recognition to the repulsive forces between ions. They gave the following equations, normalized for mole fractions:

$$\frac{G^{ex*,P}}{RT} = -(\sum_k x_k)\sqrt{\frac{1000}{M_s}}\ \frac{4A_\phi I_x}{\rho}\ \ln(1 + \rho\sqrt{I_x})$$

where:

$$A_\phi = \frac{1}{3}\sqrt{\frac{2\pi N_A d_0}{1000}}\left(\frac{e}{\sqrt{DkT}}\right)^3 \quad \text{as seen earlier}$$

$I_x = 1/2 \sum_i z_i^2 x_i$ ionic strength on a mole fraction basis

M_s - molecular weight of the solvent

ρ - closest approach parameter, set equal to 14.9 as suggested by Pitzer

The activity coefficient is then:

$$\ln f_i^P = -\sqrt{\frac{1000}{M_s}}\ A_\phi\left[\frac{2z_i^2}{\rho}\ln(1 + \rho\sqrt{I_x}) + \frac{z_i^2\sqrt{I_x} - 2I_x^{3/2}}{1 + \rho\sqrt{I_x}}\right] \quad (4.87)$$

Usage

The mean activity coefficient is defined as:

$$f_{\pm} = (f_{+}^{\nu_c} \; f_{-}^{\nu_a})^{1/\nu} \tag{2.26}$$

or

$$\ln f_{\pm}^{lc} = \frac{1}{\nu} (\nu_c \ln f_c^{lc} + \nu_a \ln f_a^{lc}) \tag{4.88}$$

for the short range salt activity coefficient from Chen's equations (4.84) and (4.85). Tables of τ values presented by Chen et al. (C4) can be found in Appendix 4.6.

Recalling equation (4.69):

$$\ln f_{\pm} = \ln f_{\pm}^{P} + \ln f_{\pm}^{lc} \tag{4.69}$$

the activity coefficient of a salt may be calculated using equations (4.87), and the results of (4.88).

Chen's equations are based on mole fractions:

$$x_i = \frac{\nu_i m}{\nu m + 1000/M_s}$$

where:

- m - molality
- M_s - molecular weight of the solvent
- ν - stoichiometric ($\nu_a + \nu_c$)

As shown by Robinson and Stokes (C6), the mole fraction based activity coefficient, f, is related to the mean molal activity coefficient, γ, by the following relationship:

$$\ln \gamma_{\pm} = \ln f_{\pm} - \ln(1. + .001\, M_s \nu m) \tag{4.89}$$

This relationship is presented for the sake of uniformity as the other methods outlined in this chapter are presented on a molality basis.

TEMPERATURE EFFECTS

The methods for calculating activity coefficients have, up to this point, been presented for solutions at 25°C. This is due to the fact that most experimental data available, used for determining the various parameters of the acitivity coefficient models, was measured at 25°C. Recognizing that activity coefficients can be strongly affected by temperature, Bromley, Meissner, Pitzer and Chen have suggested methods of adapting their models and parameters to any temperature solution.

Bromley's method

In 1973, Bromley (B3) presented two equations that he felt adequately correlated the effect of temperature on the B parameter of his equation (4.44). They are:

$$B = B^* \ln \frac{T-243}{T} + \frac{B_1}{T} + B_2 + B_3 \ln T \tag{4.90}$$

$$B = \frac{B^*}{T-230} + \frac{B_1}{T} + B_2 + B_3 \ln T \tag{4.91}$$

with T as the temperature in Kelvins. The major drawback to these equations is the need for sufficient experimental data in order to determine the constants B^*, B_1, B_2 and B_3.

Since such data is often unavailable, Bromley suggested using Meissner's technique when one good activity coefficient is known.

Meissner's method

While differing from the values at 25°C, isotherms at other temperatures also fall into the Meissner family of curves (Figure 4.6). Meissner originally indicated (M3, M4, M5) that this held for temperatures between 0° and 150°C; he later revised this range to 25° to 120° C (M10).

The reduced activity coefficient appears to be independent of temperature when the Meissner q parameter equals 1.7 and Meissner states that the closer a nonsulfate's isotherm lies to this line, the less the temperature effect is felt. The following equation was proposed to determine the temperature effect:

$$\frac{\log (\Gamma^\circ_t / \Gamma^\circ_{25^\circ C})_{I=10}}{t-25} = a \log (\Gamma^\circ_{25^\circ C})_{I=10} + b \qquad (4.92)$$

where:

t - temperature in degrees Celsius

a - equals -.005 for nonsulfates; -.0079 for sulfates

b - equals 0.0

This equation for finding the reduced activity coefficient at temperature t may only be used at ionic strength equal to 10. The equation may also be written:

$$\log (\Gamma^\circ_t)_{I=10} = (1. - .005(t-25)) \log (\Gamma^\circ_{25^\circ C})_{I=10} \qquad (4.93)$$

In 1973, Meissner (M6) suggested the following equations for finding the reduced activity at any ionic strength I, given the reduced activity coefficient at 25°C for that ionic strength:

$$\log (\Gamma^\circ_t) = (1.125 - .005t) \log (\Gamma^\circ_{25^\circ C}) - (0.125 - .005t) \log (\Gamma^\circ_{ref}) \qquad (4.94)$$

with: $$\log (\Gamma^\circ_{ref}) = - \frac{.41\sqrt{I}}{1 + \sqrt{I}} + 0.039\, I^{.92} \qquad (4.94a)$$

which is the reduced activity at 25°C where q=1.7

Meissner proposed (M7) that since $\log (\Gamma^\circ)_{I=10}$ is almost linear for values of q from -2 to 7, equation (4.92) may be expressed in terms of q:

$$\frac{q_t - q_{25}}{t - 25} = a\, q_{25} + b^* \qquad (4.95)$$

with:

a = -.0079 for sulfates;
-.005 for other electrolytes

b* = -.0029 for sulfates;
.0085 for nonsulfates

This equation is applicable for any ionic strength.

More recently, Meissner (M10) suggested the following equation as an improvement to equation (4.95):

$$q^\circ_t = q^\circ_{25^\circ C} \left[1. - \frac{0.0027(t-25)}{|z_+ z_-|} \right] \qquad (4.96)$$

The superscript "o" indicates a solution of one electrolyte resulting in one cation and one anion type.

Pitzer's method

Pitzer's most extensive investigation of temperature effects on the β_0, β_1 and C^ϕ parameters of his equations was done for NaCl (P8, P17, P19). This resulted in the following equations:

$$\beta_0 = q_1 + q_2\left(\frac{1}{T} - \frac{1}{T_r}\right) + q_3 \ln\left(\frac{T}{T_r}\right) + q_4 (T - T_r) + q_5 (T^2 - T_r^2) \qquad (4.97a)$$

$$\beta_1 = q_6 + q_9 (T - T_r) + q_{10} (T^2 - T_r^2) \qquad (4.97b)$$

$$C^\phi = q_{11} + q_{12}\left(\frac{1}{T} - \frac{1}{T_r}\right) + q_{13} \ln\left(\frac{T}{T_r}\right) + q_{14}(T - T_r) \qquad (4.97c)$$

with:

T - absolute temperature

T_r = 298.15 K

and the following q's fitted with a least-squares algorithm:

q_1 = 0.0765

q_2 = -777.03

q_3 = -4.4706

q_4 = 0.008946

q_5 = -3.3158×10^{-6}

q_6 = 0.2664

q_9 = 6.1608×10^{-5}

q_{10} = 1.0715×10^{-6}

q_{11} = 0.00127

q_{12} = 33.317

q_{13} = 0.09421

q_{14} = -4.655×10^{-5}

Pitzer noted that there was very little change in these parameters over the range of temperatures from 25 to 300°C. The Debye-Hückel paramter, A_ϕ, was affected to a much greater degree; it approximately doubled over the given temperature range. He concluded that the decrease in the dielectric constant was the dominant factor.

He defined the dielectric constant as (P8, P14, P17):

$$D = \sum_{i=0}^{4} \sum_{j=0}^{4-i} e_{ij}\, t^{i}\, d_w^{j} \qquad (4.98)$$

where:

- d_w - density of pure water
- t - temperature in degrees Celsius
- e - parameters fit via a least-squares method, given below

Table 4.1

Array of Fit Coefficients for Equation (4.98)

e_{ij}

i \ j	0	1	2	3	4
0	88.287	-78.1110993	77.6526372	0	0
1	-0.67033927	0.443465512	-0.135474228	-.334969862	
2	$1.99954913 \times 10^{-3}$	$-8.71140434 \times 10^{-4}$	$1.10849697 \times 10^{-4}$		
3	$-2.71567936 \times 10^{-6}$	$5.91444341 \times 10^{-7}$			
4	$1.40598463 \times 10^{-9}$				

This resulted in the following values for A_ϕ:

Table 4.2

Debye-Hückel Parameters

t,°C	A_ϕ	t,°C	A_ϕ
0	0.3770	140	0.5137
10	0.3820	150	0.5291
20	0.3878	160	0.5454
30	0.3944	170	0.5627
40	0.4017	180	0.5810
50	0.4098	190	0.6005
60	0.4185	200	0.6212
70	0.4279	220	0.6670
80	0.4380	240	0.7200
90	0.4488	260	0.7829
100	0.4603	280	0.8596
110	0.4725	300	0.9576
120	0.4855	325	1.1323
130	0.4992	350	1.4436

These values for the dielectric constant and resulting Debye-Hückel parameters were used by Pitzer in the work on NaCl. In 1979, Pitzer and Bradley (P12) suggested the following equations improved the calculation:

$$D = D_{1000} + C \ln \frac{B + P}{B + 1000} \tag{4.99}$$

where:

P - pressure in bars

$$D_{1000} = U_1 \exp(U_2 T + U_3 T^2) \tag{4.99a}$$

$$C = U_4 + \frac{U_5}{U_6 + T} \tag{4.99b}$$

$$B = U_7 + \frac{U_8}{T + U_9 T} \tag{4.99c}$$

T - temperature in Kelvins

U_1 = 3.4279E2
U_2 = -5.0866E-3
U_3 = 9.4690E-7
U_4 = -2.0525
U_5 = 3.1159E3
U_6 = -1.8289E2
U_7 = -8.0324E3
U_8 = 4.2142E6
U_9 = 2.1417

They suggest using this method for future work; the previous method should continue being used for the values presented in Appendices 4.4 and 4.5.

Appendix 4.5 contains the temperature derivatives of β_0, β_1, β_2 and C^ϕ for various species. Although developed for calculating molal enthalpies, Pitzer states (P14) that they can be used to convert the β_0, β_1, β_2 and C^ϕ of Appendix 4.4, which are for 25°C, to other temperatures, provided the temperature change is not too great. As noted by Silvester and Pitzer (P10), since for 1-1 electrolytes the β_0 and β_1 parameters are small (having a magnitude of one to a few tenths), the derivatives fall in the range 10^{-3}. Therefore a ten to twenty degree change in temperature causes little effect. Other authors working with Pitzer's method have determined the temperature dependence of the parameters for specific solutions and are listed in the references at the end of this chapter.

Chen's method

Chen et al. presented (C4) the following expression for calculating the Debye-Hückel parameter used in the long range contribution based on the values tabulated by Pitzer and Silvester (P17):

$$A_\phi = -61.44534 \exp\left[\frac{T-273.15}{273.15}\right] + 2.864468 \left[\exp\left(\frac{T-273.15}{273.15}\right)\right]^2 + 183.5379 \ln\left[\frac{T}{273.15}\right] - .6820223(T-273.15) + .0007875695(T^2 - (273.15)^2) + 58.95788\left[\frac{273.15}{T}\right] \quad (4.100)$$

Although they have not determined a method of readily adapting their tabulated parameters, $\tau_{m,ca}$ and $\tau_{ca,m}$, for temperature effects, they have shown (C2, C3, C4) that, for NaCl, $FeCl_2$, KCl and KBr, temperature affects these parameters very little. This suggests that, unless adequate data is available for regression, it may still be possible to get relatively good activity coefficients using the 25°C parameters and equation (4.100). This hypothesis will be tested later in this chapter.

APPLICATION

In order to test Bromley, Meissner, Pitzer and Chen's methods for calculating activity coefficients, the models were coded on the HP-85 and HP-87 desktop computers. The calculated values for various electrolytes were plotted against smoothed experimental data published by the National Bureau of Standards and others. In the first plot for each electrolyte, the maximum molality to which they were plotted is the maximum molality of the published parameters. For some of the electrolytes, when there was experimental data available, the maximum molality was extended on a second plot. These plots illustrate the wide deviation from experimental data that may occur when using the published parameters for solutions with ionic strengths greater than the noted maximum molality.

The equations used for each method are summarized on the following pages.

BROMLEY'S METHOD

For a single electrolyte solution at 25°C:

$$\log \gamma_{\pm} = -\frac{A|z_+z_-|\sqrt{I}}{1+\sqrt{I}} + \frac{(0.06+0.6B)|z_+z_-|\,I}{\left(1+\frac{1.5}{|z_+z_-|}I\right)^2} + BI \qquad (4.44)$$

with:

$\gamma_{\pm}$ - the mean molal activity coefficient

A - the Debye-Hückel parameter

$$A = \frac{1}{2.303}\left(\frac{e}{\sqrt{DkT}}\right)^3 \sqrt{\frac{2\,\pi\, d_0 N_A}{1000}} \qquad (4.31)$$

I - ionic strength

$$I = 1/2 \sum_i m_i z_i^2 \qquad (4.27)$$

z_+ - number of charges on the cation

z_- - number of charges on the anion

B - Bromley's parameter

To show temperature effects:

- regress experimental data to fit the B to one of the following equations, T in Kelvins:

$$B = B^* \ln\frac{T-243}{T} + \frac{B_1}{T} + B_2 + B_3 \ln T \qquad (4.90)$$

or

$$B = \frac{B^*}{T-230} + \frac{B_1}{T} + B_2 + B_3 \ln T \qquad (4.91)$$

- use Meissner's technique for showing temperature effects: having one good experimental value at 25°C,

$$\Gamma^{\circ} = \gamma_{\pm}^{1/z_+z_-} \qquad (4.45)$$

where: Γ - is the reduced activity coefficient

calculate the reference reduced activity coefficient:

$$\log(\Gamma^{\circ}_{ref}) = -\frac{.41\sqrt{I}}{1+\sqrt{I}} + 0.039\, I^{.92} \qquad (4.94a)$$

which can then be used in the following equation for the reduced activity coefficient at a temperature of t degrees Celsius:

$$\log(\Gamma_t^\circ) = (1.125 - .005t)\log(\Gamma_{25°C}^\circ) - (0.125 - .005t)\log(\Gamma_{ref}^\circ) \qquad (4.94)$$

Using the results of this equation, the mean activity coefficient at temperature t and ionic strength I may be calculated:

$$\log \gamma_t = z_+ z_- \log(\Gamma_t^\circ)$$

MEISSNER'S METHOD

For single electrolyte solutions at 25°C:

$$\Gamma^\circ = \{1. + B(1. + 0.1\ I)^q - B\}\ \Gamma^* \qquad (4.46)$$

with:

Γ° - reduced activity coefficient of pure solution 25°C

$$B = 0.75 - 0.065\ q \qquad (4.46a)$$

$$\log \Gamma^* = -\frac{.5107\sqrt{I}}{1 + C\sqrt{I}} \qquad (4.46b)$$

$$C = 1. + 0.055\ q\ \exp(-0.023\ I^3) \qquad (4.46c)$$

I - ionic strength $= 1/2 \sum_i m_i z_i^2$

q - Meissner's parameter

The reduced activity coefficient is converted to a mean molal activity coefficient via the following relationship:

$$\Gamma^\circ = \gamma_\pm^{1/z_+z_-} \qquad (4.45)$$

To show temperature effects:

- the reduced activity coefficient from one experimental value:

$$\log(\Gamma_t^\circ) = (1.125 - .005t)\log(\Gamma_{25°C}^\circ) - (.125 - .005t)\log(\Gamma_{ref}^\circ) \qquad (4.94)$$

where:

$\Gamma_{25°C}^\circ$ = the experimental value at 25°C

t = temperature in degrees Celsius

$$\log(\Gamma_{ref}^\circ) = -\frac{.41\sqrt{I}}{1 + \sqrt{I}} + 0.039\ I^{.92} \qquad (4.94a)$$

- to get the temperature affected q parameter from the value of the q parameter at 25°C:

$$q_t^\circ = q_{25°C}^\circ \left[1. - \frac{0.0027(t-25)}{|z_+z_-|}\right] \qquad (4.96)$$

with t = desired temperature in degrees Celsius

PITZER'S METHOD

For single electrolyte solutions at 25°C:

$$\ln \gamma_{\pm} = |z_+ z_-| f^{\gamma} + m B_{\pm}^{\gamma} + m^2 C_{\pm}^{\gamma}$$

with:

m - stoichiometric molality

z_+ - number of charges on the cation

z_- - number of charges on the anion

$$f^{\gamma} = -A_{\phi} \left[\frac{\sqrt{I}}{1 + b\sqrt{I}} + \frac{2}{b} \ln(1 + b\sqrt{I}) \right] \qquad (4.62)$$

$$b = 1.2$$

$$A_{\phi} = \frac{1}{3}\left(\frac{e}{DkT}\right)^3 \sqrt{\frac{2\pi d_0 N_A}{1000}} \qquad (4.64)$$

$$I = 1/2 \sum_i m_i z_i^2$$

$$B_{\pm}^{\gamma} = 2\beta_0 + \frac{2\beta_1}{\alpha^2 I} \{1. - (1. + \alpha\sqrt{I} - .5\,\alpha^2\, I) \exp(-\alpha\sqrt{I}\,)\} \qquad (4.63)$$

or, for 2-2 electrolytes

$$B_{\pm}^{\gamma} = 2\beta_0 + \frac{2\beta_1}{\alpha_1^2 I} \{1. - (1. + \alpha_1\sqrt{I} - .5\,\alpha_1^2\, I) \exp(-\alpha_1\sqrt{I}\,)\}$$

$$+ \frac{2\beta_2}{\alpha_2^2 I} \{1. - (1. + \alpha_2\sqrt{I} - .5\,\alpha_2^2\, I) \exp(-\alpha_2\sqrt{I}\,)\} \qquad (4.63)$$

β_0 , β_1 , β_2 - Pitzer parameters

$\alpha = 2.0$

$\alpha_1 = 1.4$

$\alpha_2 = 12.$

$$C_{\pm}^{\gamma} = \frac{3}{2} C_{\pm}^{\phi}$$

$C_{\pm}^{\phi}$ - Pitzer parameter

To show temperature effects:

- for NaCl, use equations (4.97) with the parameters given earlier in this chapter

- use the temperature dependent expression for the dielectric constant to calculate the Debye-Hückel parameter, A_ϕ:

$$D = \sum_{i=0}^{4} \sum_{j=0}^{4-i} e_{ij}\, t^i\, d_w^j \qquad (4.98)$$

when using the Pitzer published parameters, with:

d_w - density of pure water
t - temperature in degrees Celsius
e - parameters fit via a least-squares method, given in table 4.1

CHEN'S METHOD

For single electrolyte solutions at 25°C:

$$\ln f_{\pm} = \ln f_{\pm}^{P} + \ln f_{\pm}^{lc} \qquad (4.69)$$

with: f - mole fraction activity coefficient

short range:

$$\ln f_{+}^{lc} = \frac{x_m^2 \tau_{ca,m} F_{cm}}{(x_c F_{cm} + x_a F_{am} + x_m)^2} - \frac{z_- x_a \tau_{m,ca} x_m F_{ma}}{(x_c + x_m F_{ma})^2} + \frac{z_+ x_m \tau_{m,ca} F_{mc}}{x_a + x_m F_{mc}} - z_+ \tau_{m,ca} - F_{cm} \tau_{ca,m} \qquad (4.84)$$

$$\ln f_{-}^{lc} = \frac{x_m^2 \tau_{ca,m} F_{am}}{(x_c F_{cm} + x_a F_{am} + x_m)^2} - \frac{z_+ x_c \tau_{m,ca} x_m F_{mc}}{(x_a + x_m F_{mc})^2} + \frac{z_- x_m \tau_{m,ca} F_{ma}}{x_c + x_m F_{ma}} - z_- \tau_{m,ca} - F_{am} \tau_{ca,m} \qquad (4.85)$$

x_m, x_a, x_c - solvent, anion and cation mole fractions

$$x_i = \frac{\nu_i m}{m + 1000/M_s}$$

m - molality

M_s - solvent molecular weight

ν - stoichiometric ($\nu_a + \nu_c$)

$\tau_{ca,m}$, $\tau_{m,ca}$ - Chen parameters

z_+ - number of charges on the cation

z_- - number of charges on the anion

$F_{cm} = F_{am} = \exp(-\alpha \tau_{ca,m})$ (4.70b)

$F_{ma} = F_{mc} = \exp(-\alpha \tau_{m,ca})$

$\alpha = .2$

The electrolyte activity coefficient is:

$$f_{\pm} = [f_{+}^{\nu_c} f_{-}^{\nu_a}]^{1/\nu} \tag{2.26}$$

or

$$\ln f_{\pm}^{lc} = \frac{1}{\nu}(\nu_c \ln f_{+}^{lc} + \nu_a \ln f_{-}^{lc})$$

long range:

$$\ln f_{\pm}^{P} = -\sqrt{\frac{1000}{M_s}}\, A_\phi \left[\frac{2|z_+z_-|}{\rho}\ln(1+\rho\sqrt{I_x}) + \frac{|z_+z_-|\sqrt{I_x} - 2I_x^{3/2}}{1+\rho\sqrt{I_x}}\right] \tag{4.87}$$

$$A_\phi = \frac{1}{3}\left(\frac{e}{\sqrt{DkT}}\right)^3 \sqrt{\frac{2\pi d_0 N_A}{1000}} \tag{4.64}$$

$$I_x = 1/2 \sum_i x_i z_i^2$$

$$\rho = 14.9$$

The short and long range activity coefficients are combined using equation (4.69). For the sake of uniformity the result is converted to the molality based form using the following relationship:

$$\ln \gamma_{\pm} = \ln f_{\pm} - \ln(1. + .001\, M_s \nu m) \tag{4.89}$$

To show temperature effects:

- use the following equation to calculate the temperature dependent Debye-Hückel parameter for the temperature of interest, T in Kelvins:

$$A_\phi = -61.44534 \exp\left[\frac{T-273.15}{273.15}\right] + 2.864468 \left[\exp\left[\frac{T-273.15}{273.15}\right]\right]^2$$
$$+ 183.5379 \ln\left[\frac{T}{273.15}\right] - .6820223(T-273.15)$$
$$+ .0007875695(T^2 - (273.15)^2) + 58.95788\left[\frac{273.15}{T}\right] \tag{4.100}$$

NBS SMOOTHED EXPERIMENTAL DATA

From: Hamer, W.J. and Y-C Wu, "Osmotic Coefficients and Mean Activity Coefficients of Uni-univalent Electrolytes in Water at 25°C", J. Phys. Chem. Ref. Data, 1, 1047 (1972)

The National Bureau of Standards fit experimental data to the following equation:

$$\log \gamma_{\pm} = - \frac{A |z_+ z_-| \sqrt{I}}{1 + B^* \sqrt{I}} + \beta I + C I^2 + D I^3 + \ldots..$$

A = 0.5108 at 25°C

From: Goldberg, R.N., "Evaluated Activity and Osmotic Coefficients for Aqueous Solutions: Thirty-Six Uni-Bivalent Electrolytes", J. Phys. Chem. Ref. Data, 10, 671 (1981)

The data was fit to the same kind of equation, but in a log base e form:

$$\ln \gamma_{\pm} = - \frac{\alpha |z_+ z_-| \sqrt{I}}{1 + B \sqrt{I}} + C m + D m^2 + E m^3 + \ldots..$$

α = 1.176252569 at 25°C

HCl PARAMETERS

Bromley	$B = 0.1433$	maximum molality = 6.0
Meissner	$q = 6.69$	maximum molality = 4.5 to 6
Pitzer	$\beta_0 = 0.1775$	maximum molality = 6.0
	$\beta_1 = 0.2945$	
	$C^{\phi} = 0.0008$	
Chen	$\tau_{m,ca} = 10.089$	maximum molality = 6.0
	$\tau_{ca,m} = -5.212$	
NBS	$B^* = 1.525$	maximum molality = 16.
	$\beta = .10494$	
	$C = 6.5360 \times 10^{-3}$	
	$D = -4.2058 \times 10^{-4}$	
	$E = -4.07 \times 10^{-6}$	
	$F = 5.2580 \times 10^{-7}$	

HCl

COMPARISON OF ACTIVITY COEFFICIENT METHODS

molality	NBS	Bromley	Meissner	Pitzer	Chen
.002	.95241	.95214	.95205	.95232	.95215
.004	.93531	.93488	.93463	.93509	.93473
.006	.92309	.92254	.92211	.92275	.92220
.008	.91335	.91271	.91208	.91291	.91215
.01	.90517	.90447	.90363	.90464	.90366
.02	.87635	.87560	.87362	.87547	.87342
.03	.85737	.85678	.85359	.85625	.85314
.04	.84319	.84287	.83848	.84191	.83777
.05	.83193	.83195	.82636	.83054	.82541
.06	.82267	.82305	.81631	.82120	.81511
.07	.81486	.81562	.80775	.81335	.80632
.08	.80816	.80930	.80035	.80661	.79870
.09	.80234	.80385	.79386	.80078	.79200
.1	.79723	.79910	.78811	.79567	.78606
.2	.76813	.77273	.75381	.76688	.75052
.3	.75791	.76380	.73964	.75714	.73618
.4	.75604	.76235	.73457	.75575	.73175
.5	.75904	.76528	.73489	.75918	.73337
.6	.76530	.77125	.73890	.76582	.73923
.7	.77399	.77956	.74570	.77485	.74832
.8	.78461	.78978	.75474	.78578	.76004
.9	.79685	.80166	.76568	.79829	.77398
1.0	.81050	.81500	.77830	.81219	.78988
1.5	.89646	.90026	.86242	.89913	.89329
2.0	1.00872	1.01277	.97694	1.01203	1.02961
2.5	1.14752	1.15186	1.12008	1.15125	1.19439
3.0	1.31559	1.31952	1.29302	1.31967	1.38536
3.5	1.51695	1.51924	1.50029	1.52170	1.60081
4.0	1.75658	1.75570	1.75096	1.76305	1.83901
4.5	2.04033	2.03470	2.05857	2.05084	2.09810
5.0	2.37481	2.36325	2.43884	2.39372	2.37603
5.5	2.76740	2.74973	2.90666	2.80221	2.67060
6.0	3.22616	3.20403	3.47510	3.28897	2.97950
6.5	3.75979	3.73787	4.15665	3.86930	3.30034
7.0	4.37751	4.36506	4.96521	4.56166	3.63078
7.5	5.08890	5.10185	5.91734	5.38827	3.96848
8.0	5.90384	5.96740	7.03237	6.37596	4.31124
8.5	6.83225	6.98427	8.33219	7.55714	4.65694
9.0	7.88402	8.17897	9.84122	8.97089	5.00364
9.5	9.06886	9.58276	11.58637	10.66450	5.34954
10.0	10.39624	11.23238	13.59724	12.69513	5.69302
11.0	13.51565	15.44980	18.54897	18.05876	6.36710
12.0	17.31825	21.27753	24.99373	25.81006	7.01630
13.0	21.89599	29.33383	33.28420	37.04977	7.63360
14.0	27.38900	40.47537	43.83554	53.40173	8.21417
15.0	34.04291	55.88934	57.13308	77.26812	8.75502
16.0	42.30536	77.22130	73.74053	112.21240	9.25467
std. dev.		7.27395	7.94388	14.57291	9.13957

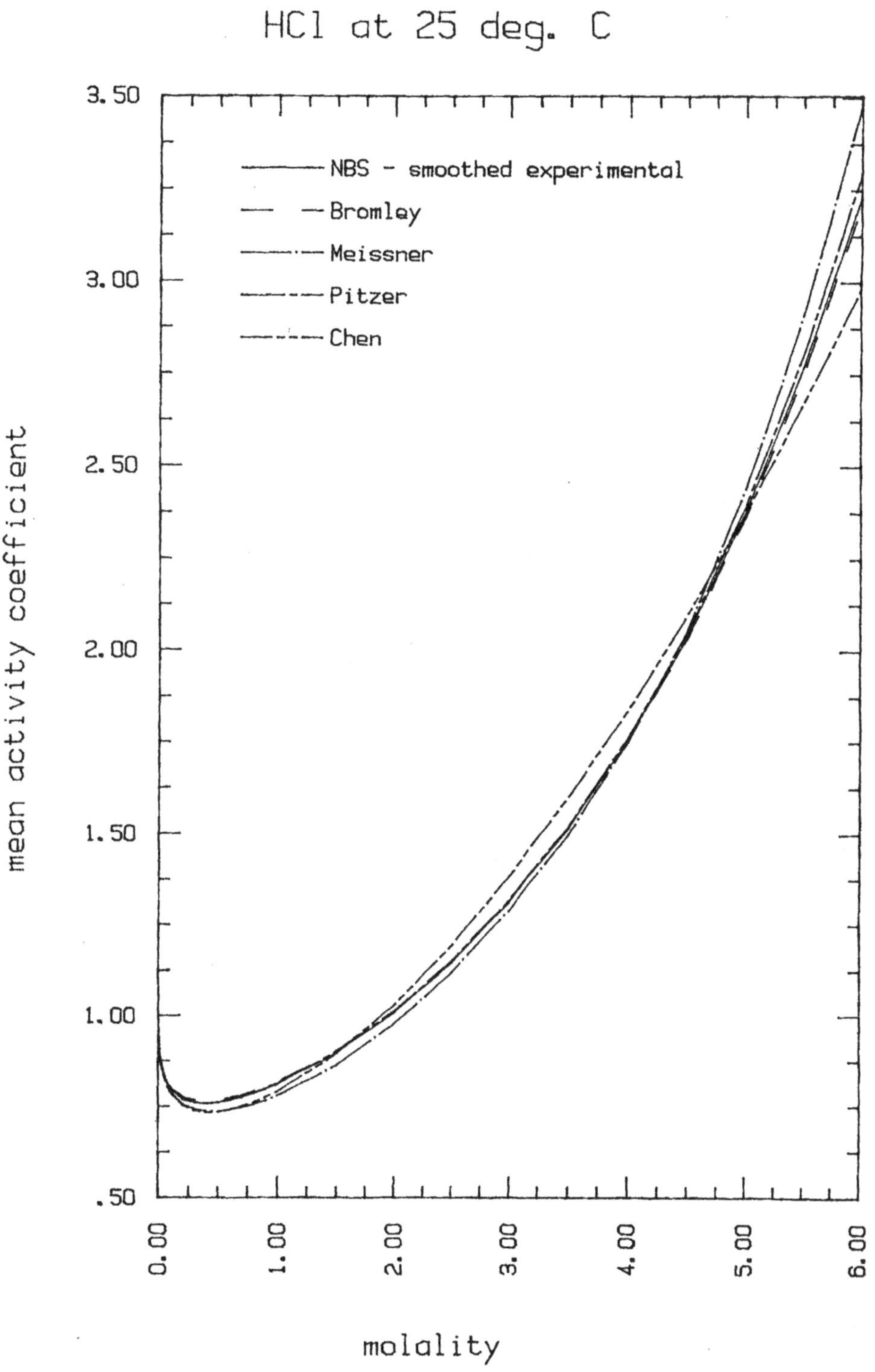
HCl at 25 deg. C
NBS - smoothed experimental
Bromley
Meissner
Pitzer
Chen
mean activity coefficient
molality
3.50
3.00
2.50
2.00
1.50
1.00
.50
0.00
1.00
2.00
3.00
4.00
5.00
6.00

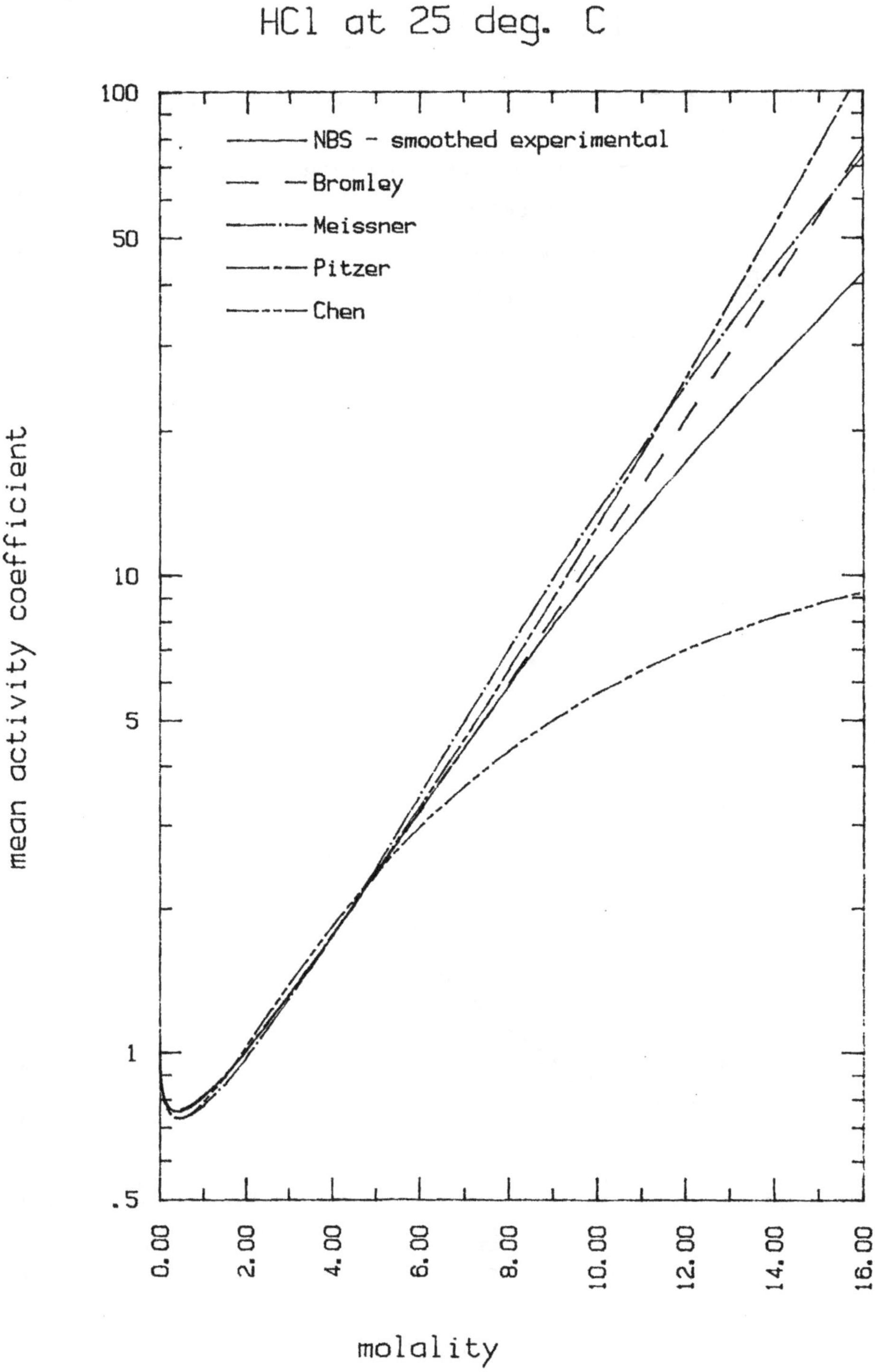
HCl at 25 deg. C
NBS - smoothed experimental
Bromley
Meissner
Pitzer
Chen
100
50
10
5
1
.5
mean activity coefficient
0.00
2.00
4.00
6.00
8.00
10.00
12.00
14.00
16.00
molality

KCl PARAMETERS

Bromley	$B = 0.0240$	maximum molality = less than 6
Meissner	$q = .92$	maximum molality = 4.5 to 6
Pitzer	$\beta_0 = 0.04835$	maximum molality = 4.8
	$\beta_1 = 0.2122$	
	$C^{\phi} = -0.00084$	
Chen	$\tau_{m,ca} = 8.064$	maximum molality = 4.5
	$\tau_{ca,m} = -4.107$	
NBS	$B^* = 1.295$	maximum molality = 5.0
	$\beta = 7.0 \times 10^{-5}$	
	$C = 3.599 \times 10^{-3}$	
	$D = -1.9540 \times 10^{-4}$	

The NBS maximum molality of 5.0 is a supersaturated solution. A KCl solution is saturated at a molality of 4.803.

KCl

COMPARISON OF ACTIVITY COEFFICIENT METHODS

molality	NBS	Bromley	Meissner	Pitzer	Chen
.001	.96490	.96479	.96471	.96494	.96506
.002	.95150	.95131	.95113	.95153	.95175
.003	.94162	.94136	.94109	.94165	.94194
.004	.93356	.93324	.93287	.93357	.93393
.005	.92664	.92627	.92581	.92664	.92707
.006	.92054	.92012	.91957	.92052	.92101
.007	.91504	.91459	.91394	.91502	.91556
.008	.91003	.90954	.90879	.90999	.91058
.009	.90541	.90488	.90405	.90536	.90599
.01	.90111	.90055	.89962	.90105	.90172
.02	.86885	.86810	.86630	.86869	.86966
.03	.84672	.84591	.84331	.84649	.84762
.04	.82959	.82879	.82546	.82931	.83051
.05	.81553	.81477	.81078	.81522	.81645
.06	.80358	.80289	.79829	.80324	.80446
.07	.79317	.79256	.78740	.79281	.79400
.08	.78394	.78343	.77775	.78358	.78472
.09	.77566	.77524	.76910	.77529	.77637
.1	.76815	.76781	.76124	.76777	.76879
.2	.71696	.71729	.70809	.71663	.71685
.3	.68653	.68703	.67707	.68627	.68579
.4	.66526	.66566	.65587	.66505	.66410
.5	.64919	.64936	.64024	.64902	.64778
.6	.63646	.63638	.62817	.63634	.63495
.7	.62607	.62577	.61858	.62599	.62457
.8	.61741	.61693	.61080	.61738	.61602
.9	.61008	.60947	.60438	.61011	.60887
1.0	.60381	.60312	.59903	.60389	.60283
1.5	.58295	.58245	.58239	.58341	.58357
2.0	.57259	.57275	.57490	.57347	.57480
2.5	.56833	.56914	.57177	.56957	.57153
3.0	.56822	.56940	.57107	.56962	.57145
3.5	.57116	.57235	.57201	.57252	.57333
4.0	.57649	.57731	.57439	.57757	.57640
4.5	.58373	.58383	.57813	.58433	.58019
5.0	.59253	.59164	.58313	.59249	.58436
std. dev.		.00427	.03789	.00392	.01476

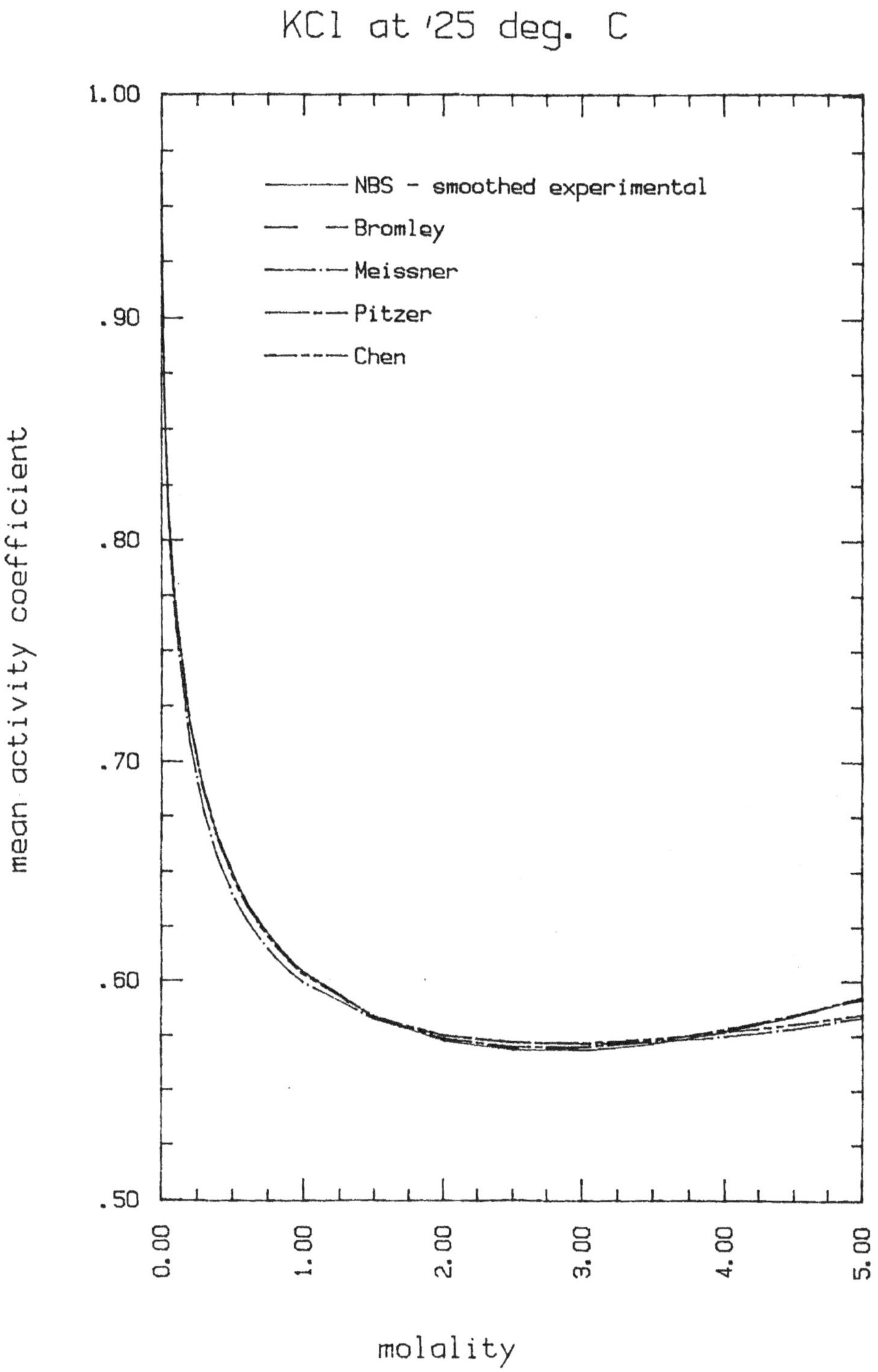
KCl at 25 deg. C
NBS - smoothed experimental
Bromley
Meissner
Pitzer
Chen
1.00
.90
.80
.70
.60
.50
mean activity coefficient
0.00
1.00
2.00
3.00
4.00
5.00
molality

KOH PARAMETERS

Bromley	$B = 0.1131$	maximum molality = 6.
Meissner	$q = 4.77$	maximum molality = 4.5 to 6
Pitzer	$\beta_0 = 0.1298$	maximum molality = 5.5
	$\beta_1 = 0.320$	
	$C^{\phi} = 0.0041$	
Chen	$\tau_{m,ca} = 9.868$	maximum molality = 6.0
	$\tau_{ca,m} = -5.059$	
NBS	$B^* = 1.15$	maximum molality = 20.
	$\beta = .1$	
	$C = 2.6270 \times 10^{-3}$	
	$D = -1.3 \times 10^{-4}$	
	$E = 1.7 \times 10^{-7}$	

KOH

COMPARISON OF ACTIVITY COEFFICIENT METHODS

molality	NBS	Bromley	Meissner	Pitzer	Chen
.001	.96497	.96511	.96508	.96530	.96519
.005	.92704	.92779	.92748	.92829	.92766
.01	.90196	.90348	.90275	.90417	.90287
.03	.84962	.85401	.85144	.85479	.85091
.05	.82064	.82757	.82322	.82806	.82181
.07	.80056	.80972	.80376	.80982	.80139
.09	.78539	.79651	.78911	.79620	.78578
.1	.77906	.79106	.78301	.79056	.77920
.2	.74003	.75830	.74567	.75628	.73743
.3	.72207	.74360	.72882	.74082	.71693
.4	.71348	.73662	.72111	.73350	.70630
.5	.71029	.73411	.71867	.73080	.70159
.6	.71065	.73462	.71971	.73113	.70094
.7	.71357	.73738	.72328	.73364	.70330
.8	.71845	.74191	.72878	.73785	.70805
.9	.72491	.74792	.73584	.74343	.71477
1.0	.73268	.75519	.74421	.75018	.72315
1.5	.78654	.80630	.80048	.79769	.78420
2.0	.85984	.87669	.87404	.86402	.86906
2.5	.95022	.96359	.96044	.94790	.97226
3.0	1.05787	1.06664	1.05823	1.05032	1.09087
3.5	1.18413	1.18660	1.16836	1.17351	1.22289
4.0	1.33101	1.32486	1.29420	1.32062	1.36664
4.5	1.50101	1.48335	1.44059	1.49575	1.52057
5.0	1.69717	1.66439	1.61215	1.70405	1.68318
5.5	1.92293	1.87079	1.81199	1.95192	1.85304
6.0	2.18227	2.10576	2.04160	2.24724	2.02872
6.5	2.47962	2.37304	2.30186	2.59971	2.20886
7.0	2.81994	2.67689	2.59391	3.02128	2.39217
7.5	3.20874	3.02219	2.91954	3.52671	2.57743
8.0	3.65204	3.41450	3.28107	4.13423	2.76351
8.5	4.15644	3.86017	3.68114	4.86643	2.94937
9.0	4.72904	4.36640	4.12259	5.75134	3.13408
9.5	5.37745	4.94140	4.60844	6.82386	3.31682
10.0	6.10974	5.59453	5.14183	8.12754	3.49694
11.0	7.85990	7.17913	6.36451	11.66009	3.84629
12.0	10.04903	9.22399	7.81870	16.97455	4.17838
13.0	12.74508	11.86338	9.53484	25.06643	4.49038
14.0	16.00624	15.27097	11.54585	37.53722	4.78057
15.0	19.87002	19.67147	13.88715	56.99130	5.04812
16.0	24.33974	25.35556	16.59681	87.71053	5.29285
17.0	29.36964	32.69937	19.71556	136.81239	5.51509
18.0	34.85075	42.18961	23.28689	216.25927	5.71554
19.0	40.60036	54.45615	27.35715	346.38041	5.89512
20.0	46.35881	70.31422	31.97558	562.11082	6.05496
std. dev.		4.41307	4.83772	98.29301	12.50386

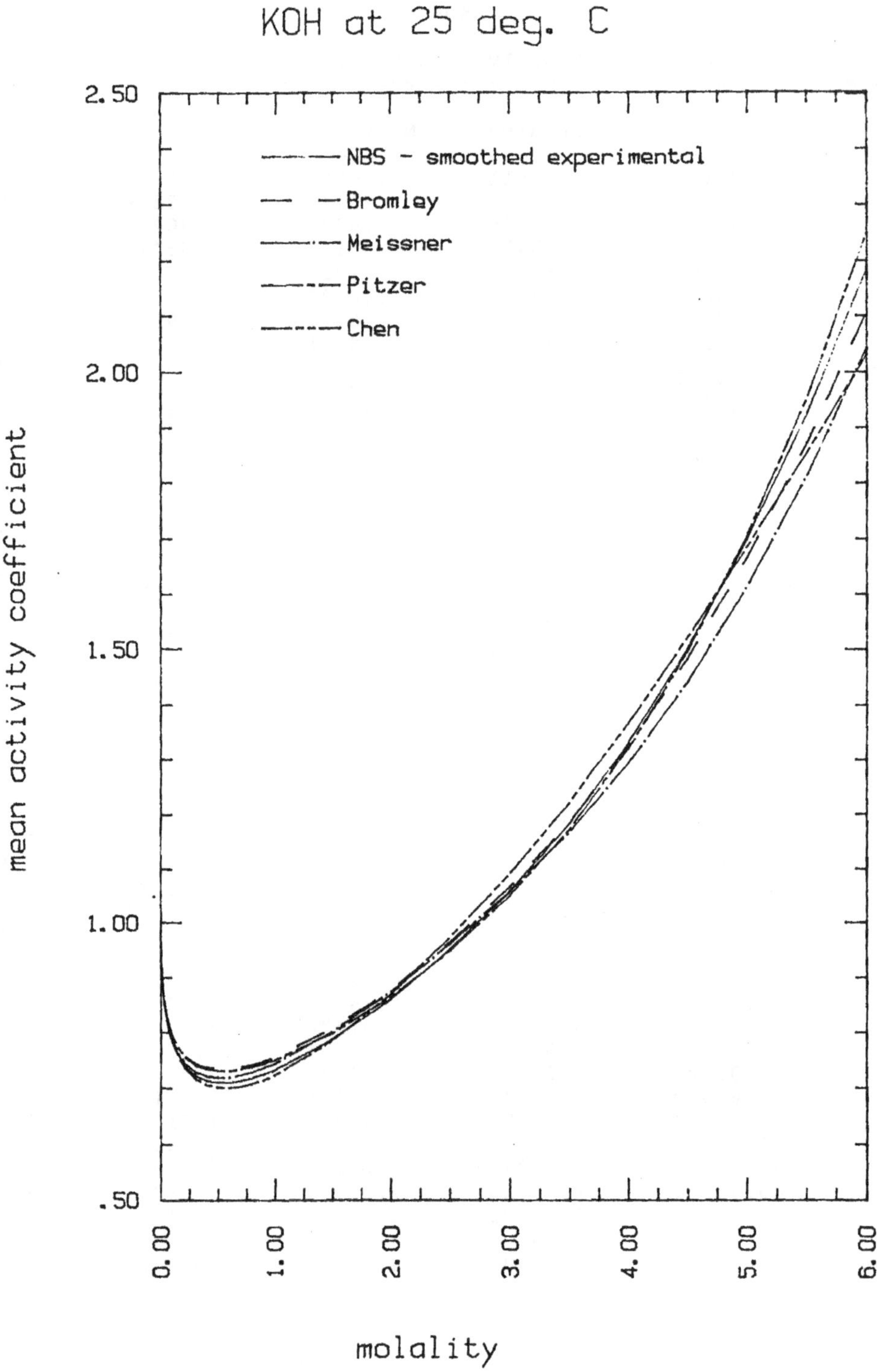
KOH at 25 deg. C
NBS - smoothed experimental
Bromley
Meissner
Pitzer
Chen
2.50
2.00
1.50
1.00
.50
mean activity coefficient
0.00
1.00
2.00
3.00
4.00
5.00
6.00
molality

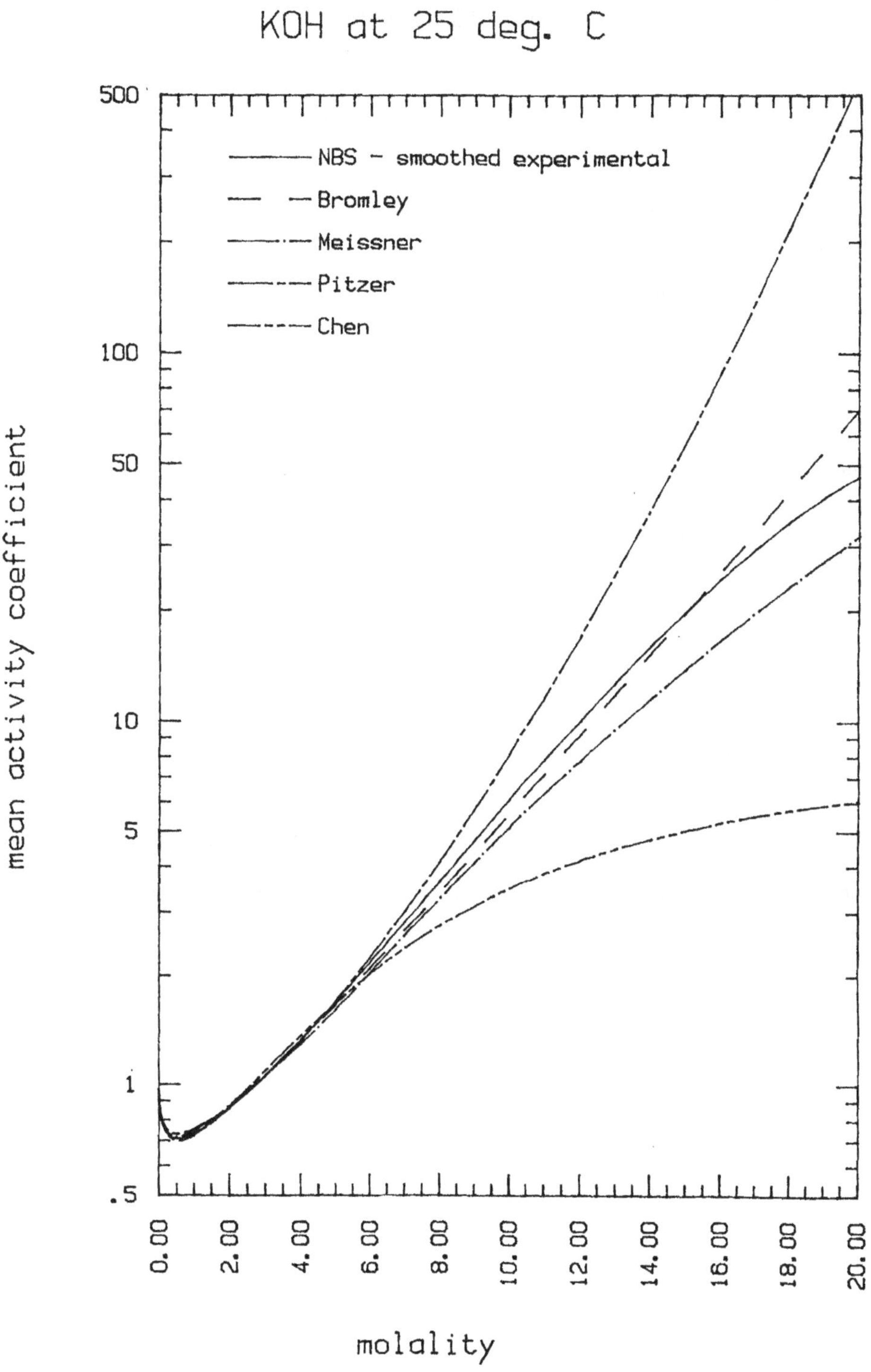
KOH at 25 deg. C
NBS - smoothed experimental
Bromley
Meissner
Pitzer
Chen
500
100
50
10
5
1
.5
mean activity coefficient
0.00
2.00
4.00
6.00
8.00
10.00
12.00
14.00
16.00
18.00
20.00
molality

NaCl PARAMETERS

Bromley	$B = 0.0574$	maximum molality = 6.
Meissner	$q = 2.23$	maximum molality = 3 to 4
Pitzer	$\beta_0 = 0.0765$	maximum molality = 6.0
	$\beta_1 = 0.2664$	
	$C^{\phi} = 0.00127$	
Chen	$\tau_{m,ca} = 8.885$	maximum molality = 6.0
	$\tau_{ca,m} = -4.549$	
NBS	$B^* = 1.4495$	maximum molality = 6.144
	$\beta = 2.0442 \times 10^{-2}$	
	$C = 5.7927 \times 10^{-3}$	
	$D = -2.8860 \times 10^{-4}$	

The maximum molality of 6.144 of the NBS is the saturation point for a NaCl solution; therefore there is only one plot for NaCl.

NaCl

COMPARISON OF ACTIVITY COEFFICIENT METHODS

molality	NBS	Bromley	Meissner	Pitzer	Chen
.001	.96511	.96491	.96486	.96510	.96512
.002	.95189	.95154	.95141	.95183	.95186
.003	.94220	.94171	.94151	.94208	.94210
.004	.93431	.93370	.93342	.93414	.93415
.005	.92756	.92684	.92648	.92735	.92733
.006	.92162	.92080	.92036	.92136	.92133
.007	.91628	.91537	.91485	.91598	.91592
.008	.91142	.91042	.90983	.91108	.91100
.009	.90695	.90587	.90519	.90657	.90646
.01	.90280	.90165	.90089	.90238	.90224
.02	.87189	.87019	.86865	.87116	.87067
.03	.85094	.84894	.84665	.85000	.84909
.04	.83489	.83271	.82972	.83378	.83245
.05	.82182	.81955	.81592	.82060	.81884
.06	.81080	.80848	.80427	.80948	.80730
.07	.80126	.79895	.79419	.79988	.79728
.08	.79288	.79059	.78533	.79144	.78844
.09	.78539	.78314	.77743	.78392	.78052
.1	.77865	.77645	.77032	.77715	.77337
.2	.73405	.73240	.72386	.73252	.72570
.3	.70907	.70771	.69876	.70768	.69891
.4	.69269	.69142	.68310	.69145	.68155
.5	.68118	.67992	.67276	.68007	.66964
.6	.67282	.67157	.66584	.67181	.66132
.7	.66667	.66548	.66129	.66576	.65556
.8	.66217	.66110	.65847	.66134	.65173
.9	.65895	.65808	.65698	.65820	.64940
1.0	.65677	.65616	.65651	.65609	.64828
1.5	.65660	.65796	.66405	.65622	.65501
2.0	.66818	.67184	.68051	.66804	.67426
2.5	.68779	.69333	.70114	.68778	.70067
3.0	.71392	.72045	.72410	.71391	.73158
3.5	.74589	.75224	.74905	.74577	.76543
4.0	.78337	.78821	.77651	.78310	.80116
4.5	.82618	.82811	.80723	.82590	.83801
5.0	.87427	.87185	.84161	.87433	.87541
5.5	.92757	.91945	.87955	.92867	.91291
6.0	.98604	.97097	.92056	.98932	.95016
std. dev.		.02454	.10214	.00681	.07016

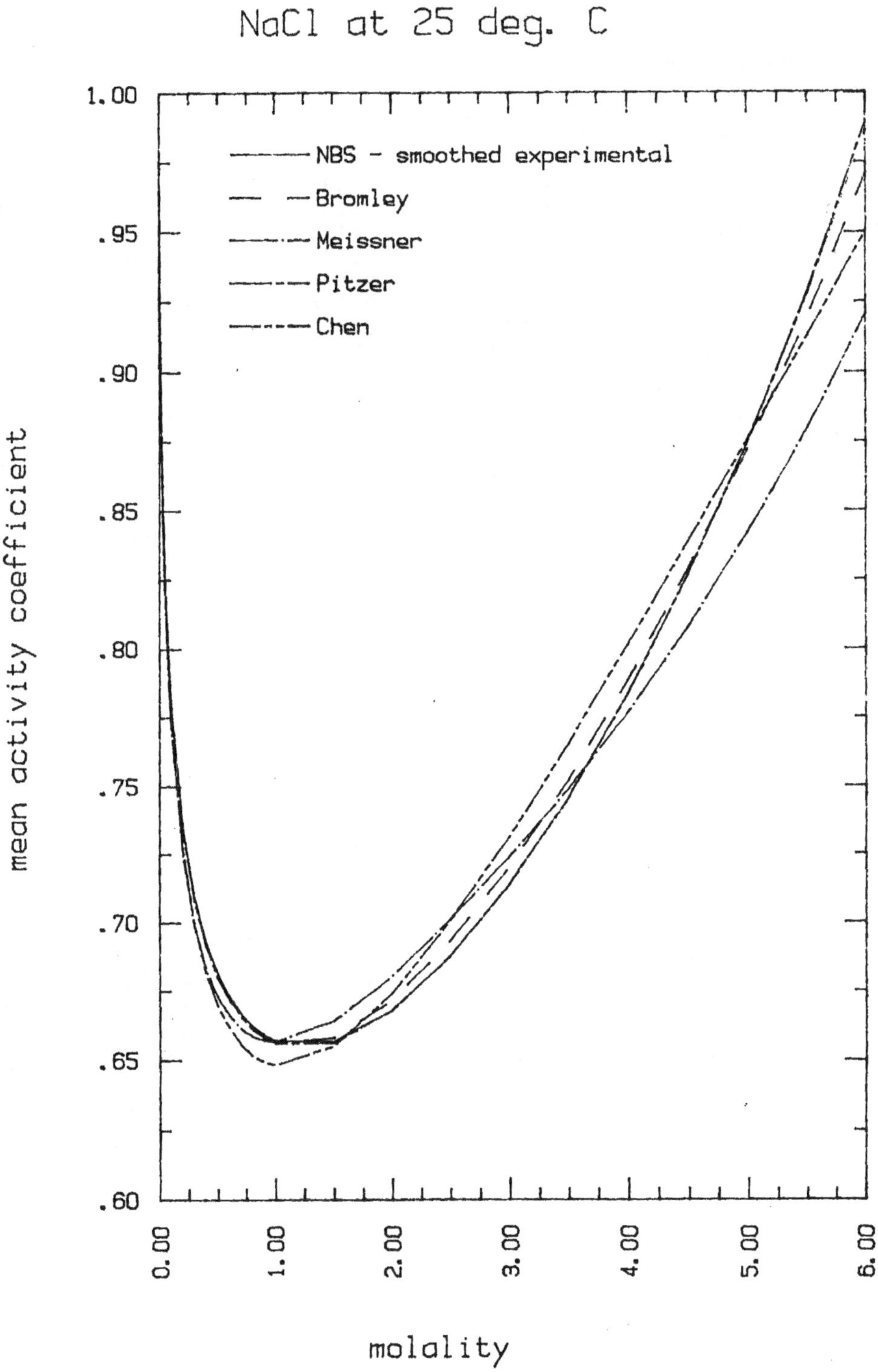
NaCl at 25 deg. C
NBS - smoothed experimental
Bromley
Meissner
Pitzer
Chen
mean activity coefficient
1.00
.95
.90
.85
.80
.75
.70
.65
.60
0.00
1.00
2.00
3.00
4.00
5.00
6.00
molality

NaOH PARAMETERS

Bromley	$B = 0.0747$	maximum molality = 6.
Meissner	$q = 3.00$	maximum molality = 3 to 4
Pitzer	$\beta_0 = 0.0864$	maximum molality = 6.0
	$\beta_1 = 0.253$	
	$C^{\phi} = 0.0044$	
Chen	$\tau_{m,ca} = 9.372$	maximum molality = 6.0
	$\tau_{ca,m} = -4.777$	
NBS	$B^* = 1.21$	maximum molality = 29
	$\beta = 6.1509 \times 10^{-2}$	
	$C = -2.9784 \times 10^{-3}$	
	$D = 1.3293 \times 10^{-3}$	
	$E = -1.1134 \times 10^{-4}$	
	$F = 3.5452 \times 10^{-6}$	
	$G = -3.9985 \times 10^{-8}$	

NaOH

COMPARISON OF ACTIVITY COEFFICIENT METHODS

molality	NBS	Bromley	Meissner	Pitzer	Chen
.001	.96495	.96497	.96493	.96509	.96512
.005	.92691	.92713	.92683	.92733	.92732
.01	.90167	.90221	.90153	.90236	.90221
.05	.81879	.82203	.81851	.82066	.81874
.1	.77499	.78096	.77486	.77748	.77330
.2	.73108	.74035	.73173	.73365	.72596
.4	.69375	.70515	.69681	.69473	.68338
.6	.67900	.69055	.68510	.67775	.66574
.8	.67379	.68521	.68328	.67031	.65957
1.0	.67394	.68544	.68698	.66847	.66029
1.5	.68778	.70085	.70947	.67880	.68030
2.0	.71388	.72974	.74214	.70342	.71638
2.5	.74934	.76796	.78016	.73905	.76274
3.0	.79367	.81383	.82168	.78477	.81641
3.5	.84736	.86664	.86662	.84065	.87550
4.0	.91140	.92617	.91606	.90734	.93869
4.5	.98712	.99247	.97149	.98593	1.00491
5.0	1.07618	1.06577	1.03397	1.07791	1.07331
5.5	1.18051	1.14642	1.10368	1.18518	1.14315
6.0	1.30231	1.23489	1.18016	1.31007	1.21380
6.5	1.44409	1.33171	1.26278	1.45543	1.28469
7.0	1.60862	1.43751	1.35111	1.62471	1.35536
7.5	1.79895	1.55300	1.44498	1.82209	1.42538
8.0	2.01838	1.67895	1.54437	2.05260	1.49441
8.5	2.27039	1.81623	1.64939	2.32233	1.56213
9.0	2.55863	1.96579	1.76012	2.63865	1.62829
9.5	2.88676	2.12867	1.87669	3.01049	1.69269
10.0	3.25839	2.30602	1.99923	3.44870	1.75515
11.0	4.14538	2.70922	2.26267	4.57995	1.87373
12.0	5.24123	3.18679	2.55148	6.17734	1.98332
13.0	6.55596	3.75228	2.86669	8.45907	2.08359
14.0	8.08298	4.42181	3.20935	11.75719	2.17450
15.0	9.79643	5.21448	3.58054	16.58238	2.25624
16.0	11.65252	6.15295	3.98132	23.72866	2.32914
17.0	13.59545	7.26410	4.41276	34.44429	2.39365
18.0	15.56699	8.57979	4.87596	50.71362	2.45026
19.0	17.51732	10.13780	5.37198	75.72683	2.49951
20.0	19.41404	11.98295	5.90192	114.67106	2.54195
21.0	21.24689	14.16833	6.46686	176.07733	2.57812
22.0	23.02646	16.75695	7.06788	274.13851	2.60854
23.0	24.77666	19.82348	7.70607	432.74167	2.63370
24.0	26.52101	23.45648	8.38252	692.56050	2.65409
25.0	28.26255	27.76097	9.09830	1123.66786	2.67015
26.0	29.95652	32.86149	9.85452	1848.21527	2.68229
27.0	31.47585	38.90575	10.65224	3081.67280	2.69088
28.0	32.57401	46.06891	11.49256	5208.66605	2.69629
29.0	32.86172	54.55874	12.37656	8924.02046	2.69883
std. dev.		7.00992	12.70134		16.70179

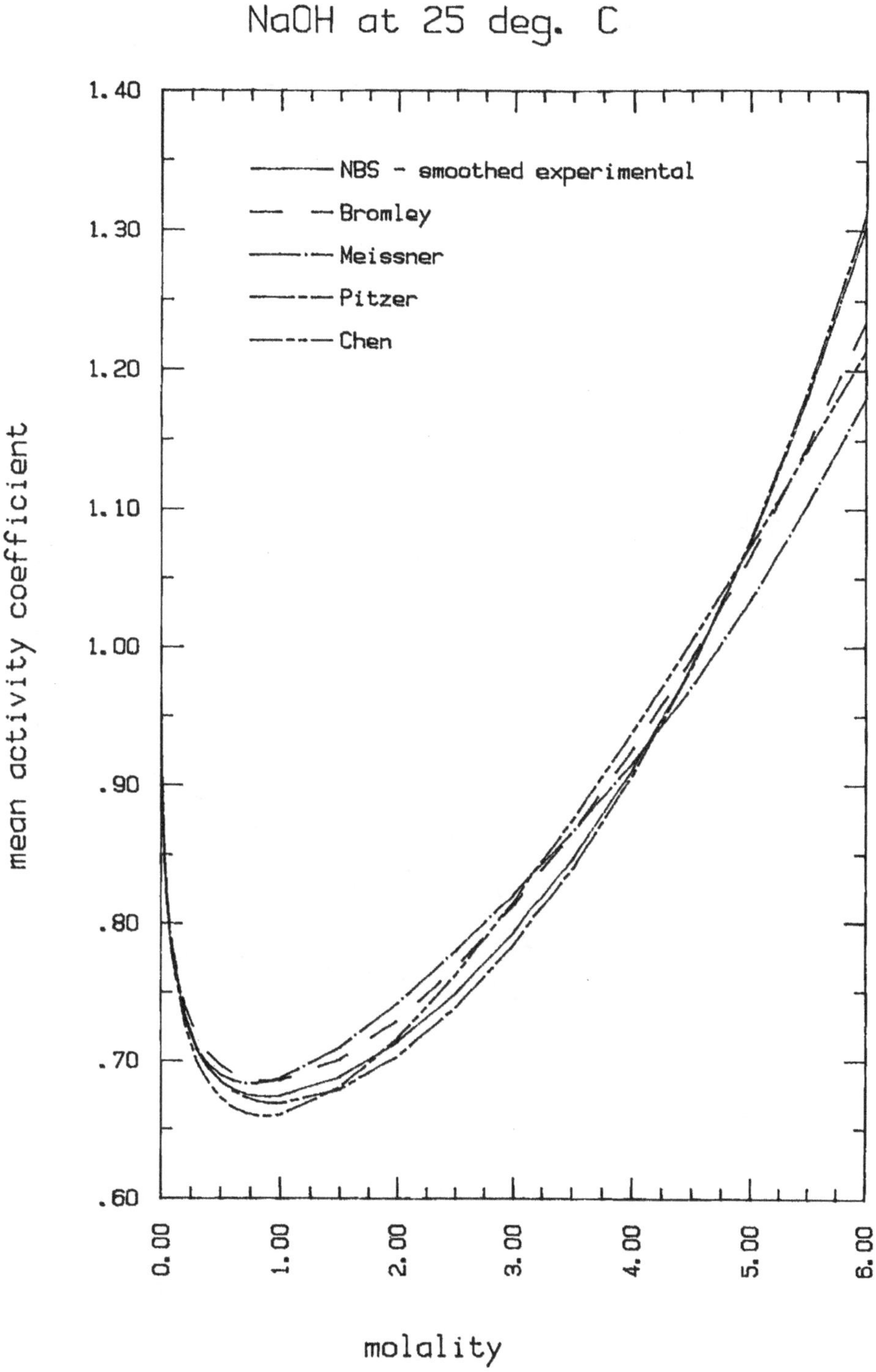
NaOH at 25 deg. C
NBS - smoothed experimental
Bromley
Meissner
Pitzer
Chen
mean activity coefficient
1.40
1.30
1.20
1.10
1.00
.90
.80
.70
.60
0.00
1.00
2.00
3.00
4.00
5.00
6.00
molality

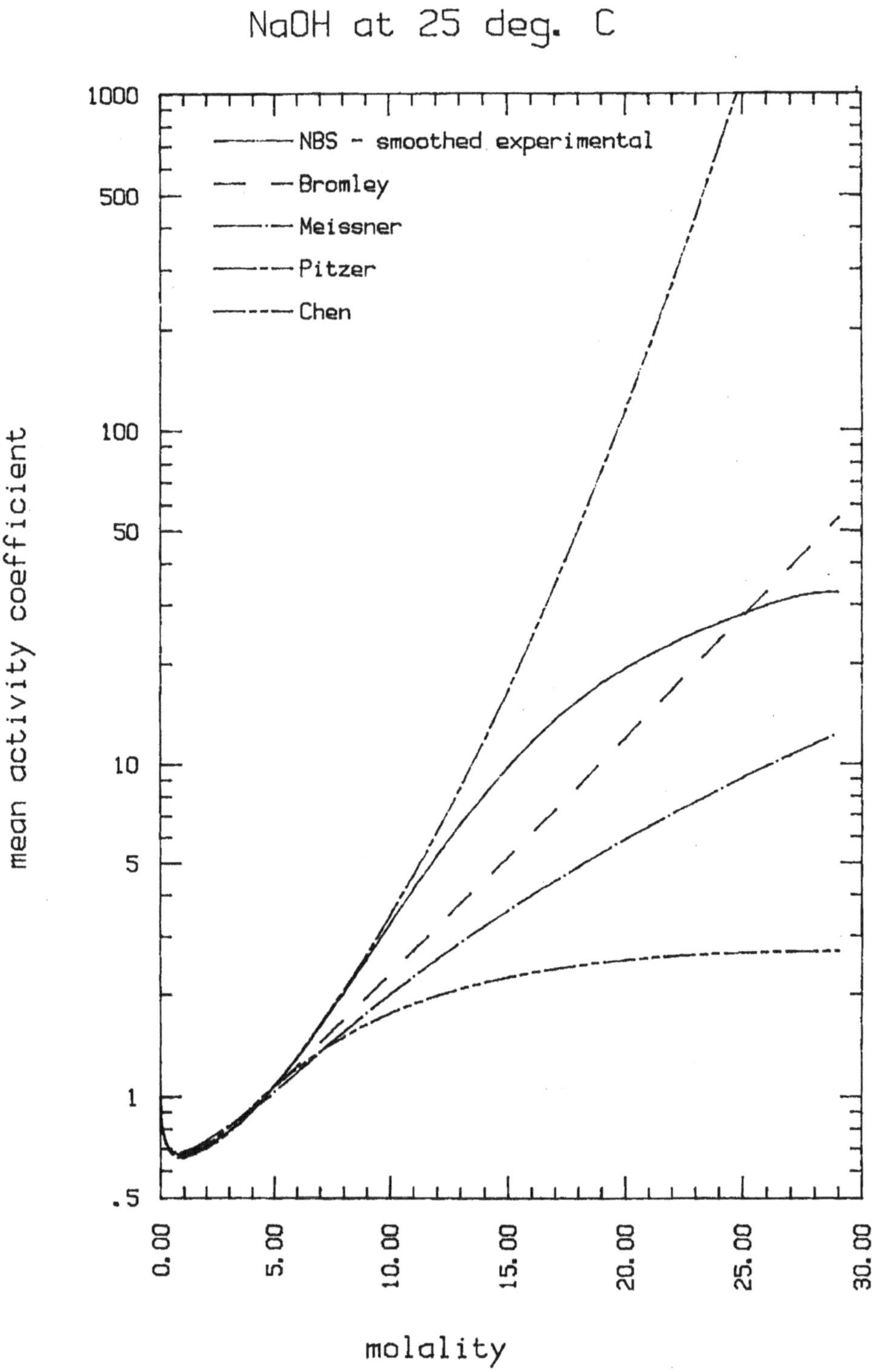

NaOH at 25 deg. C
NBS - smoothed experimental
Bromley
Meissner
Pitzer
Chen
mean activity coefficient
1000
500
100
50
10
5
1
.5
0.00
5.00
10.00
15.00
20.00
25.00
30.00
molality

$CaCl_2$ PARAMETERS

Bromley	B = .0948	maximum molality = 2.0
Meissner	q = 2.40	maximum molality = 5.0
Pitzer	$\frac{4}{3}\ \beta_0 = 0.4212$	maximum molality = 2.5
	$\frac{4}{3}\ \beta_1 = 2.152$	
	$\frac{2^{5/2}}{3}\ C^{\phi} = -0.00064$	
Chen	$\tau_{m,ca} = 11.396$	maximum molality = 6.0
	$\tau_{ca,m} = -6.218$	
NBS	B = 1.60002	maximum molality = 10.
	C = .256690	
	D = .151052	
	$E = -3.77055 \times 10^{-2}$	
	$F = 9.90578 \times 10^{-3}$	
	$G = -1.69480 \times 10^{-3}$	
	$H = 1.34960 \times 10^{-4}$	
	$I = -3.94208 \times 10^{-6}$	

From: Staples, B.R. and Nuttall, R.L., "The Activity and Osmotic Coefficients of Aqueous Calcium Chloride at 298.15 K", J. Phys. Chem. Ref. Data, 6, 385, (1977)

CaCl2

COMPARISON OF ACTIVITY COEFFICIENT METHODS

molality	NBS	Bromley	Meissner	Pitzer	Chen
.001	.88851	.88698	.88653	.88835	.88706
.002	.85077	.84818	.84723	.85043	.84789
.003	.82449	.82106	.81962	.82402	.82029
.004	.80393	.79982	.79789	.80337	.79849
.005	.78692	.78224	.77982	.78629	.78030
.006	.77237	.76720	.76430	.77168	.76462
.007	.75962	.75404	.75066	.75890	.75080
.008	.74827	.74233	.73849	.74754	.73842
.009	.73804	.73180	.72750	.73730	.72720
.01	.72872	.72222	.71747	.72798	.71693
.02	.66445	.65682	.64796	.66404	.64451
.03	.62561	.61823	.60590	.62573	.59921
.04	.59821	.59162	.57636	.59889	.56639
.05	.57734	.57179	.55406	.57856	.54087
.06	.56070	.55628	.53646	.56241	.52014
.07	.54700	.54373	.52216	.54917	.50282
.08	.53548	.53333	.51030	.53805	.48805
.09	.52562	.52454	.50031	.52855	.47525
.1	.51708	.51701	.49180	.52032	.46404
.2	.46924	.47588	.44929	.47414	.39916
.3	.45081	.46001	.43998	.45600	.37354
.4	.44419	.45414	.44379	.44927	.36514
.5	.44417	.45418	.45466	.44914	.36704
.6	.44860	.45843	.46997	.45357	.37624
.7	.45640	.46602	.48831	.46157	.39124
.8	.46702	.47642	.50890	.47259	.41123
.9	.48013	.48932	.53129	.48629	.43576
1.0	.49557	.50451	.55533	.50251	.46460
1.5	.60696	.61233	.70810	.62000	.67154
2.0	.78421	.77525	.94650	.80476	.98736
2.5	1.05288	1.00533	1.28400	1.07778	1.42017
3.0	1.45497	1.32351	1.73086	1.47410	1.97300
3.5	2.05174	1.76027	2.30935	2.04655	2.64140
4.0	2.92555	2.35824	3.04630	2.87297	3.41397
4.5	4.17627	3.17646	3.97214	4.06743	4.27431
5.0	5.90737	4.29629	5.12100	5.79688	5.20313
5.5	8.19884	5.82984	6.53087	8.30567	6.18024
6.0	11.07177	7.93145	8.24373	11.95171	7.18609
6.5	14.46054	10.81372	10.30572	17.25959	8.20280
7.0	18.21535	14.76952	12.76729	24.99873	9.21475
7.5	22.14839	20.20245	15.68338	36.29845	10.20886
8.0	26.11131	27.66885	19.11354	52.81717	11.17458
8.5	30.06689	37.93588	23.12215	76.99170	12.10373
9.0	34.11316	52.06168	27.77851	112.40410	12.99025
9.5	38.42963	71.50611	33.15704	164.32208	13.82995
10.0	43.11587	98.28386	39.33742	240.49475	14.62019
std. dev.		10.71533	3.43776	40.10493	9.12192

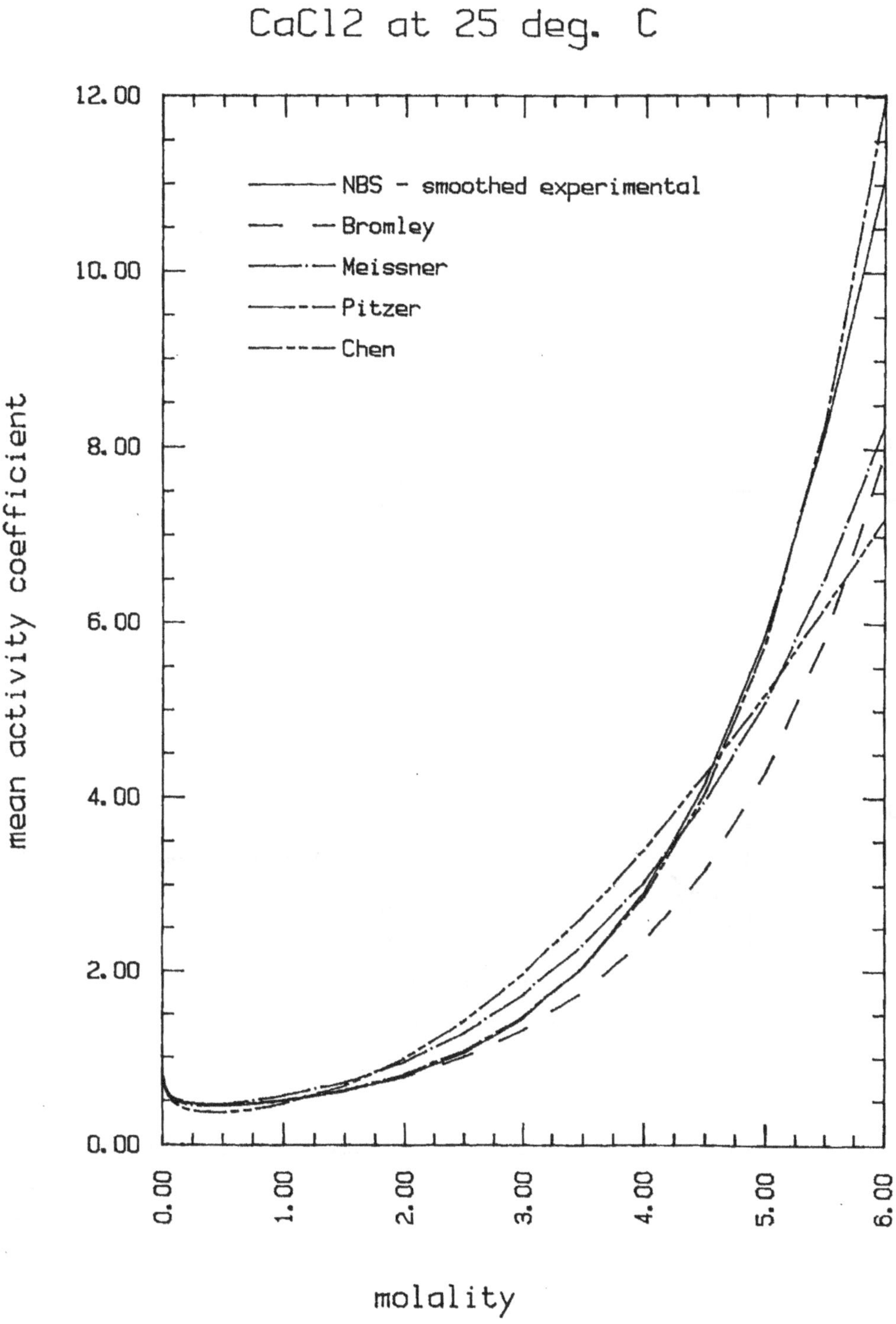
CaCl2 at 25 deg. C
12.00
10.00
8.00
6.00
4.00
2.00
0.00
mean activity coefficient
NBS - smoothed experimental
Bromley
Meissner
Pitzer
Chen
0.00
1.00
2.00
3.00
4.00
5.00
6.00
molality

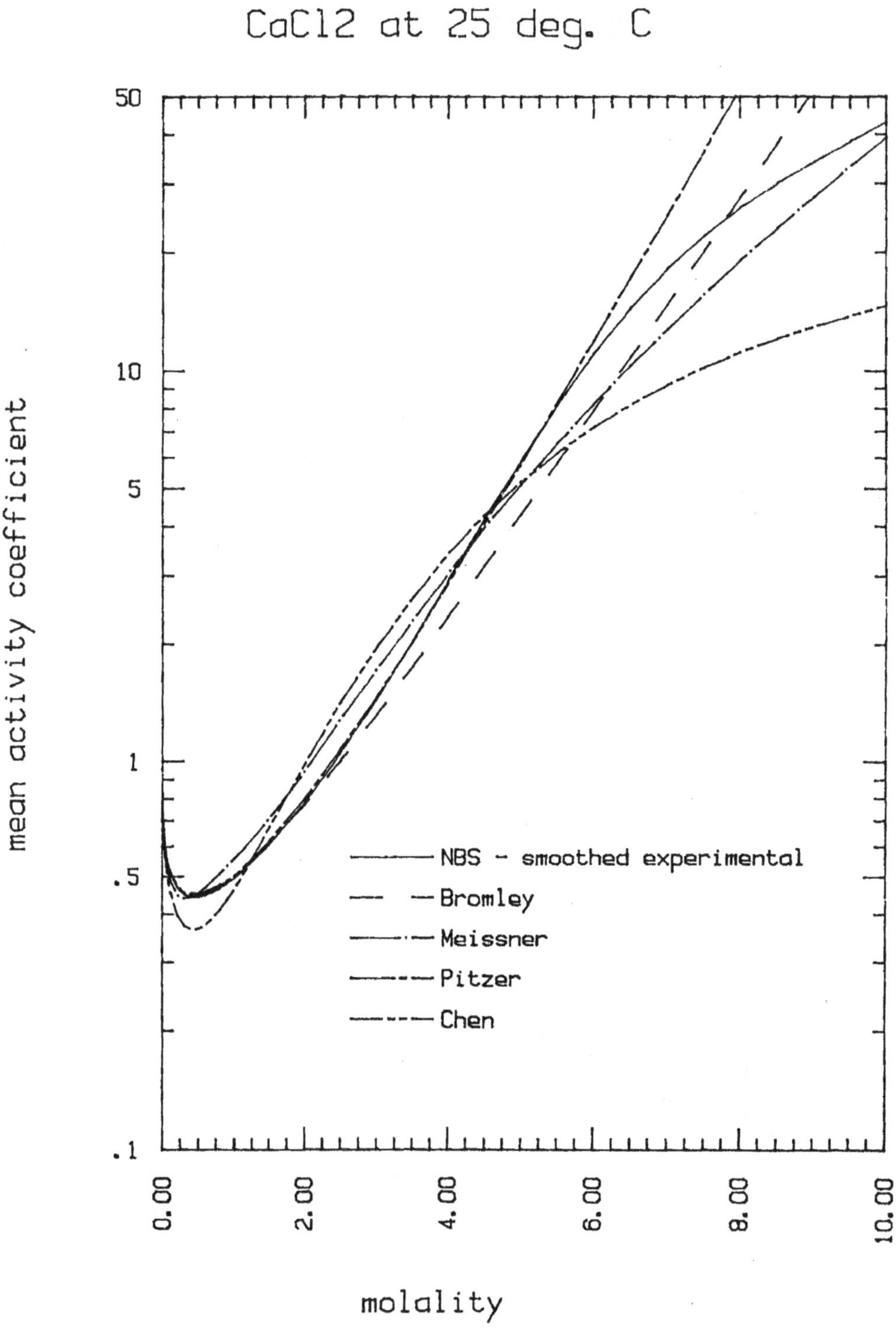
CaCl2 at 25 deg. C
50
10
5
1
.5
.1
mean activity coefficient
0.00
2.00
4.00
6.00
8.00
10.00
molality
NBS - smoothed experimental
Bromley
Meissner
Pitzer
Chen

Na_2SO_4 PARAMETERS

Bromley	B = -0.0204	maximum molality = 2.0
Meissner	q = -0.19	maximum molality = 3.0
Pitzer	$\frac{4}{3}\ \beta_0 = 0.0261$	maximum molality = 4.0
	$\frac{4}{3}\ \beta_1 = 1.484$	
	$\frac{2^{5/2}}{3}\ C^{\phi} = 0.01075$	
Chen	$\tau_{m,ca} = 8.389$	maximum molality = 4.0
	$\tau_{ca,m} = -4.539$	
NBS	B = 1.215973148	maximum molality = 4.445
	C = -.3557285519	saturation molality = 1.957
	D = .08294655619	
	E = $-.4869541257 \times 10^{-2}$	

Na2SO4

COMPARISON OF ACTIVITY COEFFICIENT METHODS

molality	NBS	Bromley	Meissner	Pitzer	Chen
.001	.88589	.88543	.88490	.88657	.88681
.002	.84599	.84523	.84419	.84710	.84741
.003	.81778	.81679	.81528	.81926	.81958
.004	.79546	.79429	.79232	.79727	.79757
.005	.77682	.77550	.77310	.77893	.77917
.006	.76072	.75930	.75648	.76312	.76329
.007	.74653	.74501	.74180	.74918	.74927
.008	.73380	.73221	.72861	.73670	.73670
.009	.72225	.72060	.71664	.72539	.72529
.01	.71168	.70998	.70566	.71504	.71482
.02	.63694	.63525	.62787	.64204	.64059
.03	.58995	.58866	.57889	.59625	.59357
.04	.55572	.55495	.54325	.56288	.55907
.05	.52888	.52869	.51549	.53668	.53187
.06	.50689	.50727	.49266	.51517	.50948
.07	.48832	.48924	.47354	.49694	.49048
.08	.47228	.47372	.45710	.48116	.47402
.09	.45820	.46011	.44273	.46725	.45952
.1	.44568	.44802	.43001	.45483	.44659
.2	.36559	.37062	.35060	.37417	.36341
.3	.32120	.32714	.30862	.32831	.31745
.4	.29101	.29716	.28119	.29672	.28666
.5	.26842	.27448	.26132	.27300	.26407
.6	.25056	.25637	.24602	.25429	.24658
.7	.23592	.24138	.23374	.23905	.23256
.8	.22363	.22866	.22360	.22636	.22102
.9	.21312	.21765	.21502	.21562	.21135
1.0	.20403	.20797	.20765	.20640	.20311
1.5	.17246	.17222	.18150	.17513	.17525
2.0	.15458	.14819	.16450	.15792	.15922
2.5	.14444	.13022	.15216	.14821	.14875
3.0	.13938	.11592	.14269	.14325	.14125
3.5	.13796	.10412	.13510	.14172	.13546
4.0	.13933	.09410	.12883	.14296	.13072
4.5	.14289	.08544	.12352	.14667	.12665
5.0	.14816	.07785	.11895	.15276	.12303
std. dev.		.29342	.12742	.05200	.08698

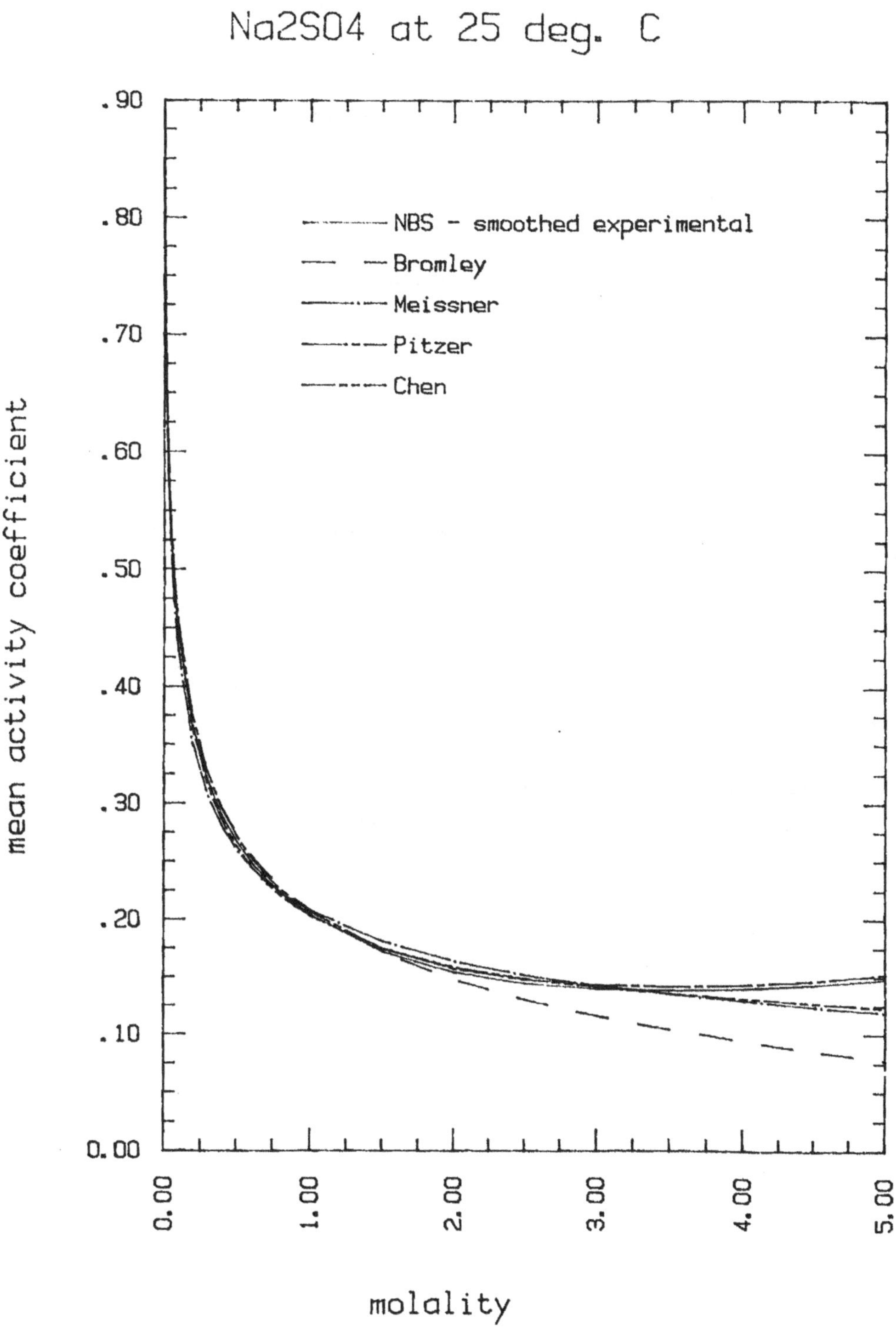
Na2SO4 at 25 deg. C
.90
.80
.70
.60
.50
.40
.30
.20
.10
0.00
NBS - smoothed experimental
Bromley
Meissner
Pitzer
Chen
mean activity coefficient
0.00
1.00
2.00
3.00
4.00
5.00
molality

$MgSO_4$ PARAMETERS

Bromley	$B = -0.0153$	maximum molality = 1.5
Meissner	$q = .15$	maximum molality = 2.25
Pitzer	$\beta_0 = .221$	maximum molality = 3.
	$\beta_1 = 3.343$	
	$\beta_2 = -37.23$	
	$C^{\phi} = 0.0250$	
Chen	$\tau_{m,ca} = 11.346$	maximum molality = 3.5
	$\tau_{ca,m} = -6.862$	

The experimental data points for $MgSO_4$ were taken from the following source:

R.A. Robinson and R.H. Stokes, Electrolyte Solutions, 2nd ed., Butterworth and Co., London, 1970, p. 502

MgSO4

COMPARISON OF ACTIVITY COEFFICIENT METHODS

molality	Robin & Stokes	Bromley	Meissner	Pitzer	Chen
.1	.15000	.18343	.16545	.16646	.17689
.2	.10700	.13150	.11326	.12003	.11868
.3	.08740	.10693	.09086	.09846	.09264
.4	.07560	.09142	.07809	.08535	.07760
.5	.06750	.08032	.06973	.07634	.06782
.6	.06160	.07182	.06379	.06973	.06101
.7	.05710	.06504	.05931	.06467	.05608
.8	.05360	.05948	.05582	.06070	.05239
.9	.05080	.05482	.05301	.05753	.04960
1.0	.04850	.05084	.05072	.05498	.04746
1.2	.04530	.04442	.04723	.05126	.04457
1.4	.04340	.03944	.04474	.04887	.04296
1.6	.04230	.03546	.04288	.04747	.04221
1.8	.04170	.03219	.04145	.04683	.04209
2.0	.04170	.02945	.04031	.04684	.04244
2.5	.04390	.02422	.03832	.04929	.04479
3.0	.04920	.02046	.03708	.05503	.04865
std. dev.		.24476	.08109	.12927	.08140

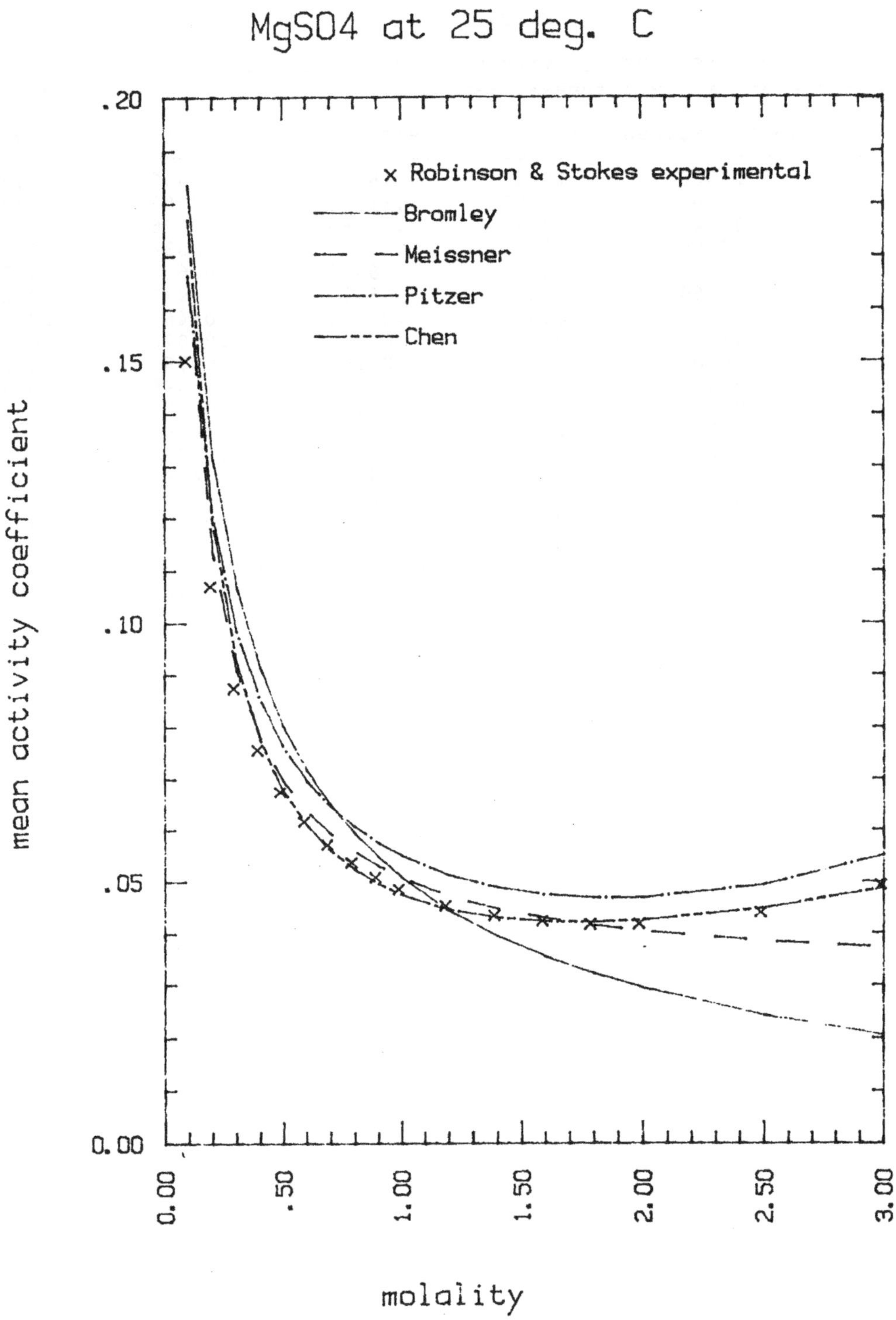
MgSO4 at 25 deg. C
x Robinson & Stokes experimental
Bromley
Meissner
Pitzer
Chen
mean activity coefficient
.20
.15
.10
.05
0.00
0.00
.50
1.00
1.50
2.00
2.50
3.00
molality

BROMLEY'S EXTENDED EQUATION

As shown earlier, Bromley (B3) presented an extended form of his equation, designed to help compensate for some of the strong ion associations that may occur, and published corresponding parameters for some bivalent metal sulfates. The following pages compare the results of using the two equations for $MgSO_4$.

The extended equation for symmetrical salts, such as the 2-2 $MgSO_4$, is:

$$\log \gamma_{\pm} = - \frac{A|z_+z_-|\sqrt{I}}{1+\sqrt{I}} + \frac{(0.06 + .6B)|z_+z_-|\, I}{\left(1 + \frac{1.5}{|z_+z_-|} I\right)^2} + B\, I$$

$$- E\alpha\sqrt{I}\ \{1 - \exp(-\alpha\sqrt{I}\,)\} \qquad (4.44a)$$

with: α - an arbitrary constant set equal to 70 for 2-2 salts

E - new parameter for the extension

For the standard Bromley equation for $MgSO_4$:

B = -.0153 maximum molality = 1.25

For the extended Bromley equation for $MgSO_4$:

α = 70. maximum molality = 4

B = .1042

E = .00464

The experimental points are once again taken from Robinson and Stokes.

MgSO4

COMPARISON OF ACTIVITY COEFFICIENT METHODS

molality	Robin & Stokes	Bromley	Extd. Bromley
.1	.15000	.18343	.15581
.2	.10700	.13150	.11476
.3	.08740	.10693	.09559
.4	.07560	.09142	.08330
.5	.06750	.08032	.07443
.6	.06160	.07182	.06769
.7	.05710	.06504	.06243
.8	.05360	.05948	.05826
.9	.05080	.05482	.05493
1.0	.04850	.05084	.05226
1.2	.04530	.04442	.04844
1.4	.04340	.03944	.04610
1.6	.04230	.03546	.04483
1.8	.04170	.03219	.04440
2.0	.04170	.02945	.04465
2.5	.04390	.02422	.04777
3.0	.04920	.02046	.05414
std. dev.		.24476	.08332

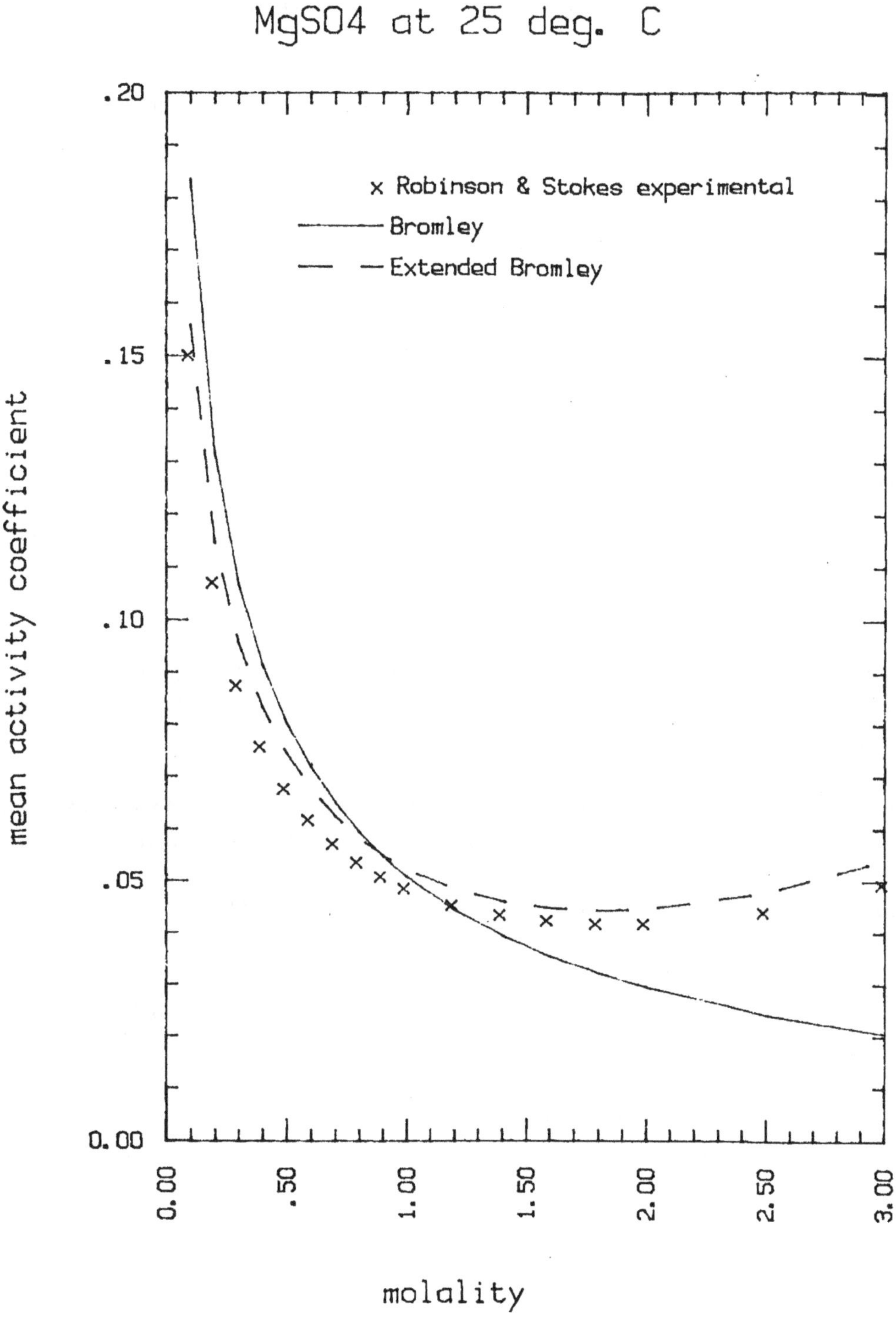
MgSO4 at 25 deg. C
x Robinson & Stokes experimental
Bromley
Extended Bromley
mean activity coefficient
.20
.15
.10
.05
0.00
0.00
.50
1.00
1.50
2.00
2.50
3.00
molality

COMPARISON OF TEMPERATURE EFFECT METHODS

As mentioned earlier, most experimental data is measured at 25°C. This limits the applicability of the published parameters for the activity coefficient models discussed. The following pages compare available experimental data and the activity coefficients calculated using Bromley, Meissner, Pitzer and Chen's models at temperatures other that 25°C.

Bromley

As suggested by Bromley, Meissner's equation was used:

$$\log(\Gamma^\circ_t) = (1.125 - .005t)\log(\Gamma^\circ_{25°C}) - (.125 - .005t)\log(\Gamma^\circ_{ref}) \qquad (4.94)$$

where: t - degrees Celsius

$$\log(\Gamma^\circ_{ref}) = -\frac{.41\sqrt{I}}{1+\sqrt{I}} + .039\, I^{.92} \qquad (4.94a)$$

$$\Gamma^\circ = \gamma_\pm^{1/z_+z_-}$$

This equation calculates the reduced activity coefficient for temperature t at a given ionic strength from the reduced activity coefficient at 25°C. The calculations were done as follows:

1) The Bromley B parameter for the species at 25°C was used to calculate the activity coefficient $\gamma_\pm$ for ionic strength I with Bromley's equation (4.44).

2) $\Gamma^\circ_{25°C}$ was calculated using this $\gamma_\pm$ with equation (4.45)

3) Calculating $\log(\Gamma^\circ_{ref})$ with equation (4.94a) and plugging it and the $\Gamma^\circ_{25°C}$ into equation (4.94), the reduced activity coefficient at t and I was determined.

4) The mean activity coefficient at this temperature and ionic strength was then calculated:

$$\log \gamma_t = z_+ z_- \log(\Gamma^\circ_t)$$

This method of calculating temperature effects is dependent upon the validity of the 25°C activity coefficient.

Meissner

Meissner suggested the following equation for adjusting his 25°C q parameter to another temperature:

$$q_t^\circ = q_{25°C}^\circ \left[1 - \frac{0.0027(t-25)}{|z_+ z_-|} \right] \qquad (4.96)$$

with t in degrees Celsius. The resulting $q_t°$ is not restricted by ionic strength.

Pitzer and Chen

As suggested by Pitzer, the dielectric constant used in calculating the Debye-Huckel constant A_ϕ is depended upon to reflect the effect of temperature on the activity coefficients. Either of the following methods may be used:

1) Use Pitzer's expression for the dielectric constant in calculating A_ϕ :

$$D = \sum_{i=0}^{4} \sum_{j=0}^{4-i} e_{ij}\, t^i\, d_w^j \qquad (4.98)$$

with: d_w - density of pure water
t - degrees Celsius
e - parameters fit via least-squares method, given in table 4.1

2) Use Chen's equation for calculating A_ϕ , regressed to fit the A_ϕ's calculated using Pitzer's dielectric constant:

$$A_\phi = -61.44534 \exp\left(\frac{T-273.15}{273.15}\right) + 2.864468 \left[\exp\left(\frac{T-273.15}{273.15}\right)\right]^2$$
$$+ 183.5379 \ln\left(\frac{T}{273.15}\right) - .6820223(T-273.15)$$
$$+ .0007875695(T^2 - (273.15)^2) + 58.95788\left(\frac{273.15}{T}\right) \qquad (4.100)$$

T is in Kelvins

Experimental data

For each species for which the comparison was done, a page describing the source of the experimental data and the parameters used for each model preceeds a listing of the calculated and experimental activity coefficients and a plot of these results.

HCl AT 50° CELSIUS

25°C parameters:

Bromley	$B = 0.1433$		maximum molality = 6.0
Meissner	$q = 6.69$		maximum molality = 4.5 to 6
Pitzer	$\beta_0 = 0.1775$		maximum molality = 6.0
	$\beta_1 = 0.2945$		
	$C^{\phi} = 0.0008$		
Chen	$\tau_{m,ca} = 10.089$		maximum molality = 6.0
	$\tau_{ca,m} = -5.212$		

Experimental data:

From: H.S. Harned and B.B. Owen, The Physical Chemistry of Electrolytic Solutions, third edition, Reinhold Publishing Corporation, New York (1958) p. 716

Harned and Owen presented tabulated values for the mean activity coefficients of HCl at temperatures from 0 to 60°C for maximum molalities from 2 to 4. The coefficients are from observed electromotive forces for molalities greater than .001; the values for molalities less than .002 were extrapolated from plots.

HCl

COMPARISON OF ACTIVITY COEFFICIENT METHODS

molality	Harned & Owen	Bromley	Meissner	Pitzer	Chen
.0001	.98790	.98876	.98848	.98795	.98795
.0002	.98310	.98424	.98385	.98311	.98311
.0005	.97380	.97551	.97490	.97376	.97374
.001	.96390	.96602	.96517	.96361	.96356
.002	.95000	.95321	.95201	.94992	.94981
.005	.92500	.92978	.92787	.92490	.92456
.01	.90000	.90627	.90347	.89977	.89900
.02	.86900	.87761	.87333	.86900	.86734
.05	.82110	.83374	.82581	.82136	.81704
.1	.78500	.79992	.78727	.78393	.77569
.2	.75080	.77143	.75258	.75203	.73791
.5	.73440	.75804	.73271	.73873	.71696
1.0	.76970	.79710	.77351	.78478	.76855
1.5	.84040	.86884	.85262	.86484	.86668
2.0	.93270	.96388	.95876	.97014	.99691
std. dev.		.07043	.03191	.05038	.07755

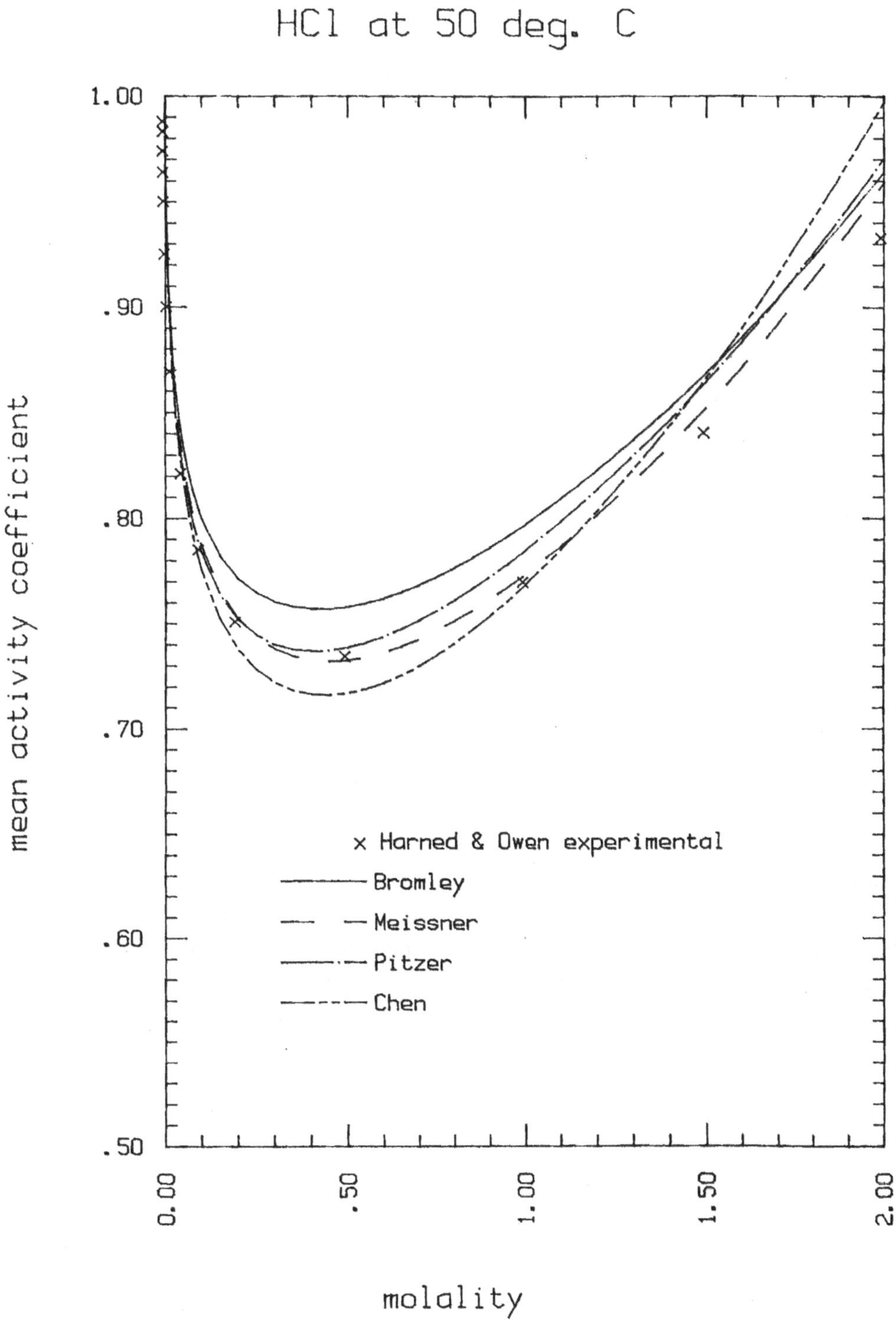
HCl at 50 deg. C
1.00
.90
.80
.70
.60
.50
mean activity coefficient
× Harned & Owen experimental
Bromley
Meissner
Pitzer
Chen
0.00
.50
1.00
1.50
2.00
molality

KCl AT 80° CELSIUS

25°C parameters:

Bromley	$B = 0.0240$		maximum molality = less than 6
Meissner	$q = .92$		maximum molality = 4.5 to 6
Pitzer	$\beta_0 = 0.04835$		maximum molality = 4.8
	$\beta_1 = 0.2122$		
	$C^{\phi} = -0.00084$		
Chen	$\tau_{m,ca} = 8.064$		maximum molality = 4.5
	$\tau_{ca,m} = -4.107$		

Experimental data:

From: H.P. Snipes, C. Manly and D.D. Ensor, "Heats of Dilution of Aqueous Electrolytes: Temperature Dependence", J. Chem. Eng. Data, v20, no. 3, 287 (1975)

Snipes et al. measured the heats of dilution of KCl over a concentration range of .005 to 2 molal for temperatures from 40 to 80°C. They used the data to fit the relative apparent molal heat content to a polynomial equation; calculated the relative partial molal heat contents of the solvent and solute from the relative apparent molal heat content data. They then fit the partial molal heat contents of the solute to a polynomial equation of the form:

$$\bar{L}_2 = f + gT + hT^2 \ldots \qquad (4.101)$$

Relating the mean activity coefficient to $\bar{L}_2$:

$$\int d \ln \gamma = \int -\bar{L}_2/\nu RT^2 \; dT \qquad (4.102)$$

replacing the $\bar{L}_2$ of equation (4.102) with (4.101) and integrating, using as a reference temperature 25°C, they obtained the following equation for calculating the activity coefficients:

$$\ln \gamma_T = \ln \gamma_{Tr} - \frac{1}{\nu R}\left[f\left(\frac{1}{Tr} - \frac{1}{T}\right) + q\left(\ln \frac{T}{Tr}\right) + h(T - Tr)\right] \tag{4.103}$$

with T and Tr in Kelvins. Not taking into account any uncertainties in the 25°C activity coefficients, they estimate the uncertainty of their derived activity coefficients for KCl to be ±0.001.

KCl

COMPARISON OF ACTIVITY COEFFICIENT METHODS

molality	Snipes et al	Bromley	Meissner	Pitzer	Chen
.1	.75200	.77806	.76020	.74071	.74457
.2	.69900	.72942	.70628	.68359	.68814
.3	.66900	.70057	.67458	.64957	.65452
.4	.64800	.68044	.65276	.62571	.63105
.5	.63200	.66530	.63655	.60761	.61338
.6	.62000	.65341	.62392	.59322	.59947
.7	.61000	.64382	.61379	.58141	.58818
.8	.60400	.63595	.60548	.57152	.57884
.9	.59600	.62941	.59855	.56310	.57099
1.0	.59200	.62394	.59270	.55585	.56433
1.2	.58300	.61546	.58343	.54403	.55370
1.4	.57700	.60946	.57650	.53491	.54575
1.6	.57300	.60529	.57122	.52778	.53974
1.8	.57100	.60251	.56715	.52219	.53519
2.0	.57000	.60085	.56399	.51782	.53177
2.5	.56900	.60032	.55880	.51091	.52676
3.0	.57200	.60348	.55615	.50814	.52511
3.5	.57700	.60924	.55522	.50829	.52552
4.0	.58200	.61695	.55571	.51063	.52722
std. dev.		.17966	.05542	.24338	.18175

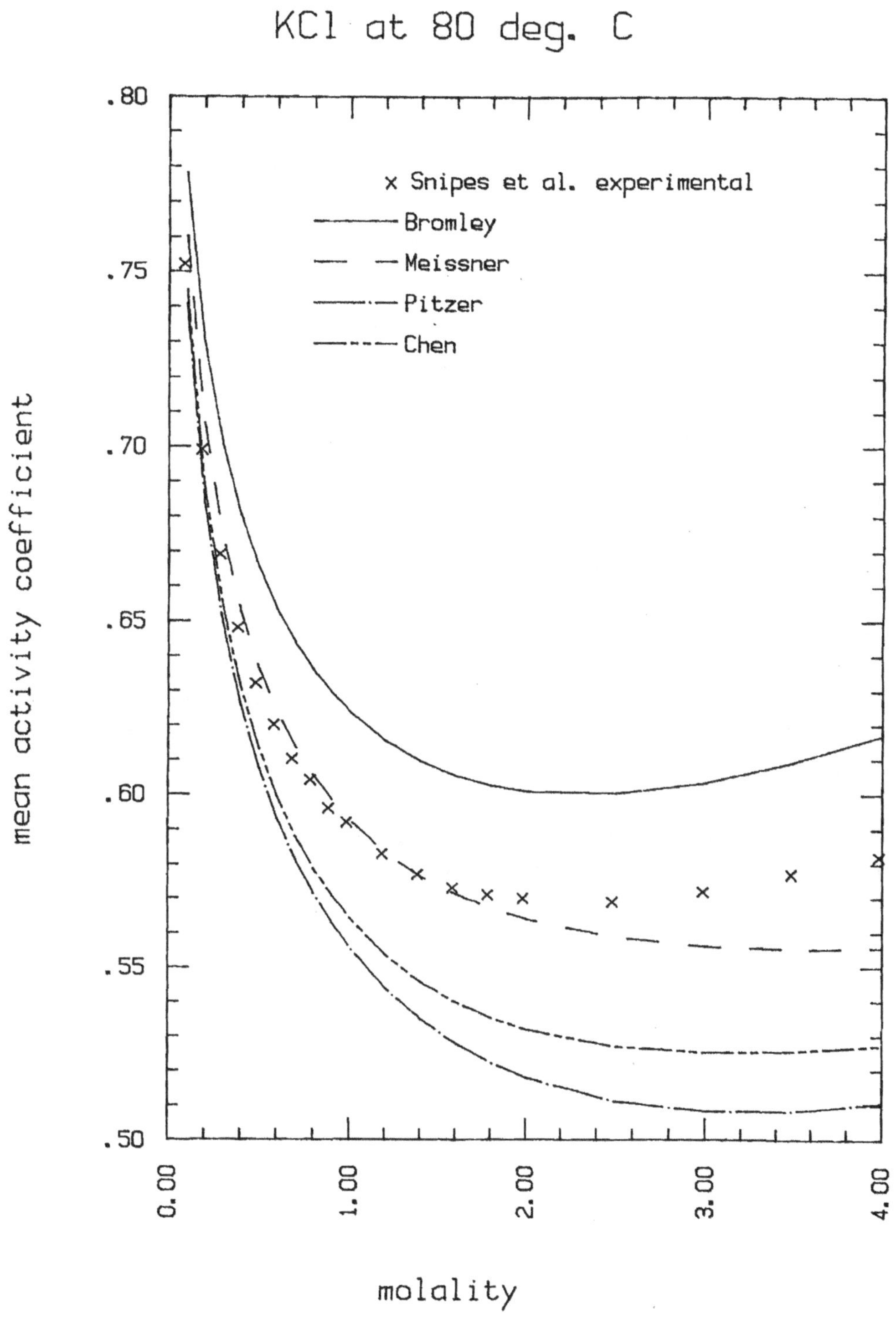
KCl at 80 deg. C
x Snipes et al. experimental
Bromley
Meissner
Pitzer
Chen
mean activity coefficient
.80
.75
.70
.65
.60
.55
.50
0.00
1.00
2.00
3.00
4.00
molality

KOH AT 80° CELSIUS

25°C parameters:

Bromley	$B = 0.1131$	maximum molality = 6.
Meissner	$q = 4.77$	maximum molality = 4.5 to 6
Pitzer	$\beta_0 = 0.1298$	maximum molality = 5.5
	$\beta_1 = 0.320$	
	$C^{\phi} = 0.0041$	
Chen	$\tau_{m,ca} = 9.868$	maximum molality = 6.
	$\tau_{ca,m} = -5.059$	

Experimental data:

From: H.S. Harned and B.B. Owen, The Physical Chemistry of Electrolytic Solutions, third edition, Reinhold Publishing Corporation, New York (1958) p. 730

Harned and Owen presented tabulated values for the mean activity coefficients of KOH for temperatures of 0 to 80°C from .1 to 17 molal.

KOH

COMPARISON OF ACTIVITY COEFFICIENT METHODS

molality	Harned & Owen	Bromley	Meissner	Pitzer	Chen
.1	.75900	.79506	.78014	.76269	.75465
.2	.71800	.75942	.74080	.72141	.70789
.4	.68300	.73229	.71266	.69011	.67115
1.0	.68000	.73442	.72376	.69050	.67696
1.5	.71000	.76862	.76689	.72621	.72906
2.0	.75500	.81810	.82375	.78017	.80401
4.0	1.04300	1.12669	1.13403	1.16756	1.25003
6.0	1.53000	1.62242	1.64700	1.96099	1.84442
8.0	1.82000	2.37692	2.43460	3.57329	2.50274
10.0	3.42000	3.51297	3.53165	6.97186	3.15892
12.0	5.11000	5.21881	5.00519	14.47002	3.76629
14.0	7.20000	7.77865	6.93177	31.82804	4.30244
17.0	11.40000	14.21175	10.85727	115.21747	4.95490
std. dev.		.97452	.53132	32.43989	2.35567

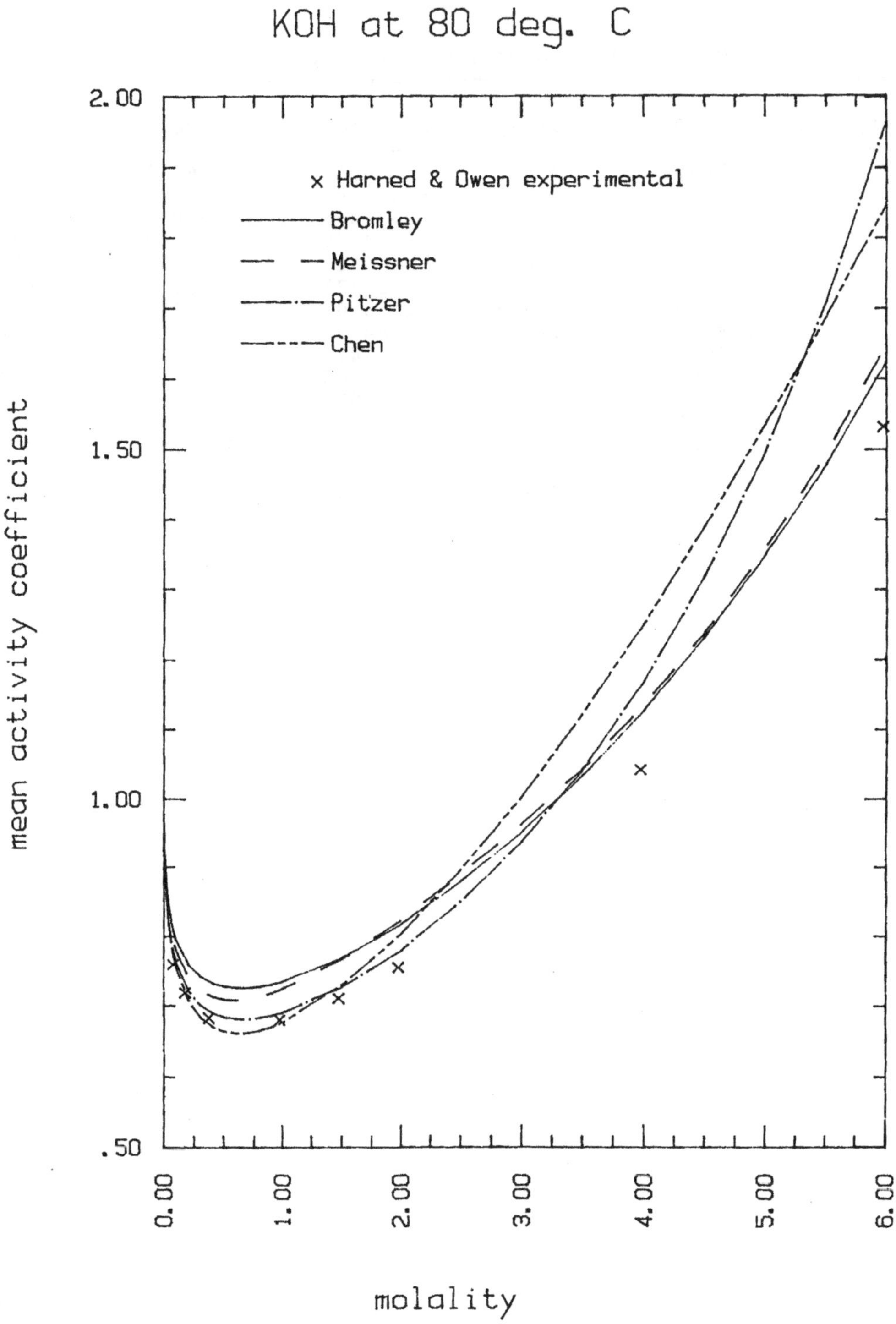
KOH at 80 deg. C
× Harned & Owen experimental
Bromley
Meissner
Pitzer
Chen
mean activity coefficient
2.00
1.50
1.00
.50
0.00
1.00
2.00
3.00
4.00
5.00
6.00
molality

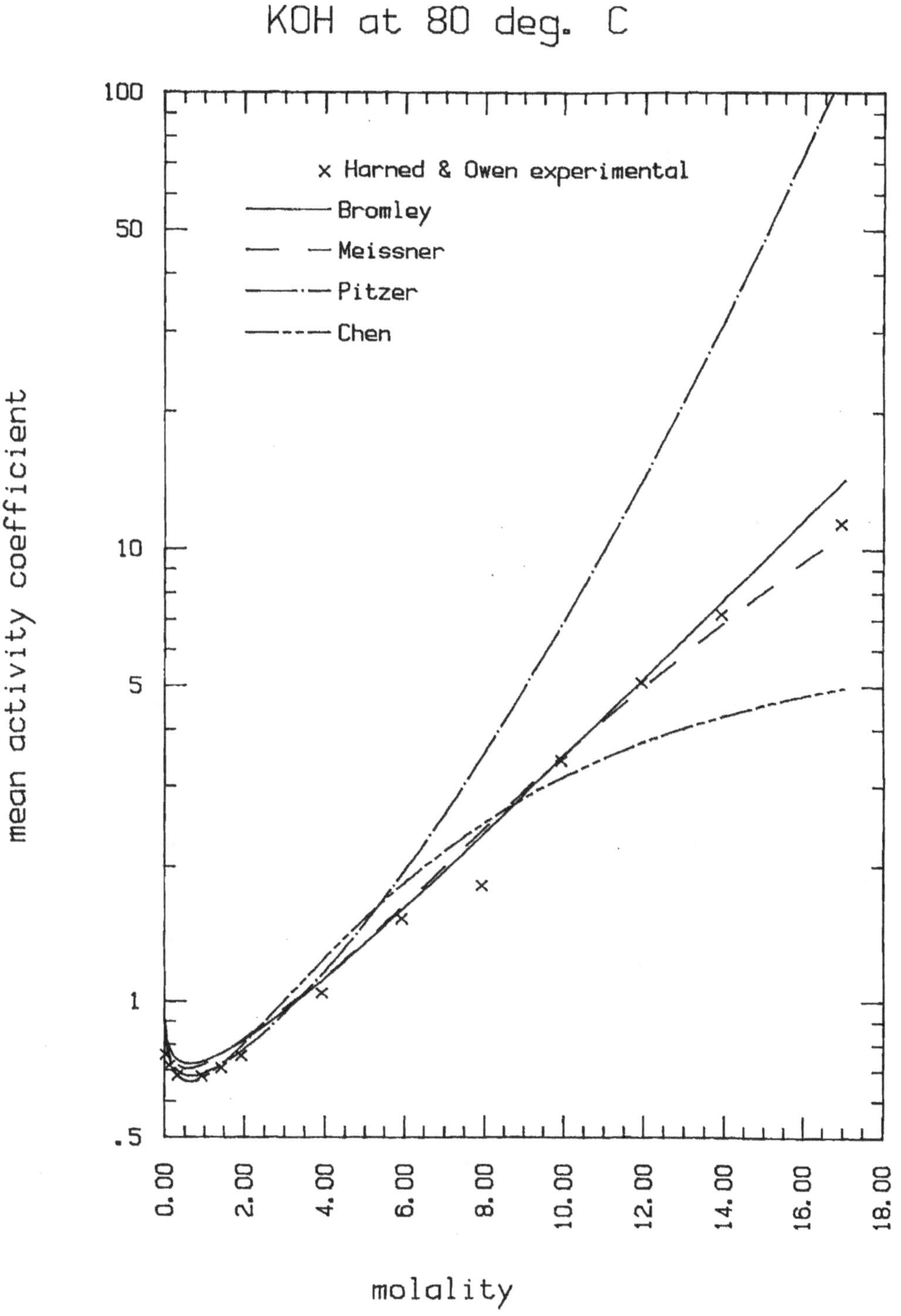
KOH at 80 deg. C
× Harned & Owen experimental
Bromley
Meissner
Pitzer
Chen
mean activity coefficient
100
50
10
5
1
.5
0.00
2.00
4.00
6.00
8.00
10.00
12.00
14.00
16.00
18.00
molality

NaCl AT 100 AND 300° CELSIUS

25°C parameters:

Bromley	$B = 0.0574$		maximum molality = 6
Meissner	$q = 2.23$		maximum molality = 3 to 4
Pitzer	$\beta_0 = 0.0765$		maximum molality = 6.
	$\beta_1 = 0.2664$		
	$C^{\phi} = 0.00127$		
Chen	$\tau_{m,ca} = 8.885$		maximum molality = 6.
	$\tau_{ca,m} = -4.459$		

Experimental data:

From: L.F. Silvester and K.S. Pitzer, "Thermodynamics of Geothermal Brines I. Thermodynamic Properties of Vapor-Saturated NaCl (aq) Solutions From 0 - 300°C", Lawrence Berkeley Laboratory Report LBL-4456, January 1976

Using various types of experimental data, (i.e. freezing point depression, isopiestic, vapor pressure lowering, emf) Silvester and Pitzer regressed the data to fit into the Pitzer parameters:

$$B_0 \equiv \frac{2\,\nu_+\,\nu_-}{\nu}\,\beta_0 \qquad B_1 \equiv \frac{2\,\nu_+\,\nu_-}{\nu}\,\beta_1 \qquad C \equiv \frac{2(\nu_+\,\nu_-)^{3/2}}{\nu}\,C^{\phi}$$

For the 1-1 salt NaCl, the stoichiometric coefficients for the B's and C are equal to unity. The temperature dependent equations for calculating these parameters are:

$$B_0 = q_1 + q_2\left(\frac{1}{T} - \frac{1}{Tr}\right) + q_3\left(\ln\frac{T}{Tr}\right) + q_4\,(T - Tr) + q_5(T^2 - Tr^2)$$

$$B_1 = q_6 + q_9\,(T - Tr) + q_{10}\,(T^2 - Tr^2)$$

$$C = q_{11} + q_{12}\left(\frac{1}{T} - \frac{1}{Tr}\right) + q_{13}\left(\ln \frac{T}{Tr}\right) + q_{14}(T - Tr)$$

with: T - desired temperature in Kelvins

Tr - reference temperature of 25°C, in Kelvins = 298.15

The fit coefficients for these equations are:

q_1 = 0.0765
q_2 = -777.03
q_3 = -4.4706
q_4 = 0.008946
q_5 = -3.3158×10^{-6}

q_6 = 0.2664
q_9 = 6.1608×10^{-5}
q_{10} = 1.0715×10^{-6}

q_{11} = 0.00127
q_{12} = 33.317
q_{13} = 0.09421
q_{14} = -4.655×10^{-5}

NaCl

COMPARISON OF ACTIVITY COEFFICIENT METHODS

molality	LBL	Bromley	Meissner	Pitzer	Chen
.001	.95908	.96742	.96481	.95900	.95912
.002	.94362	.95496	.95132	.94343	.94363
.003	.93230	.94577	.94137	.93198	.93226
.004	.92309	.93826	.93324	.92266	.92300
.005	.91523	.93183	.92626	.91468	.91508
.006	.90831	.92615	.92010	.90765	.90810
.007	.90211	.92104	.91455	.90134	.90183
.008	.89647	.91638	.90949	.89558	.89612
.009	.89128	.91209	.90482	.89028	.89086
.01	.88647	.90810	.90048	.88536	.88598
.02	.85081	.87823	.86788	.84867	.84951
.03	.82683	.85789	.84556	.82376	.82469
.04	.80859	.84225	.82834	.80464	.80559
.05	.79383	.82950	.81425	.78906	.78999
.06	.78146	.81873	.80233	.77591	.77680
.07	.77081	.80942	.79199	.76453	.76535
.08	.76149	.80122	.78287	.75451	.75525
.09	.75322	.79390	.77473	.74555	.74622
.1	.74580	.78730	.76738	.73748	.73806
.2	.69763	.74350	.71875	.68373	.68366
.3	.67152	.71876	.69173	.65310	.65287
.4	.65486	.70237	.67424	.63254	.63260
.5	.64344	.69076	.66215	.61764	.61837
.6	.63535	.68228	.65351	.60638	.60807
.7	.62957	.67601	.64726	.59767	.60056
.8	.62548	.67141	.64276	.59084	.59511
.9	.62269	.66811	.63957	.58547	.59128
1.0	.62093	.66586	.63741	.58128	.58873
1.5	.62248	.66523	.63638	.57216	.58887
2.0	.63454	.67554	.64404	.57551	.60186
2.5	.65293	.69244	.65565	.58683	.62201
3.0	.67574	.71402	.66937	.60423	.64661
3.5	.70191	.73933	.68472	.62682	.67406
4.0	.73079	.76782	.70195	.65418	.70335
4.5	.76188	.79922	.72148	.68616	.73375
5.0	.79481	.83335	.74347	.72281	.76473
5.5	.82927	.87016	.76774	.76428	.79586
6.0	.86497	.90961	.79387	.81081	.82682
std. dev.		.26577	.16979	.27385	.15370

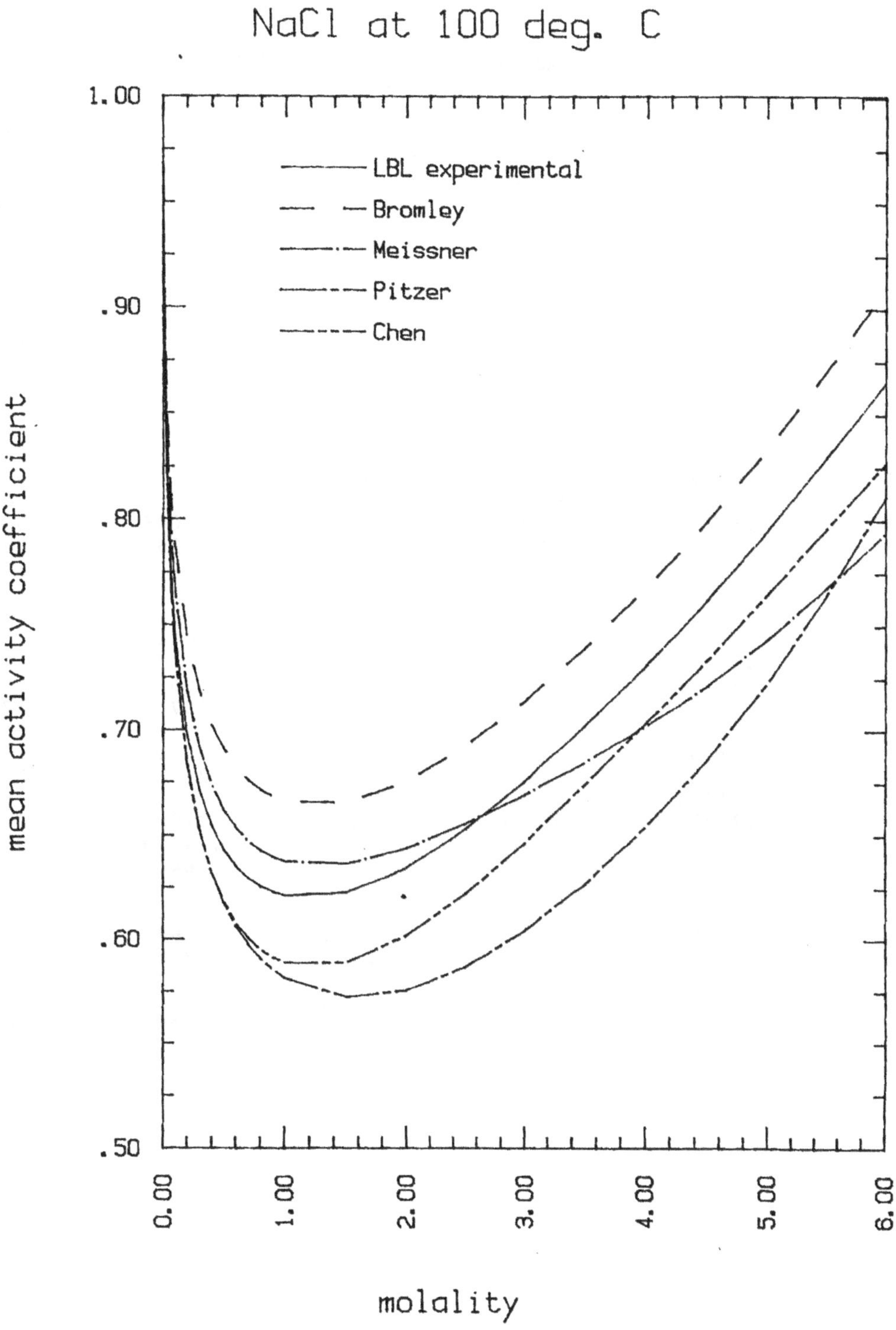
NaCl at 100 deg. C
LBL experimental
Bromley
Meissner
Pitzer
Chen
mean activity coefficient
1.00
.90
.80
.70
.60
.50
0.00
1.00
2.00
3.00
4.00
5.00
6.00
molality

NaCl

COMPARISON OF ACTIVITY COEFFICIENT METHODS

molality	LBL	Bromley	Meissner	Pitzer	Chen
.001	.91626	.97416	.96467	.91576	.91655
.002	.88534	.96412	.95105	.88441	.88587
.003	.86292	.95667	.94097	.86160	.86367
.004	.84486	.95054	.93272	.84317	.84579
.005	.82953	.94526	.92562	.82749	.83065
.006	.81614	.94058	.91934	.81376	.81741
.007	.80418	.93634	.91367	.80149	.80560
.008	.79336	.93246	.90849	.79035	.79492
.009	.78345	.92887	.90371	.78015	.78514
.01	.77429	.92552	.89926	.77071	.77611
.02	.70750	.90004	.86562	.70144	.71026
.03	.66359	.88221	.84234	.65554	.66698
.04	.63065	.86823	.82421	.62093	.63452
.05	.60429	.85663	.80927	.59314	.60854
.06	.58235	.84671	.79653	.56994	.58691
.07	.56359	.83801	.78540	.55006	.56842
.08	.54724	.83028	.77553	.53269	.55230
.09	.53276	.82331	.76664	.51731	.53805
.1	.51980	.81698	.75857	.50352	.52529
.2	.43554	.77394	.70345	.41383	.44283
.3	.38841	.74905	.67070	.36401	.39758
.4	.35651	.73243	.64792	.33070	.36778
.5	.33282	.72054	.63080	.30633	.34634
.6	.31421	.71168	.61731	.28749	.33009
.7	.29905	.70492	.60635	.27239	.31737
.8	.28635	.69969	.59724	.25997	.30716
.9	.27552	.69561	.58952	.24955	.29884
1.0	.26612	.69242	.58291	.24067	.29199
1.5	.23275	.68501	.56031	.21080	.27139
2.0	.21206	.68549	.54734	.19429	.26331
2.5	.19788	.69007	.53914	.18469	.26148
3.0	.18758	.69716	.53369	.17930	.26326
3.5	.17978	.70595	.53009	.17681	.26725
4.0	.17367	.71599	.52794	.17648	.27267
4.5	.16874	.72700	.52702	.17787	.27901
5.0	.16467	.73880	.52716	.18074	.28593
5.5	.16119	.75126	.52813	.18492	.29317
6.0	.15815	.76430	.52973	.19033	.30057
std. dev.		4.66356	3.20162	.19509	.75892

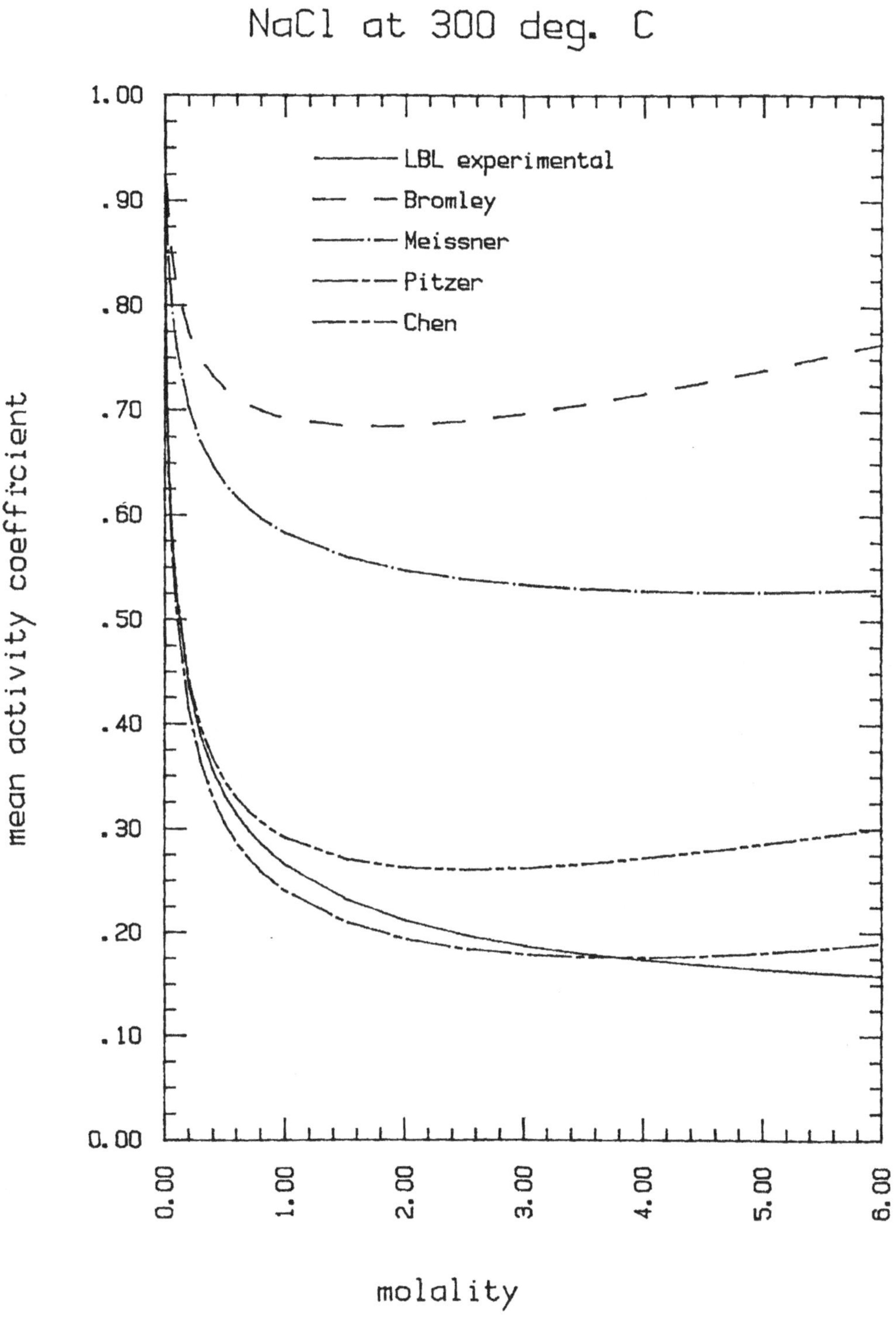
NaCl at 300 deg. C
LBL experimental
Bromley
Meissner
Pitzer
Chen
mean activity coefficient
molality
1.00
.90
.80
.70
.60
.50
.40
.30
.20
.10
0.00
0.00
1.00
2.00
3.00
4.00
5.00
6.00

NaOH AT 35° CELSIUS

25°C parameters:

Bromley	B = 0.0747		maximum molality = 6.
Meissner	q = 3.00		maximum molality = 3 to 4
Pitzer	β_0 = 0.0864		maximum molality = 6.
	β_1 = 0.253		
	C_ϕ = 0.0044		
Chen	$\tau_{m,ca}$ = 9.372		maximum molality = 6.
	$\tau_{ca,m}$ = -4.777		

Experimental data:

From: H.S. Harned and B.B. Owen, The Physical Chemistry of Electrolytic Solutions, third edition, Reinhold Publishing Corporation, New York (1958) p. 729

Harned and Owen presented tabulated values for the mean activity coefficient of NaOH for temperatures from 0 to 35°C, molalities from 0.05 to 4.

NaOH

COMPARISON OF ACTIVITY COEFFICIENT METHODS

molality	Harned & Owen	Bromley	Meissner	Pitzer	Chen
.05	.81600	.82323	.81825	.81722	.81558
.1	.76400	.78218	.77441	.77312	.76941
.25	.71200	.72913	.71817	.71448	.70635
.5	.69400	.69695	.68762	.67759	.66689
1.0	.67800	.68529	.68391	.65985	.65347
1.5	.68300	.69966	.70482	.66890	.67253
2.0	.69800	.72726	.73572	.69227	.70766
2.5	.72600	.76391	.77179	.72659	.75301
3.0	.77200	.80792	.81119	.77087	.80561
3.5	.82200	.85854	.85380	.82514	.86358
4.0	.88200	.91552	.90063	.89001	.92560
std. dev.		.09583	.09771	.03765	.09513

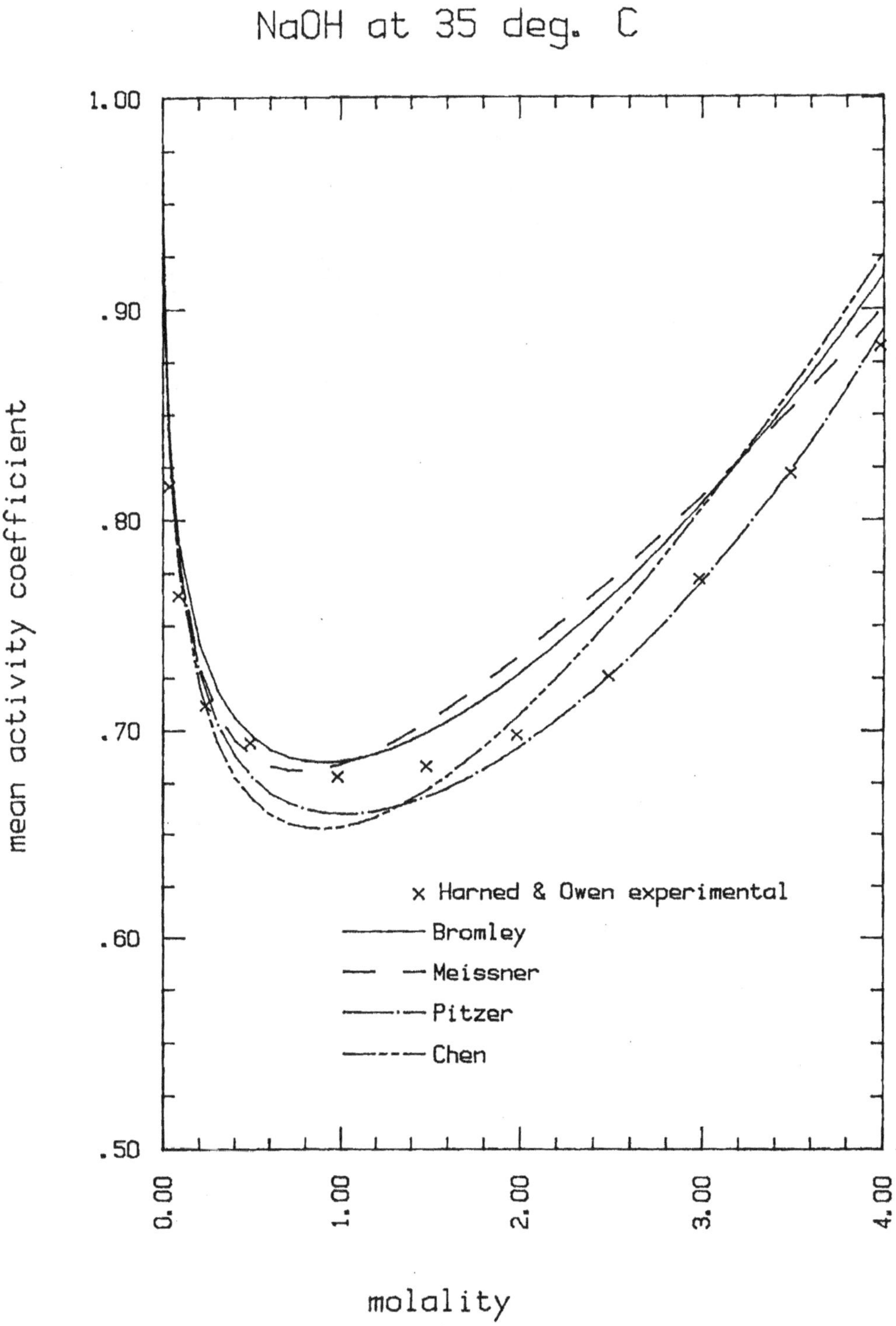
NaOH at 35 deg. C
1.00
.90
.80
.70
.60
.50
mean activity coefficient
0.00
1.00
2.00
3.00
4.00
molality
× Harned & Owen experimental
Bromley
Meissner
Pitzer
Chen

$CaCl_2$ AT 108.85 AND 201.85° CELSIUS

25°C parameters:

Bromley	B = .0948	maximum molality = 2.
Meissner	q = 2.40	maximum molality = 5.0
Pitzer	$4/3\ \beta_0$ = 0.4212	maximum molality = 2.5
	$4/3\ \beta_1$ = 2.152	
	$2^{5/2}/3\ C^{\phi}$ = -0.00064	
Chen	$\tau_{m,ca}$ = 11.396	maximum molality = 6.
	$\tau_{ca,m}$ = -6.218	

Experimental data:

From: H.F. Holmes, C.F. Baes, Jr., and R.E. Mesmer, "Isopiestic studies of aqueous solutions at elevated temperatures I. KCl, $CaCl_2$, and $MgCl_2$", J. Chem. Thermo., v 10, p. 983 (1978)

Using their experimentally determined isopiestic molalities, Mesmer et al. calculated the osmotic coefficients, ϕ, with the following relation:

$$\phi_x = \left[\frac{\nu_s m_s}{\nu_x m_x}\right] \phi_s = R\ \phi_s \qquad (4.104)$$

where: m - molality
R - isopiestic ratio
ν - number of particles per molecule
- subscript denoting the electrolyte whose properties are being calculated
s - standard solution having known properties; for this study, NaCl was used as the reference solution

With the results from equation (4.104), Mesmer et al. used a least-squares equation fitting program to regress the β_0, β_1 and C^{ϕ} of the Pitzer method equation for osmotic coefficients. These parameters may also be used in Pitzer's equation for calculating mean activity coefficients. They note that the validity of the calculated activity coefficients is dependent upon how well the equation for calculating the osmotic coefficients represents dilute solutions for which they had no experimental results.

For $CaCl_2$ they presented the following parameters:

T Kelvins/ Celsius	β_0 kg/mole	β_1 kg/mole	C^{ϕ} kg ^{2}/mole^2
382 K 108.85°C	0.30353	2.27800	-0.008680
413.8 K 140.65°C	0.29054	2.60450	-0.010430
445.4 K 172.25°C	0.28151	2.85450	-0.012690
474 K 201.85°C	0.27035	3.32150	-0.013670

CaCl2

COMPARISON OF ACTIVITY COEFFICIENT METHODS

molality	Mesmer	Bromley	Meissner	Pitzer	Chen
.001	.86788	.89548	.88638	.86650	.86580
.005	.75097	.79659	.77921	.74568	.74208
.01	.68596	.73870	.71641	.67711	.67021
.02	.61617	.67422	.64616	.60214	.58961
.04	.54662	.60787	.57339	.52572	.50461
.06	.50816	.57072	.53249	.48267	.45525
.08	.48251	.54602	.50544	.45364	.42140
.1	.46372	.52815	.48611	.43227	.39626
.2	.41262	.48204	.43974	.37420	.32844
.3	.38888	.46394	.42658	.34794	.30028
.4	.37584	.45694	.42631	.33416	.28855
.5	.36873	.45621	.43280	.32722	.28617
.6	.36557	.45970	.44338	.32474	.29009
.7	.36533	.46637	.45667	.32552	.29882
.8	.36739	.47560	.47185	.32888	.31152
.9	.37138	.48703	.48848	.33440	.32773
1.0	.37705	.50043	.50638	.34183	.34717
1.1	.38422	.51563	.52556	.35101	.36969
1.2	.39278	.53254	.54620	.36186	.39521
1.3	.40263	.55110	.56858	.37432	.42371
1.4	.41373	.57128	.59301	.38838	.45517
1.5	.42603	.59309	.61978	.40405	.48962
1.6	.43951	.61652	.64911	.42136	.52709
1.7	.45416	.64161	.68110	.44036	.56761
1.8	.46997	.66840	.71579	.46111	.61121
1.9	.48696	.69692	.75311	.48370	.65794
2.0	.50513	.72724	.79301	.50821	.70782
2.1	.52451	.75942	.83540	.53475	.76088
2.2	.54512	.79354	.88024	.56345	.81715
2.3	.56698	.82968	.92751	.59443	.87664
2.4	.59012	.86792	.97722	.62784	.93935
2.5	.61459	.90836	1.02943	.66385	1.00529
2.6	.64041	.95111	1.08419	.70261	1.07444
2.7	.66762	.99627	1.14156	.74433	1.14680
2.8	.69628	1.04395	1.20163	.78920	1.22234
2.9	.72641	1.09430	1.26448	.83746	1.30102
3.0	.75808	1.14743	1.33019	.88935	1.38280
3.1	.79132	1.20349	1.39884	.94511	1.46764
3.2	.82619	1.26263	1.47052	1.00505	1.55548
3.3	.86274	1.32500	1.54533	1.06945	1.64626
3.4	.90102	1.39078	1.62334	1.13866	1.73991
3.5	.94110	1.46014	1.70466	1.21303	1.83635
std. dev.		1.93513	2.62573	.62856	2.61780

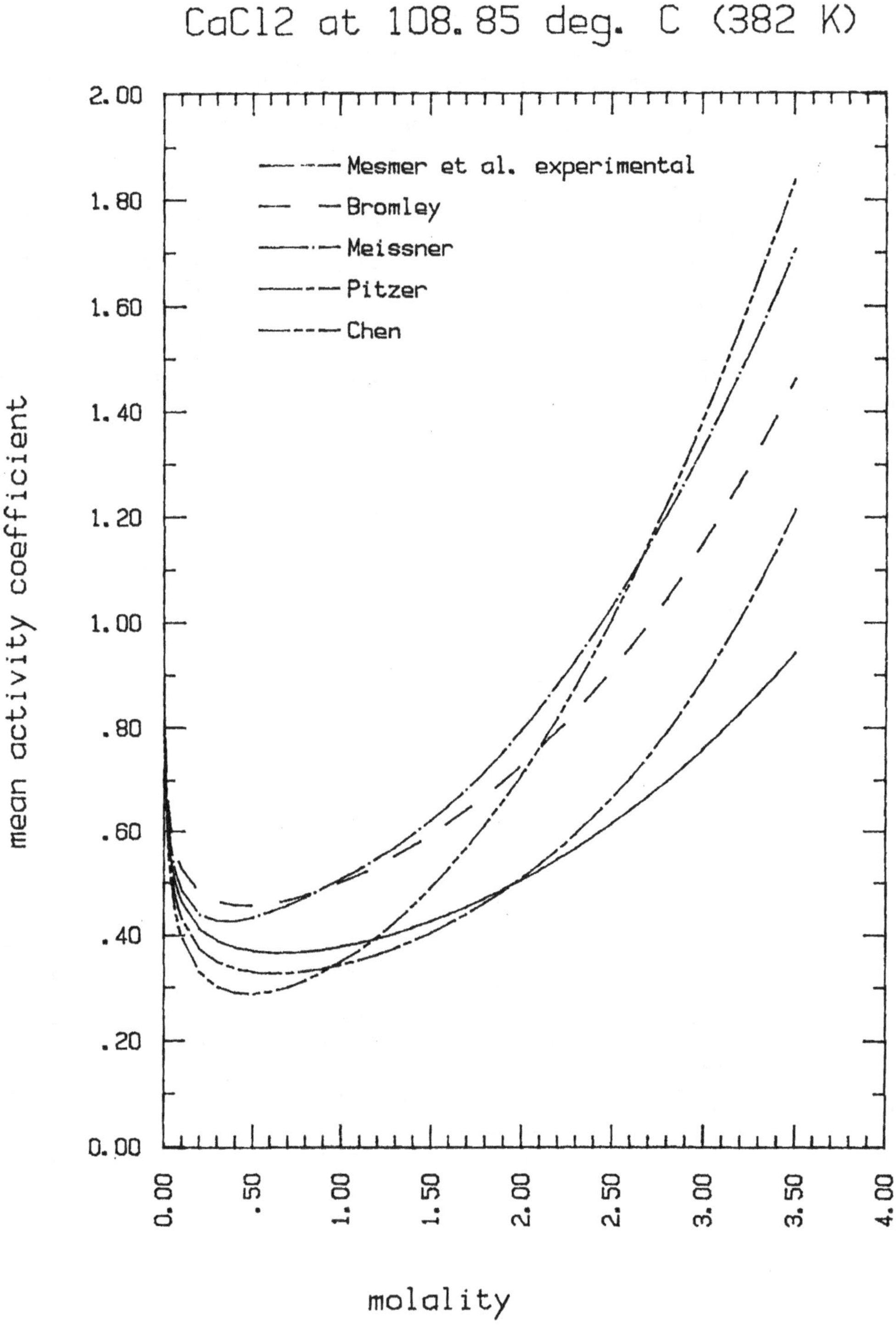
CaCl2 at 108.85 deg. C (382 K)
Mesmer et al. experimental
Bromley
Meissner
Pitzer
Chen
mean activity coefficient
molality
2.00
1.80
1.60
1.40
1.20
1.00
.80
.60
.40
.20
0.00
0.00
.50
1.00
1.50
2.00
2.50
3.00
3.50
4.00

CaCl2

COMPARISON OF ACTIVITY COEFFICIENT METHODS

molality	Mesmer	Bromley	Meissner	Pitzer	Chen
.001	.82814	.90500	.88621	.82480	.82522
.005	.68352	.81281	.77851	.67134	.67187
.01	.60629	.75741	.71518	.58660	.58652
.02	.52595	.69405	.64407	.49605	.49435
.04	.44831	.62642	.56994	.40613	.40149
.06	.40614	.58717	.52791	.35654	.34971
.08	.37810	.56045	.49985	.32353	.31512
.1	.35745	.54078	.47956	.29941	.28990
.2	.29912	.48897	.42886	.23415	.22327
.3	.26867	.46835	.41150	.20363	.19492
.4	.24878	.46008	.40689	.18592	.18108
.5	.23461	.45847	.40881	.17475	.17485
.6	.22410	.46111	.41459	.16755	.17338
.7	.21616	.46675	.42281	.16300	.17529
.8	.21014	.47469	.43269	.16039	.17980
.9	.20561	.48452	.44377	.15927	.18646
1.0	.20228	.49594	.45584	.15935	.19501
1.1	.19993	.50878	.46887	.16046	.20528
1.2	.19840	.52292	.48296	.16245	.21717
1.3	.19756	.53828	.49829	.16524	.23060
1.4	.19733	.55480	.51504	.16878	.24555
1.5	.19762	.57246	.53340	.17303	.26198
1.6	.19837	.59123	.55348	.17796	.27989
1.7	.19953	.61111	.57530	.18356	.29927
1.8	.20106	.63210	.59884	.18985	.32013
1.9	.20292	.65421	.62402	.19681	.34246
2.0	.20508	.67746	.65074	.20448	.36626
2.1	.20752	.70187	.67893	.21286	.39154
2.2	.21021	.72747	.70850	.22199	.41829
2.3	.21313	.75429	.73942	.23190	.44651
2.4	.21627	.78236	.77168	.24263	.47619
2.5	.21961	.81172	.80526	.25422	.50732
2.6	.22313	.84241	.84019	.26671	.53990
2.7	.22682	.87447	.87648	.28016	.57389
2.8	.23067	.90796	.91416	.29463	.60930
2.9	.23467	.94291	.95324	.31018	.64609
3.0	.23881	.97940	.99377	.32688	.68423
3.1	.24307	1.01746	1.03576	.34481	.72371
3.2	.24745	1.05716	1.07924	.36404	.76448
3.3	.25193	1.09856	1.12424	.38466	.80652
3.4	.25651	1.14173	1.17079	.40678	.84979
3.5	.26118	1.18674	1.21893	.43048	.89424
std. dev.		6.48670	6.33856	.82341	3.34864

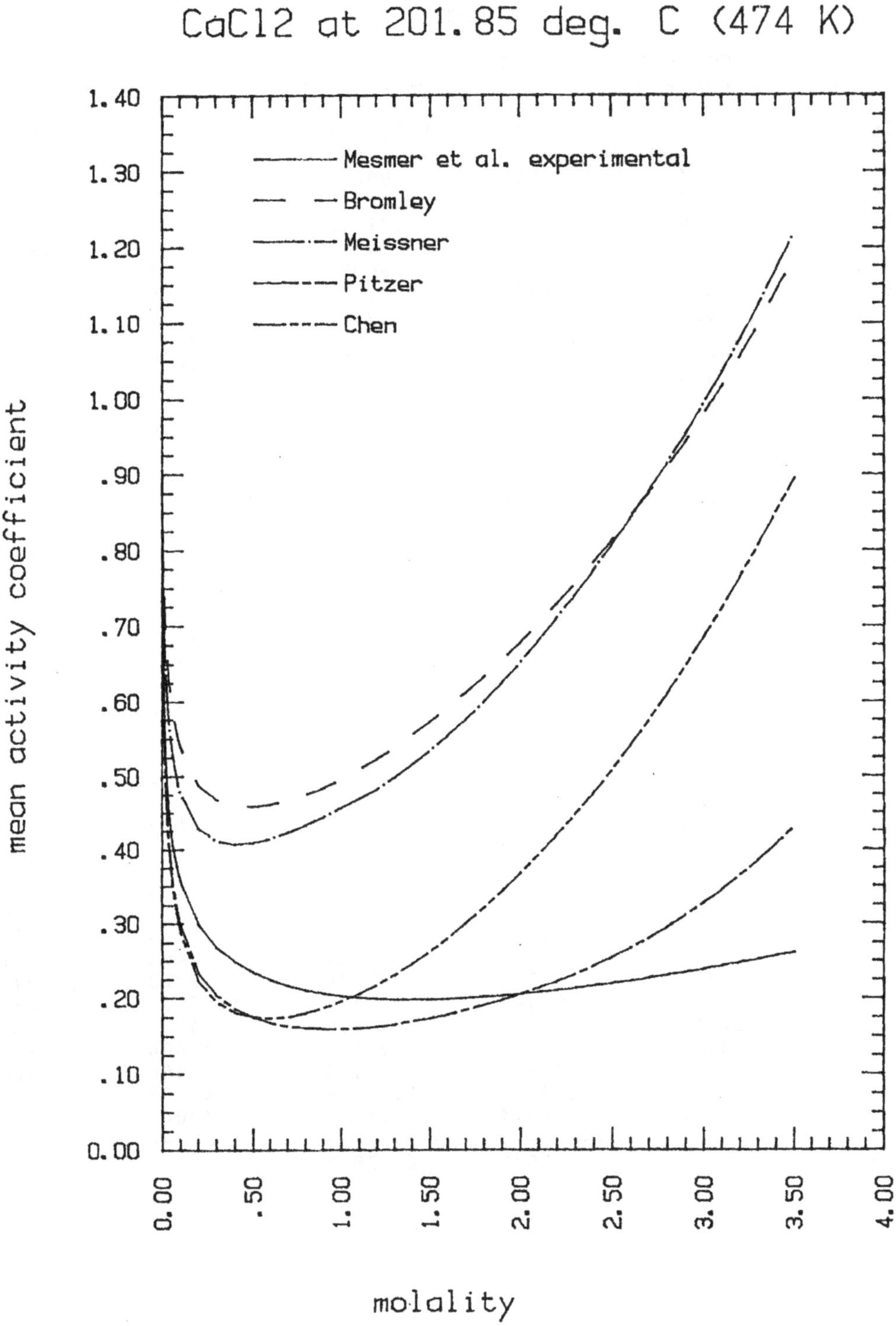
CaCl2 at 201.85 deg. C (474 K)
Mesmer et al. experimental
Bromley
Meissner
Pitzer
Chen
mean activity coefficient
1.40
1.30
1.20
1.10
1.00
.90
.80
.70
.60
.50
.40
.30
.20
.10
0.00
0.00
.50
1.00
1.50
2.00
2.50
3.00
3.50
4.00
molality

Na_2SO_4 AT 80° CELSIUS

25°C parameters:

Bromley	$B = -0.0204$	maximum molality = 2.0
Meissner	$q = -0.19$	maximum molality = 3.0
Pitzer	$4/3\ \beta_0 = 0.0261$	maximum molality = 4.
	$4/3\ \beta_1 = 1.484$	
	$2^{5/2}/3\ C^{\phi} = 0.01075$	
Chen	$\tau_{m,ca} = 8.389$	maximum molality = 4.0
	$\tau_{ca,m} = -4.539$	

Experimental data:

From: H.P. Snipes, C. Manly and D.D. Ensor, "Heats of Dilution of Aqueous Electrolytes: Temperature Dependence", J. Chem. Eng. Data, v 20, no. 3, 287 (1975)

As for KCl, Snipes et al. used the measured heats of dilution of Na_2SO_4 over a concentration range of .005 to 2 molal and a temperature range of 40 to 80°C to fit the partial molal heat contents to a polynomial equation:

$$\bar{L}_2 = f + qT + hT^2 \quad \ldots \qquad (4.101)$$

The resulting parameters were then used in the following equation to calculate the mean activity coefficients which were presented in tabular form:

$$\ln \gamma_T = \ln \gamma_{Tr} - \frac{1}{\nu R}\left[f\left(\frac{1}{Tr} - \frac{1}{T}\right) + q\left(\ln \frac{T}{Tr}\right) + h(T - Tr)\right] \qquad (4.103)$$

with the T and Tr in Kelvins (Tr = 298.15). They estimated the uncertainty of their results, excluding any uncertainty in the 25°C data, to be ±0.002.

Na2SO4

COMPARISON OF ACTIVITY COEFFICIENT METHODS

molality	Snipes et al	Bromley	Meissner	Pitzer	Chen
.1	.41500	.47258	.43039	.40749	.40666
.2	.34100	.40036	.35117	.32518	.32372
.3	.30000	.36130	.30933	.27967	.27889
.4	.27300	.33528	.28202	.24896	.24930
.5	.25300	.31617	.26225	.22627	.22782
.6	.23700	.30134	.24703	.20860	.21134
.7	.22500	.28939	.23483	.19435	.19821
.8	.21400	.27949	.22475	.18258	.18746
.9	.20400	.27113	.21624	.17269	.17849
1.0	.19700	.26396	.20891	.16425	.17087
1.2	.18400	.25222	.19685	.15064	.15863
1.4	.17500	.24298	.18718	.14020	.14923
1.6	.16600	.23548	.17910	.13201	.14178
std. dev.		.49234	.08618	.22262	.18394

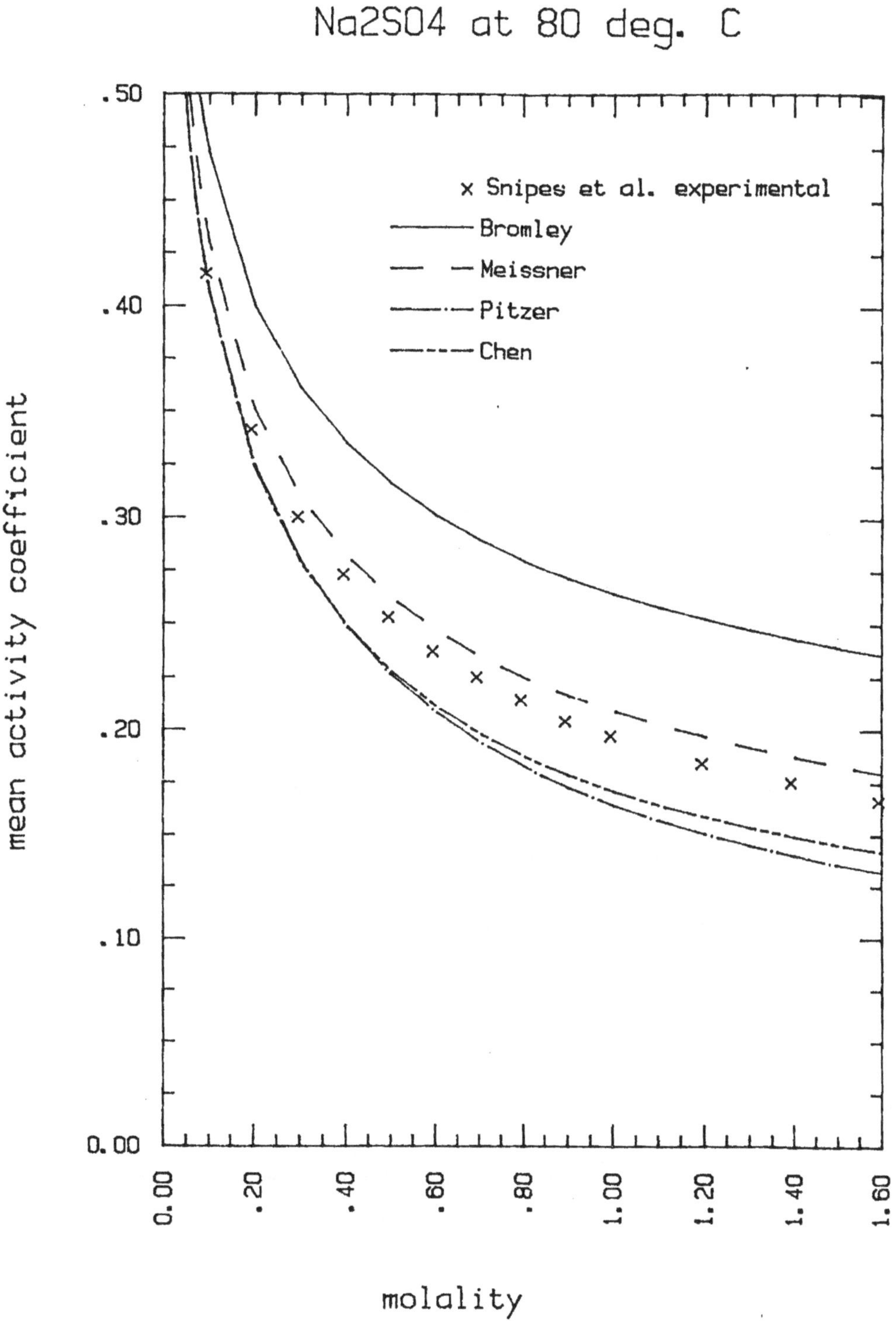
Na2SO4 at 80 deg. C
× Snipes et al. experimental
Bromley
Meissner
Pitzer
Chen
mean activity coefficient
.50
.40
.30
.20
.10
0.00
0.00
.20
.40
.60
.80
1.00
1.20
1.40
1.60
molality

$MgSO_4$ AT 80° CELSIUS

25°C parameters:

Bromley	$B = -0.0153$	maximum molality = 1.5
Meissner	$q = .15$	maximum molality = 2.25
Pitzer	$\beta_0 = .221$	maximum molality = 3.
	$\beta_1 = 3.343$	
	$\beta_2 = -37.23$	
	$C^{\phi} = .0250$	
Chen	$\tau_{m,ca} = 11.346$	maximum molality = 3.5
	$\tau_{ca,m} = -6.862$	

Experimental data:

From: H.P. Snipes, C. Manly and D.D. Ensor, "Heats of Dilution of Aqueous Electrolytes: Temperature Dependence", J. Chem. Eng. Data, v 20, no. 3, 287 (1975)

Snipes et al. used the measured heats of dilution of $MgSO_4$ over a concentration range of .005 to 2 molal and a temperature range of 40 to 80°C to fit the partial molal heat contents to a polynomial equation

$$\bar{L}_2 = f + qT + hT^2 \ \ldots \qquad (4.101)$$

The resulting parameters were then used in the following equation to calculate the mean activity coefficients which were presented in tabular form:

$$\ln \gamma_T = \ln \gamma_{Tr} - \frac{1}{\nu R}\left[f\left(\frac{1}{Tr} - \frac{1}{T}\right) + q\left(\ln \frac{T}{Tr}\right) + h(T - Tr)\right] \qquad (4.103)$$

with the T and Tr in Kelvins (Tr = 298.15). They estimated the uncertainty of their results, excluding any uncertainty in the 25°C data, to be ± 0.0009.

MgSO4

COMPARISON OF ACTIVITY COEFFICIENT METHODS

molality	Snipes et al	Bromley	Meissner	Pitzer	Chen
.1	.12900	.20406	.16530	.13043	.14404
.2	.08910	.15249	.11308	.08815	.09227
.3	.07030	.12914	.09066	.06932	.06996
.4	.06010	.11508	.07787	.05821	.05737
.5	.05350	.10543	.06950	.05075	.04931
.6	.04740	.09833	.06354	.04537	.04376
.8	.04020	.08855	.05555	.03814	.03675
1.0	.03570	.08221	.05043	.03359	.03273
1.2	.03280	.07789	.04692	.03060	.03031
1.4	.03090	.07488	.04441	.02859	.02887
1.6	.02940	.07277	.04253	.02729	.02809
1.8	.02890	.07133	.04108	.02651	.02776
2.0	.02850	.07039	.03992	.02614	.02777
std. dev.		.85125	.27857	.03773	.05990

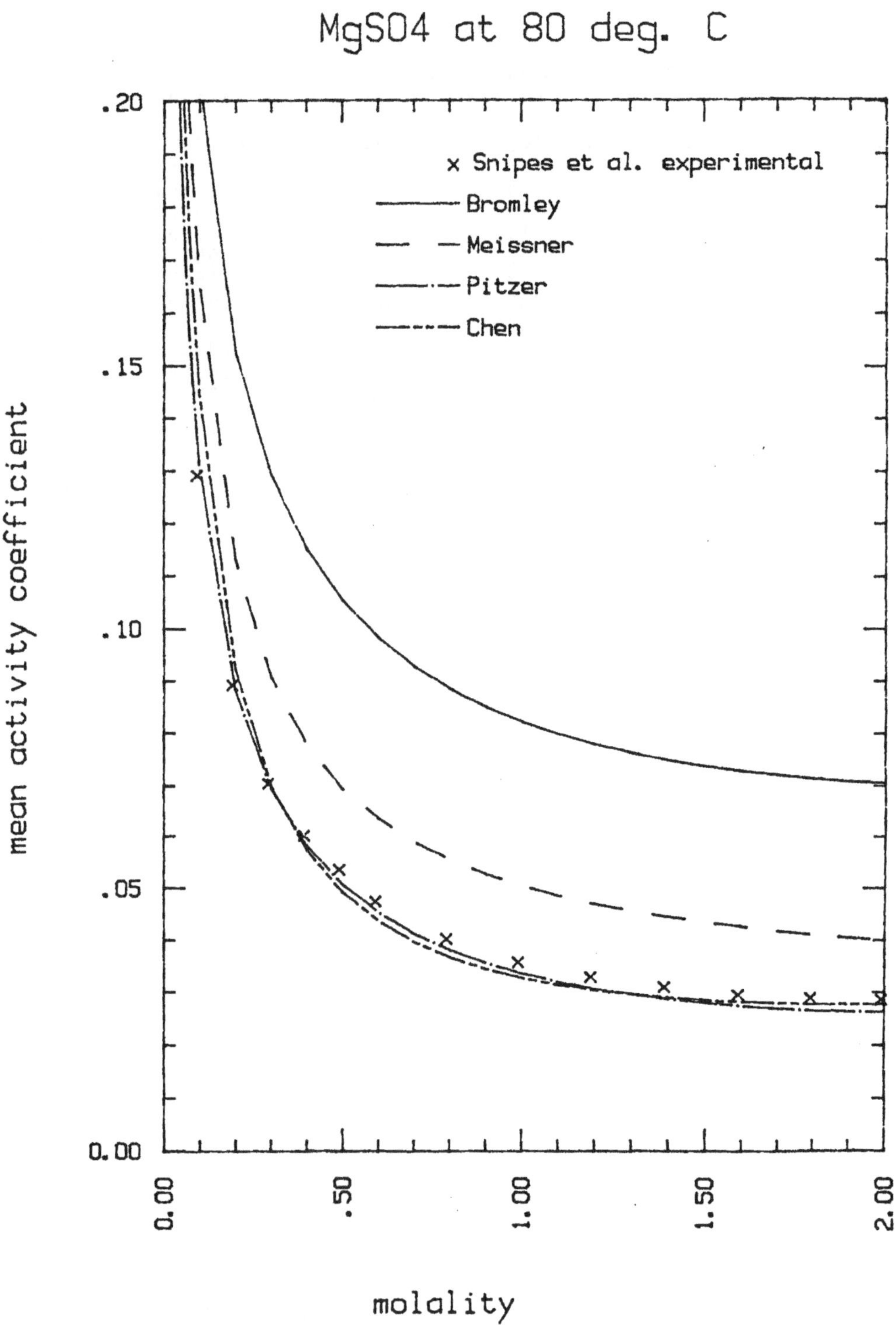
MgSO4 at 80 deg. C
x Snipes et al. experimental
Bromley
Meissner
Pitzer
Chen
mean activity coefficient
.20
.15
.10
.05
0.00
0.00
.50
1.00
1.50
2.00
molality

APPENDIX 4.1

VALUES FOR GUGGENHEIM'S β PARAMETER

From: E.A. Guggenheim and J.C. Turgeon, "Specific Interaction of Ions", Trans. Faraday Soc., v. 51, pp. 747-761 (1955)

Columns (a) - 0°C values from freezing point measurements
Columns (b) - 25°C values from e.m.f. measurements
Columns (c) - values from isopiestic measurements relative to NaCl

Table 1: β Values for Uni-univalent Electrolytes

	(a)		(b)		(c)
	0°C		25°C		25°C
electrolyte	β kg/mole	ref.	β kg/mole	ref.	β kg/mole
HCl	0.25	a1	0.27	b1	
HBr			0.33	b2	
HI					0.36
$HClO_4$					0.30
HNO_3	0.16	a2			
LiCl	0.20	a3			0.22
LiBr	0.30	a3			0.26
LiI					0.35
$LiClO_3$	0.25	a4			
$LiClO_4$	0.35	a4			0.34
$LiNO_3$	0.23	a5			0.21
LiO_2CH	0.11	a6			
LiOAc	0.19	a6			0.18
NaF					0.07
NaCl	0.11	a3	0.15	b3	(0.15)
	0.11	a7			
NaBr	0.20	a3			0.17
NaI					0.21
$NaClO_3$	0.	a4			0.10
$NaClO_4$					
$NaBrO_3$					0.01
$NaIO_3$	-0.41	a8			
$NaNO_3$	-0.04	a5			0.04
NaO_2CH	0.13	a6			
NaOAc	0.26	a6			0.23
NaCNS					0.20
NaH_2PO_4					-0.06
KF	0.05	a9			0.13
KCl	0.04	a3,a10	0.10	b4	0.10
	0.00	a11			

	(a)		(b)		(c)
	0°C		25°C		25°C
electrolyte	β kg/mole	ref.	β kg/mole	ref.	β kg/mole
KBr	0.06	a3			0.11
KI					0.15
$KClO_3$	-0.19	a4			-0.04
	-0.30	a9			
$KClO_4$	-0.55	a4			
	-0.55	a9			
$KBrO_3$					-0.07
KIO_3	-0.43	a8			-0.07
	-0.43	a9			
KNO_3	-0.26	a5			-0.11
	-0.31	a10			
KO_2CH	0.15	a6			
KOAc	0.30	a6			0.26
KCNS					0.09
KH_2PO_4					-0.16
RbCl					0.06
RbBr					0.05
RbI					0.04
$RbNO_3$					-0.14
RbOAc					0.26
CsCl					0
CsBr					0
CsI					-0.01
$CsNO_3$					-0.15
CsOAc					0.28
$AgNO_3$			-0.14	b5	-0.14
$TlClO_4$					-0.17
$TlNO_3$					-0.36
TlOAc					-0.04

Table 2: β and B Values of Bi-univalent and Uni-bivalent Electrolytes from Freezing Points

electrolyte	β kg/mole	B kg/mole	ref.
$BaCl_2$	0.7	0.8	a8
$Ba(NO_3)_2$	-0.5	-0.6	a12
$CoCl_2$	1.1	1.3	a8
K_2SO_4	-0.1	-0.1	a8

DATA SOURCES FOR TABLES 1 AND 2

a1. Randall and Vanselow, J. Amer. Chem. Soc., 46, 2418 (1924)
a2. Hartmann and Rosenfeld, Z. physik. Chem. A, 164, 377 (1933)
a3. Scatchard and Prentiss, J. Amer. Chem. Soc., 55, 4355 (1933)
a4. Scatchard, Prentiss and Jones, J. Amer. Chem. Soc., 56, 805 (1934)
a5. Scatchard, Prentiss and Jones, J. Amer. Chem. Soc., 54, 2690 (1932)
a6. Scatchard and Prentiss, J. Amer. Chem. Soc., 56, 807 (1934)
a7. Harkins and Roberts, J. Amer. Chem. Soc., 38, 2676 (1916)
a8. Hall and Harkins, J. Amer. Chem. Soc., 38, 2658 (1916)
a9. Lange and Herre, Z. physik. Chem. A., 181, 329 (1938)
a10. Adams, J. Amer. Chem. Soc., 37, 481 (1915)
a11. Lange, Z. physik. Chem. A., 168, 147 (1934)
a12. Randall and Scott, J. Amer. Chem. Soc., 49, 647 (1927)

b1. Hills and Ives, J. Chem. Soc., 305, 311, 318 (1951)
b2. Harned, Keston and Donelson, J. Amer. Chem. Soc., 58, 989 (1936)
b3. Brown and MacInnes, J. Amer. Chem. Soc., 57, 1356 (1935)
b4. Shedlovsky and MacInnes, J. Amer. Chem. Soc., 59, 503 (1937)
b5. MacInnes and Brown, Chem. Rev., 18, 335 (1936)

METHODS FOR CALCULATING β

From freezing point data:

Using the freezing point depression Θ, the osmotic coefficient ϕ can be calculated:

$$\Theta(1 + b\Theta) = (\nu_+ + \nu_-)\, m\, \lambda\, \phi$$

where: b - cryoscopic constant determined by Guggenheim and Turgeon = 4.8×10^{-4} deg^{-1}

λ - cryoscopic constant determined by Guggenheim and Turgeon = 1.860 deg. $mole^{-1}kg$

ν_+ - number of cations per molecule of electrolyte

ν_- - number of anions per molecule of electrolyte

m - electrolyte molality

They then define a quantity ϕ^{st}:

$$\phi^{st} = 1 - \frac{1}{3}\alpha|z_+z_-|\sqrt{I}\,\sigma(\sqrt{I})$$

where: α - Debye-Hückel constant, log e basis

z_+ - number of positive charges on cation

z_- - number of negative charges on anion

I - molality scale ionic strength

$\sigma(\sqrt{I})$ - function defined as:

$$\sigma(y) \equiv \frac{3}{y^3}\{1 + y - \frac{1}{1+y} - 2\ln(1+y)\}$$

A table of values for σ at values for y from 0. to 4.5 is available in Herbert S. Harned and Benton B. Owen's The Physical Chemistry of Electrolyte Solutions, 3rd ed., Reinhold Publishing Co., New York, 1958, p176.

They continue:

$$\Theta(1 + b\Theta) - (\nu_+ + \nu_-)\lambda m\phi^{st} = (\nu_+ + \nu_-)\lambda m(\phi - \phi^{st})$$

$$= (\nu_+ + \nu_-)\underline{\nu}\beta\lambda m^2 = 2\nu_+\nu_-\beta\lambda m^2$$

By plotting $\Theta(1 + b\Theta) - (\nu_+ + \nu_-)\lambda m \phi^{st}$ versus m^2, they expect to get a straight line with slope $2\nu_+\nu_-\beta\lambda$ from which they can calculate β.

From e.m.f. measurements:

Of cells without transference:

Since the increase in free energy of the cell is

$$\Delta\bar{G} = 2RT\ln\frac{\gamma m}{\gamma' m'} \quad \text{and} \quad EF = -\Delta\bar{G}$$

using the measured potential E, a quantity $E^{o\prime}$ is defined:

$$E^{o\prime} \equiv E + 2\frac{RT}{F}\ln m - 2\frac{RT}{F}\frac{\alpha\sqrt{m}}{1+\sqrt{m}}$$

where: E - e.m.f. measurements
R - gas law constant
T - temperature
F - Faraday
m - electrolyte molality
α - Debye-Hückel constant, log e basis

Plotting $E^{o\prime}$ versus m results in a straight line of slope $2\beta RT/F$ and y intercept of E from which β can be calculated.

Of cells with transference:

Since the e.m.f. E is:

$$E = -2 \frac{RT}{F} \int_{m''}^{m'} t^+ \, d \ln \gamma m$$

then:

$$E = -2\, \overline{t^+} \frac{RT}{F} \ln \frac{m'\gamma'}{m''\gamma''}$$

where: $\overline{t^+}$ - the value of t^+ averaged over the range of m" to m'
t^+ - transport number

Using e.m.f. and t^+ measurements, Guggenheim and Turgeon find $\overline{t^+}$ by using rough estimates or successive substitution for γ' and γ''. With this $\overline{t^+}$ they can calculate values for $\ln(\gamma'/\gamma'')$ and hence values of $k + \log\gamma$ for each m. Then by plotting:

$$k + \log\gamma + \frac{\alpha}{\ln 10} \frac{\sqrt{m}}{1+\sqrt{m}}$$

versus m to get a straight line of slope $2\beta/\ln 10$ with intercept k on the m=0 axis. For the example they show in their paper, they have adjusted to ordinate scale so that the arbitrarily set unknown constant k = 0.

From isopiestic measurements:

Guggenheim and Turgeon estimate β for 1-1 electrolytes by using the relation:

$$\beta = 10\phi - 9.19$$

where ϕ is the osmotic coefficient value at 25°C, .1 molal ionic strength. As isopiestic measurements give relative values, they use the NaCl ϕ as their standard substance to obtain absolute values. The need to use .1 molal values is dictated by the fact that their method is valid only to ionic strengths of .1. Unfortunately, the isopiestic measurements at this level carry possible experimental errors of up to .001 for ϕ, affecting the β at least .01 kg/mole. Guggenheim and Turgeon also differed with Robinson and Stokes' ϕ for NaCl by .002. They obtained ϕ_{NaCl} = .934.

As 1-2 and 2-1 measurements go down to .1 molal, which is .3 molal ionic strength, Guggenheim and Turgeon claim the β values estimated would have an uncertainty of ± .1 kg/mole and so have not tabulated them. They state that β is obtainable through the following relationship:

$$\beta = 7.5(\phi - .782) \quad \text{at molality .1}$$

APPENDIX 4.2

BROMLEY INTERACTION PARAMETERS

From: Leroy A. Bromley, "Thermodynamic Properties of Strong Electrolytes in Aqueous Solutions", AIChE Journal, v. 19, #2, pp. 313-320 (1973)

Table 1: B Values at 25°C Determined by the Method of Least Squares on Log γ to I=6.0 (or less if limited data)

σ is the standard deviation in log γ

1-1 salts	B kg/mole	σ	1-1 salts	B kg/mole	σ
$AgNO_3$	-0.0828	0.027	LiAc	0.0722	0.005
CsAc	0.1272	0.010	LiBr	0.1527	0.017
CsBr	-0.0039	0.015	LiCl	0.1283	0.009
CsCl	0.0025	0.017	$LiClO_3$	0.1442	0.004
CsF	0.0906	0.004	$LiClO_4$	0.1702	0.006
CsI	-0.0188	0.007	LiI	0.1815	0.007
$CsNO_3$	-0.1173	0.004	$LiNO_3$	0.0938	0.014
CsOH	0.1299	0.003	LiOH	-0.0097	0.022
HBr	0.1734	0.004	LiTol	0.0230	0.007
HCl	0.1433	0.003	NaAc	0.1048	0.009
$HClO_4$	0.1639	0.019	NaBr	0.0749	0.0009
HI	0.2054	0.004	$NaBrO_3$	-0.0278	0.002
HNO_3	0.0776	0.019	Na Butyrate	0.1474	0.023
KAc	0.1188	0.009	Na Caprate	-0.4786	0.074
KBr	0.0296	0.003	Na Caproate	0.0480	0.097
$KBrO_3$	-0.0884	0.003	Na Caprylate	-0.1419	0.060
KCl	0.0240	0.0005	NaCl	0.0574	0.002
$KClO_3$	-0.0739	0.003	$NaClO_3$	0.0127	0.005
$KClO_4$	-0.1637	0.000	$NaClO_4$	0.0330	0.007
KCNS	0.0137	0.007	NaCNS	0.0758	0.011
KF	0.0565	0.006	NaF	0.0041	0.0006
KH_2AsO_4	-0.0798	0.003	Na Formate	0.0519	0.008
$K_2H_2PO_4$	-0.1124	0.008	NaH_2AsO_4	-0.0291	0.006
KH Adipate	0.0286	0.001	NaH_2PO_4	-0.0460	0.033
KH Malonate	-0.0227	0.006	NaH Adipate	0.0461	0.001
KH Succinate	-0.0035	0.007	NaH Malonate	-0.0011	0.002
KI	0.0428	0.008	NaH Succinate	0.0131	0.005
KNO_3	-0.0862	0.013	Na Heptylate	-0.0467	0.086
KOH	0.1131	0.011	NaI	0.0994	0.002
KTol	-0.0559	0.012	$NaNO_3$	-0.0128	0.002

1-1 salts	B kg/mole	σ
NaOH	0.0747	0.010
Na Pelargonate	-0.3040	0.080
Na Propionate	0.1325	0.008
NaTol	-0.0200	0.011
Na Valerate	0.1222	0.057
NH_4Br	-0.0066	0.013
NH_4Cl	0.0200	0.005
NH_4ClO_4	-0.0640	0.011
NH_4I	0.0210	0.004
NH_4NO_3	-0.0358	0.011
RbAc	0.1239	0.008
RbBr	0.0111	0.004
RbCl	0.0157	0.005
RbF	0.0650	0.006
RbI	-0.0108	0.005
$RbNO_3$	-0.0869	0.023
TlAc	-0.0224	0.014
TlCl	0.0372	0.002
$TlClO_4$	-0.1288	0.0005
$TlNO_3$	-0.2340	0.003
1-2 salts I=3m		
Cs_2SO_4	-0.0012	0.004
H_2SO_4	0.0606	0.175
K_2CO_3	0.0372	0.015
K_2CrO_4	-0.0003	0.003
K_2HAsO_4	0.0296	0.015
K_2HPO_4	-0.0096	0.006
K_2SO_4	-0.0320	0.005
Li_2SO_4	0.0207	0.003
Na_2CO_3	0.0089	0.006
Na_2CrO_4	0.0096	0.006
Na_2 Fumarate	0.0366	0.013
Na_2HAsO_4	0.0022	0.016
Na_2HPO_4	-0.0265	0.011
Na_2 Maleate	-0.0029	0.033
Na_2SO_4	-0.0204	0.008
$Na_2S_2O_3$	-0.0005	0.005
$(NH_4)_2SO_4$	-0.0287	0.024
Rb_2SO_4	-0.0091	0.005

1-3 salts I=6m	B kg/mole	σ
K_3AsO_4	0.0551	0.033
$K_3Fe(CN)_6$	0.0195	0.003
K_3PO_4	0.0344	0.031
Na_3AsO_4	0.0159	0.033
Na_3PO_4	0.0043	0.037
1-4 salts I=10m		
$K_4Fe(CN)_6$	0.0085	0.009
$K_4Mo(CN)_8$	0.0110	0.018
2-1 salts I=3m		
$BaAc_2$	0.0357	0.015
$BaBr_2$	0.0852	0.004
$BaCl_2$	0.0638	0.007
$Ba(ClO_4)_2$	0.0936	0.012
BaI_2	0.1254	0.012
$Ba(NO_3)_2$	-0.0545	0.001
$Ba(OH)_2$	-0.0240	0.003
$CaBr_2$	0.1179	0.009
$CaCl_2$	0.0948	0.005
$Ca(ClO_4)_2$	0.1457	0.003
CaI_2	0.1440	0.007
$Ca(NO_3)_2$	0.0410	0.007
$CdBr_2$	-0.1701	0.237
$CdCl_2$	-0.1448	0.180
CdI_2	-0.2497	0.383
$Cd(NO_3)_2$	0.0719	0.017
$CoBr_2$	0.1361	0.012
$CoCl_2$	0.1016	0.004
CoI_2	0.1683	0.013
$Co(NO_3)_2$	0.0912	0.002
$CuCl_2$	0.0654	0.017
$Cu(NO_3)_2$	0.0797	0.005
$FeCl_2$	0.0961	0.004
$MgAc_2$	0.0339	0.015
$MgBr_2$	0.1419	0.012
$MgCl_2$	0.1129	0.010
$Mg(ClO_4)_2$	0.1760	0.009
MgI_2	0.1695	0.014
$Mg(NO_3)_2$	0.1014	0.004
$MnCl_2$	0.0869	0.005

2-1 salts I=3m	B kg/mole	σ
$NiCl_2$	0.1039	0.007
$Pb(ClO_4)_2$	0.0987	0.003
$Pb(NO_3)_2$	-0.0606	0.034
$SrBr_2$	0.1038	0.005
$SrCl_2$	0.0847	0.002
$Sr(ClO_4)_2$	0.1254	0.008
SrI_2	0.1339	0.005
$Sr(NO_3)_2$	0.0138	0.012
UO_2Cl_2	0.1157	0.015
$UO_2(ClO_4)_2$	0.2267	0.022
$UO_2(NO_3)_2$	0.1296	0.010
$ZnBr_2$	0.0911	0.057
$ZnCl_2$	0.0364	0.043
$Zn(ClO_4)_2$	0.1755	0.019
ZnI_2	0.1341	0.060
$Zn(NO_3)_2$	0.1002	0.006
2-2 salts I=4m		
$BeSO_4$	-0.0301	0.110
$CdSO_4$	-0.0371	0.064
$CuSO_4$	-0.0364	0.066
$MgSO_4$	-0.0153	0.051
$NiSO_4$	-0.0296	0.052
$ZnSO_4$	-0.0240	0.046
3-1 salts I=6m		
$AlCl_3$	0.1089	0.018
$CeCl_3$	0.0815	0.010
$Co(EN)_3Cl_3$	-0.0251	0.016
$CrCl_3$	0.1026	0.015
$Cr(NO_3)_3$	0.0919	0.010
$EuCl_3$	0.0867	0.009
$Ga(ClO_4)_3$	0.1607	0.019
$LaCl_3$	0.0818	0.011
$LaNO_3$	0.0868	0.034
$NdCl_3$	0.0815	0.011

3-1 salts I=6m	B kg/mole	σ
$PrCl_3$	0.0805	0.009
$ScCl_3$	0.0969	0.016
$SmCl_3$	0.0848	0.010
YCl_3	0.0882	0.014
3-2 salts I=15m		
$Al_2(SO_4)_3$	-0.0044	0.041
$Cr_2(SO_4)_3$	0.0122	0.014
4-1 salts I=10m		
$ThCl_4$	0.1132	0.056
$Th(NO_3)_4$	0.0894	0.079

Table 2: Individual Ion Values of B and δ in Aqueous Solutions at 25°C

$B = B_+ + B_- + \delta_+ \delta_-$ the units of B are kg/mole

Cation	B_+	δ_+	Anion	B_-	δ_-
H^+	0.0875	0.103	F^-	0.0295	-0.93
Li^+	0.0691	0.138	Cl^-	0.0643	-0.067
Na^+	0.0000*	0.028	Br^-	0.0741	0.064
K^+	-0.0452	-0.079	I^-	0.0890	0.196
Rb^+	-0.0537	-0.100	ClO_3^-	0.005	0.45
Cs^+	-0.0710	-0.138	ClO_4^-	0.002	0.79
NH_4^+	-0.042	-0.02	BrO_3^-	-0.032	0.14
Tl^+	-0.135	-0.02	IO_3^-	(-0.04)	(0)
Ag^+	-0.058	(0)	NO_3^-	-0.025	0.27
Be^{++}	(0.1)	(0.2)	$H_2PO_4^-$	-0.052	0.20
Mg^{++}	0.0570	0.157	$H_2AsO_4^-$	-0.030	0.05
Ca^{++}	0.0374	0.119	CNS^-	0.071	0.16
Sr^{++}	0.0245	0.110	OH^-	0.076	-1.00*
Ba^{++}	0.0022	0.098	Formate(C_1)	0.072	(-0.7)
Mn^{++}	0.037	(0.21)	Acetate(C_2)	0.104	-0.73
Fe^{++}	0.046	(0.21)	Propionate(C_3)	0.152	(-0.7)
Co^{++}	0.0490	0.210	Butyrate(C_4)	0.167	(-0.7)
Ni^{++}	0.054	(0.21)	Valerate(C_5)	0.142	(-0.7)
Cu^{++}	0.022	0.30	Caproate(C_6)	0.068	(-0.7)
Zn^{++}	0.101	0.09	Heptylate(C_7)	-0.027	(-0.7)
Cd^{++}	0.072	(0.09)	Caprylate(C_8)	-0.122	(-0.7)
Pb^{++}	-0.104	0.25	Pelargonate(C_9)	-0.284	(-0.7)
UO_2^{++}	0.079	0.19	Caprate(C_{10})	-0.459	(-0.7)
Cr^{+++}	0.066	0.15	H Malonate(C_3)	0.005	-0.22
Al^{+++}	0.052	0.12	H Succinate(C_4)	0.021	-0.27
Sc^{+++}	0.046	(0.2)	H Adipate(C_5)	0.053	-0.26
Y^{+++}	0.037	(0.2)	Toluate	-0.022	-0.16
La^{+++}	0.036	0.27	CrO_4^{--}	0.019	-0.33
Ce^{+++}	0.035	(0.27)	SO_4^{--}	0.000	-0.40
Pr^{+++}	0.034	(0.27)	$S_2O_3^{--}$	0.019	(-0.7)

Cation	B_+	δ_+	Anion	B_-	δ_-
Nd^{+++}	0.035	(0.27)	HPO_4^{---}	-0.010	-0.57
Sm^{+++}	0.039	(0.27)	$HAsO_4^{--}$	0.021	-0.67
Eu^{+++}	0.041	(0.27)	CO_3^{--}	0.028	-0.67
Ga^{+++}	0.000	(0.2)	Fumarate (Trans C_4)	0.056	(-0.7)
$Co(en)^{+++}$ †	-0.089	(0.)	Maleate(Cis C_4)	0.017	(-0.7)
Th^{++++}	0.062	0.19	PO_4^{---}	0.024	-0.70
			AsO_4^{---}	0.038	-0.78
			$Fe(CN)_6^{---}$	0.065	(0)
			$Fe(CN)_6^{---}$	0.054	(0)
			$Mo(CN)_8^{---}$	0.056	(0)

* For convenience the value of B_{Na^+} was set equal to zero and the δ_{OH^-} set equal to -1.0.

† $Co(en)_3^{+++}$ denotes tris-(ethylenediamine) cobalt III.

Values in parentheses are estimated.

Table 3: Bivalent Metal Sulfates at 25°C

$\alpha = 70$; σ = standard deviation in $\log\gamma$; maximum molality = 4

salt	B	E	σ
$BeSO_4$	0.1385	0.00654	0.045
$CaSO_4$	0.4463	0.01143	0.002
$CdSO_4$	0.0988	0.00526	0.014
$CoSO_4$	0.1244	0.00498	0.006
$CuSO_4$	0.1285	0.00640	0.008
$MgSO_4$	0.1042	0.00464	0.009
$MnSO_4$	0.1226	0.00599	0.027
$NiSO_4$	0.1056	0.00524	0.008
$ZnSO_4$	0.1008	0.00483	0.009

APPENDIX 4.3

MEISSNER PARAMETERS

From: C.L. Kusik and H.P. Meissner, "Electrolyte Activity Coefficicients in Inorganic Processing", AIChE Symposium Series 173, 74, 16 (1978)

Table: Average Values of Parameter q in Equation (4.46) for Selected Electrolytes

1-1 electrolyte	q	maximum I
$AgNO_3$	-2.55	b
Cs Acetate	5.59	a
CsBr	-0.06	b
CsCl	0.16	b
CsI	-0.41	a
CsOH	7.34	u
$CsNO_3$	-2.62	u
HCl	6.69	b
HNO_3	3.66	a
K Acetate	5.05	a
KBr	1.15	b
$KBrO_3$	-2.00	u
KCl	0.92	b
$KClO_3$	-1.70	u
KCNS	0.61	b
KF	2.13	a
KH Maleate	-0.72	b
KH_2PO_4	-2.54	a
KH Succinate	0.02	b
KI	1.62	b
KNO_3	-2.33	a
KOH	4.77	b
K p-toluene sulphonate	-1.75	a
Li Acetate	2.81	a
LiBr	7.27	b
LiCl	5.62	b
LiOH	-0.08	a
$LiNO_3$	3.80	b
Li p-toluene sulphonate	0.84	b
Na Acetate	4.20	a

1-1 electrolyte	q	maximum I
NaBr	2.98	a
$NaBrO_3$	-0.68	u
NaCl	2.23	a
$NaClO_3$	0.41	a
$NaClO_4$	1.30	b
NaCNS	2.94	a
NaF	0.37	u
Na Formate	1.83	a
NaH Maleate	0.01	b
NaH Succinate	0.60	b
NaH_2PO_4	-1.59	a
NaI	4.06	a
$NaNO_3$	-0.39	b
NaOH	3.00	a
Na Propionate	5.54	a
Na p-toluene sulphonate	-0.80	a
NH_4Cl	0.82	b
NH_4NO_3	-1.15	b
Rb Acetate	5.39	a
RbBr	0.46	b
RbCl	0.62	b
RbI	0.45	b
$RbNO_3$	-2.49	b
Tl Acetate	-0.73	b

1-2 electrolyte	q	maximum I
Cs_2SO_4	0.16	b
K_2CrO_4	0.16	c
K_2SO_4	-0.25	u
Li_2SO_4	0.57	c
Na_2CrO_4	0.41	c
Na_2 Fumarate	0.88	b
Na_2 Maleate	0.12	b
$Na_2S_2O_3$	0.18	c
Na_2SO_4	-0.19	c
$(NH_4)_2SO_4$	-0.25	c
Rb_2SO_4	0.007	b

2-1 electrolyte	q	maximum I
$BaBr_2$	1.92	b
$BaCl_2$	1.48	b
$Ba(ClO_4)_2$	1.90	c
BaI_2	2.84	b
$Ba(NO_3)_2$	-0.52	c

2-1 electrolyte	q	maximum I
$CaCl_2$	2.40	d
CaI_2	3.27	b
$Ca(NO_3)_2$	0.93	d
$Cd(NO_3)_2$	1.53	b
$CoBr_2$	3.08	d
$CoCl_2$	2.25	b
CoI_2	3.87	d
$Co(NO_3)_2$	2.08	d
$CuCl_2$	1.40	b
$Cu(NO_3)_2$	1.83	d
$FeCl_2$	2.16	b
Mg $Acetate_2$	0.83	c
$MgBr_2$	3.50	d
$MgCl_2$	2.90	d
$Mg(ClO_4)_2$	4.15	b
MgI_2	4.04	b
$Mg(NO_3)_2$	2.32	d
$MnCl_2$	1.60	d
$NiCl_2$	2.33	d
$Pb(ClO_4)_2$	2.25	d
$Pb(NO_3)_2$	-0.97	b
$SrBr_2$	2.34	b
$SrCl_2$	1.95	c
$Sr(ClO_4)_2$	2.84	d
SrI_2	3.03	b
$Sr(NO_3)_2$	0.30	c
UO_2Cl_2	2.40	b
$UO_2(ClO_4)_2$	5.64	d
$UO_2(NO_4)_2$	2.90	b
$Zn(ClO_4)_2$	4.30	a
$Zn(NO_3)_2$	2.28	d

2-2 electrolyte	q	maximum I
$CdSO_4$	0.016	b
$CuSO_4$	0.00	c
$MgSO_4$	0.15	c
$MnSO_4$	0.14	b
$NiSO_4$	0.025	c
UO_2SO_4	0.066	b
$ZnSO_4$	0.05	c

3-1 electrolyte	q	maximum I
$AlCl_3$	1.92	d
$CeCl_3$	1.41	c
$CrCl_3$	1.72	b
$Cr(NO_3)_3$	1.51	c
$EuCl_3$	1.49	c
$LaCl_3$	1.41	c
$NdCl_3$	1.42	c
$PrCl_3$	1.40	c
$ScCl_3$	1.68	c
$SmCl_3$	1.47	c
YCl_3	1.55	c

3-2 electrolyte	q	maximum I
$Al_2(SO_4)_3$	0.36	d
$Cr_2(SO_4)_3$	0.43	d

Maximum I Key:

a - 3 to 4 molal
b - 4.5 to 6 molal
c - 9 molal
d - 15 molal
u - limited data, q determined usually from Γ° at highest reported concentration

APPENDIX 4.4

PITZER PARAMETERS

From: K.S. Pitzer, "Theory: Ion Interaction Approach", Activity Coefficients in Electrolyte Solutions, vol. 1, Ricardo M. Pytkowicz, ed., CRC Press, Boca Raton, Fla. (1979)

σ is the standard deviation of fit

key: a - high; b - intermediate; c - low — accuracy in cases where some or all of the experimental data are other than isopiestic measurements

Table 1: Inorganic Acids, Bases and Salts of 1-1 Type

species	β_0	β_1	C^ϕ	Max m	σ
$AgNO_3$	-0.0856	0.0025	0.00591	6	0.001
CsBr	0.0279	0.0139	0.00004	5	0.002
CsCl	0.0300	0.0558	0.00038	5	0.002
CsF	0.1306	0.2570	-0.0043	3.2	0.002
CsI	0.0244	0.0262	-0.00365	3	0.001
$CsNO_2$	0.0427	0.060	-0.0051	6	0.004
$CsNO_3$	-0.0758	-0.0669	---	1.4	0.002
CsOH	0.150	0.30	---	-	---
HBr	0.1960	0.3564	0.00827	3	a
HCl	0.1775	0.2945	0.00080	6	a
$HClO_4$	0.1747	0.2931	0.00819	5.5	0.002
HI	0.2362	0.392	0.0011	3	b
HNO_3	0.1119	0.3206	0.0010	3	0.001
KBr	0.0569	0.2212	-0.00180	5.5	0.001
$KBrO_3$	-0.1290	0.2565	---	0.5	0.001
KCl	0.04835	0.2122	-0.00084	4.8	0.0005
$KClO_3$	-0.0960	0.2481	---	0.7	0.001
KCNS	0.0416	0.2302	-0.00252	5	0.001
KF	0.08089	0.2021	0.00093	2	0.001
KH_2AsO_4	-0.0584	0.0626	---	1.2	0.003
KH_2PO_4	-0.0678	-0.1042	---	1.8	0.003
KI	0.0746	0.2517	-0.00414	4.5	0.001
KNO_2	0.0151	0.015	0.0007	5	0.003
KNO_3	-0.0816	0.0494	0.00660	3.8	0.001
KOH	0.1298	0.320	0.0041	5.5	b

species	β_0	β_1	C^{ϕ}	Max m	σ
KPF_6	-0.163	-0.282	---	0.5	0.001
LiBr	0.1748	0.2547	0.0053	2.5	0.002
LiCl	0.1494	0.3074	0.00359	6	0.001
$LiClO_4$	0.1973	0.3996	0.0008	3.5	0.002
LiI	0.2104	0.373	---	1.4	0.006
$LiNO_2$	0.1336	0.325	-0.0053	6	0.003
$LiNO_3$	0.1420	0.2780	-0.00551	6	0.001
LiOH	0.015	0.14	---	4	c
NH_4Br	0.0624	0.1947	-0.00436	2.5	0.001
NH_4Cl	0.0522	0.1918	-0.00301	6	0.001
NH_4ClO_4	-0.0103	-0.0194	---	2	0.004
NH_4NO_3	-0.0154	0.1120	-0.00003	6	0.001
$NaBF_4$	-0.0252	0.1824	0.0021	6	0.006
$NaBO_2$	-0.0526	0.1104	0.0154	4.5	0.004
NaBr	0.0973	0.2791	0.00116	4	0.001
$NaBrO_3$	-0.0205	0.1910	0.0059	2.5	0.001
NaCl	0.0765	0.2664	0.00127	6	0.001
$NaClO_3$	0.0249	0.2455	0.0004	3.5	0.001
$NaClO_4$	0.0554	0.2755	-0.00118	6	0.001
NaCNS	0.1005	0.3582	-0.00303	4	0.001
NaF	0.0215	0.2107	---	1	0.001
NaH_2AsO_4	-0.0442	0.2895	---	1.2	0.001
NaH_2PO_4	-0.0533	0.0396	0.00795	6	0.003
NaI	0.1195	0.3439	0.0018	3.5	0.001
$NaNO_2$	0.0641	0.1015	-0.0049	5	0.005
$NaNO_3$	0.0068	0.1783	-0.00072	6	0.001
NaOH	0.0864	0.253	0.0044	6	b
RbBr	0.0396	0.1530	-0.00144	5	0.001
RbCl	0.0441	0.1483	-0.00101	5	0.001
RbF	0.1141	0.2842	-0.0105	3.5	0.002
RbI	0.0397	0.1330	-0.00108	5	0.001
$RbNO_2$	0.0269	-0.1553	-0.00366	5	0.002
$RbNO_3$	-0.0789	-0.0172	0.00529	4.5	0.001
$TlClO_4$	-0.087	-0.023	---	0.5	0.001
$TlNO_3$	-0.105	-0.378	---	0.4	0.001

Table 2: Salts of Carboxylic Acids (1-1 Type)

species	β_0	β_1	C^ϕ	Max m	σ
Cs Acetate	0.1628	0.3605	-0.00555	3.5	0.001
K Acetate	0.1587	0.3251	-0.00660	3.5	0.001
KH Adipate	0.0419	0.2523	---	1	0.001
KH Malonate	-0.0095	0.1423	0.00167	5	0.004
KH Succinate	0.0111	0.1564	0.00274	4.5	0.002
Li Acetate	0.1124	0.2483	-0.00525	4	0.001
Na Acetate	0.1426	0.3237	-0.00629	3.5	0.001
Na Formate	0.0820	0.2872	-0.00523	3.5	0.001
Na Propionate	0.1875	0.2789	-0.01277	3	0.001
NaH Adipate	0.0472	0.3168	---	0.7	0.001
NaH Malonate	0.0229	0.1600	-0.00106	5	0.002
NaH Succinate	0.0354	0.1606	0.00040	5	0.001
Rb Acetate	0.1622	0.3353	-0.00551	3.5	0.001
Tl Acetate	0.0082	0.0131	-0.00127	6	0.001

Table 3: Tetraalkylammonium Halides

species	β_0	β_1	C^ϕ	Max m	σ
Bu_4NBr	-0.0277	-0.525	0.0011	4.5	0.007
Bu_4NCl	0.2339	-0.410	-0.0567	2.5	0.001
Bu_4NF	0.6092	0.402	-0.0281	1.7	0.005
Et_4NBr	-0.0176	-0.394	0.0156	4	0.001
Et_4NCl	0.0617	-0.099	0.0105	3	0.002
Et_4NF	0.3113	0.6155	0.0349	2	0.002
Et_4NI	-0.179	-0.571	0.0412	2	0.007
Me_4NBr	-0.0082	-0.147	0.0105	3.5	0.004
Me_4NCl	0.0430	-0.029	0.0078	3.4	0.005
Me_4NF	0.2677	0.2265	0.0013	3	0.002
Me_4NI	0.0345	-0.585	---	0.3	0.003
Pr_4NBr	0.0390	-0.772	0.0099	3.5	0.003
Pr_4NCl	0.1346	-0.300	0.0119	2.5	0.002
Pr_4NF	0.4463	0.4090	0.0537	2	0.002
Pr_4NI	-0.2839	-0.863	---	0.5	0.005

Table 4: Sulfonic Acids and Salts (1-1 Type) SA = Sulfonic Acid; S = Sulfonate

species	β_0	β_1	C^{ϕ}	Max m	σ
Benzene SA	0.0526	0.445	0.0036	5	0.002
Bu_4N ethane S	0.1827	0.445	-0.0374	4	---
Bu_4N methane S	0.2145	0.235	-0.0392	4	---
Ethane SA	0.1536	0.341	-0.0056	4	---
Et_4N ethane S	0.1805	0.075	-0.0040	4	---
Et_4N methane S	0.1548	0.090	-0.0034	4	---
K ethane S	0.0965	0.250	-0.0074	4	---
K methane S	0.0581	0.165	-0.0046	4	---
K p-toluene S	-0.0985	0.453	0.0122	3.5	0.002
Li benzene S	0.1134	0.466	-0.0075	4.5	0.002
Li ethane S	0.1799	0.319	-0.0118	4	---
Li2,5 Me_2benzene S	-0.0098	0.361	0.0039	3.5	---
Li Mesitylene S	-0.1998	0.871	0.0456	2	0.004
Li methane S	0.1320	0.271	-0.0030	4	---
Li p-Et benzene S	-0.1438	0.804	0.0317	5	0.01
Li p-toluene S	0.0189	0.399	0.0046	4.5	0.004
2,5 Me_2benzene SA	-0.0965	0.141	0.0210	4.5	0.01
Me_4N ethane S	0.1796	0.083	-0.0116	4	---
Me_4N methane S	0.1458	0.168	-0.0043	4	---
Mesitylene SA	-0.2209	0.248	0.0432	2	0.01
Methane SA	0.1298	0.629	0.0052	4	---
NH_4 ethane S	0.1142	0.179	-0.0114	4	---
NH_4 methane S	0.0661	0.191	-0.0041	4	---
Na benzene S	0.0842	0.351	-0.0181	2.5	0.001
Na ethane S	0.1316	0.374	-0.0082	4	---
Na2,5 Me_2 benzene S	-0.0277	0.228	---	1	0.005
Na Mesitylene S	-0.2018	0.767	---	1	0.003
Na methane S	0.0787	0.274	-0.0024	4	---
Na p-Et benzene S	-0.2240	0.895	0.0355	2.5	0.01
Na p-toluene S	-0.0344	0.396	0.0043	4	0.003
p-Et benzene SA	-0.1736	0.435	0.0383	2	0.007
p-toluene SA	-0.0366	0.281	0.0137	5	0.002

Table 5: Additional 1-1 Type Organic Salts

species	β_0	β_1	C^ϕ	Max m	σ
Bu_3SBr	-0.0803	-0.616	0.0053	6	0.01
Bu_3SCl	0.0726	-0.245	-0.0099	6	0.01
Choline Br	-0.0066	-0.227	0.0036	6	0.004
Choline Cl	0.0457	-0.196	0.0008	6	0.004
$(HOC_2H_4)_4NBr$	-0.0474	-0.259	0.0106	3	0.002
$(HOC_2H_4)_4NF$	0.0938	0.128	-0.0030	4	0.001
Me_3BzNBr	-0.1517	-0.545	0.0187	3	0.01
Me_3BzNCl	-0.0821	-0.178	0.0162	3.5	0.01
$Me_2OEtBzNBr$	-0.1518	-0.778	0.0177	3	0.01
$Me_2OEtBzNCl$	-0.0879	-0.343	0.0134	4	0.01
Me_3SBr	-0.0228	-0.245	0.0044	6	0.004
Me_3SCl	0.0314	-0.184	0.0023	6	0.005
Me_3SI	-0.0601	-0.604	0.0006	3	0.01

Table 6: Inorganic Compounds of 2-1 Type

species	$4/3\ \beta_0$	$4/3\ \beta_1$	$2^{5/2}/3\ C^\phi$	Max m	σ
$BaBr_2$	0.4194	2.093	-0.03009	2	0.001
$BaCl_2$	0.3504	1.995	-0.03654	1.8	0.001
$Ba(ClO_4)_2$	0.4819	2.101	-0.05894	2	0.003
BaI_2	0.5625	2.249	-0.03286	1.8	0.003
$Ba(NO_3)_2$	-0.043	1.07	---	0.4	0.001
$Ba(OH)_2$	0.229	1.60	---	0.1	---
$CaBr_2$	0.5088	2.151	-0.00485	2	0.002
$CaCl_2$	0.4212	2.152	-0.00064	2.5	0.003
$Ca(ClO_4)_2$	0.6015	2.342	-0.00943	2	0.005
CaI_2	0.5839	2.409	-0.00158	2	0.001
$Ca(NO_3)_2$	0.2811	1.879	-0.03798	2	0.002
$Cd(NO_3)_2$	0.3820	2.224	-0.04836	2.5	0.002
$CoBr_2$	0.5693	2.213	-0.00127	2	0.002
$CoCl_2$	0.4857	1.967	-0.02869	3	0.004
CoI_2	0.695	2.23	-0.0088	2	0.01
$Co(NO_3)_2$	0.4159	2.254	-0.01436	5.5	0.003
Cs_2SO_4	0.1184	1.481	-0.01131	1.8	0.001
$CuCl_2$	0.3955	1.855	-0.06792	2	0.002
$Cu(NO_3)_2$	0.4224	1.907	-0.04136	2	0.002
$FeCl_2$	0.4479	2.043	-0.01623	2	0.002
K_2CrO_4	0.1011	1.652	-0.00147	3.5	0.003
K_2HAsO_4	0.1728	2.198	-0.0336	1	0.001
K_2HPO_4	0.0330	1.699	0.0309	1	0.002
$K_2Pt(CN)_4$	0.0881	3.164	0.0247	1	0.005
K_2SO_4	0.0666	1.039	---	0.7	0.002

species	$4/3\ \beta_0$	$4/3\ \beta_1$	$2^{5/2}/3\ C^{\phi}$	Max m	σ
Li_2SO_4	0.1817	1.694	-0.00753	3	0.002
$MgBr_2$	0.5769	2.337	0.00589	5	0.004
$MgCl_2$	0.4698	2.242	0.00979	4.5	0.003
$Mg(ClO_4)_2$	0.6615	2.678	0.01806	2	0.002
MgI_2	0.6536	2.4055	0.01496	5	0.003
$Mg(NO_3)_2$	0.4895	2.113	-0.03889	2	0.003
$MnCl_2$	0.4363	2.067	-0.03865	2.5	0.003
$(NH_4)_2SO_4$	0.0545	0.878	-0.00219	5.5	0.004
Na_2CO_3	0.2530	1.128	-0.09057	1.5	0.001
Na_2CrO_4	0.1250	1.826	-0.00407	2	0.002
$Na_2HA_5O_4$	0.0407	2.173	0.0034	1	0.001
Na_2HPO_4	-0.0777	1.954	0.0554	1	0.002
$Na_2S_2O_3$	0.0882	1.701	0.00705	3.5	0.002
Na_2SO_4	0.0261	1.484	0.01075	4	0.003
$NiCl_2$	0.4639	2.108	-0.00702	2.5	0.002
$Pb(ClO_4)_2$	0.4443	2.296	-0.01667	6	0.004
$Pb(NO_3)_2$	-0.0482	0.380	0.01005	2	0.002
Rb_2SO_4	0.0772	1.481	-0.00019	1.8	0.001
$SrBr_2$	0.4415	2.282	0.00231	2	0.001
$SrCl_2$	0.3810	2.223	-0.00246	4	0.003
$Sr(ClO_4)_2$	0.5692	2.089	-0.02472	2.5	0.003
SrI_2	0.5350	2.480	-0.00501	2	0.001
$Sr(NO_3)_2$	0.1795	1.840	-0.03757	2	0.002
UO_2Cl_2	0.5698	2.192	-0.06951	2	0.001
$UO_2(ClO_4)_2$	0.8151	2.859	0.04089	2.5	0.003
$UO_2(NO_3)_2$	0.6143	2.151	-0.05948	2	0.002
$ZnBr_2$	0.6213	2.179	-0.2035	1.6	0.007
$ZnCl_2$	0.3469	2.190	-0.1659	1.2	0.006
$Zn(ClO_4)_2$	0.6747	2.396	0.02134	2	0.003
ZnI_2	0.6428	2.594	-0.0269	0.8	0.002
$Zn(NO_3)_2$	0.4641	2.255	-0.02955	2	0.001
cis$[Co(en)_2NH_3NO_2](NO_3)_2$	-0.0928	0.271	---	0.6	0.002
trans$[Co(en)_2NH_3NO_2](NO_3)_2$	-0.0901	0.249	---	0.8	0.002
cis$[Co(en)_2NH_3NO_2]Cl_2$	-0.0327	0.684	0.0121	2.8	0.005
trans$[Co(en)_2NH_3NO_2]Cl_2$	0.0050	0.695	0.0066	2.4	0.005
cis$[Co(en)_2NH_3NO_2]Br_2$	-0.1152	0.128	0.0158	1	0.004
trans$[Co(en)_2NH_3NO_2]Br_2$	-0.0912	0.424	0.0223	2.4	0.005
cis$[Co(en)_2NO_3NO_2]I_2$	-0.1820	0.594	---	0.6	0.004
trans$[Co(en)_2NO_3NO_2]I_2$	-0.1970	1.003	---	0.3	0.003

Table 7: Organic Electrolytes of 2-1 Type (SA = Sulfonic Acid: S = Sulfonate)

species	$4/3\ \beta_0$	$4/3\ \beta_1$	$2^{5/2}/3\ C^{\phi}$	Max m	σ
m-Benzenedi SA	0.5611	2.637	-0.0463	1.6	0.004
Li_2 *m*-Benzenedi S	0.5464	2.564	-0.0622	2.5	0.004
Na_2 *m*-Benzenedi S	0.3411	2.698	-0.0419	3.0	0.004
4,4'bibenzyldi SA	0.1136	2.432	0.0705	2.0	0.01
Li_2 4,4-bibenzyldi S	0.1810	1.755	0.0462	1.2	0.007
Na_2 4,4'bibenzyldi S	0.0251	1.969	---	0.4	0.01
Na_2 fumarate	0.3082	1.203	-0.0378	2.0	0.003
Na_2 maleate	0.1860	0.575	-0.0170	3.0	0.004

Table 8: 3-1 Electrolytes

species	$3/2\ \beta_0$	$3/2\ \beta_1$	$3^{5/2}/2\ C^{\phi}$	Max m	σ
$AlCl_3$	1.0490	8.767	0.0071	1.6	0.005
$CeCl_3$	0.9072	8.40	-0.0746	1.8	0.007
$Co(en)_3Cl_3$	0.2603	3.563	-0.0916	1.0	0.003
$Co(en)_3(ClO_4)_3$	0.1619	5.395	---	0.6	0.007
$Co(en)_3(NO_3)_3$	0.1882	3.935	---	0.3	0.01
$Co(pn)_3(ClO_4)_3$	0.2022	3.976	---	0.3	0.003
$CrCl_3$	1.1046	7.883	-0.1172	1.2	0.005
$Cr(NO_3)_3$	1.0560	7.777	-0.1533	1.4	0.004
$DyCl_3$	0.9290	8.40	-0.0456	3.6	0.005
$Dy(ClO_4)_3$	1.201	9.80	0.0142	2.0	0.006
$ErCl_3$	0.9285	8.40	-0.0389	3.7	0.006
$Er(ClO_4)_3$	1.202	9.80	0.0144	1.8	0.004
$Er(NO_3)_3$	0.938	7.70	-0.226	1.5	0.006
$EuCl_3$	0.9115	8.40	-0.0547	3.6	0.006
$Ga(ClO_4)_3$	1.2381	9.794	0.0904	2.0	0.008
$GdCl_3$	0.9139	8.40	-0.0494	3.6	0.006
$Gd(ClO_4)_3$	1.173	9.80	0.0140	2.0	0.007
$Gd(NO_3)_3$	0.776	7.70	-0.170	1.4	0.005
$HoCl_3$	0.9376	8.40	-0.0450	3.7	0.006
$Ho(ClO_4)_3$	1.198	9.80	0.0132	2.0	0.004
$InCl_3$	-1.68	-3.85	---	0.01	---
K_3AsO_4	0.7491	6.511	-0.3376	0.7	0.001
$K_3Co(CN)_6$	0.5603	5.815	-0.1603	1.4	0.008
$K_3Fe(CN)_6$	0.5035	7.121	-0.1176	1.4	0.003
K_3PO_4	0.5594	5.958	-0.2255	0.7	0.001

species	$3/2\ \beta_0$	$3/2\ \beta_1$	$3^{3/2}/2\ C^{\phi}$	Max m	σ
$K_3P_3O_9$	0.4867	8.349	-0.0886	0.8	0.004
$LaCl_3$	0.8834	8.40	-0.0619	3.9	0.006
$La(ClO_4)_3$	1.158	9.80	0.0016	2.0	0.009
$La(NO_3)_3$	0.740	7.70	-0.199	1.3	0.009
$LuCl_3$	0.9228	8.40	-0.0332	4.1	0.005
$Lu(ClO_4)_3$	1.186	9.80	0.0290	2.0	0.006
Na_3AsO_4	0.3582	5.895	-0.1240	0.7	0.001
Na_3PO_4	0.2672	5.777	-0.1339	0.7	0.003
$NdCl_3$	0.8784	8.40	-0.0493	3.9	0.006
$Nd(ClO_4)_3$	1.131	9.80	0.0194	2.0	0.007
$PrCl_3$	0.8838	8.40	-0.0549	3.9	0.006
$Pr(ClO_4)_3$	1.132	9.80	0.0163	2.0	0.006
$Pr(NO_3)_3$	0.737	7.70	-0.188	1.1	0.006
$SmCl_3$	0.9000	8.40	-0.0535	3.6	0.007
$Sm(ClO_4)_3$	1.146	9.80	0.0140	2.0	0.005
$Sm(NO_3)_3$	0.701	7.70	-0.131	1.5	0.007
$SrCl_3$	1.0500	7.978	-0.0840	1.8	0.005
$TbCl_3$	0.9229	8.40	-0.0468	3.6	0.004
$Tb(ClO_4)_3$	1.193	9.80	0.0123	2.0	0.006
$Tb(NO_3)_3$	0.838	7.70	-0.202	1.4	0.005
$TmCl_3$	0.9262	8.40	-0.0362	3.7	0.005
$Tm(ClO_4)_3$	1.193	9.80	0.0245	2.0	0.005
$Tm(NO_3)_3$	0.952	7.70	-0.222	1.5	0.006
YCl_3	0.9367	8.40	-0.0407	3.5	0.007
$YbCl_3$	0.9235	8.40	-0.0335	3.7	0.005
$Yb(ClO_4)_3$	1.206	9.80	0.0137	1.8	0.004
$Yb(NO_3)_3$	0.948	7.70	-0.208	1.5	0.006

Table 9: 4-1 Electrolytes

species	$8/5\ \beta_0$	$8/5\ \beta_1$	$16/5\ C^{\phi}$	Max m	σ
$K_4Fe(CN)_6$	1.021	16.23	-0.5579	0.9	0.008
$K_4Mo(CN)_8$	0.854	18.53	-0.3499	0.8	0.01
$K_4P_2O_7$	0.977	17.88	-0.2418	0.5	0.01
$K_4W(CN)_8$	1.032	18.49	-0.4937	1.0	0.005
$Me_4NMo(CN)_8$	0.938	15.91	-0.3330	1.4	0.01
$Na_4P_2O_7$	0.699	17.16	---	0.2	0.01
$ThCl_4$	1.622	21.33	-0.3309	1.0	0.006
$Th(NO_3)_4$	1.546	18.22	-0.5906	1.0	0.01

Table 10: 5-1 Electrolytes

species	5/3 β_0	5/3 β_1	$5^{3/2}/3\ C^{\phi}$	Max M	σ
$K_5P_3O_{10}$	1.939	39.64	-1.055	0.5	0.015
$Na_5P_3O_{10}$	1.869	36.10	-1.630	0.4	0.01

Table 11: 2-2 Electrolytes (b = 1.2, α_1 = 1.4, α_2 = 12.)

species	β_0	β_1	β_2	C^{ϕ}	range	σ
$BeSO_4$	0.317	2.914	?	0.0062	.1-4.	0.004
$CaSO_4$	0.20	2.65	-55.7	--	.004-.011	0.003
$CdSO_4$	0.2053	2.617	-48.07	0.0114	.005-3.5	0.002
$CoSO_4$	0.20	2.70	-30.7	--	.006-0.1	0.003
$CuSO_4$	0.2340	2.527	-48.33	0.0044	.005-1.4	0.003
$MgSO_4$	0.2210	3.343	-37.23	0.0250	.006-3.0	0.004
$MnSO_4$	0.201	2.980	?	0.0182	0.1-4.0	0.003
$NiSO_4$	0.1702	2.907	-40.06	0.0366	.005-2.5	0.005
UO_2SO_4	0.322	1.827	?	-0.0176	0.1-5.0	0.003
$ZnSO_4$	0.1949	2.883	-32.81	0.0290	.005-3.5	0.004

APPENDIX 4.5

PITZER PARAMETER DERIVATIVES

From: K.S. Pitzer, "Theory: Ion Interaction Approach", Activity Coefficients in Electrolyte Solutions, vol. 1, Ricardo M. Pytkowicz, ed., CRC Press, Boca Raton, Fla. (1979)

Table 1: Temperature Derivatives of Parameters for 1-1 Electrolytes Evaluated from Calorimetric Data

species	$\left(\frac{\partial \beta_0}{\partial T}\right)_P$ x10^4	$\left(\frac{\partial \beta_1}{\partial T}\right)_P$ x10^4	$\left(\frac{\partial C^\phi}{\partial T}\right)_P$ x10^5	Max. m
Bu_4NBr	-116.2	167.2	284.1	3.0
Bu_4NCl	-122.8	163.6	258.5	2.5
Bu_4NF	-117.8	105.3	43.5	1.9
CsBr	7.80	28.44	---	1.0
CsCl	8.28	15.0	-12.25	3.0
CsF	0.95	5.97	---	1.1
CsI	9.75	34.77	---	0.7
Et_4NBr	3.81	73.4	-11.37	4.6
Et_4NCl	2.00	61.4	-13.1	5.3
Et_4NF	-16.4	43.4	---	0.5
Et_4NI	-1.97	92.0	-36.3	2.0
HBr	-2.049	4.467	-5.685	6.0
HCl	-3.081	1.419	6.213	4.5
$HClO_4$	4.905	19.31	-11.77	6.0
HI	-0.230	8.86	-7.32	6.0
KBr	7.39	17.40	-7.004	5.2
KCl	5.794	10.71	-5.095	4.5
$KClO_3$	19.87	31.8	---	0.1
$KClO_4$	0.60	100.7	---	0.1
KCNS	6.87	37.0	0.43	3.1
KF	2.14	5.44	-5.95	5.9
KH_2PO_4	6.045	28.6	-10.11	1.8
KI	9.914	11.86	-9.44	7.
KNO_3	2.06	64.5	39.7	2.4
LiBr	-1.819	6.636	-2.813	6.0
LiCl	-1.685	5.366	-4.520	6.4
$LiClO_4$	0.386	7.009	-7.712	4.0
MeH_3NCl	1.13	10.8	---	0.5
Me_2H_2NCl	0.023	18.2	---	0.5
Me_3HNCl	0.22	35.3	---	0.5

species	$\left(\frac{\partial \beta_0}{\partial T}\right)_P$ x10^4	$\left(\frac{\partial \beta_1}{\partial T}\right)_P$ x10^4	$\left(\frac{\partial C^\phi}{\partial T}\right)_P$ x10^5	Max. m
Me_4NBr	6.91	60.2	-6.69	5.5
Me_4NCl	5.93	49.0	-7.66	8.1
Me_4NF	-0.82	16.0	-9.27	3.0
Me_4NI	-7.06	100.9	---	0.3
NH_4Cl	0.779	12.58	2.10	4.0
$NH_4H_2PO_4$	1.51	22.8	-2.84	3.4
NaBr	7.692	10.79	-9.30	9.0
$NaBrO_3$	5.59	34.37	---	0.1
NaCl	7.159	7.005	-10.54	6.0
$NaClO_3$	10.35	19.07	-9.29	6.4
$NaClO_4$	12.96	22.97	-16.23	6.0
NaCNS	7.80	20.0	---	0.1
NaF	5.361	8.70	---	0.7
NaI	8.355	8.28	-8.35	6.0
$NaIO_3$	20.66	60.57	---	0.1
$NaNO_3$	12.66	20.60	-23.16	2.2
NaOH	7.00	1.34	-18.94	4.2
Pr_4NBr	-31.0	109.0	26.5	3.0
Pr_4NCl	-32.2	85.1	11.3	4.4
Pr_4NF	-39.1	41.6	---	0.8
Pr_4NI	-23.4	107.0	---	0.5
RbBr	6.780	20.35	---	1.0
RbCl	5.522	15.06	---	0.8
RbF	-0.76	14.7	---	1.0
RbI	8.578	23.83	---	0.7

Table 2: Temperature Derivatives of Parameters for 2-1 and 1-2 Electrolytes Evaluated from Calorimetric Data

species	$\frac{4}{3}\left(\frac{\partial \beta_0}{\partial T}\right)_P$ x10^3	$\frac{4}{3}\left(\frac{\partial \beta_1}{\partial T}\right)_P$ x10^3	$\frac{2}{3}\left(\frac{\partial C^\phi}{\partial T}\right)_P$ x10^4	Max. m
$BaBr_2$	-0.451	9.04	---	0.1
$BaCl_2$	0.854	4.31	-2.9	1.8
$Ba(NO_3)_2$	-3.88	38.8	---	0.1
$CaBr_2$	-0.697	8.05	---	0.6
$CaCl_2$	-0.230	5.20	---	0.1
$Ca(ClO_4)_2$	1.106	6.77	-5.83	4.0
$Ca(NO_3)_2$	0.706	12.25	---	0.1
$Co(ClO_4)_2$	0.727	7.15	-6.77	4.0
Cs_2SO_4	-1.19	19.31	---	0.1
$CuCl_2$	-3.62	11.3	---	0.6

species	$\frac{4}{3}\left(\frac{\partial\beta_0}{\partial T}\right)_P$ x10^3	$\frac{4}{3}\left(\frac{\partial\beta_1}{\partial T}\right)_P$ x10^3	$\frac{2^{5/2}}{3}\left(\frac{\partial C^\phi}{\partial T}\right)_P$ x10^4	Max. m
K_2SO_4	1.92	8.93	---	0.1
Li_2SO_4	0.674	1.88	-4.40	3.0
$MgBr_2$	-0.075	5.15	---	0.1
$MgCl_2$	-0.572	4.87	---	0.1
$Mg(ClO_4)_2$	0.697	6.00	-6.65	3.2
$Mg(NO_3)_2$	0.687	5.99	---	0.1
$Mn(ClO_4)_2$	0.529	6.70	-6.29	4.0
Na_2SO_4	3.156	7.51	-9.20	3.0
$Ni(ClO_4)_2$	0.888	6.35	-7.21	4.0
Rb_2SO_4	1.25	11.52	---	0.1
$SrBr_2$	-0.437	8.71	---	0.1
$SrCl_2$	0.956	3.79	---	0.1
$Sr(ClO_4)_2$	1.524	7.19	-5.86	3.0
$Sr(NO_3)_2$	0.236	16.63	---	0.2
$Zn(ClO_4)_2$	0.795	6.79	-7.27	4.0

Table 3: Temperature Derivatives of Parameters for 3-1 and 2-2 Electrolytes Evaluated from Calorimetric Data

species	$\frac{\partial\beta_0}{\partial T}$ x10^3	$\frac{\partial\beta_1}{\partial T}$ x10^2	$\frac{\partial\beta_2}{\partial T}$ x10	$\frac{\partial C^\phi}{\partial T}$ x10^3	Max. m
$CaSO_4$	---	5.46	-5.16	---	0.02
$CdSO_4$	-2.79	1.71	-5.22	2.61	1.0
$CuSO_4$	-4.4	2.38	-4.73	4.80	1.0
$K_3Fe(CN)_6$	-0.87	3.15	---	---	0.1
$K_4Fe(CN)_6$	4.74	3.92	---	---	0.2
$LaCl_3$	0.253	0.798	---	-0.371	3.6
$La(ClO_4)_3$	0.152	1.503	---	-0.672	2.1
$La(NO_3)_3$	0.173	1.095	---	-0.451	2.2
$MgSO_4$	-0.69	1.53	-2.53	0.523	2.0
$Na_3Fe(CN)_6$	3.05	1.52	---	---	0.1
$ZnSO_4$	-3.66	2.33	-3.33	3.97	1.0

Pitzer et al. give parameters for other rare earths in Pitzer reference P9.

APPENDIX 4.6

CHEN PARAMETERS

From: C-C Chen; H.I. Britt; J.F. Boston and L.B. Evans, "Local Composition Model for Excess Gibbs Energy of Electrolyte Systems Part I: Single Solvent, Single Completely Dissociated Electrolyte Systems", AIChE J., v. 28, #4, p. 592 (1982)

Table: τ Values Fit for Molality Mean Ionic Activity Coefficient Data of Aqueous Electrolytes at 298.15 K

$$\sigma = \sum_i (\ln \gamma_{\pm m,i}^{calc} - \ln \gamma_{\pm m,i}^{exp})^2$$

1-1 electrolyte	$\tau_{m,ca}$	$\tau_{ca,m}$	σ	Max m
$AgNO_3$	7.420	-3.285	0.010	6.0
CsAc	8.596	-4.626	0.010	3.5
CsBr	8.510	-4.188	0.006	5.0
CsCl	8.530	-4.210	0.006	6.0
CsI	8.337	-4.087	0.007	3.0
$CsNO_3$	8.996	-4.153	0.003	1.4
HBr	9.823	-5.173	0.015	3.0
HCl	10.089	-5.212	0.035	6.0
$HClO_4$	10.601	-5.422	0.063	6.0
HI	9.577	-5.150	0.018	3.0
HNO_3	8.429	-4.463	0.008	3.0
KAc	8.590	-4.602	0.008	3.5
KBr	8.093	-4.143	0.004	5.5
KCl	8.064	-4.107	0.003	4.5
KCNS	7.516	-3.846	0.002	5.0
KF	8.792	-4.500	0.005	4.0
KH Adipate	5.538	-2.986	0.002	1.0
KH Malonate	7.489	-3.661	0.005	5.0
KH Succinate	8.108	-4.022	0.003	4.5
KH_2AsO_4	8.004	-3.735	0.003	1.2
KH_2PO_4	8.932	-4.116	0.004	1.8
KI	7.800	-4.069	0.005	4.5
KNO_3	7.673	-3.479	0.008	3.5
KOH	9.868	-5.059	0.023	6.0
LiAc	8.441	-4.417	0.005	4.0

1-1 electrolyte	$\tau_{m,ca}$	$\tau_{ca,m}$	σ	Max m
LiBr	10.449	-5.348	0.050	6.0
LiCl	10.031	-5.154	0.040	6.0
$LiClO_4$	9.579	-5.088	0.023	4.0
LiI	9.257	-4.987	0.024	3.0
$LiNO_3$	8.987	-4.707	0.013	6.0
LiOH	9.008	-4.400	0.026	4.0
LiTol	7.567	-3.895	0.013	4.5
NaAc	8.392	-4.489	0.007	3.5
NaBr	8.793	-4.562	0.009	4.0
$NaBrO_3$	7.593	-3.709	0.002	2.5
Na Butyrate	7.526	-4.300	0.007	3.5
Na Caprate	11.729	-4.836	0.024	1.8
NaCl	8.885	-4.549	0.018	6.0
$NaClO_3$	7.225	-3.692	0.005	3.5
$NaClO_4$	8.021	-4.131	0.010	6.0
NaCNS	7.941	-4.238	0.010	4.0
NaF	7.540	-3.800	0.000	1.0
Na Formate	7.437	-3.942	0.004	5.0
NaH Malonate	7.694	-3.850	0.002	5.0
NaH Succinate	8.226	-4.133	0.002	5.0
NaH_2PO_4	8.277	-3.892	0.003	6.0
NaI	8.862	-4.652	0.010	3.5
$NaNO_3$	7.300	-3.620	0.002	6.0
NaOH	9.372	-4.777	0.029	6.0
Na Pelargonate	10.931	-4.657	0.042	2.5
Na Propionate	8.400	-4.559	0.006	3.0
NH_4Cl	7.842	-4.005	0.001	6.0
NH_4NO_3	7.359	-3.526	0.012	6.0
RbAc	8.728	-4.666	0.009	3.5
RbBr	8.078	-4.063	0.002	5.0
RbCl	8.239	-4.148	0.002	5.0
RbI	8.203	-4.115	0.003	5.0
$RbNO_3$	7.718	-3.464	0.012	4.5
TlAc	7.862	-3.821	0.012	6.0
1-2 electrolyte				
Cs_2SO_4	7.254	-4.113	0.009	1.8
K_2CrO_4	8.155	-4.542	0.022	3.5
K_2SO_4	9.247	-4.964	0.008	0.7
Li_2SO_4	8.416	-4.771	0.023	3.0
Na_2CrO_4	8.857	-4.912	0.057	4.0

1-2 electrolyte	$\tau_{m,ca}$	$\tau_{ca,m}$	σ	Max m
Na_2 Fumarate	8.834	-5.020	0.003	2.0
Na_2 Maleate	9.375	-5.078	0.020	3.0
Na_2SO_4	8.389	-4.539	0.024	4.0
$Na_2S_2O_3$	8.510	-4.705	0.030	3.5
$(NH_4)_2SO_4$	8.623	-4.602	0.017	4.0
Rb_2SO_4	7.312	-4.086	0.009	1.8
2-1 electrolyte				
$BaBr_2$	8.292	-5.022	0.026	2.0
$Ba(ClO_4)_2$	9.709	-5.592	0.072	5.0
BaI_2	9.340	-5.564	0.034	2.0
$CaBr_2$	12.175	-6.553	0.351	6.0
$CaCl_2$	11.396	-6.218	0.205	6.0
$Ca(ClO_4)_2$	12.058	-6.579	0.272	6.0
CaI_2	9.861	-5.809	0.046	2.0
$Ca(NO_3)_2$	8.982	-5.098	0.060	6.0
$CdBr_2$	11.656	-5.498	0.258	4.0
$CdCl_2$	10.621	-5.066	0.214	6.0
CdI_2	13.510	-6.179	0.374	2.5
$CoBr_2$	11.422	-6.339	0.141	5.0
$CoCl_2$	10.111	-5.771	0.055	4.0
CoI_2	12.350	-6.731	0.242	6.0
$Co(NO_3)_2$	10.460	-5.866	0.108	5.0
$CuCl_2$	7.888	-4.771	0.038	6.0
$Cu(NO_3)_2$	10.346	-5.782	0.113	6.0
$FeCl_2$	9.118	-5.377	0.029	2.0
$MgAc_2$	9.148	-5.141	0.013	4.0
$MgBr_2$	11.978	-6.547	0.241	5.0
$MgCl_2$	11.579	-6.338	0.202	5.0
$Mg(ClO_3)_2$	11.948	-6.620	0.208	4.0
MgI_2	12.467	-6.775	0.316	5.0
$Mg(NO_3)_2$	10.655	-5.969	0.125	5.0
$MnCl_2$	9.554	-5.508	0.047	6.0
$NiCl_2$	10.751	-6.013	0.092	5.0
$Pb(ClO_4)_2$	10.853	-6.030	0.147	6.0
$Pb(NO_3)_2$	9.143	-4.695	0.022	2.0
$SrBr_2$	9.305	-5.475	0.036	2.0
$SrCl_2$	10.285	-5.784	0.088	4.0
$Sr(ClO_4)_2$	11.415	-6.305	0.168	6.0
SrI_2	9.655	-5.704	0.046	2.0
$Sr(NO_3)_2$	6.774	-4.000	0.029	4.0
UO_2Cl_2	8.923	-5.386	0.040	3.0
$UO_2(ClO_4)_2$	13.393	-7.184	0.447	5.5

2-1 electrolyte	$\tau_{m,ca}$	$\tau_{ca,m}$	σ	Max m
$UO_2(NO_3)_2$	9.216	-5.542	0.041	5.5
$ZnCl_2$	7.997	-4.660	0.119	6.0
$Zn(ClO_4)_2$	12.118	-6.675	0.211	4.0
$Zn(NO_3)_2$	10.750	-5.995	0.148	6.0
2-2 electrolyte				
$BeSO_4$	11.556	-6.984	0.039	4.0
$CdSO_4$	11.228	-6.755	0.037	3.5
$CuSO_4$	11.703	-6.993	0.037	1.4
$MgSO_4$	11.346	-6.862	0.036	3.5
$MnSO_4$	11.294	-6.805	0.037	4.0
$NiSO_4$	11.378	-6.837	0.031	2.5
UO_2SO_4	11.201	-6.764	0.050	6.0
$ZnSO_4$	11.476	-6.888	0.038	3.5
3-1 electrolyte				
$AlCl_3$	10.399	-6.255	0.115	1.8
$CeCl_3$	9.398	-5.775	0.084	2.0
$Co(en)_3Cl_3$	7.534	-4.278	0.013	1.0
$CrCl_3$	6.567	-4.916	0.069	1.2
$Cr(NO_3)_3$	6.276	-4.777	0.054	1.4
$EuCl_3$	9.555	-5.856	0.091	2.0
$Ga(ClO_4)_3$	11.302	-6.740	0.228	2.0
$K_3Fe(CN)_6$	6.459	-3.806	0.019	1.4
$LaCl_3$	9.290	-5.737	0.082	2.0
$NdCl_3$	9.524	-5.822	0.083	2.0
$PrCl_3$	9.420	-5.779	0.082	2.0
$ScCl_3$	9.664	-5.943	0.078	1.8
$SmCl_3$	9.533	-5.840	0.087	2.0
YCl_3	9.906	-5.991	0.093	2.0
3-2 electrolyte				
$Al_2(SO_4)_3$	10.646	-7.116	0.075	1.0
$Cr_2(SO_4)_3$	6.705	-5.568	0.129	1.2

REFERENCES

1. Arrhenius, S., Z. physik. Chem., 1, 631 (1887)

2. Debye, P. and E. Hückel, Physik Z., 24, 185 (1923)

3. Lewis, G.N. and J.M. Randall, J. Am. Chem. Soc., 43, 1112 (1921)

4. Robinson, R.A. and R.H. Stokes, J. Am. Chem. Soc., 70, 1870 (1948)

5. Robinson, R.A. and R.H. Stokes, Electrolyte Solutions, 2nd ed., Butterworth and Co., London, (1970), p. 236

6. Güntelberg, E., Z. physik. Chem., 123, 199 (1926)

7. Guggenheim, E.A., Phil. Mag., 19, 588 (1926)

8. Guggenheim, E.A. and J.C. Turgeon, Trans. Faraday Soc., 51, 747 (1955)

9. Davies, C.W., J. Chem Soc., 2093 (1938)

10. Davies, C.W., Ion Association, Butterworths Scientific Publications, London (1962)

11. Guggenheim, E.A., Thermodynamics, 5th ed., North-Holland Publishing Company, Amsterdam (1967)

12. Friedman, H.L., Ionic Solution Theory, Interscience Publishers, New York (1962)

13. Goldberg, R.N., "Evaluated Activity and Osmotic Coefficients for Aqueous Solutions: Thirty-Six Uni-Bivalent Electrolytes", J. Phys. Chem. Ref. Data, 10, 735 (1981)

BROMLEY REFERENCES

B1. Bromley, L.A., "Approximate individual ion values of β (or B) in extended Debye-Hückel theory for uni-univalent aqueous solutions at 298.15 K", J. Chem. Thermo., 4, 669 (1972)

B2. Lewis, G.N., M. Randall, K.S. Pitzer and L. Brewer, Thermodynamics, 2nd ed., McGraw-Hill: New York (1961)

B3. Bromley, L.A., "Thermodynamic Properties of Strong Electrolytes in Aqueous Solutions", AIChE J., 19, 313 (1973)

B4. Bromley, L.A., "Thermodynamic Properties of Sea Salt Solutions", AIChE J., 20, 326 (1974)

OTHER AUTHORS USING BROMLEY'S METHOD

OB1. Rastogi, A. and D. Tassios, "Estimation of Thermodynamic Properties of Binary Aqueous Electrolytic Solutions in the Range 25 - 100°C", Ind. Eng. Chem. Process Des. Dev., 19, 477 (1980)

MEISSNER REFERENCES

M1. Meissner, H.P. and J.W. Tester, "Activity Coefficients of Strong Electrolytes in Aqueous Solutions", Ind. Eng. Chem. Proc. Des. Dev., 11, 128 (1972)

M2. Meissner, H.P. and C.L. Kusik, "Activity Coefficients of Strong Electrolytes in Multicomponent Aqueous Solutions", AIChE J., 18, 294 (1972)

M3. Meissner, H.P., C.L. Kusik and J.W. Tester, "Activity Coefficients of Strong Electrolytes in Aqueous Solution - Effect of Temperature", AIChE J., 18, 661 (1972)

M4. Meissner, H.P. and C.L. Kusik, "Vapor Pressures of Water Over Aqueous Solutions of Strong Electrolytes", Ind. Eng. Chem. Proc. Des. Dev., 12, 112 (1973)

M5. Meissner, H.P. and C.L. Kusik, "Aqueous Solutions of Two or More Strong Electrolytes - Vapor Pressures and Solubilities", Ind. Eng. Chem. Proc. Des. Dev., 12, 205 (1973)

M6. Meissner, H.P. and N.A. Peppas, "Activity Coefficients - Aqueous Solutions of Polybasic Acids and Their Salts", AIChE J., 19, 806 (1973)

M7. Meissner, H.P. and C.L. Kusik, "Electrolyte Activity Coefficients in Inorganic Processing", AIChE Symposium Series 173, 74, 14 (1978)

M8. Meissner, H.P. and C.L. Kusik, "Double Salt Solubilities", Ind. Eng. Chem. Proc. Des. Dev., 18, 391 (1979)

M9. Meissner, H.P., C.L. Kusik and E.L. Field, "Estimation of Phase Diagrams and Solubilities for Aqueous Multi-ion Systems", AIChE J., 25, 759 (1979)

M10. Meissner, H.P. "Prediction of Activity Coefficients of Strong Electrolytes in Aqueous Systems", Thermodynamics of Aqueous Systems with Industrial Applications, ACS Symposium Series 133, 496 (1980)

PITZER REFERENCES

P1. Pitzer, K. S., "Thermodynamics of Electrolytes. I. Theoretical Basis and General Equations", J. Phys. Chem., 77, 268 (1973)

P2. Pitzer, K.S. and G. Mayorga, "Thermodynamics of Electrolytes. II. Activity and Osmotic Coefficients for Strong Electrolytes with One or Both Ions Univalent", J. Phys. Chem., 77, 2300 (1973)

P3. Pitzer, K.S. and G. Mayorga, "Thermodynamics of Electrolytes. III. Activity and Osmotic Coefficients for 2-2 Electrolytes", J. Sol. Chem., 3, 539 (1974)

P4. Pitzer, K.D. and J.J. Kim, "Thermodynamics of Electrolytes. IV. Activity and Osmotic Coefficients for Mixed Electrolytes", J. A.C.S., 96, 5701 (1974)

P5. Pitzer, K.S., "Thermodynamics of Electrolytes. V. Effects of Higher-Order Electrostatic Terms", J. Sol. Chem., 4, 249 (1975)

P6. Pitzer, K.S. and L.F. Silvester, "Thermodynamics of Electrolytes. VI. Weak Electrolytes Including H_3PO_4", J. Sol. Chem., 5, 269 (1976)

P7. Pitzer, K.S., R.N. Roy and L.F. Silvester, "Thermodynamics of Electrolytes. 7. Sulfuric Acid", J. A.C.S., 99, 4930 (1977)

P8. Pitzer, K.S. and L.F. Silvester, "Thermodynamics of Electrolytes. 8. High-Temperature Properties, Including Enthalpy and Heat Capacity, with Application to Sodium Chloride", J. Phys. Chem., 81, 1822 (1977)

P9. Pitzer, K.S., J.F. Peterson and L.F. Silvester, "Thermodynamics of Electrolytes. IX. Rare Earth Chlorides, Nitrates, and Perchlorates", J. Sol. Chem., 7, 45 (1978)

P10. Pitzer, K.S. and L.F. Silvester, "Thermodynamics of Electrolytes. X. Enthalpy and the Effect of Temperature on the Activity Coefficients", J. Sol. Chem., 7, 327 (1978)

P11. Pitzer, K.S. and L.F. Silvester, "Thermodynamics of Electrolytes. 11. Properties of 3:2, 4:2, and Other High-Valence Types", J. Phys. Chem., 82, 1239 (1978)

P12. Pitzer, K.S. and D.J. Bradley, "Thermodynamics of Electrolytes. 12. Dielectric Properties of Water and Debye-Hückel Parameters to 350°C and 1 kbar", J. Phys. Chem., 83, 1599 (1979)

P13. Pitzer, K.S. and C.J. Downes, "Thermodynamics of Electrolytes. Binary Mixtures Formed from Aqueous NaCl, Na_2SO_4, $CuCl_2$, and $CuSO_4$ at 25°C", J. Sol. Chem., 5, 389 (1976)

P14. Pitzer, K.S., "Theory: Ion Interaction Approach", Activity Coefficients in Electrolyte Solutions, vol. 1, Ricardo M. Pytkowicz, ed., CRC Press, Boca Raton, Fla. (1979)

P15. Pitzer, K.S., "Thermodynamics of Aqueous Electrolytes at Various Temperatures, Pressures, and Compositions", Thermodynamics of Aqueous Systems with Industrial Applications, Stephen A. Newman, ed., ACS Symposium Series 133, p. 451 (1980)

P16. Pitzer, K.S. and J.C. Peiper, "Activity Coefficient of Aqueous $NaHCO_3$", J. Phys. Chem., 84, 2396 (1980)

P17. Pitzer, K.S. and L.F. Silvester, "Thermodynamics of Geothermal Brines I. Thermodynamic Properties of Vapor-Saturated NaCl(aq) Solutions From 0 - 300°C", Lawrence Berkeley Laboratory Report 4456 (1976)

P18. Pitzer, K.S., "Electrolyte Theory - Improvement since Debye and Hückel", Accounts of Chem. Res., 10, 371 (1977)

P19. Pitzer, K.S. and P.S.Z. Rogers, "Volumetric Properties of Aqueous Sodium Chloride Solutions", J. Phys. Chem. Ref. Data, 11, 15 (1982)

P20. Pitzer, K.S., M. Conceicão and P. de Lima, "Thermodynamics of Saturated Aqueous Solutions Including Mixtures of NaCl, KCl, and CsCl", J. Sol. Chem., 12, 171 (1983)

P21. Pitzer, K.S., M. Conceicão and P. de Lima, "Thermodynamics of Saturated Electrolyte Mixtures of NaCl with Na_2SO_4 and with $MgCl_2$", J. Sol. Chem., 12, 187 (1983)

P22. Pitzer, K.S., and R.C. Phutela, "Thermodynamics of Aqueous Calcium Chloride", J. Sol. Chem., 12, 201 (1983)

OTHER AUTHORS USING PITZER'S METHOD

OP1. Edwards, T.J., G. Maurer, J. Newman and J.M. Prausnitz, "Vapor-Liquid Equilibria in Multicomponent Aqueous Solutions of Volatile Weak Electrolytes", AIChE J., 24, 966 (1978)

OP2. Harvie, C.E. and J.H. Weare, "The prediction of mineral solubilities in natural waters: the Na-K-Mg-Ca-Cl-SO_4-H_2O system from zero to high concentration at 25°C", Geochimica et Cosmochimica Acta, 44, 981 (1980)

OP3. Rosenblatt, G., "Use of Pitzer's Equations to Estimate Strong-Electrolyte Activity Coefficients in Aqueous Flue Gas Desulfurization Processes", Flue Gas Desulfurization, J.L. Hudson and G.T. Rochelle, eds., ACS Symposium Series 188, p57 (1982)

OP4. Holmes, H.F., C.F. Baes, Jr. and R.E. Mesmer, "Isopiestic studies of aqueous solutions at elevated temperatures I. KCl, $CaCl_2$ and $MgCl_2$", J. Chem. Thermo., 10, 983 (1973)

OP5. Holmes, H.F., C.F. Baes, Jr. and R.E. Mesmer, "Isopiestic studies of aqueous solutions at elevated temperatures II. NaCl + KCl mixtures", J. Chem. Thermo., 11, 1035 (1979)

OP6. Holmes, H.F., C.F. Baes, Jr. and R.E. Mesmer, "Isopiestic studies of aqueous solutions at elevated temperatures III. (1-y)NaCl + y$CaCl_2$", J. Chem. Thermo., 13, 101 (1981)

OP7. Holmes, H.F. and R.E. Mesmer, "Isopiestic studies of aqueous solutions at elevated temperatures IV. $NiCl_2$ and $CoCl_2$", J. Chem. Thermo., 13, 131 (1981)

OP8. Holmes, H.F. and R.E. Mesmer, "Isopiestic studies of aqueous solutions at elevated temperatures V. $SrCl_2$ and $BaCl_2$", J. Chem. Thermo., 13, 1025 (1981)

OP9. Holmes, H.F. and R.E. Mesmer, "Isopiestic studies of aqueous solutions at elevated temperatures VI. LiCl and CsCl", J. Chem. Thermo., 13, 1035 (1981)

OP10. Fritz, J.J., "Chloride Complexes of CuCl in Aqueous Solution", J. Phys. Chem., 84, 2241 (1980)

OP11. Fritz, J.J., "Representation of the Solubility of CuCl in Solutions of Various Aqueous Chlorides", J. Phys. Chem., 85, 890 (1981)

OP12. Fritz, J.J., "Solubility of Cuprous Chloride in Various Soluble Aqueous Chlorides", J. Chem. and Eng. Data, 27, 188 (1982)

OP13. Pawlikowski, E,M., J. Newman and J.M. Prausnitz, "Phase Equilibria for Aqueous Solutions of Ammonia and Carbon Dioxide", Ind. Eng. Chem. Process Des. Dev., 21, 764 (1982)

OP14. Beutier, D. and H. Renon, "Representation of NH_3-H_2S-H_2O, NH_3-CO_2-H_2O and NH_3-SO_2-H_2O Vapor-Liquid Equilibria", Ind. Eng. Chem. Process Des. Dev., 17, 220 (1978)

OP15. Fürst, W. and H. Renon, "Effect of the Various Parameters in the Application of Pitzer's Model to Solid-Liquid Equilibrium. Preliminary Study for Strong 1-1 Electrolytes", Ind. Eng. Chem. Process Des. Dev., 21, 396 (1982)

OP16. Rogers, P.S.Z., "Thermodynamics of Geothermal Fluids", Doctoral Dissertation, U. Cal. at Berkeley (1981)

CHEN REFERENCES

C1. Chen, C-C; H.I. Britt, J.F. Boston and L.B. Evans, "Extension and Application of the Pitzer Equation for Vapor-Liquid Equilibrium of Aqueous Electrolyte Systems with Molecular Solutes", AIChE J., 25, 820 (1979)

C2. Chen, C-C, H.I. Britt, J.F. Boston and L.B. Evans, "Two New Activity Coefficient Models for the Vapor-Liquid Equilibrium of Electrolyte Systems", Thermodynamics of Aqueous Systems with Industrial Applications, ACS Symposium Series 133, 61 (1980)

C3. Chen, C-C, "Computer Simulation of Chemical Processes with Electrolytes", Sc. D. thesis, Dept. of Chem. E., M.I.T. (1980)

C4. Chen, C-C; H.I. Britt, J.F. Boston and L.B. Evans, "Local Composition Model for Excess Gibbs Energy of Electrolyte Systems Part I: Single Solvent, Single Completely Dissociated Electrolyte Systems", AIChE J., 28, 588 (1982)

C5. Renon, H. and J.M. Prausnitz, "Local Compositions in Thermodynamic Excess Functions of Liquid Mixtures", AIChE J., 14, 135 (1968)

C6. Robinson, R.A. and R.H. Stokes, Electrolyte Solutions, 2nd ed., Butterworth and Co., London, (1970), p.32

V:

Activity Coefficients of Multicomponent Strong Electrolytes

ACTIVITY COEFFICIENTS OF MULTICOMPONENT STRONG ELECTROLYTES

This chapter considers five methods for calculating activity coefficients in multicomponent strong electrolyte mixtures. Generally, these methods are built upon the calculation of activity coefficients of single salt electrolyte solutions together with the application of certain "mixing rules". The methods considered in this chapter are:

1. Guggenheim - an equation using one parameter per ion pair
2. Bromley - an equation using one parameter per ion pair
3. Meissner - an equation using one parameter per ion pair
4. Pitzer - an equation using three to four parameters per ion pair, one parameter for each like charged ion pair and one ternary parameter interaction pair
5. Chen - a method using two parameters per water-salt pair and two parameters for each salt-salt pair

The chapter is composed of three principal sections:

1. Mathematical presentation - A development of the underlying equations.

2. Application - A presentation of the specific predictive equations for activity coefficients which result from each of the five methods presented.

3. Examples - A series of computer calculated runs and resultant comparative plots for specific multicomponent systems. Systems considered include:

   a. H_2O-NaCl-KCl
   b. H_2O-NaCl-HCl
   c. H_2O-NaCl-NaOH
   d. H_2O-KCl-HCl
   e. H_2O-NaCl-$CaCl_2$
   f. H_2O-NaCl-$MgCl_2$
   g. H_2O-NaCl-Na_2SO_4
   h. H_2O-KCl-$CaCl_2$
   i. H_2O-NaOH-Na_2SO_4
   j. H_2O-NaCl-$CaSO_4$

k. H_2O-HCl-$CaSO_4$
l. H_2O-$CaCl_2$-$CaSO_4$
m. H_2O-Na_2SO_4-$CaSO_4$
n. H_2O-$MgSO_4$-$CaSO_4$

In addition to the computer results and plots, tables of the parameters used for each activity coefficient method used for each electrolyte are presented.

GUGGENHEIM'S METHOD FOR MULTICOMPONENT SOLUTIONS

Guggenheim's (1, 2) expression for the mean activity coefficient is applicable to multicomponent solutions. For the mean activity coefficient of an electrolyte of cation c and anion a, the equation is:

$$\ln \gamma_{ca} = -\frac{\alpha |z_+ z_-| \sqrt{I}}{1+\sqrt{I}} + \frac{2\nu_+}{\nu_+ + \nu_-} \sum_{a'} \beta_{ca'} m_{a'} + \frac{2\nu_-}{\nu_+ + \nu_-} \sum_{c'} \beta_{c'a} m_{c'} \qquad (5.1)$$

where:
- α - log base e Debye-Hückel constant
- ν_+ - number of cations per molecule of electrolyte ca
- ν_- - number of anions per molecule of electrolyte ca
- I - ionic strength in molality scale; $I = \frac{1}{2} \sum_i m_i z_i^2$
- z_+ - number of charges on cation c
- z_- - number of charges on anion a
- a' - all anions
- c' - all cations
- m - ionic molalities
- β - interaction coefficient; tabulated in Appendix 4.1

Converted to log base 10, this equation becomes:

$$\log \gamma_{ca} = -\frac{A |z_+ z_-| \sqrt{I}}{1+\sqrt{I}} + \frac{\nu_+}{\nu_+ + \nu_-} \sum_{a'} B_{ca'} m_{a'} + \frac{\nu_-}{\nu_+ + \nu_-} \sum_{c'} B_{c'a} m_{c'} \qquad (5.1a)$$

with: A - log base 10 Debye-Hückel constant, equation (4.31)

$$B = \frac{2\beta}{\ln 10}$$

Expressions for the mean activity coefficients of ions can be derived from equations (5.1) and (5.1a). From (5.1a), the activity coefficient of cation c is:

$$\log \gamma_c = -\frac{A z_+^2 \sqrt{I}}{1+\sqrt{I}} + \sum_{a'} B_{ca'} m_{a'} \qquad (5.2)$$

and for the anion a:

$$\log \gamma_a = -\frac{A z_-^2 \sqrt{I}}{1+\sqrt{I}} + \sum_{c'} B_{c'a} m_{c'} \qquad (5.3)$$

As indicated in the previous chapter, Guggenheim's method for single electrolytes is accurate only to about an ionic strength of .1 molal; this limitation cannot be expected to improve for multicomponent solutions.

BROMLEY'S METHOD FOR MULTICOMPONENT SOLUTIONS

Bromley defined (B3) the mean molal activity coefficient for a single salt in solution:

$$\log \gamma^{o}_{\pm} = -\frac{A\left|z_+ z_-\right| \sqrt{I}}{1+\sqrt{I}} + \frac{(0.06 + 0.6B)\left|z_+ z_-\right| I}{\left(1 + \frac{1.5}{\left|z_+ z_-\right|} I\right)^2} + BI \qquad (4.44)$$

or: $$\log \gamma^{o}_{12} = -\frac{A\left|z_1 z_2\right| \sqrt{I}}{1+\sqrt{I}} + \dot{B}_{12}\, I \qquad (5.4)$$

where:
- A - Debye-Hückel constant, equation (4.31)
- I - ionic strength, molality basis
- z_+ - number of charges on the cation
- z_- - number of charges on the anion
- B - Bromley interaction parameter, tabulated in Appendix 4.2

In the same vein, he defined the activity coefficient for a single ion

$$\log \gamma_i = -\frac{A z_i^2 \sqrt{I}}{1+\sqrt{I}} + F_i \qquad (5.5)$$

Bromley felt that treating a multicomponent solution as a single complex salt solution would be the simplest approach towards calculating the activity coefficients of electrolytes in solution. The F_i terms would then be based on the ionic interactions of this "complex salt". Using the convention that odd number subscripts denote cations and even number subscripts indicate anions, he proposed for a cation:

$$F_1 = \dot{B}_{12} Z_{12}^2 m_2 + \dot{B}_{14} Z_{14}^2 m_4 + \dot{B}_{16} Z_{16}^2 m_6 + \ldots \qquad (5.6)$$

where: $Z_{12} = \frac{z_1 + z_2}{2}$

m_i = ionic molality

For an anion, the expression would be:

$$F_2 = \dot{B}_{12} Z_{12}^2 m_1 + \dot{B}_{32} Z_{32}^2 m_3 + \dot{B}_{52} Z_{52}^2 m_5 + \ldots \qquad (5.7)$$

It can be seen, if equations (4.44) and (5.4) are approximately equal, then:

$$\dot{B}_{12} = \frac{(0.06 + .6B)\left|z_1 z_2\right|}{\left(1 + \frac{1.5}{\left|z_1 z_2\right|} I\right)^2} + B \qquad (5.8)$$

Equations (5.6) and (5.7) ignore any possible cation-cation or anion-anion interaction and any higher order interactions. The activity coefficient of an electrolyte in a multicomponent solution is, by combining equations (5.5), (5.6) and (5.7) and remembering the definition of a mean activity coefficient, equation (2.26):

$$\log \gamma_{12} = -\frac{A|z_1 z_2|\sqrt{I}}{1+\sqrt{I}} + \frac{\nu_1 F_1}{\nu} + \frac{\nu_2 F_2}{\nu} \qquad (5.9)$$

with: $|z_1 z_2| = \frac{\nu_i z_i^2}{\Sigma \nu_i}$; absolute value of the charge product for a multi-ion salt

This equation can be expressed in a more general fashion:

$$\log \gamma_{\pm} = -\frac{A \frac{\Sigma \nu_i z_i^2}{\nu} \sqrt{I}}{1+\sqrt{I}} + \frac{1}{\nu} \sum_i \nu_i F_i \qquad (5.10)$$

with: $\nu = \sum_i \nu_i$

i - representing the ions present in the solution due to the salt for which the activity is being calculated

These F_i terms may also be expressed as functions of the activity coefficients of single salt solutions:

$$F_1 = Y_{21} \log \gamma^\circ_{12} + Y_{41} \log \gamma^\circ_{14} + Y_{61} \log \gamma^\circ_{16} + \ldots$$
$$+ \frac{A\sqrt{I}}{1+\sqrt{I}} (z_1 z_2 Y_{21} + z_1 z_4 Y_{41} + z_1 z_6 Y_{61} + \ldots) \qquad (5.11)$$

and:

$$F_2 = X_{12} \log \gamma^\circ_{12} + X_{32} \log \gamma^\circ_{32} + X_{52} \log \gamma^\circ_{52} + \ldots$$
$$+ \frac{A\sqrt{I}}{1+\sqrt{I}} (z_1 z_2 X_{12} + z_3 z_2 X_{32} + z_5 z_2 X_{52} + \ldots) \qquad (5.12)$$

where: $$Y_{21} = \frac{Z_{12}^2 m_2}{I} = \left(\frac{z_1 + z_2}{2}\right)^2 \frac{m_2}{I} \qquad (5.13a)$$

$$X_{12} = \frac{Z_{12}^2 m_1}{I} = \left(\frac{z_1 + z_2}{2}\right)^2 \frac{m_1}{I}$$

Activity Coefficients of Trace Components

In describing the activity coefficient of a trace concentration, Bromley used reduced activity coefficients. This was in keeping with work done the previous year by Meissner and Kusik (M2) which will be described in the next section. The equation Bromley presented was more complex, designed to allow for solutions containing ions with different charges:

$$\Gamma_{32}^{tr} = \Gamma_{12}^{tr} = \left[\left(\Gamma_{12}^{o\;\frac{z_1 + z_2}{z_2 + z_3}} \quad \Gamma_{32}^{o\;\frac{z_3 + z_2}{z_1 + z_2}}\right)^{\frac{1}{2}}\right] \times$$

$$\left[10^{\frac{A\sqrt{I}}{1 + \sqrt{I}}\left(\frac{z_1 + z_2}{2(z_2 + z_3)} + \frac{z_3 + z_2}{2(z_1 + z_2)} - 1\right)}\right] \qquad (5.14)$$

which reduces to Meissner's equation:

$$\Gamma_{32}^{tr} = \Gamma_{12}^{tr} = (\Gamma_{12}^{o} \; \Gamma_{32}^{o})^{\frac{1}{2}}$$

when $z_1 = z_3$.

The reduced activity coefficient can be converted to a mean molal activity coefficient via the following relationship:

$$\Gamma_{ca} = \gamma_{ca}^{\frac{1}{z_+ z_-}} \qquad (4.45)$$

or:

$$\log \gamma_{ca} = |z_+ z_-| \log \Gamma_{ca} \qquad (5.15)$$

MEISSNER'S METHOD FOR MULTICOMPONENT SOLUTIONS

Meissner and Kusik (M2) presented, in 1972, a method of calculating the reduced activity coefficients of strong electrolytes in a multicomponent solution. They based their method on Brønsted's proposal that, in multicomponent solutions, the activity coefficient of an electrolyte will be influenced most by the interaction of it's cation with all the anions in solution and the interaction of it's anion with all the cations in solution. Ignoring the possible interactions between like charged ions was felt to be valid as such interactions would be very small. Meissner and Kusik proposed that the activity coefficient could then be defined as:

$$\log \Gamma_{12} = F_1 + F_2 \tag{5.16}$$

Once again using the notation that odd number subscripts indicate cations and even number subscripts indicate anions, F_1 and F_2 are expressions for calculating the interaction of cation with anions and anion with cations. They defined these as:

$$F_1 = \tfrac{1}{2}\,(Y_2 \log \Gamma^\circ_{12} + Y_4 \log \Gamma^\circ_{14} + Y_6 \log \Gamma^\circ_{16} + \ldots) \tag{5.17}$$

and

$$F_2 = \tfrac{1}{2}\,(X_1 \log \Gamma^\circ_{12} + X_3 \log \Gamma^\circ_{32} + X_5 \log \Gamma^\circ_{52} + \ldots) \tag{5.18}$$

where: Γ° indicates the pure solution reduced activity coefficient at the total ionic strength of the solution

$X_i = \dfrac{m_i z_i^2}{I_c}$ with i = odd numbers, expresses the ionic strength fraction of cation i

$I_c = \tfrac{1}{2} \sum_i m_i z_i^2$ with i = odd numbers to calculate the cationic ionic strength

m_i - ionic molality

z_i - number of charges on the ion

$Y_i = \dfrac{m_i z_i^2}{I_a}$ with i = even numbers for the anion ionic strength fraction

$I_a = \tfrac{1}{2} \sum_i m_i z_i^2$ with i = even numbers for the anionic ionic strength

For a three ion solution of two cations and one anion, the activity coefficients are:

$$\log \Gamma_{12} = \tfrac{1}{2}\,(X_1 \log \Gamma_{12}^{\circ} + X_3 \log \Gamma_{32}^{\circ} + Y_2 \log \Gamma_{12}^{\circ}) \qquad (5.19)$$

and

$$\log \Gamma_{32} = \tfrac{1}{2}\,(X_1 \log \Gamma_{12}^{\circ} + X_3 \log \Gamma_{32}^{\circ} + Y_2 \log \Gamma_{32}^{\circ}) \qquad (5.20)$$

Since there is only one anion, the ionic strength fraction, Y_2, for it equals one. The relationship between the ionic strength fractions of the cations is:

$$X_1 = 1 - X_3 \qquad (5.21)$$

Equations (5.19) and (5.20) are therefore:

$$\log \Gamma_{12} = \log \Gamma_{12}^{\circ} + .5\, X_3 \log(\Gamma_{32}^{\circ} / \Gamma_{12}^{\circ}) \qquad (5.22)$$

$$\log \Gamma_{32} = \log \Gamma_{32}^{\circ} - .5\, X_1 \log(\Gamma_{32}^{\circ} / \Gamma_{12}^{\circ}) \qquad (5.23)$$

Meissner and Kusik noted that these equations fulfill the requirement that $\Gamma = \Gamma^{\circ}$ in a pure solution due to the fact that the X_3 of equation (5.22) and the X_1 of equation (5.23) become zero for pure solutions of electrolytes 12 and 32.

They also proposed that, for a solution of electrolyte 32 with a trace amount of 12, the ionic strength fraction for cation 1 equals zero, and the cation and anion fractions Y_2 and X_3 equal one. Therefore, the activity coefficient of the trace electrolyte is, from equation (5.19):

$$\Gamma_{12}^{tr} = (\Gamma_{12}^{\circ} \cdot \Gamma_{32}^{\circ})^{\frac{1}{2}} \qquad (5.24)$$

Accordingly, for a trace amount of electrolyte 32 in a solution of electrolyte 12, equation (5.20) becomes:

$$\Gamma_{32}^{tr} = (\Gamma_{12}^{\circ} \cdot \Gamma_{32}^{\circ})^{\frac{1}{2}} \qquad (5.25)$$

Therefore, when the solutions are of equal ionic strength:

$$\Gamma_{12}^{tr} = \Gamma_{32}^{tr}$$

In 1973, Bromley (B3) disputed Meissner and Kusik's method for multicomponent solutions as follows:

> Meissner and Kusik defined the reduced activity coefficient of electrolyte 12 as:
>
> $$\log \Gamma_{12} = F_1 + F_2 \qquad (5.16)$$

as shown earlier. The mean activity coefficient is then:

$$\log \gamma_{12} = z_1 z_2 F_1 + z_1 z_2 F_2 \qquad (5.27a)$$

due to the definition of the reduced activity coefficient:

$$\Gamma_{12} = \gamma_{12}^{\frac{1}{z_1 z_2}} \qquad (4.45)$$

In Chapter II, the mean activity coefficient is defined in terms of the ionic activity coefficients:

$$\log \gamma_{12} = \log (\gamma_1^{\nu_1/\nu} \cdot \gamma_2^{\nu_2/\nu}) \qquad (2.26)$$

with: ν_i = number of ions per molecule

$\nu = \sum_i \nu_i$

Assuming electrical neutrality:

$$\nu_1 |z_1| = \nu_2 |z_2|$$

so that equation (2.26) may be written:

$$\log \gamma_{12} = \frac{z_2}{z_1 + z_2} \log \gamma_1 + \frac{z_1}{z_1 + z_2} \log \gamma_2 \qquad (5.27b)$$

Equations (5.27a) and (5.27b) must then be equal:

$$z_1 z_2 F_1 + z_1 z_2 F_2 = \frac{z_2}{z_1 + z_2} \log \gamma_1 + \frac{z_1}{z_1 + z_2} \log \gamma_2 \qquad (5.27c)$$

For a solution of four ions there will be equations for the activity coefficients of electrolytes 12, 14, 32 and 34. By expanding equation (5.27c) for the four ion solution and replacing the ionic activity coefficients with expressions in terms of F, the following relationship is found:

$$(F_1 - F_3)(z_4 - z_2) = (F_4 - F_2)(z_3 - z_1) \qquad (5.27d)$$

Bromley notes that, since the F values are usually different, it follows that:

$$z_4 - z_2 = 0 = z_3 - z_1$$

for Meissner's equation to be true. Therefore, Meissner's equation is valid for multicomponent solutions only if the cations have the same charge and the anions have the same charge: $z_2 = z_4$ and $z_1 = z_3$.

Meissner and Kusik (M7) revised their method for calculating the activity coefficients of electrolytes in multicomponent solutions in 1978. They extended equation (5.16) in order to avert the problem pointed out by Bromley. For an electrolyte of cation i and anion j, the reduced activity coefficient is:

$$\log \Gamma_{ij} = \frac{z_i}{z_i + z_j} (V_{i2}\ I_2 \log \Gamma^\circ_{i2} + V_{i4}\ I_4 \log \Gamma^\circ_{i4} + \ldots) / I$$

$$+ \frac{z_j}{z_i + z_j} (V_{j1}\ I_1 \log \Gamma^\circ_{j1} + V_{j3}\ I_3 \log \Gamma^\circ_{j3} + \ldots) / I \qquad (5.28)$$

where: z_i - number of charges on ion i

$V_{ij} = \dfrac{.5(z_i + z_j)^2}{z_i\ z_j}$ a weighting factor

$I_i = .5\ m_i\ z_i^2$ ionic strength of ion i

$I = .5 \sum_i m_i\ z_i^2$ total ionic strength of the solution

Γ°_{ij} - the reduced activity coefficient of electrolyte ij in a pure solution

In 1980, Meissner (M10) suggested a method for calculating the q parameter of his equation for calculating reduced activity coefficients:

$$q_{ij,mix} = (I_1\ q^\circ_{1j} + I_3\ q^\circ_{3j} + \ldots) / I + (I_2\ q^\circ_{i2} + I_4\ q^\circ_{i4} + \ldots) / I \qquad (5.29)$$

where: I_i - ionic strength of ion i

I - total ionic strength

q°_{ij} - Meissner q parameter for electrolyte ij in pure solution

This $q_{ij,mix}$ can then be used to calculate the reduced activity coefficient of electrolyte ij in the multicomponent solution:

$$\Gamma_{ij} = \{1 + B(1 + 0.1I)^{q_{ij,mix}} - B\}\ \Gamma^* \qquad (5.30)$$

with: $B = 0.75 - 0.065\ q_{ij,mix}$

$\log \Gamma^* = - \dfrac{0.5107 \sqrt{I}}{1 + C \sqrt{I}}$

$C = 1. + 0.055\ q_{ij,mix} \exp(-0.023\ I^3)$

Meissner and Kusik (M7) also suggested a similar method for calculating the water activity of a multicomponent solution from pure solution water activities:

$$\log (a_w)_{mix} = \sum_i \sum_j W_{ij} \log (a_w^o)_{ij} + r \qquad (5.31)$$

where: $(a_w^o)_{ij}$ - is the water activity of a pure solution of electrolyte ij

i - odd numbers denoting cations

j - even numbers denoting anions

$W_{ij} = X_i Y_j \left[\frac{(z_i + z_j)^2}{z_i z_j}\right]\left[\frac{I_c}{I}\right]\left[\frac{I_a}{I}\right]$

$X_i = I_i/I_c$, cationic fraction

$Y_j = I_j/I_a$, anionic fraction

$I_c = \frac{1}{2} \sum_i m_i z_i^2$, cationic strength

$I_a = \frac{1}{2} \sum_j m_j z_j^2$, anionic strength

$I_n = \frac{1}{2} m_n z_n^2$, individual ionic strengths

$I = \frac{1}{2} \sum_n m_n z_n^2$, total ionic strength

$r = 0.0156\ I \left[\frac{W_{12} X_1 Y_2}{z_1 z_2} + \frac{W_{23} X_3 Y_2}{z_2 z_3} + \ldots\right] - 0.0156 \left[\frac{I_1}{z_1^2} + \frac{I_2}{z_2^2} + \ldots\right]$

residue term

For solutions of similarly charged electrolytes the residue term is zero; Meissner and Kusik found that, for solutions of electrolytes with dissimilar charge, the r term is small and they stated that it usually can be ignored.

PITZER'S METHOD FOR MULTICOMPONENT SOLUTIONS

As shown in the previous chapter, Pitzer began his derivation of thermodynamic properties with the following equation for Gibb's excess energy:

$$\frac{G^{ex}}{RT} = n_w f(I) + \frac{1}{n_w} \sum_i \sum_j \lambda_{ij}(I)\, n_i n_j + \frac{1}{n_w^2} \sum_i \sum_j \sum_k \Lambda_{ijk}\, n_i n_j n_k \qquad (4.57)$$

where:
- n_w - kilograms of solvent
- n - moles of solutes i, j and k
- $f(I)$ - function describing the long range electrostatic effects as a function of temperature
- $\lambda_{ij}(I)$ - term for describing the short range inter-ionic effects as a function of ionic strength to display the type of behavior caused by the hard core effects
- Λ_{ijk} - term for triple ion interactions which ignores any ionic strength dependence

By taking the derivative of this expression and converting to a molality basis, the following expression was found for the mean activity coefficient of electrolyte CA:

$$\ln \gamma_{CA} = \frac{|z_+ z_-|}{2} f' + \frac{2\,\nu_+}{\nu_+ + \nu_-} \sum_j \lambda_{Cj} m_j + \frac{2\,\nu_-}{\nu_+ + \nu_-} \sum_j \lambda_{jA} m_j +$$
$$\frac{|z_+ z_-|}{2} \sum_j \sum_k \lambda'_{jk}\, m_j m_k + \frac{3\,\nu_+}{\nu_+ + \nu_-} \sum_j \sum_k \Lambda_{Cjk}\, m_j m_k + \frac{3\nu_-}{\nu_+ + \nu_-} \sum_j \sum_k \Lambda_{AjK}\, m_j m_k \qquad (4.59)$$

where: $f' = \frac{df}{dI}$

$$\lambda'_{ij} = \frac{d\lambda_{ij}}{dI}$$

The interaction, or virial, coefficients λ and Λ are not individually measurable but may be observable in certain combinations. For single electrolyte solutions these were the B^{γ} and C^{γ} terms of equation (4.61). Using a similar method of combining these coefficients, the equation for the mean activity coefficient of electrolyte CA in a multicomponent solution is:

$$\ln \gamma_{CA} = |z_+z_-| f^\gamma + \frac{2\nu_+}{\nu} \sum_a m_a \left[B_{Ca} + (\Sigma mz)C_{Ca} + \frac{\nu_-}{\nu_+} \Theta_{Aa} \right]$$

$$+ \frac{2\nu_-}{\nu} \sum_c m_c \left[B_{cA} + (\Sigma mz)C_{cA} + \frac{\nu_+}{\nu_-} \Theta_{Cc} \right]$$

$$+ \sum_c \sum_a m_c m_a \left[|z_+z_-| B'_{ca} + \frac{1}{\nu} (2\nu_+ z_+ C_{ca} + \nu_+ \psi_{Cca} + \nu_- \psi_{caA}) \right]$$

$$+ \tfrac{1}{2} \sum_c \sum_{c'} m_c m_{c'} \left[\frac{\nu_-}{\nu} \psi_{cc'A} + |z_+z_-| \Theta'_{cc'} \right]$$

$$+ \tfrac{1}{2} \sum_a \sum_{a'} m_a m_{a'} \left[\frac{\nu_+}{\nu} \psi_{Caa'} + |z_+z_-| \Theta'_{aa'} \right] \qquad (5.32)$$

with: $f^\gamma = \tfrac{1}{2} f' = -A_\phi \left[\frac{\sqrt{I}}{1 + b\sqrt{I}} + \frac{2}{b} \ln (1 + b\sqrt{I}) \right]$ with b = 1.2

$$B_{ij} = \lambda_{ij} + \left|\frac{z_j}{2z_i}\right| \lambda_{ii} + \left|\frac{z_i}{2z_j}\right| \lambda_{jj}$$

$$= \beta_0 + \frac{2\beta_1}{\alpha^2 I} \{1 - (1 + \alpha\sqrt{I}) \exp(-\alpha\sqrt{I})\}$$ for ions of different charge

$$B'_{ij} = \lambda'_{ij} + \left|\frac{z_j}{2z_i}\right| \lambda'_{ii} + \left|\frac{z_i}{2z_j}\right| \lambda'_{jj}$$

$$= \frac{2\beta_1}{\alpha^2 I^2} \{-1 + (1 + \alpha\sqrt{I} + \tfrac{1}{2}\alpha^2 I) \exp(-\alpha\sqrt{I})\}$$ for ions of different charge

$$C_{ij} = \frac{C^\phi_{ij}}{2\sqrt{z_+z_-}}$$ for ions of different charge

$$C^\phi_{ij} = \frac{3}{\sqrt{\nu_+\nu_-}} \left[\nu_+ \Lambda_{Iij} + \nu_- \Lambda_{iJj} \right]$$

$$\Theta_{ij} = \lambda_{ij} - \frac{z_j}{2z_i} \lambda_{ii} - \frac{z_i}{2z_j} \lambda_{jj}$$ for ions of the same charge

$$\Theta'_{ij} = \frac{\partial \Theta_{ij}}{\partial I}$$

$$\psi_{ijk} = 6\,\Lambda_{ijk} - \frac{3z_j}{z_i}\,\Lambda_{iij} - \frac{3z_i}{z_j}\,\Lambda_{jjk}$$ with two of the ions having the same charge

These last three terms are double and triple ion interactions possible in multi-component solutions. For a single electrolyte solution they would be zero.

The α term of the B and B' coefficient is set to 2.0 for 1-1, 2-1, 1-2, 3-1, 4-1 and 5-1 electrolytes as it was for single electrolyte solutions. The 2-2 electrolytes require an expanded form of the second virial coefficient:

$$B_{ij} = \beta_0 + \frac{2\beta_1}{\alpha_1^2 I}\{1 - (1 + \alpha_1\sqrt{I})\exp(-\alpha_1\sqrt{I})\} +$$

$$\frac{2\beta_2}{\alpha_2^2 I}\{1 - (1 - \alpha_2\sqrt{I})\exp(-\alpha_2\sqrt{I})\}$$

$$B'_{ij} = \frac{2\beta_1}{\alpha_1^2 I^2}\{-1 + (1 + \alpha_1\sqrt{I} + \tfrac{1}{2}\alpha_1^2 I)\exp(-\alpha_1\sqrt{I})\} +$$

$$\frac{2\beta_2}{\alpha_2^2 I^2}\{-1 + (1 + \alpha_2\sqrt{I} + \tfrac{1}{2}\alpha_2^2 I)\exp(-\alpha_2\sqrt{I})\}$$

with: $\alpha_1 = 1.4$ and $\alpha_2 = 12.0$

Although equation (5.32) appears rather formidable, actual usage is relatively straightforward. Pitzer noted (P4) that the β_0, β_1 and C^{ϕ} terms have the greatest effect on activity coefficient calculations, so that the ψ and Θ terms may contribute little or nothing to the calculations. In 1975, Pitzer (P5) showed the coefficient Θ to be defined as:

$$\Theta_{ij} = \theta_{ij} + {}^{E}\theta_{ij}(I) \tag{5.33a}$$

$$\Theta'_{ij} = {}^{E}\theta'_{ij}(I) \tag{5.33b}$$

with θ_{ij} being the only adjustable parameter for each pair of cations or each pair of anions. The E terms, which are dependent only on the ion charges and total ionic strength, account for the electrostatic effects of unsymmetrical mixing and equal zero when the ions are of the same charge. Pitzer states (P14) that these electrostatic terms seem to be unneccesary for 2-1 systems and therefore the Θ' term can be dropped. He further recommends omitting this term in all but the most extremely unsymmetrical mixes. A brief description of his method for calculating these electrostatic effects is in Appendix 5.1.

The appearance of equation (5.32) can be simplified by calculating the ionic activity coefficients separately:

$$\ln \gamma_C = z_+^2 f^\gamma + \sum_a m_a \{2 B_{Ca} + (2 \sum_c m_c z_c) C_{Ca}\} + \sum_c m_c (2 \Theta_{Cc} + \sum_a m_a \psi_{Cca}) +$$

$$\sum_c \sum_a m_c m_a (z_+^2 B'_{ca} + |z_+| C_{ca}) + \tfrac{1}{2} \sum_a \sum_{a'} m_a m_{a'} \psi_{Caa'} \qquad (5.34a)$$

$$\ln \gamma_A = z_-^2 f^\gamma + \sum_c m_c \{2 B_{cA} + (2 \sum_a m_a z_a) C_{cA}\} + \sum_a m_a (2 \Theta_{Aa} + \sum_c m_c \psi_{Aac}) +$$

$$\sum_c \sum_a m_c m_a (z_-^2 B'_{ca} + |z_-| C_{ca}) + \tfrac{1}{2} \sum_c \sum_{c'} m_c m_{c'} \psi_{cc'A} \qquad (5.34b)$$

The sums over c or over a are over all cations or all anions. The f^γ, B, B' and C terms are calculated as defined earlier. It must be noted that the stoichiometric coefficients of the B and C parameters ($2\nu_+\nu_-/\nu$ and $2(\nu_+\nu_-)^{1.5}/\nu$) found in the pure solution equation (4.61) are absent from equations (5.34a) and (5.34b) since ionic activity coefficients are being calculated. The β_0, β_1 and C^ϕ values tabulated in Appendix 4.4 include these coefficients. Therefore, when using the tabulated parameters in equations (5.34a) and (5.34b), the coefficient must be removed. For example, the 1-2 salt $CaCl_2$ has tabulated:

$$\frac{4}{3} \beta_0 = .4212$$

The value to be used in calculating the B parameter for this interaction is:

$$\beta_0 = .3159$$

Some values for ψ and Θ were calculated and tabulated by Pitzer and can be found in Appendix 5.1.

Once the ionic activity coefficients have been calculated, the activity coefficient for electrolyte ca may be calculated:

$$\gamma_\pm = \left[\gamma_c^{\nu_c} \cdot \gamma_a^{\nu_a} \right]^{1/\nu} \qquad (2.26)$$

or:

$$\ln \gamma_\pm = \frac{1}{\nu} (\nu_c \ln \gamma_c + \nu_a \ln \gamma_a)$$

CHEN'S METHOD FOR MULTICOMPONENT SOLUTIONS

As discussed in the previous chapter, Chen's method is based on expressing activity coefficients in terms of a long range interaction model, such as a Debye-Hückel or Pitzer equation, and a local composition model. In his thesis (C3), Chen expanded the local composition model so that it could be used for calculating the short range contributions in a multicomponent system using the parameters for single component solutions. Chen felt that these components need not be strong electrolytes, a limitation of other models; the method should be applicable to multicomponent solutions containing strong or weak electrolytes, molecular solvents and solutes, and multi-solvent solutions.

Recalling the equations for defining the local mole fractions x_{ji} and x_{ii} with central species i:

$$\frac{x_{ji}}{x_{ii}} = \frac{x_j}{x_i} F_{ji} \tag{4.70a}$$

$$F_{ji} = \exp(-\alpha\ \tau_{ji}) \tag{4.70b}$$

$$\tau_{ji} = \frac{G_{ji} - G_{ii}}{RT} \tag{4.70c}$$

with: x_i, x_j - overall mole fractions

x_{ji}, x_{ii} - local mole fractions

F_{ji} - NRTL parameter

α - nonrandomness factor, fixed equal to 0.2

τ_{ji} - NRTL binary interaction energy parameter

G_{ji}, G_{ii} - energies of interaction between j-i and i-i

For other local mole fractions, the following equations were presented:

$$\frac{x_{ji}}{x_{ji}} = \frac{x_j}{x_k} F_{ji,ki} \tag{4.71a}$$

$$F_{ji,ki} = \exp(-\alpha\ \tau_{ji,ki}) \tag{4.71b}$$

$$\tau_{ji,ki} = \frac{G_{ji} - G_{ki}}{RT} \tag{4.71c}$$

for species j and k around central species i. Once again, by assuming like-ion repulsion, the local mole fractions $x_{aa} = x_{cc} = 0$. The local compositions are then the sums of the local mole fractions:

around central molecule m:

$$\sum_{c'} x_{c'm} + \sum_{a'} x_{a'm} + \sum_{m'} x_{m'} = 1 \tag{5.35a}$$

around central cation c:

$$\sum_{a'} x_{a'c} + \sum_{m'} x_{m'c} = 1 \tag{5.35b}$$

around central anion a:

$$\sum_{c'} x_{c'a} + \sum_{m'} x_{m'a} = 1 \tag{5.35c}$$

The local mole fractions may then be defined in terms of the overall mole fractions:

'around central molecule m:

$$x_{im} = \frac{x_i F_{im}}{\sum_{a'} x_{a'} F_{a'm} + \sum_{c'} x_{c'} F_{c'm} + \sum_{m'} x_{m'} F_{m'm}} \tag{5.36a}$$

around central cation c:

$$x_{ac} = \frac{x_a}{\sum_{a'} x_{a'} F_{a'c,ac} + \sum_{m'} x_{m'} F_{m'c,ac}} \tag{5.36b}$$

around central anion a:

$$x_{ca} = \frac{x_c}{\sum_{c'} x_{c'} F_{c'a,ca} + \sum_{m'} x_{m'} F_{m'a,ca}} \tag{5.36c}$$

As for single electrolyte solutions, the residual Gibbs energies per mole of cation, anion and molecule were then expressed in terms of these local mole fractions:

$$G^{(c)} = z_c \left(\sum_{m'} x_{m'c} G_{m'c} + \sum_{a'} x_{a'c} G_{a'c} \right) \tag{5.37a}$$

$$G^{(a)} = z_a \left(\sum_{m'} x_{m'a} G_{m'a} + \sum_{c'} x_{c'a} G_{c'a} \right) \tag{5.37b}$$

$$G^{(m)} = \sum_{c'} x_{c'm} G_{c'm} + \sum_{a'} x_{a'm} G_{a'm} + \sum_{m'} x_{m'm} G_{m'm} \tag{5.37c}$$

with the absolute values of the ionic charges, z_c and z_a, to relate the number of molecules and ions around the central ion.

The reference state for molecules was chosen to be the pure component; for electrolytes the chosen reference state was the hypothetical completely ionized pure solution. Therefore, the reference Gibbs energies per mole were expressed as:

$$G_{ref}^{(c)} = z_c \sum_{a'} (x_{a'} G_{a'c}) / \sum_{a''} x_{a''} \tag{5.38a}$$

$$G_{ref}^{(a)} = z_a \sum_{c'} (x_{c'} G_{c'a}) / \sum_{c''} x_{c''} \tag{5.38b}$$

$$G_{ref}^{(m)} = G_{mm} \tag{5.38c}$$

The molar excess Gibbs free energy is then defined as:

$$G^{ex,lc} = \sum_m x_m (G^{(m)} - G_{ref}^{(m)}) + \sum_c x_c (G^{(c)} - G_{ref}^{(c)}) + \sum_a x_a (G^{(a)} - G_{ref}^{(a)}) \tag{5.39}$$

which expresses the change in residual Gibbs energy that results when x_m, x_c and x_a moles of molecules, cations and anions are moved from their reference states into their respective cells in the mixture. The residual Gibbs energy can also be expressed in terms of Chen's τ parameter by substituting equations (5.37) and (5.38) in (5.39) and recalling the definitions of τ, (4.70c) and (4.71c):

$$\frac{G^{ex,lc}}{RT} = \sum_m x_m (\sum_j x_{jm} \tau_{jm}) + \sum_c x_c z_c (\sum_{a'} x_{a'} z_{a'} (\sum_j x_{jc} \tau_{jc,a'c}) / \sum_{a''} x_{a''} z_{a''}) + \sum_a x_a z_a (\sum_{c'} x_{c'} z_{c'} (\sum_j x_{ja} \tau_{ja,c'a}) / \sum_{c''} x_{c''} z_{c''}) \tag{5.40}$$

As with single electrolyte solutions, this expression for the excess Gibbs free energy must be normalized to the infinite dilution reference state to get the unsymmetric expression for the excess Gibbs free energy in order to combine the short range contribution with the Pitzer expression for the long range contribution. For a solution with water as the only solvent, the unsymmetric Gibbs energy equation is:

$$\frac{G^{ex*,lc}}{RT} = \frac{G^{ex,lc}}{RT} - \sum_c x_c \ln f_c^\infty - \sum_a x_a \ln f_a^\infty - \sum_{m \neq w} x_m \ln f_m^\infty \tag{5.41}$$

The expressions for the activity coefficients of the cations, anions and molecular solutes, at infite dilution, with water as the only solvent, are:

$$\ln f_c^\infty = z_c \frac{\sum_{a'} x_{a'} z_{a'} \tau_{wc,a'c}}{\sum_{a''} x_{a''} z_{a''}} + F_{cw} \tau_{cw} \tag{5.42a}$$

$$\ln f_a^\infty = z_a \frac{\sum_{c'} x_{c'} z_{c'} \tau_{wa,c'a}}{\sum_{c''} x_{c''} z_{c''}} + F_{aw} \tau_{aw} \tag{5.42b}$$

$$\ln f_m^\infty = \tau_{wm} + F_{mw} \tau_{mw} \tag{5.42c}$$

The mole fraction based activity coefficients can thus be derived from the combined equations (5.40), (5.41) and (5.42). For the cation, anion and molecular solute they are:

$$\ln f_c^{lc} = z_c \sum_{a'} x_{a'} \left[\frac{\sum_j x_{jc,a'c} \tau_{jc,a'c}}{\sum_{a''} x_{a''}} \right] + \sum_m x_m F_{cm} \left[\frac{\tau_{cm}}{\sum_j F_{jm} x_j} - \frac{\sum_k x_k \tau_{km} F_{km}}{(\sum_j F_{jm} x_j)^2} \right] + \sum_a z_a \sum_{c'} x_{c'} \left[\frac{x_a F_{ca,c'a}}{\sum_{c''} x_{c''}} \left(\frac{\tau_{ca,c'a}}{\sum_j F_{ja,c'a} x_j} - \frac{\sum_k x_k \tau_{ka,c'a} F_{ka,c'a}}{(\sum_j F_{ja,c'a} x_j)^2} \right) \right] - z_c \left[\frac{\sum_{a'} x_{a'} z_{a'} \tau_{wc,a'c}}{\sum_{a''} x_{a''} z_{a''}} \right] - F_{cw} \tau_{cw} \tag{5.43a}$$

$$\ln f_a^{lc} = z_a \sum_{c'} x_{c'} \left[\frac{\sum_j x_{ja,c'a} \tau_{ja,c'a}}{\sum_{c''} x_{c''}} \right] + \sum_m x_m F_{am} \left[\frac{\tau_{am}}{\sum_j F_{jm} x_j} - \frac{\sum_k x_k \tau_{km} F_{km}}{(\sum_j F_{jm} x_j)^2} \right] + \sum_c z_c \sum_{a'} x_{a'} \left[\frac{x_c F_{ac,a'c}}{\sum_{a''} x_{a''}} \left(\frac{\tau_{ac,a'c}}{\sum_j F_{jc,a'c} x_j} - \frac{\sum_k x_k \tau_{kc,a'c} F_{kc,a'c}}{(\sum_j F_{jc,a'c} x_j)^2} \right) \right] - z_a \left[\frac{\sum_{c'} x_{c'} z_{c'} \tau_{wa,c'a}}{\sum_{c''} x_{c''} z_{c''}} \right] - F_{aw} \tau_{aw} \quad (5.43b)$$

$$\ln f_m^{lc} = \frac{\sum_j F_{jm} x_j \tau_{jm}}{\sum_k F_{km} x_k} + \sum_{m'} x_{m'} F_{mm'} \left[\frac{\tau_{mm'}}{\sum_j F_{jm'} x_j} - \frac{\sum_k x_k \tau_{km'} F_{km'}}{(\sum_j F_{jm'} x_j)^2} \right]$$
$$+ \sum_c z_c \sum_{a'} x_{a'} z_{a'} \left[\frac{x_c F_{mc}}{\sum_{a''} x_{a''} z_{a''}} \left(\frac{\tau_{mc,a'c}}{\sum_j F_{jc} x_j} - \frac{\sum_k x_k \tau_{kc,a'c} F_{kc}}{(\sum_j F_{jc} x_j)^2} \right) \right]$$
$$+ \sum_a z_a \sum_{c'} x_{c'} z_{c'} \left[\frac{x_a F_{ma}}{\sum_{c''} x_{c''} z_{c''}} \left(\frac{\tau_{ma,c'a}}{\sum_j F_{ja} x_j} - \frac{\sum_k x_k \tau_{ka,c'a} F_{ka}}{(\sum_j F_{ja} x_j)^2} \right) \right]$$
$$- \tau_{wm} - F_{mw} \tau_{mw} \quad (5.43c)$$

As was shown in Chapter IV, Chen applied the local electroneutrality assumption to the single electrolyte solution resulting in the following relationship between the interaction parameters:

$\tau_{am} = \tau_{cm} = \tau_{ca,m}$ for central molecule cells

$\tau_{mc,ac} = \tau_{ma,ca} = \tau_{m,ca}$ for central ion cells so that ion-molecule interaction parameters are equal to salt-molecule binary parameters. This, he noted, does not generally hold true in multicomponent solutions. He presented the following for calculating ion-molecule interaction parameters from salt-molecule binary parameters.

By applying a local electroneutrality constraint to central molecule cells:

$$\sum_c x_c z_c F_{cm} = \sum_a x_a z_a F_{am} \tag{5.44}$$

it can be seen that, for a multicomponent solution of salts c'a, c"a and solvent m:

$$x_{c'} z_{c'} F_{c'm} + x_{c''} z_{c''} F_{c''m} = x_a z_a F_{am} \tag{5.45}$$

Since like-ion repulsion is assumed, the local compositions around each cation should be the same as in a single electrolyte solution:

$$F_{c'm} = F_{c'a,m} \tag{5.46a}$$

$$F_{c''m} = F_{c''a,m} \tag{5.46b}$$

So by combining equations (5.45) and (5.46):

$$F_{am} = (x_{c'} z_{c'} F_{c'a,m} + x_{c''} z_{c''} F_{c''a,m}) \,/\, x_a z_a$$

or

$$F_{am} = (x_{c'} z_{c'} F_{c'a,m} + x_{c''} z_{c''} F_{c''a,m}) \,/\, (x_{c'} z_{c'} + x_{c''} z_{c''}) \tag{5.47}$$

Similar results are obtained by considering a ternary system containing electrolytes sharing the same anion or electrolytes of dissimilar ions. Therefore, equation (5.47) is generalized for any multicomponent system:

$$F_{cm} = \sum_a x_a z_a F_{ca,m} \,/\, \sum_{a'} x_{a'} z_{a'} \tag{5.48a}$$

$$F_{am} = \sum_c x_c z_c F_{ca,m} \,/\, \sum_{c'} x_{c'} z_{c'} \tag{5.48b}$$

With these multicomponent system F_{im}'s the ion molecule τ_{im} parameters can be calculated.

The molecule-ion interaction parameters may also be expressed in terms of the salt-molecule binary parameters:

$$\tau_{ma,ca} = \frac{G_{ma} - G_{ca}}{RT} = \frac{G_{ma} - G_{mm}}{RT} + \frac{G_{mm} - G_{ca}}{RT} = \tau_{am} - \tau_{ca,m} + \tau_{m,ca} \tag{5.49a}$$

$$\tau_{mc,ac} = \tau_{cm} - \tau_{ca,m} + \tau_{m,ca} \tag{5.49b}$$

Chen also showed the ion-ion interaction parameters to be related to salt-salt binary parameters, which only exist between salts sharing a common ion. Thus, for a ternary system of two similar salts, c'a and c"a:

$$\tau_{c'a,c''a} = \frac{G_{c'a} - G_{c''a}}{RT} = - \frac{G_{c''a} - G_{c'a}}{RT} = - \tau_{c''a,c'a} \tag{5.50}$$

since $G_{c'a}$ and $G_{c''a}$ are supposed to be pure electrolyte properties. Unsymmetrical mixing, such as mixing electrolytes of different coordination number, changes the lattice structure and may invalidate this relationship as the $G_{c'a}$ and $G_{c''a}$ will then differ from the pure solution values.

For a ternary aqueous solution of uni-univalent electrolytes c'a and c"a, there are then five necessary parameters: the four water-electrolyte parameters and, applying equation (5.50), the electrolyte-electrolyte, or salt-salt, parameter. For a ternary solution of electrolytes not sharing a common ion, c'a' and c"a", there are eight water-electrolyte parameters for the pairs H_2O-c'a', H_2O-c'a", H_2O-c"a' and H_2O-c"a", and for the salt-salt pairs c'a'-c'a", c'a'-c"a', c"a'-c"a" and c'a"-c"a" there are eight more parameters, unless equation (5.50) could be applied, in which case there would be a total of twelve parameters necessary.

The solvent-salt parameters may be obtained from single salt solution parameters. The salt-salt parameters must be obtained from data regression of ternary solution data.

For the long range contribution Chen used, in this thesis, the Debye-Hückel equation which was seen in Chapter IV:

$$\ln f_i^{dh} = - \frac{\alpha \; z_i^2 \sqrt{I}}{1 + \beta \; a \sqrt{I}}$$

$$\ln f_{ca}^{dh} = - \frac{\alpha \; z_c \; z_a \sqrt{I}}{1 + \beta \; a \sqrt{I}}$$

where: $I = .5 \sum_i m_i z_i^2$

$$\alpha^2 = \frac{2 \pi N_A d_0}{1000\ DkT} \left(\frac{e^2}{DkT}\right)^3$$

$$\beta^2 = \frac{8 \pi e^2 N_A d_0}{1000\ DkT} \qquad (4.35)$$

z - ionic charge

a - distance of closest approach, set to 4 Å by Chen

N_A - Avagadro's number = 6.0232×10^{23} mole^{-1}

d_0 - water density

D - dielectric constant

e - electronic charge

T - absolute temperature

k - Boltzmann's constant = R/N_A = 1.38045×10^{-16} erg/deg

It is at this point that a problem appears in Chen's method as outlined in his thesis. In order to duplicate his results, the long range contribution was calculated on a molality basis, and the local composition model was calculated on a mole fraction basis. This discrepancy was corrected in the paper published in the AIChE Journal (C4) for single electrolyte solutions. As can be seen in Chapter IV, the Pitzer-Debye-Hückel expression used for the long range contribution was normalized to the mole fraction basis. Due to these differences, the interaction parameters presented in the paper for use with the Pitzer-Debye-Hückel version differ from those presented in the thesis. The AIChE Journal paper notes that a paper on multicomponent solutions was being prepared so this section will be updated later. The Chen method cannot be used in the multicomponent test systems until then.

APPLICATION

A rigorous test of multicomponent solution activity coefficient prediction methods is the calculation of the mutual solubilities of salts and the calculation of salt solubilities in aqueous electrolyte solutions. The salt solubilities are affected by the solution composition. In order to calculate the saturation molalities, the activity coefficients must be adequately predicted.

This test of activity coefficient prediction methods is also recommended due to the wealth of good experimental solubility data available.

In the pages that follow, the predictive methods presented in this chapter are summarized. An outline of the program using these methods to calculate solubilities and the parameters used are included. Finally, the program results and plots of these results with experimental data for a number of multicomponent solutions are presented.

GUGGENHEIM'S METHOD

For a multicomponent electrolyte solution at 25°C:

$$\ln \gamma_{ca} = - \frac{\alpha \left| z_+ z_- \right| \sqrt{I}}{1 + \sqrt{I}} + \frac{2 \nu_+}{\nu_+ + \nu_-} \sum_{a'} \beta_{ca'} m_{a'} + \frac{2 \nu_-}{\nu_+ + \nu_-} \sum_{c'} \beta_{c'a} m_{c'} \quad (5.1)$$

where: γ_{ca} - mean activity coefficient of the electrolyte of cation c and anion a

ν_+ - number of cations per molecule of electrolyte ca

ν_- - number of anions per molecule of electrolyte ca

α - log base e Debye-Hückel constant

I - ionic strength; $I = .5 \sum_i m_i z_i^2$

z_+ - number of charges on cation c

z_- - number of charges on anion a

a' - all anions

c' - all cations

m - ionic molalities

β - interaction coefficient; tabulated in Appendix 4.1

The log base 10 expression:

$$\log \gamma_{ca} = - \frac{A \left| z_+ z_- \right| \sqrt{I}}{1 + \sqrt{I}} + \frac{\nu_+}{\nu_+ + \nu_-} \sum_{a'} B_{ca'} m_{a'} + \frac{\nu_-}{\nu_+ + \nu_-} \sum_{c'} B_{c'a} m_{c'} \quad (5.1a)$$

where: A - log base 10 Debye-Hückel constant

$$B = \frac{2\beta}{\ln 10}$$

For ionic activity coefficients:

$$\log \gamma_c = - \frac{A z_+^2 \sqrt{I}}{1 + \sqrt{I}} + \sum_{a'} B_{ca'} m_{a'} \quad (5.2)$$

and

$$\log \gamma_a = - \frac{A z_-^2 \sqrt{I}}{1 + \sqrt{I}} + \sum_{c'} B_{c'a} m_{c'} \quad (5.3)$$

Due to a lack of parameters and the low ionic strengths that this method is applicable for, Guggenheim's method was used only for the following solutions:

1. H_2O-NaCl-KCl
2. H_2O-NaCl-HCl
3. H_2O-KCl-HCl

BROMLEY'S METHOD

For the ionic activity coefficient of ion i in a multicomponent solution at 25°C:

$$\log \gamma_i = - \frac{A z_i^2 \sqrt{I}}{1 + \sqrt{I}} + F_i$$

where:
- A - Debye-Huckel constant
- I - ionic strength; $I = .5 \sum_i m_i z_i^2$
- i - odd numbers indicate cations, even numbers indicate anions
- z_i - number of charges on ion i

The F_i term is a summation of interaction parameters. For cation i in a multicomponent solution:

$$F_i = \sum_j \dot{B}_{ij} Z_{ij}^2 m_j$$

where:
- j - indicates all anions in the solution
- $Z_{ij} = \frac{z_i + z_j}{2}$
- m_j - molality of ion j

$$\dot{B}_{ij} = \frac{(0.06 + .6B) \left| z_i z_j \right|}{\left(1 + \frac{1.5}{\left| z_i \; z_j \right|} I\right)^2} + B \qquad (5.8)$$

For the activity coefficient of an anion, the subscript i indicates that anion and the subscript j then indicates the cations in the solution.

The mean molal activity coefficient of electrolyte CA in the multicomponent solution can be calculated by combining the ionic activity coefficients:

$$\gamma_\pm = \left(\gamma_C^{\ \nu_C} \cdot \gamma_A^{\ \nu_A}\right)^{1/\nu} \qquad (2.26)$$

or:

$$\log \gamma_\pm = \frac{1}{\nu} \{\nu_C \log \gamma_C + \nu_A \log \gamma_A\}$$

MEISSNER'S METHOD

The reduced activity coefficient of electrolyte CA in a multicomponent solution at 25°C can be calculated:

$$\Gamma_{CA} = \{1. + B\,(1. + 0.1\ I)^{q_{CA,mix}} - B\}\ \Gamma^* \qquad (5.30)$$

where:

$$B = 0.75 - 0.065\ q_{CA,mix}$$

$$\log\ \Gamma^* = -\frac{0.5107\ \sqrt{I}}{1 + C\sqrt{I}}$$

$$C = 1. + 0.055\ q_{CA,mix}\ \exp(-0.023\ I^3)$$

$$I = .5\ \sum_i m_i\ z_i^2$$ total ionic strength of the solution

The Meissner interaction parameter q for the solution containing cations i and anions j is:

$$q_{CA,mix} = (\sum_j I_j\ q^o_{Cj})/I \ + \ (\sum_i I_i\ q^o_{iA})/I$$

where: q^o_{mn} - Meissner q parameter for electrolyte mn in pure solution

$I_n = .5\ m_n\ z_n^2$ ionic strength of ion n

The reduced activity coefficient is related to the mean molal activity coefficient as follows:

$$\Gamma_{\pm} = \gamma_{\pm}^{1/z_+z_-}$$

PITZER'S METHOD

The activity coefficients for cation C and anion A in a multicomponent solution at 25°C can be calculated with the following equations:

$$\ln \gamma_C = z_+^2 f^\gamma + \sum_a m_a \{2 B_{Ca} + (2 \sum_c m_c z_c) C_{Ca}\} + \sum_c m_c (2 \Theta_{Cc} + \sum_a m_a \psi_{Cca}) + \sum_c \sum_a m_c m_a (z_+^2 B'_{ca} + |z_+| C_{ca}) + .5 \sum_a \sum_{a'} m_a m_{a'} \psi_{Caa'} \quad (5.34a)$$

$$\ln \gamma_A = z_-^2 f^\gamma + \sum_c m_c \{2 B_{cA} + (2 \sum_a m_a z_a) C_{cA}\} + \sum_a m_a (2 \Theta_{Aa} + \sum_c m_c \psi_{Aac}) + \sum_c \sum_a m_c m_a (z_-^2 B'_{ca} + |z_-| C_{ca}) + .5 \sum_c \sum_{c'} m_c m_{c'} \psi_{cc'A} \quad (5.34b)$$

where: $f^\gamma = -A_\phi \left[\frac{\sqrt{I}}{1 + b\sqrt{I}} + \frac{2}{b} \ln (1 + b\sqrt{I})\right]$

A_ϕ - the Debye-Hückel constant for osmotic coefficients on a log e basis; equation (4.64)

$b = 1.2$

$I = .5 \sum_i m_i z_i^2$; ionic strength

z_i - ionic charge

m_i - ionic molality

a - subscript denoting anions

c - subscript denoting cations

$$B_{ij} = \beta_0 + \frac{2\beta_1}{\alpha_1^2 I} \{1. - (1. + \alpha_1\sqrt{I}) \exp(-\alpha_1\sqrt{I})\} + \frac{2\beta_2}{\alpha_2^2 I} \{1. - (1. - \alpha_2\sqrt{I}) \exp(-\alpha_2\sqrt{I})\}$$

$$B'_{ij} = \frac{2\beta_1}{\alpha_1^2 I^2} \{-1. + (1. + \alpha_1\sqrt{I} + .5\, \alpha_1^2 I) \exp(-\alpha_1\sqrt{I})\} + \frac{2\beta_2}{\alpha_2^2 I^2} \{-1. + (1. + \alpha_2\sqrt{I} + .5\, \alpha_2^2 I) \exp(-\alpha_2\sqrt{I})\}$$

α_1 = 2.0 for 1-1, 2-1, 1-2, 3-1, 4-1 and 5-1 electrolytes
= 1.4 for 2-2 electrolytes

α_2 = 0.0 for 1-1, 2-1, 1-2, 3-1, 4-1 and 5-1 electrolytes; therefore the last part of the B and B' equations drop out
= 12.0 for 2-2 electrolytes

β_0 - Pitzer parameter, tabulated in Appendices 4.4 and 5.1

β_1 - Pitzer parameter, tabulated in Appendices 4.4 and 5.1

β_2 - Pitzer parameter for 2-2 electrolytes; tabulated in Appendices 4.4 and 5.1

C^{ϕ} - Pitzer parameter; tabulated in Appendices 4.4 and 5.1

Θ - Pitzer interaction parameter for like charged ions; some values tabulated in Appendix 5.1

ψ - Pitzer ternary interaction parameter; some values tabulated in Appendix 5.1

The mean molal activity coefficient can be calculated by combining the ionic activity coefficients:

$$\gamma_{\pm} = \left(\gamma_C^{\nu_C} \quad \gamma_A^{\nu_A} \right)^{1/\nu} \tag{2.26}$$

or:

$$\ln \gamma_{\pm} = \frac{1}{\nu} (\nu_C \ln \gamma_C + \nu_A \ln \gamma_A)$$

WATER ACTIVITIES

Some of the electrolytes to be tested precipitate in hydrated forms at 25°C. It is therefore necessary to calculate the water activities in addition to the electrolyte activity coefficients in order to model the solutions. The water activity may be calculated using an equation presented specifically for it's calculation, or from the results of an osmotic coefficient equation.

Bromley's Water Activity

In 1973, Bromley (B3) presented the following equation for the calculation of the osmotic coefficient of a single electrolyte solution at 25°C:

$$1. - \phi = 2.303\, A_\gamma \left| z_+ z_- \right| \frac{\sqrt{I}}{3} \sigma(\rho\sqrt{I}) - 2.303\,(0.06 + 0.6\,B) \left| z_+ z_- \right| \frac{I}{2} \Psi(aI) - 2.303\, B \frac{I}{2} \qquad (5.51)$$

where: ρ = 1.0

$a = 1.5 / \left| z_+ z_- \right|$

B - Bromley's parameter, tabulated in Appendix 4.2

$I = .5 \sum_i m_i z_i^2$; ionic strength

A_γ - the Debye-Hückel constant, log e basis

ϕ - the osmotic coefficient

$$\sigma(\rho\sqrt{I}) = \frac{3}{(\rho\sqrt{I})^3} \left\{1. + \rho\sqrt{I} - \frac{1}{1 + \rho\sqrt{I}} - 2\ln(1 + \rho\sqrt{I})\right\}$$

$$\Psi(aI) = \frac{2}{aI}\left[\frac{1 + 2aI}{(1 + aI)^2} - \frac{\ln(1 + aI)}{aI}\right]$$

The resulting osmotic coefficient can then be used to calculate the water activity, a°_w, of the single electrolyte solution:

$$\ln a^\circ_w = -\frac{M_s \nu m}{1000} \phi \qquad (2.30)$$

where: M_s - the molecular weight of the solvent, 18.02 for water
m - the molality of the electrolyte
ν - the number of ions the electrolyte dissociates into

In order to calculate the water activity of a multicomponent solution, it is suggested that the hypothetical pure solution water activities for the electrolytes in solution are calculated as above. The mixed solution water activity could then be calculated using a method suggested by Meissner and Kusik (M7):

$$\log (a_w)_{mix} = \sum_i \sum_j W_{ij} \log (a_w^\circ)_{ij} + r \qquad (5.31)$$

where: $(a_w^\circ)_{ij}$ - is the hypothetical pure solution water activity for electrolyte ij

i - odd number subscripts denoting cations

j - even number subscripts denoting anions

$W_{ij} = X_i\, Y_j \left[\frac{(z_i + z_j)^2}{z_i\, z_j}\right] \left[\frac{I_c}{I}\right] \left[\frac{I_a}{I}\right]$

$X_i = I_i / I_c$; cationic fraction

$Y_j = I_j / I_a$; anionic fraction

$I_c = .5 \sum_i m_i\, z_i^2$; cationic strength

$I_a = .5 \sum_j m_j\, z_j^2$; anionic strength

$I_n = .5\, m_n\, z_n^2$; individual ionic strengths

$I = .5 \sum_n m_n\, z_n^2$; total ionic strength

$r = 0.0156\; I \left[\frac{W_{12}\, X_1\, Y_2}{z_1\, z_2} + \frac{W_{32}\, X_3\, Y_2}{z_3\, z_2} + \ldots\right] -$
$0.0156 \left[\frac{I_1}{z_1^2} + \frac{I_2}{z_2^2} + \ldots\right]$; residue term

The residue term equals zero for solutions of like charged electrolytes.

Meissner's Water Activity

Another method of calculating water activities was presented by Meissner (M10) in 1980:

$$-55.51 \log a_w^o = \frac{2I^o}{z_1 z_2} + 2\int_1^{\Gamma^o} I_{12}^o \, d \ln \Gamma_{12}^o \qquad (5.52)$$

Here again the superscript "o" denotes pure solution, and the odd numbered subscripts are cations, the even are anions. The integral portion of the equation is easily solvable with a computer program.

$$\Gamma^o = \{1. + B(1. + 0.1I)^q - B\} \, \Gamma^* \qquad (4.46)$$

$$B = 0.75 - 0.065q \qquad (4.46a)$$

$$\log \Gamma^* = -\frac{0.5107\sqrt{I}}{1. + C\sqrt{I}} \qquad (4.46b)$$

$$C = 1. + 0.055\, q \exp(-0.023I^3) \qquad (4.46c)$$

The water activities calculated in this manner could then be combined using equation (5.31).

The mixed solution water activity could also be calculated by substituting the q in equations (4.46) with $q_{ij,mix}$:

$$q_{ij,mix} = (I_1 q_{1j}^o + I_3 q_{3j}^o + \ldots)/I + (I_2 q_{i2}^o + I_4 q_{i4}^o + \ldots)/I \qquad (5.29)$$

where: I_i - ionic strength of ion i

I - total ionic strength

q_{ij}^o - Meissner q parameter for electrolyte ij in pure solution

Used in conjunction with equation (5.52), for each electrolyte ij in the solution, a hypothetical water activity, A_{ij}, would be calculated. These would then be combined:

$$a_{w,mix} = A_{12}^{R_{12}} \cdot A_{23}^{R_{23}} \cdot A_{34}^{R_{34}} \ldots$$

where: $R_{12} = \dfrac{N_{12}}{N_{12} + N_{23} + N_{34} + \ldots}$

N_{ij} = gram moles of electrolyte ij added to water to make the solution

Pitzer's Water Activity

Pitzer (P14) presented two equations for calculating osmotic coefficients. One is for single electrolyte solutions, the other is for multicomponent solutions. For pure solutions of electrolyte ca:

$$\phi - 1. = |z_c z_a| \; f^{\phi} + m \left[\frac{2\,\nu_c\,\nu_a}{\nu}\right] B^{\phi}_{ca} + m^2 \left[\frac{2(\nu_c\,\nu_a)^{3/2}}{\nu}\right] C^{\phi}_{ca} \qquad (5.53)$$

where: $f^{\phi} = -A_{\phi} \dfrac{\sqrt{I}}{1. + 1.2\,\sqrt{I}}$

A_{ϕ} - the osmotic coefficient Debye-Hückel constant

I - the ionic strength

$B^{\phi}_{ca} = \beta_0 + \beta_1 \exp(-\alpha_1 \sqrt{I}) + \beta_2 \exp(-\alpha_2 \sqrt{I})$

α_1 = 1.4 for 2-2 electrolytes, = 2.0 for others

α_2 = 12. for 2-2 electrolytes, = 0.0 for others

ν_i - number of ions i in electrolyte ca; $\nu = \nu_c + \nu_a$

z_i - charge on ion i

C^{ϕ} - Pitzer parameter, tabulated in Appendix 4.4

β - Pitzer parameters, tabulated in Appendix 4.4

m - molality

The equation for a multicomponent solution is of a similar form:

$$\phi - 1. = (\sum_i m_i)^{-1} \{2\,I\,f^{\phi} + 2 \sum_c \sum_a m_c\,m_a \,(B^{\phi}_{ca} + \frac{(\Sigma\,mz)}{\sqrt{z_c z_a}} C^{\phi}_{ca}) +$$
$$\sum_c \sum_{c'} m_c\,m_{c'}\,(\Theta_{cc'} + I\,\Theta'_{cc'} + \sum_a m_a\,\psi_{cc'a}) +$$
$$\sum_a \sum_{a'} m_a\,m_{a'}\,(\Theta_{aa'} + I\,\Theta'_{aa'} + \sum_c m_c\,\psi_{caa'})\} \qquad (5.54)$$

The summations are over all cations and anions in the solution.

$$(\Sigma\,mz) = \sum_a m_a\,|z_a| = \sum_c m_c\,z_c$$

The Θ and ψ parameters for some interactions are tabulated in Appendix 5.1. The Θ' term can be neglected for solutions of electrolytes of similar or not too different charges.

PHASE DIAGRAM CALCULATIONS

A rigorous test of the applicability of activity coefficient estimation techniques is to predict the solubility of a strong electrolyte in an aqueous solution of another strong electrolyte. Varying the solution concentration, a portion of the solid-liquid phase diagram can be predicted by calculating the saturation concentration for the electrolyte in question. This test is recommended due to the fact that a wealth of experimental data is available. An excellent source of such data is the compilation Solubilities of Inorganic and Metal-Organic Compounds by W.F. Linke and A. Seidell (3).

In order to facilitate such testing, a modular program was written to allow for simple variation of activity coefficient methods and solution composition. There are two approaches to the calculation of solubility products and both have been used for the tests in this chapter. As seen in Chapter III, the solubility product, K_{sp}, can be defined in terms of the activities:

$$K_{sp} = a_c^{\nu_c} \, a_a^{\nu_a} \tag{3.13}$$

$$= (\gamma_c \, m_c)^{\nu_c} (\gamma_a \, m_a)^{\nu_a} \tag{3.14}$$

or the Gibbs free energy of the reaction:

$$K_{sp} = \exp(- \Delta G^o_{K_{sp}} \; / \; RT) \tag{3.15}$$

where, i representing the products and j the reactants:

$$\Delta G^o_{K_{sp}} = \sum_i \nu_i \, \Delta G^o_i - \sum_j \nu_j \, \Delta G^o_j \tag{3.9}$$

The calculation of an electrolyte's solubility product can therefore be based on the saturation molality of a pure solution of that electrolyte or on tabulated values of Gibbs free energies. Ideally, the solubility product should be the same no matter what the method of calculation. In reality, this is rarely the case.

Basic flow of the testing program:

1. Choose the electrolytes in solution and the activity coefficient method to be used.

2. Do the calculations for single electrolyte solution saturation points. This is either calculating the solubility product of the electrolyte based on an experimental saturation molality and the calculated activity coefficients as suggested by equation (3.14), or, calculating the solubility product with the Gibbs free energies via equation (3.15) and using the Newton method to solve for the saturation molality.

3. For a system containing two or more possible precipitates, calculate the invariant point(s), or mutual saturation point(s) using the Newton-Raphson technique.

4. Calculate the solubility of one electrolyte while varying the solution concentration of the other. This entails using the Newton-Raphson technique to solve the system model for the saturation concentrations at even increments of the solution concentration. For a solution of two or more possible precipitates, the points are found between the pure solution saturation point and the invariant point. For a solution of one possible precipitate, the points are found from the pure solution saturation point to a set maximum molality of the other electrolyte.

5. For solutions containing three possible precipitates, the saturation molality of the third salt is found between the two invariant points.

6. At the completion of these calculations, the program can either return to step 1, or end.

For example, one of the solutions tested here is the ternary system H_2O-$NaCl$-Na_2SO_4. At 25°C there are three possible precipitates. At the saturation concentration of a pure solution of Na_2SO_4, the precipitate is $Na_2SO_4 \cdot 10H_2O$, which will be called point a. The pure saturation point of NaCl, point b, has anhydrous NaCl as its precipitate. The third possible precipitate is the anhydrous Na_2SO_4. There is a

Figure 5.1

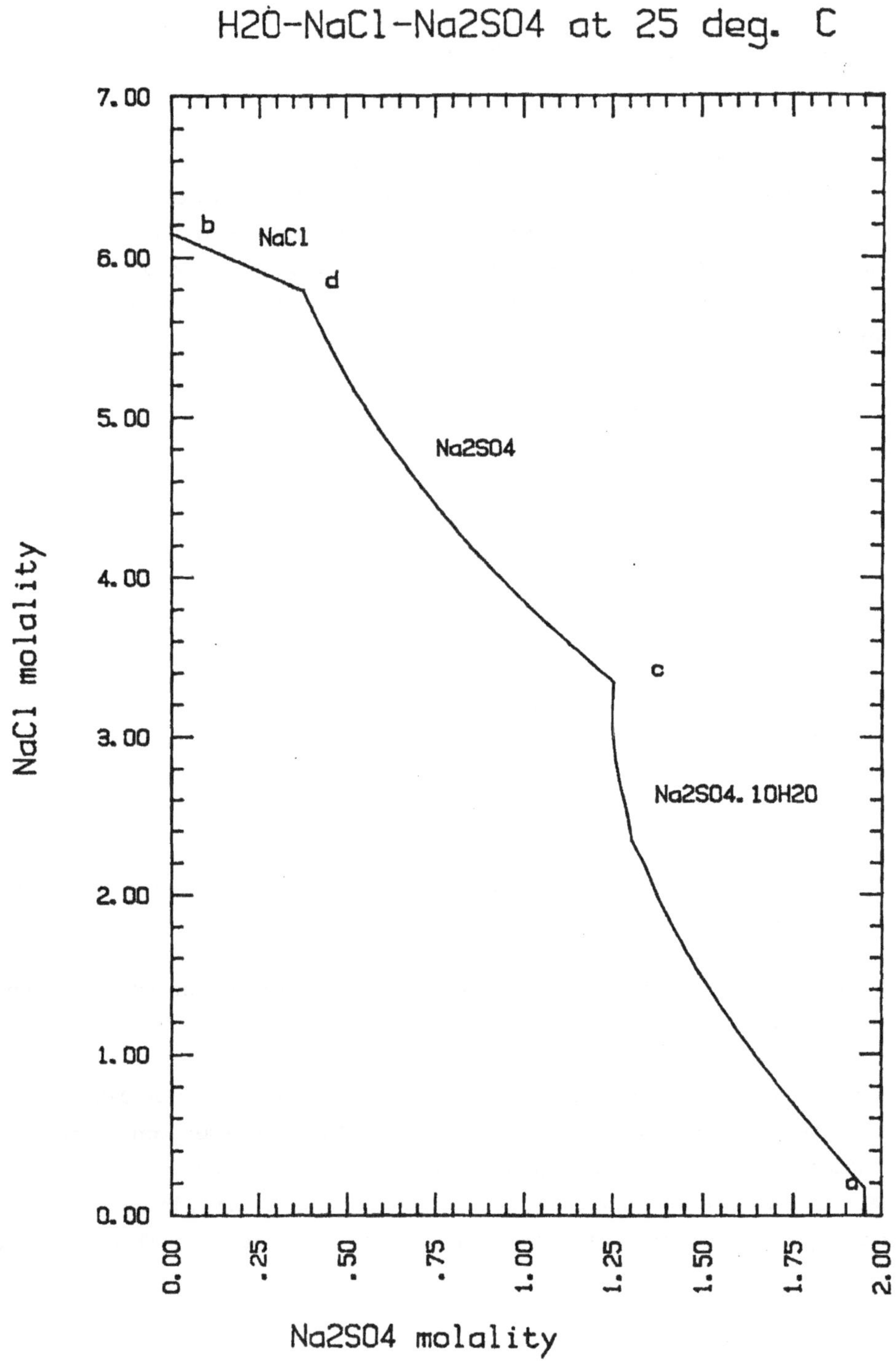
H2O-NaCl-Na2SO4 at 25 deg. C
7.00
6.00
5.00
4.00
3.00
2.00
1.00
0.00
NaCl molality
0.00
.25
.50
.75
1.00
1.25
1.50
1.75
2.00
Na2SO4 molality
b
NaCl
d
Na2SO4
c
Na2SO4. 10H2O
a

point of mutual saturation between Na $SO_4.10H_2O$ and Na_2SO_4, point c, and one between NaCl and Na_2SO_4, point d. These invariant points indicate the solution composition at which the two salts will begin to precipitate. In reference to the program steps described above:

2. The saturation point calculations for pure solutions of NaCl and Na_2SO_4, points b and a, are done.

3. The NaCl + Na_2SO_4 invariant point, point d, and Na_2SO_4 + $Na_2SO_4.10H_2O$, point c, are calculated.

4. The solubility of Na_2SO_4 in an increasing solution concentration of NaCl is calculated. These are points along the line between points a and c where $Na_2SO_4.10H_2O$ is the only precipitate. This is then done for points between the pure saturation concentration of NaCl, point b, and the other invariant point d.

5. Values along the line between the two invariant points are now found. Na_2SO_4 is the possible precipitate for this region.

Program block descriptions:

Single electrolyte saturation point calculations:

1. Based upon experimental saturation molalities:

   a) The activity coefficient for the single electrolyte solution is calculated at the given saturation molality using the pure solution equation for the chosen method as described in Chapter IV. If the precipitate is a hydrate, the water activity is also calculated.

   b) The solubility product for the precipitate is calculated based on the saturation molality, calculated activity coefficients and, if a hydrated precipitate, the water activity. The solubility product definition for precipitate ca.hH_2O is:

$$K_{sp} = a_c^{\nu_c} a_a^{\nu_a} a_w^h = (\gamma_c m_c)^{\nu_c} (\gamma_a m_a)^{\nu_a} a_w^h \qquad (5.55)$$

Since the saturation molality of the electrolyte is related to the ionic molalities:

$$m_i = \nu_i \ m$$

and the mean molal activity coefficient is:

$$\gamma_{\pm}^{\nu} = (\gamma_c^{\nu_c} \ \gamma_a^{\nu_a}) \tag{2.26}$$

the solubility product may be defined:

$$K_{sp} = (Q \ \gamma_{\pm} \ m)^{\nu} \ a_w^h \tag{5.56}$$

where $Q = (\nu_c^{\nu_c} \ \nu_a^{\nu_a})^{1/\nu}$

This Q term equals unity for 1-1 electrolytes and 2-2 electrolytes, and equals $4^{1/3}$ for 1-2 and 2-1 electrolytes.

The water activity term, a_w, drops out for anhydrous precipitates.

2. Based on Gibbs free energies:

a) Using the tabulated values of ΔG_f° in the NBS Tech Note 270 series, the Gibbs free energy of reaction is calculated:

$$\Delta G^\circ_{K_{sp}} = \sum_i \nu_i \ \Delta G^\circ_i - \sum_j \nu_j \ \Delta G^\circ_j \tag{3.9}$$

with the subscript i denoting the products and j the reactants. For example, the ΔG_{Ksp} of $Na_2SO_4.10H_2O$ would be:

$$\Delta G^\circ_{K_{sp}} = 2 * \Delta G^\circ_{Na^+} + \Delta G^\circ_{SO_4^=} + 10 * \Delta G^\circ_{H_2O} - \Delta G^\circ_{Na_2SO_4.10H_2O}$$

This value is then used to calculate the solubility product:

$$K_{sp} = \exp(-\Delta G^\circ_{K_{sp}} / RT) \tag{3.15}$$

b) The saturation molality is then solved for using Newton's method. Newton's method attempts to find a value of x such that:

$$f(x) = 0.0$$

Beginning with a guess value of x_1, a new value for x is calculated:

$$x_2 = x_1 - \frac{f(x_1)}{f'(x_1)}$$

This procedure is repeated until convergence is achieved:

$$\left| \frac{x_{n+1} - x_n}{x_n} \right| \leq \varepsilon$$

The attempt to find a solution is terminated if convergence has not been reached after one hundred iterations.

The equation to be solved by Newton's method is:

$$f(m) = 1. - (Q^{\nu} \gamma_{\pm}^{\nu} m^{\nu} a_w^{h}) / K_{sp}$$

Invariant point calculations:

Some of the multiple electrolyte solutions tested here had two or more possible precipitates. For such solutions there is a concentration at which two of the possible precipitates will begin to form. A set of equations modeling the solution was solved using the Newton-Raphson technique for solving a set of simultaneous equations. The ionic molalities were the unknowns. The two equilibrium equations were:

$$f(1) = 1. - \{(m_{c_1} \gamma_{c_1})^{\nu_{c_1}} (m_{a_1} \gamma_{a_1})^{\nu_{a_1}} a_w^{h_1}\} / K_{sp_1}$$

$$f(2) = 1. - \{(m_{c_2} \gamma_{c_2})^{\nu_{c_2}} (m_{a_2} \gamma_{a_2})^{\nu_{a_2}} a_w^{h_2}\} / K_{sp_2}$$

The solubility constants are those calculated for the pure solution saturation points. A third equation was the electroneutrality equation:

$$f(3) = z_{c_1} m_{c_1} + z_{c_2} m_{c_2} - z_{a_1} m_{a_1} - z_{a_2} m_{a_2}$$

where z is the absolute value of the ionic charge. These three equations were sufficient for the systems tested as invariant point calculations were necessary only for those solutions in which the electrolytes shared a common ion.

For a solution which has a third possible precipitate, such as H_2O-Na_2SO_4-$NaCl$, the calculations must be done twice. A third solubility product was calculated for the anhydrous precipitate, Na_2SO_4, using the Gibbs free energy. One set of calculations was done using the pure solution solubility product, K_{sp} of $Na_2SO_4.10H_2O$, and the solubility product of the anhydrous salt. The second invariant point was found using the anhydrous solubility product and the solubility product of the other pure saturation point, NaCl.

Solubility of one electrolyte in a solution of another:

The solubility of one electrolyte in varied solution concentrations of the other was done for all of the systems tested. The calculations were done at equidistant steps between the pure saturation molality and a maximum molality of the other. For systems with one possible precipitate, the maximum molality was set with regard to available experimental data. The maximum molality of multiprecipitate solutions was set using the invariant point molalities.

Here again, a Newton-Raphson technique was used for solving the system model. Since the molality of one of the electrolytes was set, there were only two unknowns and only two equations were necessary:

$$f(1) = 1. - \{(m_c \gamma_c)^{\nu_c} \; (m_a \gamma_a)^{\nu_a} \; a_w^h\} / K_{sp}$$

$$f(2) = z_c m_c + z_{cs} m_{cs} - z_a m_a - z_{as} m_{as}$$

The subscripts c and a denote the cation and anion of the precipitate whose molality is being solved for; the cs and as subscripts indicate the cation and anion of the electrolyte of set molality.

For a single precipitate solution, such as H_2O-HCl-$CaSO_4$, the HCl molality was set and the equations were solved for the molalities of Ca^{++} and $SO_4^=$ at the point where $CaSO_4.2H_2O$ would begin to precipitate. The set molality of HCl was increased until it reached a maximum of ten.

The system H_2O-NaCl-KCl, with two possible precipitates, used the above set of equations twice. The saturation molality of KCl was solved for at increasing concentrations of NaCl. The maximum NaCl molality was set to the invariant point molality. The saturation molality of NaCl was then found at increasing KCl concentrations until the maximum molality of the invariant point was reached.

Triple precipitate systems also found the saturation molalities between the pure solution and invariant point concentrations, but a third set of calculations was done for the precipitate between the two invariant points. So for the system H_2O-NaCl-Na_2SO_4, after the saturation molality of NaCl was found between the pure solution concentration and the invariant point where NaCl and Na_2SO_4 precipitate, and the saturation molality of $Na_2SO_4.10H_2O$ was found to the invariant point for $Na_2SO_4.10H_2O$ and Na_2SO_4, the saturation molality for Na_2SO_4 was found for concentrations between the two invariant points.

H_2O - NaCl - KCl EXPERIMENTAL DATA

From: W.F. Linke, A. Seidell, Solubilities of Inorganic and Metal-Organic Compounds, Vol. II, American Chemical Society, Washington, D.C. (1965) p. 147

Results at 25°C:

Sat. sol. weight percent

KCl	NaCl	Solid Phase
26.52	0.0	KCl
16.58	12.3	KCl
15.71	13.45	KCl
11.41	20.42	KCl + NaCl
8.16	22.11	NaCl
3.34	24.58	NaCl
0.0	26.48	NaCl

H_2O - NaCl - KCl PARAMETERS

Guggenheim

c	a	β_{ca}
Na	Cl	.15
K		.1

Bromley

c	a	B_{ca}
Na	Cl	.0574
K		.024

Meissner

c	a	q_{ca}
Na	Cl	2.23
K		.92

Pitzer

c	a	β_0	β_1	C^{ϕ}
Na	Cl	.0765	.2664	.00127
K		.04835	.2122	-.00084

i	j	k	Θ_{ij}	ψ_{ijk}
Na	K	Cl	-.012	-.0018

SYSTEM: H2O - NaCl - KCl

Temperature: 25.0 degrees Celsius

Method: Guggenheim

PURE SALT SOLUBILITIES:

solid phase	saturation molality	activity coefficient	solubility product
NaCl	0.402509E+01	0.152752E+01	0.376680E+02
KCl	0.327252E+01	0.902674E+00	0.870401E+01

INVARIANT POINT:

Solid phase: NaCl + KCl

	molalities	weight percents
NaCl	0.3267987E+01	14.935
KCl	0.1177911E+01	6.867
H2O		78.198

THE SOLUBILITY OF KCl IN A SOLUTION OF NaCl:

molality NaCl	molality KCl
0.3104588E+01	0.1256486E+01
0.2941188E+01	0.1338226E+01
0.2777789E+01	0.1423084E+01
0.2614390E+01	0.1511007E+01
0.2450990E+01	0.1601933E+01
0.2287591E+01	0.1695792E+01
0.2124192E+01	0.1792512E+01
0.1960792E+01	0.1892014E+01
0.1797393E+01	0.1994217E+01
0.1633994E+01	0.2099038E+01
0.1470594E+01	0.2206391E+01
0.1307195E+01	0.2316191E+01
0.1143796E+01	0.2428351E+01
0.9803962E+00	0.2542786E+01
0.8169968E+00	0.2659410E+01
0.6535974E+00	0.2778140E+01
0.4901981E+00	0.2898895E+01
0.3267987E+00	0.3021594E+01
0.1633994E+00	0.3146161E+01

THE SOLUBILITY OF NaCl IN A SOLUTION OF KCl:

molality NaCl	molality KCl
0.3304668E+01	0.1119015E+01
0.3341480E+01	0.1060120E+01
0.3378422E+01	0.1001224E+01
0.3415494E+01	0.9423286E+01
0.3452693E+01	0.8834331E+00
0.3490018E+01	0.8245376E+00
0.3527468E+01	0.7656420E+00
0.3565041E+01	0.7067465E+00
0.3602736E+01	0.6478509E+00
0.3640552E+01	0.5889554E+00
0.3678488E+01	0.5300599E+00
0.3716542E+01	0.4711643E+00
0.3754712E+01	0.4122688E+00
0.3792999E+01	0.3533732E+00
0.3831400E+01	0.2944777E+00
0.3869914E+01	0.2355822E+00
0.3908540E+01	0.1766866E+00
0.3947277E+01	0.1177911E+00
0.3986124E+01	0.5889554E-01

SYSTEM: H2O - NaCl - KCl

Temperature: 25.0 degrees Celsius

Method: Bromley

PURE SALT SOLUBILITIES:

solid phase	saturation molality	activity coefficient	solubility product
NaCl	0.619136E+01	0.992017E+00	0.376680E+02
KCl	0.499021E+01	0.591377E+00	0.870401E+01

INVARIANT POINT:

Solid phase: NaCl + KCl

	molalities	weight percents
NaCl	0.5090503E+01	20.523
KCl	0.2040451E+01	10.494
H2O		68.983

THE SOLUBILITY OF KCl IN A SOLUTION OF NaCl:

molality NaCl	molality KCl
0.4835977E+01	0.2148176E+01
0.4581452E+01	0.2260164E+01
0.4326927E+01	0.2376446E+01
0.4072402E+01	0.2497039E+01
0.3817877E+01	0.2621950E+01
0.3563352E+01	0.2751177E+01
0.3308827E+01	0.2884706E+01
0.3054302E+01	0.3022514E+01
0.2799776E+01	0.3164569E+01
0.2545251E+01	0.3310827E+01
0.2290726E+01	0.3461239E+01
0.2036201E+01	0.3615746E+01
0.1781676E+01	0.3774283E+01
0.1527151E+01	0.3936778E+01
0.1272626E+01	0.4103152E+01
0.1018101E+01	0.4273324E+01
0.7635754E+00	0.4447207E+01
0.5090503E+00	0.4624710E+01
0.2545251E+00	0.4805741E+01

THE SOLUBILITY OF NaCl IN A SOLUTION OF KCl:

molality NaCl	molality KCl
0.5143300E+01	0.1938428E+01
0.5196342E+01	0.1836406E+01
0.5249628E+01	0.1734383E+01
0.5303154E+01	0.1632360E+01
0.5356921E+01	0.1530338E+01
0.5410927E+01	0.1428315E+01
0.5465170E+01	0.1326293E+01
0.5519650E+01	0.1224270E+01
0.5574365E+01	0.1122248E+01
0.5629313E+01	0.1020225E+01
0.5684493E+01	0.9182028E+00
0.5739904E+01	0.8161802E+00
0.5795545E+01	0.7141577E+00
0.5851414E+01	0.6121352E+00
0.5907511E+01	0.5101126E+00
0.5963833E+01	0.4080901E+00
0.6020380E+01	0.3060676E+00
0.6077150E+01	0.2040451E+00
0.6134142E+01	0.1020225E+00

SYSTEM: H2O - NaCl - KCl
========================

Temperature: 25.0 degrees Celsius

Method: Meissner

PURE SALT SOLUBILITIES:
=======================

solid phase	saturation molality	activity coefficient	solubility product
NaCl	0.641588E+01	0.958516E+00	0.376680E+02
KCl	0.505413E+01	0.583890E+00	0.870401E+01

INVARIANT POINT:
================

Solid phase: NaCl + KCl

	molalities	weight percents
NaCl	0.5333751E+01	21.237
KCl	0.2093983E+01	10.636
H2O		68.127

THE SOLUBILITY OF KCl IN A SOLUTION OF NaCl:
==

molality NaCl	molality KCl
0.5067064E+01	0.2198484E+01
0.4800376E+01	0.2307500E+01
0.4533688E+01	0.2421093E+01
0.4267001E+01	0.2539313E+01
0.4000313E+01	0.2662196E+01
0.3733626E+01	0.2789767E+01
0.3466938E+01	0.2922034E+01
0.3200251E+01	0.3058995E+01
0.2933563E+01	0.3200633E+01
0.2666876E+01	0.3346923E+01
0.2400188E+01	0.3497826E+01
0.2133500E+01	0.3653299E+01
0.1866813E+01	0.3813289E+01
0.1600125E+01	0.3977735E+01
0.1333438E+01	0.4146573E+01
0.1066750E+01	0.4319735E+01
0.8000627E+00	0.4497147E+01
0.5333751E+00	0.4678733E+01
0.2666876E+00	0.4864416E+01

THE SOLUBILITY OF NaCl IN A SOLUTION OF KCl:
==

molality NaCl	molality KCl
0.5385157E+01	0.1989284E+01
0.5436852E+01	0.1884585E+01
0.5488836E+01	0.1779886E+01
0.5541109E+01	0.1675186E+01
0.5593669E+01	0.1570487E+01
0.5646514E+01	0.1465788E+01
0.5699646E+01	0.1361089E+01
0.5753061E+01	0.1256390E+01
0.5806760E+01	0.1151691E+01
0.5860741E+01	0.1046991E+01
0.5915004E+01	0.9422923E+00
0.5969546E+01	0.8375932E+00
0.6024367E+01	0.7328940E+00
0.6079466E+01	0.6281949E+00
0.6134842E+01	0.5234957E+00
0.6190492E+01	0.4187966E+00
0.6246416E+01	0.3140974E+00
0.6302612E+01	0.2093983E+00
0.6359079E+01	0.1046991E+00

SYSTEM: H2O - NaCl - KCl
========================

Temperature: 25.0 degrees Celsius

Method: Pitzer

PURE SALT SOLUBILITIES:
=======================

solid phase	saturation molality	activity coefficient	solubility product
NaCl	0.612742E+01	0.100493E+01	0.376680E+02
KCl	0.499931E+01	0.590319E+00	0.870401E+01

INVARIANT POINT:
================

Solid phase: NaCl + KCl

	molalities	weight percents
NaCl	0.5017383E+01	20.037
KCl	0.2283136E+01	11.631
H2O		68.332

THE SOLUBILITY OF KCl IN A SOLUTION OF NaCl:
==

molality NaCl	molality KCl
0.4766514E+01	0.2389376E+01
0.4515644E+01	0.2498908E+01
0.4264775E+01	0.2611723E+01
0.4013906E+01	0.2727805E+01
0.3763037E+01	0.2847133E+01
0.3512168E+01	0.2969680E+01
0.3261299E+01	0.3095415E+01
0.3010430E+01	0.3224304E+01
0.2759560E+01	0.3356306E+01
0.2508691E+01	0.3491378E+01
0.2257822E+01	0.3629472E+01
0.2006953E+01	0.3770539E+01
0.1756084E+01	0.3914526E+01
0.1505215E+01	0.4061378E+01
0.1254346E+01	0.4211035E+01
0.1003477E+01	0.4363440E+01
0.7526074E+00	0.4518532E+01
0.5017383E+00	0.4676248E+01
0.2508691E+00	0.4836525E+01

THE SOLUBILITY OF NaCl IN A SOLUTION OF KCl:
===

molality NaCl	molality KCl
0.5070015E+01	0.2168979E+01
0.5122955E+01	0.2054822E+01
0.5176201E+01	0.1940665E+01
0.5229752E+01	0.1826508E+01
0.5283609E+01	0.1712352E+01
0.5337768E+01	0.1598195E+01
0.5392231E+01	0.1484038E+01
0.5446994E+01	0.1369881E+01
0.5502058E+01	0.1255725E+01
0.5557422E+01	0.1141568E+01
0.5613084E+01	0.1027411E+01
0.5669043E+01	0.9132542E+00
0.5725298E+01	0.7990975E+00
0.5781849E+01	0.6849407E+00
0.5838693E+01	0.5707839E+00
0.5895831E+01	0.4566271E+00
0.5953261E+01	0.3424703E+00
0.6010981E+01	0.2283136E+00
0.6068992E+01	0.1141568E+00

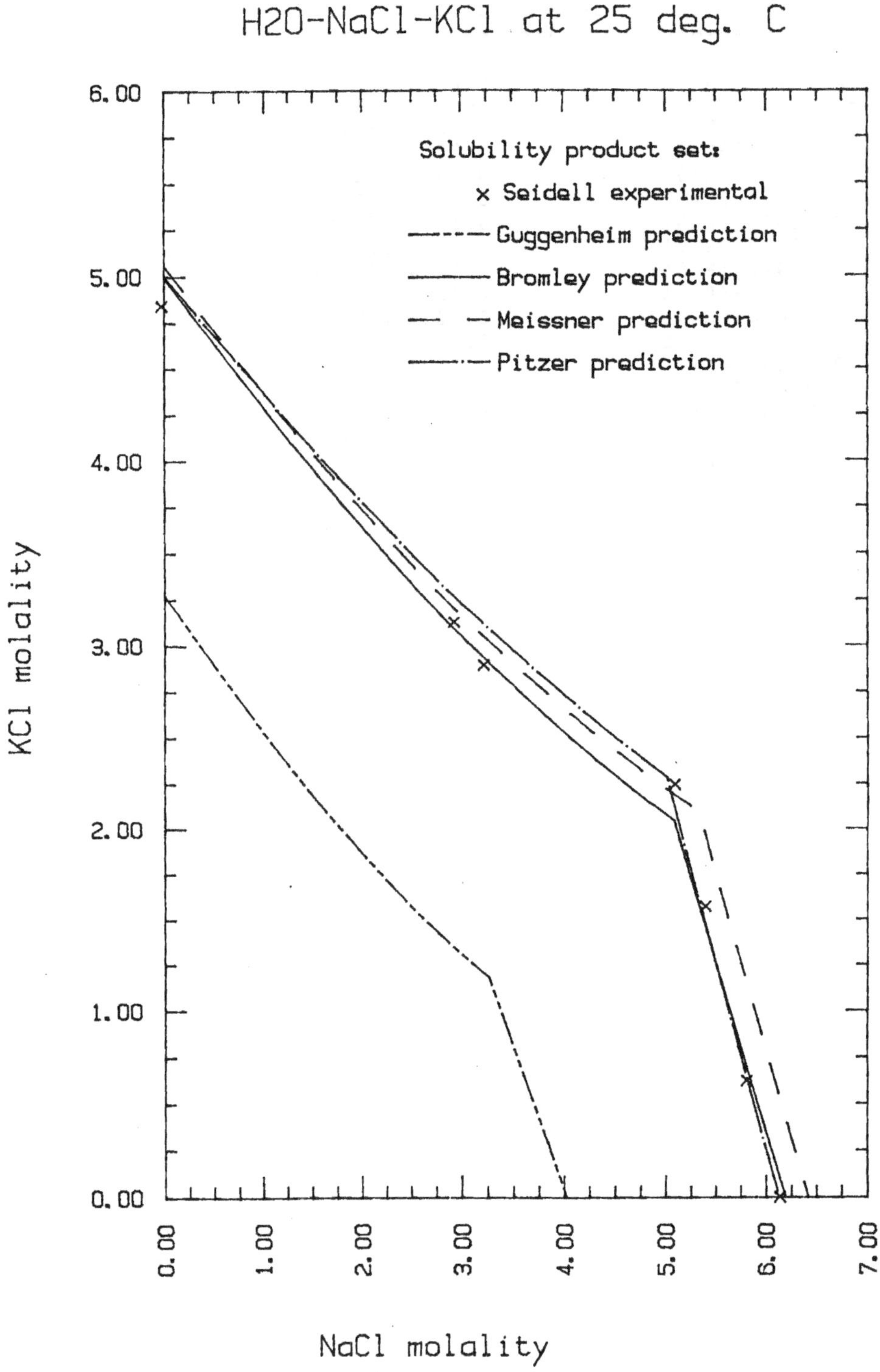
H2O-NaCl-KCl at 25 deg. C
Solubility product set:
× Seidell experimental
Guggenheim prediction
Bromley prediction
Meissner prediction
Pitzer prediction
KCl molality
NaCl molality
0.00
1.00
2.00
3.00
4.00
5.00
6.00
7.00

H_2O - NaCl - HCl EXPERIMENTAL DATA

From: W.F. Linke, A. Seidell, Solubilities of Inorganic and Metal-Organic Compounds, Vol. II, American Chemical Society, Washington, D.C. (1965) p. 962

Solubility of NaCl in hydrochloric acid solutions at 25°C:

(Ingham, 1928)

sol. density	gm. moles per liter sat. sol. HCl	NaCl
1.1981	0.0	5.4325
1.1867	.503	4.880
1.1781	.886	4.483
1.1511	2.265	3.149
1.1352	3.185	2.31
1.1319	3.487	2.079
1.1282	3.83	1.797
1.12	4.5	1.333
1.116	5.253	.907
1.1158	6.101	.544
1.1213	7.073	.293
1.1302	7.976	.158
1.1458	9.236	.091
1.1970	13.41	.017

H_2O - NaCl - HCl PARAMETERS

Guggenheim

c	a	β_{ca}
Na	Cl	.15
H		.27

Bromley

c	a	B_{ca}
Na	Cl	.0574
H		.1433

Meissner

c	a	q_{ca}
Na	Cl	2.23
H		6.69

Pitzer

c	a	β_0	β_1	C^{ϕ}
Na	Cl	.0765	.2664	.00127
H		.1775	.2945	.0008

i	j	k	Θ_{ij}	ψ_{ijk}
Na	H	Cl	.036	-.004

SYSTEM: H2O - NaCl - HCl

Temperature: 25.0 degrees Celsius

Method: Guggenheim

PURE SALT SOLUBILITIES:

solid phase	saturation molality	activity coefficient	solubility product
NaCl	0.402509E+01	0.152752E+01	0.376680E+02

THE SOLUBILITY OF NaCl IN A SOLUTION OF HCl:

molality NaCl	molality HCl
0.3536934E+01	0.5000000E+00
0.3066138E+01	0.1000000E+01
0.2615865E+01	0.1500000E+01
0.2190000E+01	0.2000000E+01
0.1793162E+01	0.2500000E+01
0.1430623E+01	0.3000000E+01
0.1107956E+01	0.3500000E+01
0.8302635E+00	0.4000000E+01
0.6009576E+00	0.4500000E+01
0.4204101E+00	0.5000000E+01
0.2852107E+00	0.5500000E+01
0.1886841E+00	0.6000000E+01
0.1225131E+00	0.6500000E+01
0.7854629E-01	0.7000000E+01
0.4996500E-01	0.7500000E+01
0.3164590E-01	0.8000000E+01
0.2000276E-01	0.8500000E+01
0.1263625E-01	0.9000000E+01
0.7985143E-02	0.9500000E+01
0.5050094E-02	0.1000000E+02

SYSTEM: H2O - NaCl - HCl
========================

Temperature: 25.0 degrees Celsius

Method: Bromley

PURE SALT SOLUBILITIES:
=======================

solid phase	saturation molality	activity coefficient	solubility product
NaCl	0.619136E+01	0.992017E+00	0.376680E+02

THE SOLUBILITY OF NaCl IN A SOLUTION OF HCl:
==

molality NaCl	molality HCl
0.5664119E+01	0.5000000E+00
0.5151923E+01	0.1000000E+01
0.4656309E+01	0.1500000E+01
0.4178982E+01	0.2000000E+01
0.3721824E+01	0.2500000E+01
0.3286876E+01	0.3000000E+01
0.2876294E+01	0.3500000E+01
0.2492271E+01	0.4000000E+01
0.2136923E+01	0.4500000E+01
0.1812132E+01	0.5000000E+01
0.1519355E+01	0.5500000E+01
0.1259422E+01	0.6000000E+01
0.1032367E+01	0.6500000E+01
0.8373314E+00	0.7000000E+01
0.6725718E+00	0.7500000E+01
0.5355950E+00	0.8000000E+01
0.4233785E+00	0.8500000E+01
0.3326363E+00	0.9000000E+01
0.2600703E+00	0.9500000E+01
0.2025699E+00	0.1000000E+02

SYSTEM: H2O - NaCl - HCl

Temperature: 25.0 degrees Celsius

Method: Meissner

PURE SALT SOLUBILITIES:

solid phase	saturation molality	activity coefficient	solubility product
NaCl	0.641588E+01	0.958516E+00	0.376680E+02

THE SOLUBILITY OF NaCl IN A SOLUTION OF HCl:

molality NaCl	molality HCl
0.5838130E+01	0.5000000E+00
0.5280053E+01	0.1000000E+01
0.4743263E+01	0.1500000E+01
0.4229542E+01	0.2000000E+01
0.3740784E+01	0.2500000E+01
0.3278984E+01	0.3000000E+01
0.2846214E+01	0.3500000E+01
0.2444567E+01	0.4000000E+01
0.2076065E+01	0.4500000E+01
0.1742498E+01	0.5000000E+01
0.1445194E+01	0.5500000E+01
0.1184750E+01	0.6000000E+01
0.9607803E+00	0.6500000E+01
0.7717713E+00	0.7000000E+01
0.6151309E+00	0.7500000E+01
0.4874267E+00	0.8000000E+01
0.3847494E+00	0.8500000E+01
0.3030955E+00	0.9000000E+01
0.2386790E+00	0.9500000E+01
0.1881272E+00	0.1000000E+02

SYSTEM: H2O - NaCl - HCl

Temperature: 25.0 degrees Celsius

Method: Pitzer

PURE SALT SOLUBILITIES:

solid phase	saturation molality	activity coefficient	solubility product
NaCl	0.612742E+01	0.100493E+01	0.376680E+02

THE SOLUBILITY OF NaCl IN A SOLUTION OF HCl:

molality NaCl	molality HCl
0.5575051E+01	0.5000000E+00
0.5036595E+01	0.1000000E+01
0.4514226E+01	0.1500000E+01
0.4010514E+01	0.2000000E+01
0.3528317E+01	0.2500000E+01
0.3070744E+01	0.3000000E+01
0.2641078E+01	0.3500000E+01
0.2242619E+01	0.4000000E+01
0.1878463E+01	0.4500000E+01
0.1551198E+01	0.5000000E+01
0.1262561E+01	0.5500000E+01
0.1013114E+01	0.6000000E+01
0.8020506E+00	0.6500000E+01
0.6271890E+00	0.7000000E+01
0.4851957E+00	0.7500000E+01
0.3719715E+00	0.8000000E+01
0.2831006E+00	0.8500000E+01
0.2142536E+00	0.9000000E+01
0.1614745E+00	0.9500000E+01
0.1213386E+00	0.1000000E+02

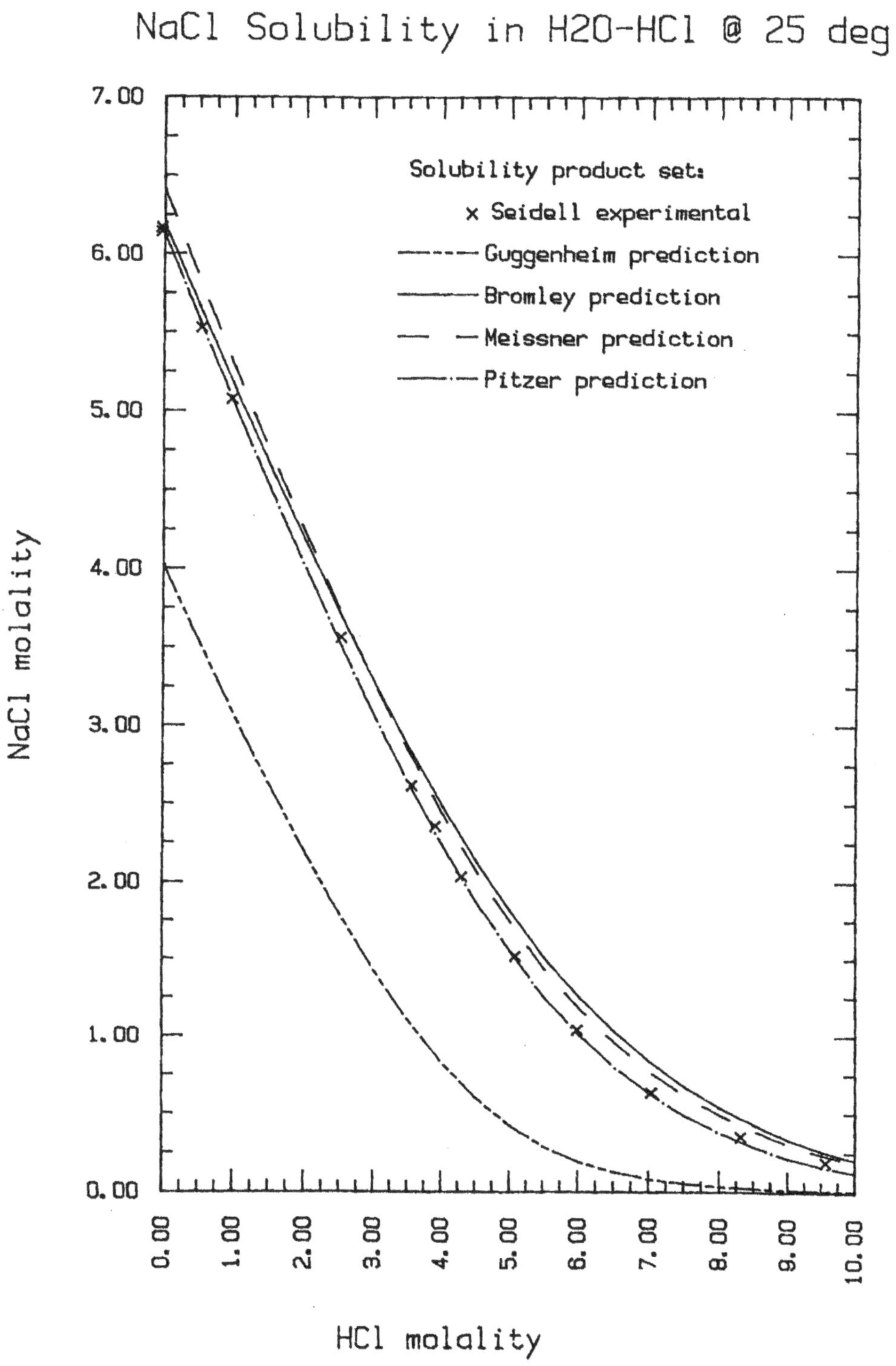
NaCl Solubility in H2O-HCl @ 25 deg
Solubility product set:
x Seidell experimental
Guggenheim prediction
Bromley prediction
Meissner prediction
Pitzer prediction
NaCl molality
0.00
1.00
2.00
3.00
4.00
5.00
6.00
7.00
0.00
1.00
2.00
3.00
4.00
5.00
6.00
7.00
8.00
9.00
10.00
HCl molality

H_2O - NaCl - NaOH EXPERIMENTAL DATA

From: W.F. Linke, A. Seidell, Solubilities of Inorganic and Metal-Organic Compounds, Vol. II, American Chemical Society, Washington, D.C. (1965) p. 966

Solubility of NaCl in aqueous sodium hydroxide at 25°C:

(Ackerlof and Short, 1937)

gm. moles per 1000 gms. H_2O

NaOH	NaCl
.920	5.613
1.858	5.048
2.714	4.672
3.07	4.486
3.812	4.048
4.798	3.61
5.721	3.206
6.688	2.635
7.141	2.442
8.031	2.173
9.824	1.558
11.437	1.28

H_2O - NaCl - NaOH PARAMETERS

Bromley

c	a	B_{ca}
Na	Cl	.0574
	OH	.0747

Meissner

c	a	q_{ca}
Na	Cl	2.23
	OH	3.0

Pitzer

c	a	β_0	β_1	C^{ϕ}
Na	Cl	.0765	.2664	.00127
	OH	.0864	.2530	.00440

i	j	k	Θ_{ij}	ψ_{ijk}
Cl	OH	Na	-.05	-.006

SYSTEM: H2O - NaCl - NaOH
========================

Temperature: 25.0 degrees Celsius

Method: Bromley

PURE SALT SOLUBILITIES:
======================

solid phase	saturation molality	activity coefficient	solubility product
NaCl	0.614618E+01	0.986332E+00	0.367500E+02

THE SOLUBILITY OF NaCl IN A SOLUTION OF NaOH:
==

molality NaCl	molality NaOH
0.5762196E+01	0.5000000E+00
0.5386495E+01	0.1000000E+01
0.5019695E+01	0.1500000E+01
0.4662451E+01	0.2000000E+01
0.4315457E+01	0.2500000E+01
0.3979437E+01	0.3000000E+01
0.3655144E+01	0.3500000E+01
0.3343344E+01	0.4000000E+01
0.3044806E+01	0.4500000E+01
0.2760283E+01	0.5000000E+01
0.2490491E+01	0.5500000E+01
0.2236080E+01	0.6000000E+01
0.1997608E+01	0.6500000E+01
0.1775513E+01	0.7000000E+01
0.1570075E+01	0.7500000E+01
0.1381400E+01	0.8000000E+01
0.1209394E+01	0.8500000E+01
0.1053759E+01	0.9000000E+01
0.9139903E+00	0.9500000E+01
0.7893950E+00	0.1000000E+02

SYSTEM: H2O - NaCl - NaOH

Temperature: 25.0 degrees Celsius

Method: Meissner

PURE SALT SOLUBILITIES:

solid phase	saturation molality	activity coefficient	solubility product
NaCl	0.614618E+01	0.933045E+00	0.328863E+02

THE SOLUBILITY OF NaCl IN A SOLUTION OF NaOH:

molality NaCl	molality NaOH
0.5768890E+01	0.5000000E+00
0.5400533E+01	0.1000000E+01
0.5041486E+01	0.1500000E+01
0.4692441E+01	0.2000000E+01
0.4354353E+01	0.2500000E+01
0.4027806E+01	0.3000000E+01
0.3713526E+01	0.3500000E+01
0.3412227E+01	0.4000000E+01
0.3124597E+01	0.4500000E+01
0.2851274E+01	0.5000000E+01
0.2592819E+01	0.5500000E+01
0.2349699E+01	0.6000000E+01
0.2122254E+01	0.6500000E+01
0.1910677E+01	0.7000000E+01
0.1715016E+01	0.7500000E+01
0.1535090E+01	0.8000000E+01
0.1370601E+01	0.8500000E+01
0.1221058E+01	0.9000000E+01
0.1085820E+01	0.9500000E+01
0.9641184E+00	0.1000000E+02

SYSTEM: H2O - NaCl - NaOH

Temperature: 25.0 degrees Celsius

Method: Pitzer

PURE SALT SOLUBILITIES:

solid phase	saturation molality	activity coefficient	solubility product
NaCl	0.614618E+01	0.100413E+01	0.380880E+02

THE SOLUBILITY OF NaCl IN A SOLUTION OF NaOH:

molality NaCl	molality NaOH
0.5868961E+01	0.5000000E+00
0.5594324E+01	0.1000000E+01
0.5322454E+01	0.1500000E+01
0.5053633E+01	0.2000000E+01
0.4788130E+01	0.2500000E+01
0.4526240E+01	0.3000000E+01
0.4268287E+01	0.3500000E+01
0.4014624E+01	0.4000000E+01
0.3765632E+01	0.4500000E+01
0.3521722E+01	0.5000000E+01
0.3283329E+01	0.5500000E+01
0.3050915E+01	0.6000000E+01
0.2824900E+01	0.6500000E+01
0.2605913E+01	0.7000000E+01
0.2394399E+01	0.7500000E+01
0.2190873E+01	0.8000000E+01
0.1995846E+01	0.8500000E+01
0.1809808E+01	0.9000000E+01
0.1633222E+01	0.9500000E+01
0.1466505E+01	0.1000000E+02

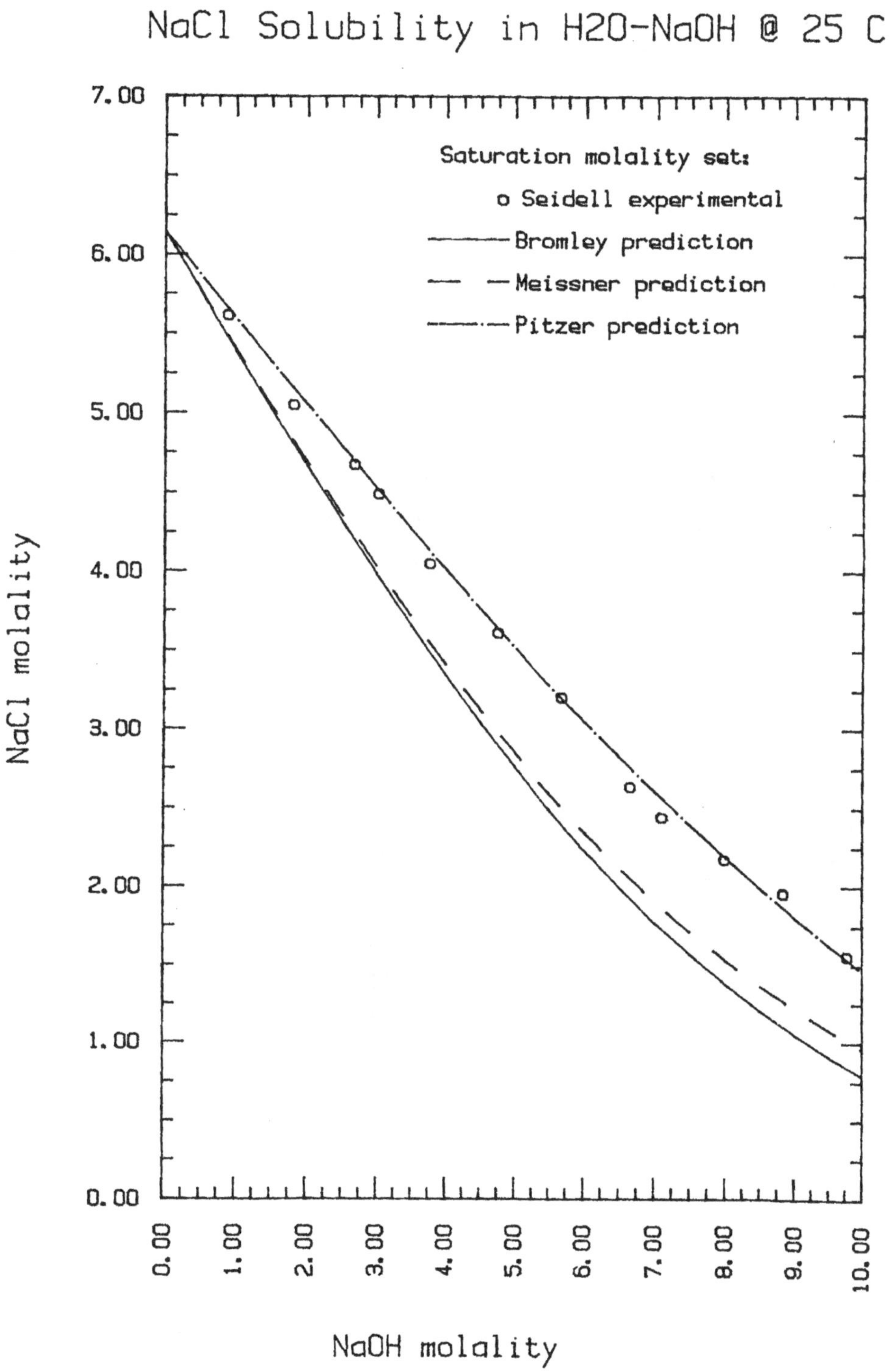
NaCl Solubility in H2O-NaOH @ 25 C
Saturation molality set:
o Seidell experimental
Bromley prediction
Meissner prediction
Pitzer prediction
NaCl molality
NaOH molality
7.00
6.00
5.00
4.00
3.00
2.00
1.00
0.00
0.00
1.00
2.00
3.00
4.00
5.00
6.00
7.00
8.00
9.00
10.00

SYSTEM: H2O - NaCl - NaOH
==========================

Temperature: 25.0 degrees Celsius

Method: Bromley

PURE SALT SOLUBILITIES:
=======================

solid phase	saturation molality	activity coefficient	solubility product
NaCl	0.619136E+01	0.992017E+00	0.376680E+02

THE SOLUBILITY OF NaCl IN A SOLUTION OF NaOH:
==

molality NaCl	molality NaOH
0.5806681E+01	0.5000000E+00
0.5430179E+01	0.1000000E+01
0.5062522E+01	0.1500000E+01
0.4704339E+01	0.2000000E+01
0.4356316E+01	0.2500000E+01
0.4019173E+01	0.3000000E+01
0.3693657E+01	0.3500000E+01
0.3380530E+01	0.4000000E+01
0.3080559E+01	0.4500000E+01
0.2794498E+01	0.5000000E+01
0.2523065E+01	0.5500000E+01
0.2266916E+01	0.6000000E+01
0.2026622E+01	0.6500000E+01
0.1802632E+01	0.7000000E+01
0.1595248E+01	0.7500000E+01
0.1404597E+01	0.8000000E+01
0.1230611E+01	0.8500000E+01
0.1073017E+01	0.9000000E+01
0.9313385E+00	0.9500000E+01
0.8049074E+00	0.1000000E+02

SYSTEM: H2O - NaCl - NaOH

Temperature: 25.0 degrees Celsius

Method: Meissner

PURE SALT SOLUBILITIES:

solid phase	saturation molality	activity coefficient	solubility product
NaCl	0.641588E+01	0.958516E+00	0.376680E+02

THE SOLUBILITY OF NaCl IN A SOLUTION OF NaOH:

molality NaCl	molality NaOH
0.6034590E+01	0.5000000E+00
0.5661943E+01	0.1000000E+01
0.5298219E+01	0.1500000E+01
0.4944191E+01	0.2000000E+01
0.4600520E+01	0.2500000E+01
0.4267789E+01	0.3000000E+01
0.3946901E+01	0.3500000E+01
0.3638439E+01	0.4000000E+01
0.3343076E+01	0.4500000E+01
0.3061454E+01	0.5000000E+01
0.2794154E+01	0.5500000E+01
0.2541682E+01	0.6000000E+01
0.2304436E+01	0.6500000E+01
0.2082718E+01	0.7000000E+01
0.1876601E+01	0.7500000E+01
0.1686086E+01	0.8000000E+01
0.1510977E+01	0.8500000E+01
0.1350909E+01	0.9000000E+01
0.1205364E+01	0.9500000E+01
0.1073684E+01	0.1000000E+02

SYSTEM: H2O - NaCl - NaOH
==========================

Temperature: 25.0 degrees Celsius

Method: Pitzer

PURE SALT SOLUBILITIES:
=======================

solid phase	saturation molality	activity coefficient	solubility product
NaCl	0.612742E+01	0.100493E+01	0.376680E+02

THE SOLUBILITY OF NaCl IN A SOLUTION OF NaOH:
===

molality NaCl	molality NaOH
0.5850309E+01	0.5000000E+00
0.5575795E+01	0.1000000E+01
0.5304141E+01	0.1500000E+01
0.5035562E+01	0.2000000E+01
0.4770328E+01	0.2500000E+01
0.4508733E+01	0.3000000E+01
0.4251103E+01	0.3500000E+01
0.3997793E+01	0.4000000E+01
0.3749184E+01	0.4500000E+01
0.3505686E+01	0.5000000E+01
0.3267738E+01	0.5500000E+01
0.3035800E+01	0.6000000E+01
0.2810293E+01	0.6500000E+01
0.2591847E+01	0.7000000E+01
0.2380907E+01	0.7500000E+01
0.2177985E+01	0.8000000E+01
0.1983591E+01	0.8500000E+01
0.1798214E+01	0.9000000E+01
0.1622311E+01	0.9500000E+01
0.1456296E+01	0.1000000E+02

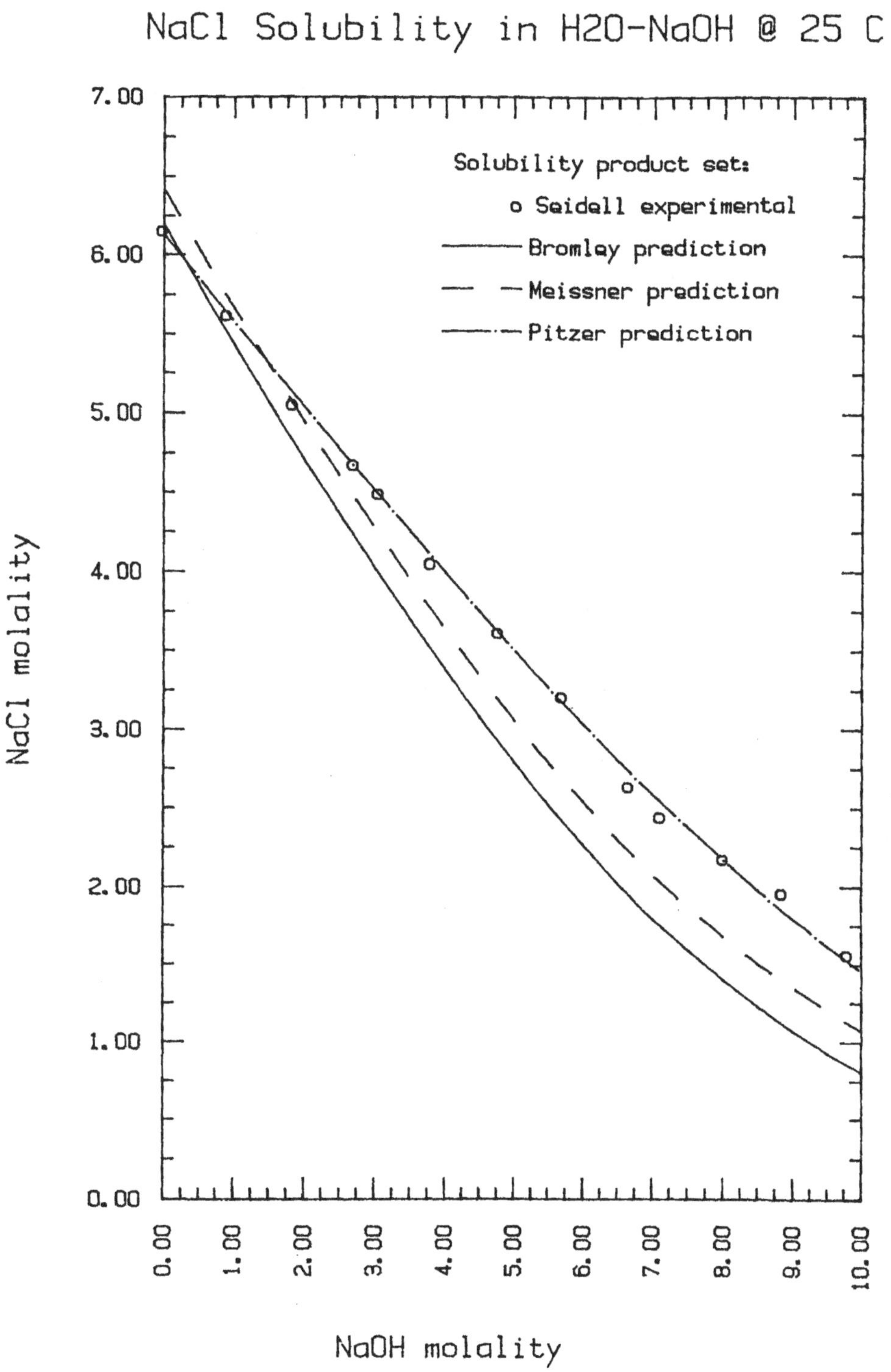
NaCl Solubility in H2O-NaOH @ 25 C
Solubility product set:
o Seidell experimental
Bromley prediction
Meissner prediction
Pitzer prediction
NaCl molality
NaOH molality
7.00
6.00
5.00
4.00
3.00
2.00
1.00
0.00
0.00
1.00
2.00
3.00
4.00
5.00
6.00
7.00
8.00
9.00
10.00

H_2O - KCl - HCl EXPERIMENTAL DATA

From: W.F. Linke, A. Seidell, Solubilities of Inorganic and Metal-Organic Compounds, Vol. II, American Chemical Society, Washington, D.C. (1965) p. 116

Solubility of KCl in hydrochloric acid solutions at 25°C:

(Armstrong, Eyre, Hussey and Paddinson, 1907; Armstrong and Eyre, 1910; Herz, 1911; Ingham, 1928; Malquori, 1928)

gms. per 100 gms. sat. sol.

HCl	KCl
0.	26.4
2.	23.2
4.	20.
6.	16.6
8.	13.7
10.	11.4
12.	9.3
14.	7.4
16.	5.8
18.	4.6
20.	3.6
22.	2.9
24.	2.4
26.	2.0
28.	1.7

H_2O - KCl - HCl PARAMETERS

Guggenheim

c	a	β_{ca}
K	Cl	.1
H		.27

Bromley

c	a	B_{ca}
K	Cl	.024
H		.1433

Meissner

c	a	q_{ca}
K	Cl	.92
H		6.69

Pitzer

c	a	β_0	β_1	C^{ϕ}
K	Cl	.04835	.2122	-.00084
H		.1775	.2945	-.0008

i	j	k	Θ_{ij}	ψ_{ijk}
K	H	Cl	.005	-.007

SYSTEM: H2O - KCl - HCl
======================

Temperature: 25.0 degrees Celsius

Method: Guggenheim

PURE SALT SOLUBILITIES:
======================

solid phase	saturation molality	activity coefficient	solubility product
KCl	0481115E+01	0.116604E+01	0.314722E+02

THE SOLUBILITY OF KCl IN A SOLUTION OF HCl:
==

molality KCl	molality HCl
0.4234906E+01	0.5000000E+00
0.3681337E+01	0.1000000E+01
0.3153891E+01	0.1500000E+01
0.2656677E+01	0.2000000E+01
0.2194463E+01	0.2500000E+01
0.1772528E+01	0.3000000E+01
0.1396257E+01	0.3500000E+01
0.1070365E+01	0.4000000E+01
0.7977987E+00	0.4500000E+01
0.5786220E+00	0.5000000E+01
0.4094442E+00	0.5500000E+01
0.2838500E+00	0.6000000E+01
0.1937125E+00	0.6500000E+01
0.1307366E+00	0.7000000E+01
0.8759551E-01	0.7500000E+01
0.5843649E-01	0.8000000E+01
0.3889585E-01	0.8500000E+01
0.2586666E-01	0.9000000E+01
0.1720205E-01	0.9500000E+01
0.1144611E-01	0.1000000E+02

SYSTEM: H2O - KCl - HCl

Temperature: 25.0 degrees Celsius

Method: Bromley

PURE SALT SOLUBILITIES:

solid phase	saturation molality	activity coefficient	solubility product
KCl	0.481115E+01	0.588291E+00	0.801092E+01

THE SOLUBILITY OF KCl IN A SOLUTION OF HCl:

molality KCl	molality HCl
0.4252994E+01	0.5000000E+00
0.3727492E+01	0.1000000E+01
0.3236968E+01	0.1500000E+01
0.2783747E+01	0.2000000E+01
0.2369949E+01	0.2500000E+01
0.1997213E+01	0.3000000E+01
0.1666401E+01	0.3500000E+01
0.1377349E+01	0.4000000E+01
0.1128739E+01	0.4500000E+01
0.9181380E+00	0.5000000E+01
0.7422087E+00	0.5500000E+01
0.5970320E+00	0.6000000E+01
0.4784558E+00	0.6500000E+01
0.3824021E+00	0.7000000E+01
0.3050841E+00	0.7500000E+01
0.2431394E+00	0.8000000E+01
0.1936746E+00	0.8500000E+01
0.1542615E+00	0.9000000E+01
0.1228984E+00	0.9500000E+01
0.9795719E-01	0.1000000E+02

SYSTEM: H2O - KCl - HCl

Temperature: 25.0 degrees Celsius

Method: Meissner

PURE SALT SOLUBILITIES:

solid phase	saturation molality	activity coefficient	solubility product
KCl	0.481115E+01	0.581101E+00	0.781631E+01

THE SOLUBILITY OF KCl IN A SOLUTION OF HCl:

molality KCl	molality HCl
0.4191918E+01	0.5000000E+00
0.3619407E+01	0.1000000E+01
0.3096161E+01	0.1500000E+01
0.2623637E+01	0.2000000E+01
0.2202064E+01	0.2500000E+01
0.1830498E+01	0.3000000E+01
0.1507033E+01	0.3500000E+01
0.1229052E+01	0.4000000E+01
0.9934396E+00	0.4500000E+01
0.7966784E+00	0.5000000E+01
0.6348599E+00	0.5500000E+01
0.5037106E+00	0.6000000E+01
0.3987409E+00	0.6500000E+01
0.3155124E+00	0.7000000E+01
0.2499284E+00	0.7500000E+01
0.1984173E+00	0.8000000E+01
0.1580008E+00	0.8500000E+01
0.1262683E+00	0.9000000E+01
0.1013070E+00	0.9500000E+01
0.8161789E-01	0.1000000E+02

SYSTEM: H2O - KCl - HCl

Temperature: 25.0 degrees Celsius

Method: Pitzer

PURE SALT SOLUBILITIES:

solid phase	saturation molality	activity coefficient	solubility product
KCl	0.481115E+01	0.586953E+00	0.797453E+01

THE SOLUBILITY OF KCl IN A SOLUTION OF HCl:

molality KCl	molality HCl
0.4292985E+01	0.5000000E+00
0.3799383E+01	0.1000000E+01
0.3333511E+01	0.1500000E+01
0.2898494E+01	0.2000000E+01
0.2497212E+01	0.2500000E+01
0.2132032E+01	0.3000000E+01
0.1804520E+01	0.3500000E+01
0.1515210E+01	0.4000000E+01
0.1263472E+01	0.4500000E+01
0.1047543E+01	0.5000000E+01
0.8647159E+00	0.5500000E+01
0.7116328E+00	0.6000000E+01
0.5846179E+00	0.6500000E+01
0.4799673E+00	0.7000000E+01
0.3941798E+00	0.7500000E+01
0.3240887E+00	0.8000000E+01
0.2669277E+00	0.8500000E+01
0.2203415E+00	0.9000000E+01
0.1823619E+00	0.9500000E+01
0.1513657E+00	0.1000000E+02

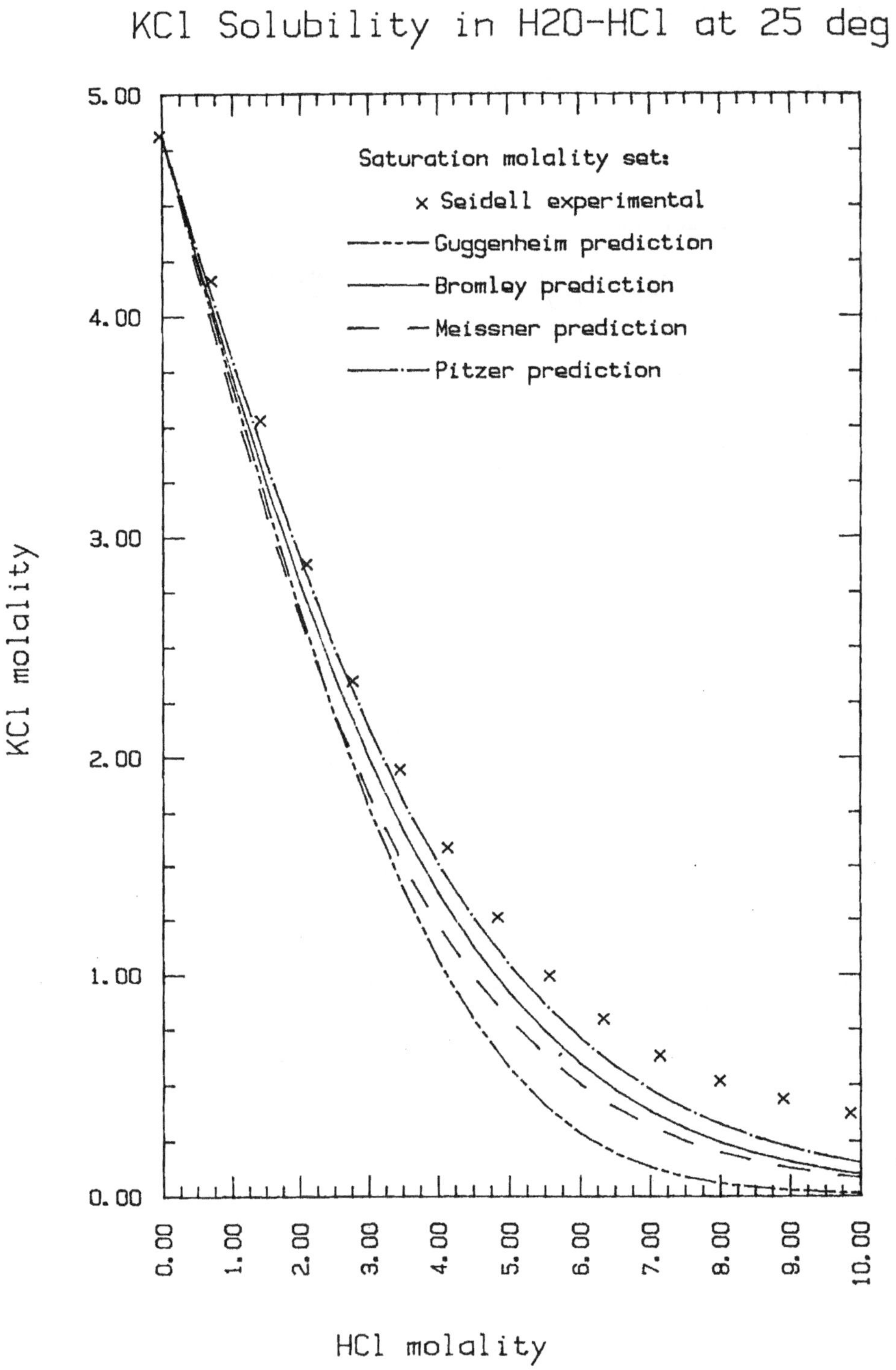
KCl Solubility in H2O-HCl at 25 deg
Saturation molality set:
× Seidell experimental
Guggenheim prediction
Bromley prediction
Meissner prediction
Pitzer prediction
KCl molality
HCl molality
0.00
1.00
2.00
3.00
4.00
5.00
6.00
7.00
8.00
9.00
10.00

SYSTEM: H2O - KCl - HCl

Temperature: 25.0 degrees Celsius

Method: Guggenheim

PURE SALT SOLUBILITIES:

solid phase	saturation molality	activity coefficient	solubility product
KCl	0.327252E+01	0.902674E+00	0.870401E+01

THE SOLUBILITY OF KCl IN A SOLUTION OF HCl:

molality KCl	molality HCl
0.2768275E+01	0.5000000E+00
0.2295779E+01	0.1000000E+01
0.1860609E+01	0.1500000E+01
0.1468796E+01	0.2000000E+01
0.1126260E+01	0.2500000E+01
0.8375569E+00	0.3000000E+01
0.6043559E+00	0.3500000E+01
0.4243371E+00	0.4000000E+01
0.2913121E+00	0.4500000E+01
0.1966582E+00	0.5000000E+01
0.1312613E+00	0.5500000E+01
0.8701009E-01	0.6000000E+01
0.5746857E-01	0.6500000E+01
0.3790334E-01	0.7000000E+01
0.2499873E-01	0.7500000E+01
0.1650112E-01	0.8000000E+01
0.1090598E-01	0.8500000E+01
0.7218920E-02	0.9000000E+01
0.4786005E-02	0.9500000E+01
0.3178108E-02	0.1000000E+02

SYSTEM: H2O - KCl - HCl
========================

Temperature: 25.0 degrees Celsius

Method: Bromley

PURE SALT SOLUBILITIES:
======================

solid phase	saturation molality	activity coefficient	solubility product
KCl	0.499021E+01	0.591377E+00	0.870401E+01

THE SOLUBILITY OF KCl IN A SOLUTION OF HCl:
==

molality KCl	molality HCl
0.4421888E+01	0.5000000E+00
0.3885880E+01	0.1000000E+01
0.3384433E+01	0.1500000E+01
0.2919838E+01	0.2000000E+01
0.2494231E+01	0.2500000E+01
0.2109344E+01	0.3000000E+01
0.1766214E+01	0.3500000E+01
0.1464932E+01	0.4000000E+01
0.1204475E+01	0.4500000E+01
0.9827098E+00	0.5000000E+01
0.7965470E+00	0.5500000E+01
0.6422340E+00	0.6000000E+01
0.5156921E+00	0.6500000E+01
0.4128335E+00	0.7000000E+01
0.3297989E+00	0.7500000E+01
0.2631162E+00	0.8000000E+01
0.2097654E+00	0.8500000E+01
0.1671902E+00	0.9000000E+01
0.1332693E+00	0.9500000E+01
0.1062678E+00	0.1000000E+02

SYSTEM: H2O - KCl - HCl
========================

Temperature: 25.0 degrees Celsius

Method: Meissner

PURE SALT SOLUBILITIES:
=======================

solid phase	saturation molality	activity coefficient	solubility product
KCl	0.505413E+01	0.583890E+00	0.870401E+01

THE SOLUBILITY OF KCl IN A SOLUTION OF HCl:
===

molality KCl	molality HCl
0.4420322E+01	0.5000000E+00
0.3832078E+01	0.1000000E+01
0.3291966E+01	0.1500000E+01
0.2801671E+01	0.2000000E+01
0.2361788E+01	0.2500000E+01
0.1971824E+01	0.3000000E+01
0.1630338E+01	0.3500000E+01
0.1335147E+01	0.4000000E+01
0.1083497E+01	0.4500000E+01
0.8721417E+00	0.5000000E+01
0.6973468E+00	0.5500000E+01
0.5549194E+00	0.6000000E+01
0.4403584E+00	0.6500000E+01
0.3491338E+00	0.7000000E+01
0.2769891E+00	0.7500000E+01
0.2201616E+00	0.8000000E+01
0.1754733E+00	0.8500000E+01
0.1403262E+00	0.9000000E+01
0.1126426E+00	0.9500000E+01
0.9078450E-01	0.1000000E+02

SYSTEM: H2O - KCl - HCl

Temperature: 25.0 degrees Celsius

Method: Pitzer

PURE SALT SOLUBILITIES:

solid phase	saturation molality	activity coefficient	solubility product
KCl	0.499931E+01	0.590319E+00	0.870401E+01

THE SOLUBILITY OF KCl IN A SOLUTION OF HCl:

molality KCl	molality HCl
0.4473062E+01	0.5000000E+00
0.3970603E+01	0.1000000E+01
0.3495039E+01	0.1500000E+01
0.3049490E+01	0.2000000E+01
0.2636884E+01	0.2500000E+01
0.2259716E+01	0.3000000E+01
0.1919768E+01	0.3500000E+01
0.1617863E+01	0.4000000E+01
0.1353712E+01	0.4500000E+01
0.1125889E+01	0.5000000E+01
0.9319801E+00	0.5500000E+01
0.7688359E+00	0.6000000E+01
0.6328941E+00	0.6500000E+01
0.5204743E+00	0.7000000E+01
0.4280311E+00	0.7500000E+01
0.3523075E+00	0.8000000E+01
0.2904236E+00	0.8500000E+01
0.2399029E+00	0.9000000E+01
0.1986598E+00	0.9500000E+01
0.1649642E+00	0.1000000E+02

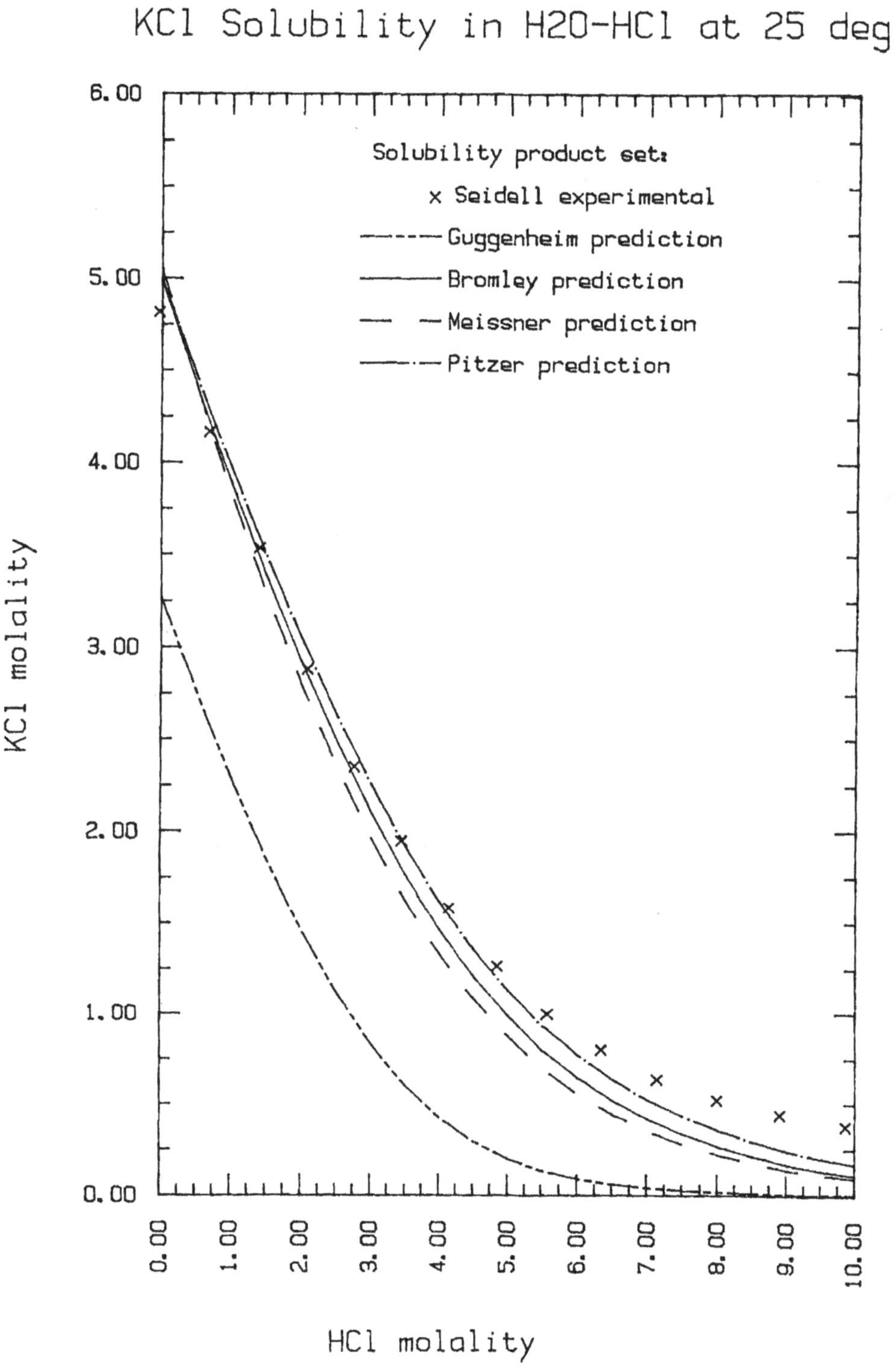
KCl Solubility in H2O-HCl at 25 deg
Solubility product set:
x Seidell experimental
Guggenheim prediction
Bromley prediction
Meissner prediction
Pitzer prediction
KCl molality
HCl molality
6.00
5.00
4.00
3.00
2.00
1.00
0.00
0.00
1.00
2.00
3.00
4.00
5.00
6.00
7.00
8.00
9.00
10.00

H_2O - NaCl - $CaCl_2$ EXPERIMENTAL DATA

From: W.F. Linke, A. Seidell, Solubilities of Inorganic and Metal-Organic Compounds, Vol. I, American Chemical Society, Washington, D.C. (1958) p. 579

Results at 25°C:

(Cameron, Bell and Robinson, 1907)

$d\frac{25}{25}$ Sat. Sol	gms. per 100 gms. of H_2O $CaCl_2$	NaCl	Solid Phase
...	84.	0.0	$CaCl_2 \cdot 6H_2O$
1.4441	78.49	1.846	$CaCl_2 \cdot 6H_2O$ + NaCl
1.3651	48.58	1.637	NaCl
1.3463	53.47	1.799	NaCl
1.2831	36.80	7.77	NaCl
1.2653	30.08	10.70	NaCl
1.2367	19.53	18.85	NaCl
1.2080	3.92	32.48	NaCl
1.2030	0.0	35.80	NaCl

(Yanatieva, 1946)

gms. per 100 gms. sat. sol. $CaCl_2$	NaCl	Solid Phase
45.60	0.0	$CaCl_2 \cdot 6H_2O$
0.0	26.80	NaCl
27.30	4.70	NaCl
32.40	2.80	NaCl
35.60	1.40	NaCl
43.50	1.00	NaCl + $CaCl_2 \cdot 6H_2O$

H_2O - NaCl - $CaCl_2$ PARAMETERS

Bromley

c	a	B_{ca}
Na	Cl	.0574
Ca		.0948

Meissner

c	a	q_{ca}
Na	Cl	2.23
Ca		2.40

Pitzer

c	a	β_0	β_1	C^{ϕ}
Na	Cl	.0765	.2664	.00127
Ca		.3159	1.6140	-.00034

i	j	k	Θ_{ij}	ψ_{ijk}
Na	Ca	Cl	.07	-.014

SYSTEM: H2O - NaCl - CaCl2
==========================

Temperature: 25.0 degrees Celsius

Method: Bromley

PURE SALT SOLUBILITIES:
=======================

solid phase	saturation molality	activity coefficient	water activity	solubility product
NaCl	0.614618E+01	0.986332E+00		0.367500E+02
CaCl2.6H2O	0.746178E+01	0.197025E+02	0.277177E+00	0.576359E+04

INVARIANT POINT:
================

Solid phase: NaCl + CaCl2.6H2O

	molalities	weight percents
NaCl	0.3199004E-01	0.098
CaCl2	0.8170584E+01	47.510
H2O		52.392

THE SOLUBILITY OF CaCl2 IN A SOLUTION OF NaCl:
==

Solid phase: CaCl2.6H2O

molality NaCl	molality CaCl2
0.3039054E-01	0.8171494E+01
0.2879104E-01	0.8172404E+01
0.2719154E-01	0.8173315E+01
0.2559203E-01	0.8174226E+01
0.2399253E-01	0.8175137E+01
0.2239303E-01	0.8176049E+01
0.2079353E-01	0.8176961E+01
0.1919402E-01	0.8177873E+01
0.1759452E-01	0.8178785E+01
0.1599502E-01	0.8179697E+01
0.1439552E-01	0.8180610E+01
0.1279602E-01	0.8181523E+01
0.1119651E-01	0.8182437E+01
0.9597012E-02	0.8183350E+01
0.7997510E-02	0.8184264E+01
0.6398008E-02	0.8185179E+01
0.4798506E-02	0.8186093E+01
0.3199004E-02	0.8187008E+01
0.1599502E-02	0.8187923E+01

THE SOLUBILITY OF NaCl IN A SOLUTION OF CaCl2:

molality NaCl	molality CaCl2
0.4520365E-01	0.7762055E+01
0.6388665E-01	0.7353526E+01
0.9025637E-01	0.6944996E+01
0.1273532E+00	0.6536467E+01
0.1792589E+00	0.6127938E+01
0.2512844E+00	0.5719409E+01
0.3500401E+00	0.5310880E+01
0.4832501E+00	0.4902350E+01
0.6591896E+00	0.4493821E+01
0.8857189E+00	0.4085292E+01
0.1169108E+01	0.3676763E+01
0.1513040E+01	0.3268234E+01
0.1918187E+01	0.2859704E+01
0.2382474E+01	0.2451175E+01
0.2901843E+01	0.2042646E+01
0.3471166E+01	0.1634117E+01
0.4085042E+01	0.1225588E+01
0.4738362E+01	0.8170584E+00
0.5426639E+01	0.4085292E+00

SYSTEM: H2O - NaCl - CaCl2

Temperature: 25.0 degrees Celsius

Method: Meissner

PURE SALT SOLUBILITIES:

solid phase	saturation molality	activity coefficient	water activity	solubility product
NaCl	0.614618E+01	0.933045E+00		0.328863E+02
CaCl2.6H2O	0.746178E+01	0.154431E+02	0.317055E+00	0.621736E+04

INVARIANT POINT:

Solid phase: NaCl + CaCl2.6H2O

	molalities	weight percents
NaCl	0.1473920E+00	0.466
CaCl2	0.7570374E+01	45.446
H2O		54.089

THE SOLUBILITY OF CaCl2 IN A SOLUTION OF NaCl:

Solid phase: CaCl2.6H2O

molality NaCl	molality CaCl2
0.1400224E+00	0.7564788E+01
0.1326528E+00	0.7559256E+01
0.1252832E+00	0.7553442E+01
0.1179136E+00	0.7547802E+01
0.1105440E+00	0.7542053E+01
0.1031744E+00	0.7536348E+01
0.9580479E-01	0.7530832E+01
0.8843519E-01	0.7525489E+01
0.8106559E-01	0.7520282E+01
0.7369599E-01	0.7514729E+01
0.6632639E-01	0.7509053E+01
0.5895679E-01	0.7503822E+01
0.5158719E-01	0.7498210E+01
0.4421759E-01	0.7493052E+01
0.3684800E-01	0.7487775E+01
0.2947840E-01	0.7482392E+01
0.2210880E-01	0.7477429E+01
0.1473920E-01	0.7472364E+01
0.7369599E-02	0.7466819E+01

THE SOLUBILITY OF NaCl IN A SOLUTION OF CaCl2:

molality NaCl	molality CaCl2
0.1789701E+00	0.7191856E+01
0.2184494E+00	0.6813337E+01
0.2679916E+00	0.6434818E+01
0.3303328E+00	0.6056300E+01
0.4088865E+00	0.5677781E+01
0.5078169E+00	0.5299262E+01
0.6320378E+00	0.4920743E+01
0.7870935E+00	0.4542225E+01
0.9788402E+00	0.4163706E+01
0.1212942E+01	0.3785187E+01
0.1494220E+01	0.3406669E+01
0.1826019E+01	0.3028150E+01
0.2209779E+01	0.2649631E+01
0.2644943E+01	0.2271112E+01
0.3129228E+01	0.1892594E+01
0.3659094E+01	0.1514075E+01
0.4230287E+01	0.1135556E+01
0.4838262E+01	0.7570374E+00
0.5478429E+01	0.3785187E+00

SYSTEM: H2O - NaCl - CaCl2

Temperature: 25.0 degrees Celsius

Method: Pitzer

PURE SALT SOLUBILITIES:

solid phase	saturation molality	activity coefficient	water activity	solubility product
NaCl	0.614618E+01	0.100413E+01		0.376680E+02
CaCl2.6H2O	0.746178E+01	0.348856E+02	0.238699E+00	0.130507E+05

INVARIANT POINT:

Solid phase: NaCl + CaCl2.6H2O

	molalities	weight percents
NaCl	0.5039100E-01	0.162
CaCl2	0.7309413E+01	44.717
H2O		55.121

THE SOLUBILITY OF CaCl2 IN A SOLUTION OF NaCl:

Solid phase: CaCl2.6H2O

molality NaCl	molality CaCl2
0.4787145E-01	0.7309419E+01
0.4535190E-01	0.7309427E+01
0.4283235E-01	0.7309435E+01
0.4031280E-01	0.7309444E+01
0.3779325E-01	0.7309453E+01
0.3527370E-01	0.7309463E+01
0.3275415E-01	0.7309474E+01
0.3023460E-01	0.7309486E+01
0.2771505E-01	0.7309498E+01
0.2519550E-01	0.7309511E+01
0.2267595E-01	0.7309525E+01
0.2015640E-01	0.7309539E+01
0.1763685E-01	0.7309554E+01
0.1511730E-01	0.7309570E+01
0.1259775E-01	0.7309586E+01
0.1007820E-01	0.7309603E+01
0.7558650E-02	0.7309621E+01
0.5039100E-02	0.7309639E+01
0.2519550E-02	0.7309658E+01

THE SOLUBILITY OF NaCl IN A SOLUTION OF CaCl2:

molality NaCl	molality CaCl2
0.6703078E-01	0.6943942E+01
0.8967584E-01	0.6578472E+01
0.1205840E+00	0.6213001E+01
0.1628278E+00	0.5487530E+01
0.2205168E+00	0.5482060E+01
0.2989975E+00	0.5116589E+01
0.4049463E+00	0.4751118E+01
0.5462086E+00	0.4385648E+01
0.7312353E+00	0.4020177E+01
0.9680466E+00	0.3654706E+01
0.1262886E+01	0.3289236E+01
0.1618996E+01	0.2923765E+01
0.2035999E+01	0.2558294E+01
0.2510130E+01	0.2192824E+01
0.3035101E+01	0.1827353E+01
0.3603202E+01	0.1461883E+01
0.4206235E+01	0.1096412E+01
0.4836129E+01	0.7309413E+00
0.5485210E+01	0.3654706E+00

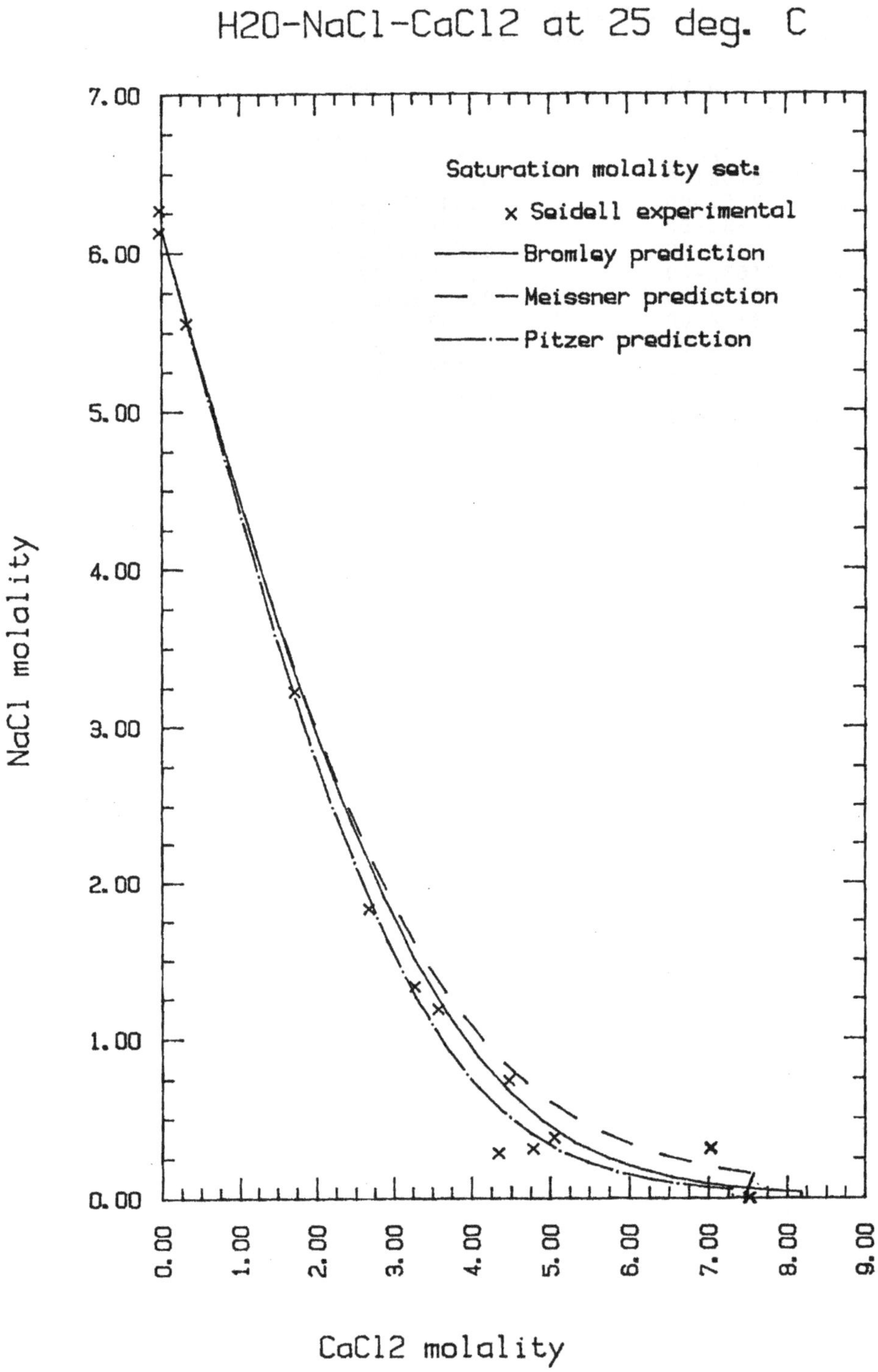
H2O-NaCl-CaCl2 at 25 deg. C
Saturation molality set:
× Seidell experimental
Bromley prediction
Meissner prediction
Pitzer prediction
7.00
6.00
5.00
4.00
3.00
2.00
1.00
0.00
NaCl molality
0.00
1.00
2.00
3.00
4.00
5.00
6.00
7.00
8.00
9.00
CaCl2 molality

H_2O - NaCl - $MgCl_2$ EXPERIMENTAL DATA

From: W.F. Linke, A. Seidell, Solubilities of Inorganic and Metal-Organic Compounds, Vol. II, American Chemical Society, Washington, D.C. (1965) p. 489.

Results at 25°C:

(Keitel, 1923)

gms. per 100 gms. H_2O

$MgCl_2$	NaCl	Solid Phase
35.6	0.0	$MgCl_2.6H_2O$
35.	.97	$MgCl_2.6H_2O$ + NaCl
30.87	1.39	NaCl
24.28	4.01	NaCl
9.68	16.38	NaCl
8.01	17.74	NaCl

(Takegami, 1921)

35.86	0.0	$MgCl_2.6H_2O$
35.73	.21	$MgCl_2.6H_2O$
35.68	.48	$MgCl_2.6H_2O$ + NaCl
31.65	1.12	NaCl
24.4	4.11	NaCl
18.79	7.80	NaCl
9.15	16.62	NaCl
6.24	19.77	NaCl
.5	26.49	NaCl

H_2O - NaCl - $MgCl_2$ PARAMETERS

Bromley

c	a	B_{ca}
Na	Cl	.0574
Mg		.1129

Meissner

c	a	q_{ca}
Na	Cl	2.23
Mg		2.90

Pitzer

c	a	β_0	β_1	C^{ϕ}
Na	Cl	.0765	.2664	.00127
Mg		.35235	1.6815	.00519

i	j	k	Θ_{ij}	ψ_{ijk}
Na	Mg	Cl	.07	-.012

SYSTEM: H2O - NaCl - MgCl2

Temperature: 25.0 degrees Celsius

Method: Bromley

PURE SALT SOLUBILITIES:

solid phase	saturation molality	activity coefficient	water activity	solubility product
NaCl	0.614618E+01	0.986332E+00		0.367500E+02
MgCl2.6H2O	0.578071E+01	0.143203E+02	0.394868E+00	0.860146E+04

INVARIANT POINT:

Solid phase: NaCl + MgCl2.6H2O

	molalities	weight percents
NaCl	0.1168651E+00	0.433
MgCl2	0.5995714E+01	36.183
H2O		63.384

THE SOLUBILITY OF MgCl2 IN A SOLUTION OF NaCl:

Solid phase: MgCl2.6H2O

molality NaCl	molality MgCl2
0.1110218E+00	0.5997859E+01
0.1051786E+00	0.6000003E+01
0.9933533E-01	0.6002148E+01
0.9349207E-01	0.6004293E+01
0.8764882E-01	0.6006438E+01
0.8180557E-01	0.6008583E+01
0.7596231E-01	0.6012876E+01
0.7011906E-01	0.6012876E+01
0.6427580E-01	0.6015023E+01
0.5843255E-01	0.6017170E+01
0.5258929E-01	0.6019318E+01
0.4674604E-01	0.6021466E+01
0.4090278E-01	0.6023614E+01
0.3505953E-01	0.6025763E+01
0.2921627E-01	0.6027913E+01
0.2337302E-01	0.6030062E+01
0.1752976E-01	0.6032213E+01
0.1168651E-01	0.6034363E+01
0.5843255E-02	0.6036514E+01

THE SOLUBILITY OF NaCl IN A SOLUTION OF MgCl2:

molality NaCl	molality MgCl2
0.1548537E+00	0.5695929E+01
0.2046993E+00	0.5396143E+01
0.2696482E+00	0.5096357E+01
0.3535029E+00	0.4796572E+01
0.4605052E+00	0.4496786E+01
0.5951013E+00	0.4197000E+01
0.7615836E+00	0.3897214E+01
0.9636630E+00	0.3597429E+01
0.1204064E+01	0.3297643E+01
0.1484256E+01	0.2997857E+01
0.1804383E+01	0.2698072E+01
0.2163405E+01	0.2398286E+01
0.2559380E+01	0.2098500E+01
0.2989789E+01	0.1798714E+01
0.3451855E+01	0.1498929E+01
0.3942780E+01	0.1199143E+01
0.4459918E+01	0.8993572E+00
0.5000876E+01	0.5995714E+00
0.5563557E+01	0.2997857E+00

SYSTEM: H2O - NaCl - MgCl2

Temperature: 25.0 degrees Celsius

Method: Meissner

PURE SALT SOLUBILITIES:

solid phase	saturation molality	activity coefficient	water activity	solubility product
NaCl	0.614618E+01	0.933045E+00		0.328863E+02
MgCl2.6H2O	0.578071E+01	0.175650E+02	0.396151E+00	0.161851E+05

INVARIANT POINT:

Solid phase: NaCl + MgCl2.6H2O

	molalities	weight percents
NaCl	0.2243353E+00	0.838
MgCl2	0.5793927E+01	35.254
H2O		63.908

THE SOLUBILITY OF MgCl2 IN A SOLUTION OF NaCl:

Solid phase: MgCl2.6H2O

molality NaCl	molality MgCl2
0.2131185E+00	0.5793477E+01
0.2019018E+00	0.5792828E+01
0.1906850E+00	0.5792162E+01
0.1794682E+00	0.5791538E+01
0.1682515E+00	0.5790829E+01
0.1570347E+00	0.5790184E+01
0.1458179E+00	0.5789501E+01
0.1346012E+00	0.5788857E+01
0.1233844E+00	0.5788177E+01
0.1121677E+00	0.5787522E+01
0.1009509E+00	0.5786859E+01
0.8973412E-01	0.5786180E+01
0.7851736E-01	0.5785493E+01
0.6730059E-01	0.5784841E+01
0.5608383E-01	0.5784146E+01
0.4486706E-01	0.5783452E+01
0.3365030E-01	0.5782786E+01
0.2243353E-01	0.5782107E+01
0.1121677E-01	0.5781418E+01

THE SOLUBILITY OF NaCl IN A SOLUTION OF MgCl2:

molality NaCl	molality MgCl2
0.2721795E+00	0.5504231E+01
0.3313090E+00	0.5214535E+01
0.4043488E+00	0.4924838E+01
0.4943976E+00	0.4635142E+01
0.6049803E+00	0.4345445E+01
0.7399275E+00	0.4055749E+01
0.9031769E+00	0.3766053E+01
0.1098451E+01	0.3476356E+01
0.1328907E+01	0.3186660E+01
0.1596801E+01	0.2896964E+01
0.1903257E+01	0.2607267E+01
0.2248209E+01	0.2317571E+01
0.2630512E+01	0.2027875E+01
0.3048186E+01	0.1738178E+01
0.3498702E+01	0.1448482E+01
0.3979255E+01	0.1158785E+01
0.4486981E+01	0.8690891E+00
0.5019093E+01	0.5793927E+00
0.5572967E+01	0.2896964E+00

SYSTEM: H2O - NaCl - MgCl2

Temperature: 25.0 degrees Celsius

Method: Pitzer

PURE SALT SOLUBILITIES:

solid phase	saturation molality	activity coefficient	water activity	solubility product
NaCl	0.614618E+01	0.100413E+01		0.380880E+02
MgCl2.6H2O	0.578071E+01	0.302625E+02	0.334958E+00	0.302455E+05

INVARIANT POINT:

Solid phase: NaCl + MgCl2.6H2O

	molalities	weight percents
NaCl	0.2574263E-01	0.093
MgCl2	0.6514690E+01	38.246
H2O		61.661

THE SOLUBILITY OF MgCl2 IN A SOLUTION OF NaCl:

Solid phase: MgCl2.6H2O

molality NaCl	molality MgCl2
0.2445550E-01	0.6514829E+01
0.2316836E-01	0.6514965E+01
0.2188123E-01	0.6515101E+01
0.2059410E-01	0.6515236E+01
0.1930697E-01	0.6515371E+01
0.1801984E-01	0.6515508E+01
0.1673271E-01	0.6515643E+01
0.1544558E-01	0.6515780E+01
0.1415844E-01	0.6515917E+01
0.1287131E-01	0.6516053E+01
0.1158418E-01	0.6516190E+01
0.1029705E-01	0.6516326E+01
0.9009919E-02	0.6516463E+01
0.7722788E-02	0.6516600E+01
0.6435657E-02	0.6516736E+01
0.5148525E-02	0.6516875E+01
0.3861394E-02	0.6517010E+01
0.2574263E-02	0.6517147E+01
0.1287131E-02	0.6517284E+01

THE SOLUBILITY OF NaCl IN A SOLUTION OF MgCl2:

molality NaCl	molality MgCl2
0.3741132E-01	0.6188955E+01
0.5437885E-01	0.5863221E+01
0.7899800E-01	0.5537486E+01
0.1145757E+00	0.5211752E+01
0.1656486E+00	0.4886017E+01
0.2382201E+00	0.4560283E+01
0.3398250E+00	0.4234548E+01
0.4792307E+00	0.3908814E+01
0.6655976E+00	0.3583079E+01
0.9071112E+00	0.3257345E+01
0.1209442E+01	0.2931610E+01
0.1574638E+01	0.2605876E+01
0.2000921E+01	0.2280141E+01
0.2483371E+01	0.1954407E+01
0.3015076E+01	0.1628672E+01
0.3588255E+01	0.1302938E+01
0.4195077E+01	0.9772035E+00
0.4828129E+01	0.6514690E+00
0.5480585E+01	0.3257345E+00

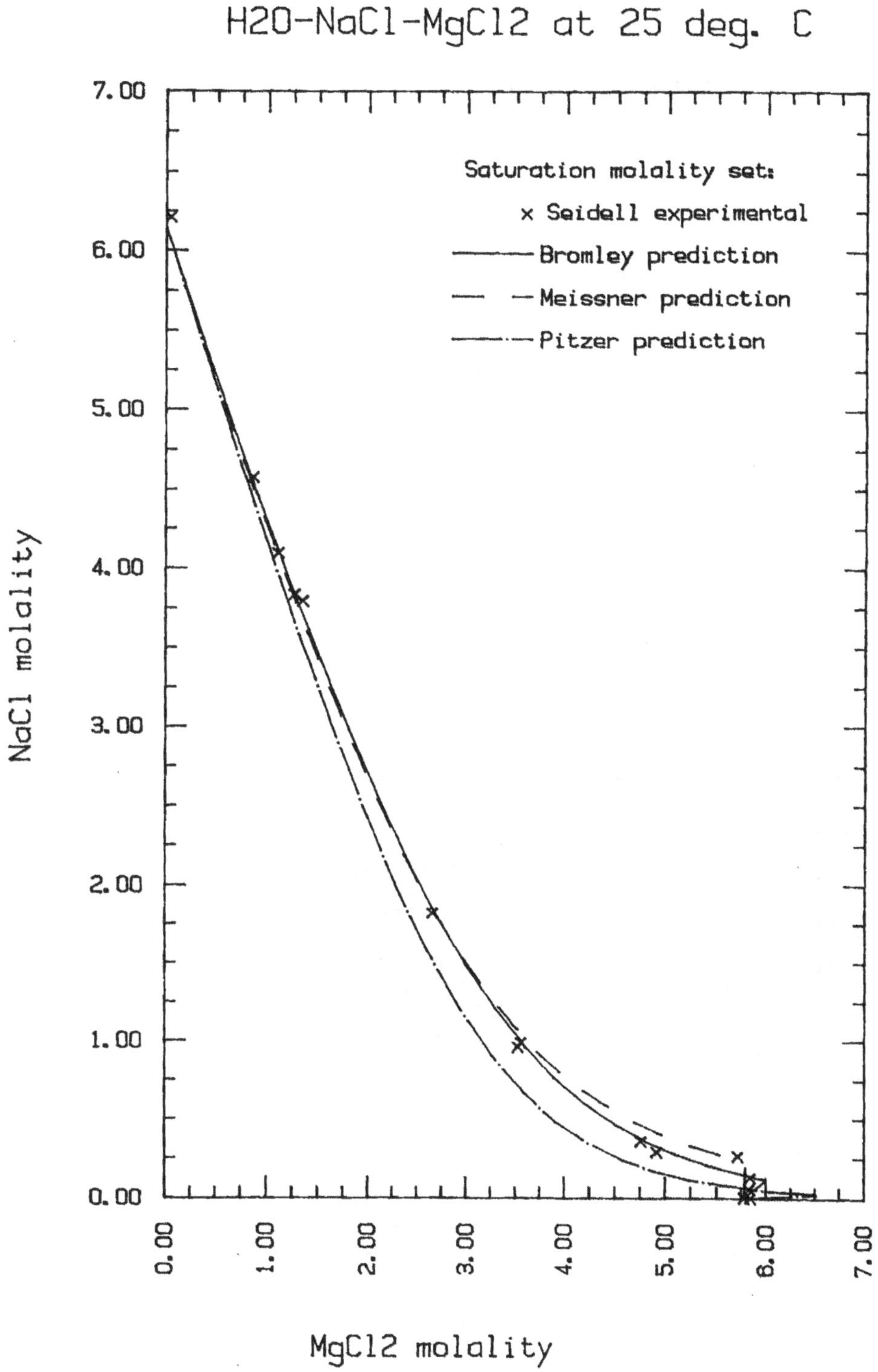
H2O-NaCl-MgCl2 at 25 deg. C
Saturation molality set:
× Seidell experimental
Bromley prediction
Meissner prediction
Pitzer prediction
NaCl molality
MgCl2 molality
7.00
6.00
5.00
4.00
3.00
2.00
1.00
0.00
0.00
1.00
2.00
3.00
4.00
5.00
6.00
7.00

SYSTEM: H2O - NaCl - MgCl2

Temperature: 25.0 degrees Celsius

Method: Bromley

PURE SALT SOLUBILITIES:

solid phase	saturation molality	activity coefficient	water activity	solubility product
NaCl	0.619136E+01	0.992017E+00		0.376680E+02
MgCl2.6H2O	0.728632E+01	0.442990E+02	0.244109E+00	0.284296E+05

INVARIANT POINT:

Solid phase: NaCl + MgCl2.6H2O

	molalities	weight percents
NaCl	0.2175919E-01	0.073
MgCl2	0.7796573E+01	42.574
H2O		57.353

THE SOLUBILITY OF MgCl2 IN A SOLUTION OF NaCl:

Solid phase: MgCl2.6H2O

molality NaCl	molality MgCl2
0.2067123E-01	0.7797075E+01
0.1958327E-01	0.7797578E+01
0.1849531E-01	0.7798082E+01
0.1740735E-01	0.7798585E+01
0.1631939E-01	0.7799088E+01
0.1523143E-01	0.7799591E+01
0.1414347E-01	0.7800094E+01
0.1305551E-01	0.7800598E+01
0.1196755E-01	0.7801101E+01
0.1087959E-01	0.7801604E+01
0.9791635E-02	0.7802108E+01
0.8703676E-02	0.7802611E+01
0.7615716E-02	0.7803115E+01
0.6527757E-02	0.7803618E+01
0.5439797E-02	0.7804122E+01
0.4351838E-02	0.7804626E+01
0.3263878E-02	0.7805129E+01
0.2175919E-02	0.7805633E+01
0.1087959E-02	0.7806137E+01

THE SOLUBILITY OF NaCl IN A SOLUTION OF MgCl2:

molality NaCl	molality MgCl2
0.3147002E-01	0.7406744E+01
0.4553848E-01	0.7016915E+01
0.6589949E-01	0.6627087E+01
0.9529830E-01	0.6237258E+01
0.1375629E+00	0.5847429E+01
0.1978909E+00	0.5457601E+01
0.2830650E+00	0.5067772E+01
0.4014391E+00	0.4677944E+01
0.5624897E+00	0.4288115E+01
0.7757927E+00	0.3898286E+01
0.1049546E+01	0.3508458E+01
0.1389111E+01	0.3118629E+01
0.1796195E+01	0.2728800E+01
0.2268998E+01	0.2338972E+01
0.2803128E+01	0.1949143E+01
0.3392801E+01	0.1559315E+01
0.4031874E+01	0.1169486E+01
0.4714557E+01	0.7796573E+00
0.5435786E+01	0.3898286E+00

SYSTEM: H2O - NaCl - MgCl2
==========================

Temperature: 25.0 degrees Celsius

Method: Meissner

PURE SALT SOLUBILITIES:
=======================

solid phase	saturation molality	activity coefficient	water activity	solubility product
NaCl	0.641588E+01	0.958516E+00		0.376680E+02
MgCl2.6H2O	0.656091E+01	0.272805E+02	0.327809E+00	0.284296E+05

INVARIANT POINT:
================

Solid phase: NaCl + MgCl2.6H2O

	molalities	weight percents
NaCl	0.1531772E+00	0.547
MgCl2	0.6592430E+01	38.352
H2O		61.101

THE SOLUBILITY OF MgCl2 IN A SOLUTION OF NaCl:
==

Solid phase: MgCl2.6H2O

molality NaCl	molality MgCl2
0.1455184E+00	0.6590840E+01
0.1378595E+00	0.6589238E+01
0.1302007E+00	0.6587657E+01
0.1225418E+00	0.6586064E+01
0.1148829E+00	0.6584478E+01
0.1072241E+00	0.6582903E+01
0.9956521E-01	0.6581306E+01
0.9190635E-01	0.6579712E+01
0.8424748E-01	0.6578149E+01
0.7658862E-01	0.6576571E+01
0.6892976E-01	0.6574995E+01
0.6127090E-01	0.6573420E+01
0.5361204E-01	0.6571852E+01
0.4595317E-01	0.6570284E+01
0.3829431E-01	0.6568717E+01
0.3063545E-01	0.6567158E+01
0.2297659E-01	0.6565591E+01
0.1531772E-01	0.6564032E+01
0.7658862E-02	0.6562469E+01

THE SOLUBILITY OF NaCl IN A SOLUTION OF MgCl2:

molality NaCl	molality MgCl2
0.1884726E+00	0.6262809E+01
0.2330287E+00	0.5933187E+01
0.2894150E+00	0.5603566E+01
0.3608486E+00	0.5273944E+01
0.4512677E+00	0.4944323E+01
0.5653369E+00	0.4614701E+01
0.7083366E+00	0.4285080E+01
0.8858672E+00	0.3955458E+01
0.1103370E+01	0.3625837E+01
0.1365509E+01	0.3296215E+01
0.1675565E+01	0.2966594E+01
0.2035020E+01	0.2636972E+01
0.2443486E+01	0.2307351E+01
0.2898946E+01	0.1977729E+01
0.3398219E+01	0.1648108E+01
0.3937475E+01	0.1318486E+01
0.4512671E+01	0.9888645E+00
0.5119857E+01	0.6592430E+00
0.5755352E+01	0.3296215E+00

SYSTEM: H2O - NaCl - MgCl2

Temperature: 25.0 degrees Celsius

Method: Pitzer

PURE SALT SOLUBILITIES:

solid phase	saturation molality	activity coefficient	water activity	solubility product
NaCl	0.612742E+01	0.100493E+01		0.376680E+02
MgCl2.6H2O	0.573435E+01	0.290883E+02	0.339833E+00	0.284296E+05

INVARIANT POINT:

Solid phase: NaCl + MgCl2.6H2O

	molalities	weight percents
NaCl	0.2831122E-01	0.103
MgCl2	0.6422201E+01	37.906
H2O		61.992

THE SOLUBILITY OF MgCl2 IN A SOLUTION OF NaCl:

Solid phase: MgCl2.6H2O

molality NaCl	molality MgCl2
0.2689566E-01	0.6422379E+01
0.2548010E-01	0.6422556E+01
0.2406454E-01	0.6422733E+01
0.2264898E-01	0.6422909E+01
0.2123341E-01	0.6423086E+01
0.1981785E-01	0.6423264E+01
0.1840229E-01	0.6423441E+01
0.1698673E-01	0.6423618E+01
0.1557117E-01	0.6423795E+01
0.1415561E-01	0.6423973E+01
0.1274005E-01	0.6424151E+01
0.1132449E-01	0.6424328E+01
0.9908927E-02	0.6424506E+01
0.8493366E-02	0.6424684E+01
0.7077805E-02	0.6424862E+01
0.5662244E-02	0.6425040E+01
0.4246683E-02	0.6425219E+01
0.2831122E-02	0.6425397E+01
0.1415561E-02	0.6425575E+01

THE SOLUBILITY OF NaCl IN A SOLUTION OF MgCl2:

molality NaCl	molality MgCl2
0.4093271E-01	0.6101091E+01
0.5918273E-01	0.5779981E+01
0.8550628E-01	0.5458871E+01
0.1233084E+00	0.5137761E+01
0.1772102E+00	0.4816650E+01
0.2532505E+00	0.4495540E+01
0.3588969E+00	0.4174430E+01
0.5026881E+00	0.3853320E+01
0.6933618E+00	0.3532210E+01
0.9385245E+00	0.3211100E+01
0.1243231E+01	0.2889990E+01
0.1609036E+01	0.2568880E+01
0.2033907E+01	0.2247770E+01
0.2512921E+01	0.1926660E+01
0.3039358E+01	0.1605550E+01
0.3605732E+01	0.1284440E+01
0.4204540E+01	0.9633301E+00
0.4828679E+01	0.6422201E+00
0.5471602E+01	0.3211100E+00

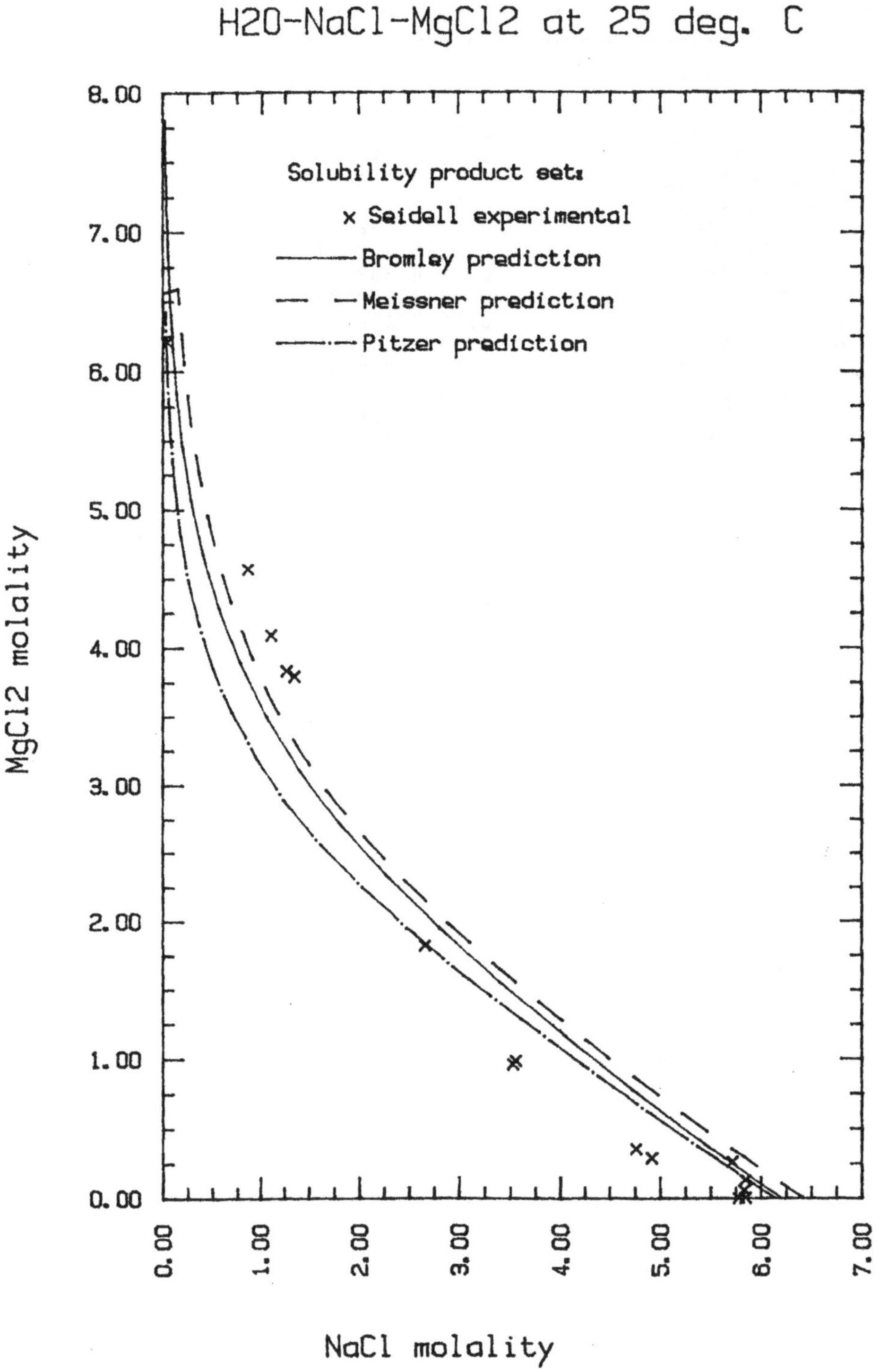
H2O-NaCl-MgCl2 at 25 deg. C
Solubility product set:
× Seidell experimental
Bromley prediction
Meissner prediction
Pitzer prediction
MgCl2 molality
NaCl molality
8.00
7.00
6.00
5.00
4.00
3.00
2.00
1.00
0.00
0.00
1.00
2.00
3.00
4.00
5.00
6.00
7.00

H_2O - NaCl - Na_2SO_4 EXPERIMENTAL DATA

From: W.F. Linke, A. Seidell, Solubilities of Inorganic and Metal-Organic Compounds, Vol. II, American Chemical Society, Washington, D.C. (1965) pp. 982, 983, 984

(Chretien, 1926, 1929; Foote and Schairer, 1930; Takejami, 1920, 1921; Schreinemakers and deBaat, 1909; Bogoyaulenskii and Sukmanskaya, 1955)

density	gms. per 100 gms. sat. sol. NaCl	Na_2SO_4	Solid Phase
1.207	7.66	16.0	$Na_2SO_4 .10H_2O$
	14.5	14.5	$Na_2SO_4 .10H_2O$ + Na_2SO_4
	14.12	14.93	$Na_2SO_4 .10H_2O$ + Na_2SO_4
	14.02	15.22	$Na_2SO_4 .10H_2O$ + Na_2SO_4
	13.61	15.18	$Na_2SO_4 .10H_2O$ + NaCl
1.239	18.4	10.4	Na_2SO_4
1.249	22.7	7.06	Na_2SO_4 + NaCl
	22.77	6.02	Na_2SO_4 + NaCl
	22.98	6.80	Na_2SO_4 + NaCl
1.2453	23.1	6.89	Na_2SO_4 + NaCl
	23.23	7.20	Na_2SO_4 + NaCl
1.216	24.6	3.35	NaCl

(Seidell, 1902)

density	gms. per 100 gms. H_2O NaCl	Na_2SO_4	Solid Phase
1.2173	2.96	26.60	$Na_2SO_4 .10H_2O$
1.2162	5.79	24.32	$Na_2SO_4 .10H_2O$
1.2150	9.90	21.41	$Na_2SO_4 .10H_2O$
1.2275	13.43	19.62	$Na_2SO_4 .10H_2O$
1.2385	15.82	19.64	$Na_2SO_4 .10H_2O$
1.2571	19.13	20.73	$Na_2SO_4 .10H_2O$ + Na_2SO_4
1.2476	23.22	16.28	Na_2SO_4
1.2429	26.54	12.64	Na_2SO_4
1.2438	31.06	9.98	Na_2SO_4
1.2451	32.41	9.93	Na_2SO_4
1.2453	33.00	9.84	Na_2SO_4 + NaCl
1.2309	33.81	6.66	NaCl
1.2162	34.60	3.38	NaCl
1.2002	35.80	0.0	NaCl

(Pel'sh, 1949)

sat. sol. weight percent NaCl	Na_2SO_4	Solid Phase
22.85	6.89	Na_2SO_4 + NaCl
14.08	14.92	$Na_2SO_4 .10H_2O$ + Na_2SO_4

H_2O - NaCl - Na_2SO_4 PARAMETERS

Bromley

c	a	B_{ca}
Na	Cl	.0574
	SO_4	-.0204

Meissner

c	a	q_{ca}
Na	Cl	2.23
	SO_4	-.19

Pitzer

c	a	β_0	β_1	C^{ϕ}
Na	Cl	.0765	.2664	.000127
	SO_4	.01058	1.113	.00497

i	j	k	Θ_{ij}	ψ_{ijk}
Cl	SO_4	Na	.02	.0014

SYSTEM: H2O - NaCl - Na2SO4

Temperature: 25.0 degrees Celsius

Method: Bromley

PURE SALT SOLUBILITIES:

solid phase	saturation molality	activity coefficient	water activity	solubility product
NaCl	0.614618E+01	0.986332E+00		0.367500E+02
Na2SO4.10H2O	0.195688E+01	0.149838E+00	0.940547E+00	0.546299E-01

INVARIANT POINTS:

Solid phase: NaCl + Na2SO4

	molalities	weight percents
NaCl	0.5550143E+01	22.286
Na2SO4	0.9229246E+00	9.007
H2O		68.707

THE SOLUBILITY OF NaCl IN A SOLUTION OF Na2SO4:

molality NaCl	molality Na2SO4
0.5579062E+01	0.8767783E+00
0.5608077E+01	0.8306321E+00
0.5637189E+01	0.7844859E+00
0.5666396E+01	0.7383397E+00
0.5695699E+01	0.6921934E+00
0.5725097E+01	0.6460472E+00
0.5754589E+01	0.5999010E+00
0.5784174E+01	0.5537547E+00
0.5813852E+01	0.5076085E+00
0.5843623E+01	0.4614623E+00
0.5873484E+01	0.4153161E+00
0.5903437E+01	0.3691698E+00
0.5933478E+01	0.3230236E+00
0.5963609E+01	0.2768774E+00
0.5993826E+01	0.2307311E+00
0.6024130E+01	0.1845849E+00
0.6054519E+01	0.1384387E+00
0.6084992E+01	0.9229246E-01
0.6115546E+01	0.4614623E-01

THE SOLUBILITY OF Na2SO4 IN A SOLUTION OF NaCl:

Solid phase: Na2SO4

molality NaCl	molality Na2SO4
0.5272635E+01	0.1030520E+01
0.4995128E+01	0.1153199E+01
0.4717621E+01	0.1293417E+01
0.4440114E+01	0.1454160E+01
0.4162607E+01	0.1639198E+01
0.3885100E+01	0.1853543E+01
0.3607593E+01	0.2104386E+01
0.3330086E+01	0.2402968E+01
0.3052578E+01	0.2769597E+01
0.2775071E+01	0.3249715E+01
0.2497564E+01	0.4012992E+01

SYSTEM: H2O - NaCl - Na2SO4
===========================

Temperature: 25.0 degrees Celsius

Method: Meissner

PURE SALT SOLUBILITIES:
=======================

solid phase	saturation molality	activity coefficient	water activity	solubility product
NaCl	0.614618E+01	0.933045E+00		0.328863E+02
Na2SO4.10H2O	0.195688E+01	0.165749E+00	0.931964E+00	0.674682E-01

INVARIANT POINTS:
=================

Solid phase: NaCl + Na2SO4

	molalities	weight percents
NaCl	0.5977305E+01	25.230
Na2SO4	0.2482861E+00	2.547
H2O		72.223

Solid phase: Na2SO4 + Na2SO4.10H2O

	molalities	weight percents
NaCl	0.3004872E+01	6.267
Na2SO4	0.1173266E+01	22.547
H2O		70.834

THE SOLUBILITY OF Na2SO4 IN A SOLUTION OF NaCl:

Solid phase: Na2SO4.10H2O

molality NaCl	molality Na2SO4
0.1438135E+01	0.2234159E+01
0.1362443E+01	0.2200927E+01
0.1286752E+01	0.2173735E+01
0.1211061E+01	0.2150980E+01
0.1135370E+01	0.2131581E+01
0.1059678E+01	0.2114763E+01
0.9839869E+00	0.2099947E+01
0.9082956E+00	0.2086684E+01
0.8326043E+00	0.2074616E+01
0.7569130E+00	0.2063451E+01
0.6812217E+00	0.2052937E+01
0.6055304E+00	0.2042831E+01
0.5298391E+00	0.2033063E+01
0.4541478E+00	0.2023281E+01
0.3784565E+00	0.2013406E+01
0.3027652E+00	0.2003280E+01
0.2270739E+00	0.1992737E+01
0.1513826E+00	0.1981608E+01
0.7569130E-01	0.1969720E+01

THE SOLUBILITY OF NaCl IN A SOLUTION OF Na2SO4:

molality NaCl	molality Na2SO4
0.5985607E+01	0.2358718E+00
0.5993913E+01	0.2234575E+00
0.6002234E+01	0.2110432E+00
0.6010572E+01	0.1986289E+00
0.6018927E+01	0.1862146E+00
0.6027298E+01	0.1738003E+00
0.6035686E+01	0.1613860E+00
0.6044089E+01	0.1489717E+00
0.6052510E+01	0.1365574E+00
0.6060946E+01	0.1241431E+00
0.6069399E+01	0.1117288E+00
0.6077868E+01	0.9931445E-01
0.6086354E+01	0.8690014E-01
0.6094855E+01	0.7448584E-01
0.6103382E+01	0.6207153E-01
0.6111906E+01	0.4965722E-01
0.6120455E+01	0.3724292E-01
0.6129020E+01	0.2482861E-01
0.6129020E+01	0.1241431E-01

THE SOLUBILITY OF Na2SO4 IN A SOLUTION OF NaCl:

Solid phase: Na2SO4

molality NaCl	molality Na2SO4
0.5754131E+01	0.2778126E+00
0.5530957E+01	0.3113068E+00
0.5307783E+01	0.3493219E+00
0.5084609E+01	0.3924758E+00
0.4861435E+01	0.4414496E+00
0.4638261E+01	0.4969809E+00
0.4415087E+01	0.5598495E+00
0.4191913E+01	0.6308524E+00
0.3968740E+01	0.7107702E+00
0.3745566E+01	0.8003237E+00
0.3522392E+01	0.9001280E+00
0.3299218E+01	0.1010650E+01
0.3076044E+01	0.1132177E+01
0.2852879E+01	0.1264796E+01
0.2629696E+01	0.1408399E+01
0.2406522E+01	0.1562689E+01
0.2183348E+01	0.1727211E+01
0.1960174E+01	0.1901383E+01
0.1737000E+01	0.2084539E+01

SYSTEM: H2O - NaCl - Na2SO4
=============================

Temperature: 25.0 degrees Celsius

Method: Pitzer

PURE SALT SOLUBILITIES:
=======================

solid phase	saturation molality	activity coefficient	water activity	solubility product
NaCl	0.614618E+01	0.100413E+01		0.380880E+02
Na2SO4.10H2O	0.195688E+01	0.156509E+00	0.936222E+00	0.594509E-01

INVARIANT POINTS:
=================

Solid phase: NaCl + Na2SO4

	molalities	weight percents
NaCl	0.5786473E+01	24.299
Na2SO4	0.3769484E+00	3.847
H2O		71.854

Solid phase: Na2SO4 + Na2SO4.10H2O

	molalities	weight percents
NaCl	0.3338436E+01	14.209
Na2SO4	0.1253449E+01	12.966
H2O		72.826

THE SOLUBILITY OF Na2SO4 IN A SOLUTION OF NaCl:

Solid phase: Na2SO4.10H2O

molality NaCl	molality Na2SO4
0.3171515E+01	0.1249830E+01
0.3004593E+01	0.1252478E+01
0.2837671E+01	0.1260693E+01
0.2670749E+01	0.1274002E+01
0.2503827E+01	0.1292076E+01
0.2336905E+01	0.1314674E+01
0.2169984E+01	0.1341612E+01
0.2003062E+01	0.1372744E+01
0.1836140E+01	0.1407939E+01
0.1669218E+01	0.1447074E+01
0.1502296E+01	0.1490025E+01
0.1335375E+01	0.1536659E+01
0.1168453E+01	0.1586835E+01
0.1001531E+01	0.1640396E+01
0.8346091E+00	0.1697177E+01
0.6676873E+00	0.1757001E+01
0.5007655E+00	0.1819685E+01
0.3338436E+00	0.1885039E+01
0.1669218E+00	0.1952876E+01

THE SOLUBILITY OF NaCl IN A SOLUTION OF Na2SO4:

molality NaCl	molality Na2SO4
0.5804651E+01	0.3581010E+00
0.5822733E+01	0.3392535E+00
0.5840807E+01	0.3204061E+00
0.5858872E+01	0.3015587E+00
0.5876927E+01	0.2827113E+00
0.5894972E+01	0.2638639E+00
0.5913007E+01	0.2450164E+00
0.5931031E+01	0.2261690E+00
0.5949043E+01	0.2073216E+00
0.5967044E+01	0.1884742E+00
0.5985032E+01	0.1696268E+00
0.6003007E+01	0.1507793E+00
0.6020968E+01	0.1319319E+00
0.6038915E+01	0.1130845E+00
0.6056847E+01	0.9423709E-01
0.6074763E+01	0.7538967E-01
0.6092664E+01	0.5654226E-01
0.6110548E+01	0.3769484E-01
0.6128415E+01	0.1884742E-01

THE SOLUBILITY OF Na2SO4 IN A SOLUTION OF NaCl:
==

Solid phase: Na2SO4

molality NaCl	molality Na2SO4
0.5664071E+01	0.4022058E+00
0.5541669E+01	0.4290249E+00
0.5419267E+01	0.4574846E+00
0.5296866E+01	0.4876631E+00
0.5174464E+01	0.5196383E+00
0.5052062E+01	0.5534861E+00
0.4929660E+01	0.5892801E+00
0.4807258E+01	0.6270899E+00
0.4684856E+01	0.6669803E+00
0.4562455E+01	0.7090098E+00
0.4440053E+01	0.7532297E+00
0.4317651E+01	0.7996826E+00
0.4195249E+01	0.8484018E+00
0.4072847E+01	0.8994104E+00
0.3950446E+01	0.9527205E+00
0.3828044E+01	0.1008333E+01
0.3705642E+01	0.1066239E+01
0.3583240E+01	0.1126416E+01
0.3460838E+01	0.1188833E+01

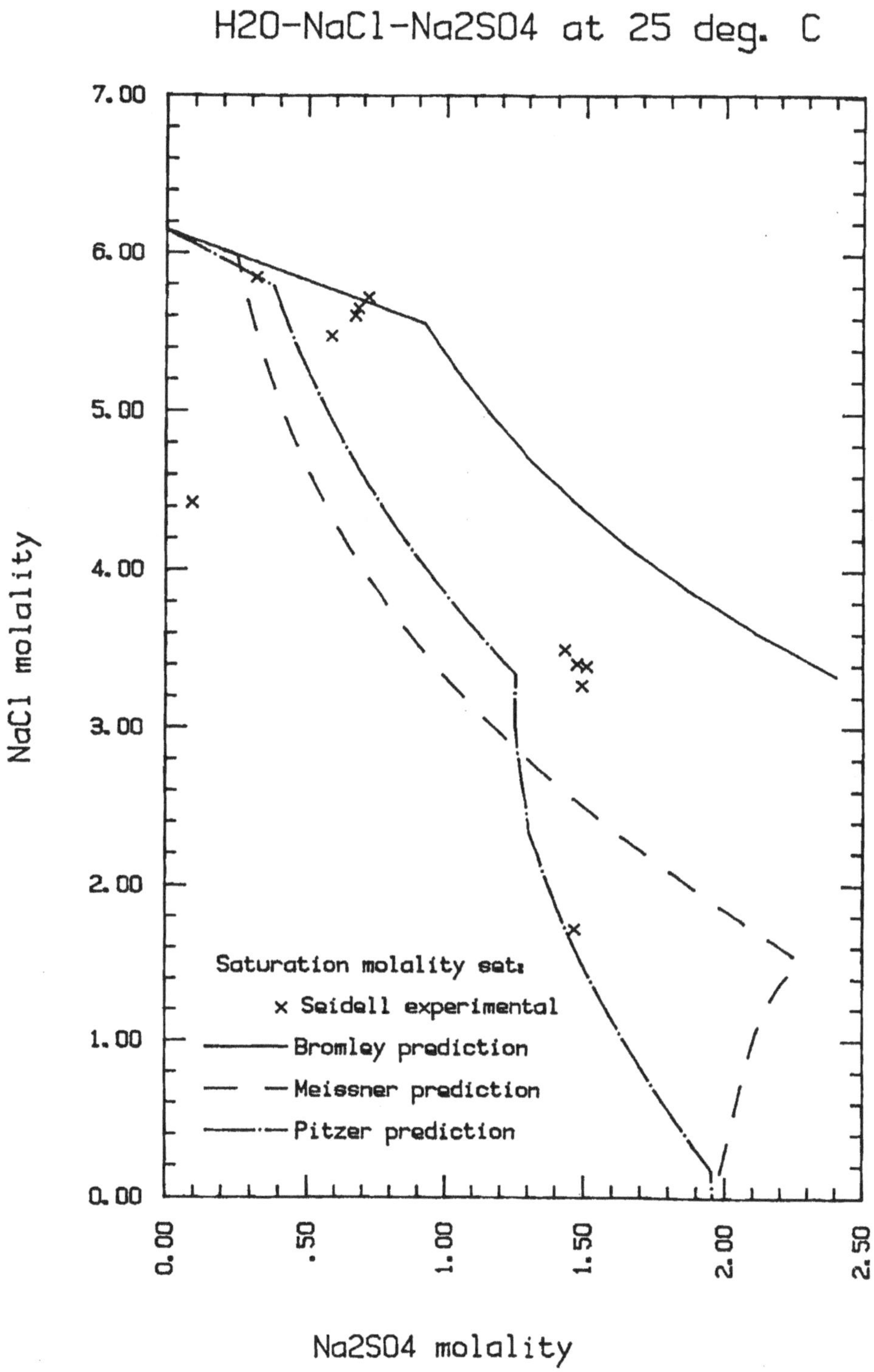
H2O-NaCl-Na2SO4 at 25 deg. C
7.00
6.00
5.00
4.00
3.00
2.00
1.00
0.00
NaCl molality
0.00
.50
1.00
1.50
2.00
2.50
Na2SO4 molality
Saturation molality set:
× Seidell experimental
Bromley prediction
Meissner prediction
Pitzer prediction

H_2O - KCl - $CaCl_2$ EXPERIMENTAL DATA

From: W.F. Linke, A. Seidell, Solubilities of Inorganic and Metal-Organic Compounds, Vol. I, American Chemical Society, Washington, D.C. (1958) p. 576

Results at 25°C:

(Lee and Egerton, 1923)

$d\frac{25}{25}$ Sat. Sol	gms. per 100 gms. sat. sol. $CaCl_2$	KCl	Solid Phase
1.47	45.06	0.0	$CaCl_2.6H_2O$
...	44.73	3.28	$CaCl_2.6H_2O$
1.485	44.72	3.06	$CaCl_2.6H_2O$
1.485	44.66	3.05	$CaCl_2.6H_2O$ + KCl
1.402	37.82	3.15	KCl
1.349	32.34	3.72	KCl
1.273	23.15	7.52	KCl
1.236	16.55	11.64	KCl
1.204	8.53	17.63	KCl
1.182	0.0	26.74	KCl

H_2O - KCl - $CaCl_2$ PARAMETERS

Bromley

c	a	B_{ca}
K	Cl	.024
Ca		.0948

Meissner

c	a	q_{ca}
K	Cl	.92
Ca		2.40

Pitzer

c	a	β_0	β_1	C^{ϕ}
K	Cl	.04835	.2122	-.00084
Ca		.3159	1.6140	-.00034

i	j	k	θ_{ij}	ψ_{ijk}
K	Ca	Cl	.032	-.025

SYSTEM: H2O - KCl - CaCl2
=========================

Temperature: 25.0 degrees Celsius

Method: Bromley

PURE SALT SOLUBILITIES:
=======================

solid phase	saturation molality	activity coefficient	water activity	solubility product
KCl	0.481115E+01	0.588291E+00		0.801092E+01
CaCl2.6H2O	0.746178E+01	0.197025E+02	0.277177E+00	0.576359E+04

INVARIANT POINT:
================

Solid phase: KCl + CaCl2.6H2O

	molalities	weight percents
KCl	0.2441876E-01	0.095
CaCl2	0.8185093E+01	47.556
H2O		52.349

THE SOLUBILITY OF CaCl2 IN A SOLUTION OF KCl:
===

Solid phase: CaCl2.6H2O

molality KCl	molality CaCl2
0.2319782E-01	0.8185280E+01
0.2197688E-01	0.8185467E+01
0.2075594E-01	0.8185654E+01
0.1953501E-01	0.8185841E+01
0.1831407E-01	0.8186028E+01
0.1709313E-01	0.8186216E+01
0.1587219E-01	0.8186403E+01
0.1465125E-01	0.8186590E+01
0.1343032E-01	0.8186777E+01
0.1220938E-01	0.8186965E+01
0.1098844E-01	0.8187152E+01
0.9767503E-02	0.8187339E+01
0.8546565E-02	0.8187527E+01
0.7325627E-02	0.8187714E+01
0.6104690E-02	0.8187901E+01
0.4883752E-02	0.8188089E+01
0.3662814E-02	0.8188276E+01
0.2441876E-02	0.8188463E+01
0.1220938E-02	0.8188651E+01

THE SOLUBILITY OF KCl IN A SOLUTION OF CaCl2:

molality KCl	molality CaCl2
0.3251292E-01	0.7775838E+01
0.4335557E-01	0.7366584E+01
0.5790267E-01	0.6957329E+01
0.7744741E-01	0.6548074E+01
0.1037364E+00	0.6138820E+01
0.1391197E+00	0.5729565E+01
0.1867382E+00	0.5320310E+01
0.2507411E+00	0.4911056E+01
0.3365109E+00	0.4501801E+01
0.4508359E+00	0.4092546E+01
0.6019315E+00	0.3683292E+01
0.7991508E+00	0.3274037E+01
0.1052256E+01	0.2864782E+01
0.1370239E+01	0.2455528E+01
0.1759949E+01	0.2046273E+01
0.2225025E+01	0.1637019E+01
0.2765801E+01	0.1227764E+01
0.3379727E+01	0.8185093E+00
0.4062890E+01	0.4092546E+00

SYSTEM: H2O - KCl - CaCl2

Temperature: 25.0 degrees Celsius

Method: Meissner

PURE SALT SOLUBILITIES:

solid phase	saturation molality	activity coefficient	water activity	solubility product
KCl	0.481115E+01	0.581101E+00		0.781631E+01
CaCl2.6H2O	0.746178E+01	0.154431E+02	0.317055E+00	0.621736E+04

INVARIANT POINT:

Solid phase: KCl + CaCl2.6H2O

	molalities	weight percents
KCl	0.9016195E-01	0.365
CaCl2	0.7520565E+01	45.328
H2O		54.307

THE SOLUBILITY OF CaCl2 IN A SOLUTION OF KCl:

Solid phase: CaCl2.6H2O

molality KCl	molality CaCl2
0.8565386E-01	0.7517919E+01
0.8114576E-01	0.7514838E+01
0.7663766E-01	0.7511976E+01
0.7212956E-01	0.7509033E+01
0.6762146E-01	0.7505448E+01
0.6311337E-01	0.7502902E+01
0.5860527E-01	0.7499684E+01
0.5409717E-01	0.7496875E+01
0.4958907E-01	0.7493668E+01
0.4508098E-01	0.7490884E+01
0.4057288E-01	0.7488077E+01
0.3606478E-01	0.7485072E+01
0.3155668E-01	0.7482060E+01
0.2704859E-01	0.7479138E+01
0.2254049E-01	0.7476141E+01
0.1803239E-01	0.7473306E+01
0.1352429E-01	0.7470541E+01
0.9016195E-02	0.7467649E+01
0.4508098E-02	0.7464685E+01

THE SOLUBILITY OF KCl IN A SOLUTION OF CaCl2:

molality KCl	molality CaCl2
0.1065457E+00	0.7144536E+01
0.1266172E+00	0.6768508E+01
0.1513640E+00	0.6392480E+01
0.1820726E+00	0.6016452E+01
0.2204253E+00	0.5640424E+01
0.2686254E+00	0.5264395E+01
0.3295520E+00	0.4888367E+01
0.4069440E+00	0.4512339E+01
0.5055905E+00	0.4136311E+01
0.6314846E+00	0.3760282E+01
0.7918451E+00	0.3384254E+01
0.9948853E+00	0.3008226E+01
0.1249172E+01	0.2632198E+01
0.1562573E+01	0.2256169E+01
0.1941000E+01	0.1880141E+01
0.2387373E+01	0.1504113E+01
0.2901228E+01	0.1128085E+01
0.3479240E+01	0.7520565E+00
0.4116949E+01	0.3760282E+00

SYSTEM: H2O - KCl - CaCl2
==========================

Temperature: 25.0 degrees Celsius

Method: Pitzer

PURE SALT SOLUBILITIES:
=======================

solid phase	saturation molality	activity coefficient	water activity	solubility product
KCl	0.481115E+01	0.586953E+00		0.797453E+01
CaCl2.6H2O	0.746178E+01	0.348856E+02	0.238699E+00	0.130507E+05

INVARIANT POINT:
================

Solid phase: KCl + CaCl2.6H2O

	molalities	weight percents
KCl	0.2275742E+00	0.926
CaCl2	0.7353829E+01	44.523
H2O		54.551

THE SOLUBILITY OF CaCl2 IN A SOLUTION OF KCl:
===

Solid phase: CaCl2.6H2O

molality KCl	molality CaCl2
0.2161955E+00	0.7351491E+01
0.2048168E+00	0.7349169E+01
0.1934381E+00	0.7346857E+01
0.1820594E+00	0.7344565E+01
0.1706807E+00	0.7342283E+01
0.1593019E+00	0.7340013E+01
0.1479232E+00	0.7337762E+01
0.1365445E+00	0.7335519E+01
0.1251658E+00	0.7333259E+01
0.1137871E+00	0.7331080E+01
0.1024084E+00	0.7328883E+01
0.9102969E-01	0.7326695E+01
0.7965097E-01	0.7324521E+01
0.6827226E-01	0.7322362E+01
0.5689355E-01	0.7320215E+01
0.4551484E-01	0.7318082E+01
0.3413613E-01	0.7315964E+01
0.2275742E-01	0.7313854E+01
0.1137871E-01	0.7311760E+01

THE SOLUBILITY OF KCl IN A SOLUTION OF CaCl2:
===

molality KCl	molality CaCl2
0.2411881E+00	0.6986137E+01
0.2596471E+00	0.6618446E+01
0.2838980E+00	0.6250754E+01
0.3152322E+00	0.5883063E+01
0.3553811E+00	0.5515371E+01
0.4066394E+00	0.5147680E+01
0.4720168E+00	0.4779989E+01
0.5554058E+00	0.4412297E+01
0.6617395E+00	0.4044606E+01
0.7970766E+00	0.3676914E+01
0.9685089E+00	0.3309223E+01
0.1183748E+01	0.2941531E+01
0.1450261E+01	0.2573840E+01
0.1774006E+01	0.2206149E+01
0.2157988E+01	0.1838457E+01
0.2601206E+01	0.1470766E+01
0.3098390E+01	0.1103074E+01
0.3640577E+01	0.7353829E+00
0.4216069E+01	0.3676914E+00

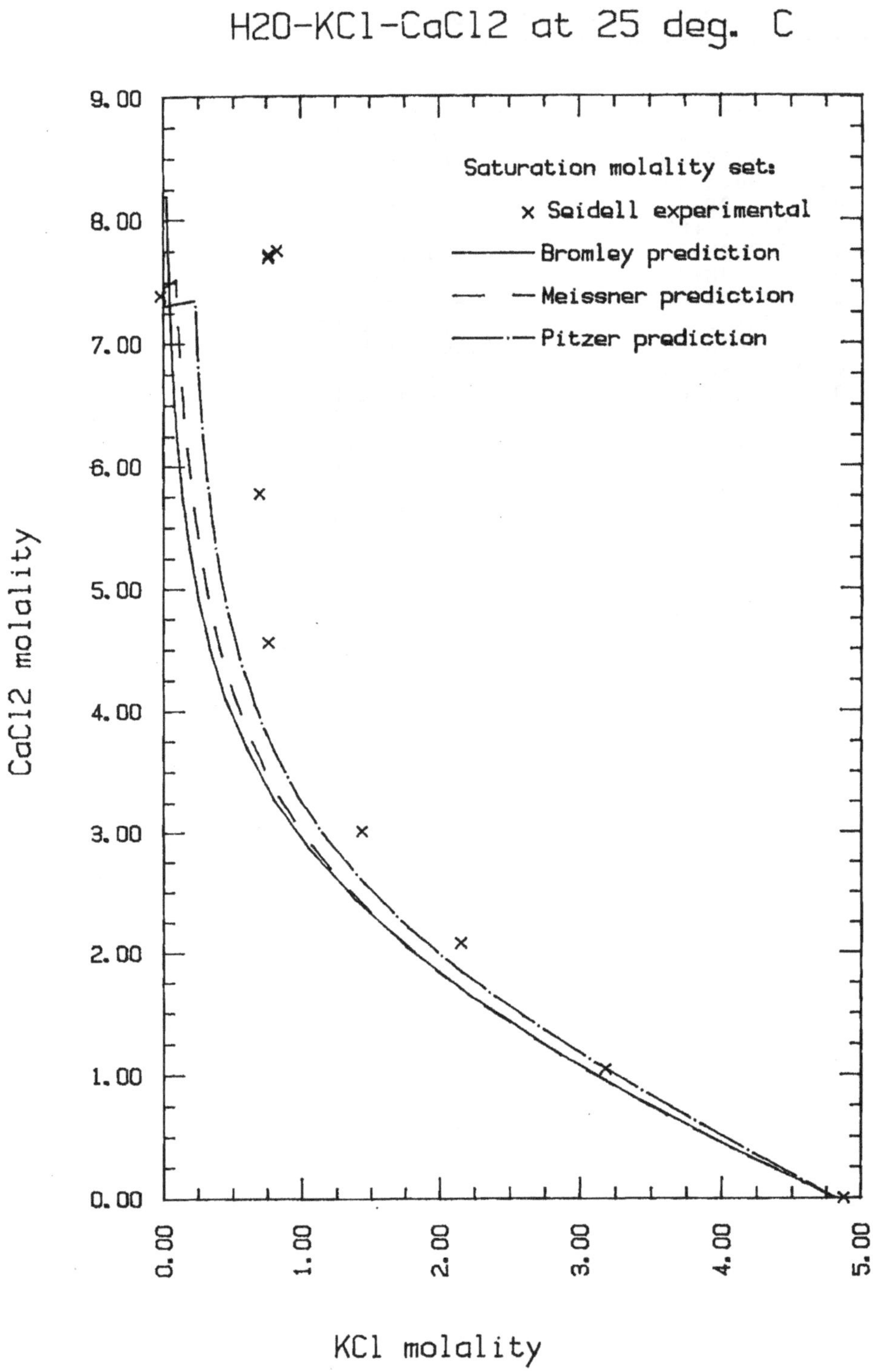
H2O-KCl-CaCl2 at 25 deg. C
Saturation molality set:
× Seidell experimental
Bromley prediction
Meissner prediction
Pitzer prediction
CaCl2 molality
KCl molality
9.00
8.00
7.00
6.00
5.00
4.00
3.00
2.00
1.00
0.00
0.00
1.00
2.00
3.00
4.00
5.00

H_2O - NaOH - Na_2SO_4 EXPERIMENTAL DATA

From: W.F. Linke, A. Seidell, Solubilities of Inorganic and Metal-Organic Compounds, Vol. II, American Chemical Society, Washington, D.C. (1965) p. 1130

(Windmaisser and Stockl, 1950)

sat. sol. weight percent

Na_2SO_4	NaOH	Solid Phase
21.48	0.0	$Na_2SO_4.10H_2O$
20.42	1.0	$Na_2SO_4.10H_2O$
18.61	3.68	$Na_2SO_4.10H_2O$
17.86	5.83	$Na_2SO_4.10H_2O$
17.86	7.79	$Na_2SO_4.10H_2O$
18.49	8.6	$Na_2SO_4.10H_2O$
18.61	8.92	$Na_2SO_4.10H_2O$
18.69	8.9	$Na_2SO_4.10H_2O$ + Na_2SO_4
18.4	9.03	Na_2SO_4
17.88	19.46	Na_2SO_4
15.22	11.42	Na_2SO_4
11.05	14.36	Na_2SO_4
7.9	27.23	Na_2SO_4
2.96	33.8	Na_2SO_4
.7	41.17	Na_2SO_4
.22	53.5	Na_2SO_4
.24	51.16	Na_2SO_4 + $NaOH.H_2O$
0.0	1.95	$NaOH.H_2O$

H_2O - NaOH - Na_2SO_4 PARAMETERS

Bromley

c	a	B_{ca}
Na	OH	.0747
	SO_4	-.0204

Meissner

c	a	q_{ca}
Na	OH	3.0
	SO_4	-.19

Pitzer

c	a	β_0	β_1	C^{ϕ}
Na	OH	.0864	.2530	.00440
	SO_4	.01958	1.113	.00497

i	j	k	Θ_{ij}	ψ_{ijk}
OH	SO_4	Na	0.0	0.0

SYSTEM: H2O - NaOH - Na2SO4

Temperature: 25.0 degrees Celsius

Method: Bromley

PURE SALT SOLUBILITIES:

solid phase	saturation molality	activity coefficient	water activity	solubility product
NaOH,H2O	0.285352E+02	0.504053E+02	0.321016E-01	0.664112E+05
Na2SO4.10H2O	0.195688E+01	0.149838E+00	0.940547E+00	0.546299E-01

INVARIANT POINTS:

Solid phase: NaOH.H2O + Na2SO4

	molalities	weight percents
NaOH	0.2874966E+02	53.486
Na2SO4	0.2287223E-03	0.002
H2O		46.513

Solid phase: Na2SO4 + Na2SO4.10H2O

	molalities	weight percents
NaOH	0.2559265E+01	6.944
Na2SO4	0.2617461E+01	25.220
H2O		67.836

THE SOLUBILITY OF Na2SO4 IN A SOLUTION OF NaOH:
==

Solid phase: Na2SO4.10H2O

molality NaOH	molality Na2SO4
0.2431302E+01	0.2548484E+01
0.2303339E+01	0.2513935E+01
0.2175375E+01	0.2501821E+01
0.2047412E+01	0.2507277E+01
0.1919449E+01	0.2527323E+01
0.1791486E+01	0.2560655E+01
0.1663522E+01	0.2606719E+01
0.1535559E+01	0.2665505E+01
0.1407596E+01	0.2737505E+01
0.1279633E+01	0.2823746E+01
0.1151669E+01	0.2925933E+01
0.1023706E+01	0.3046762E+01
0.8957428E+00	0.3190583E+01
0.7677795E+00	0.3365266E+01
0.6398163E+00	0.3585510E+01
0.5118530E+00	0.3891215E+01
0.3838898E+00	0.7504014E+01
0.2559265E+00	0.7486903E+01
0.1279633E+00	0.7503739E+01

THE SOLUBILITY OF NaOH IN A SOLUTION OF Na2SO4:
==

Solid phase: NaOH.H2O

molality NaOH	molality Na2SO4
0.2874967E+02	0.2172862E-03
0.2874969E+02	0.2058501E-03
0.2874971E+02	0.1944140E-03
0.2874972E+02	0.1829778E-03
0.2874974E+02	0.1715417E-03
0.2874975E+02	0.1601056E-03
0.2874977E+02	0.1486695E-03
0.2874979E+02	0.1372334E-03
0.2874980E+02	0.1257973E-03
0.2874982E+02	0.1143612E-03
0.2874983E+02	0.1029250E-03
0.2874985E+02	0.9148892E-04
0.2874987E+02	0.8005281E-04
0.2874988E+02	0.6861669E-04
0.2874990E+02	0.5718058E-04
0.2874992E+02	0.4574446E-04
0.2874993E+02	0.3430835E-04
0.2874985E+02	0.2287223E-04
0.2874986E+02	0.1143612E-04

THE SOLUBILITY OF Na2SO4 IN A SOLUTION OF NaOH:
==

Solid phase: Na2SO4

molality NaOH	molality Na2SO4
0.2744014E+02	0.3350706E-03
0.2613062E+02	0.4923339E-03
0.2482110E+02	0.7257757E-03
0.2351158E+02	0.1073761E-02
0.2220206E+02	0.1594935E-02
0.2089254E+02	0.2379619E-02
0.1958302E+02	0.3568112E-02
0.1827350E+02	0.5380532E-02
0.1696398E+02	0.8166278E-02
0.1565446E+02	0.1248772E-01
0.1434494E+02	0.1926533E-01
0.1303542E+02	0.3003699E-01
0.1172590E+02	0.4743879E-01
0.1041638E+02	0.7613799E-01
0.9106863E+01	0.1247461E+00
0.7797344E+01	0.2100012E+00
0.6487824E+01	0.3664868E+00
0.5178304E+01	0.6695739E+00
0.3868785E+01	0.1284125E+01

SYSTEM: H2O - NaOH - Na2SO4
===========================

Temperature: 25.0 degrees Celsius

Method: Meissner

PURE SALT SOLUBILITIES:
======================

solid phase	saturation molality	activity coefficient	water activity	solubility product
NaOH.H2O	0.285352E+02	0.119602E+02	0.860151E-01	0.100187E+05
Na2SO4.10H2O	0.195688E+01	0.165749E+00	0.931964E+00	0.674682E-01

INVARIANT POINTS:
================

Solid phase: NaOH.H2O + Na2SO4

	molalities	weight percents
NaOH	0.2853522E+02	53.300
Na2SO4	0.2022725E-04	0.000
H2O		46.700

Solid phase: Na2SO4 + Na2SO4.10H2O

	molalities	weight percents
NaOH	0.1737149E+01	5.254
Na2SO4	0.1781355E+01	19.132
H2O		75.614

THE SOLUBILITY OF Na2SO4 IN A SOLUTION OF NaOH:

Solid phase: Na2SO4.10H2O

molality NaOH	molality Na2SO4
0.1650291E+01	0.1803855E+01
0.1563434E+01	0.1824415E+01
0.1476576E+01	0.1843242E+01
0.1389719E+01	0.1860507E+01
0.1302861E+01	0.1876327E+01
0.1216004E+01	0.1890794E+01
0.1129147E+01	0.1903979E+01
0.1042289E+01	0.1915927E+01
0.9554318E+00	0.1926666E+01
0.8685743E+00	0.1936118E+01
0.7817169E+00	0.1944526E+01
0.6948595E+00	0.1951574E+01
0.6080020E+00	0.1957476E+01
0.5211446E+00	0.1962041E+01
0.4342872E+00	0.1965212E+01
0.3474297E+00	0.1966937E+01
0.2605723E+00	0.1967102E+01
0.1737149E+00	0.1965592E+01
0.8685743E-01	0.1962257E+01

THE SOLUBILITY OF NaOH IN A SOLUTION OF Na2SO4:

Solid phase: NaOH.H2O

molality NaOH	molality Na2SO4
0.2853522E+02	0.1921589E-04
0.2853522E+02	0.1820453E-04
0.2853522E+02	0.1719317E-04
0.2853522E+02	0.1618180E-04
0.2853422E+02	0.1517044E-04
0.2853521E+02	0.1415908E-04
0.2853521E+02	0.1314772E-04
0.2853521E+02	0.1213635E-04
0.2853521E+02	0.1112499E-04
0.2853521E+02	0.1011363E-04
0.2853521E+02	0.9102264E-05
0.2853520E+02	0.8090902E-05
0.2853520E+02	0.7079539E-05
0.2853520E+02	0.6068176E-05
0.2853520E+02	0.5056814E-05
0.2853520E+02	0.4045451E-05
0.2853519E+02	0.3034088E-05
0.2853519E+02	0.2022725E-05
0.2853519E+02	0.1011363E-05

THE SOLUBILITY OF Na2SO4 IN A SOLUTION OF NaOH:

Solid phase: Na2SO4

molality NaOH	molality Na2SO4
0.2719532E+02	0.2869668E-04
0.2585542E+02	0.4121743E-04
0.2451551E+02	0.5999524E-04
0.2317561E+02	0.8858427E-04
0.2183571E+02	0.1328363E-03
0.2049580E+02	0.2025684E-03
0.1915590E+02	0.3146133E-03
0.1781599E+02	0.4985276E-03
0.1647609E+02	0.8075861E-03
0.1513619E+02	0.1340646E-02
0.1379628E+02	0.2287216E-02
0.1245638E+02	0.4024209E-02
0.1111648E+02	0.7333578E-02
0.9776572E+01	0.1391897E-01
0.8436668E+01	0.2771105E-01
0.7096764E+01	0.5839798E-01
0.5756860E+01	0.1314474E+00
0.4416956E+01	0.3161023E+00
0.3077052E+01	0.7897837E+00

SYSTEM: H2O - NaOH - Na2SO4

Temperature: 25.0 degrees Celsius

Method: Pitzer

PURE SALT SOLUBILITIES:

solid phase	saturation molality	activity coefficient	water activity	solubility product
NaOH.H2O	0.285352E+02	0.689654E+04	0.953259E-03	0.369177E+08
Na2SO4.10H2O	0.195688E+01	0.156509E+00	0.936222E+00	0.594509E-01

INVARIANT POINTS:

Solid phase: Na2SO4 + Na2SO4.10H2O

	molalities	weight percents
NaOH	0.2779832E+01	8.475
Na2SO4	0.1413735E+01	15.305
H2O		76.220

THE SOLUBILITY OF Na2SO4 IN A SOLUTION OF NaOH:

Solid phase: Na2SO4.10H2O

molality NaOH	molality Na2SO4
0.2640841E+01	0.1395379E+01
0.2501849E+01	0.1387648E+01
0.2362857E+01	0.1388064E+01
0.2223866E+01	0.1395129E+01
0.2084874E+01	0.1407857E+01
0.1945883E+01	0.1425562E+01
0.1806891E+01	0.1447740E+01
0.1667899E+01	0.1474000E+01
0.1528908E+01	0.1504029E+01
0.1389916E+01	0.1537566E+01
0.1250925E+01	0.1574381E+01
0.1111933E+01	0.1614268E+01
0.9729413E+00	0.1657036E+01
0.8339497E+00	0.1702508E+01
0.6949581E+00	0.1750511E+01
0.5559665E+00	0.1800884E+01
0.4169748E+00	0.1853468E+01
0.2779832E+00	0.1908112E+01
0.1389916E+00	0.1964671E+01

THE SOLUBILITY OF Na2SO4 IN A SOLUTION OF NaOH:
==

Solid phase: Na2SO4

molality NaOH	molality Na2SO4
0.4067600E+01	0.7368550E+00
0.5355368E+01	0.3399332E+00
0.6643136E+01	0.1449355E+00
0.7930904E+01	0.5774811E-01
0.9218671E+01	0.2147100E-01
0.1050644E+02	0.7447224E-02
0.1179421E+02	0.2411778E-02
0.1308197E+02	0.7297399E-03
0.1436974E+02	0.2063300E-03
0.1565751E+02	0.5450807E-04
0.1694528E+02	0.1345097E-04
0.1823305E+02	0.3099756E-05
0.1952081E+02	0.6669336E-06
0.2080858E+02	0.1339479E-06
0.2209635E+02	0.2510844E-07
0.2338412E+02	0.4392159E-08
0.2467189E+02	0.7169101E-09
0.2595965E+02	0.1091794E-09
0.2724742E+02	0.1551221E-10

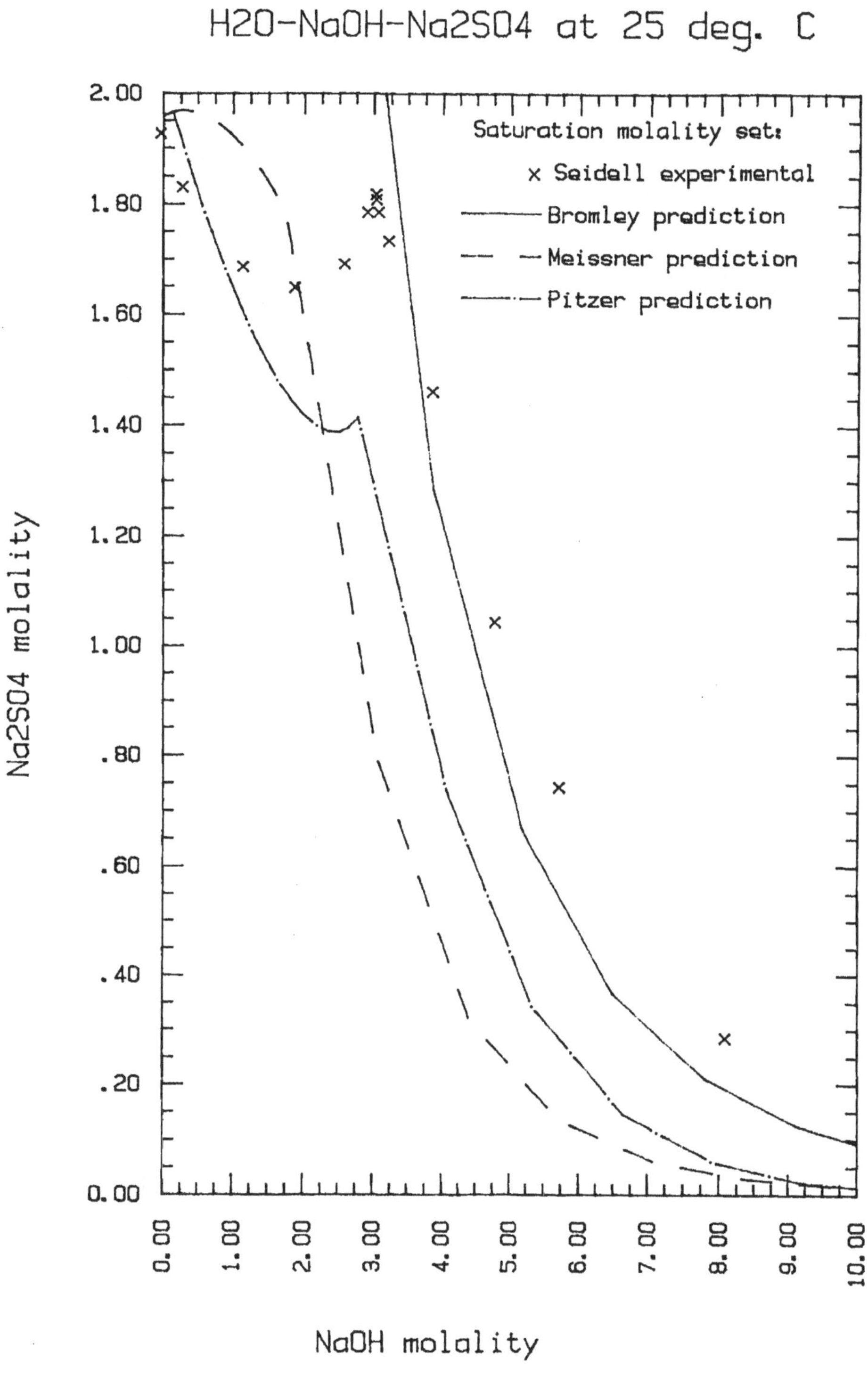
H2O-NaOH-Na2SO4 at 25 deg. C
Saturation molality set:
× Seidell experimental
Bromley prediction
Meissner prediction
Pitzer prediction
Na2SO4 molality
NaOH molality
2.00
1.80
1.60
1.40
1.20
1.00
.80
.60
.40
.20
0.00
0.00
1.00
2.00
3.00
4.00
5.00
6.00
7.00
8.00
9.00
10.00

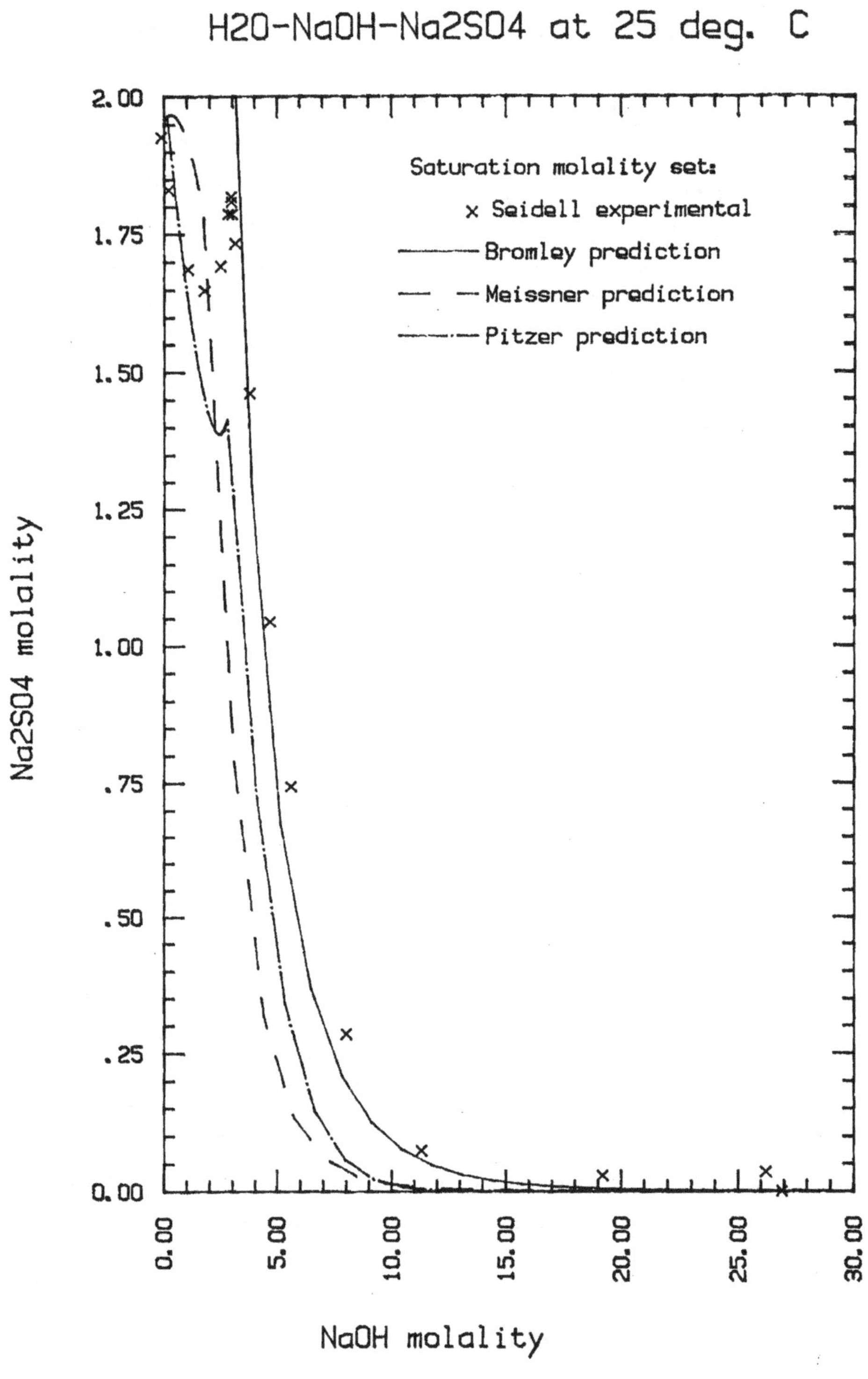
H2O-NaOH-Na2SO4 at 25 deg. C
Saturation molality set:
Seidell experimental
Bromley prediction
Meissner prediction
Pitzer prediction
Na2SO4 molality
2.00
1.75
1.50
1.25
1.00
.75
.50
.25
0.00
5.00
10.00
15.00
20.00
25.00
30.00
NaOH molality

H_2O - NaCl - $CaSO_4$ EXPERIMENTAL DATA

From: W.L. Marshall and R. Slusher, "Thermodynamics of Calcium Sulfate Dihydrate in Aqueous Sodium Chloride Solutions, 0 - 110°", J. Phys. Chem., v. 70, #12 (1966) p 4017

Solubility of $CaSO_4$ in aqueous solutions of NaCl at 25°C:

NaCl molality	$CaSO_4$ molality
0.0	0.0151
0.0117	0.0162
0.0257	0.0175
0.0513	0.0194
0.1147	0.0231
0.1921	0.0266
0.2319	0.0281
0.548	0.0372
0.689	0.0388
0.834	0.0430
1.005	0.0457
1.024	0.0452
2.024	0.0540
2.870	0.0560
4.125	0.0560
6.13	0.0489
6.22	0.0481*

* solution saturated with two solid phases, $CaSO_4 \cdot 2H_2O$ and NaCl

H_2O - NaCl - $CaSO_4$ PARAMETERS

Bromley

c	a	B_{ca}
Na	Cl	.0574
Ca		.0948
Na	SO_4	-.0204
Ca		-.0816

Meissner

c	a	q_{ca}
Na	Cl	2.23
Ca		2.40
Na	SO_4	-.19
Ca		0.0

Pitzer

c	a	β_0	β_1	β_2	C^{ϕ}
Na	Cl	.0765	.2664	0.0	.000127
Ca		.3159	1.6142	0.0	-.00034
Na	SO_4	.01958	1.113	0.0	.00497
Ca		.2	2.65	-55.7	0.0

i	j	k	Θ_{ij}	ψ_{ijk}
Na	Ca	Cl	.07	-.014
		SO_4		-.023
Cl	SO_4	Na	.02	.0014
		Ca		0.0

SYSTEM: H2O - NaCl - CaSO4

Temperature: 25.0 degrees Celsius

Method: Bromley

PURE SALT SOLUBILITIES:

solid phase	saturation molality	activity coefficient	water activity	solubility product
NaCl	0.614618E+01	0.986332E+00		0.367500E+02
CaSO4.2H2O	0.149938E-01	0.393840E+00	0.999610E+00	0.348435E-04

THE SOLUBILITY OF CaSO4 IN A SOLUTION OF NaCl:

Solid phase: CaSO4.2H2O

molality NaCl	molality CaSO4
0.3000000E+00	0.3324969E-01
0.6000000E+00	0.4193581E-01
0.9000000E+00	0.4832143E-01
0.1200000E+01	0.5339758E-01
0.1500000E+01	0.5750696E-01
0.1800000E+01	0.6082855E-01
0.2100000E+01	0.6348347E-01
0.2400000E+01	0.6556623E-01
0.2700000E+01	0.6715560E-01
0.3000000E+01	0.6831908E-01
0.3300000E+01	0.6911528E-01
0.3600000E+01	0.6959529E-01
0.3900000E+01	0.6980397E-01
0.4200000E+01	0.6977973E-01
0.4500000E+01	0.6955755E-01
0.4800000E+01	0.6916633E-01
0.5100000E+01	0.6863281E-01
0.5400000E+01	0.6797950E-01
0.5700000E+01	0.6722623E-01
0.6000000E+01	0.6639029E-01

SYSTEM: H2O - NaCl - CaSO4

Temperature: 25.0 degrees Celsius

Method: Meissner

PURE SALT SOLUBILITIES:

solid phase	saturation molality	activity coefficient	water activity	solubility product
NaCl	0.614618E+01	0.933045E+00		0.328863E+02
CaSO4.2H2O	0.149938E-01	0.396404E+00	0.999602E+00	0.352980E-04

THE SOLUBILITY OF CaSO4 IN A SOLUTION OF NaCl:

Solid phase: CaSO4.2H2O

molality NaCl	molality CaSO4
0.3000000E+00	0.3477347E-01
0.6000000E+00	0.4395287E-01
0.9000000E+00	0.4949033E-01
0.1200000E+01	0.5296573E-01
0.1500000E+01	0.5516124E-01
0.1800000E+01	0.5652912E-01
0.2100000E+01	0.5735050E-01
0.2400000E+01	0.5780166E-01
0.2700000E+01	0.5798856E-01
0.3000000E+01	0.5796781E-01
0.3300000E+01	0.5776524E-01
0.3600000E+01	0.5738798E-01
0.3900000E+01	0.5683878E-01
0.4200000E+01	0.5612369E-01
0.4500000E+01	0.5525759E-01
0.4800000E+01	0.5426435E-01
0.5100000E+01	0.5317391E-01
0.5400000E+01	0.5201791E-01
0.5700000E+01	0.5082557E-01
0.6000000E+01	0.4962100E-01

SYSTEM: H2O - NaCl - CaSO4

Temperature: 25.0 degrees Celsius

Method: Pitzer

PURE SALT SOLUBILITIES:

solid phase	saturation molality	activity coefficient	water activity	solubility product
NaCl	0.614618E+01	0.100413E+01		0.380880E+02
CaSO4.2H2O	0.149938E-01	0.332037E+00	0.999627E+00	0.247669E-04

THE SOLUBILITY OF CaSO4 IN A SOLUTION OF NaCl:

Solid phase: CaSO4.2H2O

molality NaCl	molality CaSO4
0.3000000E+00	0.2649644E-01
0.6000000E+00	0.3048823E-01
0.9000000E+00	0.3305114E-01
0.1200000E+01	0.3468505E-01
0.1500000E+01	0.3565569E-01
0.1800000E+01	0.3612488E-01
0.2100000E+01	0.3620453E-01
0.2400000E+01	0.3597819E-01
0.2700000E+01	0.3551172E-01
0.3000000E+01	0.3485797E-01
0.3300000E+01	0.3406022E-01
0.3600000E+01	0.3315409E-01
0.3900000E+01	0.3216902E-01
0.4200000E+01	0.3112931E-01
0.4500000E+01	0.3005499E-01
0.4800000E+01	0.2896249E-01
0.5100000E+01	0.2786518E-01
0.5400000E+01	0.2677392E-01
0.5700000E+01	0.2569737E-01
0.6000000E+01	0.2464241E-01

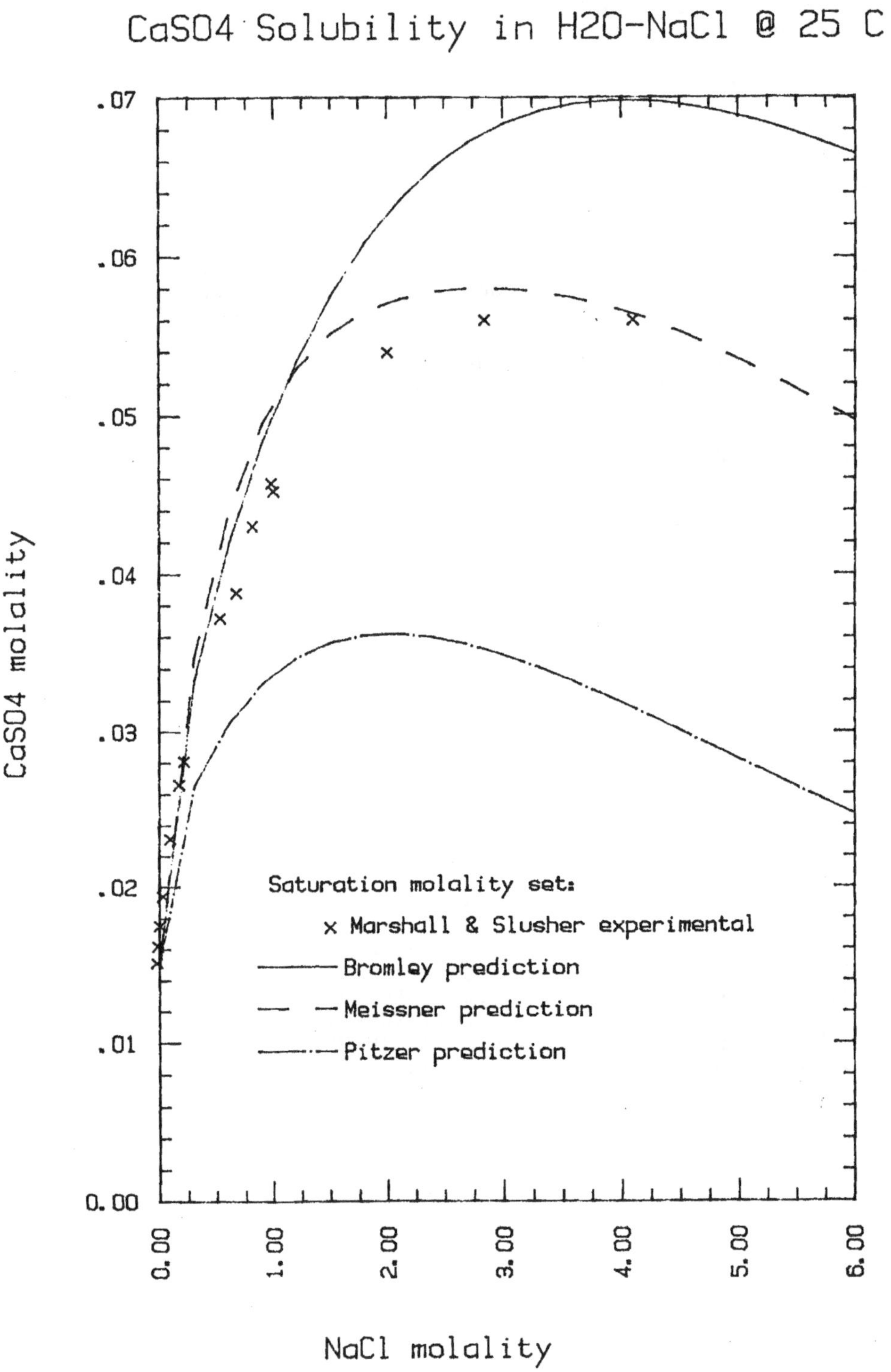
CaSO4 Solubility in H2O-NaCl @ 25 C
.07
.06
.05
.04
.03
.02
.01
0.00
CaSO4 molality
0.00
1.00
2.00
3.00
4.00
5.00
6.00
NaCl molality
Saturation molality set:
× Marshall & Slusher experimental
Bromley prediction
Meissner prediction
Pitzer prediction

SYSTEM: H2O - NaCl - CaSO4

Temperature: 25.0 degrees Celsius

Method: Bromley

PURE SALT SOLUBILITIES:

solid phase	saturation molality	activity coefficient	water activity	solubility product
NaCl	0.619136E+01	0.992017E+00		0.376680E+02
CaSO4.2H2O	0.178316E-01	0.367996E+00	0.999547E+00	0.430715E-04

THE SOLUBILITY OF CaSO4 IN A SOLUTION OF NaCl:

Solid phase: CaSO4.2H2O

molality NaCl	molality CaSO4
0.3000000E+00	0.3805874E-01
0.6000000E+00	0.4779341E-01
0.9000000E+00	0.5495356E-01
0.1200000E+01	0.6063440E-01
0.1500000E+01	0.6522078E-01
0.1800000E+01	0.6891527E-01
0.2100000E+01	0.7185604E-01
0.2400000E+01	0.7415113E-01
0.2700000E+01	0.7589047E-01
0.3000000E+01	0.7715103E-01
0.3300000E+01	0.7799947E-01
0.3600000E+01	0.7849381E-01
0.3900000E+01	0.7868479E-01
0.4200000E+01	0.7861614E-01
0.4500000E+01	0.7832713E-01
0.4800000E+01	0.7785023E-01
0.5100000E+01	0.7721578E-01
0.5400000E+01	0.7644898E-01
0.5700000E+01	0.7557210E-01
0.6000000E+01	0.7460451E-01

SYSTEM: H2O - NaCl - CaSO4

Temperature: 25.0 degrees Celsius

Method: Meissner

PURE SALT SOLUBILITIES:

solid phase	saturation molality	activity coefficient	water activity	solubility product
NaCl	0.641588E+01	0.958516E+00		0.376680E+02
CaSO4.2H2O	0.176093E-01	0.372611E+00	0.999543E+00	0.430715E-04

THE SOLUBILITY OF CaSO4 IN A SOLUTION OF NaCl:

Solid phase: CaSO4.2H2O

molality NaCl	molality CaSO4
0.3000000E+00	0.3928705E-01
0.6000000E+00	0.4939201E-01
0.9000000E+00	0.5542461E-01
0.1200000E+01	0.5918881E-01
0.1500000E+01	0.6155169E-01
0.1800000E+01	0.6301184E-01
0.2100000E+01	0.6387744E-01
0.2400000E+01	0.6434036E-01
0.2700000E+01	0.6451506E-01
0.3000000E+01	0.6446162E-01
0.3300000E+01	0.6420770E-01
0.3600000E+01	0.6375960E-01
0.3900000E+01	0.6312178E-01
0.4200000E+01	0.6230154E-01
0.4500000E+01	0.6131640E-01
0.4800000E+01	0.6019360E-01
0.5100000E+01	0.5896669E-01
0.5400000E+01	0.5767062E-01
0.5700000E+01	0.5633730E-01
0.6000000E+01	0.5499291E-01

SYSTEM: H2O - NaCl - CaSO4

Temperature: 25.0 degrees Celsius

Method: Pitzer

PURE SALT SOLUBILITIES:

solid phase	saturation molality	activity coefficient	water activity	solubility product
NaCl	0.612742E+01	0.100493E+01		0.376680E+02
CaSO4.2H2O	0.236208E-01	0.277887E+00	0.999438E+00	0.430715E-04

THE SOLUBILITY OF CaSO4 IN A SOLUTION OF NaCl:

Solid phase: CaSO4.2H2O

molality NaCl	molality CaSO4
0.3000000E+00	0.3924835E-01
0.6000000E+00	0.4349319E-01
0.9000000E+00	0.4635829E-01
0.1200000E+01	0.4813941E-01
0.1500000E+01	0.4911187E-01
0.1800000E+01	0.4946541E-01
0.2100000E+01	0.4933803E-01
0.2400000E+01	0.4883492E-01
0.2700000E+01	0.4803977E-01
0.3000000E+01	0.4701969E-01
0.3300000E+01	0.4582945E-01
0.3600000E+01	0.4451392E-01
0.3900000E+01	0.4310994E-01
0.4200000E+01	0.4164776E-01
0.4500000E+01	0.4015210E-01
0.4800000E+01	0.3864309E-01
0.5100000E+01	0.3713704E-01
0.5400000E+01	0.3564702E-01
0.5700000E+01	0.3418339E-01
0.6000000E+01	0.3275426E-01

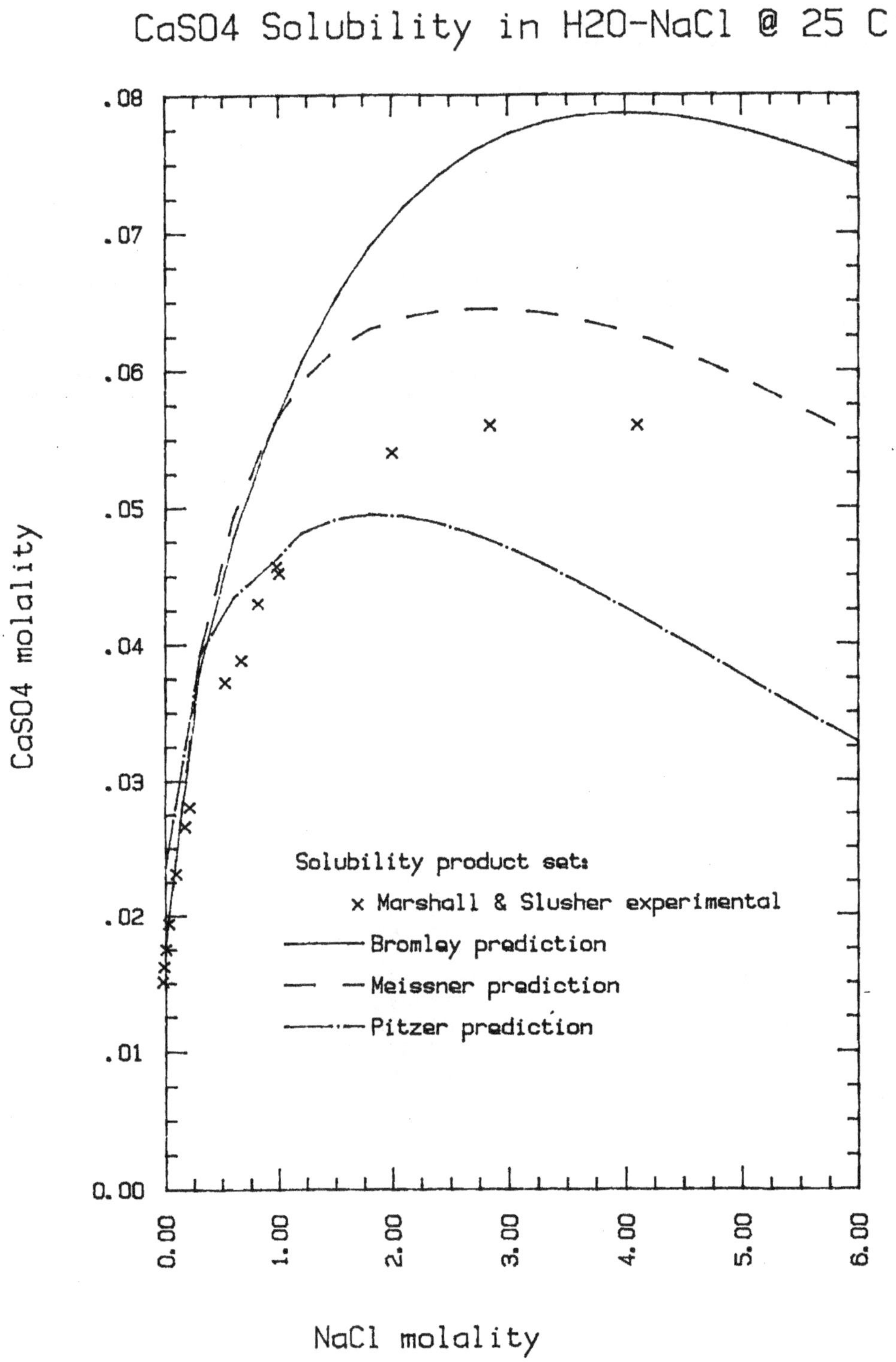
CaSO4 Solubility in H2O-NaCl @ 25 C
.08
.07
.06
.05
.04
.03
.02
.01
0.00
CaSO4 molality
0.00
1.00
2.00
3.00
4.00
5.00
6.00
NaCl molality
Solubility product set:
× Marshall & Slusher experimental
Bromley prediction
Meissner prediction
Pitzer prediction

H_2O - HCl - $CaSO_4$ EXPERIMENTAL DATA

From: W.F. Linke, A. Seidell, Solubilities of Inorganic and Metal-Organic Compounds, Vol. I, American Chemical Society, Washington, D.C. (1958) p. 664

Solubility of $CaSO_4$ in hydrochloric acid solutions at 25°C:

(Ryss and Nilus, 1955)

gms. per 100 gms. sat. sol.

HCl	$CaSO_4$	Solid Phase
3.56	1.50	$CaSO_4 . 2H_2O$
6.48	1.80	$CaSO_4 . 2H_2O$
9.15	1.85	$CaSO_4 . 2H_2O$
12.08	1.73	$CaSO_4 . 2H_2O$
15.09	1.56	$CaSO_4 . 2H_2O$
18.33	1.27	$CaSO_4 . 2H_2O$
21.15	1.04	$CaSO_4 . 2H_2O$
24.26	0.88	$CaSO_4 . 2H_2O$
26.73	0.73	$CaSO_4 . 2H_2O$

H_2O - HCl - $CaSO_4$ PARAMETERS

Bromley

c	a	B_{ca}
H	Cl	.1433
Ca		.0948
H	SO_4	.0606
Ca		-.0816

Meissner

c	a	q_{ca}
H	Cl	6.69
Ca		2.40
H	SO_4	0.0
Ca		0.0

Pitzer

c	a	β_0	β_1	β_2	C^{ϕ}
H	Cl	.1775	.2945	0.0	.0008
Ca		.3159	1.614	0.0	-.00034
H	SO_4	0.0	0.0	0.0	0.0
Ca		.2	2.65	-55.7	0.0

i	j	k	Θ_{ij}	Ψ_{ijk}
H	Ca	Cl	0.0	0.0
		SO_4		0.0
Cl	SO_4	H	.02	0.0
		Ca		0.0

SYSTEM: H2O - HCl - CaSO4

Temperature: 25.0 degrees Celsius

Method: Bromley

PURE SALT SOLUBILITIES:

solid phase	saturation molality	activity coefficient	water activity	solubility product
CaSO4.2H2O	0.178316E-01	0.367996E+00	0.999547E+00	0.430715E-04

THE SOLUBILITY OF CaSO4 IN A SOLUTION OF HCl:

Solid phase: CaSO4.2H2O

molality HCl	molality CaSO4
0.5000000E+00	0.3682898E-01
0.1000000E+01	0.4094136E-01
0.1500000E+01	0.4187926E-01
0.2000000E+01	0.4109280E-01
0.2500000E+01	0.3927926E-01
0.3000000E+01	0.3688341E-01
0.3500000E+01	0.3420185E-01
0.4000000E+01	0.3143071E-01
0.4500000E+01	0.2869636E-01
0.5000000E+01	0.2607692E-01
0.5500000E+01	0.2361749E-01
0.6000000E+01	0.2134086E-01
0.6500000E+01	0.1925503E-01
0.7000000E+01	0.1735848E-01
0.7500000E+01	0.1564354E-01
0.8000000E+01	0.1409969E-01
0.8500000E+01	0.1271401E-01
0.9000000E+01	0.1147312E-01
0.9500000E+01	0.1036378E-01
0.1000000E+02	0.9372898E-02

SYSTEM: H2O - HCl - CaSO4

Temperature: 25.0 degrees Celsius

Method: Meissner

PURE SALT SOLUBILITIES:

solid phase	saturation molality	activity coefficient	water activity	solubility product
CaSO4.2H2O	0.176093E-01	0.372611E+00	0.999543E+00	0.430715E-04

THE SOLUBILITY OF CaSO4 IN A SOLUTION OF HCl:

Solid phase: CaSO4.2H2O

molality HCl	molality CaSO4
0.5000000E+00	0.4582685E-01
0.1000000E+01	0.5533696E-01
0.1500000E+01	0.5949075E-01
0.2000000E+01	0.6133180E-01
0.2500000E+01	0.6214664E-01
0.3000000E+01	0.6247382E-01
0.3500000E+01	0.6246304E-01
0.4000000E+01	0.6209935E-01
0.4500000E+01	0.6137040E-01
0.5000000E+01	0.6034610E-01
0.5500000E+01	0.5916066E-01
0.6000000E+01	0.5795062E-01
0.6500000E+01	0.5681141E-01
0.7000000E+01	0.5579228E-01
0.7500000E+01	0.5491168E-01
0.8000000E+01	0.5417342E-01
0.8500000E+01	0.5357584E-01
0.9000000E+01	0.5311558E-01
0.9500000E+01	0.5278897E-01
0.1000000E+02	0.5259255E-01

SYSTEM: H2O - HCl - CaSO4

Temperature: 25.0 degrees Celsius

Method: Pitzer

PURE SALT SOLUBILITIES:

solid phase	saturation molality	activity coefficient	water activity	solubility product
CaSO4.2H2O	0.236208E-01	0.277887E+00	0.999438E+00	0.430715E-04

THE SOLUBILITY OF CaSO4 IN A SOLUTION OF HCl:

Solid phase: CaSO4.2H2O

molality HCl	molality CaSO4
0.5000000E+00	0.5752802E-01
0.1000000E+01	0.7411045E-01
0.1500000E+01	0.8585106E-01
0.2000000E+01	0.9394974E-01
0.2500000E+01	0.9923480E-01
0.3000000E+01	0.1023369E+00
0.3500000E+01	0.1037531E+00
0.4000000E+01	0.1038802E+00
0.4500000E+01	0.1030354E+00
0.5000000E+01	0.1014724E+00
0.5500000E+01	0.9939318E-01
0.6000000E+01	0.9695840E-01
0.6500000E+01	0.9429513E-01
0.7000000E+01	0.9150332E-01
0.7500000E+01	0.8866114E-01
0.8000000E+01	0.8582915E-01
0.8500000E+01	0.8305387E-01
0.9000000E+01	0.8037058E-01
0.9500000E+01	0.7780564E-01
0.1000000E+02	0.7537839E-01

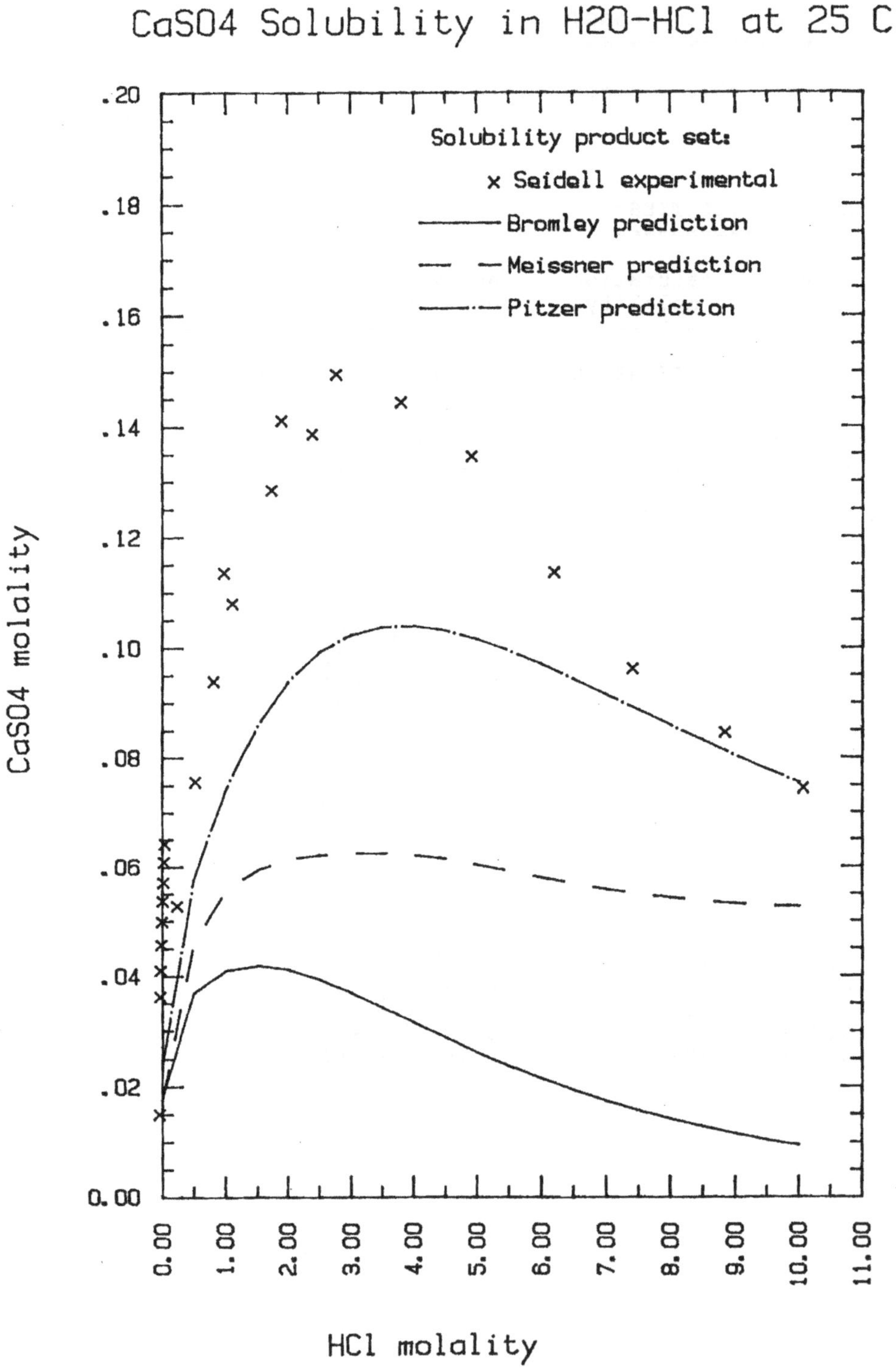
CaSO4 Solubility in H2O-HCl at 25 C
Solubility product set:
× Seidell experimental
Bromley prediction
Meissner prediction
Pitzer prediction
CaSO4 molality
HCl molality
0.00
.02
.04
.06
.08
.10
.12
.14
.16
.18
.20
0.00
1.00
2.00
3.00
4.00
5.00
6.00
7.00
8.00
9.00
10.00
11.00

H_2O - $CaCl_2$ - $CaSO_4$ EXPERIMENTAL DATA

From: W.F. Linke, A. Seidell, Solubilities of Inorganic and Metal-Organic Compounds, Vol. I, American Chemical Society, Washington, D.C. (1958) p. 669

Solubility of $CaSO_4$ in aqueous solutions of calcium chloride at 25°C:

(Sveshnikova, 1949; d'Ans, 1933*)

gms. per 100 gms. sat. sol.		
$CaCl_2$	$CaSO_4$	Solid Phase
0.16*	0.16	$CaSO_4 . 2H_2O$
0.27*	0.14	$CaSO_4 . 2H_2O$
0.44*	0.132	$CaSO_4 . 2H_2O$
2.99	0.112	$CaSO_4 . 2H_2O$
4.98*	0.103	$CaSO_4 . 2H_2O$
9.24*	0.086	$CaSO_4 . 2H_2O$
10.36*	0.091	$CaSO_4 . 2H_2O$
20.30	0.031	$CaSO_4 . 2H_2O$
26.51	0.018	$CaSO_4 . 2H_2O$
27.53	0.021	$CaSO_4 . 2H_2O$
33.87	0.011	$CaSO_4 . 2H_2O$
38.91	0.008	$CaSO_4 . 2H_2O$

H_2O - $CaCl_2$ - $CaSO_4$ PARAMETERS

Bromley

c	a	B_{ca}
Ca	Cl	.0948
	SO_4	-.0816

Meissner

c	a	q_{ca}
Ca	Cl	2.40
	SO_4	0.0

Pitzer

c	a	β_0	β_1	β_2	C^{ϕ}
Ca	Cl	.3159	1.614	0.0	-.00034
	SO_4	.2	2.65	-55.7	0.0

i	j	k	Θ_{ij}	Ψ_{ijk}
Cl	SO_4	Ca	.02	0.0

SYSTEM: H2O - CaCl2 - CaSO4

Temperature: 25.0 degrees Celsius

Method: Bromley

PURE SALT SOLUBILITIES:

solid phase	saturation molality	activity coefficient	water activity	solubility product
CaCl2.6H2O	0.746178E+01	0.197025E+02	0.277177E+00	0.576359E+04
CaSO4.2H2O	0.149938E-01	0.393840E+00	0.999610E+00	0.348435E-04

THE SOLUBILITY OF CaSO4 IN A SOLUTION OF CaCl2:

Solid phase: CaSO4.2H2O

molality CaCl2	molality CaSO4
0.3000000E+00	0.8858302E-02
0.6000000E+00	0.9330257E-02
0.9000000E+00	0.9715796E-02
0.1200000E+00	0.9969717E-02
0.1500000E+00	0.1013592E-01
0.1800000E+00	0.1025656E-01
0.2100000E+01	0.1036311E-01
0.2400000E+01	0.1047819E-01
0.2700000E+01	0.1061833E-01
0.3000000E+01	0.1079623E-01
0.3300000E+01	0.1102236E-01
0.3600000E+01	0.1130615E-01
0.3900000E+01	0.1165687E-01
0.4200000E+01	0.1208424E-01
0.4500000E+01	0.1259912E-01
0.4800000E+01	0.1321398E-01
0.5100000E+01	0.1394351E-01
0.5400000E+01	0.1480528E-01
0.5700000E+01	0.1582047E-01
0.6000000E+01	0.1701484E-01

SYSTEM: H2O - CaCl2 - CaSO4

Temperature: 25.0 degrees Celsius

Method: Meissner

PURE SALT SOLUBILITIES:

solid phase	saturation molality	activity coefficient	water activity	solubility product
CaCl2.6H2O	0.746178E+01	0.154431E+02	0.317055E+00	0.621736E+04
CaSO4.2H2O	0.149938E-01	0.396404E+00	0.999602E+00	0.352980E-04

THE SOLUBILITY OF CaSO4 IN A SOLUTION OF CaCl2:

Solid phase: CaSO4.2H2O

molality CaCl2	molality CaSO4
0.3000000E+00	0.7394397E-02
0.6000000E+00	0.5809538E-02
0.9000000E+00	0.4608385E-02
0.1200000E+00	0.3742726E-02
0.1500000E+00	0.3057656E-02
0.1800000E+00	0.2482420E-02
0.2100000E+01	0.2016681E-02
0.2400000E+01	0.1655903E-02
0.2700000E+01	0.1378357E-02
0.3000000E+01	0.1163159E-02
0.3300000E+01	0.9944247E-03
0.3600000E+01	0.8606737E-03
0.3900000E+01	0.7535700E-03
0.4200000E+01	0.6669991E-03
0.4500000E+01	0.5964302E-03
0.4800000E+01	0.5384702E-03
0.5100000E+01	0.4905521E-03
0.5400000E+01	0.4507154E-03
0.5700000E+01	0.4174480E-03
0.6000000E+01	0.3895741E-03

SYSTEM: H2O - CaCl2 - CaSO4
============================

Temperature: 25.0 degrees Celsius

Method: Pitzer

PURE SALT SOLUBILITIES:
=======================

solid phase	saturation molality	activity coefficient	water activity	solubility product
CaCl2.6H2O	0.746178E+01	0.348856E+02	0.238699E+00	0.130507E+05
CaSO4.2H2O	0.149938E-01	0.332037E+00	0.999627E+00	0.247669E-04

THE SOLUBILITY OF CaSO4 IN A SOLUTION OF CaCl2:
===

Solid phase: CaSO4.2H2O

molality CaCl2	molality CaSO4
0.3000000E+00	0.7870573E-02
0.6000000E+00	0.6693788E-02
0.9000000E+00	0.5493934E-02
0.1200000E+00	0.4351194E-02
0.1500000E+00	0.3357324E-02
0.1800000E+00	0.2543428E-02
0.2100000E+01	0.1902705E-02
0.2400000E+01	0.1411447E-02
0.2700000E+01	0.1041461E-02
0.3000000E+01	0.7661597E-03
0.3300000E+01	0.5629486E-03
0.3600000E+01	0.4137096E-03
0.3900000E+01	0.3044229E-03
0.4200000E+01	0.2244897E-03
0.4500000E+01	0.1660210E-03
0.4800000E+01	0.1232059E-03
0.5100000E+01	0.9179363E-04
0.5400000E+01	0.6868825E-04
0.5700000E+01	0.5164041E-04
0.6000000E+01	0.3901757E-04

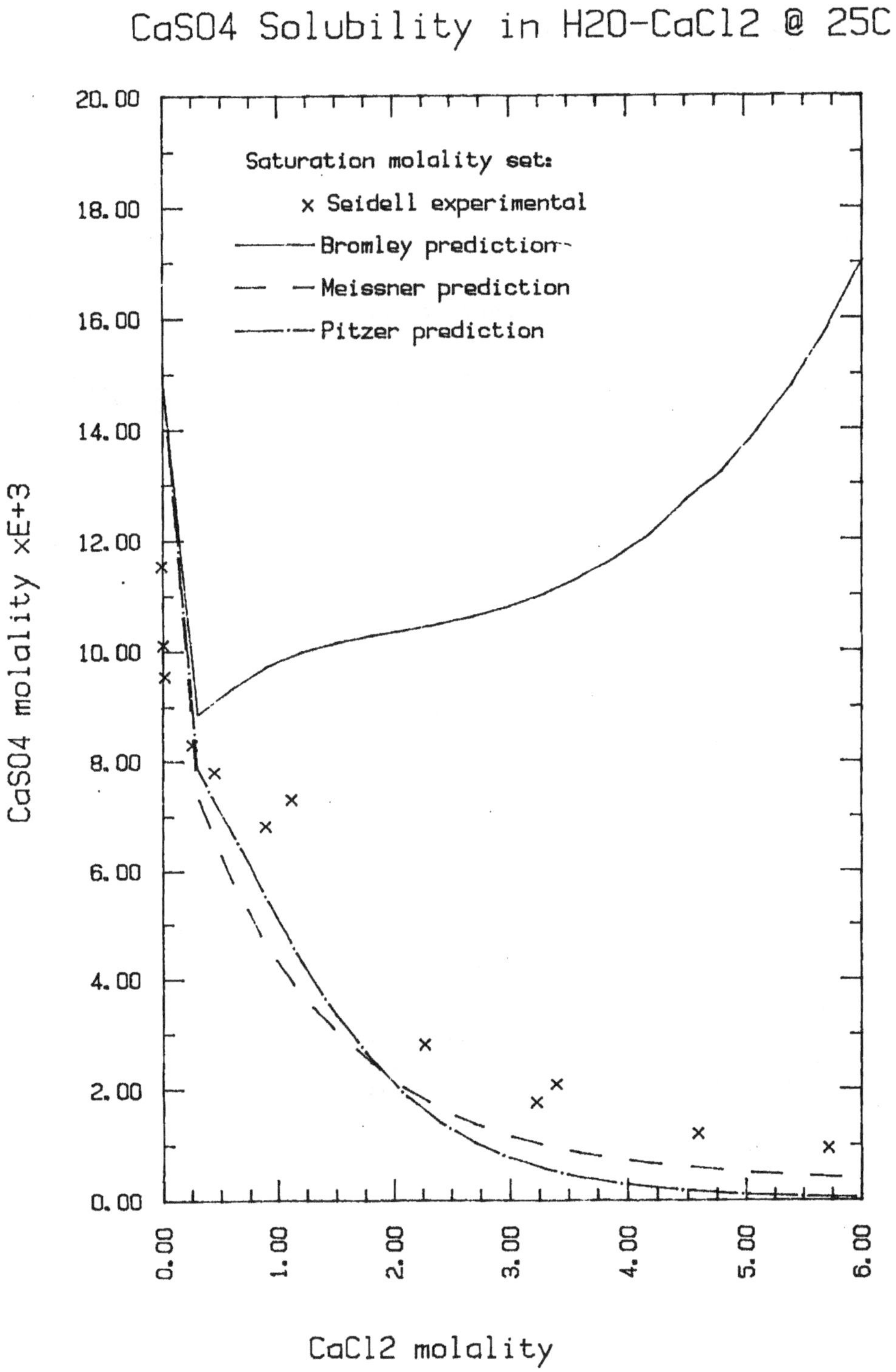
CaSO4 Solubility in H2O-CaCl2 @ 25C
Saturation molality set:
x Seidell experimental
Bromley prediction
Meissner prediction
Pitzer prediction
CaSO4 molality xE+3
CaCl2 molality
20.00
18.00
16.00
14.00
12.00
10.00
8.00
6.00
4.00
2.00
0.00
0.00
1.00
2.00
3.00
4.00
5.00
6.00

H_2O - Na_2SO_4 - $CaSO_4$ EXPERIMENTAL DATA

From: W.F. Linke, A. Seidell, Solubilities of Inorganic and Metal-Organic Compounds, Vol. I, American Chemical Society, Washington, D.C. (1958) p. 676

(Hill and Wills, 1938)

gms. per 100 gms. sat. sol.

Na_2SO_4	$CaSO_4$	Solid Phase
0.0	0.209	$CaSO_4.2H_2O$
0.595	0.148	$CaSO_4.2H_2O$
1.601	0.139	$CaSO_4.2H_2O$
3.200	0.144	$CaSO_4.2H_2O$
6.251	0.161	$CaSO_4.2H_2O$
10.93	0.181	$CaSO_4.2H_2O$
15.23	0.194	$CaSO_4.2H_2O$
18.09	0.197	$CaSO_4.2H_2O$
20.15	0.198	$CaSO_4.2H_2O$
21.75	0.197	$CaSO_4.2H_2O$ + $Na_2SO_4.10H_2O$
21.70	0.0	$Na_2SO_4.10H_2O$
21.72	0.120	$Na_2SO_4.10H_2O$
0.0	0.210	$CaSO_4.2H_2O$
0.521	0.154	$CaSO_4.2H_2O$
0.989	0.149	$CaSO_4.2H_2O$
1.521	0.148	$CaSO_4.2H_2O$
1.981	0.149	$CaSO_4.2H_2O$
3.908	0.164	$CaSO_4.2H_2O$
7.496	0.185	$CaSO_4.2H_2O$
14.05	0.214	$CaSO_4.2H_2O$
18.28	0.224	$CaSO_4.2H_2O$
20.07	0.228	$CaSO_4.2H_2O$
24.40	0.230	$CaSO_4.2H_2O$
26.93	0.230	$CaSO_4.2H_2O$
29.75	0.224	$CaSO_4.2H_2O$
31.77	0.216	$CaSO_4.2H_2O$ + Na_2SO_4
31.73	0.0	Na_2SO_4

H_2O - Na_2SO_4 - $CaSO_4$ PARAMETERS

Bromley

c	a	B_{ca}
Na	SO_4	-.0204
Ca		-.0816

Meissner

c	a	q_{ca}
Na	SO_4	-.19
Ca		0.0

Pitzer

c	a	β_0	β_1	β_2	C^{ϕ}
Na	SO_4	.01958	1.113	0.0	-.00497
Ca		.2	2.65	-55.7	0.0

i	j	k	Θ_{ij}	Ψ_{ijk}
Na	Ca	SO_4	.07	-.023

SYSTEM: H2O - Na2SO4 - CaSO4
==============================

Temperature: 25.0 degrees Celsius

Method: Bromley

PURE SALT SOLUBILITIES:
=======================

solid phase	saturation molality	activity coefficient	water activity	solubility product
Na2SO4.10H2O	0.195105E+01	0.149757E+00	0.940502E+00	0.544867E-01
CaSO4.2H2O	0.178316E-01	0.367996E+00	0.999547E+00	0.430715E-04

THE SOLUBILITY OF CaSO4 IN A SOLUTION OF Na2SO4:
==

Solid phase: CaSO4.2H2O

molality Na2SO4	molality CaSO4
0.1000000E+00	0.1406393E-01
0.2000000E+00	0.1609469E-01
0.3000000E+00	0.1862927E-01
0.4000000E+00	0.2149460E-01
0.5000000E+00	0.2469213E-01
0.6000000E+00	0.2825341E-01
0.7000000E+00	0.3222220E-01
0.8000000E+00	0.3665140E-01
0.9000000E+00	0.4160325E-01
0.1000000E+01	0.4715071E-01
0.1100000E+01	0.5337955E-01
0.1200000E+01	0.6039133E-01
0.1300000E+01	0.6830738E-01
0.1400000E+01	0.7727414E-01
0.1500000E+01	0.8747071E-01
0.1600000E+01	0.9911934E-01
0.1700000E+01	0.1125009E+00
0.1800000E+01	0.1279780E+00
0.1900000E+01	0.1460317E+00
0.2000000E+01	0.1673217E+00

SYSTEM: H2O - Na2SO4 - CaSO4

Temperature: 25.0 degrees Celsius

Method: Meissner

PURE SALT SOLUBILITIES:

solid phase	saturation molality	activity coefficient	water activity	solubility product
Na2SO4.10H2O	0.166697E+01	0.175298E+00	0.941675E+00	0.544867E-01
CaSO4.2H2O	0.176093E-01	0.372611E+00	0.999543E+00	0.430715E-04

INVARIANT POINT:

Solid phase: Na2SO4.10H2O + CaSO4.2H2O

	molalities	weight percents
Na2SO4	0.1675595E+01	19.175
CaSO4	0.2279859E-01	0.256
H2O		80.569

THE SOLUBILITY OF CaSO4 IN A SOLUTION OF Na2SO4:

Solid phase: CaSO4.2H2O

molality Na2SO4	molality CaSO4
0.1591815E+01	0.2252667E-01
0.1508036E+01	0.2224664E-01
0.1424256E+01	0.2195731E-01
0.1340476E+01	0.2165698E-01
0.1256696E+01	0.2134338E-01
0.1172917E+01	0.2101366E-01
0.1089137E+01	0.2066431E-01
0.1005357E+01	0.2029129E-01
0.9215773E+00	0.1988996E-01
0.8377975E+00	0.1945508E-01
0.7540178E+00	0.1898069E-01
0.6702380E+00	0.1845990E-01
0.5864583E+00	0.1788450E-01
0.5026785E+00	0.1724442E-01
0.4188988E+00	0.1652710E-01
0.3351190E+00	0.1571721E-01
0.2513393E+00	0.1479895E-01
0.1675595E+00	0.1377455E-01
0.8377975E-01	0.1282606E-01

THE SOLUBILITY OF Na2SO4 IN A SOLUTION OF CaSO4:
==

Solid phase: Na2SO4.10H2O

molality Na2SO4	molality CaSO4
0.1675166E+01	0.2165866E-01
0.1674734E+01	0.2051873E-01
0.1674306E+01	0.1937881E-01
0.1673876E+01	0.1823888E-01
0.1673446E+01	0.1709895E-01
0.1673016E+01	0.1595902E-01
0.1672583E+01	0.1481909E-01
0.1672154E+01	0.1367916E-01
0.1671723E+01	0.1253923E-01
0.1671292E+01	0.1139930E-01
0.1670863E+01	0.1025937E-01
0.1670432E+01	0.9119438E-02
0.1670000E+01	0.7979508E-02
0.1669568E+01	0.6839578E-02
0.1669136E+01	0.5699649E-02
0.1668703E+01	0.4559719E-02
0.1668271E+01	0.3419789E-02
0.1667838E+01	0.2279859E-02
0.1667405E+01	0.1139930E-02

SYSTEM: H2O - Na2SO4 - CaSO4

Temperature: 25.0 degrees Celsius

Method: Pitzer

PURE SALT SOLUBILITIES:

solid phase	saturation molality	activity coefficient	water activity	solubility product
Na2SO4.10H2O	0.182718E+01	0.160313E+00	0.940375E+00	0.544867E-01
CaSO4.2H2O	0.236208E-01	0.277887E+00	0.999438E+00	0.430715E-04

INVARIANT POINT:

Solid phase: Na2SO4.10H2O + CaSO4.2H2O

	molalities	weight percents
Na2SO4	0.1897005E+01	21.172
CaSO4	0.2293324E-01	0.251
H2O		78.577

THE SOLUBILITY OF CaSO4 IN A SOLUTION OF Na2SO4:

Solid phase: CaSO4.2H2O

molality Na2SO4	molality CaSO4
0.1802155E+01	0.2314265E-01
0.1707305E+01	0.2332771E-01
0.1612454E+01	0.2348463E-01
0.1517604E+01	0.2360933E-01
0.1422754E+01	0.2369742E-01
0.1327904E+01	0.2374422E-01
0.1233053E+01	0.2374478E-01
0.1138203E+01	0.2369395E-01
0.1043353E+01	0.2358649E-01
0.9485025E+00	0.2341725E-01
0.8536523E+00	0.2318137E-01
0.7588020E+00	0.2287495E-01
0.6639518E+00	0.2249588E-01
0.5691015E+00	0.2204583E-01
0.4742513E+00	0.2153441E-01
0.3794010E+00	0.2098982E-01
0.2845508E+00	0.2049149E-01
0.1897005E+00	0.2030571E-01
0.9485025E-01	0.2185690E-01

THE SOLUBILITY OF Na2SO4 IN A SOLUTION OF CaSO4:
==

Solid phase: Na2SO4.10H2O

molality Na2SO4	molality CaSO4
0.1895963E+01	0.2178658E-01
0.1894918E+01	0.2063992E-01
0.1893871E+01	0.1949326E-01
0.1892821E+01	0.1834659E-01
0.1891769E+01	0.1719993E-01
0.1890715E+01	0.1605327E-01
0.1889659E+01	0.1490661E-01
0.1888599E+01	0.1375995E-01
0.1887538E+01	0.1261328E-01
0.1886474E+01	0.1146662E-01
0.1885407E+01	0.1031996E-01
0.1884338E+01	0.9173297E-02
0.1883267E+01	0.8026635E-02
0.1882193E+01	0.6879973E-02
0.1881116E+01	0.5733311E-02
0.1880037E+01	0.4586649E-02
0.1878956E+01	0.3439986E-02
0.1877871E+01	0.2293324E-02
0.1876785E+01	0.1146662E-02

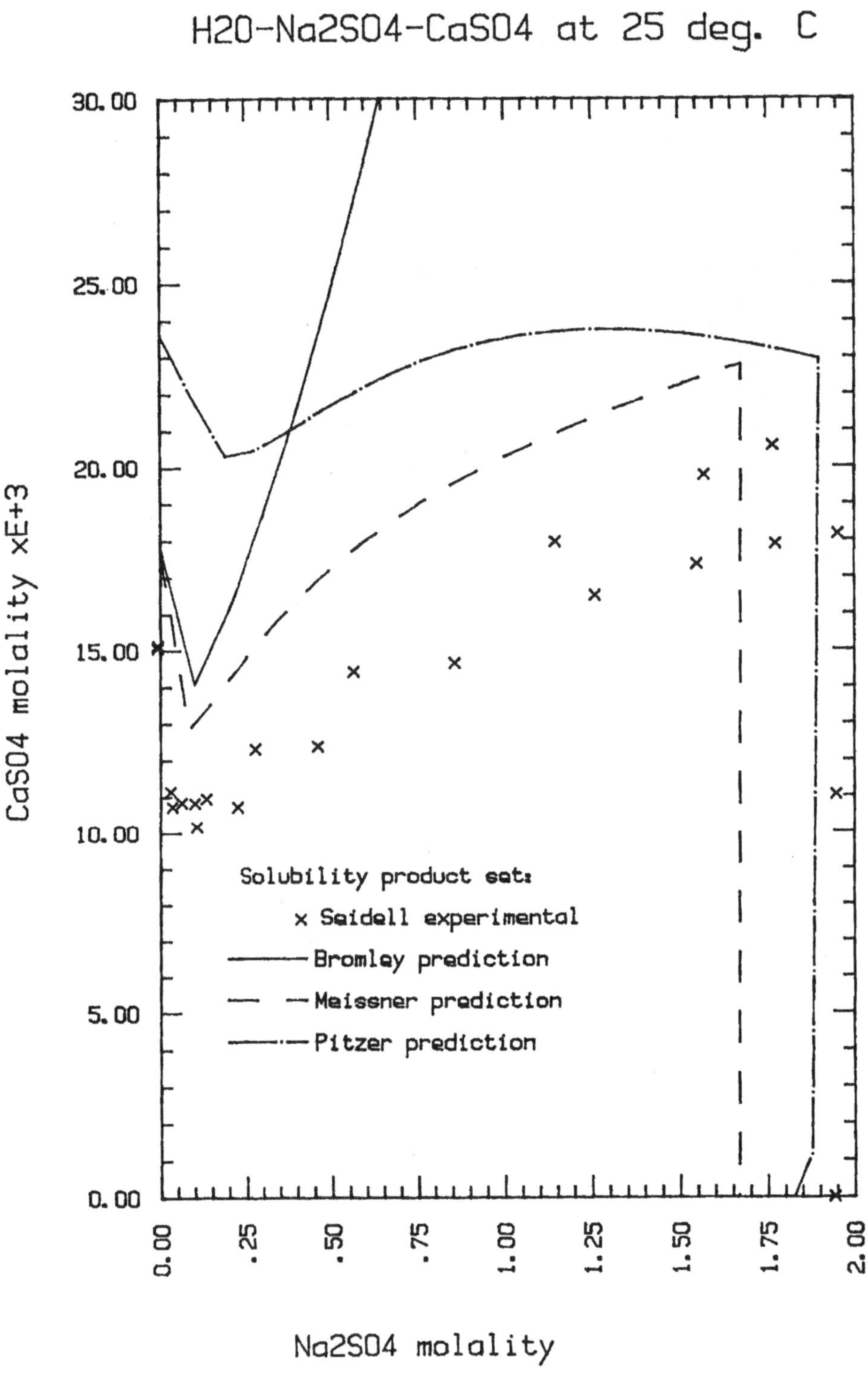
H2O-Na2SO4-CaSO4 at 25 deg. C
CaSO4 molality xE+3
30.00
25.00
20.00
15.00
10.00
5.00
0.00
0.00
.25
.50
.75
1.00
1.25
1.50
1.75
2.00
Na2SO4 molality
Solubility product set:
x Seidell experimental
Bromley prediction
Meissner prediction
Pitzer prediction

H_2O - $MgSO_4$ - $CaSO_4$ EXPERIMENTAL DATA

From: W.F. Linke, A. Seidell, Solubilities of Inorganic and Metal-Organic Compounds, Vol. I, American Chemical Society, Washington, D.C. (1958) p. 673

Solubility of $CaSO_4$ in aqueous solution of magnesium sulfate at 25°C:

(Cameron and Bell, 1906)

sp. gr. of sol. at $\frac{25°}{25}$	gms. per liter of sol. $MgSO_4$	$CaSO_4$	Solid Phase
1.0032	0.0	2.046	$CaSO_4.2H_2O$
1.0055	3.20	1.620	$CaSO_4.2H_2O$
1.0090	6.39	1.507	$CaSO_4.2H_2O$
1.0118	10.64	1.471	$CaSO_4.2H_2O$
1.0226	21.36	1.478	$CaSO_4.2H_2O$
1.0419	42.68	1.558	$CaSO_4.2H_2O$
1.0626	64.14	1.608	$CaSO_4.2H_2O$
1.0833	85.67	1.617	$CaSO_4.2H_2O$
1.1190	128.28	1.627	$CaSO_4.2H_2O$
1.1377	149.67	1.597	$CaSO_4.2H_2O$
1.1479	165.7	1.549	$CaSO_4.2H_2O$
1.1537	171.2	1.474	$CaSO_4.2H_2O$
1.1813	198.9	1.422	$CaSO_4.2H_2O$
1.2095	232.1	1.254	$CaSO_4.2H_2O$
1.2382	265.6	1.070	$CaSO_4.2H_2O$
1.2624	298.0	0.860	$CaSO_4.2H_2O$
1.2877	330.6	0.647	$CaSO_4.2H_2O$
1.3023	355.0	0.501	$CaSO_4.2H_2O$

(Harkins and Paine, 1919)

$d\frac{25}{4}$ of sat. sol.	gms. per 100 gms. sat. sol. $MgSO_4$	$CaSO_4$	Solid Phase
0.99911	0.0	0.20854	$CaSO_4.2H_2O$
0.99960	0.060294	0.19565	$CaSO_4.2H_2O$
1.0001	0.12167	0.1848	$CaSO_4.2H_2O$
1.00067	0.18339	0.1777	$CaSO_4.2H_2O$

H_2O - $MgSO_4$ - $CaSO_4$ PARAMETERS

Bromley

c	a	B_{ca}
Mg	SO_4	-.0153
Ca		-.0816

Meissner

c	a	q_{ca}
Mg	SO_4	0.15
Ca		0.0

Pitzer

c	a	β_0	β_1	β_2	C^{ϕ}
Mg	SO_4	.221	3.434	-37.25	.025
Ca		.2	2.65	-55.7	0.0

i	j	k	Θ_{ij}	ψ_{ijk}
Mg	Ca	SO_4	.007	0.5

SYSTEM: H2O - MgSO4 - CaSO4

Temperature: 25.0 degrees Celsius

Method: Bromley

PURE SALT SOLUBILITIES:

solid phase	saturation molality	activity coefficient	water activity	solubility product
MgSO4.7H2O	0.168665E+02	0.145558E-02	0.142639E+01	0.722565E-02
CaSO4.2H2O	0.178316E-01	0.367996E+00	0.999547E+00	0.430715E-04

THE SOLUBILITY OF CaSO4 IN A SOLUTION OF MgSO4:

Solid phase: CaSO4.2H2O

molality MgSO4	molality CaSO4
0.1500000E+00	0.1815975E-01
0.3000000E+00	0.2149233E-01
0.4500000E+00	0.2552271E-01
0.6000000E+00	0.3031177E-01
0.7500000E+00	0.3594680E-01
0.9000000E+00	0.4254791E-01
0.1050000E+01	0.5027423E-01
0.1200000E+01	0.5933150E-01
0.1350000E+01	0.6998446E-01
0.1500000E+01	0.8257644E-01
0.1650000E+01	0.9756032E-01
0.1800000E+01	0.1155489E+00
0.1950000E+01	0.1374021E+00
0.2100000E+01	0.1643797E+00
0.2250000E+01	0.1984612E+00
0.2400000E+01	0.2430678E+00
0.2550000E+01	0.3051290E+00
0.2700000E+01	0.4039055E+00

SYSTEM: H2O - MgSO4 - CaSO4
==============================

Temperature: 25.0 degrees Celsius

Method: Meissner

PURE SALT SOLUBILITIES:
========================

solid phase	saturation molality	activity coefficient	water activity	solubility product
MgSO4.7H2O	0.291189E+01	0.372594E-01	0.932559E+00	0.722565E-02
CaSO4.2H2O	0.176093E-01	0.372611E+00	0.999543E+00	0.430715E-04

THE SOLUBILITY OF CaSO4 IN A SOLUTION OF MgSO4:
==

Solid phase: CaSO4.2H2O

molality MgSO4	molality CaSO4
0.1500000E+00	0.1727295E-01
0.3000000E+00	0.1888455E-01
0.4500000E+00	0.1972418E-01
0.6000000E+00	0.2014703E-01
0.7500000E+00	0.2033373E-01
0.9000000E+00	0.2037149E-01
0.1050000E+01	0.2030290E-01
0.1200000E+01	0.2015433E-01
0.1350000E+01	0.1994963E-01
0.1500000E+01	0.1971061E-01
0.1650000E+01	0.1945391E-01
0.1800000E+01	0.1919003E-01
0.1950000E+01	0.1892508E-01
0.2100000E+01	0.1866271E-01
0.2250000E+01	0.1840528E-01
0.2400000E+01	0.1815434E-01
0.2550000E+01	0.1791090E-01
0.2700000E+01	0.1767562E-01
0.2850000E+01	0.1744883E-01
0.3000000E+01	0.1723071E-01

SYSTEM: H2O - MgSO4 - CaSO4

Temperature: 25.0 degrees Celsius

Method: Pitzer

PURE SALT SOLUBILITIES:

solid phase	saturation molality	activity coefficient	water activity	solubility product
MgSO4.7H2O	0.215986E+01	0.475779E-01	0.947195E+00	0.722565E-02
CaSO4.2H2O	0.236208E-01	0.277887E+00	0.999438E+00	0.430715E-04

THE SOLUBILITY OF CaSO4 IN A SOLUTION OF MgSO4:

Solid phase: CaSO4.2H2O

molality MgSO4	molality CaSO4
0.1500000E+00	0.3516163E-01
0.3000000E+00	0.3592368E-01
0.4500000E+00	0.3712192E-01
0.6000000E+00	0.3809403E-01
0.7500000E+00	0.3865498E-01
0.9000000E+00	0.3874520E-01
0.1050000E+01	0.3836376E-01
0.1200000E+01	0.3754318E-01
0.1350000E+01	0.3633562E-01
0.1500000E+01	0.3480375E-01
0.1650000E+01	0.3301437E-01
0.1800000E+01	0.3103387E-01
0.1950000E+01	0.2892512E-01
0.2100000E+01	0.2674535E-01
0.2250000E+01	0.2454496E-01
0.2400000E+01	0.2236688E-01
0.2550000E+01	0.2024648E-01
0.2700000E+01	0.1821176E-01
0.2850000E+01	0.1628390E-01
0.3000000E+01	0.1447783E-01

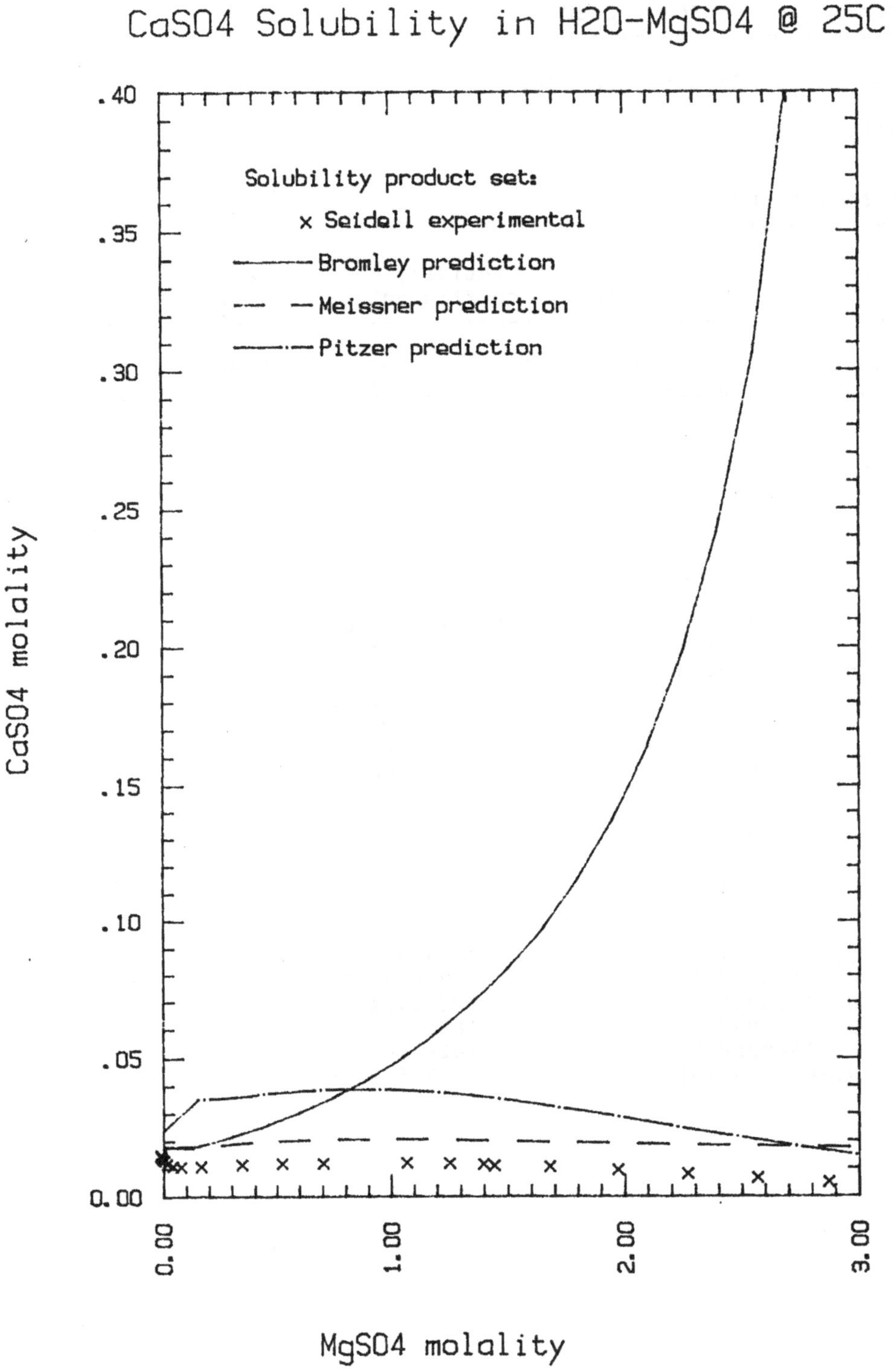
CaSO4 Solubility in H2O-MgSO4 @ 25C
Solubility product set:
x Seidell experimental
Bromley prediction
Meissner prediction
Pitzer prediction
CaSO4 molality
MgSO4 molality
.40
.35
.30
.25
.20
.15
.10
.05
0.00
0.00
1.00
2.00
3.00

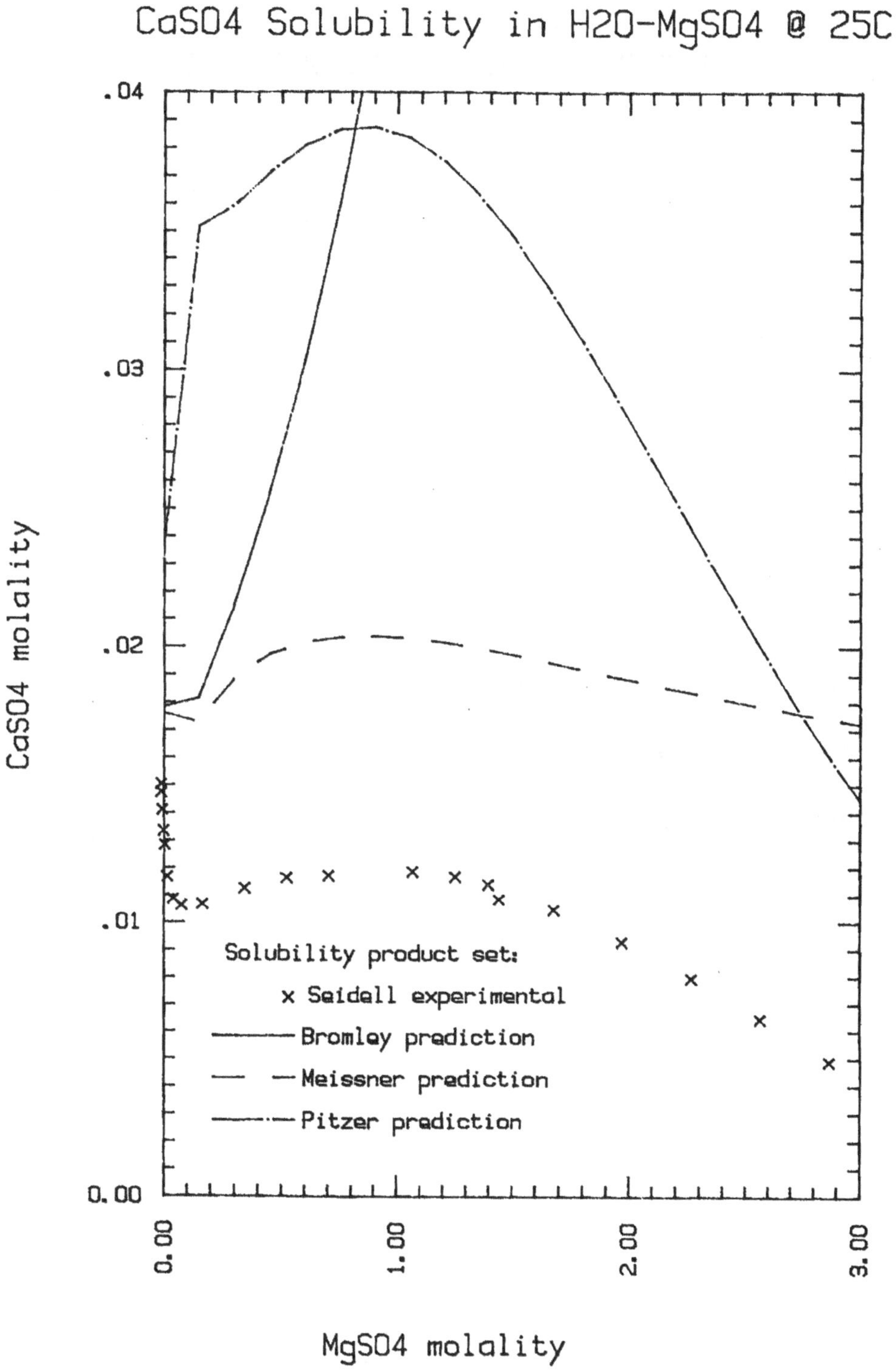
CaSO4 Solubility in H2O-MgSO4 @ 25C
.04
.03
.02
.01
0.00
CaSO4 molality
0.00
1.00
2.00
3.00
MgSO4 molality
Solubility product set:
x Seidell experimental
Bromley prediction
Meissner prediction
Pitzer prediction

APPENDIX 5.1

VALUES FOR PITZER'S θ AND ψ PARAMETERS

From: K.S. Pitzer, "Thermodynamics of Aqueous Electrolytes at Various Temperatures, Pressures, and Compositions", Thermodynamics of Aqueous Systems with Industrial Applications, Stephen A. Newman, ed., ACS Symposium Series 133 (1980)

Table 1: Parameters for mixed electrolytes with the virial coefficient equations (at 25°C)

i	j	k	Θ_{ij}	ψ_{ijk}
Na	K	Cl	-.012	-.0018
		SO_4		-.010
Na	Mg	Cl	.07	-.012
		SO_4		-.015
Na	Ca	Cl	.07	-.014
		SO_4		-.023
K	Mg	Cl	.0	-.022
		SO_4		-.048
K	Ca	Cl	.032	-.025
		SO_4		0
Mg	Ca	Cl	.007	-.012
		SO_4		.05
Cl	SO_4	Na	.02	.0014
		K		0
		Mg		-.004
		Ca		0

Table 2: Parameters for the virial coefficient equations at 25°C

C	A	β_0	β_1	β_2	C^{ϕ}
Ca	Cl	.31590	1.6140	--	-.00034
Ca	SO_4	.20000	2.650	-57.70	0
K	Cl	.04835	.2122	--	-.00084
K	SO_4	.04995	.7793	--	0
Mg	Cl	.35235	1.6815	--	.00519
Mg	SO_4	.22100	3.3430	-37.25	.025
Na	Cl	.07650	.2264	--	.00127
Na	SO_4	.01958	1.1130	--	.00497

EFFECTS OF HIGHER-ORDER ELECTROSTATIC TERMS

As shown in Chapter V, Pitzer included terms to account for the electrostatic effects of unsymmetrical mixing in his method. He stated (P5) that, while the effect on 1-2 mixing is small and may be safely omitted, it should be included when doing calculations for 3-1 and greater unsymmetrical types of mixing.

H.L. Friedman (12), in 1962, proposed that there is a higher-order electrostatic effect, or limiting law, even when mixing ions of the same charge. Pitzer presented Friedmans' equations for Gibbs excess free energy in a molality basis and omitted terms occurring due to differences in concentration, molality, etc. as these are dealt with by Pitzer's virial coefficients:

$$\frac{G^{ex}}{RT} = -4A_\phi I^{3/2} + \sum_i \sum_j m_i m_j \frac{z_i z_j}{4I} J_{ij} \qquad \text{(A.5.1)}$$

where: A_ϕ - Debye-Hückel coefficient for the osmotic coefficient

$$J_{ij} = \frac{K^2}{z_i z_j l} \int_0^\infty \left(\exp q_{ij} - \frac{u_{ij}}{kT} - 1 - q_{ij} - \tfrac{1}{2} q_{ij}^2\right) r^2\, dr$$

$$q_{ij} = -\frac{z_i z_j l}{r} \exp(-Kr)$$

$$l = \frac{e^2}{DkT}$$

e - electronic charge
D - dielectric constant
k - Boltzmann's constant
T - temperature

$$K^2 = \frac{8\pi e^2 N_A d_0}{1000\, DkT} I \qquad \text{(4.26a)}$$

r - interionic distance

u_{ij} - short range potential

$$\text{potential } v_{ij} = u_{ij} + \frac{z_i z_j l}{r}$$

Pitzer used the following expression to define the excess free energy:

$$\frac{G^{ex}}{RT} = f(I) + \sum_i \sum_j m_i m_j \lambda_{ij}(I) \qquad \text{(4.57)}$$

$$f(I) = -A_\phi \frac{4I}{b} \ln(1 + b\sqrt{I})$$

The first terms on the right of equations (A.5.1) and (4.57) are expressions based on the Debye-Hückel limiting law, and Pitzer notes that the coefficients λ_{ij} correspond to the J_{ij} terms.

The θ parameter of Pitzer's equation was defined in terms of λ:

$$\theta_{ij} = \lambda_{ij} - \frac{z_j}{2z_i}\lambda_{ii} - \frac{z_i}{2z_j}\lambda_{jj} \qquad \text{(A.5.2)}$$

for ions i and j of the same charge. Since they are of the same charge the ions are assumed to seldom approach one another due to electrostatic repulsion; thus the short range potential, u_{ij}, is negligible.

Pitzer also defined the Θ parameters:

$$\Theta_{ij} = \theta_{ij} + {}^E\theta_{ij}(I) \qquad \text{(5.33a)}$$

$$\Theta'_{ij} = {}^E\theta'_{ij}(I) \qquad \text{(5.33b)}$$

The E term is then:

$$^E\theta_{ij} = {}^E\lambda_{ij} - \frac{z_j}{2z_i}\,{}^E\lambda_{ii} - \frac{z_i}{2z_j}\,{}^E\lambda_{jj}$$

$$E\lambda_{ij} = \frac{z_i z_j}{4I} J_{ij}$$

This J_{ij}, with $u_{ij} = 0$, is then:

$$J_{ij} = \frac{K^2}{z_i z_j I} \int_0^\infty \{1 + q_{ij} + \tfrac{1}{2} q_{ij}^2 - \exp(q_{ij})\}\, r^2\, dr \qquad \text{(A.5.3)}$$

The appearance of this expression for J was simplified to:

$$J(x) = x^{-1} \int_0^\infty \{1 - q + \tfrac{1}{2}q^2 - \exp(q)\}\; y^2\, dy \qquad \text{(A.5.4)}$$

by substituting:

$y = Kr$

$x = z_i z_j I$

$q = -(x/y)\exp(-y)$

By integrating the second and third terms of (A.5.4):

$$J(x) = \tfrac{1}{4}x - 1 + J_2(x) \qquad \text{(A.5.5)}$$

with: $J_2(x) = x^{-1} \int_0^\infty (1 - \exp(q))\, y^2\, dy$ (A.5.5a)

Defining:

$$x_{ij} = 6\ z_i\ z_j\ A_\phi\ \sqrt{I}$$

E_θ and $E_{\theta'}$ are then:

$$E_{\theta_{ij}} = \frac{z_i z_j}{4I} \{J(x_{ij}) - \tfrac{1}{2}J(x_{ii}) - \tfrac{1}{2}J(x_{jj})\} \qquad \text{(A.5.6a)}$$

$$E_{\theta'_{ij}} = -(E_{\theta_{ij}}/I) + \frac{z_i z_j}{8I^2} \{x_{ij}\ J'(x_{ij}) - \tfrac{1}{2}x_{ii}\ J'(x_{ii}) - \tfrac{1}{2}x_{jj}\ J'(x_{jj})\} \qquad \text{(A.5.6b)}$$

where: $J'(x) = \frac{dJ}{dx} = \tfrac{1}{4} - \frac{J_2}{x} + J_3$

$$J_3 = x^{-1} \int_0^\infty \exp(-y)\ \exp(q)\ y\ dy$$

for any ionic strength.

For discussion in greater depth about these higher-order electrostatic terms, refer to the following source:

K.S. Pitzer, "Thermodynamics of Electrolytes. V. Effects of Higher-Order Electrostatic Terms", J. Solution Chem., 4, 249 (1975)

From: K.S. Pitzer, "Theory: Ion Interaction Approach", <u>Activity Coefficients in Electrolyte Solutions</u>, Ricardo M. Pytkowicz, ed., CRC Press, Boca Raton, Fla. (1979) pp. 191, 192

Table 3: Parameters for binary mixtures with a common ion at 25°C

i	j	k	θ_{ij}	ψ_{ijk}	Max. I
Ca	Co	Cl	-.055	.013	7.5
Cs	Ba	Cl	-.150	0.0	4.
H	Ba	Cl	-.036	0.024	3.
H	Bu_4N	Br	-.22	---	1.
H	Cs	Cl	-.044	-.019	3.
H	Et_4N	Cl	0.0	---	.1
H	K	Br	.005	-.021	3.
		Cl		-.007	3.5
H	Li	Br	.015	0.0	2.5
		Cl		0.0	5.
		ClO_4		-.0017	4.5
H	Me_4N	Cl	0.0	---	.1
H	Mn	Cl	0.0	0.0	3.
H	Na	Br	.036	-.012	3.
		Cl		-.004	3.
		ClO_4		-.016	5.
H	NH_4	Br	-.019	0.0	3.
		Cl		0.0	2.
H	Pr_4N	Br	-.170	-.15	2.
H	Sr	Cl	-.02	.018	8.
K	Ba	Cl	-.072	0.0	5.
K	Ca	Cl	-.040	-.015	5
K	Cs	Cl	0.0	-.0013	5
Li	Ba	Cl	-.070	0.019	4.3
Li	Cs	Cl	-.095	-.0094	5
Li	K	Cl	-.022	-.010	4.8

i	j	k	Θ_{ij}	ψ_{ijk}	Max. I
Li	Na	Cl	.012	-.003	6.
		ClO_4		-.008	2.6
		NO_3		-.0072	6.
		OAc		-.0043	3.5
Mg	Ca	Cl	.01	0.0	7.7
Na	Ba	Cl	-.003	0.0	5.
Na	Ca	Cl	0.0	0.0	8.
Na	Co	Cl	-.016	-.001	4.
Na	Cs	Cl	-.033	-.003	5.
Na	Cu	Cl	0.0	-.014	7.3
		SO_4		-.011	5.5
Na	K	Br	-.012	-.0022	4.
		Cl		-.0018	4.8
		NO_3		-.0012	3.3
		SO_4		-.010	3.6
Na	Mg	Cl	0.0	0.0	5.9
		SO_4		0.0	9.
Na	Mn	Cl	0.0	-.003	5.5
Na	Zn	Br	0.0	---	.4
Br	OH	K	-.065	-.014	3.
		Na		-.018	3.
Cl	Br	K	0.0	0.0	4.4
		Na		0.0	4.4
Cl	F	Na	---	---	1.
Cl	H_2PO_4	K	.1	-.01	1.
		Na		0.0	1.
Cl	NO_3	Ca	.016	-.017	6.
		K		-.006	4.
		Li		-.003	6.
		Mg		0.0	4.
		Na		-.006	5.
Cl	OH	K	-.050	-.008	3.5
		Na		-.006	3.

i	j	k	Θ_{ij}	ψ_{ijk}	Max. I
Cl	SO_4	Cu	-.02	.043	7.2
		K		-.007	2.3
		Mg		-.007	7.
		Na		.004	9.

REFERENCES

1. Guggenheim, E.A. and J.C. Turgeon, Trans. Faraday Soc., 51, 747 (1955)

2. Guggenheim, E.A., Thermodynamics, 5th ed., North-Holland Publishing Company, Amsterdam (1967)

3. Linke, W.F. and A. Seidell, Solubilities of Inorganic and Metal-Organic Compounds, Am. Chem. Soc., Washington, D.C., Vol. I (1958), Vol. II (1965)

BROMLEY REFERENCES

B1. Bromley, L.A., "Approximate individual ion values of β (or B) in extended Debye-Hückel theory for uni-univalent aqueous solutions at 298.15 K", J. Chem. Thermo., 4, 669 (1972)

B2. Lewis, G.N., M. Randall, K.S. Pitzer and L. Brewer, Thermodynamics, 2nd ed., McGraw-Hill: New York (1961)

B3. Bromley, L.A., "Thermodynamic Properties of Strong Electrolytes in Aqueous Solutions", AIChE J., 19, 313 (1973)

B4. Bromley, L.A., "Thermodynamic Properties of Sea Salt Solutions", AIChE J., 20, 326 (1974)

MEISSNER REFERENCES

M1. Meissner, H.P. and J.W. Tester, "Activity Coefficients of Strong Electrolytes in Aqueous Solutions", Ind. Eng. Chem. Proc. Des. Dev., 11, 128 (1972)

M2. Meissner, H.P. and C.L. Kusik, "Activity Coefficients of Strong Electrolytes in Multicomponent Aqueous Solutions", AIChE J., 18, 294 (1972)

M3. Meissner, H.P., C.L. Kusik and J.W. Tester, "Activity Coefficients of Strong Electrolytes in Aqueous Solution - Effect of Temperature", AIChE J., 18, 661 (1972)

M4. Meissner, H.P. and C.L. Kusik, "Vapor Pressures of Water Over Aqueous Solutions of Strong Electrolytes", Ind. Eng. Chem. Proc. Des. Dev., 12, 112 (1973)

M5. Meissner, H.P. and C.L. Kusik, "Aqueous Solutions of Two or More Strong Electrolytes - Vapor Pressures and Solubilities", Ind. Eng. Chem. Proc. Des. Dev., 12, 205 (1973)

M6. Meissner, H.P. and N.A. Peppas, "Activity Coefficients - Aqueous Solutions of Polybasic Acids and Their Salts", AIChE J., 19, 806 (1973)

M7. Meissner, H.P. and C.L. Kusik, "Electrolyte Activity Coefficients in Inorganic Processing", AIChE Symposium Series 173, 74, 14 (1978)

M8. Meissner, H.P. and C.L. Kusik, "Double Salt Solubilities", Ind. Eng. Chem. Proc. Des. Dev., 18, 391 (1979)

M9. Meissner, H.P., C.L. Kusik and E.L. Field, "Estimation of Phase Diagrams and Solubilities for Aqueous Multi-ion Systems", AIChE J., 25, 759 (1979)

M10. Meissner, H.P. "Prediction of Activity Coefficients of Strong Electrolytes in Aqueous Systems", Thermodynamics of Aqueous Systems with Industrial Applications, ACS Symposium Series 133, 496 (1980)

PITZER REFERENCES

P1. Pitzer, K. S., "Thermodynamics of Electrolytes. I. Theoretical Basis and General Equations", J. Phys. Chem., 77, 268 (1973)

P2. Pitzer, K.S. and G. Mayorga, "Thermodynamics of Electrolytes. II. Activity and Osmotic Coefficients for Strong Electrolytes with One or Both Ions Univalent", J. Phys. Chem., 77, 2300 (1973)

P3. Pitzer, K.S. and G. Mayorga, "Thermodynamics of Electrolytes. III. Activity and Osmotic Coefficients for 2-2 Electrolytes", J. Sol. Chem., 3, 539 (1974)

P4. Pitzer, K.D. and J.J. Kim, "Thermodynamics of Electrolytes. IV. Activity and Osmotic Coefficients for Mixed Electrolytes", J. A.C.S., 96, 5701 (1974)

P5. Pitzer, K.S., "Thermodynamics of Electrolytes. V. Effects of Higher-Order Electrostatic Terms", J. Sol. Chem., 4, 249 (1975)

P6. Pitzer, K.S. and L.F. Silvester, "Thermodynamics of Electrolytes. VI. Weak Electrolytes Including H_3PO_4", J. Sol. Chem., 5, 269 (1976)

P7. Pitzer, K.S., R.N. Roy and L.F. Silvester, "Thermodynamics of Electrolytes. 7. Sulfuric Acid", J. A.C.S., 99, 4930 (1977)

P8. Pitzer, K.S. and L.F. Silvester, "Thermodynamics of Electrolytes. 8. High-Temperature Properties, Including Enthalpy and Heat Capacity, with Application to Sodium Chloride", J. Phys. Chem., 81, 1822 (1977)

P9. Pitzer, K.S., J.F. Peterson and L.F. Silvester, "Thermodynamics of Electrolytes. IX. Rare Earth Chlorides, Nitrates, and Perchlorates", J. Sol. Chem., 7, 45 (1978)

P10. Pitzer, K.S. and L.F. Silvester, "Thermodynamics of Electrolytes. X. Enthalpy and the Effect of Temperature on the Activity Coefficients", J. Sol. Chem., 7, 327 (1978)

P11. Pitzer, K.S. and L.F. Silvester, "Thermodynamics of Electrolytes. 11. Properties of 3:2, 4:2, and Other High-Valence Types", J. Phys. Chem., 82, 1239 (1978)

P12. Pitzer, K.S. and D.J. Bradley, "Thermodynamics of Electrolytes. 12. Dielectric Properties of Water and Debye-Hückel Parameters to 350°C and 1 kbar", J. Phys. Chem., 83, 1599 (1979)

P13. Downes, C.J. and K.S. Pitzer, "Thermodynamics of Electrolytes. Binary Mixtures Formed from Aqueous NaCl, Na_2SO_4, $CuCl_2$, and $CuSO_4$ at 25°C", J. Sol. Chem., 5, 389 (1976)

P14. Pitzer, K.S., "Theory: Ion Interaction Approach", Activity Coefficients in Electrolyte Solutions, vol. 1, Ricardo M. Pytkowicz, ed., CRC Press, Boca Raton, Fla. (1979)

P15. Pitzer, K.S., "Thermodynamics of Aqueous Electrolytes at Various Temperatures, Pressures, and Compositions", Thermodynamics of Aqueous Systems with Industrial Applications, Stephen A. Newman, ed., ACS Symposium Series 133, p. 451 (1980)

P16. Pitzer, K.S. and J.C. Peiper, "Activity Coefficient of Aqueous $NaHCO_3$ ", J. Phys. Chem., 84, 2396 (1980)

P17. Pitzer, K.S. and L.F. Silvester, "Thermodynamics of Geothermal Brines I. Thermodynamic Properties of Vapor-Saturated NaCl(aq) Solutions From 0 - 300°C", Lawrence Berkeley Laboratory Report 4456 (1976)

P18. Pitzer, K.S., "Electrolyte Theory - Improvement since Debye and Hückel", Accounts of Chem. Res., 10, 371 (1977)

P19. Pitzer, K.S. and P.S.Z. Rogers, "Volumetric Properties of Aqueous Sodium Chloride Solutions", J. Phys. Chem. Ref. Data, 11, 15 (1982)

P20. Pitzer, K.S., M. Conceicão and P. de Lima, "Thermodynamics of Saturated Aqueous Solutions Including Mixtures of NaCl, KCl, and CsCl", J. Sol. Chem., 12, 171 (1983)

P21. Pitzer, K.S., M. Conceicão and P. de Lima, "Thermodynamics of Saturated Electrolyte Mixtures of NaCl with Na_2SO_4 and with $MgCl_2$", J. Sol. Chem., 12, 187 (1983)

P22. Pitzer, K.S., and R.C. Phutela, "Thermodynamics of Aqueous Calcium Chloride", J. Sol. Chem., 12, 201 (1983)

CHEN REFERENCES

C1. Chen, C-C; H.I. Britt, J.F. Boston and L.B. Evans, "Extension and Application of the Pitzer Equation for Vapor-Liquid Equilibrium of Aqueous Electrolyte Systems with Molecular Solutes", AIChE J., 25, 820 (1979)

C2. Chen, C-C, H.I. Britt, J.F. Boston and L.B. Evans, "Two New Activity Coefficient Models for the Vapor-Liquid Equilibrium of Electrolyte Systems", Thermodynamics of Aqueous Systems with Industrial Applications, ACS Symposium Series 133, 61 (1980)

C3. Chen, C-C, "Computer Simulation of Chemical Processes with Electrolytes", Sc. D. thesis, Dept. of Chem. E., M.I.T. (1980)

C4. Chen, C-C; H.I. Britt, J.F. Boston and L.B. Evans, "Local Composition Model for Excess Gibbs Energy of Electrolyte Systems Part I: Single Solvent, Single Completely Dissociated Electrolyte Systems", AIChE J., 28, 588 (1982)

C5. Renon, H. and J.M. Prausnitz, "Local Compositions in Thermodynamic Excess Functions of Liquid Mixtures", AIChE J., 14, 135 (1968)

C6. Robinson, R.A. and R.H. Stokes, Electrolyte Solutions, 2nd ed., Butterworth and Co., London, (1970), p.32

VI:

Activity Coefficients of Strongly Complexing Compounds

ACTIVITY COEFFICIENTS OF STRONGLY COMPLEXING COMPOUNDS

The subject of this chapter is ions which form complexes. As contrasted with strong electrolytes, discussed in Chapters IV and V, complexing species form ionic and/or molecular intermediates. A classic example is orthophosphoric acid which dissociates in water to form, among others:

1) $H_3PO_4(aq)$
2) $H_2PO_4^-$
3) HPO_4^{2-}
4) PO_4^{3-}

This phenomena of complexing has a profound effect upon the solution behavior of these electrolytes in that:

1) The ionic strength, $I = .5 \Sigma m_i z_i^2$, has been seen to be the key variable in electrolyte calculations and is greatly affected by the distribution among the various ionic charge levels. For example, a 2 molal H_3PO_4-H_2O solution, if fully dissociated to three hydrogen ions and the PO_4^{3-} ion, would have an ionic strength of $.5(1^2\ (6) + 3^2(2)) = 12.0$, but if dissociated to H^+ and $H_2PO_4^-$ would have $I = .5\ (1^2(2) + 1^2\ (2)) = 2$.

2) The activity coefficient of each ion is determined by interactions with each oppositely charged ion as well as each molecule. Thus the formulations for activity coefficients of the ions, molecules and the activity of water are all affected by the proliferation of different ionic forms.

3) The concentration of the free ion can be significantly reduced by forming complexes. The free ion concentration may be so low that certain of its chemical reactions do not occur. For example, Ni^{2+} can be complexed with NH_3 to form $Ni(NH_3)_4^{2+}$. The nickel can be maintained in solution, as the complex, even at pH 10 where the formation of $Ni(OH)_2$ precipitate lowers the free Ni^{2+} concentration to 10^{-8} molal. The nature of the ion being complexed is often totally changed as if the complexing specie masks the ion. For example, the F^- ion is very corrosive in acidic solutions. It can be complexed with Al^{3+} to form AlF^{2+} which is not corrosive.

It is thus clear that identification of salts which complex and techniques for proper formulation of electrolyte models and activity coefficients is a very important task particularly since many systems of commercial interest exhibit this behavior. The high concentrations found in many industrial applications tend to promote significant formation of complexes.

In this chapter we will present the following subsections:

1. Identification of complexing species:
   A methodology, in the form of a set of tests and/or checks, which enable identification of complexing species.

2. Evaluation of certain common electrolytes:
   An application, using one or more of the tests for complexing electrolyte identification, to certain familiar compounds, including:
   - H_3PO_4
   - H_2SO_4
   - $ZnCl_2$
   - $FeCl_3$
   - $CuCl$
   - $CaSO_4$
   - Na_2SO_4
   - other uni-bivalent chlorides

3. Activity coefficient formulations which accomodate complexors:
   A review of some of the methods discussed in Chapter V in terms of their applicability in representing activity coefficients in complexing systems. The multicomponent formulations of Chapter V are needed, inasmuch as they accomodate the presence of more than one principal anion and cation.

In order to simplify the calculation of activity coefficients, it has been common practice to assume that electrolytes dissolved in water completely dissociate into ions. With the advent of the high speed digital computer and the coinciding development of more sophisticated activity coefficient model equations in the last decade, the mathematical modeling of solutions has become more precise. It is now becoming apparent that this assumption of complete dissociation may be the cause of some of the modeling problems previously encountered.

Some compounds are well known complexors with many studies having been done on them. These are often treated as completely dissociated, having the equilibrium constant adjusted to compensate for this. For example a solution of zinc chloride, $ZnCl_2$, is known to result in multiple complexes. Among those suggested are $ZnCl^+$, $ZnCl_2$ (aq), $ZnCl_3^-$ and $ZnCl_4^{2-}$. Yet zinc chloride is often modeled:

$$ZnCl_2 = Zn^{2+} + 2Cl^-$$

This is the case with the evaluated mean activity coefficients tabulated by Goldberg (Z5). The activity coefficients were listed to 23.193 molal and values were given for the parameters of the equation to which the activity coefficients were fit:

$$\ln \gamma_{\pm} = -A_1 \sqrt{I} + \sum_i B_i\, m_i^{(i+1)/2} \tag{6.1}$$

The solubility product of $ZnCl_2$, assuming complete dissociation, is the product of the activities:

$$K_{sp} = (a_{Zn^{2+}})\,(a_{Cl^-})^2$$

or:

$$K_{sp} = (\gamma_{Zn^{2+}}\; m_{Zn^{2+}})\,(\gamma_{Cl^-}\; m_{Cl^-})^2 \tag{6.2}$$

Recalling the definitions of the mean activity coefficient and mean molality from Chapter II:

$$\gamma_{\pm}^{\nu} = \gamma_c^{\nu_c}\, \gamma_a^{\nu_a} \tag{2.26}$$

$$m_{\pm}^{\nu} = m_c^{\nu_c}\, m_a^{\nu_a} \tag{2.27}$$

equation (6.2) can be restated:

$$K_{sp} = \gamma_{\pm}^3\; m_{\pm}^3 \tag{6.3}$$

The saturation molality at 25°C is 31.7 molal (9). Continuing the assumption of complete dissociation, the ionic strength is:

$$I = .5 \sum_i z_i^2\, m_i$$

$$= .5\,[2^2\, m_{\pm} + 1^2\,(2\, m_{\pm})]$$

$$= 3m_{\pm} = 3\ (31.7)$$
$$= 95.1 \text{ molal}$$

The activity coefficient at the saturation point, calculated with equation (6.1) and the parameters given by Goldberg (Z5) is:

$$\gamma_{\pm} = 0.0448171$$

The solubility product may therefore be calculated using equation (6.3):

$$K_{sp} = \gamma_{\pm}^3\ m_{\pm}^3 = (.0448171)^3\ \ (31.7)^3$$
$$= 2.86754$$

The solubility product may also be calculated using the Gibbs free energies. Assuming complete dissociation and using the ΔG's tabulated by the National Bureau of Standards (1), at 25°C:

$$K_{sp} = \exp\left[\frac{-\Delta G_{RXN}}{RT}\right]$$
$$= \exp\left[\frac{-[-35140. + 2.\ (-31372.) - (-88296.)]}{1.98719\ (298.15)}\right]$$
$$= 1.0668496 \times 10^7$$

This differs considerably from the solubility product calculated with the activity coefficient and saturation molality. Rearranging equation (6.3) using this value for the solubility product and the saturation molality of 31.7, the activity coefficient needed for use with the K_{sp} can be calculated:

$$\gamma_{\pm}^3 = K_{sp}\ /\ m_{\pm}^3$$
$$\gamma_{\pm} = \sqrt[3]{K_{sp}\ /\ m_{\pm}^3} \qquad (6.4a)$$
$$= 6.9445$$

This problem is no doubt created by using the activity coefficient equation past its maximum molality range.

A large difference can also be seen if the following reaction is considered:

$$ZnCl_2 = ZnCl^+ + Cl^-$$

Once again using values tabulated by the National Bureau of Standards (1) the solubility product at 25°C is:

$$K_{sp} = \exp\left[\frac{-\Delta G_{RXN}}{RT}\right]$$
$$= \left[\frac{-(-65800. - 31372. + 88296.)}{1.98719(298.15)}\right]$$
$$= 3.2077449 \times 10^6$$

The solubility product calculated by the product of the activities would be for this reaction:

$$K_{sp} = (\gamma_{ZnCl^+} \; m_{ZnCl^+}) \; (\gamma_{Cl^-} \; m_{Cl^-})$$
$$= \gamma_{\pm}^2 \; m_{\pm}^2$$

The activity coefficient needed for this equation to hold true would be:

$$\gamma_{\pm} = \sqrt{K_{sp} \; / \; m_{\pm}^2} \quad (6.4b)$$
$$= 56.49889$$

This should point out how crucial it is to be consistent in the use of thermodynamic data and tabulated activity coefficient values.

Compensating for these differences by adjusting the K_{sp} values or activity coefficients can result in a good prediction of the saturation point of the salt in binary solution. Difficulties are apt to arise though in multicomponent solutions, particularly if the ions in solution are present as the result of dissolving two different salts or if more than one salt contributes to a particular ion. In the case of dissolving two different salts, MY and NX:

$$MY = M^{2+} + Y^{2-}$$
$$NX = N^{+} + X^{-}$$

failing to account for the possible association of these two ions:

$$M^{2+} + X^{-} = MX^{+}$$

could result in an incorrect prediction of the solution behavior.

Successful prediction of the system, ignoring possible ion association, could be made by regressing ternary experimental data for the activity coefficients and/or equilibrium constants. However, these regression results would be applicable only to that ternary system due to the compensating effects. Regression results are most "portable" when based on binary solution data.

One special form of the MX^+ complex results from the phenomena of cation hydrolysis. This phenomena is generally ignored when modeling solutions. Hydrolysis, in inorganic solutions, is often described as the creation of new ionic species, such as oxide and hydroxide complexes, or precipitates. The hydroxide complexing reaction can be expressed in general terms:

$$xM^{z+} + yH_2O \rightleftharpoons M_x(OH)_y^{(xz-y)+} + yH^+$$

or:

$$xM^{z+} + yOH^- \rightleftharpoons M_x(OH)_y^{(xz-y)+}$$

A narrow 1 to 2 pH units is often the range in which cation hydrolysis begins and reaches the precipitation point; knowledge of the range can help avoid unwanted precipitation. The hydroxide ion concentration, particularly when the result of the dissociation of water, can be very low. Cation hydrolysis apparently increases with the cation charge and decreases with cation size. It is also found to occur more with post-transition metals than with pre-transition metals.

Cation hydrolysis can be of particular importance in modeling natural water systems with trace metal content and when attempting to avoid or cause hydrolytic precipitation of metals. Problems arise in that the true nature of many of the complexes resulting from cation hydrolysis are not known due to the tendency towards polynuclear species, that is complexes containing multiple metal ions, and the difficulty in their experimental determination. For example, Baes and Mesmer (5) note that chromic ion, Cr^{3+}, reacts in water to form $CrOH^{2+}$, $Cr(OH)_2^+$ and $Cr(OH)_4^-$ ions. The aqueous molecule $Cr(OH)_3$ is also likely to be present. Other complexes forming may be $Cr_2(OH)_2^{4+}$ and $Cr_3(OH)_4^{5+}$.

In comparison with other areas of chemical knowledge, cation hydrolysis is a topic which could be said to be still in its infancy. The amount of research has increased greatly over the last thirty years, a good deal of which was tabulated by Sillén (4). It is impossible for us to discuss this topic in great depth here. Instead, for more detailed information we would suggest, in addition to Sillén's work, the excellent book on cation hydrolysis by Baes and Mesmer (5) and their more recent article (6).

Identification of Complexing Electrolytes

Given the crucial roles complexes play in modeling solution behavior, it is quite necessary to determine whether or not a given species will form complexes and what those complexes will be. There is, at this point, no certain or best method. Of the methods suggested here at least one should be applicable to any given species. Applications of these methods will be seen later in this chapter when discussing specific compounds. The three methods are:

- Literature Review

Complexing behavior can be checked for via a thorough literature review. Consulting the extensive compilations by the National Bureau of Standards (1) and the Russian Academy of Science (2) is a good first step. Other sources to check are the Smith and Martell (3) and Sillén (4) compilations. These references provide values for the Gibbs free energies and heats of formation for precipitate, ionic, vapor, aqueous molecule and complex species which can be used to calculate the various equilibrium constants needed for modeling aqueous solutions. There are also values for some of the cation hydrolysis complexes.

This review should cover more than one of these sources as they do not always present the same complexes. In those cases where there are differences, a choice of values will have to be made. This decision can be aided by a literature search which often will yield articles detailing work done to identify the ionic and molecular species in solution. Articles may be found describing modeling efforts similar to the one desired. Such articles are becoming increasingly common and they often use one of the activity coefficient modeling equations presented in Chapters IV and V. The Pitzer formalism, in particular, is frequently used.

- Use of Meissner's Curves

Use of the Meissner family of curves as presented in Chapter IV, Figure (4.6), in order to correlate the reduced activity coefficients of strong electrolytes. Meissner found that the reduced activity coefficients of strong electrolytes fell into a pattern which he formalized. If it can be assumed that the reduced activity coefficients of strong electrolytes follow this pattern, then deviation from these curves may be said to indicate that the species in question does not completely dissociate or may be forming complexes. If activity coefficients from experimental data are available they may be plotted on the Meissner chart. Severe deviation from the curves, such as the crossing of lines, may be taken to indicate complex formation. However, there would be no indication of what the complexes might be.

- Comparison of Osmotic Coefficients

A technique based upon comparison of osmotic coefficients in order to predict the likelihood of complexing was presented in 1945 by Stokes and Stokes (7).

They note that the Debye-Hückel theory relates osmotic behavior to the effective ionic diameter. It would be expected then that species containing the same anions and having cations of similar diameter would have nearly equal osmotic coefficients unless there was a "chemical" difference between the two. This difference could be interpreted as incomplete dissociation or complex formation.

There are drawbacks to this method. One is the need for osmotic coefficients for the species of interest as well as for the species of similar diameter. It is also advisable to compare species that would not be expected to form complexes, such as perchlorates or nitrates. And of course, while it may suggest the occurrence of complexes, it gives no clue as to what they may be.

These methods will now be used to examine specific compounds for complexing behavior. Not all methods will be applicable to each compound. Some will, in fact, yield contradictory results.

Phosphoric Acid

Orthophosphoric acid, H_3PO_4, is a polyprotic acid, that is it contains more than one acid hydrogen per molecule. Its stepwise dissociation is usually expressed:

$$H_3PO_4 = H^+ + H_2PO_4^-$$
$$H_2PO_4^- = H^+ + HPO_4^{2-}$$
$$HPO_4^{2-} = H^+ + PO_4^{3-}$$

More recently however it has been suggested that there are other complexes forming in an H_3PO_4-H_2O solution.

The first dissociation constant at 25°C is given similar values by most investigators:

	K_1
Robinson and Stokes (8)	.007112135
Barta and Bradley (P1)	.0071425
Elmore et al. (P3)	.007107
Pitzer and Silvester (P7)	.0071425
Farr (P4)	.007156
NBS thermodynamic data (1)	.0071167

The second dissociation constant is also fairly close:

	K_2
Robinson and Stokes (8)	6.3387E-8
Farr (P4)	6.339E-8
NBS thermodynamic data (1)	6.2303E-8

The third dissociation:

$$HPO_4^{2-} = H^+ + PO_4^{3-}$$

is given a dissociation constant K_3 = 4.73E-13 by Farr (P4) and K_3 = 4.5301E-13 from NBS thermodynamic data.

Some of the additional suggested equilibria in the solution are:

$$H_2PO_4^- + H_3PO_4 \rightleftharpoons H_5(PO_4)_2^-$$

$$H_2PO_4^- + H_2PO_4^- \rightleftharpoons H_4(PO_4)_2^{2-}$$
$$H_2PO_4^- + HPO_4^{2-} \rightleftharpoons H_3(PO_4)_2^{3-}$$
$$H_3PO_4 + H_3PO_4 = H_6(PO_4)_2$$
$$H_3P_2O_7^- + H_2O = H_3PO_4 + H_2PO_4^-$$
$$H_2P_2O_7^{2-} + H_2O = 2H_2PO_4^-$$

The equilibrium constants for these reactions are varied; indeed, investigators disagree as to which reactions actually occur.

In 1965, Elmore et al. (P3) presented the results of their study of the degree of dissociation of phosphoric acid based on vapor pressure, conductance and pH data at 25°C. They concluded that, over a concentration range of .1 to 10 molal, the equilibria that best describe the system are:

$$H_3PO_4 = H^+ + H_2PO_4^- \qquad K_1 = .007107$$

$$H_3PO_4 + H_2PO_4^- = H_5P_2O_8^- \qquad K_0 = \frac{a_{H_5P_2O_8^-}}{a_{H_3PO_4}\, a_{H_2PO_4^-}} = 1.263$$

$$H_6P_2O_8 = H^+ + H_5P_2O_8^- \qquad K_4 = \frac{a_{H^+}\, a_{H_5P_2O_8^-}}{a_{H_6P_2O_8}} = .3$$

They assumed that the activity coefficients of the $H_2PO_4^-$ and $H_5P_2O_8^-$ were equal; this assumption was also made for the H_3PO_4 and $H_6P_2O_8$ molecules.

When the hydrogen ion concentration is set as:

$[H^+] = \alpha m$

where: α = degree of dissociation

m = H_3PO_4 input molality

and the activity of the undissociated H_3PO_4, a_u, is calculated from vapor pressure data, the final equations for solving for the system concentrations are:

$$[H_2PO_4^-] = \frac{[H^+]}{1 + K_0\, a_u}$$

$$[H_5P_2O_8^-] = \frac{K_0\, a_u}{1 + K_0\, a_u}[H^+]$$

$$[H_3PO_4] = \frac{m - \dfrac{1 + 2K_0\, a_u}{1 + K_0\, a_u}[H^+]}{1 + \dfrac{2K_1 K_0}{K_4}\, a_u}$$

$$[H_6P_2O_8] = \frac{K_1 K_0}{K_4} a_u \left[\frac{m - \frac{1 + 2K_0 a_u}{1 + K_0 a_u}[H^+]}{1 + \frac{2K_1 K_0}{K_4} a_u} \right]$$

$$= \frac{K_1 K_0}{K_4} a_u [H_3PO_4]$$

C.W. Childs (P2) presented the results of a study of dilute solutions of orthophosphoric acid at 37°C. In addition to the standard first three dissociation constants, it was proposed that the complexes $H_3(PO_4)_2^{3-}$, $H_4(PO_4)_2^{2-}$ and $H_5(PO_4)_2^-$ would be present. For the first three dissociation constants, the following values were chosen:

$pK_1 = 1.95 \pm .25$

$pK_2 = 6.70 \pm .05$

$pK_3 = 11.30 \pm .15$

The other equilibria were set:

$$H_2PO_4^- + HPO_4^{2-} = H_3(PO_4)_2^{3-}$$

$$K_4 = \frac{[H_3(PO_4)_2^{3-}]}{[H_2PO_4^-][HPO_4^{2-}]} = 2.7 \pm 2.5 \text{ l/mole}$$

$$H_2PO_4^- + H_2PO_4^- = H_4(PO_4)_2^{2-}$$

$$K_5 = \frac{[H_4(PO_4)_2^{2-}]}{[H_2PO_4^-][H_2PO_4^-]} = 5.6 \pm 2. \text{ l/mole}$$

$$H_2PO_4^- + H_3PO_4 = H_5(PO_4)_2^-$$

$$K_6 = \frac{[H_5(PO_4)_2^-]}{[H_2PO_4^-][H_3PO_4]} < 100 \text{ l/mole}$$

The experimental data used was pH titrations.

Mesmer and Baes (P6) studied the dissociation equilibria potentiometrically over a wide temperature range. Using a hydrogen-electrode concentration cell measurements were made to 300°C. The equilibria were expressed as neutralization reactions:

$$H_xPO_4^{(3-x)-} + OH^- = H_{x-1}PO_4^{(4-x)-} + H_2O$$

The neutralization quotient, Q_x, was expressed:

$$\log Q_x = \log\left[\frac{\bar{n} + x - 3}{(4 - x - \bar{n})\,[OH^-]}\right]$$

where: $\bar{n}$ is the average number of protons neutralizated

$$= \frac{[H^+] + m_{OH^-} - [OH^-]}{m_P}$$

$[H^+]$ is the hydrogen ion concentration

$[OH^-]$ is the hydroxyl ion concentration

m_{OH^-} is the stoichiometric concentration of the base

m_P is the stoichiometric concentration of phosphoric acid

They found values only for the first two neutralization constants.

In a study of the NaH_2PO_4-$NaClO_4$-H_2O system, Wood and Platford (P10) found large free energies of mixing. They theorized that the association:

$$H_2PO_4^- + H_2PO_4^- = (H_2PO_4)_2^{2-}$$

may be the cause and the association constant would be K_5 = .25 ±.1 kg/mole at 25°C. They note that when a correction to the excess free energies of NaH_2PO_4 solution is made for the dimer formation, the behavior is then more like that of strong electrolytes such as KCl. They felt that this lent strong credence to this association.

Figure 6.1

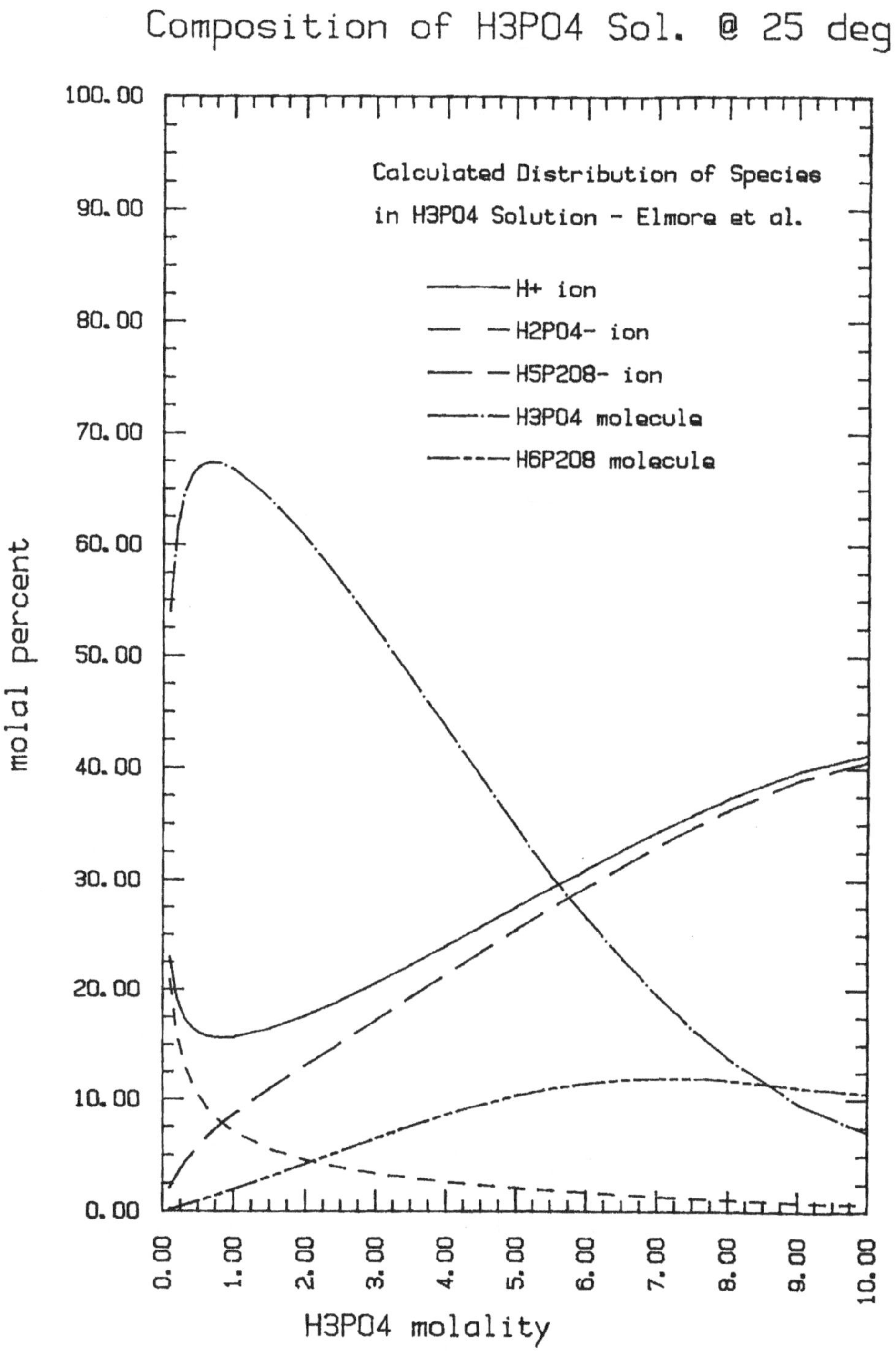
Composition of H3PO4 Sol. @ 25 deg
Calculated Distribution of Species
in H3PO4 Solution - Elmore et al.
H+ ion
H2PO4- ion
H5P2O8- ion
H3PO4 molecule
H6P2O8 molecule
molal percent
100.00
90.00
80.00
70.00
60.00
50.00
40.00
30.00
20.00
10.00
0.00
0.00
1.00
2.00
3.00
4.00
5.00
6.00
7.00
8.00
9.00
10.00
H3PO4 molality

Figure 6.2

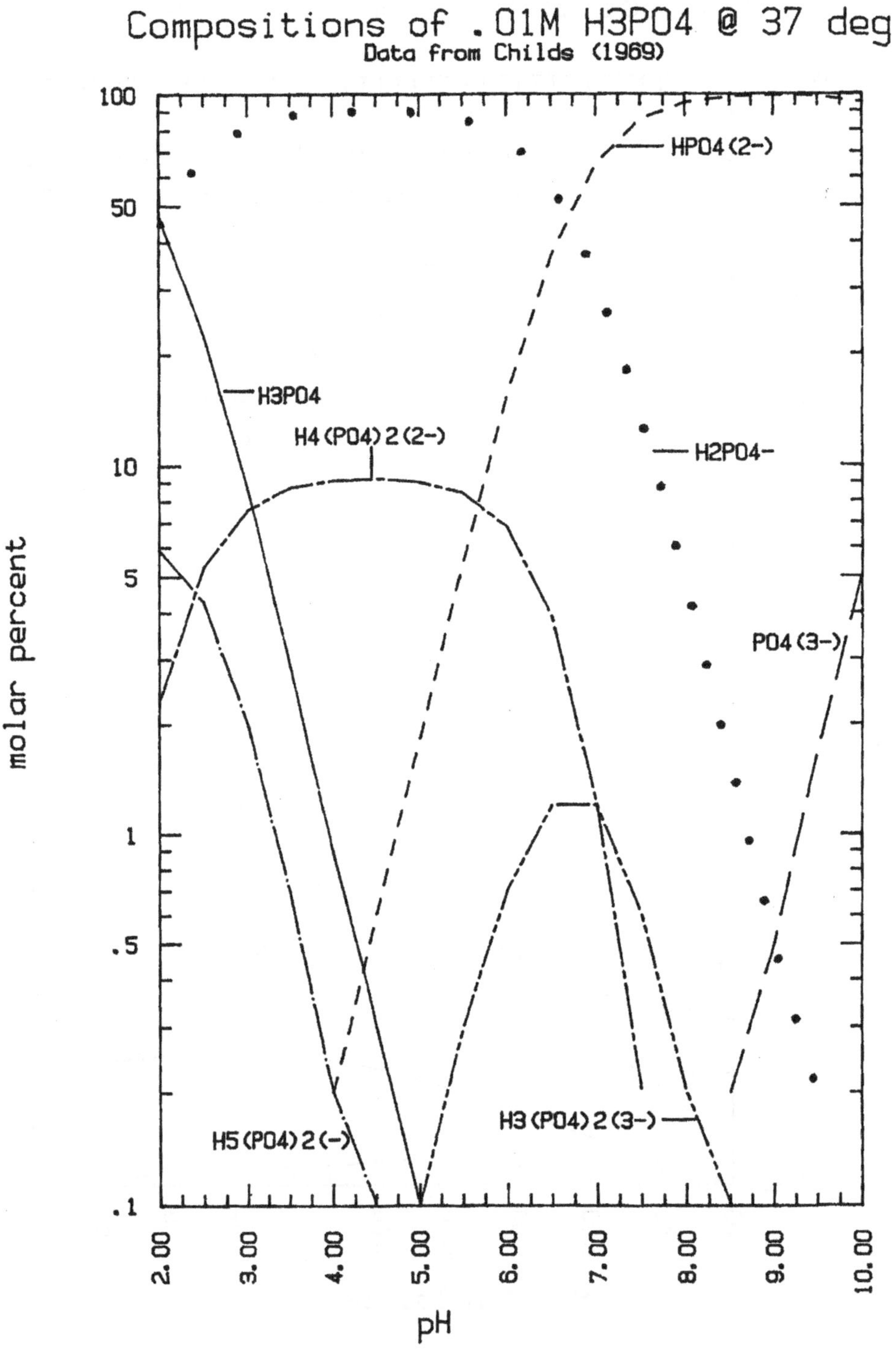
Compositions of .01M H3PO4 @ 37 deg
Data from Childs (1969)
molar percent
pH
100
50
10
5
1
.5
.1
2.00
3.00
4.00
5.00
6.00
7.00
8.00
9.00
10.00
HPO4(2-)
H3PO4
H4(PO4)2(2-)
H2PO4-
PO4(3-)
H3(PO4)2(3-)
H5(PO4)2(-)

Sulfuric Acid

In its pure liquid state, sulfuric acid has a high electrical conductivity and high viscosity. Self-dissociation occurs in the pure liquid; the reaction:

$$2H_2SO_4 \rightleftharpoons H_3SO_4^+ + HSO_4^-$$

has been proposed as the cause of the high conductivity. A second reaction has been suggested as occurring simultaneously:

$$2H_2SO_4 \rightleftharpoons H_3O^+ + H_2S_2O_7^-$$

It has been estimated that the total amount of self-dissociation products is .043 molal, in comparison with the $2.\mathrm{x}10^{-7}$ molal concentration of H^+ and OH^- ions in water.

In solution with water this polyprotic acid dissociates:

$$H_2SO_4 = H^+ + HSO_4^-$$
$$HSO_4^- = H^+ + SO_4^{2-}$$

The first dissociation is usually treated as being complete; however, at high concentrations of sulfuric acid, the acid may be said to be the solvent and the water the solute. The second dissociation prevails only in very low sulfuric acid concentrations. In 1949, Young and Blatz (H15) presented the following diagram to describe the concentrations of the SO_4^{2-} and HSO_4^- ions and undissociated acid ($\cdot HHSO_4$) based on Raman spectrum measurements:

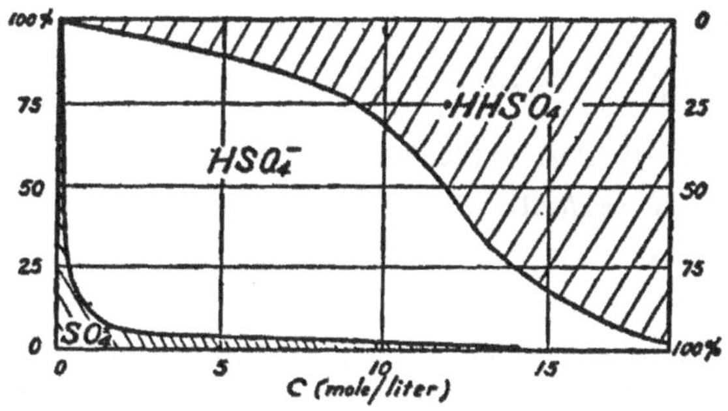

Figure 6.3 Fractions of sulfuric acid present as undissociated acid, as bisulfate, and as sulfate ion, represented by vertical distances between adjacent lines. *(Reprinted with permission from T.F. Young & L.A. Blatz, Chem. Rev., v44, 102 (1949) Copyright American Chemical Society).*

The second dissociation constant has been the subject of much study. Hamer (H5) estimated a value of .0120 at 25° in 1934 based on his e.m.f. measurements of a cell containing $NaHSO_4$, Na_2SO_4 and NaCl. The data was recalculated to take into consideration the $NaSO_4^-$ complex in 1952 by Davies et al. (H1) which resulted in K_2 = .0102. Robinson and Stokes (8) presented a temperature fit for the second ionization constant based on values presented by Singleterry (H13):

$$\log K_2 = 5.0435 - \frac{475.14}{T} - .018222\, T$$

T - Kelvins

The resulting thermodynamic properties at 25° of the second dissociation are:

ΔH^o = -5237. cal/mole

ΔC_p^o = -49.7 cal/deg mole

ΔS^o = -26.6 cal/deg mole

In Helgeson's (H6) investigation of the effect of temperature on complex dissociations, a value of $\log K_2$ = -1.99, or K_2 = .010233, at 25°C was presented. Wirth (H14) calculated values for the mean ionic activity coefficients for H^+, H^+, SO_4^{2-} and H^+, HSO_4^- from experimental activity coefficients. After examining the literature, he decided on a value of K_2 = .0102. Pitzer et al. (H8) chose K_2 = .0105 in their study of sulfuric acid.

Using the NBS (1) tabulated values of the free energies of the constituent ions:

$$HSO_4^- = H^+ + SO_4^{2-}$$

$$\Delta G_{RXN} = \Delta G_{H^+} + \Delta G_{SO_4^{2-}} - \Delta G_{HSO_4^-}$$

$$= 0. -177970. + 180690.$$

$$= 2720 \text{ cal/mole}$$

$$K_2 = .010144$$

Tabulated values for the mean molal activity coefficient of H_2SO_4 presented by Rard et al. (H9) and Parker et al. (H7) are based on a stoichiometric coefficient, ν, of three. That means they assume complete dissociation:

$$H_2SO_4 = 2H^+ + SO_4^{2-}$$

Figure 6.4 shows the activity coefficients presented by Parker et al. (H7) plotted on the Meissner chart.

Figure 6.4

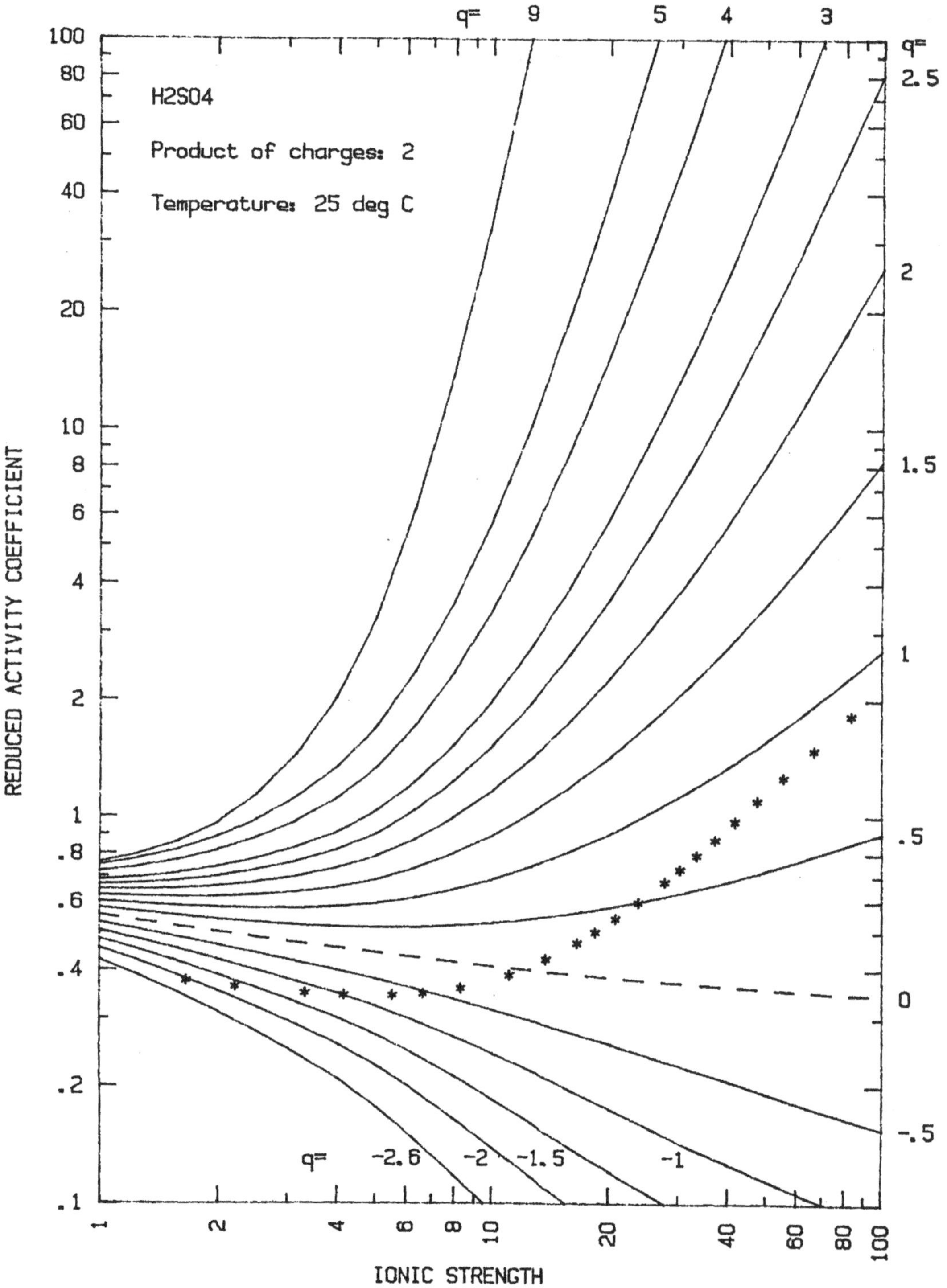
q= 9
5
4
3
q= 2.5
2
1.5
1
.5
0
-.5
H2SO4
Product of charges: 2
Temperature: 25 deg C
REDUCED ACTIVITY COEFFICIENT
100
80
60
40
20
10
8
6
4
2
1
.8
.6
.4
.2
.1
q= -2.6 -2 -1.5 -1
1
2
4
6
8
10
20
40
60
80
100
IONIC STRENGTH

In 1977, Pitzer et al. (H8) published values for the interaction parameters of Pitzer's activity coefficient equation fit to a concentration of 6 molal. The interaction parameters are for a solution containing H^+, HSO_4^- and SO_4^{2-} ions.

Zinc Chloride

Zinc chloride has long been known to form complexes in solution. In 1940, Robinson and Stokes (Z14) presented e.m.f. measurements for various concentrations at five degree temperature intervals from 10 to 40°C which they fit to the equation:

$$E = E_{25} + a(t-25) + b(t-25)^2$$

They then encountered difficulty in extrapolating back to infinite dilution which led them to reject as inaccurate a number of points below .008 m. They hypothesized that the deviation of these points from their linear extrapolation might point to incomplete dissociation:

$$ZnCl^+ = Zn^{2+} + Cl^-$$

but they felt that this was unlikely at such low concentrations.

In a similar study of zinc bromide in 1945, Stokes and Stokes (7) noted similar behavior. They suggested the osmotic coefficient ratio plots described earlier as a method of testing for complexing behavior. With ionic radii of .65 Å and .74 Å for the Mg and Zn ions, they felt that the osmotic coefficients should be similar. A plot of the osmotic coefficient ratios:

$$\frac{\phi_{ZnX_2}}{\phi_{MgX_2}}$$

versus molality was presented. The maximum molality was 1. m and the X anions were nitrates, iodides, bromides and chlorides. The nitrate ratio held close to one over the molality range; the iodides began deviating at about .7 m. The bromides and chlorides dropped away from unity rapidly. This can be seen in Figure 6.5 which has the maximum molality increased to 5 and includes perchlorates as well. The osmotic coefficients were taken from Robinson and Stokes (8). Stokes and Stokes interpreted their plot as indicating that $ZnBr_2$ may not be treated as completely dissociated at as low as .1 m; this is also true for $ZnCl_2$ which shows a greater departure from unity.

The activity coefficients for $ZnCl_2$ presented by Goldberg in 1981 (Z5) allow for the Meissner plot test. As can be seen in Figure 6.6, when plotted as the reduced activity coefficient, $\Gamma = \gamma^{1/z_+z_-}$, versus ionic strength, the activity coefficients show severe deviation from the expected strong electrolyte behavior. The activity coefficients of zinc perchlorate (Z5) show much better behavior in Figure 6.7, although some deviation can be seen at the higher concentrations.

Figure 6.5

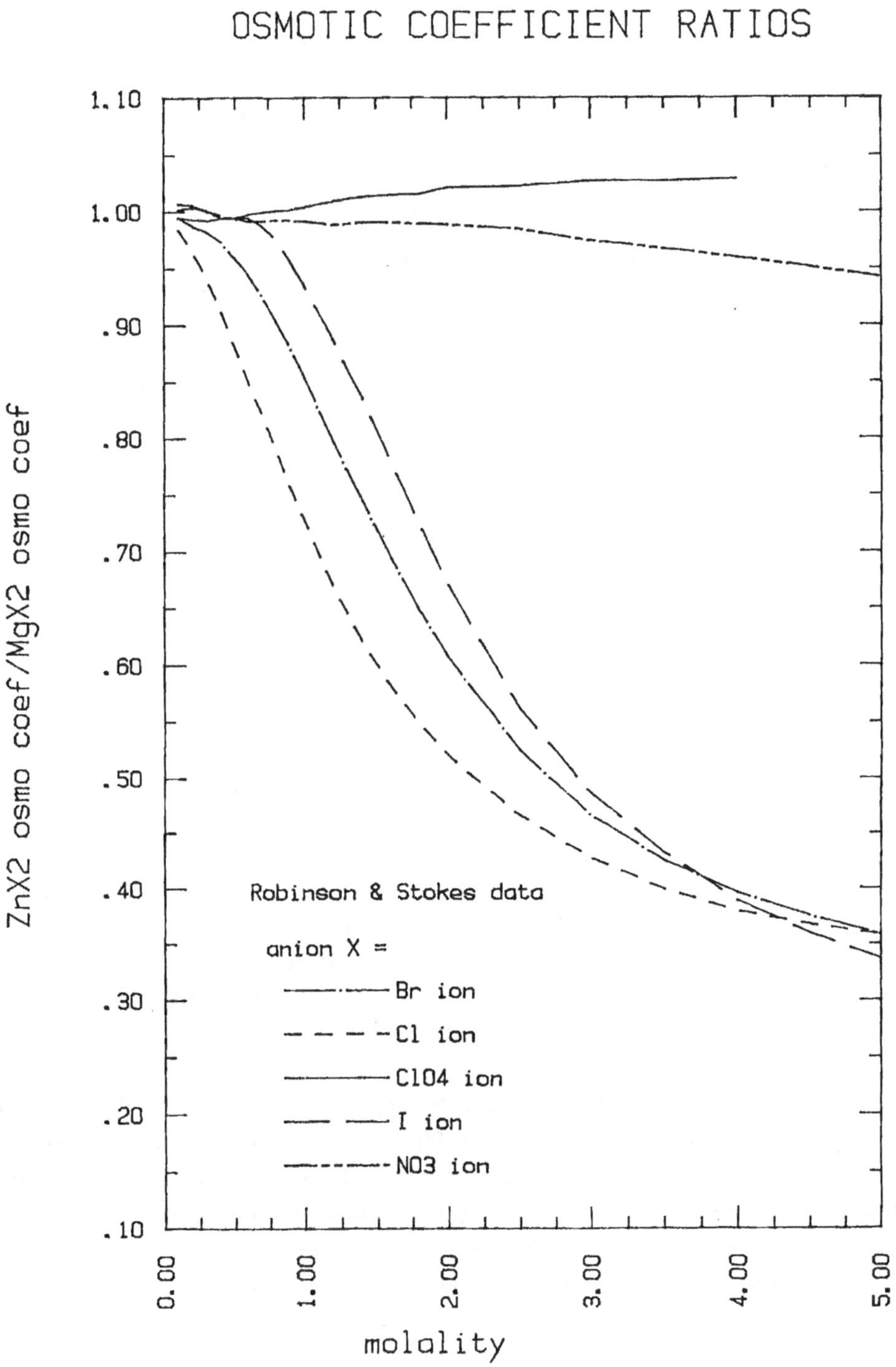
OSMOTIC COEFFICIENT RATIOS
1.10
1.00
.90
.80
.70
.60
.50
.40
.30
.20
.10
ZnX2 osmo coef/MgX2 osmo coef
0.00
1.00
2.00
3.00
4.00
5.00
molality
Robinson & Stokes data
anion X =
Br ion
Cl ion
ClO4 ion
I ion
NO3 ion

Figure 6.6

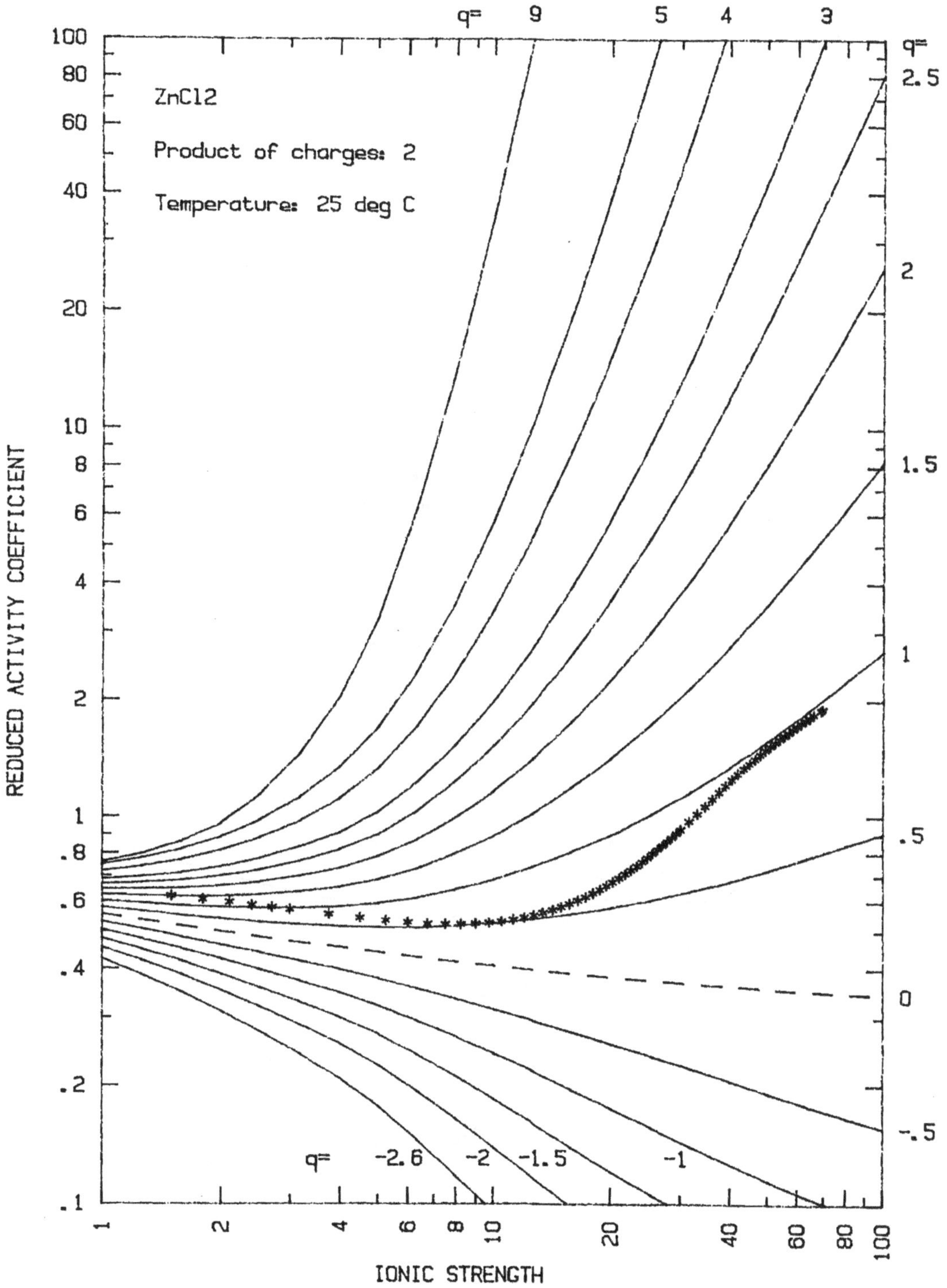
q= 9 5 4 3
q= 2.5 2 1.5 1 .5 0 -.5
ZnCl2
Product of charges: 2
Temperature: 25 deg C
100 80 60 40 20 10 8 6 4 2 1 .8 .6 .4 .2 .1
REDUCED ACTIVITY COEFFICIENT
q= -2.6 -2 -1.5 -1
1 2 4 6 8 10 20 40 60 80 100
IONIC STRENGTH

Figure 6.7

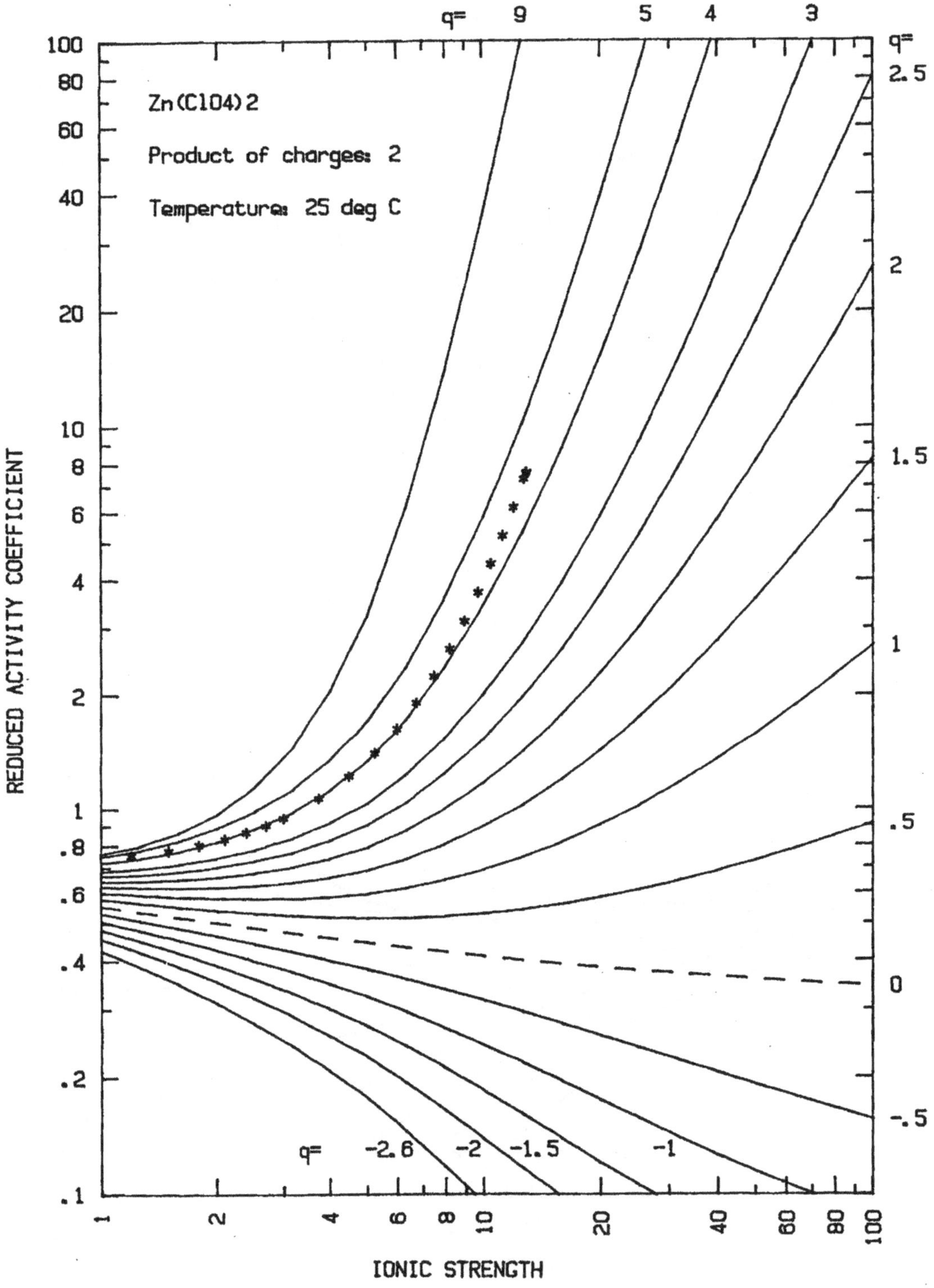
q= 9 5 4 3
q= 2.5
2
1.5
1
.5
0
-.5
q= -2.6 -2 -1.5 -1
Zn(ClO4)2
Product of charges: 2
Temperature: 25 deg C
100
80
60
40
20
10
8
6
4
2
1
.8
.6
.4
.2
.1
REDUCED ACTIVITY COEFFICIENT
1
2
4
6
8
10
20
40
60
80
100
IONIC STRENGTH

A literature review reveals attempts to determine the complexes present in a zinc chloride solution. In 1956, Shchukarev et al. (Z16) presented values for the ΔG, ΔH and $T \Delta S$ of formation of the $ZnCl^+$, $ZnCl_2$, $ZnCl_3^-$ and $ZnCl_4^{2-}$ complexes. Morris et al. (Z13) used a solvent extraction method and found stability constants for $ZnCl^+$, $ZnCl_2$ and $ZnCl_3^-$. While aware of the possibility of the $ZnCl_4^{2-}$ complex, they were unable to find it. Spivakovskii and Moisa (Z17) used a potentiometric method to determine stability constants for complex formation of zinc ions in $CaCl_2$ solutions. They found values for $ZnCl^+$, $ZnCl_2$, $ZnCl_3^-$ and $ZnCl_4^{2-}$ and found evidence of $ZnCl_2(H_2O)_4$, $ZnCl_3(H_2O)^-$, and $ZnCl_4(H_2O)_2^{2-}$ complexes. Aksel'rud and Spivakovskii (Z1) in 1958, found the complexes $Zn(OH)_2$ and $Zn(OH)_{1.5}Cl_{.5}$ or $Zn_2(OH)_3Cl$. In 1975, Belousov and Alovyainikov (Z2) and Federov et al. (Z3) presented values for the stability constants of $ZnCl^+$, $ZnCl_2$, $ZnCl_3^-$ and $ZnCl_4^{2-}$. That same year, Lutfullah et al. (Z11) presented e.m.f. measurements and considered only the first complex, $ZnCl^+$. Libuś and Tialowska (Z10) used potentiometric and spectrophotometric methods to study the $ZnCl^+$ complex. Latysheva and Andreeva (Z9) concluded that the highest complex formed would be $ZnCl_4^{2-}$.

The NBS (1) and Russian (2) thermodynamic tables present data for the $ZnCl^+$, $ZnCl_2(aq)$ and $ZnCl_3^-$ complexes. The NBS also includes the $ZnCl_4^{2-}$. There appears to be sufficient evidence of its existence that it should be included in attempts to model the zinc chloride system. It can be expected that this complex will increase in importance with an increase in concentration.

Ferric Chloride

The highly charged transition metals have a marked tendency to form complexes. The ferric ion, Fe^{3+}, is one for which numerous studies have been done in an attempt to define its complexing behavior with anions such as hydroxyl, cyanide, bromide and chloride. The complexes occurring between ferric and chloride ions will be discussed here.

Early estimates of the first association constant were made in the 1930's by Bray and Hershey (F2) and Möller (F8) using potentiometric measurements. In 1942, Rabinowitch and Stockmayer (F9) presented the results of their spectroscopic study under controlled acidity, temperature and ionic strengths. From their results they calculated estimates of the first three association constants:

$$Fe^{3+} + Cl^- = FeCl^{2+}$$

$$K_1 = \frac{[FeCl^{2+}]}{[Fe^{3+}][Cl^-]}$$

$$FeCl^{2+} + Cl^- = FeCl_2^+$$

$$K_2 = \frac{[FeCl_2^+]}{[FeCl^{2+}][Cl^-]}$$

$$FeCl_2^+ + Cl^- = FeCl_3$$

$$K_3 = \frac{[FeCl_3]}{[FeCl_2^+][Cl^-]}$$

While they note that there is evidence of at least one more complex in concentrated chloride solutions, $FeCl_4^-$, they chose not to consider it. Their calculations included a term to compensate for the association of ferric and hydroxyl ions, even though this term is small due to the low hydroxyl concentration. The results, in which they felt only K_1 could be considered as close to exact are:

at 26.7°C, ionic strength = 1

$K_1 = 4.2 \pm 0.2$

$K_2 = 1.3 \pm 0.4$

$K_3 = 0.04 \pm .02$

Extrapolated back to zero ionic strength and 25°C:

$K_1^o = 30. \pm 5$

$K_2^o = 4.5 \pm 2$

$K_3^o = 0.1 \pm .05$

K_1 was also presented as a function of ionic strength:

$$\log K_1 = 1.51 - \frac{3.\sqrt{I}}{1. + 1.5\sqrt{I}} + .295\ I$$

The constant 1.51 is the log of the association constant at 26.7°C extrapolated back to infinite dilution ($K_1^o = 32.$). Using this equation, they calculated values for log K at temperatures from 20 to 50°C which were then plotted versus 1/T. From its slope, a value for the heat of association was determined. Values for ΔG^o and ΔS^o of the reaction were derived using the value of 30 for K_1^o at 25°C. The results presented are:

$\Delta H = 8.5 \pm .2$ kcal/mole

$\Delta G^o = -2.0 \pm .2$

$\Delta S^o = 35. \pm 2$

Gamlen and Jordan (F6) used spectrophotometric measurements to determine K_3 and the next association constant:

$$FeCl_3 + Cl^- = FeCl_4^-$$

$$K_4 = \frac{[FeCl_4^-]}{[FeCl_3][Cl^-]}$$

They concluded that this was the highest chloro-complex possible after reviewing the spectrophotometric measurements and conductometric titration results of other authors (F3, F4, F5). Gamlen and Jordan define the association constants in terms of activities:

$$K_3 = \frac{a_{FeCl_3}}{a_{FeCl_2^+}\ a_{Cl^-}} = \frac{[FeCl_3]}{[FeCl_2^+][Cl^-]}\ \frac{f_{FeCl_3}}{f_{FeCl_2^+}\ f_{Cl^-}}$$

$$K_4 = \frac{a_{FeCl_4^-}}{a_{FeCl_3}\ a_{Cl^-}} = \frac{[FeCl_4^-]}{[FeCl_3][Cl^-]}\ \frac{f_{FeCl_4^-}}{f_{FeCl_3}\ f_{Cl^-}}$$

or, in general terms:

$$K_n = \frac{[FeCl_n^{3-n}]}{[FeCl_{n-1}^{4-n}][Cl^-]} \frac{f_{FeCl_n^{3-n}}}{f_{FeCl_{n-1}^{4-n}} f_{Cl^-}}$$

Since these activity coefficients are unknown, the activity coefficients are lumped into one term, F:

$$\frac{1}{F} = \frac{f_{FeCl^{2+}}}{f_{Fe^{3+}} f_{Cl^-}} = \frac{f_{FeCl_2^+}}{f_{FeCl^{2+}} f_{Cl^-}} = \frac{f_{FeCl_3}}{f_{FeCl_2^+} f_{Cl^-}} = \frac{f_{FeCl_4^-}}{f_{FeCl_3} f_{Cl^-}}$$

from which it can be assumed that the complex activity coefficient ratios remain constant:

$$C = \frac{f_{FeCl_n^{3-n}}}{f_{FeCl_{n-1}^{4-n}}}$$

They assume that C = 1 and $f_{HCl} = f_{Cl^-}$:

$$K'_n = \frac{[FeCl_n^{3-n}]}{[FeCl_{n-1}^{4-n}][Cl^-]} \frac{1}{f_{HCl}}$$

Noting that Bjerrum (F1) expressed the lumped activity coefficient as:

$$\log F = A + B[Cl^-]$$

they express the HCl activity coefficient as

$$\log f_{HCl} = -.42 + .18[Cl^-]$$

Their final results presented for an ionic strength of zero and temperature of 20°C are:

$$K'_3 = .73$$
$$K'_4 = .0105$$

In 1960, Marcus (F7) presented the formation constants:

$$\log K_2 \geq -.7$$
$$\log K_3 = -1.40 \pm .06$$
$$\log K_4 = -1.92 \pm .08$$

This value for log K_4 is close to that presented by Gamlen and Jordan, and Rabinowitch and Stockmeyer's K_3 is similar to Marcus'.

The thermodynamic tables of the National Bureau of Standards (1) and Russian Academy of Science (2) present values for the Gibbs free energies for the $FeCl^{2+}$, $FeCl_2^+$ and $FeCl_3$(aq) complexes. The Russian tables also include a value for the $FeCl_4^-$ ion:

	$\Delta G°$, kcal/mole					
	Fe^{3+}	Cl^-	$FeCl^{2+}$	$FeCl_2^+$	$FeCl_3$	$FeCl_4^-$
NBS	-1.1	-31.372	-34.4	-66.7	-96.7	
Russian	-1.108	-31.373	-34.35	-66.91	-96.74	-125.5

The formation constants at 25°C can be calculated:

$$K_n = \exp\left[\frac{-\Delta G_{RXN}}{R\,T}\right]$$

- for $Fe^{3+} + Cl^- = FeCl^{2+}$
  K_1(NBS) = 25.8967
  K_1(Russian) = 23.4421

- for $FeCl^{2+} + Cl^- = FeCl_2^+$
  K_2(NBS) = 4.7889
  K_2(Russian) = 7.4145

- for $FeCl_2^+ + Cl^- = FeCl_3$
  K_3(NBS) = .098698
  K_3(Russian) = .073955

- for $FeCl_3 + Cl^- = FeCl_4^-$
  K_4(Russian) = .012152

Cuprous Chloride

Characterizing and modeling cuprous chloride presents a challenge to investigators in two ways. Experimental data is quite difficult to obtain due to the rapid oxidation of the cuprous to cupric ion in air. For the investigator attempting to model this species in solution, there is disagreement in the experimental data and uncertainty in thermodynamic data primarily because of the possibility of numerous complex forms of Cu^+.

The NBS (1) and Russian (2) thermodynamic data compilations present values for the solid CuCl, and the aqueous complexes $CuCl_2^-$ and $CuCl_3^{2-}$. Sukhova et al. (C8) report the possibility of nineteen complexes up to $Cu_5Cl_9^{4-}$. Hakita et al. (C6) suggested $CuCl_2^-$, $CuCl_3^{2-}$ and $CuCl_4^{3-}$ and gave approximate equilibrium constants for these complexes. Ahrland and Rawsthorne (C1) suggested values for the CuCl solubility product and $CuCl_2^-$, $CuCl_3^{2-}$ and $Cu_2Cl_4^{2-}$ formation constants based on solubility and potentiometric data of the NaCl-CuCl-H_2O system at 25°C. These values fit their data to above 1M chloride. Smith and Martell (3) present values for CuCl(aq), $CuCl_2^-$, $CuCl_3^{2-}$ and the dimer $Cu_2Cl_4^{2-}$.

J.J. Fritz (C2-C5) has proposed that the species present in a cuprous chloride solution are CuCl(aq), $CuCl_2^-$, $CuCl_3^{2-}$, $Cu_2Cl_4^{2-}$ and triply charged ions represented by $Cu_3Cl_6^{3-}$. Fritz used Pitzer's activity coefficient equations to model cuprous chloride solubility in HCl-$HClO_4$, NaCl, KCl, NH_4Cl, $CuCl_2$ and $FeCl_2$ solutions. His general expression for complex formation was:

$$mCuCl(s) + nCl^- = Cu_mCl_{m+n}^{n-}$$

with the resulting equilibrium constants $K_{s,m+n}$ and equilibrium quotients $K'_{s,m+n}$. For complexes arising from the addition of chlorine ions to a complex, the notation $K_{m,m+n}$ was used. The formation of the $CuCl_2^-$ was therefore defined:

$$CuCl(s) + Cl^- = CuCl_2^-$$

$$K_{s2} = \frac{a^2_{HCuCl_2}}{a^2_{HCl}} = \frac{CuCl_2^-}{Cl^-}\,\frac{\gamma^2_{HCuCl_2}}{\gamma^2_{HCl}} = K'_{s2}\,\frac{\gamma^2_{HCuCl_2}}{\gamma^2_{HCl}}$$

$$\ln K'_{s2} = \ln K_{s2} = \ln \gamma_2$$

The term γ_2 is used to denote the activity coefficient ratio.

For cuprous chloride in $HCl-HClO_4$ solutions, the solubility data of Hikita et al. (C6) met the requirement for data taken at multiple ionic strengths and chlorine concentrations that Fritz needed in order to solve for the stability constants and activity coefficient equation parameters. Unfortunately, this data, like many sets of solubility data, was presented as molarities without the solution densities needed to convert them to the molalities required by Pitzer's equations. Consequently Fritz replaced the molality terms of the equations with molarities. He presented the following justifications (C2):

"1. Each solubility point represents a combination of a set of thermodynamic equilibrium constants and the corresponding activity coefficients. Correlation of a set of measurements requires only that the expressions for the activity coefficients be adequate to represent them within the accuracy of the data. Provided that the expressions have a reasonable functional form, selection of suitable parameters will accomplish this purpose.

2. For the data used, the ratio of solubility to (total) chloride concentration varied only slowly with concentration. Thus, conversion from one concentration scale to another left the functional form virtually unchanged. It should be observed that the change of concentration units has only a trivial effect on the thermodynamic equilibrium constants. Their determination requires only that the expressions provide suitable functions for extrapolation to infinite dilution."

Fritz emphasizes that the activity coefficient parameters regressed are good only for predictions done on a molarity basis. Using Pitzer's values for the HCl and $HClO_4$ parameters, Fritz regressed this data to get the equilibrium constants at 15, 25 and 35°C and the activity coefficient parameters for the ion pairs. With the cuprous ion concentration being negligible at all solution concentrations, the only cation present would be the hydrogen ion. As a result, the β_1, β_0 and C parameters regressed were for the following ion pairs: $H^+-CuCl_2^-$, $H^+-CuCl_3^{2-}$, $H^+-Cu_2Cl_4^{2-}$ and $H^+-Cu_3Cl_6^{3-}$.

A similar procedure was followed for the other multicomponent solutions. The resultant parameters and equilibrium constants are tabulated in Appendix 6.1.

Figure 6.8

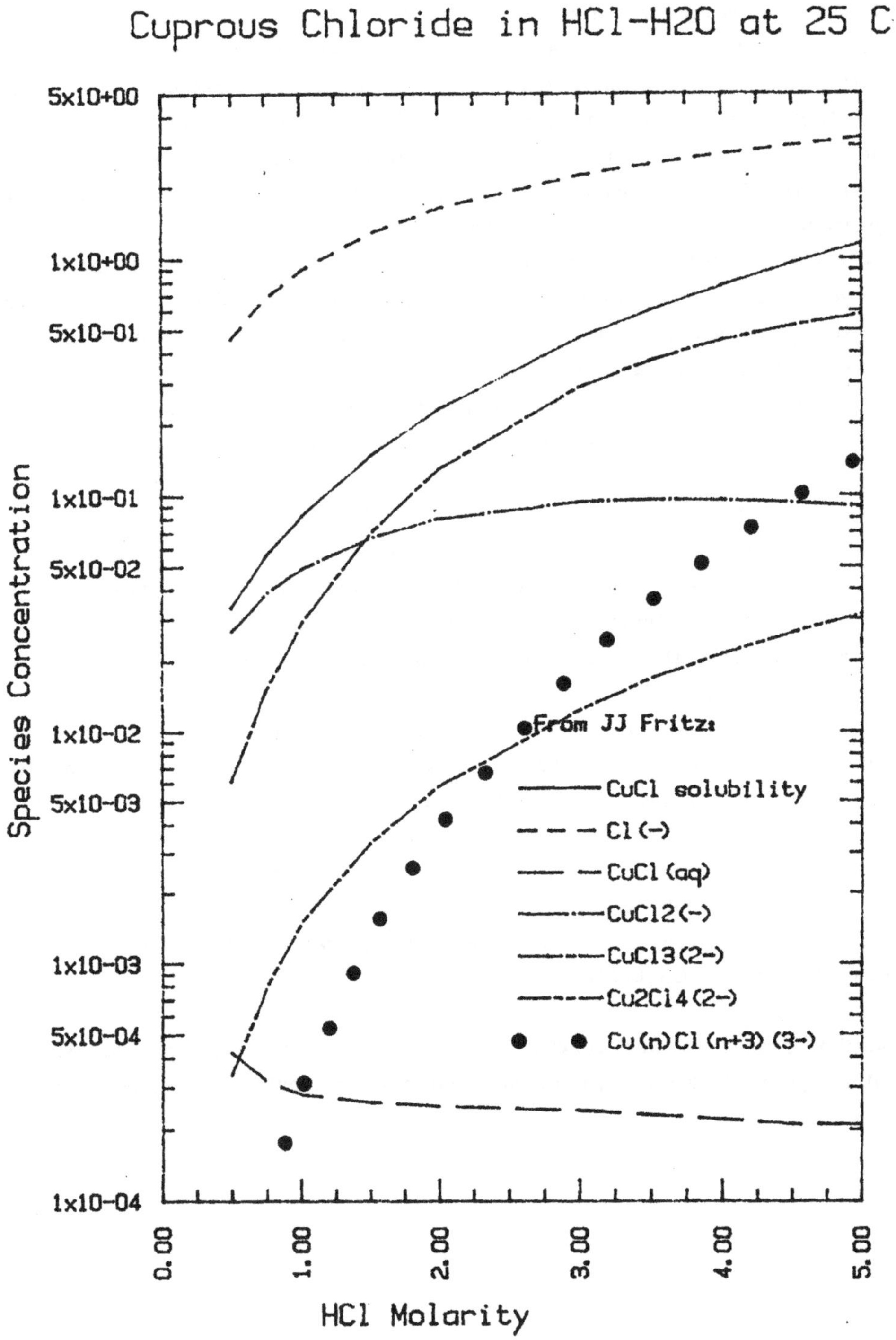
Cuprous Chloride in HCl-H2O at 25 C
5x10+00
1x10+00
5x10-01
1x10-01
5x10-02
1x10-02
5x10-03
1x10-03
5x10-04
1x10-04
Species Concentration
0.00
1.00
2.00
3.00
4.00
5.00
HCl Molarity
From JJ Fritz:
CuCl solubility
Cl(-)
CuCl(aq)
CuCl2(-)
CuCl3(2-)
Cu2Cl4(2-)
Cu(n)Cl(n+3)(3-)

Calcium Sulfate

Gypsum, $CaSO_4.2H_2O$, solubility in pure water and multicomponent solutions has been the topic of many studies. The reference list at the end of this chapter bears witness to this even though it does not list all the work reported. This body of work includes experimental measurements and attempts to predict gypsum and anhydrite precipitation.

Despite the large body of experimental data for calcium sulfate, there is still disagreement about its behavior in solution and how to best model this behavior. One major area of confusion is the transition temperature of gypsum to anhydrite, with values ranging from 38°C (G68) to 97°C (G50). The disparity in mathematical models of the compound is based upon whether or not the aqueous molecule $CaSO_4$ has been considered.

The transition temperature is most often stated to be about 42°C (G52). This value has been calculated from the intersections of solubility curves, thermodynamic calculations and vapor pressure measurements. Part of the difficulty lies in the extremely slow transformation rate between the gypsum and anhydrite phases. Power et al. (G53) note that in their studies of gypsum solubility between 25 and 100°C, the characteristic solubility of gypsum was reached in less than a minute and the solid phase was found to be the dihydrate at all temperatures. When they dissolved insoluble anhydrite they found that, after an initial rapid rate of solubility, the solute concentration continued increasing over a period of days, particularly at temperatures up to 65°C. Checking the solid after 160 hours of contact with water at 85°C with X-ray diffraction indicated only the presence of the insoluble anhydrite. Any hydrated form of calcium sulfate would be less than 1%, but the method was insensitive to amounts this small. This seeming stability of the two forms in this temperature range no doubt lead to the difficulty in determining the transition temperature.

In 1967, Hardie (G24) reported on this extremely slow rate of reaction. The first experimental technique he tried involved the additions of anhydrite or gypsum to sodium sulfate solutions in tubes that were sealed and placed in a water bath for temperature maintenance. After test periods of several days to several months, the tubes were opened and the solids present were examined under the microscope and by X-ray diffraction. Although the solution volumes were too small for analysis, an even greater drawback of the method was that, even after many months, equilibrium had not been attained in many of the tubes. Hardie then chose to use a method incorporating constant agitation. Even with agitation the reaction rates were very slow. In one run at 70°C, a solution of 95% gypsum and 5% anhydrite remained unchanged after 359 days. It is noted that this does not indicate the stability of the phase as it is common in studies of mineral equilibria for metastable starting phases to remain. In runs begun with equal amounts of gypsum and anhydrite it would be expected that any phase changes would be in the direction of the more stable phase. Combining the results of his test between 20 and 70°C, begun with varying gypsum and anhydrite concentrations in pure water or Na_2SO_4 or H_2SO_4 solutions, Hardie determined a transition temperature of 58° ± 2°C. While most of his test results were consistent, Hardie did have some results at 50° that differed with the main body of data. In a 15% solution of Na_2SO_4 and a starting phase of gypsum, after 94 days anhydrite had formed so that the solid phase composition was 95% gypsum and 5% anhydrite. The change stopped at this point and no further change was found after 10 months. At the same sodium sulfate concentration and temperature, a starting phase of anhydrite remained unchanged after 155 days. A few other differences from the main body of data were noted around this temperature.

The insoluble anhydrite mentioned in the report by Power and co-workers' is one of three possible anhydrites. Kelley et at. (G28) found the only unique and reproducible forms of calcium sulfate to be $CaSO_4.2H_2O$, α-$CaSO_4.\frac{1}{2}H_2O$, α-$CaSO_4$, β-$CaSO_4.\frac{1}{2}H_2O$, β-$CaSO_4$, and insoluble anhydrite. The dehydration of gypsum saturated steam at one atmosphere results in the α-hemihydrate. Dehydration at a lower water vapor pressure forms the more soluble β-hemihydrate. Further dehydration of the α- and β-hemihydrates results in the formation of the α- and β-anhydrite. These anhydrites readily rehydrate to gypsum as opposed to the insoluble anhydrite which seems to avoid hydration above 40°C. In their studies of the transient solubilities of calcium sulfate, Power et al. (G53) conclude that β-anhydrite hydrates to β-hemihydrate at a much more rapid rate than the further hydration to gypsum due to the identical, within experimental error, behavior of both. They also note that the difference in crystalline size between the α- and β-hemihydrates (.3 micron and .09 micron respectively, determined by X-ray diffraction), does not account for the greater instability of the β-hemihydrate, but is more likely the result of a higher energy content. Some of the discrepancies in the solubility data no doubt arise from confusion over which of the hemihydrates or anhydrites was used or was precipitated.

Blount and Dickson (G9) have also suggested possible reasons for some of the disparities in the data. They note that very fine grained or poorly crystalline forms of gypsum may have resulted in higher apparent solubilities in some experiments. Anhydrite solubility results may also vary due to poorly crystallized anhydrite or hemihydrate contamination. Most reported saturation curves were the result of approaching saturation from undersaturated solutions. Few experimenters have presented results from both undersaturated and supersaturated solutions.

Other than the disagreements between the various sets of data just discussed, the person attempting to predict the behavior of calcium sulfate in aqueous solution and in solution with other compounds is faced with further problems. One is that, no matter which of the forms of calcium sulfate being considered, the saturation molality of the pure salt in water is very low. For example, Marshall and Slusher (G38) reported the solubility of $CaSO_4.2H_2O$ to be .0151 molal $CaSO_4$ at 25°C. Due to such low saturation molalities, if calcium sulfate behavior is to be modeled in solutions containing any other compounds, parameters developed using binary solution data may be expected to be invalid at multicomponent solution concentrations. This is very likely the cause of the bad results seen in Chapter V when attempting to predict the phase diagrams of multicomponent solutions containing calcium sulfate.

Another problem is deciding how to model the system. In checking the two most complete sources of thermodynamic data, the National Bureau of Standards (1) and Russian Academy of Sciences (2) compilations, two different pictures of the solution behavior emerge. The NBS gives the following data for a gypsum solution:

	ΔG^o	ΔH^o	S^o	C^o_p
	kcal/mole		cal/deg mole	
Ca^{2+} aqueous ion	-132.30	-129.74	-12.7	--
SO_4^{2-} aqueous ion	-177.97	-217.32	4.8	-70.0
$CaSO_4.\ 2H_2O$ solid	-429.60	-483.42	46.4	44.46
H_2O liquid	-56.687	-68.315	16.71	17.995

This leads to modeling the solution as:

$$CaSO_4.2H_2O = Ca^{2+} + SO_4^{2-} + 2H_2O$$

The solubility product at 25°C would be solved for as follows:

$$\Delta G_{RXN} = \Delta G_{Ca^{2+}} + \Delta G_{SO_4^{2-}} + 2\,\Delta G_{H_2O} - \Delta G_{CaSO_4.2H_2O}$$

$$= -132300. - 177970. + 2.(-56687.) + 429600.$$

$$= 5956 \text{ cal/mole}$$

$$K_{CaSO_4.2H_2O} = \exp\left\{\frac{-\Delta G_{RXN}}{R\,T}\right\} = \exp\left\{\frac{-5956}{1.98719 * 298.15}\right\}$$

$$= 4.307154E-5$$

It has been known for years though that calcium and sulfate ions in solution associate to form the aqueous molecule $CaSO_4$. Many studies have been done in an attempt to determine the effect of this association on the system (G2, G6, G13, G16, G17, G20, G23, G30, G32, G33, G44, G45, G46, G47, G55, G67, G72, G73). The Russian thermodynamic data compilation reflects this:

	ΔG^o (kcal/mole)	ΔH^o (kcal/mole)	S^o (cal/deg mole)	C_p^o (cal/deg mole)
Ca^{2+} aqueous ion	-132.112	-129.80	-13.5	--
SO_4^{2-} aqueous ion	-178.215	-217.73	4.3	--
$CaSO_4$ aqueous molecule	-313.48	-345.7	7.514	--
$CaSO_4.2H_2O$ solid	-430.102	-483.94	46.4	44.25
H_2O liquid	-56.703	-68.3149	16.75	17.997

This results in the descriptive equations:

$$CaSO_4.2H_2O = Ca^{2+} + SO_4^{2-} + 2H_2O$$

$$CaSO_4(aq) = Ca^{2+} + SO_4^{2-}$$

The gypsum solubility product at 25°C is:

$$\Delta G_{RXN} = \Delta G_{Ca^{2+}} + \Delta G_{SO_4^{2-}} + 2\,\Delta G_{H_2O} - \Delta G_{CaSO_4.2H_2O}$$

$$= -132112. - 178215. + 2.(-56703.) + 430102.$$

$$= 6369. \text{ cal/mole}$$

$$K_{CaSO_4.2H_2O} = \exp\left\{\frac{-\Delta G_{RXN}}{RT}\right\} = \exp\left\{\frac{-6369.0}{1.98719 * 298.15}\right\}$$

$$= 2.14515E-5$$

This is about half the value of the solubility product calculated using the NBS data that assumes complete dissociation. Since the amount of calcium and sulfate ions in solution are so low and the association of the two to form an aqueous molecule would have little effect on the ionic strength, it may be decided to disregard the interaction. The large difference in the solubility products points out that care must be taken in matching the thermodynamic data to the modeling method chosen.

Sodium Sulfate

Although there is ample evidence of its existence, the $NaSO_4^-$ ion is generally ignored when calculating activity coefficients in solutions containing sodium and sulfate ions. Sodium sulfate is treated as a completely dissociating electrolyte. As early as 1930, Righellato and Davies (S34) stated that, even in dilute solutions, most uni-bivalent salts are incompletely dissociated. Based on conductance measurements at 18°C, they presented dissociation constants for a number of intermediate ions. For the salt M_2X the dissociations were defined:

$$M_2X = M^+ + MX^-$$

$$MX^- = M^+ + X^{2-}$$

and the dissociation constant of the intermediate ion MX^-:

$$K = \frac{c_M\, c_X}{c_{MX}}\ \frac{y_M\, y_X}{y_{MX}}$$

where: c_i - gram moles of ion i per liter

y_i - activity coefficient of ion i

This activity coefficient was given as:

$$\log y = -A\, z^2 \sqrt{I} + BI$$

with the Debye-Hückel value A equal to .5 at 18°C. When fit back into the K equation:

$$\log K = \left(\log \frac{c_M\, c_X}{c_{MX}} - 2\ \sqrt{I}\right) + \Sigma BI$$

The term ΣB is set as 1.30. For the $NaSO_4^-$ ion they found a value of K = .198 at 18°C.

Jenkins and Monk (S19) similarly calculated a value for this dissociation constant in 1950. Based on conductance measurements at 25°C, they arrived at a value of K = .19.

More recent work includes that of Izatt et al. (S16) who presented a calorimetrically determined value of K = .2239 at 25°C in 1969. The $\Delta H°$ value for the reaction did not compare well with that determined by Austin and Mair (S1):

	Izatt et al.	Austin and Mair
$\Delta H°$	-.49 ± .05 kcal/mole	1.12 ± .8 kcal/mole

Martynova et al. (S21) used a calcium-selective electrode to study the behavior of a $CaSO_4$-$NaCl$-H_2O solution at five temperatures. They concluded that the $NaSO_4^-$ complex must be considered as well as the aqueous molecule $CaSO_4$ in order to adequately describe the system. The resulting dissociation constants for $NaSO_4^-$ are:

t°C	$K_{NaSO_4^-}$
20	.206
40	.164
60	.126
80	.103
98	.078

In 1975, Santos et al. (S41) determined the association constant of $NaSO_4^-$ at 25°C to be $\beta_1 = 2.5 \pm .2$ at .5M ionic strength using a sodium-selective electrode. That same year, Reardon (S33) presented a value of $pK = .82 \pm .05$ ($K \simeq .1514$) at 25°C, .1 molal from stoichiometric activity coefficients.

The pressure effects on $NaSO_4^-$ were the subject of papers by Fisher (S10) and Fisher and Fox (S9, S11). The following equation for the dissociation constant resulted:

$$\log K_{NaSO_4^-} = (-1.02 + 9.6x10^{-5}P - 4.3x10^{-9}\ P^2) \pm .03$$

where P is the pressure in atmospheres. The dissociation constant at 1 atm. is .094 using this equation and is considerably smaller than those presented by the other authors.

The NBS (1) and Russian (2) thermodynamic data compilations both present values of $\Delta G°$ for the $NaSO_4^-$ complex. The resulting dissociation constants for the reaction $NaSO_4^- = Na^+ + SO_4^{2-}$ are:

- NBS

$$\Delta G_{RXN} = \Delta G_{Na^+} + \Delta G_{SO_4^{2-}} - \Delta G_{NaSO_4^-}$$

$$= -62593. - 177970. + 241560.$$

$$= 997 \text{ cal/mole}$$

$$K = \exp\left(\frac{-\Delta G_{RXN}}{R\,T}\right) = .18586$$

- Russian

$$\Delta G_{RXN} = \Delta G_{Na^+} + \Delta G_{SO_4^{2-}} - \Delta G_{NaSO_4^-}$$

$$= -62647. - 178215. + 241820.$$

$$= 958 \text{ cal/mole}$$

$$K = \exp\left(\frac{-\Delta G_{RXN}}{R\,T}\right) = .19851$$

Using the Meissner plot test gives an indication of possible complexing. The activity coefficients tabulated by Goldberg (S13) were plotted as reduced activity coefficients versus ionic strength in Figure 6.9. The activity coefficients presented were fit past the saturation molality of 1.957. The upward curve at the end of the plotted values points towards ion association of some sort.

Figure 6.9

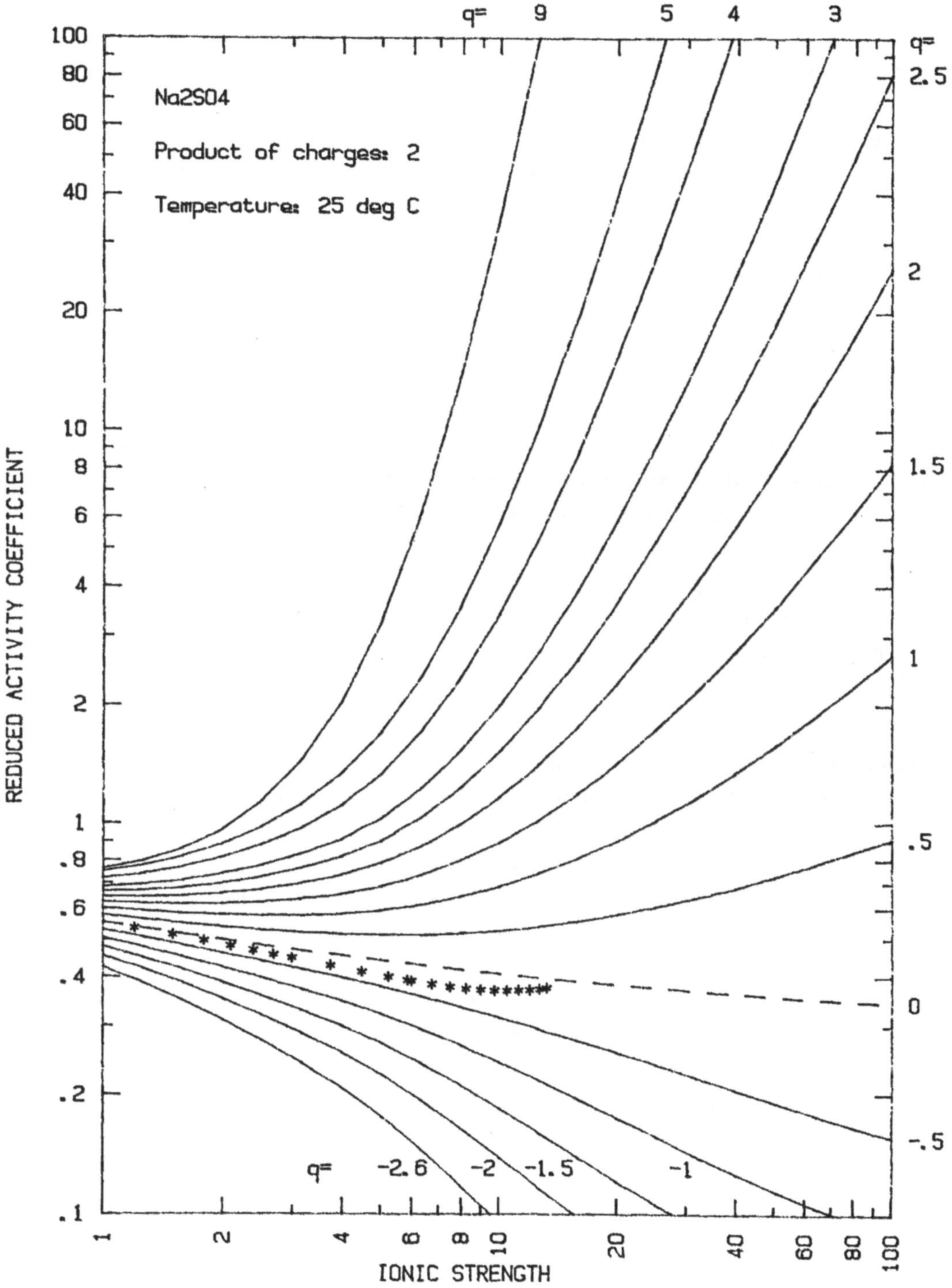
q=
9
5
4
3
q=
2.5
2
1.5
1
.5
0
-.5
Na2SO4
Product of charges: 2
Temperature: 25 deg C
100
80
60
40
20
10
8
6
4
2
1
.8
.6
.4
.2
.1
REDUCED ACTIVITY COEFFICIENT
q= -2.6 -2 -1.5 -1
1
2
4
6
8
10
20
40
60
80
100
IONIC STRENGTH

Other Chloride Complexes

Righellato and Davies (S34) postulated in 1930 that most uni-bivalent salts must be considered as incompletely dissociated even in dilute solutions. It may be expected then that ion association will occur for many bi-univalent salts. This would seem to be particularly true of the doubly charged transition metals. Zinc chloride is one such salt which has been shown to form many complexes. The "complexing tests" described earlier in this chapter are applied to manganous chloride, cobalt chloride, nickel chloride, and cupric chloride.

Manganous chloride

When plotted on the Meissner chart, Figure 6.10, the activity coefficients presented by Goldberg (O6) can be seen to deviate from the q line pattern. The activity coefficient of manganous perchlorate, which would not be expected to complex, can be seen in Figure 6.11 to follow the q lines quite well. $MnCl_2$ may therefore be suspected of complex formation. Libuś and Tialowska (O10) studied a $MnCl_2$ solution potentiometrically and spectrophotometrically in order to determine the nature of the complex and its stability constant at 25°C. The stability constants:

$$\beta_1 = \frac{m_{MnCl^+}}{m_{Mn^{2+}}\ m_{Cl^-}}$$

calculated from potentiometric measurements of a cell using a $Mn(ClO_4)_2$ solution as the reaction medium, were found to be:

	molality of $Mn(ClO_4)_2$					
	1.0	1.5	2.0	2.5	3.0	0.0
β_1	0.14	0.19	0.28	0.38	0.50	0.72

According to the thermodynamic data tables, $MnCl^+$ is not the only complex occurring in a manganous chloride solution. Both the NBS (1) and Russians (2) report an undissociated aqueous molecule, $MnCl_2$, and the ion $MnCl_3^-$ as well as the first complex $MnCl^+$. The Gibbs free energies given and the resulting stepwise formation constants at 25°C are:

Figure 6.10

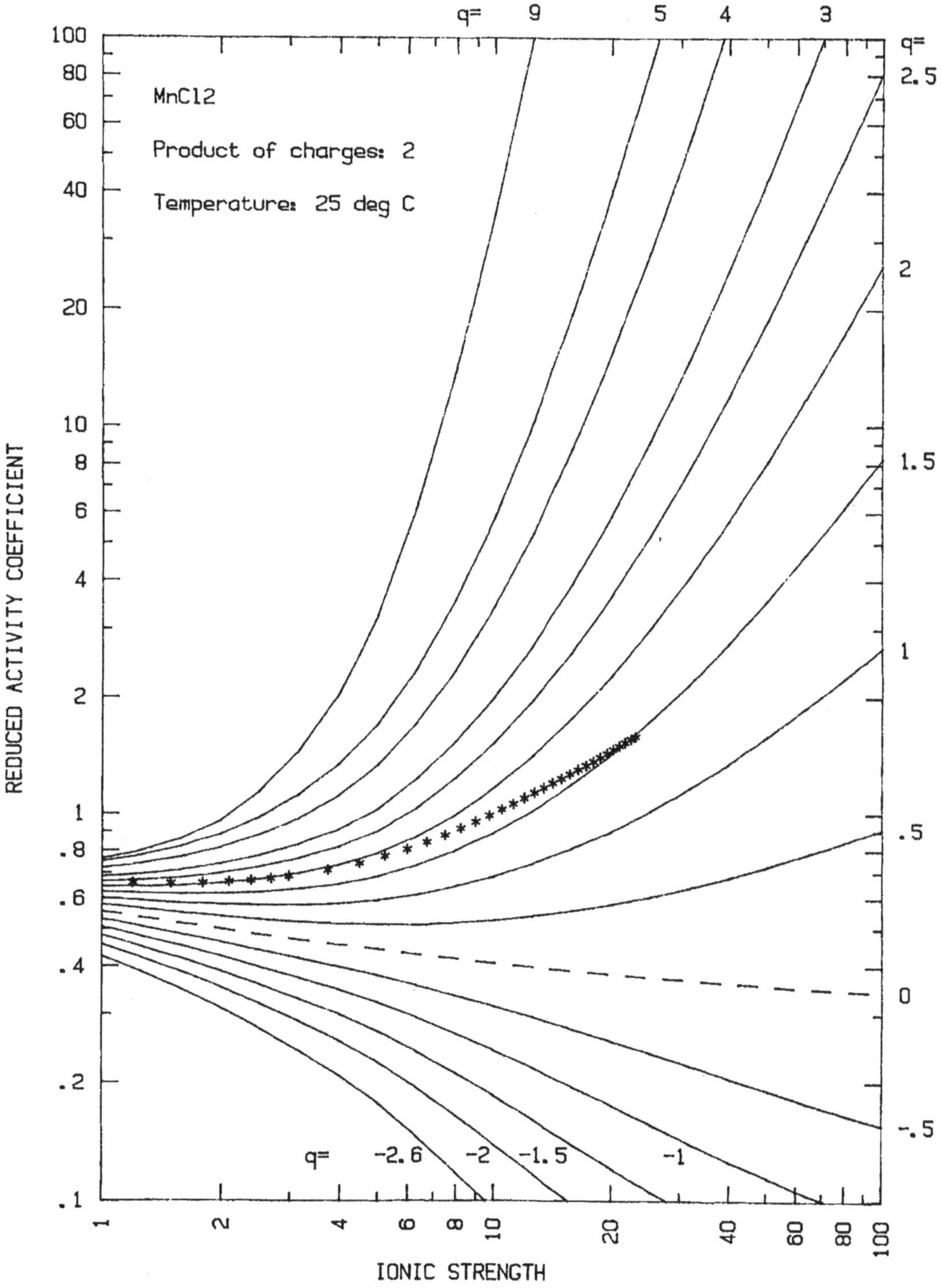
q= 9
5
4
3
q=
2.5
2
1.5
1
.5
0
-.5
MnCl2
Product of charges: 2
Temperature: 25 deg C
q= -2.6
-2
-1.5
-1
100
80
60
40
20
10
8
6
4
2
1
.8
.6
.4
.2
.1
REDUCED ACTIVITY COEFFICIENT
1
2
4
6
8
10
20
40
60
80
100
IONIC STRENGTH

Figure 6.11

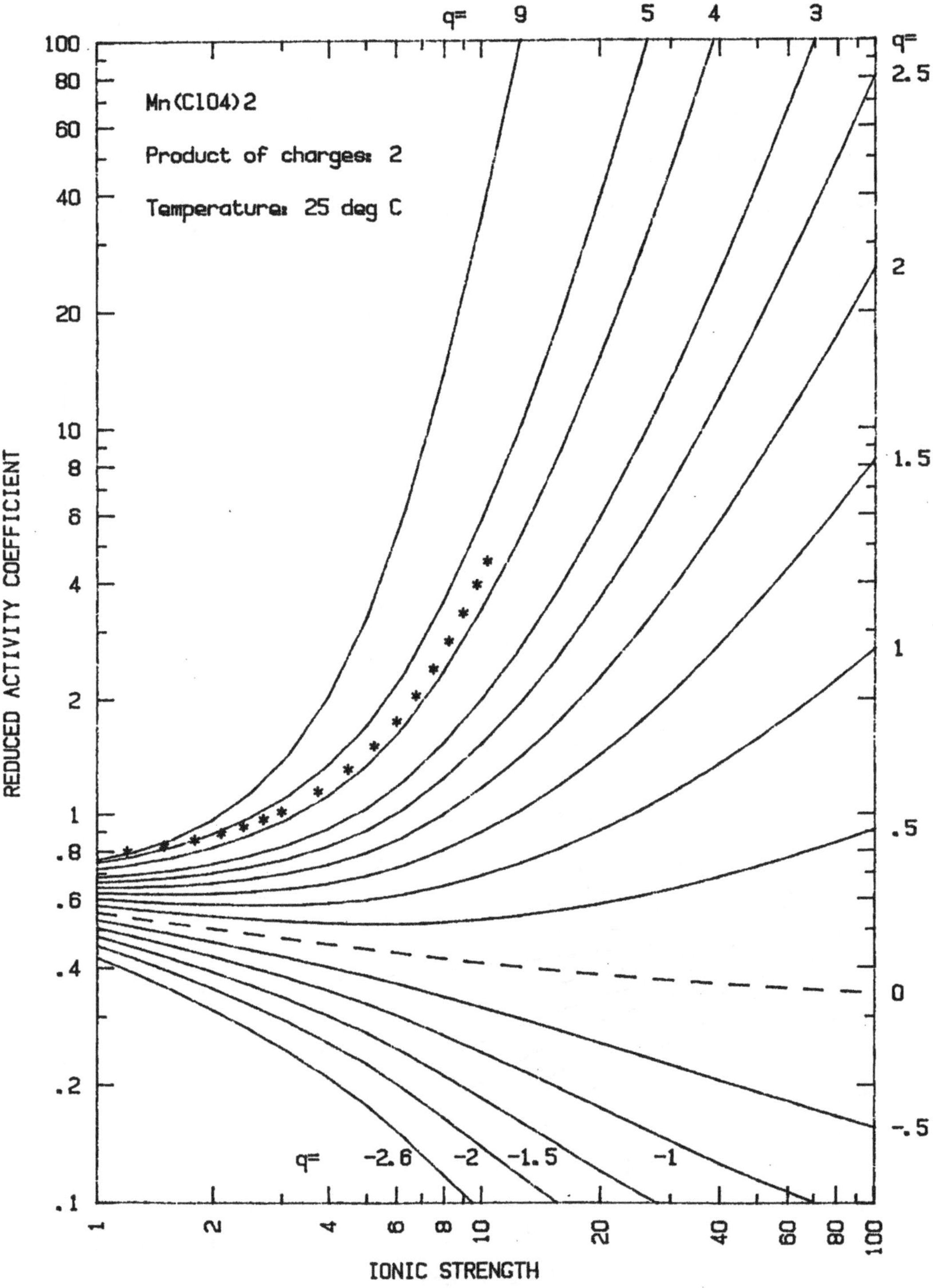
q= 9 5 4 3
q= 2.5 2 1.5 1 .5 0 -.5
Mn(ClO4)2
Product of charges: 2
Temperature: 25 deg C
REDUCED ACTIVITY COEFFICIENT
100 80 60 40 20 10 8 6 4 2 1 .8 .6 .4 .2 .1
q= -2.6 -2 -1.5 -1
1 2 4 6 8 10 20 40 60 80 100
IONIC STRENGTH

	ΔG°, kcal/mole	
	NBS	Russian
Mn^{2+}	-54.5	-55.2
Cl^-	-31.372	-31.373
$MnCl^+$	-86.7	-88.4
$MnCl_2(aq)$	-117.6	-119.82
$MnCl_3^-$	-148.2	-150.36

- for $Mn^{2+} + Cl^- = MnCl^+$

  K_1 (NBS) = 4.045

  K_1 (Russian) = 21.838

- for $MnCl^+ + Cl^- = MnCl_2(aq)$

  K_2 (NBS) = .45084

  K_2 (Russian) = 1.08256

- for $MnCl_2(aq) + Cl^- = MnCl_3^-$

  K_3 (NBS) = .27172

  K_3 (Russian) = .24513

Cobalt Chloride

The osmotic coefficient ratio test, seen in Figure 6.12, shows the increased deviation from unity for the chloride anion indicating possible complexing. The activity coefficients for $CoCl_2$ and $Co(ClO_4)_2$, tabulated by Goldberg et al. (O5), have been plotted on the Meissner chart, Figures 6.13 and 6.14. The fit to the q line pattern is quite good and does not appear to indicate complexing behavior.

Libuś and Tialowska (O10) presented the following stability constants at 25°C:

$$\beta_1 = \frac{m_{CoCl^-}}{m_{Co^{2+}}\ m_{Cl^-}}$$

	molality of $Co(ClO_4)_2$					
	1.0	1.5	2.0	2.5	3.0	0.0
β_1	0.10	0.10	0.14	0.18	0.22	0.45

Figure 6.12

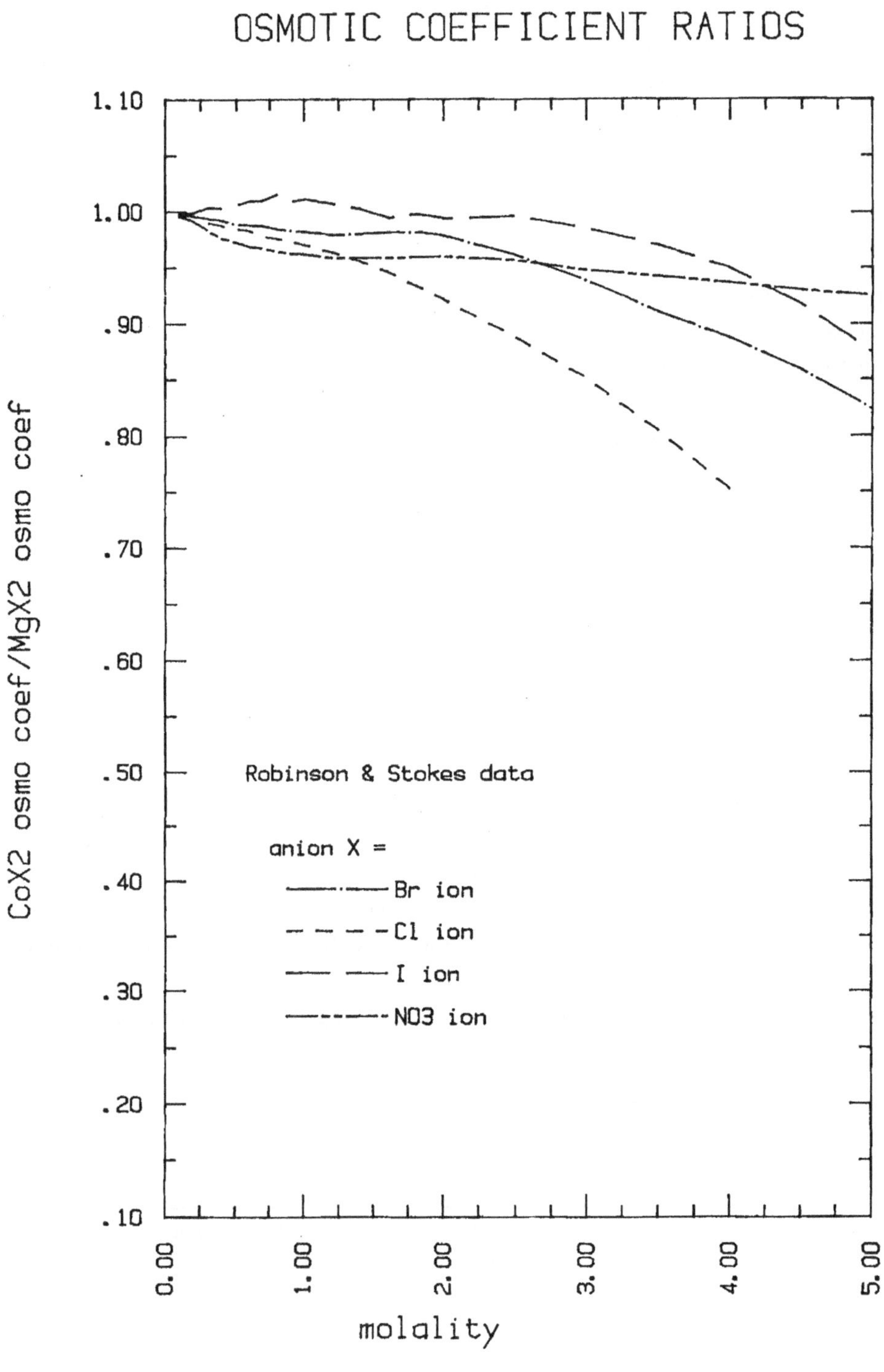
OSMOTIC COEFFICIENT RATIOS
1.10
1.00
.90
.80
.70
.60
.50
.40
.30
.20
.10
CoX2 osmo coef/MgX2 osmo coef
Robinson & Stokes data
anion X =
Br ion
Cl ion
I ion
NO3 ion
0.00
1.00
2.00
3.00
4.00
5.00
molality

Figure 6.13

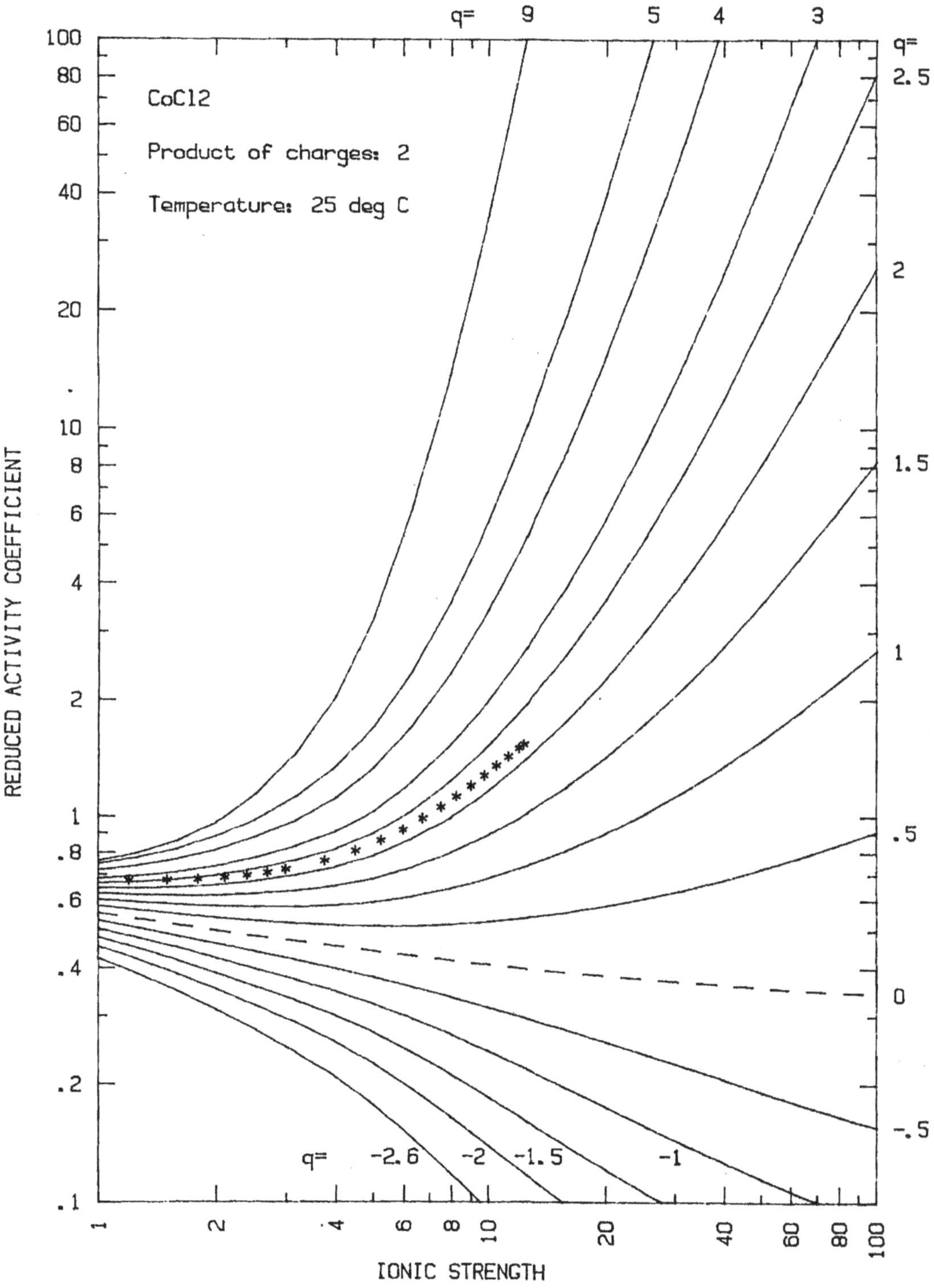
q= 9 5 4 3
q=
2.5
2
1.5
1
.5
0
-.5
CoCl2
Product of charges: 2
Temperature: 25 deg C
100
80
60
40
20
10
8
6
4
2
1
.8
.6
.4
.2
.1
REDUCED ACTIVITY COEFFICIENT
q= -2.6 -2 -1.5 -1
1
2
4
6
8
10
20
40
60
80
100
IONIC STRENGTH

Figure 6.14

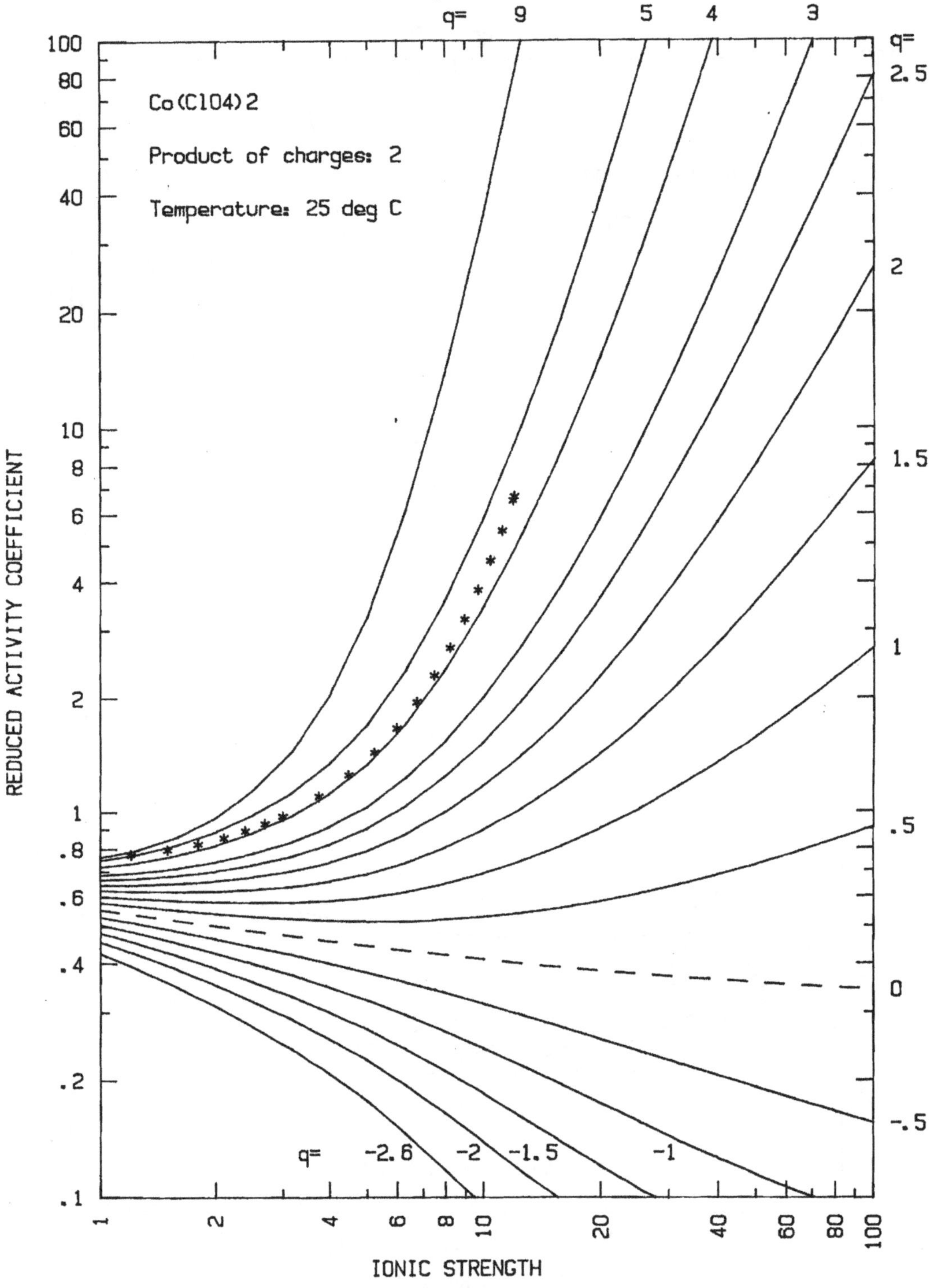
q= 9
5
4
3
q=
2.5
2
1.5
1
.5
0
-.5
Co(ClO4)2
Product of charges: 2
Temperature: 25 deg C
100
80
60
40
20
10
8
6
4
2
1
.8
.6
.4
.2
.1
REDUCED ACTIVITY COEFFICIENT
q= -2.6
-2
-1.5
-1
1
2
4
6
8
10
20
40
60
80
100
IONIC STRENGTH

They found that the activity coefficient quotient dependence on the perchlorate molality differed for the $CoCl^+$ complex from the dependence shown by the $CuCl^+$, $MnCl^+$ and $ZnCl^+$ complexes. As a careful spectrophotometric study they did indicated that the $CuCl^+$ complex existed primarily as the inner-sphere complex $[CuCl(OH_2)_5]^+$, they speculated that the difference was that the $CoCl^+$ complex was an outer-sphere complex $\{[Co(OH_2)_6]^{2+}Cl^-\}^+$.

An examination of the thermodynamic data tables of the National Bureau of Standards (1) and Russian Academy of Science (2) present different pictures of the solution composition. The NBS table does not point to any ion pairing while the Russian table indicates the existence of the $CoCl^+$ complex. The free energy values and resulting formation constant for 25°C from the Russian table are:

	$\Delta G°$, kcal/mole
Co^{2+}	-12.82
Cl^-	-31.373
$CoCl^+$	-44.74

$Co^{2+} + Cl^- = CoCl^+$

$K_1 = 2.5174$

<u>Nickel Chloride</u>

No clear picture of complexing behavior in aqueous nickel chloride solutions has yet been presented. As with cobalt chloride, Libuś and Tialowska (O10) suggest the $NiCl^+$ complex exists as an outer-sphere ion pair $[Ni(OH_2)_6]^{2+}Cl^-$ $^+$. They presented the following values for the stability constant at 25°C:

	molality of $Ni(ClO_4)_2$					
	1.0	1.5	2.0	2.5	3.0	0.0
β_1	0.08	0.08	0.10	0.12	0.16	0.37

In 1980, Lagarde et al. (O8) presented the results of their EXAFS (extended x-ray absorption fine structure) study of $NiCl_2$ solutions. Previous studies, including Raman experiments (O3, O4, O12), x-ray diffraction (O1, O9, O11), neutron diffraction (O7, O13, O14) and inelastic neutron scattering (O2) presented conflicting reports of ion pairing behavior. Lagarde et al. were unable to conclude whether there was complex formation or not.

Figure 6.15

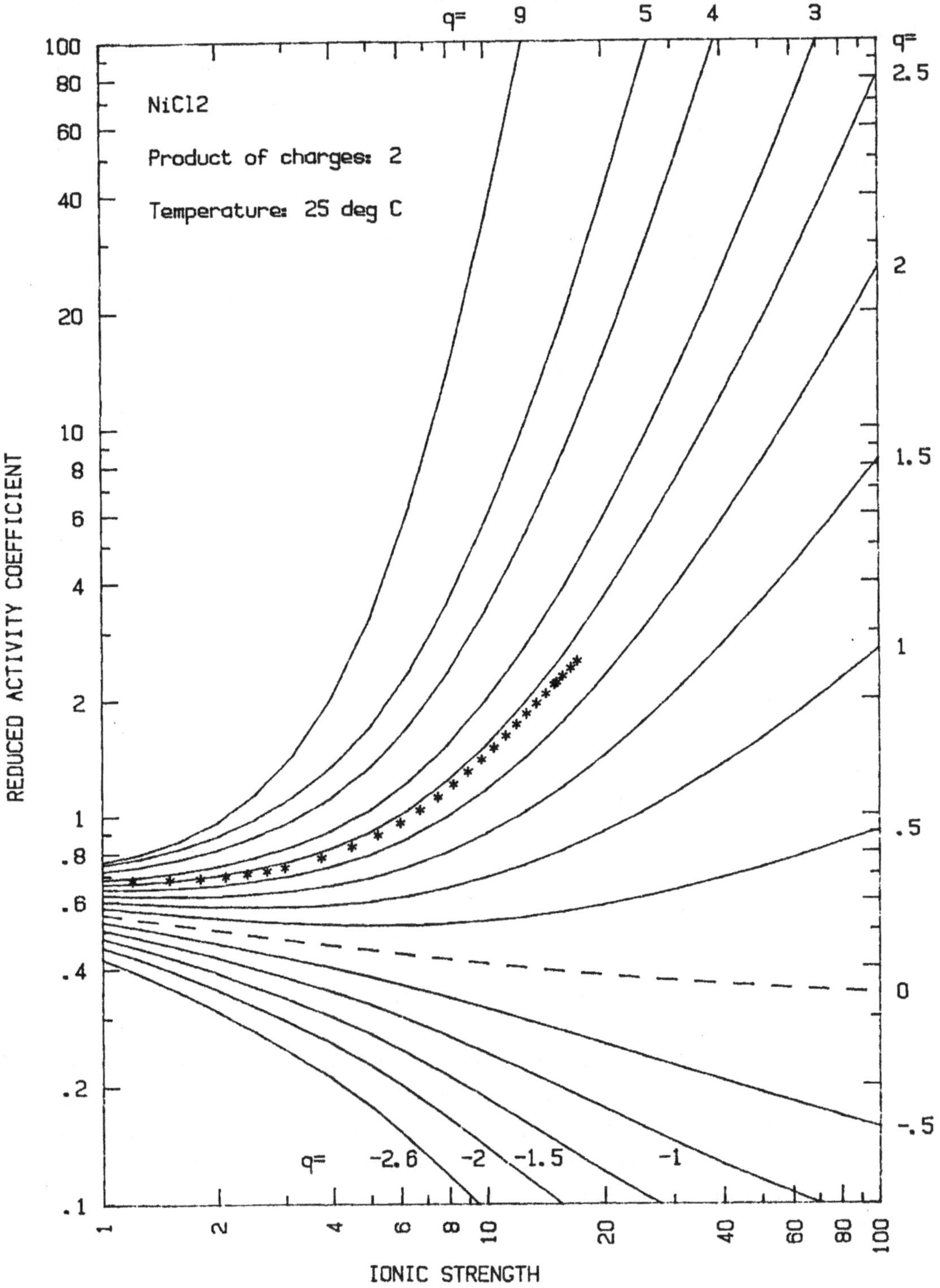
q= 9 5 4 3
q= 2.5 2 1.5 1 .5 0 -.5
NiCl2
Product of charges: 2
Temperature: 25 deg C
REDUCED ACTIVITY COEFFICIENT
100 80 60 40 20 10 8 6 4 2 1 .8 .6 .4 .2 .1
q= -2.6 -2 -1.5 -1
1 2 4 6 8 10 20 40 60 80 100
IONIC STRENGTH

The activity coefficients tabulated by Goldberg et al. (O5) behave well when plotted on a Meissner chart, Figure 6.15. The thermodynamic compilations of the National Bureau of Standards (1) and Russian Academy of Science (2) also disagree as to possible ion pairing. While the NBS presents no data for any complexes of the nickel and chlorine ions, the Russian table has data for not only the intermediate $NiCl^+$ ion, but also for the $NiCl_2(aq)$ undissociated molecule:

	ΔG°, kcal/mole
Ni^{2+}	-10.89
Cl^-	-31.373
$NiCl^+$	-42.71
$NiCl_2(aq)$	-74.69

For the intermediate $NiCl^+$:

$$Ni^{2+} + Cl^- = NiCl^+$$

$$K_1 = 2.1265$$

The formation constant for the aqueous molecule is:

$$Ni^{2+} + 2Cl^- = NiCl_2(aq)$$

$$K_2 = 5.9237$$

Cupric Chloride

A more definite picture is presented in the literature for cupric chloride solutions. Lagarde et al. (O8) found evidence of Cu-Cl pairing in their EXAFS study, as did Libuš (O9) in spectrophotometric measurements which resulted in the following stability constants at 25°C:

	molality of $Cu(ClO_4)_2$					
	1.0	1.5	2.0	2.5	3.0	0.0
β_1	0.34	0.43	0.56	0.76	1.13	1.63

The activity coefficients tabulated by Goldberg (O6) cut across the q lines of Meissner's chart, Figure 6.16, as opposed to the perchlorate activity coefficients which hold to the pattern, Figure 6.17.

Figure 6.16

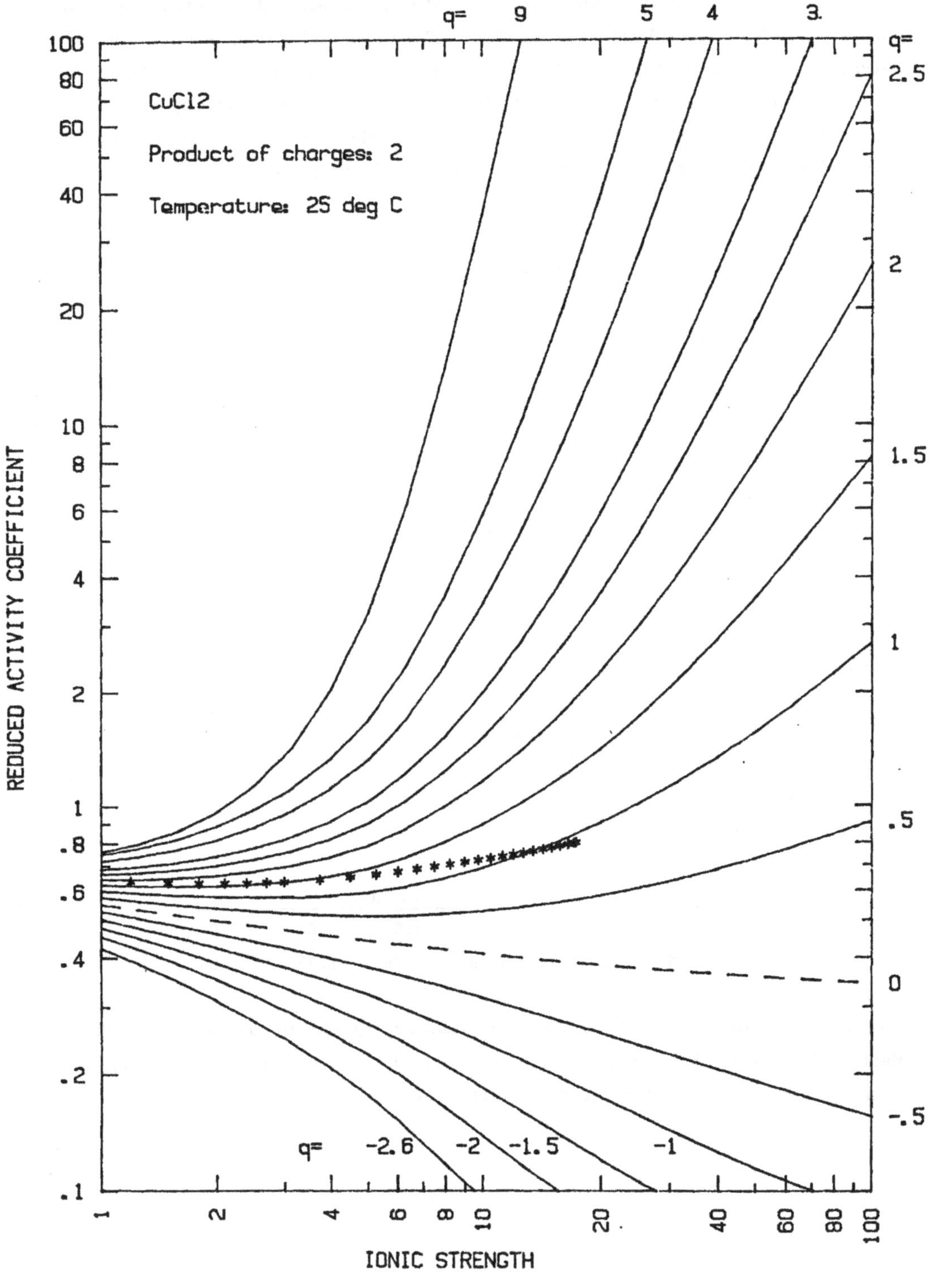
q= 9
5
4
3.
q=
2.5
2
1.5
1
.5
0
-.5
100
80
60
40
20
10
8
6
4
2
1
.8
.6
.4
.2
.1
CuCl2
Product of charges: 2
Temperature: 25 deg C
q= -2.6
-2
-1.5
-1
REDUCED ACTIVITY COEFFICIENT
1
2
4
6
8
10
20
40
60
80
100
IONIC STRENGTH

Figure 6.17

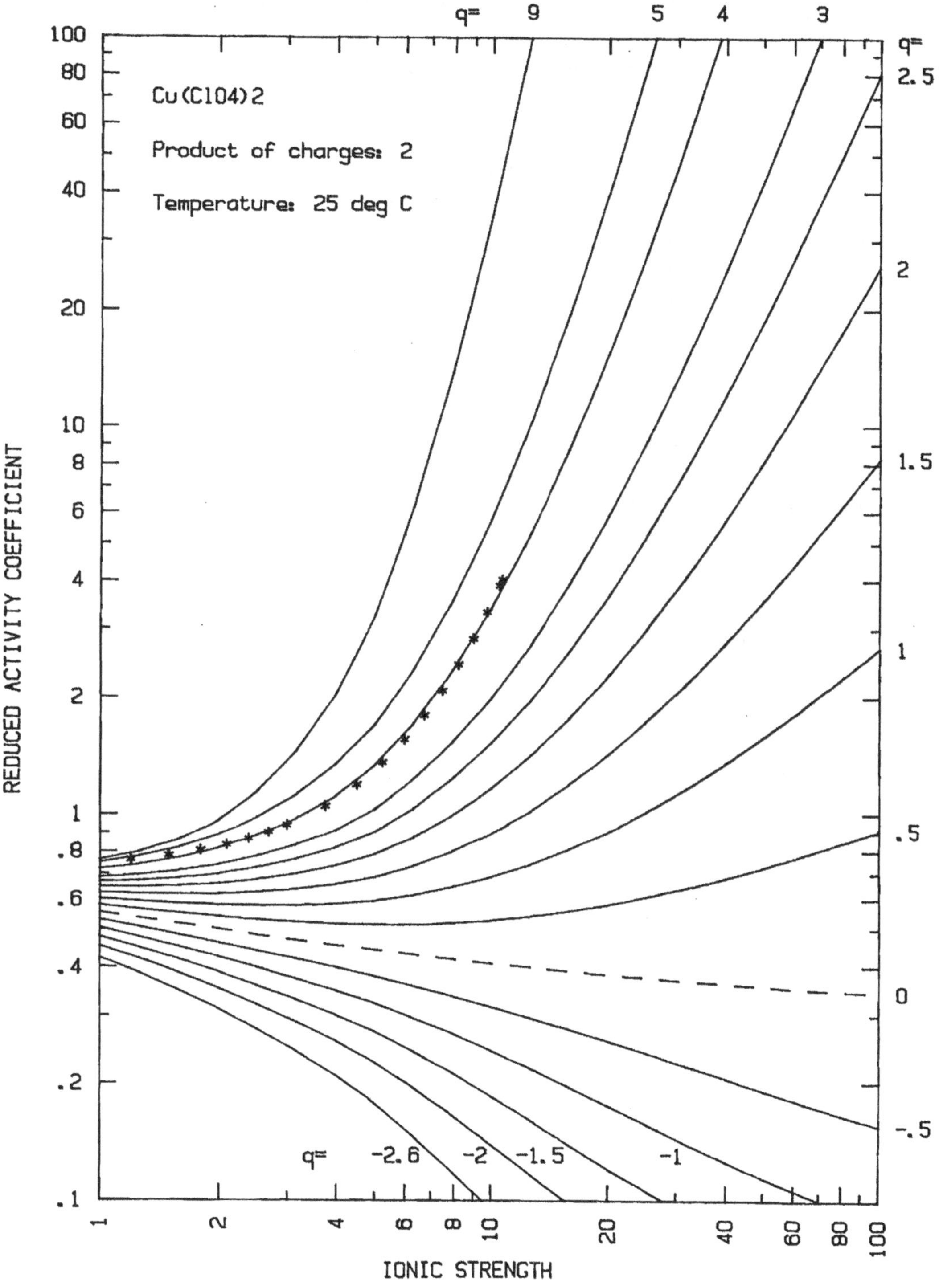
q= 9 5 4 3
q= 2.5
2
1.5
1
.5
0
-.5
Cu(C104)2
Product of charges: 2
Temperature: 25 deg C
REDUCED ACTIVITY COEFFICIENT
100
80
60
40
20
10
8
6
4
2
1
.8
.6
.4
.2
.1
q= -2.6 -2 -1.5 -1
1
2
4
6
8
10
20
40
60
80
100
IONIC STRENGTH

A uniform view of complexes in a cupric chloride solution is found in the NBS (1) and Russian (2) thermodynamic data tables. Both present data for not only the $CuCl^+$ complex, but also the aqueous molecule $CuCl_2$:

	$\Delta G°$, kcal/mole	
	NBS	Russian
Cu^{2+}	15.66	15.67
Cl^-	-31.372	-31.373
$CuCl^+$	-16.3	-16.71
$CuCl_2(aq)$	-47.3	-47.62

At 25°C, the following formation constants result:

- for $Cu^{2+} + Cl^- = CuCl^+$

  K_1 (NBS) = 2.6978

  K_1 (Russian) = 5.4719

- for $CuCl^+ + Cl^- = CuCl_2(aq)$

  K_2 (NBS) = .53373

  K_2 (Russian) = .45774

Activity Coefficient Methods

Due to the fact that complexing results in multiple anions and/or cations, activity coefficients for such species must be calculated using multicomponent methods even for simple binary solutions. The methods presented in Chapter V are suitable for use.

Gugggenheim's method is applicable for use with these complexes. The ionic activity equation:

$$\log \gamma_i = \frac{-A\, z_i^2 \sqrt{I}}{1 + \sqrt{I}} + \sum_s B_{is}\, m_s \tag{5.2}$$

where:

- γ_i - activity coefficient of ion i
- A - log base 10 Debye-Hückel constant
- z_i - ionic charge
- I - solution ionic strength
- B_{is} - coefficient for the interaction between ion i and species s
- m_s - molality of species s

is applicable due the summation term. The species may be any of the complexes in the solution, including molecular species. However, since Guggenheim's method is accurate at low ionic strengths only, it would be advisable to choose one of the other methods for general use.

Bromley's expression for the ionic activity coefficient of ion i is:

$$\log \gamma_i = -\frac{A\, z_i^2 \sqrt{I}}{1 + \sqrt{I}} + F_i \tag{5.5}$$

where:

- A - Debye-Hückel constant
- I - ionic strength
- z_i - ionic charge
- F_i - interaction terms

$$F_i = \sum_j \dot{B}_{ij}\, Z_{ij}^2\, m_j$$

$$Z_{ij} = \frac{z_i + z_j}{2}$$

m_j - molality of ion j

$$\dot{B}_{ij} = \frac{(.06 + .6B)\,|z_i z_j|}{\left(1 + \frac{1.5}{|z_i z_j|}\, I\right)^2} + B \tag{5.8}$$

Due to the mixing rule's dependence on the ionic charges, this method is really only applicable for species which have no molecular complexes.

The same holds true for Meissner's method since the mixing terms are ionic strength fractions and are dependent on the ionic charge. The equation for the reduced activity coefficient of species CA is:

$$\Gamma_{CA} = \left[1. + B\,(1. + 0.1\ I)^{q_{CA,mix}} - B\right]\Gamma^* \tag{5.30}$$

where:

$B = 0.75 - 0.065\ q_{CA,mix}$

$\log \Gamma^* = -\dfrac{0.5107\ \sqrt{I}}{1 + C\sqrt{I}}$

$C = 1. + 0.055\ q_{CA,mix}\ \exp(-0.023\ I^3)$

I = ionic strength

$q_{CA,mix} = (\sum_j I_j\, q^o_{Cj})/I + (\sum_i I_i\, q^o_{iA})/I$

with: q^o_{mn} - Meissner q parameter for electrolyte mn in pure solution

$I_n = .5\ m_n\, z_n^2$ ionic strength of ion n

To date, Pitzer's method is the one most often used for complexing species. The expression for the ionic activity coefficient is:

$$\ln \gamma_i = z_i^2\, f^{\gamma} + 2 \sum_j m_j\,[B_{ij} + (\sum mz)\, C_{ij}] +$$

$$2 \sum_k m_k\, \Theta_{ik} + \sum_m \sum_n m_m m_n\,[z_i^2\, B'_{mn} + |z_i|\, C_{mn}$$

$$+ \psi_{imn}] + 0.5 \sum_e \sum_{e'} m_e\, m_{e'}\, \psi_{iee'}$$

where: $$f^{\gamma} = -A_{\phi}\left[\frac{\sqrt{I}}{1 + 1.2\sqrt{I}} + \frac{2}{1.2}\ln(1 + 1.2\sqrt{I}\,)\right] \tag{5.34}$$

A_{ϕ} - Debye-Hückel constant for osmotic coefficients, log e basis; equation (4.64)

I - ionic strength

m - molality

Θ_{ik} - interaction parameter for like-charged ions

ψ - ternary interaction term

$$B_{ij} = \beta_0 + \frac{2\beta_1}{\alpha_1^2 I}\{1. - (1. + \alpha_1\sqrt{I})\exp(-\alpha_1\sqrt{I})\} + \frac{2\beta_2}{\alpha_2^2 I}\{1. - (1. + \alpha_2\sqrt{I})\exp(-\alpha_2\sqrt{I})\}$$

$$B'_{ij} = \frac{2\beta_1}{\alpha_1^2 I}\{-1. + (1. + \alpha_1\sqrt{I} + .5\,\alpha_1^2 I)\exp(-\alpha_1\sqrt{I})\} + \frac{2\beta_2}{\alpha_2^2 I}\{-1. + (1. + \alpha_2\sqrt{I} + .5\,\alpha_2^2 I)\exp(-\alpha_2\sqrt{I})\}$$

$$C_{ij} = \frac{C^{\phi}_{ij}}{2|z_i z_j|^{1/2}}$$

The most often used interaction parameters are the β_0, β_1 and C^{ϕ}, many of which can be found in the appendices of Chapter IV. Values for the interaction parameters Θ and ψ are tabulated in the Chapter V appendices. This method can be used for systems containing aqueous molecular species; ion-ion, ion-molecule, ion-ion-ion, molecule-ion-ion and molecule-molecule-ion interaction parameters can be found for calculating ionic activity coefficients. The equation has been modified for calculating undissociated aqueous molecule activity coefficients. This will be discussed in greater depth in the next chapter.

Summary

The most important message of this chapter is that nearly all electrolytes form one or more complexes over some range of their phase space with water and/or other compounds in water. For many species the conditions over which this occurs and the attendant quantities make it unreasonable to ignore the phenomena. Trying to bend strong electrolyte models to handle species which form complexes is often as invalid as using conventional VLE formulations to handle electrolyte solutions, ie. ignoring ion formation.

As we have seen in this chapter, there are several techniques for identifying complexors as well as the actual species propagated. In addition there are several models available for activity coefficients with one, in particular, Pitzer's, able to handle both ionic and molecular intermediates.

APPENDIX 6.1
Cuprous Chloride

The following parameters for the Pitzer activity coefficient equation were presented by Fritz. The parameters are for use only when the solution concentrations are expressed in terms of molarities:

Table 1a - Interaction Parameters

ion pair	β_0	β_1	β_2	C	ref.
H^+-$CuCl_2^-$	.2072	.3215		.0107	C2
H^+-$CuCl_3^{2-}$	.2746	1.5870		.0254	C2
H^+-$Cu_2Cl_4^{2-}$	.3666	1.7740		.0076	C2
H^+-$Cu_3Cl_6^{3-}$	.4498	4.5670		.0218	C2
Na^+-$CuCl_2^-$	.0837	.1595		.0098	C3
Na^+-$CuCl_3^{2-}$	.0896	1.1780		.0329	C3
Na^+-$Cu_2Cl_4^{2-}$	-.0169	.6372		.0154	C3
Na^+-$Cu_3Cl_6^{3-}$	.1577	3.4014		.0209	C3
K^+-$CuCl_2^-$	.2837	.4180		.0098	C3
K^+-$CuCl_3^{2-}$	.0480	.9540		.0114	C3
K^+-$Cu_2Cl_4^{2-}$	-.1719	-.7000		.0076	C3
K^+-$Cu_3Cl_6^{3-}$	-.1123	2.1330		.0219	C3

Table 1b - Three Parameter Set

ion pair	β_0	β_1	β_2	C	ref.
NH_4^+-$CuCl_2^-$	-.1422	.0545		.0107	C3
NH_4^+-$CuCl_3^{2-}$	-.0677	.2610		.0254	C3
NH_4^+-$Cu_3Cl_6^{3-}$	-.1775	1.8930		.0215	C3
Cu^{2+}-$CuCl_2^-$	.4589	1.7570		-.0372	C4
Cu^{2+}-$CuCl_3^{2-}$	.0192	7.6990	-20.79	-.0396	C4
Cu^{2+}-$Cu_3Cl_6^{3-}$	.9515	9.2520	-10.57	-.1708	C4

ion pair	β_0	β_1	β_2	C	ref.
Fe^{2+}-$CuCl_2^-$	.5489	1.6713		-.00071	C4
Fe^{2+}-$CuCl_3^{2-}$	.3880	4.2740	-4.82	.03230	C4
Fe^{2+}-$Cu_3Cl_6^{3-}$	.9329	7.4660	-3.20	-.01229	C4

The parameters listed as being from the three parameter set were the result of regression done on data sets with an insufficient number of points to regress for the full twelve parameters. As a result, Fritz treated the two doubly charged complexes together.

Table 2: Equilibrium Constants and Heats of Reaction - Ref. C4

complex formed	K(298.15K)	ΔH, cal/mole
$CuCl_2^-$	0.0604	6669.
$CuCl_3^{2-}$	0.0128	3450.
$Cu_2Cl_4^{2-}$	8.24×10^{-4}	6700.
$Cu_3Cl_6^{3-}$	3.41×10^{-5}	0.0
combined^{2-} *	0.0144	4010.

* combined should be used when using the β's and C's of the three parameter set.

Table 3a - Equilibrium Constants and Changes in Thermodynamic Properties for Formation of $CuCl_2^-$ and $CuCl_3^{2-}$ from $CuCl(s) + nCl^- = CuCl_{n+1}^{n-}$ - Ref. C5

complex	°C	K	ΔH° cal/mole	ΔG° cal/mole	ΔS° cal/mole K
$CuCl_2^-$	15	.0409	6350.	1830.	15.6
	25	.0604	6530.	1660.	16.3
	35	.0869	6720.	1500.	16.9
$CuCl_3^{2-}$	15	.0105	3000.	2610.	1.4
	25	.0128	3430.	2580.	2.9
	35	.0150	3530.	2570.	3.1

Table 3b - Equilibrium Constants and Changes in Thermodynamic Properties for Formation of $CuCl_2^-$ and $CuCl_3^{2-}$ from $Cu^+ + nCl^- = CuCl_n^{(n-1)-}$ - Ref. C5

complex	°C	K	ΔH° cal/mole	ΔG° cal/mole	ΔS° cal/mole K
$CuCl_2^-$	15	3.1×10^5	-2860.	-7240.	15.2
	25	2.6×10^5	-2680.	-7390.	15.8
	35	2.3×10^6	-2500.	-7560.	16.4
$CuCl_3^{2-}$	15	7.9×10^4	-6300.	-6460.	0.6
	25	5.6×10^4	-5880.	-6480.	2.0
	35	4.0×10^4	-5780.	-6490.	2.3

REFERENCES

1. (a) Wagman, D.D.; W.H. Evans; V.B. Parker; I. Halow; S.M. Bailey; R.H. Schumm; Selected Values of Chemical Thermodynamic Properties - Tables for the First Thirty-Four Elements in the Standard Order of Arrangement, National Bureau of Standards Technical Note 270-3, 1968

(b) Wagman, D.D.; W.H. Evans; V.B. Parker; I. Halow; S.M. Bailey; R.H. Schumm; Selected Values of Chemical Thermodynamic Properties - Tables for Elements 35 through 53 in the Standard Order of Arrangement, NBS Tech Note 270-4, 1969

(c) Wagman, D.D.; W.H. Evans; V.B. Parker; I. Halow; S.M. Bailey; R.H. Schumm; K.L. Churney; Selected Values of Chemical Thermodynamic Properties - Tables for Elements 54 through 61 in the Standard Order of Arrangement, NBS Tech Note 270-5, 1971

(d) Parker, V.B.; D.D. Wagman and W.H. Evans; Selected Values of Chemical Thermodynamic Properties - Tables for the Alkaline Earth Elements (Elements 92 through 97 in the Standard Order of Arrangement), NBS Tech Note 270-6, 1971

(e) Schumm, R.H.; D.D. Wagman; S. Bailey; W.H. Evans; V.B. Parker, Selected Values of Chemical Thermodynamic Properties - Tables for the Lanthanide (Rare Earth) Elements (Elements 62 through 76 in the Standard Order of Arrangement), NBS Tech Note 270-7, 1973

(f) Wagman, D.D.; W.H. Evans; V.B. Parker; R.H. Schumm; R.L. Nuttall; Selected Values of Chemical Thermodynamic Properties - Compounds of Uranium, Protactinium, Thorium, Actinium, and the Alkali Metals, NBS Tech Note 270-8, 1981

(g) Wagman, D.D.; W.H. Evans, V.B. Parker, R.H. Schumm, I. Halow, S.M. Bailey, K.L. Churney, R.L. Nutall, The NBS tables of chemical thermodynamic properties - Selected values for inorganic and C_1 and C_2 organic substances in SI units, J. Phys. Chem. Ref. Data, v11, suppl. 2 (1982)

2. (a) Glushko, V.P., dir.; Thermal Constants of Compounds - Handbook in Ten Issues, VINITI Academy of Sciences of the U.S.S.R., Moscow

Volume I (1965) - O, H, D, T, F, Cl, Br, I, At, ^{3}He, He, Ne, Ar, Kr, Xe, Rn

(b) Volume II (1966) - S, Se, Te, Po

(c) Volume III (1968) - N, P, As, Sb, Bi

(d) Volume IV-1 (1970) - C, Si, Ge, Sn, Pb
Volume IV-2 (1971)

(e) Volume V (1971) - B, Al, Ga, In, Tl

(f) Volume VI-1 (1972) - Zn, Cd, Hg, Cu, Ag, Au, Fe, Co, Ni, Ru, Rh, Pd, Os, Ir, Pt
Volume VI-2 (1973)

(g) Volume VII-1 (1974) - Mn, Tc, Re, Cr, Mo, W, V, Nb, Ta, Ti, Zr, Hf
Volume VII-2 (1974)

(h) Volume VIII-1 (1978) - Sc, Y, La, Ce, Pr, Nd, Pm, Sm, Eu, Gd, Tb, Dy, Ho, Er, Tm, Yb, Lu, Ac, Th, Pa, U, Np, Pu, Am, Cm, Bk, Cf, Es, Fm, Md, No
Volume VIII-2 (1978)

(i) Volume IX (1979) - Be, Mg, Ca, Sr, Ba, Ra

(j) Volume X-1 (1981) - Li, Na, K, Rb, Cs, Fv
Volume X-2 (1981)

3. Smith, R.M.; A.E. Martell, Critical Stability Constants, Vol. 4: Inorganic Complexes, Plenum Press, New York (1976)

4. Sillén, L.G.; Stability Constants of Metal-Ion Complexes, Section I: Inorganic Ligands, The Chemical Society, London, Special Publication No. 17 (1964)

5. Baes, C.F. Jr.; R.E. Mesmer, The Hydrolysis of Cations, John Wiley & Sons, New York (1976)

6. Baes, C.F. Jr.; R.E. Mesmer, "The Thermodynamics of Cation Hydrolysis", Amer. J. Sci., v281, pp935-962 (1981)

7. Stokes, R.H.; J.M. Stokes, "A Thermodynamic Study of Bivalent Metal Halides in Aqueous Solution. Part XII. Electromotive Force Measurements on Zinc Bromide Solutions", Trans. Faraday Soc., v41, pp688-695 (1945)

8. Robinson, R.A.; R.H. Stokes, Electrolyte Solutions, 2nd ed., Butterworths & Co., London (1970)

9. Linke, W.F.; A. Seidell, Solubilities of Inorganic and Metal-Organic Compounds, Am. Chem. Soc., Washington, D.C., Vol. I (1958), Vol. II (1965)

PHOSPHORIC ACID REFERENCES

P1. Barta, L.; D.J. Bradley, "Interaction Model for the Volumetric Properties of Weak Electrolytes with Application to H_3PO_4", J. Sol. Chem., v12, #9, pp631-643 (1983)

P2. Childs, C.W., "Equilibria in Dilute Aqueous Solutions of Orthophosphates", J. Phys. Chem., v73, #9, pp2956-2960 (1969)

P3. Elmore, K.L.; J.D. Hatfield, R.L. Dunn, A.D. Jones, "Dissociation of Phosphoric Acid Solutions at 25°", J. Phys. Chem., v69, #10, pp3520-3525 (1965)

P4. Farr, T.D., "Phosphorus - Properties of the Element and Some of Its Compounds", TVA Chem. Eng. Rep. #8 (1950)

P5. Luff, B.B., "Heat Capacity and Enthalpy of Phosphoric Acid", J. Chem. Eng. Data, v26, pp70-74 (1981)

P6. Mesmer, R.E.; C.F. Baes Jr., "Phosphoric Acid Dissociation Equilibria in Aqueous Solutions to 300°C", J. Sol. Chem, v3, #4, pp307-322 (1974)

P7. Pitzer, K.S.; L.F. Silvester, "Thermodynamics of Electrolytes VI. Weak Electrolytes Including H_3PO_4", J. Sol. Chem., v5, #4, pp269-278 (1976)

P8. Platford, R.F.; "Thermodynamics of Aqueous Solutions of Orthophosphoric Acid from the Freezing Point to 298.15°K", J. Sol. Chem., v4, #7, pp591-598 (1975)

P9. Van Wazer, J.R. ed., Phosphorus and Its Compounds, Interscience (1961)

P10. Wood, R.H.; R.F. Platford, "Free Energies of Aqueous Mixtures of NaH_2PO_4 and $NaClO_4$: Evidence for the Species $(H_2PO_4)_2^{2-}$", J. Sol. Chem., v4, #12, pp977-982 (1975)

SULFURIC ACID REFERENCES

H1. Davies, C.W.; H.W. Jones, C.B. Monk, Trans. Faraday Soc., v48, p921 (1952)

H2. Gardner, W.L.; R.E. Mitchell, J.W. Cobble, "The thermodynamic properties of high-temperature aqueous solutions. X. The electrode potentials of sulfate ion electrodes from 0-100°. Activity coefficients and the entropy of aqueous sulfuric acid", J. Phys. Chem., v73, #6, pp2021-2024 (1969)

H3. Gmitro, J.I.; T. Vermeulen, "Vapor-liquid equilibria for aqueous sulfuric acid", AIChE J., v10, #5, pp740-746 (1964)

H4. Gurney, R.W., Ionic Processes in Solution, Dover Publications, New York (1962)

H5. Hamer, W.J., JACS, v56, p860 (1934)

H6. Helgeson, H.C., "Thermodynamics of complex dissociation in aqueous solution at elevated temperatures", J. Phys. Chem., v71, #10, pp3121-3136 (1967)

H7. Parker, V.B.; B.R. Staples, T.L. Jobe Jr., D.B. Neumann, "A report on some thermodynamic data for desulfurization processes", NBSIR 81-2345 (1981)

H8. Pitzer, K.S.; R.N. Roy, L.F. Silvester, "Thermodynamics of electrolytes 7. Sulfuric acid", JACS, v99, #5, pp4930-4936 (1977)

H9. Rard, J.A.; A. Habenschuss, F.H. Spedding, "A review of the osmotic coefficients of aqueous H_2SO_4 at 25°C", J. Chem. Eng. Data, v21, #3, pp374-379 (1976)

H10. Redlich, O., "The dissociation of strong electrolytes", Chem. Rev., v39, pp333-356 (1946)

H11. Rosenblatt, G.M., "Estimation of activity coefficients in concentrated sulfite-sulfate solutions", AIChE J., v27, #4, pp619-626 (1981)

H12. Sanders, S.J., "Organics in Aqueous Sulfuric Acid Solutions", presented at the AIChE Annual Meeting, Los Angeles, Cal., Nov. 1982, Submitted to Ind. Eng. Process Des. Dev.

H13. Singleterry, Thesis, U. Chicago (1940)

H14. Wirth, H.E., "Activity coefficients in sulfuric acid and sulfuric acid-sodium sulphate mixtures", Electrochim. Acta, v16, pp1345-1356 (1971)

H15. Young, T.F.; L.A. Blatz, "The variation of the properties of electrolytic solutions with degrees of dissociation", Chem. Rev., v44 (1949)

ZINC CHLORIDE REFERENCES

Z1. Aksel'rud, N.V.; V.B. Spivakovskii, "Study of basic salts and hydroxides of metals II. Basic chlorides and hydroxides of zinc", J. Inorg. Chem., USSR, V. III, #2, pp269-277 (1958)

Z2. Belousov, E.A.; A.A. Alovyainikov, "Determination of the stability constants of zinc and cadmium chloro-complexes in aqueous solutions by the distribution method", Russ. J. Inorg. Chem., v20, #5, pp803-804 (1975)

Z3. Fedorov, V.A.; G.E. Chernikova, M.A. Kuznechikhina, T.I. Kuznetsova, "Formation of mixed chloro(sulphato)- and bromo(sulphato)- complexes of zinc and cadmium in solutions", Russ. J. Inorg. Chem., v20, #11, pp1613-1614 (1975)

Z4. Gerding, P., "Thermochemical studies on metal complexes IX. Free energy, enthalpy, and entropy changes for stepwise formation of zinc(II) halide complexes in aqueous solution", Acta Chem. Scand., v23, #5, pp1695-1703 (1969)

Z5. Goldberg, R.N., "Evaluated activity and osmotic coefficients for aqueous solutions: Bi-univalent compounds of zinc, cadmium, and ethylene bis(trimethylammonium) chloride and iodide", J. Phys. Chem. Ref. Data, v10, #1, pp1-55 (1981)

Z6. Grauer, R.; P. Schindler, "Die loslichkeitskonstanten der zinkhydroxidchloride - ein beitrag zur kenntnis der korrosionsprodukte des zinks", Corrosion Sci., v12, pp405-414 (1972)

Z7. Latysheva, V.A.; I.N. Andreeva, "Change in heat capacity accompanying formation of chloride complexes of zinc and cadmium in aqueous solution at 25°C", Russ. J. Phys. Chem., v43, #2, pp260-261 (1969)

Z8. Latysheva, V.A., I.N. Andreeva, A.I. Gol'denberg, "Change in volume in the formation of chloro-complexes of zinc", Russ. J. Inorg. Chem., v14, #5, pp618-620 (1969)

Z9. Latysheva, V.A., I.N. Andreeva, "Forms of existence of higher chloride complexes of zinc and cadmium in aqueous solutions", Zhur. Obshchei Khim., v41, #8, pp1649-1652 (1971)

Z10. Libuś, Z.; H. Tialowska, "Stability and nature of complexes of the type MCl^+ in aqueous solution (M = Mn, Co, Ni, and Zn)", J. Sol. Chem., v4, #12, pp1011-1022 (1975)

Z11. Lutfullah; R. Paterson, H.S. Dunsmore, "Re-determination of the standard electrode potential of zinc and mean molal activity coefficients for aqueous zinc chloride at 298.15K", J. Chem. Soc. Faraday Trans., v72, pp495-503 (1976)

Z12. Mihailov, M.H., "A correlation between the overall stability constants of metal complexes - I. Calculation of the stability constants using the formation function $\bar{n}$", J. inorg. nucl. Chem., v36, pp107-113 (1974)

Z13. Morris, D.F.C.; D.T. Anderson, S.L. Waters, G.L. Reed, "Zinc chloride and zinc bromide complexes - IV. Stability constants", Electrochim. Acta, v14, pp643-650 (1969)

Z14. Robinson, R.A.; R.H. Stokes, "Part IV. The thermodynamics of zinc chloride solutions", Trans. Faraday Soc., v36, pp740-748 (1940)

Z15. Robinson, R.A.; R.O. Farrelly, "Some e.m.f. and conductance measurements in concentrated solutions of zinc and calcium chlorides", J. Phys. Chem., v51, pp704-708 (1947)

Z16. Shchukarev, S.A.; V.A. Latysheva, L.S. Lilich, "Studies on halide complexes of zinc, cadmium and mercury in aqueous solutions", J. Inorg. Chem., USSR, V.I, #2, pp225-231 (1956)

Z17. Spivakovskii, V.B.; L.P. Moisa, "Chloro-complexes of zinc in concentrated solutions of calcium chloride", Russ. J. Inorg. Chem., v14, #5, pp615-618 (1969)

FERRIC CHLORIDE REFERENCES

F1. Bjerrum, J., Kge. Danske Videnskab. Selskab., Mat.-fys. Medd., v22, #18 (1946)

F2. Bray, W.C.; A.V. Hershey, JACS, v56, p1889 (1934)

F3. Brealy, Evans, Uri, Nature, v166, p959 (1950)

F4. Brealy, Uri, J. Chem. Phys., v20, p257 (1952)

F5. Friedman, H.L.; JACS, v74, p5 (1952)

F6. Gamlen, G.A., D.O. Jordan, "A spectrophometric study of the iron(III) chloro-complexes", J. Chem. Soc., pp1435-1443 (1953)

F7. Marcus, Y., "The anion exchange of metal complexes - IV. The iron(III)-chloride system", J. Inorg. Nucl. Chem., v12, pp287-296 (1960)

F8. Möller, M., J. Phys. Chem., v41, p1123 (1937)

F9. Rabinowitch, E.; W.H. Stockmayer, "Association of ferric ions with chloride, bromide and hydroxyl ions (a spectroscopic study)", JACS, v64, pp335-347 (1942)

F10. Shimmel, F.A., "The ternary systems ferrous chloride - hydrogen chloride - water, ferric chloride - ferrous chloride - water", JACS, v74, pp4689-4691 (1952)

CUPROUS CHLORIDE REFERENCES

C1. Ahrland, S.; J. Rawsthorne, Acta Chem. Scand., v24, p157 (1974)

C2. Fritz, J.J., "Chloride Complexes of CuCl in Aqueous Solution", J. Phys. Chem., v84, #18, pp2241-2246 (1980)

C3. Fritz, J.J., "Representation of the Solubility of CuCl in Solutions of Various Aqueous Chlorides", J. Phys. Chem., v85, #7, pp890-894 (1981)

C4. Fritz, J.J., "Solubility of Cuprous Chlorides in Various Soluble Aqueous Chlorides", J. Chem. Eng. Data, v27, #2, pp188-193 (1982)

C5. Fritz, J.J., "Heats of Solution of Cuprous Chloride in Aqueous HCl-$HClO_4$ Mixtures", J. Sol. Chem., v13, #5, pp369-382 (1984)

C6. Hikita, H.; H. Ishikawa, N. Esaka, Nippon Kagaku Kaishi, v1, p13 (1973)

C7. Kale, S.S.; S.S. Tamhankar, R.V. Chaudhari, "Solubility of Cuprous Chloride in Aqueous Hydrochloric Acid Solutions", J. Chem. Eng. Data, v24, #2, pp110-111 (1979)

C8. Sukhova, T.G.; O.N. Ternkin, R.M. Flid, T.K. Kaliga, Russ. J. Inorg. Chem. (Eng. trans.), v13, p1072 (1968)

CALCIUM SULFATE REFERENCES

G1. Adler, M.S.; J. Glater, J.W. McCutchan, "Prediction of gypsum solubility and scaling limits in saline waters", J. Chem. Eng. Data, v24, #3, pp187-192 (1979)

G2. Ainsworth, R.G., "Dissociation constant of calcium sulfate from 25 to 50°C", J. Chem. Soc. Faraday Trans I, v69, pp1028-1032 (1973)

G3. Ball, M.C., "The dehydration of disodiumpentacalcium sulphate trihydrate, $Na_2Ca_5.(SO_4)_6.3H_2O$", Thermoch. Acta, v24, pp190-193 (1978)

G4. Barba, D.; V. Brandani, G. diGiacomo, "A thermodynamic model of $CaSO_4$ solubility in multicomponent aqueous solutions", Chem. Eng. J., v24, pp191-200 (1982)

G5. Barba, D.; V. Brandani, G. diGiacomo, "Solubility of calcium sulfate dihydrate in the system Na_2SO_4-$MgCl_2$-H_2O", J. Chem. Eng. Data, v29, pp42-45 (1984)

G6. Bell, R.P.; J.H.B. George, "The incomplete dissociation of some thallous and calcium salts at different temperatures", Trans. Faraday Soc., v49, pp619-627 (1953)

G7. Block, J.; O.B. Waters, Jr., "The $CaSO_4$-Na_2SO_4-NaCl-H_2O system at 25° to 100°C", J. Chem. Eng. Data, v13, #3, pp336-344 (1968)

G8. Blount, C.W.; F.W. Dickson, "The solubility of anydrite ($CaSO_4$) in NaCl-H_2O from 100 to 450°C and 1 to 1000 bars", Geo. et Cosmo. Acta, v33, pp227-245 (1969)

G9. Blount, C.W.; F.W. Dickson, "Gypsum-anhydrite equilibria in systems $CaSO_4$-H_2O and $CaCO_4$-NaCl-H_2O", Amer. Mineralogist, v58, pp323-331 (1973)

G10. Bock, E., "On the solubility of anhydrous calcium sulphate and of gypsum in concentrated solutions of sodium chloride at 25°C, 30°C, 40°C, and 50°C", Can. J. Chem., v39, pp1746-1751 (1961)

G11. Booth, H.S.; R.M. Bidwell, "Solubilities of salts in water at high temperatures", JACS, v72, pp2567-2575 (1950)

G12. Briggs, C.C.; T.H. Lilley, "Activity coefficients of calcium sulphate in water at 25°C", Proc. R. Soc. Lond., v349, pp355-368 (1976)

G13. Brown, P.G.M; J.E. Prue, "A study of ionic association in aqueous solutions of bi-bivalent electrolytes by freezing-point measurements", Proc. Roy. Soc. Lond. A, v232, pp320-336 (1955)

G14. Cameron, F.K., "Solubility of gypsum in aqueous solutions of sodium chloride", J. Phys. Chem., v5, pp556-576 (1901)

G15. Conley, R.F.; W.M. Bundy, "Mechanism of gypsification", Geo. et Cosmo. Acta, v15, pp57-72 (1958)

G16. Corti, H.R.; R. Fernandez-Prini, "Thermodynamics of solution of gypsum and anhydrite in water over a wide temperature range", Can. J. Chem., v62, pp484-488 (1984)

G17. Emara, M.M.; G. Atkinson, N.A. Farid, "An ion selective electrode study of calcium and magnesium sulfate in aqueous solution", Anal. Letters, A11, (10), pp797-811 (1978)

G18. Eugster, H.P.; C.E. Harvie, J.H. Weare, "Mineral equilibria in a six-component seawater system, Na-K-Mg-Ca-SO_4-Cl-H_2O, at 25°C", Geochimica et Cosmochim. Acta, v44, pp1335-1347 (1980)

G19. Frydman, M.; L.G. Sillen; G. Nilsson, T. Rengemo, "Some solution equilibria involving calcium sulfite and carbonate III. The acidity constants of H_2CO_3 and H_2SO_3, and $CaCO_3$ + $CaSO_3$ equilibria in $NaClO_4$ medium at 25°C", Acta Chem. Scand., v12, #5, pp878-884 (1958)

G20. Gardner, A.W.; E. Glueckauf, "Ionic association in aqueous solutions of bivalent sulphates", Proc. Roy. Soc. Lond., v313, p131-147 (1969)

G21. Glater, J.; J. Schwartz, "High-temperature solubilities of calcium sulfate hemihydrate and anhydrite in natural seawater concentrates", J. Chem. Eng. Data, v21, #1, pp47-52 (1976)

G22. Glew, D.N.: D.A. Hames, "Gypsum, disodium pentacalcium sulfate, and anhydrite solubilities in concentrated sodium choride solutions", Can. J. Chem., v48, pp3733-3738 (1970)

G23. Gordievskii, A.V.; E.L. Filippov, V.S. Sherman, A.S. Krivoshein, "Ion-exchange Membranes. III. Application of an ion-exchange membrane electrode in the study of complex-formation reactions", Russ. J. Phys. Chem., v42, #8, pp1050-1054 (1968)

G24. Hardie, L.A., "The gypsum-anhydrite equilibrium at one atm. pressure", Amer. Mineralogist, v52, pp171-200 (1967)

G25. Harvie, C.E.; J.H. Weare, "The prediction of mineral solubilities in natural waters: the Na-K-Mg-Ca-Cl-SO_4 -H_2O system from zero to high concentration at 25°C", Geochimica et Cosmochimica Acta, v44, pp981-997 (1980)

G26. Hill, A.E., "The transition temperature of gypsum to anhydrite", JACS, v59, pp2242-2244 (1937)

G27. Hill, A.E.; J.H. Wills, "Ternary systems. XXIV. Calcium sulfate, sodium sulfate and water", JACS, v60, pp1647-1655 (1938)

G28. Kelley, K.K.; J.C. Southard, and C.T. Anderson, "Thermodynamic properties of gypsum and its dehydration products", U.S. Bur. Mines Tech. Paper 625 (1941)

G29. King, G.A.; M.J. Ridge; G.S. Walker, "Location of sodium incorporated in crystals of gypsum", J. appl. Chem. Biotechnol., v25, pp721-726 (1975)

G30. Larson, J.W., "Thermodynamics of divalent metal sulfate dissociation and the structure of the solvated metal sulfate ion pair", J. Phys. Chem., v74, #18, pp3392-3396 (1970)

G31. Latimer, W.M.; J.F.G. Hicks, Jr.; P.W. Schutz, "The heat capacities and entropies of calcium and barium sulfates from 15 to 300K. The entropy and free energy of sulfate ion", J. Chem. Physics, v1, pp620-624 (1933)

G32. Lieser, V.K.H., "Radiochemische Messung der Loslichkeit von Erdalkalisulfaten in Wasser und in Natriumsulfatlosungen", Zeitschrift fur anorganishe und allgemeine Chemie, v335, pp225-231 (1965)

G33. Lilley, T.H.; C.C. Briggs, "Thermodynamic studies on aqueous solutions containing calcium chloride + calcium sulphate", J. Chem. Thermo., v8, pp151-158 (1976)

G34. Luk'yanova, N.K., "Solubility of natural gypsum and anhydrite in solutions", Zh. P. Khim., v69, #7, p1654 (1976)

G35. MacDonald, G.J.F., "Anhydrite-gypsum equilibrium relations", Amer. J Sci, v251, pp884-898 (1953)

G36. Madgin, W.M.; D.A. Swales, "Solubilities in the system $CaSO_4$-NaCl-H_2O at 25° and 35°", J appl. Chem., v6, pp482-487 (1956)

G37. Marshall, W.L., R. Slusher, E.V. Jones, "Aqueous systems at high temperature. XIV. Solubility and thermodynamic relationships for $CaSO_4$ in NaCl-H_2O solutions from 40° to 200°C, 0 to 4 molal NaCl", J. Chem. Eng. Data, v9, #2, pp187-191 (1964)

G38. Marshall, W.L.; R. Slusher, "Thermodynamics of calcium sulfate dihydrate in aqueous sodium chloride solutions, 0-110°", J. Phys. Chem., v70, #12, pp4015- 4027 (1966)

G39. Marshall, W.L.; E.V. Jones, "Second dissociation constant of sulfuric acid from 25 to 350° evaluated from solubilities of calcium sulfate in sulfuric acid solutions", J. Phys. Chem., v70, #12, pp4028-4040 (1966)

G40. Marshall, W.L.; R. Slusher, "Aqueous systems at high temperature. Solubility to 200°C of calcium sulfate and its hydrates in sea water and saline water concentrates and temperature-concentration limits", J. Chem. Eng. Data, v13, #1, pp83-93 (1968)

G41. Marshall, W.L.; R. Slusher, "Debye-Hückel correlated solubilities of calcium sulfate in water and in aqueous sodium nitrate and lithium nitrate solutions of molality 0 to 6 mole/kg and at temperatures from 398 to 623 K", J. Chem. Thermo., v5, pp189-197 (1973)

G42. Marshall, W.L., R. Slusher, "The ionization constant of nitric acid at high temperatures from solubilities of calcium sulfate in HNO_3-H_2O, 100-350°C; activity coefficients and thermodynamic functions", J. inorg. nucl. Chem., v37, pp1191-1202 (1975)

G43. Marshall, W.L., "Thermodynamic functions at saturation of several metal sulfates in aqueous sulfuric and deuterosulfuric acids at temperatures up to 350°C", J. inorg. nucl. Chem., v37, pp2155-2163 (1975)

G44. Martynova, O.I.; L.G. Vasina, S.A. Pozdnyakova, V.A. Kishnevskii, "Use of the calcium-selective electrode for determining the solubility product and dissociation constant of calcium sulfate", Doklady Akad. Nauk SSSR, v217, #4, pp862-864 (1974)

G45. Millero, F.J.; F. Gombar, J. Oster, "The partial molal volume and compressibility change for the formation of the calcium sulfate ion pair at 25°C", J. Sol. Chem., v6, #4, pp269-280 (1977)

G46. Money, R.W.; C.W. Davies, "The extent of dissociation of salts in water. Part IV: Bi-bivalent salts", Trans. Faraday Soc., v28, pp609-614 (1932)

G47. Moreno, E.C.; G. Osborn, "Solubility of gypsum and dicalcium phosphate dihydrate in the system CaO-P_2O_5-SO_3-H_2O and in soils", Soil Sci. Soc. of America Proc., v27, pp614-619 (1963)

G48. Nancollas, G.H., "The growth of crystals in solution", Adv. in Colloid and Interface Sci., v10, pp215-252 (1979)

G49. Nilsson, G.; L.G. Sillen; T. Rengemo, "Some solution equilibria involving calcium sulfite and carbonate. I. Simple solubility equilibria of CO_2, SO_2, $CaCO_3$, and $CaSO_4$", Acta Chem. Scand., v12, #5, pp868-872 (1958)

G50. Ostroff, A.G.; "Conversion of gypsum to anhydrite in aqueous salt solutions", Geo. et Cosmo. Acta, v28, pp1363-1372 (1964)

G51. Pfefferkorn, O., "The economics of the seed slurry process for scale control", Desalination, v19, pp75-82 (1976)

G52. Posnjak, E., "The system, $CaSO_4$-H_2O", Amer. J. Sci., v235-A, pp247-272 (1938)

G53. Power, W.H., B.M. Fabuss, C.N. Satterfield, "Transient solubilities in the calcium sulfate-water system", J. Chem. Eng. Data, v9 #3, pp437-442 (1964)

G54. Power, W.H.; B.M. Fabuss, C.N. Satterfield, "Transient solute concentrations and phase changes of calcium sulfate in aqueous sodium chloride", J. Chem. Eng. Data, v11, #2, pp149-154 (1966)

G55. Reardon, E.J.; D. Langmuir, "Activity coefficients of $MgCO_3$ and $CaSO_4$ ion pairs as a function of ionic strength", Geochimica et Cosmochimica Acta, v40, pp549-554 (1976)

G56. Rengemo, T.; L.G. Sillen; U. Brune, "Some solution equilibria involving calcium sulfite and carbonate II. The equilibrium between calcium sulfate and calcium sulfite in aqueous solutions", Acta Chem. Scand., v12, #5, pp873-877 (1958)

G57. Rogers, P.S.Z., "Thermodynamics of geothermal fluids", Doctoral dissertation, U. Cal. Berkeley (1981)

G58. Rogozovskaya, M.Z., N.K. Luk'yanova, T.I. Konochuk, "The properties of the calcium sulfate sediment formed in brine treatment for the production of chlorine", Khim. prom., v9, #10, pp770-772 (1977)

G59. Rogozovskaya, M.Z., N.K. Luk'yanova, T.I. Kononchuk, "Gypsum and pentasalt formation in brine purification from chlorine production", Khim. prom., v10, #5, pp362-365 (1978)

G60. Rogozovskaya, M.Z.; T.I. Kononchuk, N.K. Luk'yanova, "Influence of pH on formation of gypsum incrustations", Zhur. Prikladnoi Chim., v52, #3, pp601-605 (1979)

G61. Rogozovskaya, M.Z.; T.I. Konochuk, "Crystallization of gypsum and of "sodium penta salt" during desulfation of sodium chloride solutions", Zhur. Prik. Khim., v54, #8, pp1708-1711 (1981)

G62. Rosenblatt, G.M., "The use of Pitzer's equations to estimate activity coefficients in FGD scrubber systems", Paper presented at the FGD Meeting, Morgantown, W. Virg., Nov. 1980

G63. Rosenblatt, G.M., "Estimation of activity coefficients in concentrated sulfite-sulfate solutions", LBL-9671 preprint, November 1979

G64. Ryss, I.G.; E.L. Nilus, "Solubility of calcium sulfate in solutions of hydrochloric acid at 25°", J. Gen. Chem. of USSR (Eng. trans), v25, #6, pp1035-1038 (1955)

G65. Scrivner, N.C.; and B.R. Staples, "Equilibria in aqueous solution - Industrial applications", Paper presented at the Second World Congress of Chem. Eng., Montreal (Oct. 1982)

G66. Susarla, V.R.K.S.; J.R. Sanghavi, "The solubility data of the system $NaCl-CaSO_4-MgCl_2-H_2O$ at 35°C", Salt Res. & Ind., v15, #2, pp1-3 (1979)

G67. Templeton, C.C.; J.C. Rodgers, "Solubility of anhydrite in several aqueous salt solutions between 250 and 325°C", J. Chem Eng. Data., v12, #4, pp536-547 (1967)

G68. Toriumi, T.; R. Hara, "On the transition point of calcium sulfate in water and concentrated sea water", Tohoku Imperial Univ. Tech. Rep., v12, pp17-90 (1938)

G69. Treibus, E.B., S.V. Moshkin, T.G. Il'inskaya, "Growth kinetics of gypsum crystals", Zhur. Fizicheskoi Khim., v55, pp112-115 (1981)

G70. Weare, J.H.; C.E. Harvie; N.E. Moller-Weare, "Towards an accurate and efficient chemical model for geothermal waters", preprint of paper presented at the SPE Fifth Int. Symp. on Oilfield & Geothermal Chem., May, 1980, AIME, SPE 8993, pp179-186 (1980)

G71. Weare, J.H.; C.E. Harvie; N. Moller-Weare, "Toward an accurate and efficient chemical model for hydrothermal brines", S. Petro. Eng. J., pp699-708, Oct 1982

G72. Yeatts, L.B.; W.L. Marshall, "Apparent invariance of activity coefficients of calcium sulfate at constant ionic strength and temperature in the system $CaSO_4-Na_2SO_4-NaNO_3-H_2O$ to the critical temperature of water. Association equilibria", J. Phys. Chem., v73, #1, pp81-90 (1969)

G73. Yeatts, L.B.; W.L. Marshall, "Solubility of calcium sulfate dihydrate and association equilibria in several aqueous mixed electrolyte salt systems at 25°C", J. Chem. Eng. Data, v17, #2, pp163-168 (1972)

G74. Zen, E-An, "Solubility measurements in the system $CaSO_4$-NaCl-H_2O at 35°, 50°, and 70°C and 1 atmosphere pressure", J. Petrology, v6, pp124-164 (1965)

SODIUM SULFATE REFERENCES

S1. Austin, J.M.; A.D. Mair, "The standard enthalpy of formation of complex sulfate ions in water. I. HSO_4^-, $LiSO_4^-$, $NaSO_4^-$", J. Phys. Chem, v66, pp519-521 (1962)

S2. Bhatnagar, Om. N.; A.N. Campbell, "Osmotic and activity coefficients of sodium sulphate in water from 50 to 150°C", Can. J. Chem., v59, pp123-126 (1981)

S3. Booth, J.S.; R.M. Bidwell, "Solubilities of salts in water at high temperatures", J. Am. Chem. Soc., v72, pp2567-2575 (1950)

S4. Correla, R.J.; J. Kestin, "Viscosity and density of aqueous Na_2SO_4 and K_2SO_4 solutions in the temperature range 20-90°C and the pressure range 0-30 MPa", J. Chem. Eng. Data, v26, #1, pp43-47 (1981)

S5. deLima, M.C.P.; K.S. Pitzer, "Thermodynamics of saturated electrolyte mixtures of NaCl with Na_2SO_4 and with $MgCl_2$", J. Sol. Chem., v12, #3, pp187-199 (1983)

S6. Downes, C.J.; K.S. Pitzer, "Thermodynamics of electrolytes. Binary mixtures formed from aqueous NaCl, Na_2SO_4, $CuCl_2$, and $CuSO_4$ at 25°C", J. Sol. Chem., v5, #6, pp389-398 (1976)

S7. Dunn, L.A., "Apparent molar volumes of electrolytes: Part 1. Some 1-1, 1-2, 2-1, 3-1 electrolytes in aqueous solution at 25°C", Trans. Faraday Soc., v62, pp2348-2354 (1966)

S8. Fabuss, B.M.; A. Korosi, "Vapor pressures of binary aqueous solutions of NaCl, KCl, Na_2SO_4, and $MgSO_4$ at concentrations and temperatures of interest in desalination processes", Desalination, v1, pp139-148 (1966)

S9. Fisher, F.H.; A.P. Fox, "$NaSO_4^-$ ion pairs in aqueous solutions at pressures up to 2000 atm", J. Sol. Chem., v4, #3, pp225-236 (1975)

S10. Fisher, F.H., "Dissociation of Na_2SO_4 from ultrasonic absorption reduction in $MgSO_4$-NaCl solutions", J. Sol. Chem., v4, #3, pp237-240 (1975)

S11. Fisher, F.H.; A.P. Fox, "KSO_4^-, $NaSO_4^-$, $MgCl^+$ ion pairs in aqueous solutions up to 2000 atm", J. Sol. Chem., v6, #10, pp641-650 (1977)

S12. Gardner, W.L.; J.W. Cobble, E.C. Jekel, "The thermodynamic properties of high-temperature aqueous solutions. IX. The standard partial molal heat capacities of sodium sulfate and sulfuric acid from 0 to 100°", J. Phys. Chem., v73, #6, p2017 (1969)

S13. Goldberg, R.N., "Evaluated activity and osmotic coefficients for aqueous solutions: thirty-six uni-bivalent electrolytes", J. Phys. Chem. Ref. Data, v10, #3, pp671-764 (1981)

S14. Harvie, C.E.; J.H. Weare, "The prediction of mineral solubilities in natural waters: the Na-K-Mg-Ca-Cl-SO_4-H_2O system from zero to high concentration at 25°C", Geochim. et Cosmochim. Acta, v44, pp981-997 (1980)

S15. Humphries, W.T.; C.S. Patterson, C.F. Kohrt, "Osmotic properties of some aqueous electrolytes at 60°C", J. Chem. Eng. Data, v13, #3, pp327-330 (1968)

S16. Izatt, R.M.; D. Eatough, J.J. Christensen, C.H. Bartholomew, "Calorimetrically determined Log K, delta H and delta S values for the interaction of sulphate ion with H^+, Na^+ and K^+ in the presence of tetra-n-alkylammonium ions", J. Chem. Soc., (A) pp45-47 (1969)

S17. Izatt, R.M.; D. Eatough, J.J. Christensen, C.H. Bartholomew, "Calorimetrically determined log K, delta H and delta S values for the interaction of sulphate ion with several bi and ter-valent metal ions", J. Chem. Soc (A), pp47-53 (1969)

S18. Jakli, Gy; T.C. Chan, W.A. Van Hook, "Equilibrium isotope effects in aqueous systems. IV. The vapor pressures of NaBr, NaI, KF, Na_2SO_4 and $CaCl_2$ solutions in H_2O and D_2O (0 to 90°C). Vapor pressures of $Na_2SO_4.10H_2O$, $.10D_2O$ and $CaCl_2.2H_2O$, $.2D_2O$", J. Sol. Chem., v4, #1, pp71-90 (1975)

S19. Jenkins, I.L.; C.B. Monk, "The conductances of sodium, potassium and lanthanum sulfates at 25°", JACS, v72, pp2695-98 (1950)

S20. LoSurdo, A.; F.J. Millero; E.M. Alzola, "The (p, V, T) properties of concentrated aqueous electrolytes. I. Densities and apparent molar volumes of NaCl, Na_2SO_4, $MgCl_2$ and $MgSO_4$ solutions from .1 mol/kg to saturation and from 273.15 to 323.15K", J. Chem. Thermo., v14, pp649-662 (1982)

S21. Martynova, O.I.; L.G. Vasina, S.A. Pozdnyakova, "Determination of the dissociation constant of $NaSO_4^-$ ion pairs in the temperature interval 20-98°C", Doklady Akademii Nauk SSSR, v217, #5, pp1080-1082 (1974)

S22. Mayrath, J.E.; R.H. Wood, "Enthalpy of dilution of aqueous solutions of Na_2SO_4, K_2SO_4, and $MgSO_4$ at 373.15 and 423.65 K and of MgCl at 373.15, 423.65, and 472.95 K", J. Chem. Eng. Data, v28, pp56-59 (1983)

S23. Millero, F.J., "Effect of pressure on sulfate ion association in sea water", Geochim et Cosmo. Acta, v35, p1089 (1971)

S24. Millero, F.J.; G.K. Ward, F.K. Lepple, E.V. Hoff, "Isothermal compressibility of aqueous sodium chloride, magnesium chloride, sodium sulfate and magnesium sulfate solutions from 0 to 45° at 1 atm", J. Phys. Chem., v78, #16, p1636 (1974)

S25. Moore, J.T.; C.S. Patterson, W.T. Humphries, "Isopiestic studies of some aqueous electrolyte solutions at 80°C, J. Chem. Eng. Data, v17, #2, pp180-182 (1972)

S26. Patterson, C.S.; L.O. Gilpatrick, B.A. Soldano, "The osmotic behavior of representative aqueous salt solutions at 100°", J. Chem. Soc., pp2730-2734 (1960)

S27. Perron, G.; F.J. Millero, J.E. Desnoyers, "Apparent molal volumes and heat capacities of some sulfates and carbonates in water at 25°C", Can. J. Chem., v53, p1134 (1975)

S28. Pitzer, K.S.; J.S. Murdzek, "Thermodynamics of aqueous sodium sulfate", J. Sol. Chem., v11, #6, pp409-413 (1982)

S29. Platford, R.F., "Thermodynamics of the system H_2O-NaCl-$MgCl_2$-Na_2SO_4-$MgSO_4$ at 25°C", Marine Chem., v3, pp261-270 (1975)

S30. Platford, R.F., "Excess Gibbs energies of mixing in the system H_2O-NaCl-Na_2SO_4-$MgSO_4$ at 298.15K", J. Sol. Chem., v4, #1, pp37-43 (1975)

S31. Rard, J.A.; D.G. Miller, "Isopiestic determination of the osmotic coefficients of aqueous Na_2SO_4, $MgSO_4$, and Na_2SO_4-$MgSO_4$ at 25°C", J. Chem. Eng. Data, v26, pp33-38 (1981)

S32. Rastogi, A.; D. Tassios, "Estimation of the thermodynamic properties of binary aqueous electrolytic solutions in the range 25-100°C", Ind. Eng. Chem. Process Des. Dev., v19, pp477-482 (1980)

S33. Reardon, E.J., "Dissociation constants of some monovalent sulfate ion pairs at 25° from stoichiometric activity coefficients", J. Phys. Chem., v79, #5, pp422-425 (1975)

S34. Righellato, E.C.; C.W. Davies, "The extent of dissociation of salts in water. Part II. Uni-bivalent salts", Trans. Faraday Soc., v26, pp592-600 (1930)

S35. Robinson, R.A.; J.M. Wilson; R.H. Stokes, "The activity coefficients of lithium, sodium and potassium sulfate and sodium thiosulfate at 25° from isopiestic vapor pressure measurements", JACS, v63, pp1011 (1941)

S36. Robinson, R.A., "Excess Gibbs energy of mixing of the systems H_2O-LiCl-Na_2SO_4 and H_2O-CsCl-Na_2SO_4 at 25°C", J. Sol. Chem., v1, #1, pp71-75 (1972)

S37. Robinson, R.A.; R.F. Platford, C.W. Childs, "Thermodynamics of aqueous mixtures of sodium chloride, potassium chloride, sodium sulfate, and potassium sulfate at 25°C", J. Sol. Chem., v1, #2, pp167-172 (1972)

S38. Rogers, P.S.Z., "Thermodynamics of geothermal fluids", Doctoral dissertation, U. Cal. Berkeley, 1981

S39. Rosenblatt, G.M., "Estimation of activity coefficients in concentrated sulfite-sulfate solutions", LBL-9671 preprint, sub. to AIChE, Nov. 1979

S40. Rosenblatt, G.M.; "The use of Pitzer's equations to estimate activity coefficients in FGD scrubber systems", presented at FGD meeting, Morgantown WV, Nov. 1980

S41. Santos, M.M.; J.R.F.G. de Carvalho, R.A.G. de Carvalho, "Determination of the association constant of $NaSO_4^-$ with the sodium-selective electrode", J. Sol. Chem., v4, #1, pp25-29 (1975)

S42. Scatchard, G.; Y.C. Wu, R.M. Rush, "Osmotic and activity coefficients for binary mixtures of sodium chloride, sodium sulfate, magnesium sulfate and magnesium chloride in water at 25°C. III. Treatment with the ions as components", J. Phys. Chem., v74, #21, pp3786-3796 (1970)

S43. Snipes, H.P., C. Manly, D.D. Ensor, "Heats of dilution of aqueous electrolytes: Temperature dependence", J. Chem. Eng. Data, v20, #3, pp287-291 (1975)

S44. Soldano, B.A.; C.S. Patterson, "Osmotic behavior of aqueous salt solutions at elevated temperatures. Part II", J. Chem. Soc., pp937-940 (1962)

S45. Srna, R.F.; R.H. Wood, "Heats of mixing aqueous electrolytes XII. The reciprocal salt pair Na^+, Mg^{++} || Cl^-, $SO_4^=$, J. Phys. Chem., v79, #15, p1535 (1975)

S46. Staples, B.R.; V.B. Parker, T.L. Jobe Jr.; D.B. Neumann, "A report on some thermodynamic data for desulfurization processes", NBSIR 81-2345, Sept. 1981

S47. Tran, T-T; F. Lenzi, "Methods of estimating the water activity of supersaturated aqueous solutions", Can. J. Chem. Eng., v52, pp798-802 (1974)

S48. Wirth, H.E., "Activity coefficients in sulphuric acid and sulphuric acid-sodium sulphate mixtures", Electro. Acta, v16, pp1345-1356 (1971)

S49. Wood, R.H.; D.E. Smith, H.K.W. Chen, P.T. Thompson, "Heats of mixing aqueous electrolytes XII. The reciprocal salt pair Na^+, Li^+ || Cl^-, $SO_4^=$, J. Phys. Chem., v79, #15, p1532 (1975)

S50. Wu, Y.C.; G. Scatchard, R.M. Rush, "Osmotic and activity coefficients for binary mixtures of sodium chloride, sodium sulfate, magnesium sulfate and magnesium chloride in water at 25°C. I. Isopiestic measurements on the four systems with common ions", J. Phys. Chem., v72, #12, pp4048-4053 (1968)

S51. Wu, Y.C.; G. Scatchard, R.M. Rush, "Osmotic and activity coefficients for binary mixtures of sodium chloride, sodium sulfate, magnesium sulfate and magnesium chloride in water at 25°C. II. Isopiestic and electromotive force measurements on the two systems without common ions", J. Phys. Chem., v73, #6, pp2047-2053 (1969)

OTHER CHLORIDE REFERENCES

O1. Caminiti, R.; G. Licheri, G. Piccaluya, G. Pinna, Faraday Discuss. Chem. Soc. v64 (1977)

O2. Cubiotti, G; F. Sacchetti, M.C. Spinelli, Solid State Commun., v27, p349 (1978)

O3. Fontana, M.P.; G. Maisano, P. Migliardo, F. Wanderlingh, Solid State Comm., v23, p489 (1977)

O4. Fontana, M.P.; G. Maisano; P. Migliardo; F. Wanderlingh; J. Chem. Phys., v69, p676 (1978)

O5. Goldberg, R.N.; R.L. Nuttall, B.R. Staples, "Evaluated activity and osmotic coefficients for aqueous solutions: iron chloride, and the bi-univalent compounds of nickel and cobalt", J. Phys. Chem. Ref. Data, v8, #4, pp923-1003 (1979)

O6. Goldberg, R.N.; "Evaluated activity and osmotic coefficients for aqueous solutions: bi-univalent compounds of lead, copper, manganese, and uranium", J. Phys. Chem. Ref. Data, v8, #4, pp1005-1050 (1979)

O7. Howe, R.A.; W.S. Howells, J.R. Enderby, J. Phys. C 7, L111 (1974)

O8. Largarde, P.; A. Fontaine, D. Raoux, A. Sadoc, P. Migliardo, "EXAFS studies of strong electrolytic solutions", J. Chem. Phys., v72, #5, pp3061-3069 (1980)

O9. Libuś, Z.; Inorg. Chem., v12, p2972 (1973)

O10. Libuś, Z.; H. Tialowska, "Stability and nature of complexes of the type MCl^+ in aqueous solution (M = Mn, Co, Ni, and Zn)", J. Sol. Chem., v4, #12, pp1011-1022 (1975)

O11. Long, D.T.; E.E. Angino, "Chemical speciation of Cd, Cu, Pb and Zn in mixed freshwater, seawater, and brine solutions", Geochim. et Cosmochim. Acta, v41, pp1183-1191 (1977)

O12. Maisano, G.; P. Migliardo, F. Wanderlingh; M.P. Fontana, J. Chem. Phys., v68, p5594 (1978)

O13. Nielson, G.W.; J.E. Enderby, J. Phys. C 11, L625 (1978)

O14. Soper, A.K.; G.W. Nielson, J.E. Enderby, R.A. Howe, J. Phys., C. 10, p1793 (1977)

VII:

Activity Coefficients of Weak Electrolytes and Molecular Species

ACTIVITY COEFFICIENTS OF WEAK ELECTROLYTES AND MOLECULAR SPECIES

In Chapters IV and V considerable effort was spent describing strong electrolytes and alternative formulations for their corresponding mean and/or ionic activity coefficients. A strict definition of a strong electrolyte is a species which completely dissociates in water. In reality, very few species fit this definition of strong electrolytes. The following definitions are offered in order to provide a practical classification of electrolytes:

Strong electrolyte - A compound which nearly completely dissociates to its maximally charged constituent ions when placed in aqueous solution.

Complexing electrolyte - A compound which forms non-trival quantities of non-maximally charged complexes when placed in aqueous solution. These intermediates may be molecular or ionic.

Weak electrolyte - A compound which forms a non-trivial quantity of a molecular species when placed in aqueous solution. By this definition, weak electrolytes form a subset of complexing electrolytes.

The terms "nearly" and "non-trivial" are qualitative and are used intentionally. This is because there are probably no totally ionized electrolytes. Indeed, Helgeson (3) shows dissociation constants for molecular NaCl, KCl, HCl and NaOH. In his 1967 paper on complex dissociation in the temperature range 0 to 370°C, Helgeson presented estimates for the reactions:

$$HCl(aq) \rightleftharpoons H^+ + Cl^-$$

over the temperature range 0 to 370°C the values of $\log_{10} K$ varied from 7.25 to -4.75

$$NaOH(aq) \rightleftharpoons Na^+ + OH^-$$

over the temperature range 0 to 225°C the approximate values of $\log_{10} K$ varied from .5 to -.1

$$KCl(aq) \rightleftharpoons K^+ + Cl^-$$

over the temperature range 275 to 325°C the approximate values of log10 K varied from 0.1 to -1.

$$NaCl(aq) \rightleftharpoons Na^+ + Cl^-$$

over the temperature range 275 to 370°C the approximate values of $\log_{10}$ K varied from -.5 to -2.75

In 1981, Helgeson et al. (7) presented an approximate dissociation constant of NaCl(aq) over the full 0 to 370°C temperature range. At 0°C the approximate value for $\log_{10}$ K was .65.

Arbitrarily ignoring a molecular species over a wide range of conditions can be dangerous because:

- The dissociation constant may vary greatly with temperature and ignoring the molecular species formation a priori may be quite dangerous.

- Even a tiny amount of a molecular species can greatly influence a corresponding vapor pressure which in turn can be a pivotal element in controlling a process.

- Even if a molecular species has a large dissociation constant, significant amounts of the molecular species may form if the solubility of the species is high.

- In multicomponent solutions, a common ion from another electrolyte may force formation of an aqueous specie. For example, adding HCl to an acetic acid solution promotes the formation of the aqueous acetic acid molecule, but adding NaOH promotes the formation of acetate ions.

One good general test for a weak electrolyte is to see if a corresponding molecular vapor species exists. Thus, although aqueous molecular HCl is rarely mentioned, (in fact neither the National Bureau of Standards (1) nor the Russian Academy of Sciences (2) consider it in their data compilations), its presence is presaged by the obvious existence of the vapor HCl. A situation where a trace amount of a molecular aqueous species is in equilibrium with a vapor having a high vapor pressure implies a large Henry's law constant:

$$p_i = y_i P = H_i x_i \quad \text{or} \quad H_i = \left(\frac{y_i}{x_i}\right) P$$

where: p_i - vapor pressure of species i
y_i - vapor mole fraction
P - total vapor pressure
H_i - Henry's law constant for species i
x_i - liquid mole fraction of species i

As with the strong electrolytes and complexors, the key to the effective modeling of these systems is accurate prediction of:

- Equilibrium constants as a function of, primarily, temperature, and to a lesser extent, pressure. In particular, vapor-liquid equilibria, vapor species = aqueous species, and liquid phase molecular dissociation equilibria, aqueous species = ion + ion for example, must be dealt with.

- Activity coefficients as a function of the temperature and ionic strength.

- Fugacity coefficients for the vapor species. These values, functions of temperature, pressure and vapor phase composition, are usually unimportant until pressures reach several atmospheres.

The equilibrium constants for vapor-liquid equilibria (VLE), aqueous dissociation equilibria and solid-liquid equilibria (SLE) can be obtained either by the thermodynamic relationship involving the $\Delta G°$, $\Delta H°$ and ΔC_p of the reaction or via the numerous articles in the literature providing functions, tables or graphs of the temperature dependent constants for various species. Prominent among these are the articles by Wilhelm, Battino and Wilcock (8) for VLE and Edwards, Newman and Prausnitz (P5) for liquid phase dissociation and/or hydrolysis.

The issue of the role played by a molecular species in determining ionic and molecular activity coefficients as well as water activities has received much less attention than the role played by ionic species, particularly for strong electrolytes. In this chapter, attention will be focused on:
- the Setschênow equation
- the Pitzer based equations

The material in this chapter is divided into two principal sections, namely:

- Theoretical relationships for activity coefficients. The Setschénow equation is used to calculate the activity coefficients of aqueous molecular species in salt solutions. The Pitzer based methods may be used for binary or multicomponent solution activity coefficient calculations for all species in the solution.

- Comparison of predictions based upon the theoretical equations with experimental data. The specific systems considered are:

  a) NH_3-H_2O
  b) CO_2-H_2O
  c) NH_3-CO_2-H_2O
  d) SO_2-H_2O
  e) O_2-NaCl-H_2O

No attempt is made in this chapter to depict the actual equations or the attendant calculation procedure which requires the digital computer. Rather, the underlying detail is deferred until Chapter IX on worked examples where all of the systems of this chapter are covered.

Setschénow Equation

It has long been known that the solubility of a gas in a salt solution is less than that in pure water. The solubility decrease is called salting out. This behavior has been attributed to the greater attraction between ions and water molecules than between nonpolar or slightly polar gas molecules and water. The ions show increasing salting out influence with increasing ionic charge; the influence also increases with the decrease in ionic radius. In an attempt to empirically define this behavior, J. Setschénow (S33) proposed the following relationship:

$$\ln\left(\frac{S^{\circ}}{S}\right) = kc \tag{7.1}$$

where: S° - is the solubility of the gas in pure water

S - is the solubility of the gas in a salt solution

k - is the salt coefficient, constant for a given salt

c - is the salt concentration

The equation has been used with the concentrations in molalities, molarities and mole fractions. S and S° must be in the same units but are reported in cm^3/liter, gm/liter, moles/liter, molality, etc. The salting out coefficient, k, also changes with the different units.

The chemical potential of a species is the same at equilibrium, whether in a binary or multicomponent solution:

$$\mu_u = \mu_u^{\circ} \tag{7.2}$$

with the subscript u denoting an undissociated species and the superscript ° indicating binary solution. This relationship may also be expressed in terms of the activities,

$$a_u = a_u^{\circ} \tag{7.3}$$

The activity of a species is a function of its activity coefficients, γ, and molalities, m, as seen in earlier chapters:

$$\gamma_u m_u = \gamma_u^{\circ} m_m^{\circ} \tag{7.4}$$

By rearrangement:

$$\frac{\gamma_u}{\gamma_u^{\circ}} = \frac{m_u^{\circ}}{m_u} \tag{7.5}$$

As noted in Chapter II, the activity coefficient of a species approaches unity as its molality approaches zero due to the chosen standard state:

$$\gamma_i = a_i/m_i \qquad \gamma_i \longrightarrow 1 \text{ as } m_i \longrightarrow 0$$

Since the solubility of a gas in pure water is low, the activity coefficient $\gamma_u°$ is assumed equal to one. As a result, the activity coefficient in a multicomponent solution must therefore deviate from unity if equation (7.5) is to hold true:

$$\gamma_u = \frac{m_u^o}{m_u} \tag{7.6}$$

The right hand side of equation (7.6) is now of the same form as the Setschénow equation (7.1):

$$\ln \gamma_u = \ln\left(\frac{m_u^o}{m_u}\right) = k\, m_s \tag{7.7}$$

where: γ_u - is the activity coefficient of the undissociated gas molecule in the multicomponent system

m_u^o - is the solubility of the gas in pure water, molal units

m_u - is the solubility of the gas in the multicomponent solution, molal units

k - is the salting out parameter

m_s - is the molality of the salt in the solution

This is the equation that has been used to calculate the activity coefficient of an undissociated molecule in a salt solution.

Salting Out Parameter Determination by Randall and Failey

In 1927 a series of papers by M. Randall and C.F. Failey were published, presenting activity coefficients based on the Setschénow equation. The salt molality was expressed as ionic strength. The first paper (S28) concentrated on the activity coefficients of gases in salt solutions. The gases considered were oxygen, hydrogen, nitrogen, nitrous oxide, carbon dioxide, hydrogen sulfide, ammonia and acetylene. The tabulated results were the ionic strengths of the salts, the activity coefficients of the dissolved gases and the quotients log γ/I for each gas in different salt solutions of varying concentrations and temperatures. They plotted the quotients versus the square roots of the ionic strengths to a maximum 4 molal. The maximum ionic strength tabulated varied, with the highest being 21.9 for a solution of $Fe_2(SO_4)_3$ at 25°C into which was dissolved nitrous oxide. The quotient, which is essentially the salting coefficient k, did show some variation in their plots with the change in ionic strength. Randall and Failey felt however that it was within the range of experimental error and decided that the arithmetic mean of the term could be used as a constant k for most of the salts. Some of the quotient versus $\sqrt{I}$ plots can be seen in Figure 7.1. Here the plots were extended to the maximum ionic strengths tabulated for each of the solutions.

In the second paper of their series (S29), Randall and Failey found similar results for the nonelectrolytes iodine, phenylthiourea and o-nitrobenzaldehyde.

The last paper (S30) presented activity coefficients for the undissociated part of the following weak electrolyte acids: benzoic, ortho-toluylic, salicylic, ortho-nitrobenzoic, acetic, monochloracetic and dichloracetic acids. In a study of binary solutions of sodium and potassium acetate, Randall, McBain and White (S26) found the ionic activity coefficients to be very close in value to the ionic activity coefficients for binary sodium and potassium chloride solutions. Consequently, in Randall and Failey's paper on the activity coefficients of the undissociated part of weak electrolytes, the activity coefficients of the monobasic acids in salt solutions of varying concentrations were assumed to be equal to the activity coefficient of hydrochloric acid in the same or similar salt solution at the same concentration. This meant that:

- the activity coefficient of the dissociated portion of a weak electrolyte dissolved in any sodium salt solution was set equal to the activity coefficient of .01 M HCl in a NaCl solution of the same concentration. The values used were those presented by Harned (S10).

- the activity coefficients for .01 M HCl in KCl solutions presented by Harned (S10) were used for the activity coefficients of the dissociated portion of the weak electrolyte in any potassium salt solution of the same concentration.

- the activity coefficient of the dissociated portion of a weak electrolyte in any barium salt solution was equivilenced to the activity coefficient of HCl

Figure 7.1

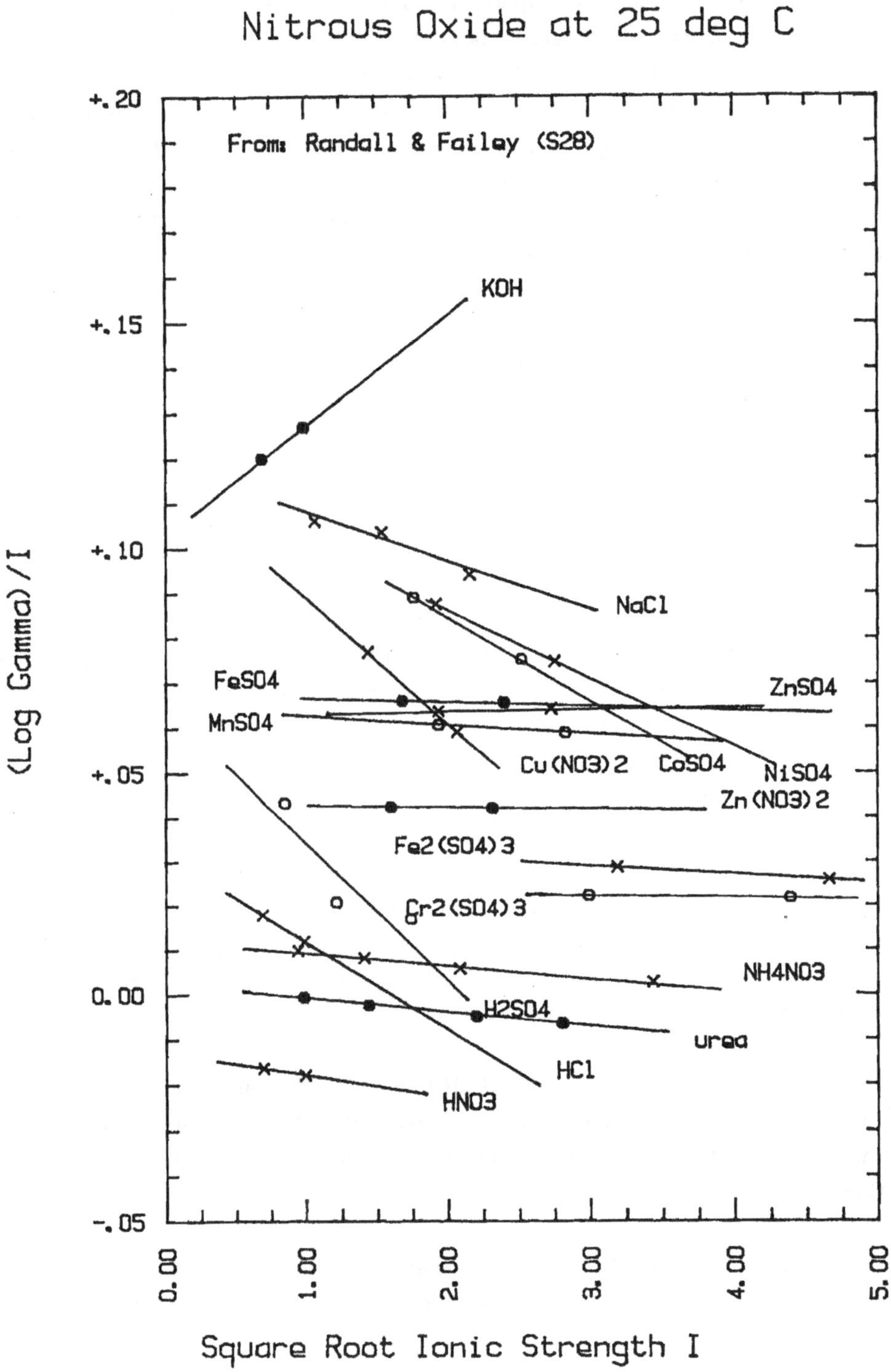
Nitrous Oxide at 25 deg C
From: Randall & Failey (S28)
KOH
NaCl
FeSO4
MnSO4
ZnSO4
Cu(NO3)2
CoSO4
NiSO4
Zn(NO3)2
Fe2(SO4)3
Cr2(SO4)3
NH4NO3
H2SO4
urea
HCl
HNO3
+.20
+.15
+.10
+.05
0.00
-.05
(Log Gamma)/I
0.00
1.00
2.00
3.00
4.00
5.00
Square Root Ionic Strength I

in barium chloride solution of the same concentration. The values used were those presented by Randall and Breckenridge (S27).

Randall and Failey felt that, while not completely accurate, such assumptions could be safely made for such dilute solutions and would not have much effect on the calculations for the activity coefficients of the undissociated portion of the weak electrolyte acid.

The activity coefficients of the undissociated portions of the weak acids were calculated in the following manner using data presented by numerous authors of the solubility of the weak acid in salt solutions of varying concentrations:

1. Since the activity of the weak acid in a solution saturated with that weak acid is constant:

$$a_2 = m_+ \, m_- \, \gamma_\pm^2 = \text{constant}$$

and the molalities of the hydrogen and acid ions are equal

$$m_+ = m_-$$

then $m_\pm \, \gamma_\pm$ is also constant. This $\gamma_\pm$ is the activity coefficient of the dissociated portion of the weak acid and is set to the activity coefficient of HCl as detailed above.

2. The following calculations were done using data for a binary solution of the weak electrolyte acid:

- Given: m° - the solubility of the weak acid in water

  K - the dissociation constant of the acid

  $\gamma_\pm^\circ$ - the activity coefficients of HCl in a binary solution from a plot of $\gamma_\pm$ versus ionic strength

the equilibrium equation is:

$$K = \frac{(m_+^\circ \, \gamma_\pm^\circ)}{m_u^\circ \, \gamma_u^\circ}$$

where m_u° and γ_u° are the molality and activity coefficient of the undissociated portion of the weak electrolyte acid in the binary solution. Making the simplifying assumption that the activity coefficient of the undissociated weak acid is unity in binary solution, and noting that

$$m_u^\circ = m^\circ - m_+^\circ \qquad (7.8)$$

the equilibrium was rewritten

$$(m_+^o \ \gamma_\pm^o)^2 = K\ (m^o - m_+^o)$$

This equation was solved via successive substitution for the hydrogen ion molality, m_+, using values for the activity coefficients, $\gamma_\pm$, read from the plot for each m_+ chosen. Once m_+ was found, the molality of the undissociated part of the weak acid was calculated using equation (7.8) and the term $m_+^o \ \gamma_\pm^o$ was set.

3. The following calculations were then done using ternary weak acid - salt - water solubility data in which the solution is saturated with the weak electrolyte to get the activity coefficient of the undissociated portion of the weak acid:

- Since the activity of the weak acid in any solution saturated with that weak acid is constant, the following is true:

  $$a_2^o = a_2$$

  That is, the activity of the weak acid in binary solution is equal to the activity of the weak acid in the ternary solution. This relationship may also be expressed in terms of the molalities and activity coefficients:

  $$m_+^o \ \gamma_\pm^o = m_+ \ \gamma_\pm \qquad (7.9)$$

  The left hand side of the equation is the term that was found using the binary solution data. The activity coefficient $\gamma_\pm$ was set equal to the activity coefficient for HCl in a salt solution of equal ionic strength using the assumptions detailed earlier. Therefore the only unknown in equation (7.9) is the molality of the hydrogen ion in the multicomponent solution:

  $$m_+ = \frac{m_+^o \ \gamma_\pm^o}{\gamma_\pm}$$

- Having the hydrogen ion molality in the ternary solution, m_+, and having the saturation molality of the weak acid in the solution, m, the molality of the undissociated portion of the weak acid, m_u, is merely the difference between the two:

  $$m_u = m - m_+$$

- Equation (7.9) equating the activities in binary and ternary solutions applies to the undissociated weak acid also; the molalities of the undissociated part of the weak acid are known for the binary and ternary solutions and since the activity coefficient in the binary solution was assumed to be unity, equation (7.9) may be rearranged in order to calculate the activity coefficient of the undissociated portion of the weak electrolyte in the ternary solution:

$$\gamma_u = \frac{m_u^o}{m_u}$$

Randall and Failey found that these activity coefficients behaved very much like those found for the nonelectrolytes in the previous papers. That is:

$$\frac{\log \gamma_u}{I} = \text{constant}$$

Salting Out Parameter Determination by Long and McDevit

In 1952, F.A. Long and W.F. McDevit (S15) presented the results of their extensive study of the activity coefficients of nonelectrolytes in aqueous salt solutions. Their results for the molar activity coefficients of undissociated nonelectrolytes in salt solutions were based mainly on solubility, distribution and vapor pressure measurements. They noted that since the activity coefficient for any species i could be expressed as a power series to show the effects of the concentrations of all solutes j in the solution:

$$\log y_i = \sum_{n,m^0}^{\infty} k_{nm} C_j^n C_i^m$$

then in their study of nonelectrolytes dissolved in salt solutions, since the concentrations are low, only the linear terms involving the nonelectrolyte and the salt would remain to calculate the activity coefficient of the nonelectrolyte:

$$\log y_u = k_s C_s + k_u C_u \qquad (7.10)$$

where: y_u -is the molar activity coefficient of the nonelectrolyte u

k_s - is the salting out parameter for salt s; also called the ion-molecule interaction term

C_s - molar concentration of salt s in the solution

k_u - the nonelectrolyte self interaction term

C_u - molar concentration of the nonelectrolyte u in the solution

Due to the low concentrations, interactions between the solutes would not be expected.

The ion-molecule interaction term, k_s, is the one that is most often calculated. Long and McDevit felt that the nonelectrolyte self interaction term could safely be ignored only when the nonelectrolyte solubility was very low as in the case of the nonpolar electrolytes hydrogen, oxygen and benzene. For the more soluble polar nonelectrolytes, such as ammonia, carbon dioxide and phenol, the self interaction term is of much greater importance and should be determined where data is available.

Long and McDevit outlined the following methods for calculating the activity coefficient of a nonelectrolyte as a function of the salt concentration of the solution:

1. From solubility data

   The data used by Long and McDevit was presented by many authors. It consisted of the saturation concentrations of nonelectrolytes in binary aqueous solutions and in salt solutions of varying concentrations. Both types of measurements can be found at different temperatures in the literature.

   a) Since at equilibrium the chemical potential of a species is the same in binary and multicomponent solutions saturated with that species, the activity coefficients and concentrations may be equivilenced as seen earlier in this chapter:

   $$y_u C_u = y_u^\circ C_u^\circ \qquad (7.4)$$

   Here equation (7.4) is based on molar units:

   y_u - molar activity coefficient of nonelectrolyte u in the salt solution

   C_u - solubility in moles per liter of the nonelectrolyte in the salt solution

   y_u° - molar activity coefficient of nonelectrolyte u in binary solution

   C_u° - solubility in moles per liter of the nonelectrolyte in binary solution

Rearranging equation (7.4) and taking the log results in the following equation for the log of the activity coefficient of the nonelectrolyte in the salt solution:

$$\log y_u = \log y_u^\circ + \log \frac{C_u^\circ}{C_u} \qquad (7.11)$$

b) Recalling the earlier equation for log y_u:

$$\log y_u = k_s C_s + k_u C_u \qquad (7.10)$$

combining equations (7.10) and (7.11) results in:

$$\log y_u^\circ + \log \frac{C_u^\circ}{C_u} = k_s C_s + k_u C_u \qquad (7.12)$$

Equation (7.10) may also be used to express the activity coefficient of the nonelectrolyte in a binary solution:

$$\log y_u^\circ = k_u^\circ C_u^\circ \qquad (7.13)$$

The self interaction term is independent of solution concentration and relates only to the nonelectrolyte; therefore it is considered equal in binary and multicomponent solutions, $k_u^\circ = k_u$. The log y_u° term in equation (7.12) is replaced by the right hand side of equation (7.13) and subtracted from both sides of the equation:

$$\log \frac{C_u^\circ}{C_u} = k_s C_s + k_u (C_u - C_u^\circ) \qquad (7.14)$$

c) The left hand side of equation (7.14) recalls the Setschénow equation:

$$\log y_u = \log \frac{C_u^\circ}{C_u} \qquad (7.7)$$

$$\log y_u = k_s C_s + k_u (C_u - C_u^\circ) \qquad (7.15)$$

2. From distribution data

The distribution data used by Long and McDevit was by other authors found in the literature. The applicable data were measurements of the solubilities of the nonelectrolyte in water, in a nonaqueous solvent that is immiscible with water and in an aqueous salt solution. The first measurements taken were the solubilities of the nonelectrolyte in the two solvents, water and di-butyl ether for example. The second system contained the two solvents to which a salt was added. The addition of salt must not create an atmosphere in which the solvents lose their immiscibility or cause one of them to begin to dissolve into the other. The salt also must dissolve only into the aqueous phase. The addition of the nonelectrolyte to the system resulted in the solubility in the aqueous phase differing from that in the system without salt, while the reference nonaqueous phase solubility remained the same. The following equation defines the relationship between the nonelectrolyte solubilities in the pure water, salt solution and the nonaqueous reference phase:

$$y_u^{\circ} \; C_u^{\circ} = b \; y_u^R \; C_u^R = y_u \; C_u \tag{7.16}$$

where:

- y_u° - is the molar activity coefficient of the nonelectrolyte in the pure water phase
- C_u° - is the solubility of the nonelectrolyte in the pure water phase in moles per liter
- b - is a constant
- y_u^R - is the activity coefficient of the nonelectrolyte in the reference phase
- C_u^R - is the solubility of the nonelectrolyte in the nonaqueous reference phase
- y_u - is the activity coefficient of the nonelectrolyte in the water - salt phase
- C_u - is the solubility of the nonelectrolyte in the water - salt phase

3. From vapor pressure data

The vapor pressure data presented in the literature is often the partial pressure of the nonelectrolyte in binary solution and the partial pressures in salt solutions of varying concentrations. The relationship between the partial vapor pressures of a nonelectrolyte in binary solution and in multicomponent solutions is described:

$$\frac{y_u \, C_u}{y_u^\circ \, C_u^\circ} = \frac{p_u}{p_u^\circ} \qquad (7.17)$$

where:

y_u - is the molar activity coefficient of the nonelectrolyte in the salt solution

C_u - is the solubility in moles per liter of the nonelectrolyte in the salt solution

y_u° - is the molar activity coefficient of the nonelectrolyte in binary solution; as usual this is treated as being the standard state and $y_u^\circ = 1$

C_u° - is the solubility in moles per liter of the nonelectrolyte in the binary solution

p_u - is the partial vapor pressure of the nonelectrolyte in the salt solution

p_u° - is the partial vapor pressure of the nonelectrolyte in binary solution

Equation (7.17) assumes ideal vapor phase behavior. By keeping the nonelectrolyte concentration the same in the binary and salt solutions, the concentration terms of equation (7.17) cancel, and since the activity coefficient in the binary solution is set to unity, equation (7.17) simplifies to:

$$y_u = \frac{p_u}{p_u^\circ}$$

The following advantages and drawbacks to these methods of calculation were noted:

Advantages:

- Solubility data are easy to take with good precision.
- Distribution data is particularly good where the nonelectrolyte is miscible in water. The experimental method is quite simple and the nonelectrolyte concentration in the aqueous phase can be kept low in order to avoid the self interaction term of equation (7.15).

- Vapor pressure measurements allow for easy calculations when the nonelectrolyte concentrations are kept the same in the binary and salt solutions, a process that is not difficult.

Drawbacks:

- Solubility data presents difficulty if the solubility is high; the self interaction term must then be determined based on separate experiments. Measurement errors are possible when the nonelectrolyte is a liquid; the salt or water may dissolve in it.
- Care must be taken during distribution experiments to avoid reference solvents which are miscible with water; the calculation of the activity coefficients can be distorted by any mutual solution of water and the nonaqueous reference solvent. The difference in distribution must also be sufficient to accurately measure.
- Vapor pressure measurements are applicable only to volatile species. In order to get accurate measurements, the nonelectrolyte concentration must also be quite high.

In addition to presenting values of k_s, and k_u for phenol, Long and McDevit presented, in the appendices, lists of studies done on the salting effects for many polar and nonpolar nonelectrolytes. Values for the salting parameters given in the paper can be found at the end of this chapter in Appendix 7.1.

Salting Out Parameter Determination by Other Authors

More recently other authors have presented results of their studies of the salting out effects on gases. Some of these are listed in the references at the end of this chapter (S1, S2, S5, S7, S8, S13, S16, S17, S19, S22, S23, S32, S34, S37). One of the most recent papers is that of E.M. Pawlikowski and J.M. Prausnitz (S24). They presented a simple method of estimating salting out constants for nonpolar gas solutions and included a temperature fit for some. In reviewing the perturbation theory method of calculating the salting out constants presented by Tiepel and Gubbins (S35) they decided that the results were not good enough to justify using the method. The calculations required a large number of parameters (salt solution density, Lennard-Jones size and energy parameters for water, for both salt ions and for the gas, the dipole moment of water and the polarizability of the gas). A computer was also required to do numerical integrations. Pawlikowski and Prausnitz proposed that for the activity coefficient of a nonpolar gas in a salt solution calculated using the Setschénow equation:

$$\ln \gamma_g = k_{s,g} m_s$$

where:

γ_g - is the activity coefficient of the gas

m_s - is the molality of the salt

the following equation was suggested by the perturbation theory for calculating the salting out parameter or salt-gas interaction parameter $k_{s,g}$:

$$k_{s,g} = a_s + b_s \left(\frac{\varepsilon_{gas}}{k}\right) \qquad (7.18)$$

where:

a_s and b_s are temperature dependent parameters for the salt and are tabulated in Appendix 7.2

ε_{gas} is the Lennard-Jones energy interaction for the gas. They used Liabastre's (S14) reported values for ε which were calculated based on pure water gas solubility data. These values are tabulated in Appendix 7.2.

k is Boltzmann's constant

$k_{s,g}$ is the salting out parameter for a solution of salt s and gas g. Values of $k_{s,g}$ in kg/mole are tabulated in Appendix 7.2.

They found the salting out constant increasing with ε/k for all salts other than $NaNO_3$, which agreed with behavior noted by other authors for nitrate salts.

In addition to presenting values for a_s and b_s for various salts, they also presented values for the following temperature fit equation

$$k_{s,g} = a_{1s} + a_{2s} T + (b_{1s} + b_{2s} T) \left(\frac{\varepsilon_{gas}}{k}\right) \qquad (7.19)$$

with T in Kelvins for temperatures from 0 to 60° Celsius. The parameters are tabulated in Appendix 7.2.

Based on Bromley's (S3) assumption that activity coefficient interaction parameters for individual ions could be added to estimate the value for a salt:

$$\beta_{CA} = \beta_{C^+} + \beta_{A^-}$$

Pawlikowski and Prausnitz presented salting out parameters for individual ions that could be added to estimate a salt's k_s:

$$k_{s(CA)} = k_{s(C^+)} + k_{s(A^-)}$$

Using the $a_{s(i)}$ and $b_{s(i)}$ values they presented for individual ions and the Liabastre value for a gas's Lennard-Jones parameter, the salt-gas interaction term can be estimated:

$$k_{s(C^+),g} = a_{s(C^+)} + b_{s(C^+)} \left(\frac{\varepsilon_{gas}}{k}\right) \qquad (7.20a)$$

$$k_{s(A^-),g} = a_{s(A^-)} + b_{s(A^-)} \left(\frac{\varepsilon_{gas}}{k}\right) \qquad (7.20b)$$

$$k_{s,g} = k_{s(C^+),g} + k_{s(A^-),g}$$

Pawlikowski and Prausnitz note that although the parameters presented are intended primarily for 1-1 electrolytes, they may be used to estimate the salting out parameter for 1-2 electrolytes if the additive equation is used as follows:

$$k_{CA_2,g} = k_{s(C^{2+}),g} + 2\, k_{s(A^-),g}$$

A similar change may also be made for estimating the salting out parameter for a 2-1 electrolyte. Temperature dependent parameters were also presented for the ions; the values for the temperature dependent and independent ion parameters are given in Appendix 7.2.

Although the Setschénow equation can be used with success in predicting the solubility of a nonelectrolyte in dilute salt solutions, it has some major drawbacks. One is the restriction of use to salt solutions. Also, due to the linearity of the equation, published values of the salting out parameter can not be expected to extrapolate well past the solution concentration at which the parameter was determined. In 1977, Yasunishi (S37) suggested an expanded version of the Setschénow equation in an attempt to avoid the linear behavior. In a study of the solubilities of nitrogen and oxygen in aqueous sodium sulfite and sulfate solutions, Yasunishi presented the solubilities in terms of the Ostwald coefficient:

$$L = \frac{V_g}{V_L}$$

where:

L - Ostwald coefficient

V_g - volume of gas absorbed

V_L - volume of the absorbing liquid

The Setschénow equation for the activity coefficient of the gas was then expressed:

$$\log y_g = \log\left(\frac{L^\circ}{L}\right) = k\, C_s \tag{7.21}$$

where:

L° - is the Ostwald coefficient for the binary gas-water system

k - is the salting out parameter

C_s - is the salt concentration in moles/liter

Yasunishi found that a plot of log (L°/L) versus C_s showed the linear behavior required for using the Setschénow equation for the nitrogen in both the sodium sulfite and sodium sulfate solutions at 25°C. This can be seen in Figure 7.2, plotted to the maximum salt concentrations of approximately 1.1 moles/liter. A similar plot of log (L°/L) versus the Na_2SO_4 concentration at 25°C however, began to deviate from the linear line resulting from equation (7.21) at approximately .5 moles/liter Na_2SO_4. This is shown in Figure 7.3. The broken line was calculated using equation (7.21). As a result, Yasunishi proposed the following equation be used:

$$\log y_g = \log\left(\frac{L^\circ}{L}\right) = \frac{k_1\, C_s}{1. + k_2 C_s} \tag{7.22}$$

With this equation, and the parameters k_1 and k_2 given in Table 7.1, the data was successfully fit to 1.5 moles/liter Na_2SO_4. The solid line in Figure 7.3 was the result of using equation (7.22).

Another disadvantage of the Setschénow equation is that the salting out parameter affects only the molecular activity coefficient; no effect is shown on the other species in the solution. These problems are addressed in the next section covering the adaptation of the Pitzer equation to solutions containing molecular solutes.

Figure 7.2

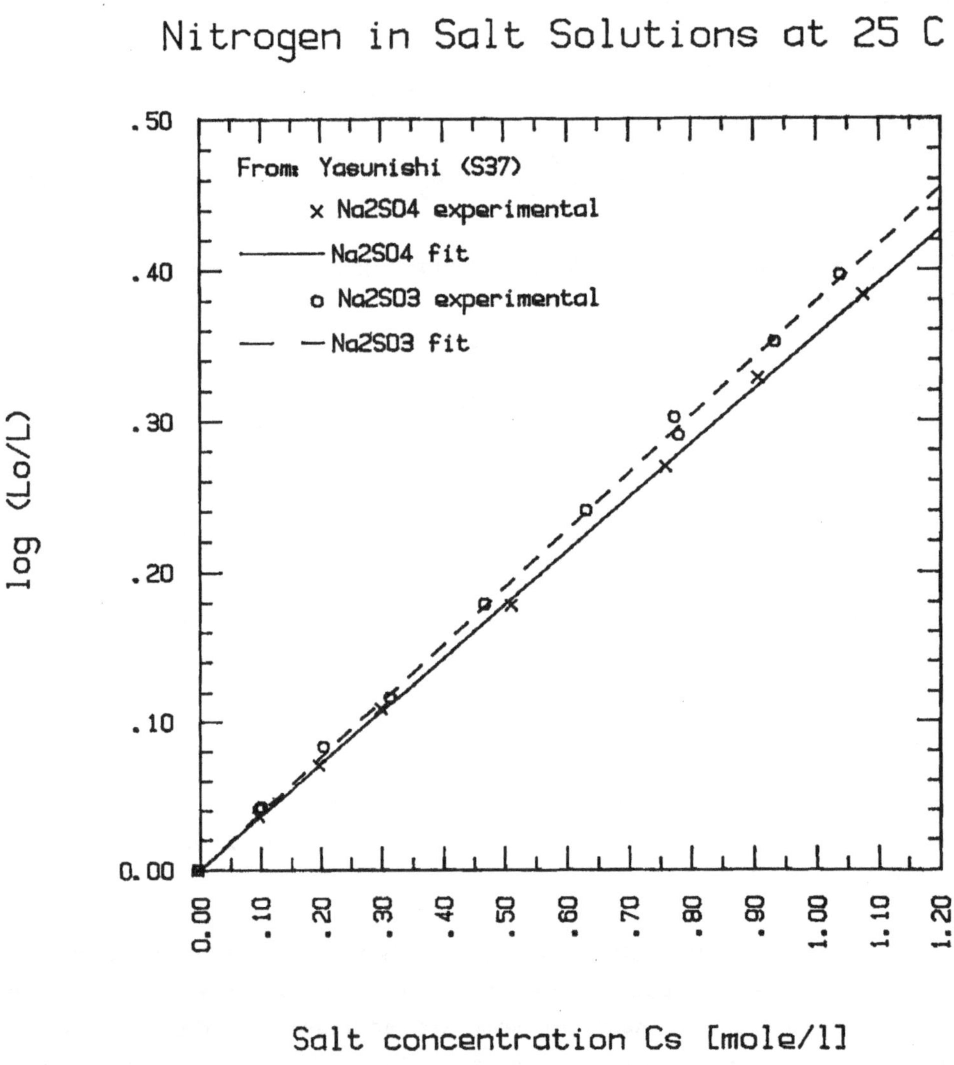
Nitrogen in Salt Solutions at 25 C
From: Yasunishi (S37)
x Na2SO4 experimental
Na2SO4 fit
o Na2SO3 experimental
Na2SO3 fit
log (Lo/L)
.50
.40
.30
.20
.10
0.00
.10
.20
.30
.40
.50
.60
.70
.80
.90
1.00
1.10
1.20
Salt concentration Cs [mole/l]

Figure 7.3

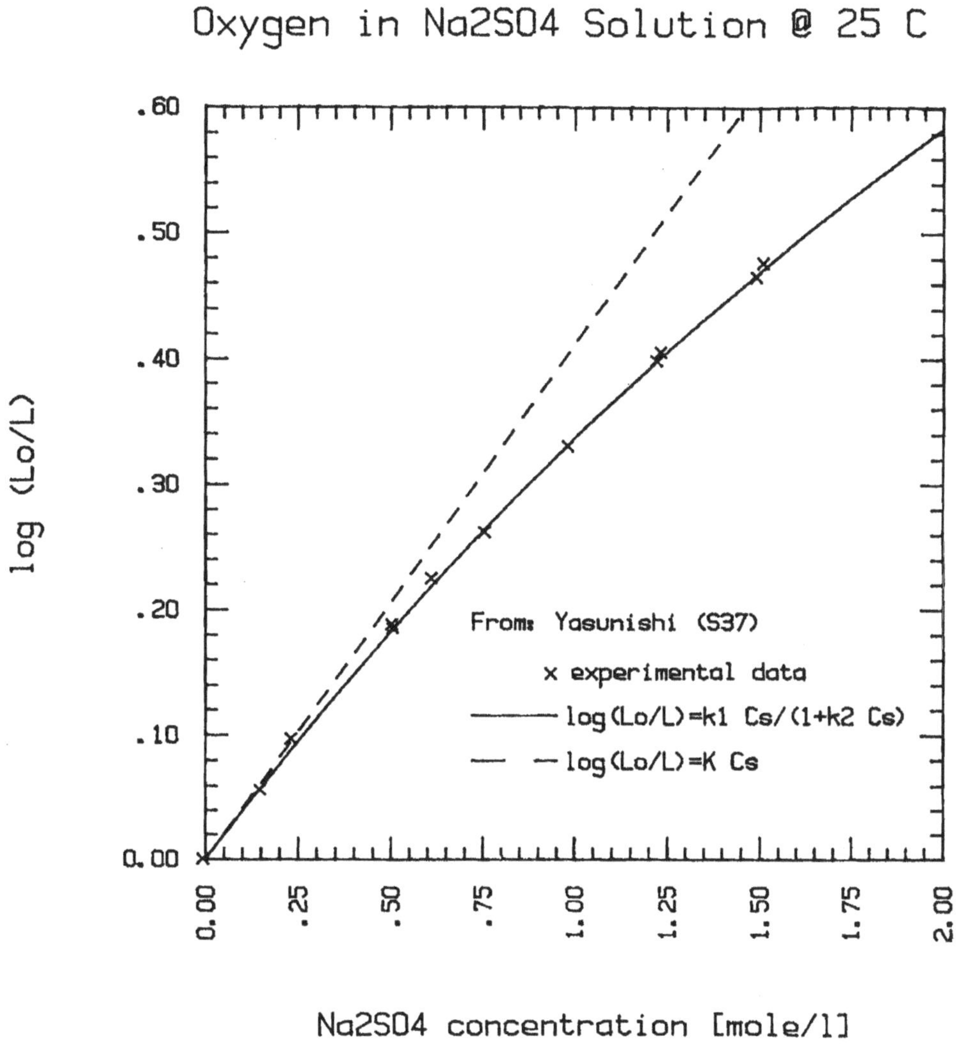
Oxygen in Na2SO4 Solution @ 25 C
log (Lo/L)
.60
.50
.40
.30
.20
.10
0.00
0.00
.25
.50
.75
1.00
1.25
1.50
1.75
2.00
From: Yasunishi (S37)
x experimental data
log(Lo/L)=k1 Cs/(1+k2 Cs)
log(Lo/L)=K Cs
Na2SO4 concentration [mole/l]

Table 7.1 - Constants of equation (7.22) for oxygen-sodium sulfate systems (S37)

temp (°C)	k_1 (l/mole)	k_2 (l/mole)
15.	0.420	0.251
25.	0.398	0.183
35.	0.421	0.231

Pitzer Based Equations

Due to the need to model the equilibria of solutions containing multiple weak electrolytes, such as the H_2O - NH_3- CO_2 system, it became necessary to go beyond the Setschénow equation for activity coefficient calculations. For such solutions to be modeled well, the ion-molecule interactions must affect not only the molecular activity coefficients, but also the ionic activity coefficients and water activities. An early attempt by Edwards, Newman and Prausnitz (P5) used the Guggenheim equation for activity coefficients and assumed the water activity to be unity. This application was felt to be good for low weak electrolyte concentrations at temperatures no higher than 80°C.

Edwards, Maurer, Newman and Prausnitz Pitzer Based Method

In 1978, Edwards, Maurer, Newman and Prausnitz (P6) used the Pitzer formalism for such solutions. They felt the resulting equations and parameters to be applicable for solutions of concentrations as high as 10 to 20 molal for 0 to 170°C.

As seen in Chapter V, the equation for the activity coefficient of a species i in a multicomponent solution can be expressed using the Pitzer equation:

$$\ln \gamma_i = z_i^2 f^\gamma + 2. \sum_{j \neq H_2O} m_j B_{ij} + z_i^2 \sum_{j \neq H_2O} \sum_{k \neq H_2O} m_j m_k B'_{jk} \qquad (7.23)$$

where:

$$f^\gamma = -A_\phi \left[\frac{\sqrt{I}}{1. + 1.2\sqrt{I}} + \frac{2.}{1.2} \ln(1. + 1.2\sqrt{I}) \right]$$

$$B_{ij} = \beta_0 + \frac{\beta_1}{2I} \left[1. - (1. + 2.\sqrt{I}) \exp(-2.\sqrt{I}) \right]$$

$$B'_{ij} = \frac{\beta_1}{4 I^2} \left[-1. + (1. + 2.\sqrt{I} + 2.\ I) \exp(-2.\sqrt{I}) \right]$$

A_ϕ = the natural log based Debye-Hückel constant for osmotic coefficients; equation (4.64)

I = $.5 \Sigma z^2 m$; ionic strength, summation over all ions in the solution

z - ionic charge

m - species molality

β_0 - interaction parameter

β_1 - interaction parameter

This equation, used by Edwards et al., ignores interaction parameters other than the binary β_0 and β_1. Like charge ionic interactions are ignored. Unlike Pitzer's definition of the interaction parameters being between ions of different charges, Edwards et al. used the terms for ion-ion, ion-molecule and molecule-molecule interactions. They ignored the β_1 term for interactions involving molecules as they felt the experimental data was inadequate. Equation (7.23) reduces to:

$$\ln \gamma_m = 2. \sum_{j \neq H_2O} m_j B_{mj} \tag{7.24}$$

for aqueous molecular species. The first and last terms of equation (7.23) drop out because the charge on a molecule is zero. For the molecular activity coefficient, the term B_{mj} is the β_0 interaction due to ignoring the β_1 term.

The water activity is expressed:

$$\ln a_w = M_w \left\{ \frac{2.\, A\phi\, I^{3/2}}{1. + 1.2\sqrt{I}} - \sum_{i \neq H_2O} \sum_{j \neq H_2O} m_i m_j \left[\beta_0 + \beta_1 \exp(-2.\sqrt{I}) \right] \right\} - M_w \sum_{i \neq H_2O} m_i \tag{7.25}$$

where the molecular weight of water is denoted by M_w.

Edwards et al. note that for a weak electrolyte in solution, equations (7.24) and (7.25) reduce to molecule-molecule interactions:

$$\ln \gamma_m = 2.\, \beta_{0(m-m)}\, m_m \tag{7.26}$$

$$\ln a_w = \left[-\beta_{0(m-m)}\, m_m^2 - m_m \right] M_w \tag{7.27}$$

when the dissociation constant is small and the resulting ionic concentrations are very low. Equation (7.26) is very similar in form to the Setschénow equation seen earlier in this chapter:

$$\ln \gamma_m = k\, m_s \tag{7.7}$$

However, equation (7.7) is based on the salt molality in a mixed solution; equation (7.26) has the advantage of being dependent on the aqueous molecule molality and therefore may be used for binary solutions.

Equations (7.23) and (7.25) have other improvements over the Setschénow formalism. In addition to their usefulness in solutions not containing salts, the presence of undissociated molecules in a solution is reflected in the water activity equation and in the ionic activity coefficient equations as well as the molecular activity coefficient equations. Another advantage is the ability to set up many of the ion-ion, ion-molecule and molecule-molecule interactions using binary solution data. These may then be used in different multicomponent solutions, unlike the Setschénow salting out parameters which are good only for the one mixed solution.

As in Pawlikowski and Prausnitz's (S24) paper on Setschénow constants, Edwards et al. (P5, P6) used Bromley's theory of the additivity of single ion interactions to estimate ion-ion and ion-molecule interactions:

$$\beta_{ca} = \beta_c + \beta_a$$

$$\beta_{mi} = \beta_m + \beta_i$$

The molecule-molecule interaction parameters between different molecules were estimated from the molecular self interaction parameters:

$$\beta_{m_1 m_2} = .5\,(\beta_{m_1 m_1} + \beta_{m_2 m_2})$$

Beutier and Renon's Pitzer Based Method

Also in 1978, Beutier and Renon (P1) presented a slightly different interpretation of the Pitzer equation for use with solutions containing molecular solutes. Noting the additive form of the equation given by Pitzer for the excess Gibbs free energy:

$$\frac{G^{ex}}{.018\, n_w\, R\, T} = f(I) + \sum_k \sum_l \lambda_{kl}\, m_k\, m_l + \sum_k \sum_l \sum_h \mu_{klh}\, m_k\, m_l\, m_h \qquad (4.57)$$

where:

n_w - kilograms of water

m - molalities

f(I) - long range electrostatic effects

$$= -\frac{A}{3}\left(\frac{4\,I}{1.2}\right) \ln\,(1. + 1.2\sqrt{I}\,)$$

λ_{kl} - short range interaction effects

μ_{klh} - triple body interaction between k, l, and h

Beutier and Renon defined the Gibbs excess free energy as a summation:

$$G^{ex} = G^{ex}_i + G^{ex}_{im} + G^{ex}_m$$

in terms of the ion-ion, ion-molecule and molecule-molecule interactions. This differs from Pitzer's definition (4.57) in that he considered k, l and h to be ions only, whereas Beutier and Renon felt that molecules should also be included.

The activity coefficient is the derivative of the Gibbs excess free energy:

$$\ln \gamma_k = \frac{\partial}{\partial n_k}\left(\frac{G^{ex}}{RT}\right)$$

which resulted in the following for the molecular and ionic activity coefficients:

$$\ln \gamma_m = (\ln \gamma_m)_{mm} + (\ln \gamma_m)_{im} \tag{7.28}$$

$$\ln \gamma_i = (\ln \gamma_i)_{ij} + (\ln \gamma_i)_{im} \tag{7.29}$$

with the subcripts denoting:

mm - molecule-molecule interactions

im - ion-molecule interactions

ij - ion-ion interactions

The derivatives of equation (4.57) for the terms of equations (7.28) and (7.29) are:

$$\ln (\gamma_m)_{mm} = 2.\ \lambda_{mm}\ m_m + 2.\ \mu_{mmm}\ m_m^2 \tag{7.30}$$

where:

λ_{mm} - is the self interaction term, equal to the $\beta_{0\,(m-m)}$ of equation (7.26)

μ_{mmm} - is the pseudo ternary self interaction term, set:

$$\mu_{mmm} = -\frac{1.}{55.5}\left(\lambda_{mm} + \frac{1.}{166.5}\right)$$

and:

$$\ln (\gamma_i)_{ij} = \frac{z_i^2}{2.}\frac{df}{dI} + 2.\ \sum_j \lambda_{ij}\ m_j + \frac{z_i^2}{2.}\sum_j\sum_k \frac{d\lambda_{jk}}{dI}\ m_j\ m_k +$$

$$3.\ \sum_j\sum_k \mu_{ijk}\ m_j\ m_k \tag{7.31}$$

where:

$$\frac{df}{dI} = -\frac{2A}{3.}\left[\frac{\sqrt{I}}{1. + 1.2\sqrt{I}} + \frac{2.}{1.2}\ln\left(1. + 1.2\sqrt{I}\right)\right]$$

A - is the Debye-Hückel constant

λ_{ij} - Pitzer ion-ion interaction

$$\lambda_{ij} = \beta_0 + \frac{2\beta_1}{2^2 I}\left[1. - (1. + 2.\sqrt{I})\exp(-2.\sqrt{I})\right]$$

$$\frac{d\lambda_{jk}}{dI} = -\frac{2\beta_1}{2^2 I^2}\left[1. - (1. + 2.\sqrt{I} + 2.\ I)\exp(-2.\sqrt{I})\right]$$

μ_{ijk} - ternary ion interactions. The following assumptions were made:

$$\mu_{MXY} = .5\ (\mu_{MXX} + \mu_{MYY})$$

$$\mu_{MMX} = \mu_{MXX}$$

and are related to Pitzer's C^{ϕ} parameter as follows:

for the 1-1 electrolyte CA:

$$\frac{3}{2} C^{\phi} = C^{\gamma} = \frac{9}{2}(\mu_{CCA} + \mu_{CAA})$$

for the 1-2 electrolyte C_2A:

$$\frac{\sqrt{2}}{3} C = \frac{2\sqrt{2}}{9} C^{\gamma} = 2\mu_{CCA} + \mu_{CAA}$$

The ion-molecule interaction terms were estimated based on the Debye-McAulay electrostatic theory which gives an equation for the electric work needed to transfer ions from a solution of dielectric constant D_i to one of dielectric constant D_f:

$$W_{el} = \left(\frac{1.}{D_f} - \frac{1.}{D_i}\right)\sum_j \frac{N_A e^2 z_j^2}{2 r_j} n_j \times 10^8 \qquad (7.32)$$

where:

- N_A - is Avagadro's number
- e - is the electron charge in esu
- z_j - is the ionic charge
- r_j - is the ionic cavity radius in Angstroms, Å
- D_i - is the dielectric constant of the solution with ionic solutes and no molecular solutes
- D_f - is the dielectric constant of the solution with ionic and molecular solutes
- n_j - number of moles of ion j

This expression of electric work was used to determine the ion-molecule portions of the activity coefficients:

$$G_{im}^{ex} = W_{el} \tag{7.33}$$

Using Pottel's (P8) explanation of Onsager's theory, the dielectric constants of equation (7.32) are defined:

$$D_i = D_w \left[\frac{1. - Y_i}{1. + .5\, Y_i} \right] \tag{7.34}$$

$$D_f = D_s \left[\frac{1. - Y_f}{1. + .5\, Y_f} \right] \tag{7.35}$$

where:

D_w - is the dielectric constant of water

$D_w = 305.7 \exp(-\exp(-12.741 + .01875\, T) - T/219.)$

T - Kelvins

D_s - dielectric constant of solution containing molecular but no ionic solutes; with the partial molal volumes assumed to be concentration independent:

$$D_s = D_w \left[1. + \sum_m \alpha_m \frac{m_m}{V_m} \right] \tag{7.36}$$

Y_i - assuming that every ion in solution is the center of a dielectrically saturated cavity which includes hydration molecules, this term is the volume fraction of the ionic cavities in an ionic solution:

$$Y_i = \frac{4}{3} \pi N_A \sum_j r_j^3 m_j \frac{1}{V_i} \times 10^{-27} \tag{7.37}$$

Y_f - volume fraction of the ionic cavities with all species, ions and molecules, in the solution:

$$Y_f = Y_i \left(\frac{V_i}{V_f} \right) \tag{7.38}$$

V_i - volume of solution with only ionic solutes, $dm^3/kg\ H_2O$:

$$V_i = \frac{1}{d_0} + \sum_i m_i \overline{v}_i \tag{7.39}$$

V_f - volume of the real solution containing both molecular and ionic solutes, $dm^3/kg\ H_2O$:

$$V_f = \frac{1}{d_0} + \sum_i m_i \overline{v}_i + \sum_m m_m \overline{v}_m \tag{7.40}$$

V_m - volume of solution with only molecular solutes, $dm^3/kg\ H_2O$:

$$V_m = \frac{1.}{d_0} + \sum_m m_j \bar{v}_m \qquad (7.41)$$

d_0 - water density

$\bar{v}_k$ - partial molal volume of k, dm^3/mole

α_m - dielectric coefficient of neutral solute m, dm^3/mole:

$$\alpha_m = \bar{v}_m \left[\frac{D_m - D_w}{D_w} \right] \qquad (7.42)$$

By substituting equations (7.34) and (7.35) for D_i and D_f in equation (7.32) and defining a term:

$$L_j = \frac{e^2\ z_j^2}{2\ r_j\ k\ T\ D_w} \times 10^8 \qquad (7.43)$$

where k is Boltzmann's constant, 1.38045 x 10^{-16} erg/K, equation (7.32) becomes:

$$\frac{W_{el}}{0.18\ n_w\ R\ T} = \sum_m L_j\ m_j \left[\frac{D_w}{D_s}\ \frac{1. + .5\ Y_f}{1. - Y_f} - \frac{1. + .5\ Y_i}{1. - Y_i} \right]$$

Using equations (7.36) and (7.37) and summing over all ions, j and k, and molecules, m, Beutier and Renon arrived at the following equation:

$$\frac{W_{el}}{.018\ n_w\ R\ T} = \sum_j \sum_m - \frac{D_w\ (1. - .5\ Y_i\ Y_f)}{D_s\ (1. - Y_i)\ (1. - Y_f)\ V_m}\ L_j\ \alpha_m\ m_m\ m_j +$$

$$\sum_j \sum_k \frac{D_w}{D_s\ (1. - Y_i)\ (1. - Y_f)} \left[\frac{1}{2V_f} - \frac{1}{V_i} + \frac{D_s}{D_w\ V_f} - \frac{D_s}{2\ D_w\ V_i} \right] \frac{4}{3}\ \pi\ N_A \times 10^{-27}\ L_j\ r_k^3\ m_k\ m_j \qquad (7.44)$$

Since G_{im}^{el} = W_{el} and the Gibbs excess free energy G^{ex} was defined as a summation of terms, equation (7.44) is then related to the Pitzer expression, equation (4.57), seen earlier. The following definitions were made by equivalencing the terms of equation (7.44) to the interaction terms, λ:

$$\lambda_{jm} = \frac{- D_w\ (1. - .5\ Y_i\ Y_f)}{D_s\ (1. - Y_i)\ (1. - Y_f)\ V_m}\ L_j\ \alpha_m$$

λ_{mj} = 0 since the z term in L equals zero

$$\lambda_{jk} = \frac{4}{3}\,\pi\,N_A \times 10^{-27}\;\frac{D_w}{D_s\,(1. - Y_i)\,(1. - Y_f)}\left[\frac{1}{2V_f} - \frac{1}{V_i} + \frac{D_s}{D_w V_f} - \frac{D_s}{2\,D_w V_i}\right] L_j\, r_k^3$$

The ion-molecule interaction parts of the activity coefficients (7.28) and (7.29) can now be determined by taking the appropriate derivatives:

$$(\ln \gamma_m)_{im} = (\sum_j L_j\, m_j)\,\frac{D_w}{D_f}\left[-\frac{D_w}{D_s}\,\frac{\alpha_m + \bar{v}_m}{V_m} + \frac{\bar{v}_m}{V_m} - \frac{D_s}{D_f}\,\frac{\bar{v}_m}{V_f}\,\frac{1.5\,Y_f}{(1. + .5\,Y_f)^2}\right] \qquad (7.45)$$

$$(\ln \gamma_i)_{im} = L_i\left(\frac{D_w}{D_f} - \frac{D_w}{D_i}\right) + \sum_j m_j\, L_j\left[\frac{D_w}{D_i^2}\,\frac{\partial D_i}{\partial m_i} - \frac{D_w}{D_f^2}\,\frac{\partial D_f}{\partial m_i}\right] \qquad (7.46)$$

where:

$$\frac{\partial D_i}{\partial m_i} = -\,D_w\,\frac{1.5}{(1. + .5\,Y_i)^2}\,\frac{1.}{V_i}\left[\frac{4}{3}\,\pi\,N_A'\, r_i^3 - \bar{v}_i\, Y_i\right]$$

$$\frac{\partial D_f}{\partial m_i} = -\,D_s\,\frac{1.5}{(1. + .5\,Y_f)^2}\,\frac{1.}{V_f}\left[\frac{4}{3}\,\pi\,N_A'\, r_i^3 - \bar{v}_i\, Y_f\right]$$

$$N_A' = N_A \times 10^{-27}$$

By defining two new terms:

V_c - volume of all the ionic cavities, $dm^3/kg\ H_2O$:

$$V_c = \sum_i v_i^c\, m_i$$

v_i^c - volume of ionic cavities for ion i, $dm^3/mole$:

$$v_i^c = \frac{4}{3}\pi N_A'\, r_i^3$$

and using previous definitions to express equations (7.45) and (7.46) in terms of the volumes, the various equations may be combined:

- for the activity coefficient of a molecular solute:

$$\ln \gamma_m = (\ln \gamma_m)_{mm} + (\ln \gamma_m)_{im} \qquad (7.28)$$

$$= 2\,\lambda_{mm}\,m_m + 3\,\mu_{mmm}\,m_m^2 + (\sum_j L_j\,m_j)\,\frac{D_w}{D_s}\left[\frac{-1.5\,\overline{V}_m\,V_c}{(V_f - V_c)^2} + \left(\frac{V_f + .5\,V_c}{V_f - V_c}\right)\left(\frac{\overline{V}_m}{V_m} - \frac{\overline{V}_m + \alpha_m}{V_m}\,\frac{D_w}{D_s}\right)\right] \qquad (7.47)$$

- for the activity coefficient of an ionic solute:

$$\ln \gamma_i = (\ln \gamma_i)_{ij} + (\ln \gamma_i)_{im} \qquad (7.29)$$

$$= \frac{z_i^2}{2.}\,\frac{df}{dI} + 2.\sum_j \lambda_{ij}\,m_j + \frac{z_i^2}{2.}\sum_j\sum_k \frac{d\lambda_{jk}}{dI}\,m_j\,m_k + 3.\sum_j\sum_k \mu_{ijk}\,m_j\,m_k + L_i\,\frac{D_w}{D_s}\sum_m m_m\left[-\frac{\alpha_m}{V_m}\,\frac{V_i + .5\,V_c}{V_i - V_c} - \frac{1.5\,\overline{V}_m\,V_c}{(V_i - V_c)(V_f - V_c)}\right] + 1.5\,(\sum_j L_j\,m_j)\left[\frac{D_w}{D_s}\,\frac{V_f\,v_i^c - V_c\,\overline{V}_i}{(V_f - V_c)^2} - \frac{V_i\,v_i^c - V_c\,\overline{V}_i}{(V_i - V_c)^2}\right] \qquad (7.48)$$

An equation for the water activity was also derived from this expression of the Gibbs free energy:

$$\ln a_w = M_w\left[-\sum_i m_i - \sum_m m_m + \frac{1.}{M_w}\,\frac{\partial}{\partial n_w}\left(\frac{G^{ex}}{RT}\right)\right]$$

$$= M_w\left[-\sum_i m_i - \sum_m m_m + f(I) - I\,\frac{df}{dI} - \sum_i\sum_j\left(\lambda_{ij} + I\,\frac{d\lambda_{ij}}{dI}\right)m_i\,m_j - 2.\sum_i\sum_j\sum_k \mu_{ijk}\,m_i\,m_j\,m_k - \sum_m \lambda_{mm}\,m_m^2 - 2.\sum_m \mu_{mmm}\,m_m^3 - (\sum_i L_i\,m_i)\left\{-\frac{\sum_m \alpha_m\,m_m}{d_0\,V_m^2}\left(\frac{D_w}{D_s}\right)^2\left(\frac{V_f + .5\,V_c}{V_f - V_c}\right) + \frac{1.5\,V_c}{d_0(V_f - V_c)^2}\,\frac{D_w}{D_s} - \frac{1.5\,V_c}{d_0\,(V_i - V_c)^2}\right\}\right] \qquad (7.49)$$

Chen's Pitzer Based Method

Chen et al. (P2, P3, P4) also presented an extended form of the Pitzer equation for use with aqueous molecular solutes. As first seen in Chapter IV, the equation for the Gibbs excess free energy from which activity coefficient equations may be derived was suggested by Pitzer:

$$\frac{G^{ex}}{n_w RT} = f(I) + \sum_i \sum_j \lambda_{ij} m_i m_j + \sum_i \sum_j \sum_k \mu_{ijk} m_i m_j m_k \qquad (4.57)$$

By combining terms and choosing an empirical formulation, Pitzer arrived at the following equation for the Gibbs excess free energy of a multicomponent solution of cations, c, and anions, a:

$$\frac{G^{ex}}{n_w RT} = f(I) + 2.\sum_c \sum_a m_c m_a \left[B_{ca} + (\Sigma mz) C_{ca}\right] +$$

$$\sum_c \sum_{c'} m_c m_{c'} \left[\Theta_{cc'} + .5 \sum_a m_a \psi_{cc'a}\right] +$$

$$\sum_a \sum_{a'} m_a m_{a'} \left[\Theta_{aa'} + .5 \sum_c m_c \psi_{caa'}\right] \qquad (7.50)$$

where:

$$f(I) = -A_\phi \frac{4\ I}{1.2} \ln(1. + 1.2\sqrt{I})$$

$$B_{ca} = \lambda_{ca} + \left(\frac{\nu_c}{2\nu_a}\right)\lambda_{cc} + \left(\frac{\nu_a}{2\nu_c}\right)\lambda_{aa}$$

$$= \beta_0 + \frac{2\ \beta_1}{2^2\ I}\left[1. - (1. + 2.\sqrt{I})\exp(-2.\sqrt{I})\right]$$

$$C_{ca} = \frac{C^\phi_{ca}}{2\sqrt{|z_c z_a|}}$$

$$C^\phi_{ca} = \frac{3}{\sqrt{\nu_c \nu_a}}(\nu_c \mu_{cca} + \nu_a \mu_{caa})$$

with triple interactions between ions all having the same sign ignored

$$\Theta_{cc'} = \lambda_{cc'} - \left(\frac{z_{c'}}{2z_c}\right) \lambda_{cc} - \left(\frac{z_c}{2z_{c'}}\right) \lambda_{c'c'}$$

$$\psi_{cc'a} = 6\ \mu_{cc'a} - \left(\frac{3z_{c'}}{z_c}\right) \mu_{cca} - \left(\frac{3z_c}{z_{c'}}\right) \mu_{c'c'a}$$

Chen et al. proposed that a similar process be followed to get interactions involving molecular solutes. If the i, j and k's of equation (4.57) can be any solute, ionic or molecular, the Gibbs excess free energy contribution of the terms involving molecules is:

$$\left(\frac{G^{ex}}{n_w\,R\,T}\right)_{molecule} = \sum_m \sum_{i \neq m}{}' m_m\, m_i\ (2.\ \lambda_{mi} + 6.\ \sum_j m_j\ \mu_{mij}) + \sum_m \sum_{m'} m_m\, m_{m'}\ (\lambda_{mm'} + \sum_{m''} m_{m''}\ \mu_{mm'm''})$$

Ignoring the ternary interactions, Chen defined the following terms:

$$D_{ca,m} = 2\left(\frac{\nu_{cm}}{\nu_a} + \frac{\nu_{am}}{\nu_c}\right)\Big/\nu_c$$

$$= D_{ac,m}\left(\frac{\nu_a}{\nu_c}\right)$$

These terms describe the interaction between salt ca and a molecular solute m in an aqueous solution of one salt and one molecular solute which Chen says is equivalent to the Setschénow salt-molecule interaction term.

$$\omega_{cc',m} = 2.\left(\frac{\lambda_{cm}}{z_c} - \frac{\lambda_{c'm}}{z_{c'}}\right)\Big/z_{c'}$$

For describing the difference in the interactions between a molecular solute and two salts with the same anion; equivalent to the difference between two Setschénow constants.

$$\omega_{aa',m} = 2.\left(\frac{\lambda_{am}}{z_a} - \frac{\lambda_{a'm}}{z_{a'}}\right)\Big/z_{a'}$$

To describe the difference in the interactions of a molecular solute with two salts having the same cation.

The final version of the extended equation for the excess Gibbs free energy was proposed by Chen to be:

$$\frac{G^{ex}}{n_w R T} = f(I) + 2. \sum_c \sum_a m_c m_a \left[B_{ca} + (\Sigma mz)\, C_{ca}\right] +$$

$$\sum_c \sum_{c'} m_c m_{c'} (\Theta_{cc'} + .5 \sum_a m_a \psi_{cc'a}) +$$

$$\sum_a \sum_{a'} m_a m_{a'} (\Theta_{aa'} + .5 \sum_c m_c \psi_{caa'}) +$$

$$\sum_m \sum_{m'} m_m m_{m'} \lambda_{mm'} + \sum_m (\sum_a D_{ca,m} m_m m_a -$$

$$\sum_{c'} \omega_{cc',m} m_m m_{c'}) \tag{7.51}$$

Chen took the ternary molecule-molecule interaction term to be zero under the assumption that its effect would be minimal in solutions of low molecular solute concentration. The disregard of the ternary ion-molecule interactions was based on the approximate additivity of the contributions made by ions from inorganic salts; Chen notes that organic ion contributions do not show this additivity so that the ternary ion-molecule interactions in organic salt solutions should not be ignored.

The appropriate derivatives of equation (7.51) were taken to arrive at the following equations:

- for the activity of water, a_w:

$$\ln a_w = \ln x_w + \ln \gamma_w$$

$$= \ln \left[\frac{55.55}{55.55 + \sum_i m_i}\right] + \frac{(\partial G^{ex} / \partial n_w)}{55.55\, R T}$$

$$= \ln \left[\frac{55.55}{55.55 + \sum_i m_i}\right] - \frac{(\phi - 1) \sum_i m_i}{55.55} \tag{7.52}$$

- for the osmotic coefficient, ϕ :

$$(\phi - 1.) \sum_i m_i = - A_\phi \left[\frac{2.\ I^{3/2}}{1. + 1.2\sqrt{I}}\right] +$$

$$2. \sum_c \sum_a m_c m_a \left[\beta_0 + \beta_1 \exp(-2.\sqrt{I}) + \frac{2. \sum_c m_c z_c}{\sqrt{|z_c z_a|}} C_{ca}\right] +$$

$$\sum_c \sum_{c'} m_c m_{c'} (\Theta_{cc'} + \sum_a m_a \psi_{cc'a}) +$$

$$\sum_a \sum_{a'} m_a m_{a'} (\Theta_{aa'} + \sum_c m_c \psi_{aa'c}) + \sum_m \sum_{m'} m_m m_{m'} \lambda_{mm'} +$$

$$\sum_m (\sum_a D_{ca,m} m_m m_a - \sum_{c'} \omega_{cc',m} m_m m_{c'}) \qquad (7.53)$$

- for the activity coefficient of an anion, γ_a:

$$\ln \gamma_a = \frac{1.}{n_w R T} \left(\frac{\partial G^{ex}}{\partial m_a}\right)$$

$$= - A_\phi z_a^2 \left[\frac{\sqrt{I}}{1. + 1.2\sqrt{I}} + \frac{2.}{1.2} \ln (1. + 1.2\sqrt{I})\right] +$$

$$2. \sum_c m_c \left(\beta_0 + \frac{\beta_1}{2I} \left[1. - (1. + 2.\sqrt{I}) \exp(-2.\sqrt{I})\right]\right) +$$

$$2. \sum_{a'} m_{a'} \Theta_{aa'} - \frac{z_a^2}{2I^2} \sum_c \sum_{a'} m_c m_{a'} \beta_1 \left[1. - (1. + 2.\sqrt{I} + 2I)\right.$$

$$\left.\exp(-2.\sqrt{I})\right] + \sum_c \sum_{a'} m_c m_{a'} \psi_{aa'c} + .5 \sum_c \sum_{c'} m_c m_{c'} \psi_{cc'a} +$$

$$\sum_c m_c \frac{2 \sum_c m_c z_c}{\sqrt{|z_c z_a|}} C_{ca} + z_a \sum_c \sum_{a'} \frac{m_c m_{a'} C_{ca'}}{\sqrt{|z_c z_{a'}|}} + \sum_m m_m D_{ca,m} \qquad (7.54)$$

A cation activity coefficient may be calculated by reversing the cation and anion subscripts.

- for the activity coefficient of a molecular solute, γ_m:

$$\ln \gamma_m = \frac{1.}{n_w R T} \left(\frac{\partial G^{ex}}{\partial m_m} \right)$$

$$= 2. \sum_{m'} m_{m'} \lambda_{mm'} + \sum_a D_{ca,m} m_a - \sum_{c'} \omega_{cc',m} m_{c'} \qquad (7.55)$$

Predictions Based Upon Theoretical Equations

The theoretical equations given by Setschénow and Pitzer can be evaluated based upon an extensive body of published data on a large number of binary and ternary systems involving weak electrolytes. The systems chosen for evaluation and illustration are:

1) NH_3-H_2O
2) CO_2-H_2O
3) NH_3-CO_2-H_2O
4) SO_2-H_2O
5) O_2-NaCl-H_2O

In the case of the first four systems, the Pitzer equation used is based upon the formulation of Edwards, Maurer, Newman and Prausnitz and of Beutier and Renon. Comparisons of predicted and experimentally measured total pressure and partial pressure(s) are presented in both tabular and graphical form.

In the case of the last example wherein NaCl is present, the emphasis will be on prediction using the Setschénow equation and comparison to resonable published data.

The tabular and graphical output will, in every case, reference both:

1) The method used
2) The source of the data

Ammonia - Water

Figures 7.4 - 7.9 show the tabular and graphical results of Pitzer predictions measured against the data of Macriss, et al (D1) and Clifford and Hunter (D2). The species considered in the predictive model are NH_3(aq), NH_4^+, OH^-, H^+ and H_2O(aq).

Figure 7.4

SYSTEM: AMMONIA - WATER

==

Method: Edwards, Maurer, Newman & Prausnitz, AIChE J (1978)

Experimental Data Source:
Macriss, et al, IGT34 (1964); Clifford & Hunter, JPC 37 (1933)

Temperature: 60.0 degrees Celsius

Experimental data:

point	Pressure (atm)	NH3 Partial Pressure (atm)	Moles NH3(aq)	Moles H2O
1	0.24000	0.04324	0.05637	5.49755
2	0.28600	0.08570	0.11567	5.44149
3	0.35500	0.14968	0.19964	5.36211
4	0.35500	0.14358	0.19964	5.36211
5	0.36600	0.16281	0.21491	5.34768
6	0.41600	0.23540	0.27539	5.29050
7	0.43000	0.24503	0.29007	5.27663
8	0.48700	0.29969	0.35466	5.21557
9	0.58400	0.41333	0.46387	5.11232
10	0.63000	0.46249	0.50850	5.07013
11	0.71443	0.53700	0.58718	4.99575
12	0.84200	0.67535	0.70403	4.88529
13	1.10400	0.93767	0.90308	4.69712
14	1.45100	1.31308	1.11035	4.50117
15	1.61257	1.45929	1.17436	4.44067
16	3.08907	2.96467	1.76154	3.88559
17	5.57937	5.46848	2.34871	3.33050
18	9.18555	9.09954	2.93589	2.77542

Calculation results:

Error function = {[pNH3(calc) - pNH3(exp)] / pNH3(exp)} +
{[P(calc) - P(exp)] / P(exp)}

point	P(exp)	P(calc)	pNH3(exp)	pNH3(calc)	error
1	0.24000	0.23482	0.04324	0.04024	-0.09102
2	0.28600	0.27719	0.08570	0.08472	-0.04233
3	0.35500	0.34090	0.14968	0.15150	-0.02762
4	0.35500	0.34090	0.14358	0.15150	0.01537
5	0.36600	0.35297	0.16281	0.16414	-0.02743
6	0.41600	0.40238	0.23540	0.21583	-0.11585
7	0.43000	0.41476	0.24503	0.22877	-0.10176
8	0.48700	0.47115	0.29969	0.28767	-0.07267
9	0.58400	0.57404	0.41333	0.39494	-0.06155
10	0.63000	0.61904	0.46249	0.44178	-0.06217
11	0.71443	0.70294	0.53700	0.52900	-0.03098
12	0.84200	0.83931	0.67535	0.67047	-0.01041
13	1.10400	1.10951	0.93767	0.94987	0.01801
14	1.45100	1.45387	1.31308	1.30446	-0.00459
15	1.61257	1.57601	1.45929	1.42991	-0.04281
16	3.08907	3.22316	2.96467	3.11033	0.09254
17	5.57937	6.77150	5.46848	6.69593	0.43813
18	9.18555	16.30672	9.09954	16.26729	1.56296

Sum [error ^ 2] = 2.695161

Figure 7.5

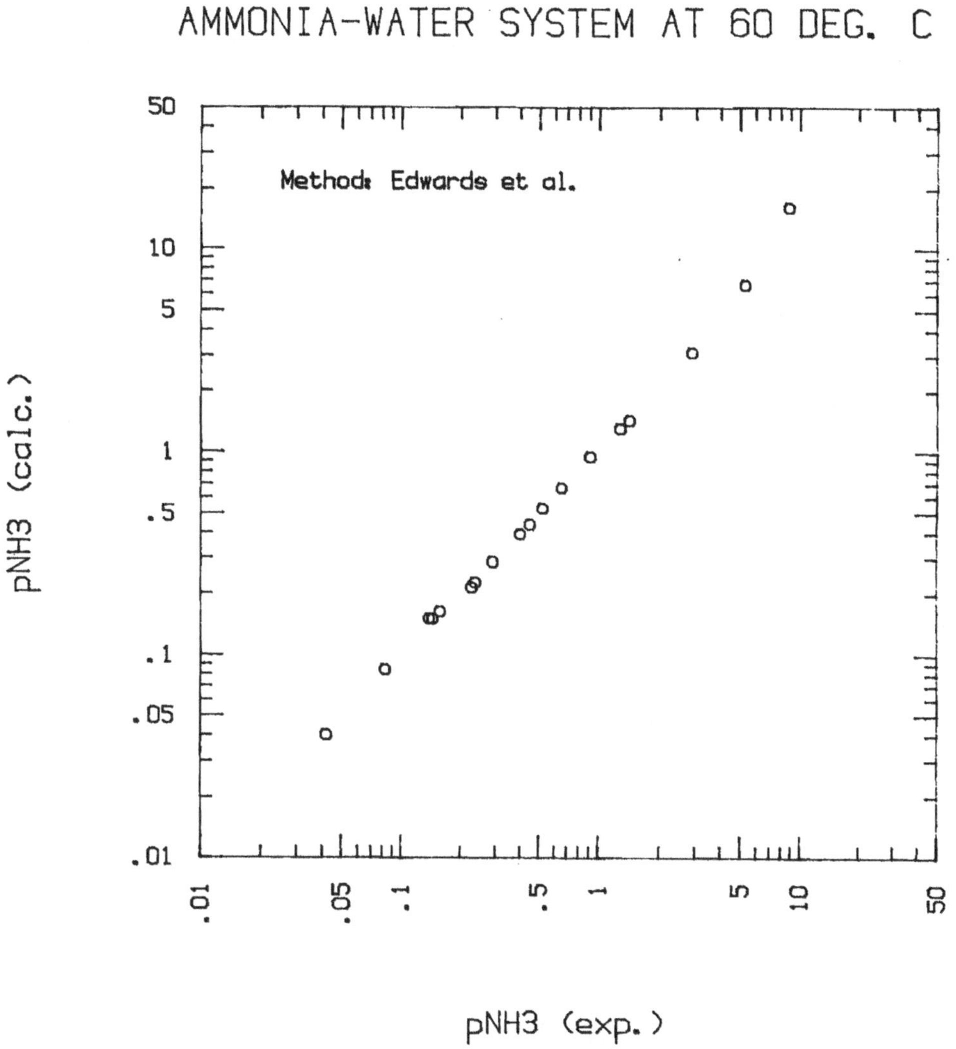
AMMONIA-WATER SYSTEM AT 60 DEG. C
Method: Edwards et al.
50
10
5
1
.5
.1
.05
.01
pNH3 (calc.)
.01
.05
.1
.5
1
5
10
50
pNH3 (exp.)

Figure 7.6

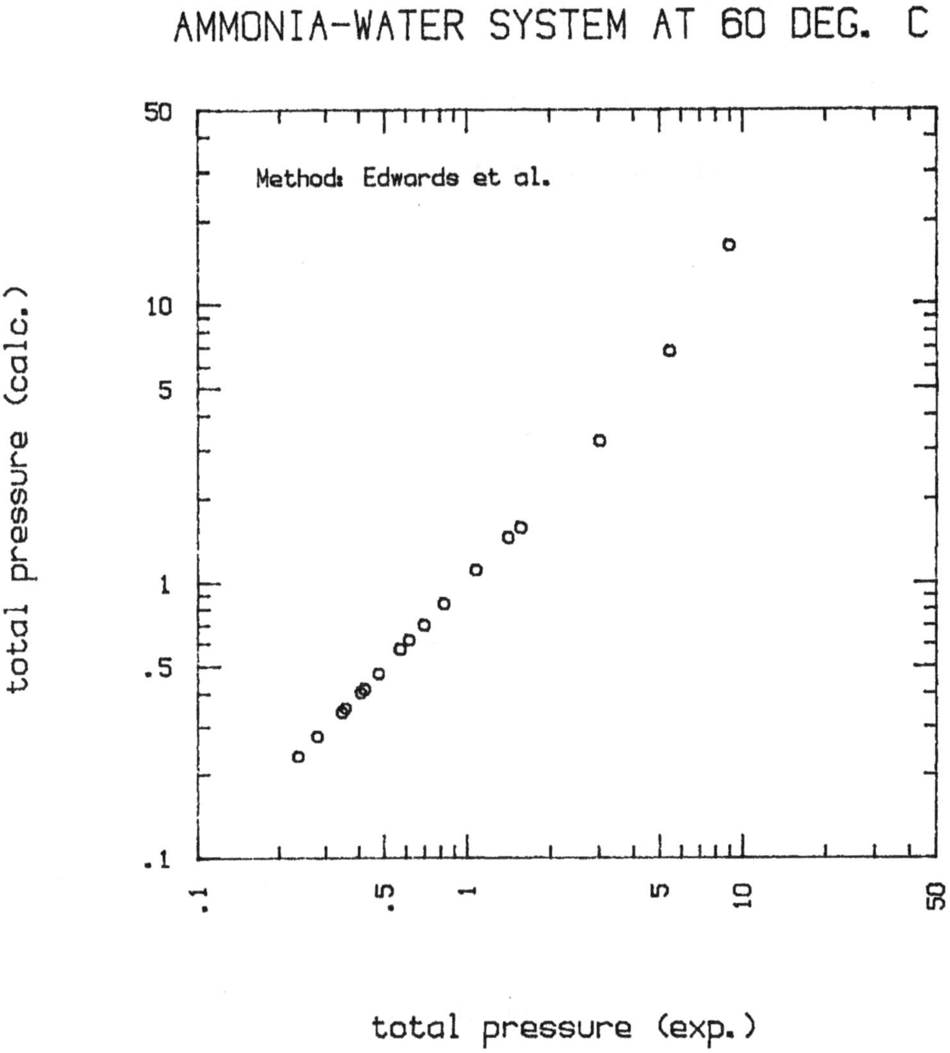
AMMONIA-WATER SYSTEM AT 60 DEG. C
Method: Edwards et al.
total pressure (calc.)
total pressure (exp.)
50
10
5
1
.5
.1
.1
.5
1
5
10
50

Figure 7.7

SYSTEM: AMMONIA - WATER

Method: Beutier & Renon, IECPDD (1978)

Experimental Data Source:
Macriss, et al, IGT34 (1964); Clifford & Hunter, JPC 37 (1933)

Temperature: 60.0 degrees Celsius

Experimental data:

point	Pressure (atm)	NH3 Partial Pressure (atm)	Moles NH3(aq)	Moles H2O
1	0.24000	0.04324	0.05637	5.49755
2	0.28600	0.08570	0.11567	5.44149
3	0.35500	0.14968	0.19964	5.36211
4	0.35500	0.14358	0.19964	5.36211
5	0.36600	0.16281	0.21491	5.34768
6	0.41600	0.23540	0.27539	5.29050
7	0.43000	0.24503	0.29007	5.27663
8	0.48700	0.29969	0.35466	5.21557
9	0.58400	0.41333	0.46387	5.11232
10	0.63000	0.46249	0.50850	5.07013
11	0.71443	0.53700	0.58718	4.99575
12	0.84200	0.67535	0.70403	4.88529
13	1.10400	0.93767	0.90308	4.69712
14	1.45100	1.31308	1.11035	4.50117
15	1.61257	1.45929	1.17436	4.44067
16	3.08907	2.96467	1.76154	3.88559
17	5.57937	5.46848	2.34871	3.33050
18	9.18555	9.09954	2.93589	2.77542

Calculation results:

Error function = {[pNH3(calc) - pNH3(exp)] / pNH3(exp)} +
{[P(calc) - P(exp)] / P(exp)}

point	P(exp)	P(calc)	pNH3(exp)	pNH3(calc)	error
1	0.24000	0.23487	0.04324	0.04023	-0.09102
2	0.28600	0.27715	0.08570	0.08462	-0.04364
3	0.35500	0.34040	0.14968	0.15092	-0.03286
4	0.35500	0.34040	0.14358	0.15092	0.00996
5	0.36600	0.35233	0.16281	0.16341	-0.03365
6	0.41600	0.40087	0.23540	0.21422	-0.12634
7	0.43000	0.41297	0.24503	0.22687	-0.11373
8	0.48700	0.46762	0.29969	0.28399	-0.09220
9	0.58400	0.56534	0.41333	0.38594	-0.09822
10	0.63000	0.60713	0.46249	0.42950	-0.10763
11	0.71443	0.68337	0.53700	0.50888	-0.09584
12	0.84200	0.80221	0.67535	0.63242	-0.11082
13	1.10400	1.01752	0.93767	0.85580	-0.16564
14	1.45100	1.25169	1.31308	1.09821	-0.30100
15	1.61257	1.32412	1.45929	1.17311	-0.37499
16	3.08907	1.86435	2.96467	1.73132	-0.81248
17	5.57937	1.74584	5.46848	1.60543	-1.39351
18	9.18555	0.95995	9.09954	0.63824	-1.82535

Sum [error ^ 2] = 6.285167

Figure 7.8

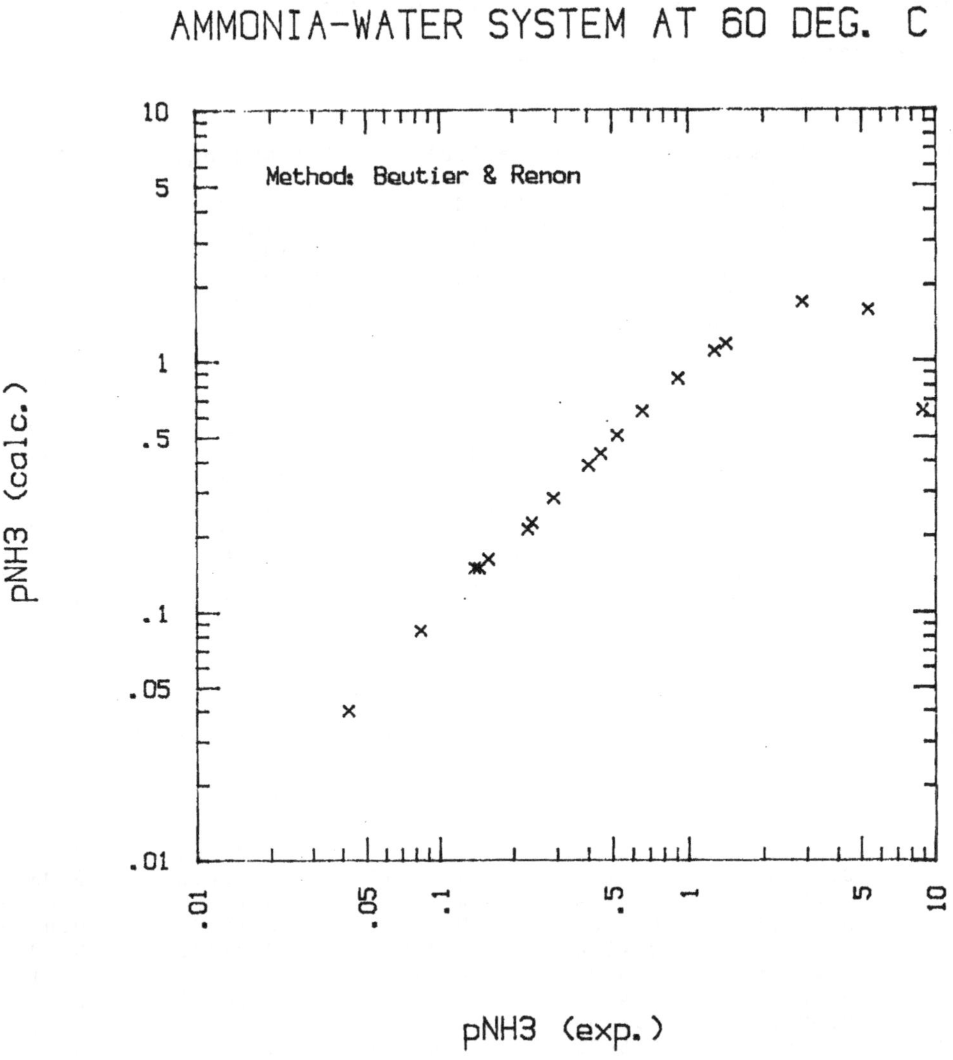
AMMONIA-WATER SYSTEM AT 60 DEG. C
Method: Beutier & Renon
pNH3 (calc.)
pNH3 (exp.)
10
5
1
.5
.1
.05
.01
.01
.05
.1
.5
1
5
10

Figure 7.9

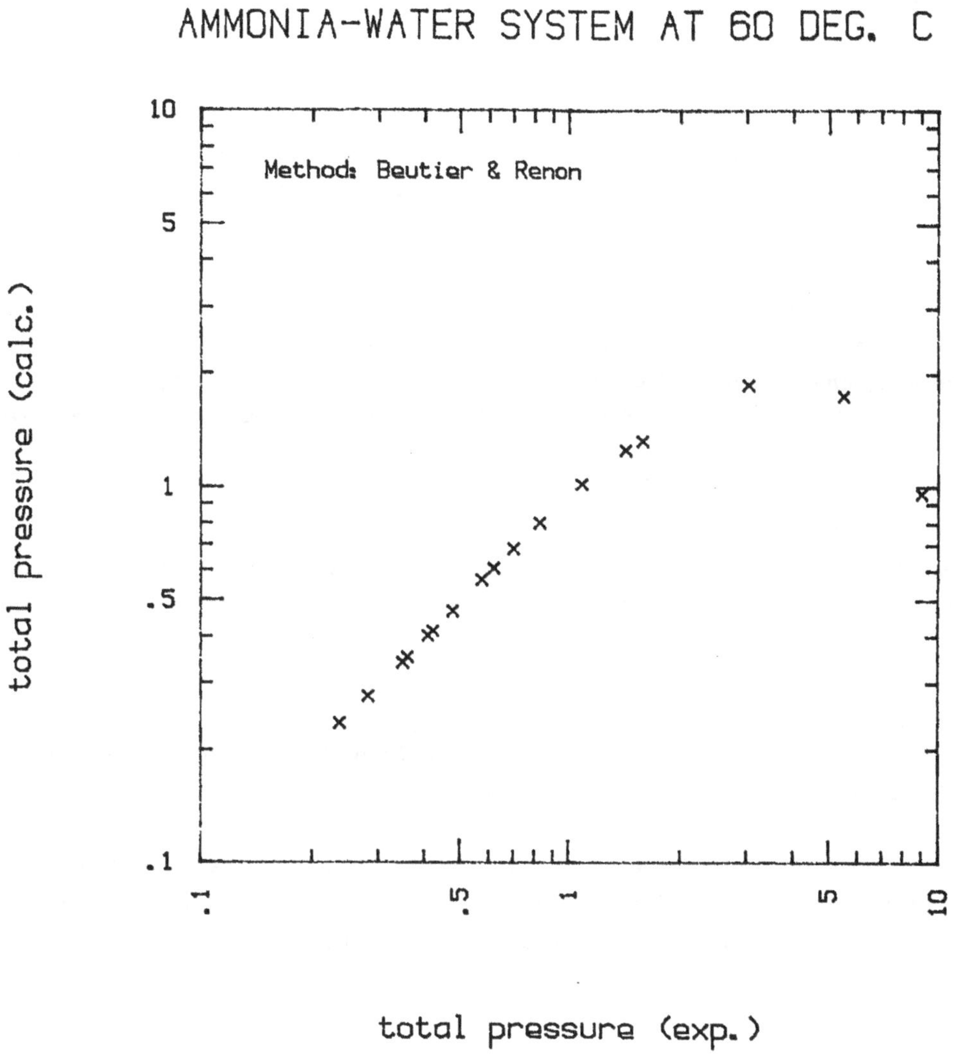
AMMONIA-WATER SYSTEM AT 60 DEG. C
Method: Beutier & Renon
total pressure (calc.)
total pressure (exp.)
10
5
1
.5
.1
.1
.5
1
5
10

Carbon Dioxide - Water

Figures 7.10 - 7.13 show the tabular and graphical results of Pitzer predictions measured against the data of Houghton, McLean and Ritchie (D3). The species considered in the predictive model are CO_2 (aq), CO_3^{2-}, HCO_3^-, H^+, OH^- and H_2O(aq).

Ammonia - Carbon Dioxide - Water

Figures 7.14 - 7.15 show the tabular and graphical results of Pitzer predictions measured against the data of Otsuku (D4). The species considered in the predictive model are NH_3 (aq), NH_4^+, H^+, OH^-, H_2O(aq), CO_2 (aq), CO_3^{2-}, HCO_3^- and $NH_2CO_2^-$.

Sulfur Dioxide - Water

Figures 7.16 - 7.19 show the tabular and graphical results of Pitzer predictions measured against the data of Sherwood (D6) and Johnstone and Leppla (D5). The species considered in the predictive model are SO_2(aq), SO_3^{2-}, HSO_3^-, H^+, OH^- and H_2O(aq).

Oxygen - Sodium Chloride - Water

Figures 7.20 - 7.21 show the tabular and graphical results of Setschénow measured against the data of Cramer (D7). The species considered in the predictive model are: O_2(aq), Na^+, Cl^-, and H_2O(aq).

Conclusions

Using the published Pitzer coefficients and applying them to the binary systems NH_3-H_2O, CO_2-H_2O and SO_2-H_2O, resulted in

1) An excellent match between the predicted points and the experimental data over the entire range for the CO_2-H_2O system. An excellent match for NH_3-H_2O and SO_2-H_2O at low concentrations of NH_3 and SO_2, but a methodical worsening of the fit at higher concentrations.

2) Very strong agreement between the two different Pitzer formulations.

One likely reason for the poor fit at high concentrations is found in the material discussed in Chapter VI. In the case of NH_3-H_2O, Helgeson (3) indicated that the molecular species NH_4OH(aq), with a dissociation constant of about

Figure 7.10

SYSTEM: CARBON DIOXIDE - WATER

Method: Edwards, Maurer, Newman & Prausnitz, AIChE J (1978)

Experimental Data Source:
Houghton, McLean & Ritchie, Chem E Sci, 6 (1957)

Temperature: 50.0 degrees Celsius

Experimental data:

point	CO2 Partial Pressure (atm)	CO2 Liquid Mole Fraction	H2O Mole Fraction
1	1.000	0.00034	0.99966
2	2.000	0.00068	0.99932
3	4.000	0.00135	0.99865
4	6.000	0.00202	0.99798
5	8.000	0.00266	0.99734
6	10.000	0.00330	0.99670
7	12.000	0.00393	0.99607
8	14.000	0.00455	0.99545
9	16.000	0.00515	0.99485
10	18.000	0.00575	0.99425
11	20.000	0.00634	0.99366
12	22.000	0.00691	0.99309
13	24.000	0.00747	0.99253
14	26.000	0.00803	0.99197
15	28.000	0.00857	0.99143
16	30.000	0.00910	0.99090
17	32.000	0.00962	0.99038
18	34.000	0.01013	0.98987
19	36.000	0.01063	0.98937

Calculation results:

Error function = [pCO2(calc) - pCO2(exp)] / pCO2(exp)

point	pCO2(exp)	pCO2(calc)	error
1	1.000	0.96913	-0.03087
2	2.000	1.94599	-0.02701
3	4.000	3.89218	-0.02695
4	6.000	5.85561	-0.02406
5	8.000	7.77332	-0.02833
6	10.000	9.72273	-0.02773
7	12.000	11.67413	-0.02716
8	14.000	13.62758	-0.02660
9	16.000	15.55081	-0.02807
10	18.000	17.50802	-0.02733
11	20.000	19.46753	-0.02662
12	22.000	21.39534	-0.02748
13	24.000	23.32441	-0.02815
14	26.000	25.29027	-0.02730
15	28.000	27.22282	-0.02776
16	30.000	29.15693	-0.02810
17	32.000	31.09272	-0.02835
18	34.000	33.03032	-0.02852
19	36.000	34.96989	-0.02861

Sum [error ^ 2] = 0.014538

Figure 7.11

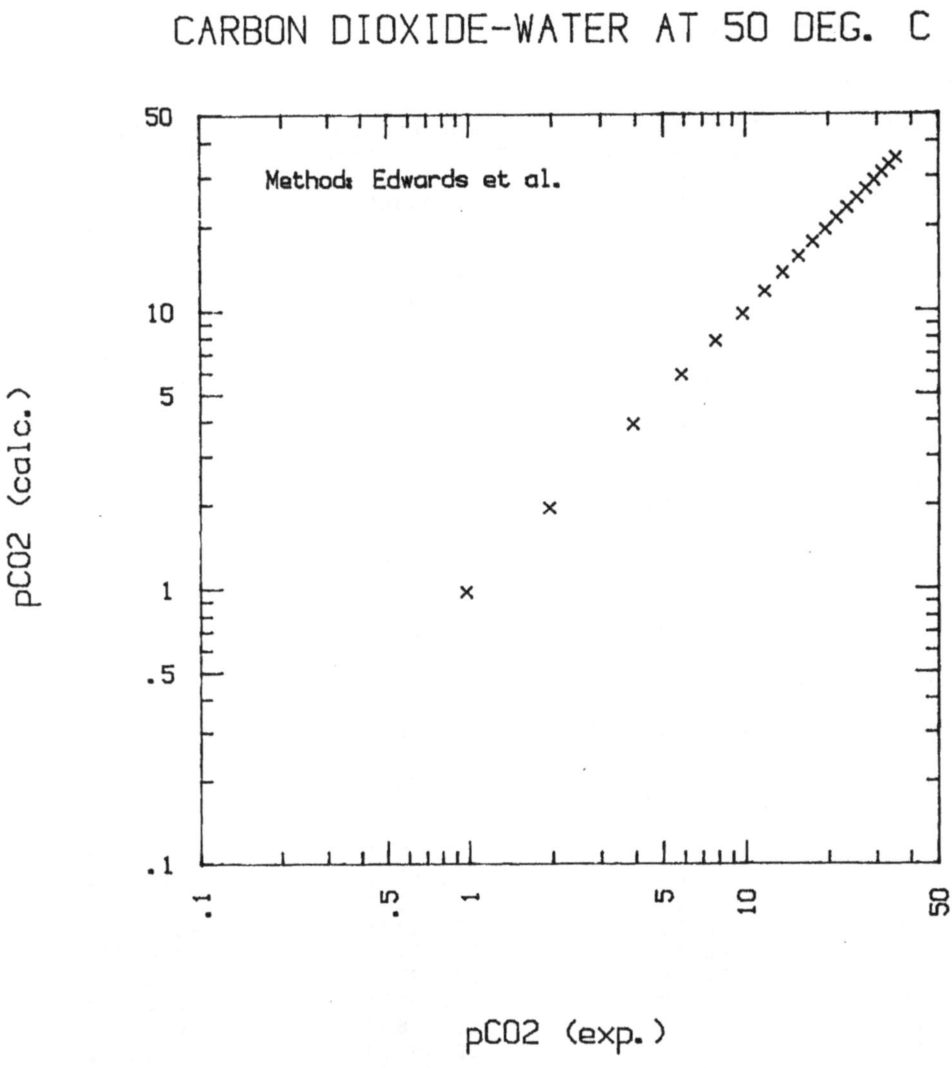
CARBON DIOXIDE-WATER AT 50 DEG. C
Method: Edwards et al.
50
10
5
1
.5
.1
.1
.5
1
5
10
50
pCO2 (calc.)
pCO2 (exp.)

Figure 7.12

SYSTEM: CARBON DIOXIDE - WATER

==

Method: Beutier & Renon, IECPDD (1978)

Experimental Data Source:
Houghton, McLean & Ritchie, Chem E Sci, 6 (1957)

Temperature: 50.0 degrees Celsius

Experimental data:

point	CO2 Partial Pressure (atm)	CO2 Liquid Mole Fraction	H2O Mole Fraction
1	1.000	0.00034	0.99966
2	2.000	0.00068	0.99932
3	4.000	0.00135	0.99865
4	6.000	0.00202	0.99798
5	8.000	0.00266	0.99734
6	10.000	0.00330	0.99670
7	12.000	0.00393	0.99607
8	14.000	0.00455	0.99545
9	16.000	0.00515	0.99485
10	18.000	0.00575	0.99425
11	20.000	0.00634	0.99366
12	22.000	0.00691	0.99309
13	24.000	0.00747	0.99253
14	26.000	0.00803	0.99197
15	28.000	0.00857	0.99143
16	30.000	0.00910	0.99090
17	32.000	0.00962	0.99038
18	34.000	0.01013	0.98987
19	36.000	0.01063	0.98937

Calculation results:

Error function = [pCO2(calc) - pCO2(exp)] / pCO2(exp)

point	pCO2(exp)	pCO2(calc)	error
1	1.000	0.96913	-0.03087
2	2.000	1.94600	-0.02700
3	4.000	3.89223	-0.02694
4	6.000	5.85574	-0.02404
5	8.000	7.77361	-0.02830
6	10.000	9.72325	-0.02767
7	12.000	11.67500	-0.02708
8	14.000	13.62894	-0.02650
9	16.000	15.55279	-0.02795
10	18.000	17.51080	-0.02718
11	20.000	19.47131	-0.02643
12	22.000	21.40029	-0.02726
13	24.000	23.33077	-0.02788
14	26.000	25.29831	-0.02699
15	28.000	27.23277	-0.02740
16	30.000	29.16908	-0.02770
17	32.000	31.10736	-0.02789
18	34.000	33.04778	-0.02801
19	36.000	34.99050	-0.02804

Sum [error ^ 2] = 0.014323

Figure 7.13

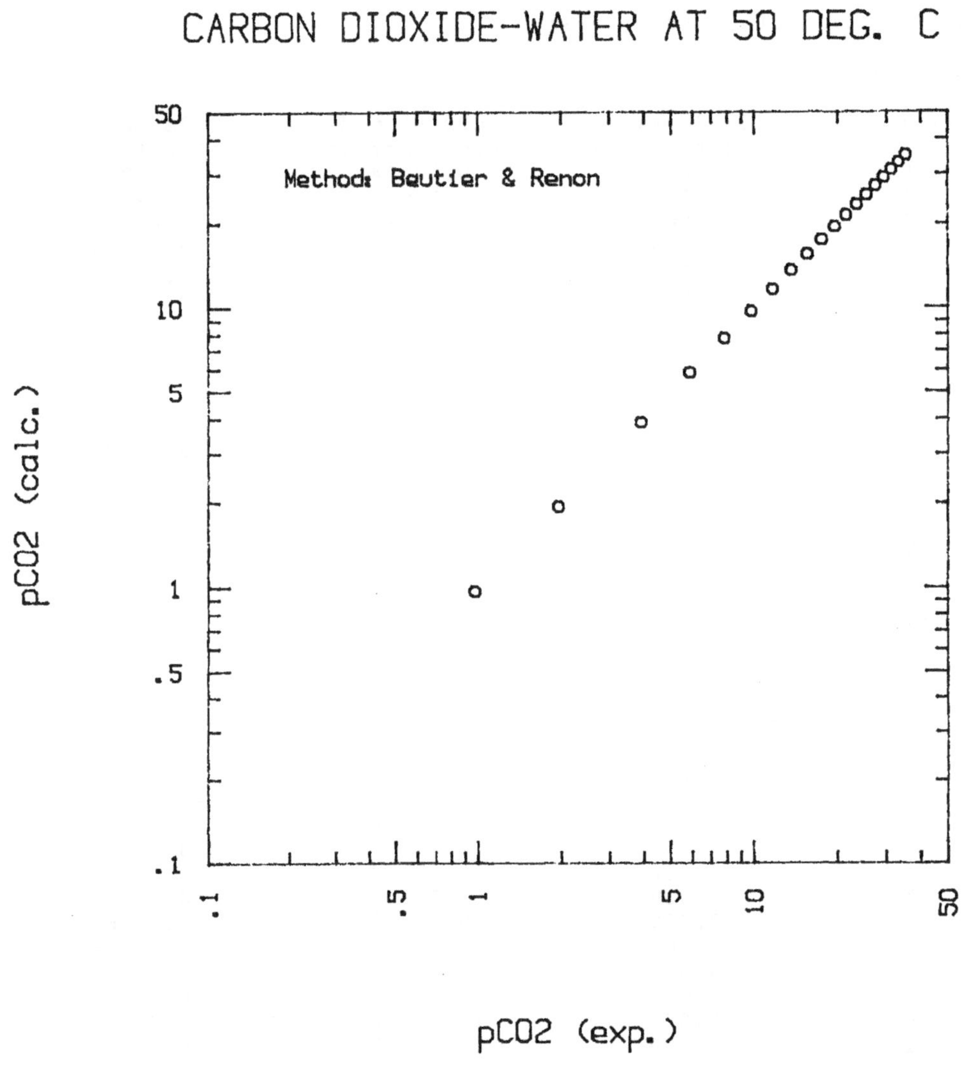
CARBON DIOXIDE-WATER AT 50 DEG. C
Method: Bautier & Renon
pCO2 (calc.)
pCO2 (exp.)
50
10
5
1
.5
.1
.1
.5
1
5
10
50

Figure 7.14

SYSTEM: CARBON DIOXIDE - AMMONIA - WATER

Method: Edwards, Maurer, Newman & Prausnitz, AIChE J (1978)

Experimental Data Source:
Otsuku, Kog. Kag. Zas., v63 pp126-130 (1960)

Temperature: 60.0 degrees Celsius

Experimental data:

point	Liquid molality NH3	CO2	Partial pressure (atm) NH3	CO2	Pressure (atm)
1	1.9866212	1.3882554	0.0300000	0.5815789	0.8065789
2	5.4475198	3.4276729	0.0788158	0.5144737	0.7798684
3	9.3666811	4.9811323	0.1332895	0.3575000	0.6861842
4	9.3227092	4.6902155	0.1842105	0.2526316	0.6263158
5	6.2184708	3.1152872	0.1368421	0.2144737	0.5394737
6	2.5361150	1.2685824	0.0671053	0.1552632	0.4164474

Calculation results:

Error function = {[pNH3(calc) - pNH3(exp)] / pNH3(exp)} +
{[pCO2(calc) - pCO2(exp)] / pCO2(exp)}

point	pNH3(exp)	pNH3(calc)	pCO2(exp)	pCO2(calc)	error
1	0.03000	0.02165	0.58158	0.43575	-0.52900
2	0.07882	0.05475	0.51447	0.37517	-0.57610
3	0.13329	0.14159	0.35750	0.09676	-0.66704
4	0.18421	0.17138	0.25263	0.07773	-0.76193
5	0.13684	0.11505	0.21447	0.12277	-0.58685
6	0.06711	0.05469	0.15526	0.11788	-0.42578

Error function = {[P(calc) - P(exp)] / P(calc)}

point	P(exp)	P(calc)	error
1	0.80658	0.64636	-0.19864
2	0.77987	0.60970	-0.21820
3	0.68618	0.41411	-0.39651
4	0.62632	0.42262	-0.32523
5	0.53947	0.41514	-0.23047
6	0.41645	0.36000	-0.13555

Sum [error ^ 2] = 2.584434

Figure 7.15

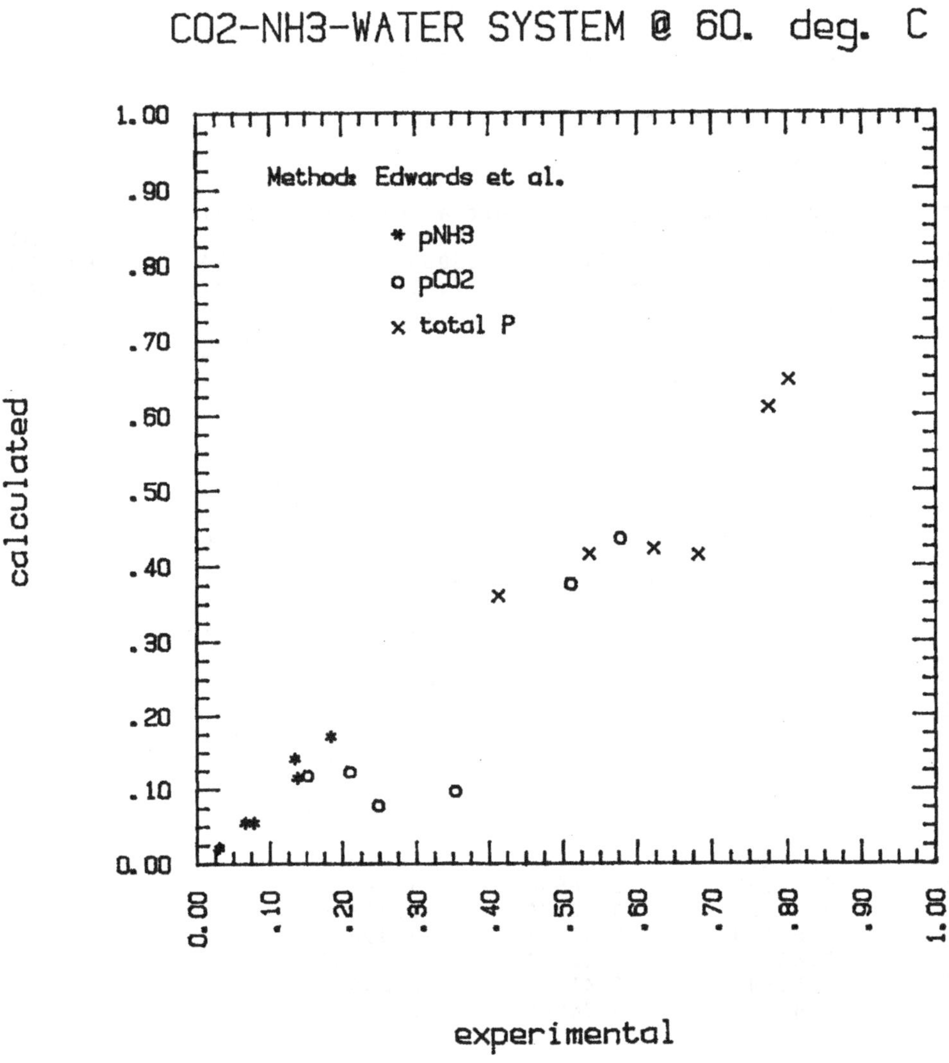
CO2-NH3-WATER SYSTEM @ 60. deg. C
Method: Edwards et al.
* pNH3
o pCO2
x total P
calculated
experimental
1.00
.90
.80
.70
.60
.50
.40
.30
.20
.10
0.00
.10
.20
.30
.40
.50
.60
.70
.80
.90
1.00

Figure 7.16

SYSTEM: SULFUR DIOXIDE - WATER

==

Method: Edwards, Maurer, Newman & Prausnitz, AIChE J (1978)

Experimental Data Source:
Sherwood, I&EC, 17 (1925); Johnstone & Leppla, JACS, 56 (1934)

Temperature: 50.0 degrees Celsius

Experimental data:

point	SO2 Partial Pressure (atm)	SO2 Liquid Molality	Moles H2O
1	0.00171	0.00312	55.51
2	0.00230	0.00467	55.51
3	0.00383	0.00637	55.51
4	0.00538	0.00838	55.51
5	0.00618	0.00780	55.51
6	0.00879	0.01057	55.51
7	0.01579	0.01561	55.51
8	0.02632	0.02341	55.51
9	0.04079	0.03122	55.51
10	0.10789	0.04683	55.51
11	0.15263	0.07805	55.51
12	0.22632	0.10927	55.51
13	0.35000	0.15610	55.51
14	0.60263	0.23415	55.51

Calculation results:

Error function = [pSO2(calc) - pSO2(exp)] / pSO2(exp)

point	pSO2(exp)	pSO2(calc)	error
1	0.00171	0.00118	-0.31138
2	0.00230	0.00222	-0.03530
3	0.00383	0.00354	-0.07471
4	0.00538	0.00529	-0.01764
5	0.00618	0.00477	-0.22857
6	0.00879	0.00734	-0.16489
7	0.01579	0.01251	-0.20787
8	0.02632	0.02128	-0.19130
9	0.04079	0.03065	-0.24858
10	0.10789	0.05049	-0.53208
11	0.15263	0.09271	-0.39260
12	0.22632	0.13692	-0.39501
13	0.35000	0.20554	-0.41274
14	0.60263	0.32397	-0.46240

Sum [error ^ 2] = 1.302569

Figure 7.17

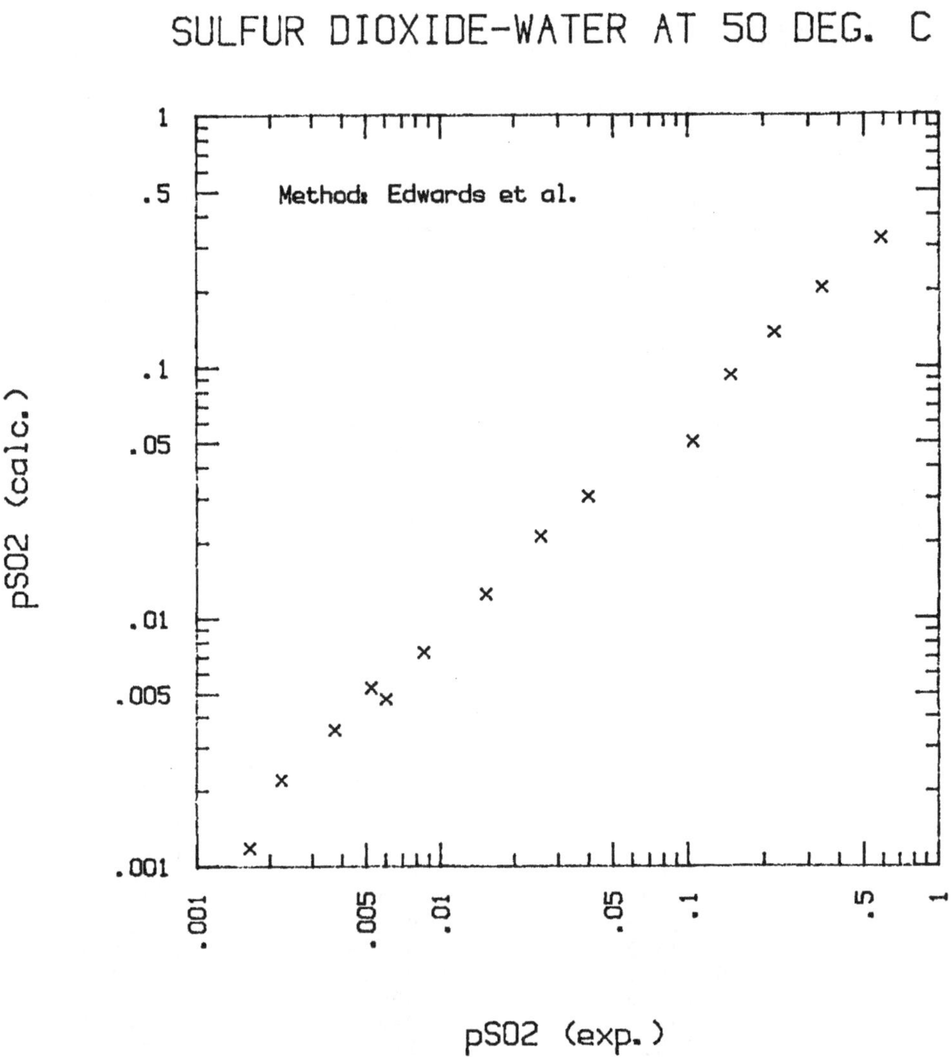
SULFUR DIOXIDE-WATER AT 50 DEG. C
Method: Edwards et al.
pSO2 (calc.)
pSO2 (exp.)
1
.5
.1
.05
.01
.005
.001
.001
.005
.01
.05
.1
.5
1

Figure 7.18

SYSTEM: SULFUR DIOXIDE - WATER
==

Method: Beutier & Renon, IECPDD (1978)

Experimental Data Source:
Sherwood, I&EC, 17 (1925); Johnstone & Leppla, JACS, 56 (1934)

Temperature: 50.0 degrees Celsius

Experimental data:

point	SO2 Partial Pressure (atm)	SO2 Liquid Molality	Moles H2O
1	0.00171	0.00312	55.51
2	0.00230	0.00467	55.51
3	0.00383	0.00637	55.51
4	0.00538	0.00838	55.51
5	0.00618	0.00780	55.51
6	0.00879	0.01057	55.51
7	0.01579	0.01561	55.51
8	0.02632	0.02341	55.51
9	0.04079	0.03122	55.51
10	0.10789	0.04683	55.51
11	0.15263	0.07805	55.51
12	0.22632	0.10927	55.51
13	0.35000	0.15610	55.51
14	0.60263	0.23415	55.51

Calculation results:

Error function = [pSO2(calc) - pSO2(exp)] / pSO2(exp)

point	pSO2(exp)	pSO2(calc)	error
1	0.00171	0.00111	-0.35262
2	0.00230	0.00211	-0.08214
3	0.00383	0.00340	-0.11146
4	0.00538	0.00512	-0.04916
5	0.00618	0.00461	-0.25482
6	0.00879	0.00715	-0.18649
7	0.01579	0.01231	-0.22066
8	0.02632	0.02113	-0.19715
9	0.04079	0.03059	-0.24993
10	0.10789	0.05071	-0.52999
11	0.15263	0.09360	-0.38678
12	0.22632	0.13841	-0.38840
13	0.35000	0.20761	-0.40684
14	0.60263	0.32570	-0.45954

Sum [error ^ 2] = 1.353709

Figure 7.19

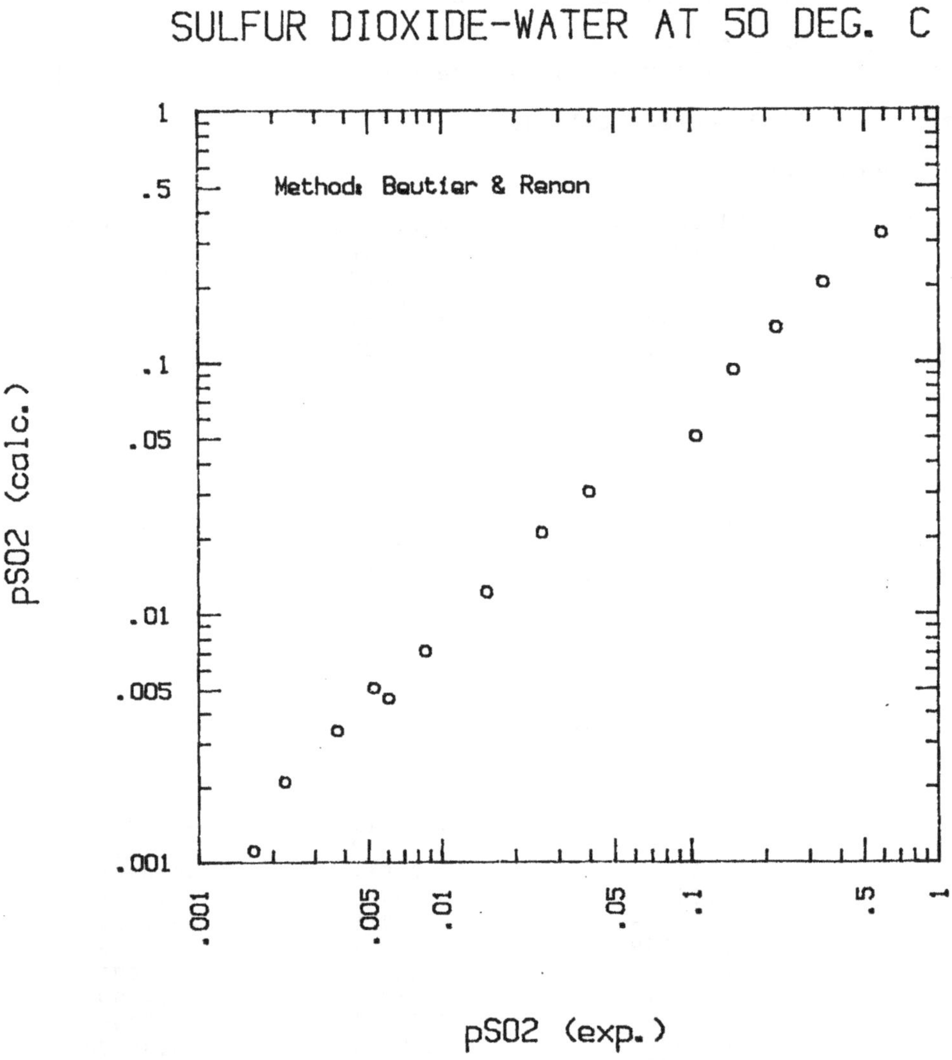
SULFUR DIOXIDE-WATER AT 50 DEG. C
Method: Beutier & Renon
pSO2 (calc.)
pSO2 (exp.)
1
.5
.1
.05
.01
.005
.001
.001
.005
.01
.05
.1
.5
1

Figure 7.20

WATER - OXYGEN - SODIUM CHLORIDE

Method: Setschenow for O2; Pitzer for NaCl

Experimental data source:
S.D. Cramer, IECPDD, v19 (1980)

Total vapor pressure = 1. atm.

Error = sum {[(calc - exp)/exp]^2}

temp deg. C	Oxygen solubility (in ppm) in sodium chloride concentrations of: 0.87 molal		2.97 molal		5.69 molal		error
	exp	calc	exp	calc	exp	calc	
0.0	9.27	7.0251	3.74	1.4167	1.61	0.2440	-0.8484409
10.0	7.34	5.7212	3.20	1.3391	1.53	0.2822	-0.8155306
20.0	6.12	4.9108	2.85	1.3173	1.44	0.3206	-0.7773472
30.0	5.29	4.3596	2.61	1.3278	1.35	0.3617	-0.7320628
40.0	4.66	3.9414	2.41	1.3536	1.26	0.4068	-0.6771500
50.0	4.12	3.5757	2.23	1.3779	1.17	0.4550	-0.6111464
60.0	3.59	3.1985	2.03	1.3793	1.07	0.5016	-0.5312114
70.0	3.01	2.7475	1.78	1.3267	0.96	0.5355	-0.4422354
80.0	2.30	2.1514	1.43	1.1737	0.81	0.5333	-0.3415736
90.0	1.37	1.3201	0.95	0.8495	0.60	0.4511	-0.2482303
100.0	0.12	0.1349	0.27	0.2474	0.29	0.2083	-0.2816433

Figure 7.21

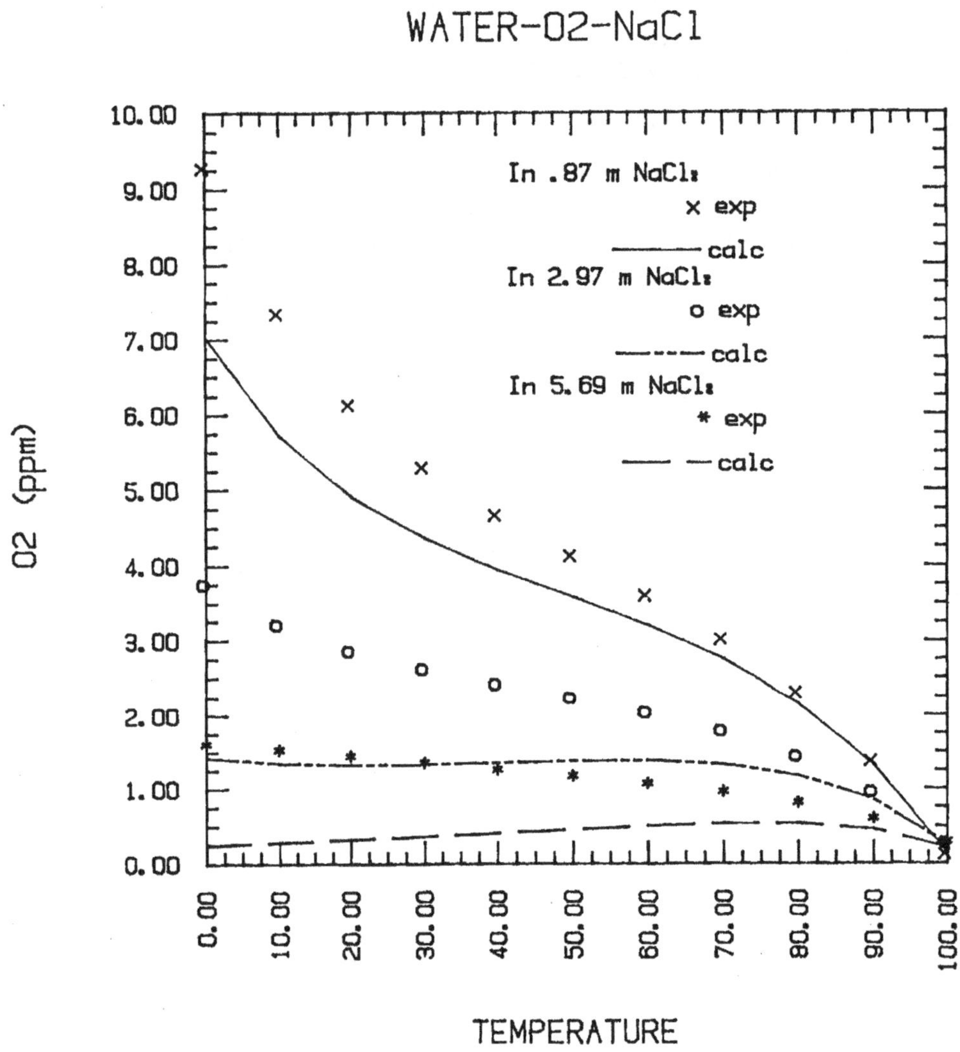
WATER-O2-NaCl
In .87 m NaCl:
x exp
calc
In 2.97 m NaCl:
o exp
calc
In 5.69 m NaCl:
* exp
calc
O2 (ppm)
10.00
9.00
8.00
7.00
6.00
5.00
4.00
3.00
2.00
1.00
0.00
0.00
10.00
20.00
30.00
40.00
50.00
60.00
70.00
80.00
90.00
100.00
TEMPERATURE

$10^{-4.74}$ at 25°C, is likely to form to a significant degree so long as the product of aNH_4^+ x aOH^- is greater than $10^{-5.0}$. Since there were no published Pitzer coefficients for the interactions between this species and others in solution, it was ignored. Once again we see that at high concentrations intermediates (complexes) are the key to accurate simulation.

Using the published Pitzer coefficients and applying them to the ternary NH_3-CO_2-H_2O yielded similar results. This is not surprising since $NH_4OH(aq)$ was not considered. In addition to the scatter, several published points could not even be simulated. It appears that from a numerical point of view, the activity coefficient formulations, based upon the published interaction coefficients, were ill-behaved. This could be due to a number of factors such as a typographical error in one of the published interaction values. Since the numerical problems did not occur for either of the constituent binary systems, it would have to be an error in one of the crossover terms such as any $NH_2CO_2^-$ interaction.

Using the Setschénow equation on the oxygen-sodium chloride-water system resulted in a fairly poor fit. Nonetheless, the trend of solubility of oxygen with increasing salt molality is consistent with the published data. The problem seems to be that the Setschénow predictions for the activity coefficient of oxygen worsens as the salt concentration increases.

APPENDIX 7.1

Salting Out Parameters for Phenol in Aqueous Salt Solutions at 25° Celsius

From: Long, F.A.; M.W. McDevit, "Activity coefficients of nonelectrolyte solutes in aqueous salt solutions", Chem. Rev., v51, pp139 (1952)

For calculating the molar activity coefficient:

$$\log y_u = k_s C_s + k_u C_u \qquad (7.10)$$

where:

y_u - molar activity coefficient of phenol

k_s - is the salting out parameter for salt s

C_s - is the molar concentration of salt s in the solution

k_u - is the phenol self interaction term

C_u - is the molar concentration of phenol in the solution

salt	salt concentration (moles/liter)	k_s	k_u	k (sol.)
NaCl	0.94	0.172	-0.15	0.229
	1.88	0.172	-0.13	0.210
	2.80		-0.13	0.203
KCl	0.93	0.133	-0.18	0.191
	1.84	0.133	-0.16	0.174
$NaNO_3$	0.93	0.113	-0.14	0.148
	1.82	0.113	-0.12	0.139
	3.49		-0.14	0.137
KNO_3	0.91	0.080	-0.14	0.107
	1.78	0.080	-0.14	0.1045
	3.38		-0.16	0.1035
average			-0.14	

The k_s term was calculated using distribution data where C_u° for phenol ranged from .05 to .15. The k_u term was then calculated:

$$k_u = \frac{1}{C_u - C_u^\circ}\left[\log\left(\frac{C_u^\circ}{C_u}\right) - k_s C_s\right]$$

The k term is the standard salting parameter from solubility data which does not give a self interaction term:

$$\log y_u = k\, C_s$$

and:

$$k = \frac{\log (C_u^\circ \ / \ C_u)}{C_s}$$

$C_u{}^\circ$ for phenol, the solubility in water, was .90 mole/liter at 25°C.

APPENDIX 7.2

Salting Out Parameters from Pawlikowski and Prausnitz for Nonpolar Gases in Common Salt Solutions at Moderate Temperatures

From: Pawlikowski, E.M.; J.M. Prausnitz, "Estimation of Setchenow Constants for Nonpolar Gases in Common Salts at Moderate Temperatures", Ind. Eng. Chem. Fundam., v22, #1, pp87, 88 (1983)
and
"Correction", Ind. Eng. Chem. Fundam., v23, #2, p270 (1984)

Table 1 - Lennard-Jones Parameters for Nonpolar Gases as Reported by Liabastre (S14)

gas	ε_{gas}/k, Kelvins
He	6.03
H_2	29.2
N_2	95.
N_2O	114.
O_2	118.
Ar	125.
CH_4	137.
Xe	217.
C_2H_6	230.

Table 2 - Salting Out Parameters for Strong Electrolytes in Equation (7.18) at 25°C (ε/k in Kelvins)

salt	a_s, kg/mole	10^4 b_s, kg/(mole K)
NH_4Br	0.0345	1.31
KI	0.075	1.31
LiCl	0.0658	3.31
NaCl	0.0805	3.56
KCl	0.0652	6.00
NaBr	0.0855	2.98

Table 2 continued

$NaNO_3$	0.0864	-1.40
NH_4Cl	0.0255	4.60
$CaCl_2$	0.162	6.95
$BaCl_2$	0.0895	8.42
$(NH_4)_2SO_4$	0.101	11.45
Na_2SO_4	0.142	13.6
HCl	0.193	0.115
H_2SO_4	0.0908	0.199
KOH	0.114	5.01

Table 3 - Temperature Dependence of the Salting Out Parameters for Equation (7.19) (Temperature range: 0 - 60°C)

salt	a_{1s} kg/mole	10^2 a_{2s} kg/(mole K)	b_{1s} kg/(mole K)	b_{2s} kg/(mole K^2)
NH_4Br	0.249	-0.0721	3.123×10^{-4}	-6.08×10^{-7}
KI	0.228	-0.0512	9.120×10^{-4}	-2.62×10^{-6}
LiCl	0.134	-0.0229	1.800×10^{-3}	-4.93×10^{-6}
NaCl	0.516	-0.146	3.740×10^{-4}	-6.30×10^{-8}
KOH	0.0679	0.0155	1.950×10^{-3}	-4.86×10^{-6}
KCl	0.0861	-0.007	3.464×10^{-3}	-9.61×10^{-6}

Table 4 - Salting Out Parameters for Individual Ions for Equation (7.20) (ϵ/k in Kelvins)

ion i	a_{si}, kg/mole	$10^4\ b_{si}$, kg/(mole K)
OH^-	0	0
K^+	0.114	5.01
Cl^-	-0.049	0.99
Na^+	0.129	2.57
Li^+	0.115	2.32
NH_4^+	0.074	3.61
Br^-	-0.042	-0.95
H^+	0.068	-0.875
NO_3^-	-0.043	-3.97
Ca^{2+}	0.260	4.97
Ba^{2+}	0.187	6.44
SO_4^{2-}	-0.070	4.88
I^-	-0.039	-3.70

Table 5 - Temperature Dependence of the Salting Out Constants for Individual Ions

ion i	a_{1i} kg/mole	$10^2\ a_{2i}$ kg/(mole K)	b_{1i} kg/(mole K)	b_{2i} kg/(mole K^2)
OH^-	0.	0.	0.	0.
K^+	0.0679	0.0155	1.95×10^{-3}	-4.86×10^{-6}
Cl^-	0.0182	-0.0225	1.514×10^{-3}	-4.75×10^{-6}
Li^+	0.116	-0.0004	2.86×10^{-4}	-1.80×10^{-7}
I^-	0.160	-0.0667	-1.038×10^{-3}	2.24×10^{-6}
Na^+	0.497	-0.123	-1.14×10^{-3}	4.68×10^{-6}

REFERENCES/FURTHER READING

1. (a) Wagman, D.D.; W.H. Evans; V.B. Parker; I. Halow; S.M. Bailey; R.H. Schumm; Selected Values of Chemical Thermodynamic Properties - Tables for the First Thirty-Four Elements in the Standard Order of Arrangement, National Bureau of Standards Technical Note 270-3, 1968

   (b) Wagman, D.D.; W.H. Evans; V.B. Parker; I. Halow; S.M. Bailey; R.H. Schumm; Selected Values of Chemical Thermodynamic Properties - Tables for Elements 35 through 53 in the Standard Order of Arrangement, NBS Tech Note 270-4, 1969

   (c) Wagman, D.D.; W.H. Evans; V.B. Parker; I. Halow; S.M. Bailey; R.H. Schumm; K.L. Churney; Selected Values of Chemical Thermodynamic Properties - Tables for Elements 54 through 61 in the Standard Order of Arrangement, NBS Tech Note 270-5, 1971

   (d) Parker, V.B.; D.D. Wagman and W.H. Evans; Selected Values of Chemical Thermodynamic Properties - Tables for the Alkaline Earth Elements (Elements 92 through 97 in the Standard Order of Arrangement), NBS Tech Note 270-6, 1971

   (e) Schumm, R.H.; D.D. Wagman; S. Bailey; W.H. Evans; V.B. Parker, Selected Values of Chemical Thermodynamic Properties - Tables for the Lanthanide (Rare Earth) Elements (Elements 62 through 76 in the Standard Order of Arrangement), NBS Tech Note 270-7, 1973

   (f) Wagman, D.D.; W.H. Evans; V.B. Parker; R.H. Schumm; R.L. Nuttall; Selected Values of Chemical Thermodynamic Properties - Compounds of Uranium, Protactinium, Thorium, Actinium, and the Alkali Metals, NBS Tech Note 270-8, 1981

   (g) Wagman, D.D.; W.H. Evans, V.B. Parker, R.H. Schumm, I. Halow, S.M. Bailey, K.L. Churney, R.L. Nutall, The NBS tables of chemical thermodynamic properties - Selected values for inorganic and C_1 and C_2 organic substances in SI units, J. Phys. Chem. Ref. Data, v11, suppl. 2 (1982)

2. (a) Glushko, V.P., dir.; Thermal Constants of Compounds - Handbook in Ten Issues, VINITI Academy of Sciences of the U.S.S.R., Moscow

   Volume I (1965) - O, H, D, T, F, Cl, Br, I, At, ^{3}He, He, Ne, Ar, Kr, Xe, Rn

   (b) Volume II (1966) - S, Se, Te, Po

   (c) Volume III (1968) - N, P, As, Sb, Bi

   (d) Volume IV-1 (1970) - C, Si, Ge, Sn, Pb
   Volume IV-2 (1971)

   (e) Volume V (1971) - B, Al, Ga, In, Tl

(f) Volume VI-1 (1972) - Zn, Cd, Hg, Cu, Ag, Au, Fe, Co, Ni, Ru, Rh, Pd, Os, Ir, Pt
Volume VI-2 (1973)

(g) Volume VII-1 (1974) - Mn, Tc, Re, Cr, Mo, W, V, Nb, Ta, Ti, Zr, Hf
Volume VII-2 (1974)

(h) Volume VIII-1 (1978) - Sc, Y, La, Ce, Pr, Nd, Pm, Sm, Eu, Gd, Tb, Dy, Ho, Er, Tm, Yb, Lu, Ac, Th, Pa, U, Np, Pu, Am, Cm, Bk, Cf, Es, Fm, Md, No
Volume VIII-2 (1978)

(i) Volume IX (1979) - Be, Mg, Ca, Sr, Ba, Ra

(j) Volume X-1 (1981) - Li, Na, K, Rb, Cs, Fv
Volume X-2 (1981)

3. Helgeson, H.C.; "Thermodynamics of Complex Dissociation in Aqueous Solution at Elevated Temperatures", J. Phys. Chem., v71, #10, pp3121-3136 (1967)

4. Helgeson, H.C.; D.H. Kirkham, "Theoretical Prediction of the Thermodynamic Behavior of Aqueous Electrolytes at High Pressures and Temperatures: I. Summary of the Thermodynamic/Electrostatic Properties of the Solvent", Am. J. Sci., v274, pp1089-1198 (1974)

5. Helgeson, H.C.; D.H. Kirkham, "Theoretical Prediction of the Thermodynamic Behavior of Aqueous Electrolytes at High Pressures and Temperatures: II. Debye-Hückel Parameters for Activity Coefficients and Relative Partial Molal Properties", Am. J. Sci., v274, pp1199-1261 (1974)

6. Helgeson, H.C.; D.H. Kirkham, "Theoretical Prediction of the Thermodynamic Behavior of Aqueous Electrolytes at High Pressures and Temperatures: III. Equation of State for Aqueous Species at Infinite Dilution", Am. J. Sci., v276, pp97-240 (1976)

7. Helgeson, H.C.; D.H. Kirkham, G.C. Flowers, "Theoretical Prediction of the Thermodynamic Behavior of Aqueous Electrolytes at High Pressures and Temperatures: IV. Calculation of Activity Coefficients, Osmotic Coefficients, and Apparent Molal and Standard and Relative Partial Molal Properties to 600°C and 5 kb", Am. J. Sci., v281, pp1249-1516 (1981)

8. Wilhelm, E.; R. Battino, R.J. Wilcock, "Low-Pressure Solubility of Gases in Liquid Water", Chem. Rev., v77, #2, pp219-262 (1977)

SETSCHÉNOW REFERENCES

S1. Aquan-Yuen, M.; D. Mackay, W.Y. Shiu, "Solubility of hexane, phenanthrene, chlorobenzene, and p-dichlorobenzene in aqueous electrolyte solutions", J. Chem. Eng. Data, v24, #1, pp30-34 (1979)

S2. Armenante, P.M.; H.T. Karisson, "Salting-out parameters for organic acids", J. Chem. Eng. Data, v27, pp155-156 (1982)

S3. Bromley, L.A., "Approximate individual ion values of β (or B) in extended Debye-Huckel theory for uni-univalent aqueous solutions at 298.15 K", J. Chem. Thermo., v4, pp669-673 (1972)

S4. Cysewski, G.R.; J.M. Prausnitz, "Estimation of gas solubilities in polar and nonpolar solvents", Ind. Eng. Chem. Fundam., v15, #4, pp304-309 (1976)

S5. Desnoyers, J.E.; M. Billon, S. Leger, G. Perron, J-P Morel, "Salting out of alcohols by alkali halides at the freezing temperature", J. Sol. Chem., v5, #10, pp681-691 (1976)

S6. Furter, W.F., "Salt effect in distillation: a literature review II", Can. J. Chem. Eng., v55, pp229-239 (1977)

S7. Gordon, J.E.; R.L. Thorne, "Salt effects on the activity coefficient of naphthalene in mixed aqueous electrolyte solutions. I. Mixtures of two salts", J. Phys. Chem., v71, #3, pp4390-4399 (1967)

S8. Gordon, J.E.; R.L. Thorne, "Salt effects on non-electrolyte activity coefficients in mixed aqueous electrolyte solutions. II. Artificial and natural sea waters", Geochim. et Cosmochim. Acta, v31, pp2433-2443 (1967)

S9. Gordon, J.E., The Organic Chemistry of Electrolyte Solutions, J. Wiley & Sons, New York (1975)

S10. Harned, H.S., JACS, v48, p326 (1926)

S11. Himmelblau, D.M.; "Partial molal heats and entropies of solution for gases dissolved in water from the freezing to near the critical point", J. Phys. Chem., v63, pp1803-1808 (1959)

S12. Khomutov, N.E.; E.I. Konnik, "Solubility of oxygen in aqueous electrolyte solutions", Russ. J. Phys. Chem., v48, #3, pp359-362 (1974)

S13. Krishnan, C.V.; H.L. Friedman, "Model calculations for Setchenow coefficients", J. Sol. Chem., v3, #9, pp727-744 (1974)

S14. Liabastre, A.A.; Ph.D. Dissertation, Georgia Inst. Tech., Aug, (1974)

S15. Long, F.A.; W.F. McDevit, "Activity coefficients of nonelectrolyte solutes in aqueous salt solutions", Chem. Rev., v51, pp119-169 (1952)

S16. Masterton, W.L.; T.P. Lee, "Salting coefficients from scaled particle theory", J. Phys. Chem., v74, pp1776-1782 (1970)

S17. Masterton, W.L.; D. Bolocofsky, T.P. Lee, "Ionic radii from scaled particle theory of the salt effect", J. Phys. Chem., v75, #18, pp2809-2814 (1971)

S18. Masterton, W.L.; D. Polizzotti, H. Welles, "The solubility of argon in aqueous solutions of a complex-ion electrolyte", J. Sol. Chem., v2, #4, pp417-423 (1973)

S19. Masterson, W.L., "Salting coefficients for gases in seawater from scaled-particle theory", J. Sol. Chem., v4, #6, pp523-534 (1975)

S20. McDevit, W.F.; F.A. Long, "The activity coefficient of benzene in aqueous salt solutions", JACS, v74, pp1773-1777 (1952)

S21. Morrison, T.J., "The salting-out of non-electrolytes. Part I. The effect of ionic size, ionic charge, and temperature", J. Chem. Soc., pp3814-3818 (1952)

S22. Morrison, T.J.; F. Billet, "The salting-out of non-electrolytes. Part II. The effect of variation in non-electrolyte", J. Chem. Soc., pp3819-3822 (1952)

S23. Onda, K.; E. Saka, T. Kobayashi, S. Kito, K. Ito, "Salting-out parameters of gas solubility in aqueous salt solutions", J. Chem. Eng. Japan, v3, #1, pp18-24 (1970)

S24. a. Pawlikowski, E.M.; J.M. Prausnitz, "Estimation of Setchenow constants for nonpolar gases in common salts at moderate temperatures", Ind. Eng. Chem. Fundam., v22, #1, pp86-90 (1983)

b. Pawlikowski, E.M.; J.M. Prausnitz, "Correction", Ind. Eng. Chem. Fundam., v23, #2, p270 (1984)

S25. Prausnitz, J.M.; F.H. Shair, "A thermodynamic correlation of gas solubilities", AIChE J., v7, #4, pp682-687 (1961)

S26. Randall, M.; J.W. McBain, White, JACS, v48, p2517 (1926)

S27. Randall, M.; G.F. Breckenridge, JACS, v49, p1435 (1927)

S28. Randall, M.; C.F. Failey, "The activity coefficients of gases in aqueous salt solutions", Chem. Rev., v4, #3, pp271-284 (1927)

S29. Randall, M.; C.F. Failey, "The activity coefficient of non-electrolytes in aqueous salt solutions from solubility measurements. The salting-out order of the ions", Chem. Rev., v4, #3, pp285-290 (1927)

S30. Randall, M.; C.F. Failey, "The activity coefficient of the undissociated part of weak electrolytes", Chem. Rev., v4, #3, pp291-318 (1927)

S31. Schumpe, A.; I. Adler, W.D. Deckwer, "Solubility of oxygen in electrolyte solutions", Biotech. & Bioeng., v20, pp145-150 (1978)

S32. Sergeeva, V.F., "Salting-out and salting-in of non-electrolytes", Russ. Chem. Rev., v34, #4, pp309-318 (1965)

S33. Setschénow, J.; "Über die Konstitution der Salzlösungen auf Grund ihres Verhaltens zu Kohlensaure", Z. physik. Chem., v4, p117 (1889)

S34. Shoor, S.K.; K.E. Gubbins, "Solubility of nonpolar gases in concentrated electrolyte solutions", J. Phys. Chem., v73, #3, pp498-505 (1969)

S35. Tiepel, E.W.; K.E. Gubbins, "Thermodynamic properties of gases dissolved in electrolyte solutions", Ind. Eng. Chem. Fundam., v12, #1, pp18-25 (1973)

S36. Wilhelm, E.; R. Battino, "Thermodynamic functions of the solubilities of gases in liquids at 25°C", Chem. Rev., v73, #1, pp1-9 (1973)

S37. Yasunishi, A., "Solubilities of sparingly soluble gases in aqueous sodium sulfate and sulfite solutions", J. Chem. Eng. Japan, v10, #2, pp89-94 (1977)

S38. Yen, L.C.; J.J. McKetta, Jr., "A thermodynamic correlation of nonpolar gas solubilities in polar, nonassociated liquids", AIChE J., v8, #4, pp501-507 (1962)

PITZER EQUATION REFERENCES

P1. a. Beutier, D.; H. Renon, "Representation of NH_3-H_2S-H_2O, NH_3-CO_2-H_2O, and NH_3-SO_2-H_2O Vapor-Liquid Equilibria", Ind. Eng. Chem. Process Des. Dev., v17, #3, pp220-230 (1978)

b. Beutier, D.; H. Renon, "Corrections", Ind. Eng. Chem. Process Des. Dev., v19, #4, p722 (1980)

P2. a. Chen, C-C; H.I. Britt, J.F. Boston, L.B. Evans, "Extension and application of the Pitzer equation for vapor-liquid equilibrium of aqueous electrolyte systems with molecular solutes", AIChE J, v25, #5, pp820-831 (1979)

b. Chen, C-C; H.I. Britt, J.F. Boston, L.B. Evans, "Errata", AIChE J, v26, #5, p879 (1980)

P3. Chen, C-C, "Computer Simulation of Chemical Processes with Electrolytes", Sc. D. thesis, MIT, Boston (1980)

P4. a. Chen, C-C; H.I. Britt, J.F. Boston, L.B. Evans, "Two New Activity Coefficient Models for the Vapor-liquid Equilibrium of Electrolyte Systems", Thermodynamics of Aqueous Systems with Industrial Applications, S.A. Newman ed., ACS Symp. Series 133, pp61-89 (1980)

b. Chen, C-C; H.I. Britt, J.F. Boston, L.B. Evans, "Appendix 2: Comparison of Calculated Partial Pressures of NH_3, CO_2 and H_2S with Experimental Data in the NH_3-CO_2-H_2S-H_2O and NH_3-CO_2-H_2O Systems: Comparison 6", Thermodynamics of Aqueous Systems with Industrial Applications, S.A. Newman ed., ACS Symp. Series 133 Suppl., pp76-94 (1980)

P5. Edwards, T.J.; J. Newman, J.M. Prausnitz, "Thermodynamics of Aqueous Solutions Containing Volatile Weak Electrolytes", AIChE J, v21, #2, pp248-259 (1975)

P6. Edwards, T.J.; G. Maurer, J. Newman, J.M. Prausnitz, "Vapor-liquid Equilibria in Multicomponent Aqueous Solutions of Volatile Weak Electrolytes", AIChE J, v24, #6, pp966-976 (1978)

P7. Pawlikowski, E.M.; J. Newman, J.M. Prausnitz, "Phase Equilibria for Aqueous Solutions of Ammonia and Carbon Dioxide", Ind. Eng. Chem. Process Des. Dev., v21, #4, pp764-770 (1982)

P8. Pottel, R., "Dielectric Properties", Water, A Comprehensive Treatise, Vol. 3, F. Franks ed., Plenum Press, New York, pp401-431 (1973)

P9. a. Renon, H., "Representation of NH_3-H_2S-H_2O, NH_3-SO_2-H_2O and NH_3-CO_2-H_2O Vapor-Liquid Equilibria", Thermodynamics of Aqueous Systems with Industrial Applications, S.A. Newman ed., ACS Symp. Series 133, pp173-186 (1980)

b. Renon, H. "Appendix 2: Comparison of Calculated Partial Pressures of NH_3, CO_2 and H_2S with Experimental Data in the NH_3-CO_2-H_2S-H_2O and NH_3-CO_2-H_2O Systems: Comparison 7", Thermodynamics of Aqueous Systems with Industrial Applications, S.A. Newman ed., ACS Symp. Series 133 Suppl., pp95-98 (1980)

DATA REFERENCES

D1. Clifford, I.L.; E. Hunter, "The system ammonia-water at temperatures up to 150°C and at pressures up to twenty atmospheres", J. Phys. Chem., v37, pp101-118 (1933)

D2. Macriss, R.A.; B.E. Eakin, R.T. Ellington, J. Huebler, "Physical and thermodynamic properties of ammonia-water mixtures", Inst. Gas. Tech. Res. Bull. 34 (1964)

D3. Houghton, G.; A.M. McLean, P.D. Ritchie, "Compressibility, fugacity, and water-solubility of carbon dioxide in the region 0-36 atm. and 0-100°C", Chem. E. Sci., v6, pp132-137 (1957)

D4. Otsuku, E., "NH_3-CO_2-H_2O", Kogyo Kagaku Zasshi, v63, pp126-131 (1960)

D5. Johnstone, H.F.; P.W. Leppla, "The solubility of sulfuric dioxide at low partial pressures. The ionization constant and heat of ionization of sulfurous acid", JACS, v56, pp2233-2238 (1934)

D6. Sherwood, T.K., "Solubilities of sulfur dioxide and ammonia in water", I & EC, v17, #7, pp745-747 (1925)

D7. Cramer, S.D.; "The solubility of oxygen in brines from 0 to 300°C", Ind. Eng. Chem. Process Des. Dev., v19, pp300-305 (1980)

VIII:

Thermodynamic Functions Derived from Activity Coefficients

THERMODYNAMIC FUNCTIONS DERIVED FROM ACTIVITY COEFFICIENTS

This chapter deals with techniques for calculating certain properties of ionic solutions. In general, this is an area which has received very little attention in the literature. Specifically, the properties to be considered are density and enthalpy. The treatment of these two distinct properties herein should serve to suggest the manner in which such properties can be approached.

In the case of density the approach is similar to that taken with respect to the calculation of activity coefficients in a multicomponent solution. Specifically, the approach is:

1) If possible, obtain good binary (water + electrolyte) data and fit this data to a reasonable function of ionic strength.

2) Invoke a reasonable mixing rule in order to predict the same property in multicomponent solutions.

The approach for enthalpy involves a more classical thermodynamic treatment. The enthalpy of a solution is composed of contributions from each constituent species in the solution. Each species contributes a pure component enthalpy term and an "excess enthalpy" term. This latter term is based upon the derivative of the species' activity coefficient with respect to temperature.

Each of these properties will now be considered in more detail.

DENSITY

This section describes a procedure for predicting densities in multicomponent electrolyte solutions comprised of strong electrolytes. The treatment is based upon a mixing rule which requires fitting binary (water + electrolyte) density data of the constituent electrolyte. To begin with we will consider a general solution made up of water plus one or more salts in solution.

The definition of the density of a multicomponent solution is

$$d = \frac{\text{mass}}{V} = \frac{\Sigma n_i \cdot M_i}{\Sigma n_i \cdot \overline{V}_i} \tag{8.1}$$

where,

V = solution volume, cm^3

n_i = moles of species i, gm-mole

M_i = molecular weight of species i, gm/gm-mole

$\overline{V}_i$ = partial molal volume of species i, cm^3/gm-mole

d = density of the mixture, gm/cm^3

The definition of $\overline{V}_i$ is,

$$\overline{V}_i = \left(\frac{\partial V}{\partial n_i}\right)_{T,\, P,\, n_{j \neq i}} \tag{8.2}$$

where,

T = temperature

P = pressure

Binary Density

Our first objective is to derive the partial molal volumes of both constituents of a binary mixture as a function of ionic strength. Once this is accomplished, a reasonable mixing rule for multicomponent systems can be proposed.

The definition for the apparent molal volume, ϕ_2, of an electrolyte in a binary system is:

$$\phi_2 = \frac{V - V_1}{n_2} = \frac{V - n_1 \overline{V}_1^{\circ}}{n_2} \tag{8.3}$$

with the subscript 1 denoting water and the subscript 2 indicating the electrolyte.

By this definition, we can see that the apparent molal volume of an electrolyte in a binary system is the difference between the total solution volume and the volume of water in the solution divided by the number of moles of the electrolyte present.

The total solution volume can now be expressed in molality units as follows. $\bar{V}_1°$ is defined as the partial molal volume of pure water at the solution temperature. This can be expressed as:

$$\bar{V}_1^o = \frac{M_1}{d_0} = \frac{18.016}{d_0} \tag{8.4}$$

where,

d_0 = density of pure water at the solution temperature.

To convert to the molal scale, it is necessary to consider 1000 grams of water, or, $n_1 = 1000\ gm/M_1$ gm-moles and, thus, $V_1 = n_1\bar{V}_1° = (1000/M_1)\ (M_1/d_0) = 1000/d_0$. Thus, given this and the fact that in 1000 grams of water the number of moles of a species equals its molality, $n_2 = m_2$, it follows that:

$$V = \frac{1000 + m_2 M_2}{d} \tag{8.5}$$

Substituting equations (8.4) and (8.5) into equation (8.3) we get

$$\phi_2 = \frac{1000\ (d_0 - d)}{m_2\ d\ d_0} + \frac{M_2}{d} \tag{8.6}$$

With this expression we can directly compute the apparent molal volume of the electrolyte if we have data on d, m_2 and d_0.

Density variations are often small and thus must be measured with considerable accuracy. Much old data is measured in gm/ml and not gm/cm . When using such data, conversion based upon temperature can be crucial to reasonable results. Reference to Kell (18) will facilitate these conversions.

Since our objective is to ultimately express $\bar{V}_1$ and $\bar{V}_2$ in terms of ionic strength, the next step is to relate $\bar{V}_2$ and $\bar{V}_1$ to ϕ_2. From equation (8.3) we have, by rearrangement:

$$V = n_2 \phi_2 + n_1 \overline{V}^\circ_1 \tag{8.7}$$

Applying the definition of the partial molal volume given in equation (8.2),

$$\overline{V}_2 = \frac{\partial}{\partial n_2} (n_2 \phi_2 + n_1 \overline{V}^\circ_1) = \phi_2 + n_2 \frac{\partial \phi_2}{\partial n_2} = \phi_2 + m_2 \frac{\partial \phi_2}{\partial m_2} \tag{8.8}$$

gives us the desired relationship between $\overline{V}_2$ and ϕ_2. The $\overline{V}_1$ and ϕ_2 relation comes by equating (8.7) with the definition of V:

$$n_1 \overline{V}_1 + n_2 \overline{V}_2 = V = n_2 \phi_2 + n_1 \overline{V}^\circ_1 \tag{8.9}$$

Substituting the definition of $\overline{V}_2$, equation (8.8), into equation (8.9) and solving for $\overline{V}_1$ yields:

$$\overline{V}_1 = \overline{V}^\circ_1 - \frac{n_2^2}{n_1} \frac{\partial \phi_2}{\partial n_2} = \overline{V}^\circ_1 - \frac{m_2^2}{55.508} \frac{\partial \phi_2}{\partial m_2} \tag{8.10}$$

Thus, equations (8.8) and (8.10) are the desired relationships for $\overline{V}_1$ and $\overline{V}_2$ expressed in terms of molality.

In order to introduce ionic strength, I, it is useful to note that for a solution involving a single strong electrolyte:

$$I = Wm \tag{8.11}$$

where,

W = 1 for 1:1 salts

W = 3 for 2:1 salts

W = 6 for 3:1 salts

W = 4 for 2:2 salts

This simple expression can be seen to be valid by noting the general definition for ionic strength:

$$I = .5 \quad \Sigma\, m_i \, z_i^2$$

where,

m_i = the molality of ion i

z_i = the charge of ion i

and since

$$m_i = \nu_i \, m$$

where,

m = the molality of the electrolyte

ν_i = the stoichiometric number of ion i

the ionic strength of a single strong electrolyte in solution is:

$$I = .5\ (\ z_c^2\ \nu_c\ m\ +\ z_a^2\ \nu_a\ m\ )$$

$$=\ .5\ (\ z_c^2\ \nu_c + z_a^2\ \nu_a\ )\ m$$

The factor W of equation (8.11) is therefore:

$$W = .5\ (\ z_c^2\ \nu_c + z_a^2\ \nu_a\ )$$

Substituting this expression for the ionic strength into equations (8.8) and (8.10) yields:

$$\overline{V}_2 = \phi_2 + I\frac{\partial \phi_2}{\partial I} \tag{8.12}$$

$$\overline{V}_1 = \overline{V}_1^o - \frac{I^2}{55.508\ W}\frac{\partial \phi_2}{\partial I} \tag{8.13}$$

The remaining step in the process of deriving $\overline{V}_1$ and $\overline{V}_2$ in terms of I is to propose a functional form for:

$$\phi_2 = \phi_2(I) \tag{8.14}$$

and thereby complete equations (8.12) and (8.13). The unknown coefficients of equation (8.14) can then be fit based upon published binary density data, utilizing equation (8.6) to calculate ϕ_2.

The function considered herein is

$$\phi_2 = \phi_2^\infty + AI^{1/2} + BI + CI^{3/2} + DI^2 + \ldots \tag{8.15}$$

where ϕ_2^∞ is the apparent molal volume at infinite dilution and at zero ionic strength, $\phi_2^\infty = \overline{V}^o$.

Differentiating and substituting into equations (8.12) and (8.13), we get:

$$\overline{V}_2 = \phi_2^\infty + 1.5\ AI^{1/2} + 2\ BI + 2.5\ CI^{3/2} + 3\ DI^2 + \ldots \tag{8.16}$$

$$\bar{V}_1 = \bar{V}_1^o - \frac{1}{55.508\ W} [\frac{A}{2} I^{3/2} + BI^2 + \frac{3}{2} CI^{5/2} + 2\ DI^3 + ...] \tag{8.17}$$

Multicomponent Density

The definition of density given in equation (8.1), recast in molality units, is:

$$d = \frac{1000 + \sum_i m_i M_i}{18.016\ \bar{\bar{V}}_W + \sum_i m_i \bar{\bar{V}}_i}$$

The mixing rule proposed here is to be based upon defining $\bar{V}_i$ and $\bar{V}_W$ as follows:

$$\bar{V}_i = \bar{V}_{2,i} \tag{8.18}$$

$$\bar{V}_W = \frac{\sum_i (\bar{V}_{1,i}) (I_i)}{I} \tag{8.19}$$

where,

$\bar{V}_{2,i}$ = equation (8.16) for the partial molal volume of electrolyte i evaluated at the ionic strength of the multicomponent solution

$\bar{V}_{1,i}$ = equation (8.17) for the partial molal volume of water in the "binary" solution of electrolyte i evaluated at the ionic strength of the multicomponent solution

I_i = the contribution of the ith electrolyte to the ionic strength, $.5\ z_i^2\ m_i$

I - ionic strength of the solution; $.5 \sum_i z_i^2\ m_i$

$\bar{V}_W$ = the partial molal volume of water in the multicomponent mixture estimated as the weighted average of the constituent binary partial molal volumes. The weighting factor is taken to be I_i.

Strong Electrolytes Which Complex

The above treatment applies directly to strong electrolytes which completely dissociate to their constituent ions. The factor W in equation (8.17) accounts for the type of salt involved. In Chapter VI, we saw that certain salts complex. For example, the strong electrolyte reaction, (Case 1)

$$CA_2(s) = C^{2+} + 2A^- \quad (8.20)$$

may not occur, but rather the reaction (Case 2):

$$CA_2(s) = CA^+ + A^- \quad (8.21)$$

or, the sequence (Case 3):

$$CA_2(s) = CA^+ + A^- \quad (8.22)$$

$$CA^+ = C^{2+} + A^- \quad (8.23)$$

may occur. The resulting ionic strengths are:

Case 1: $I = 3m$
Case 2: $I = m$
Case 3: $m \leq I \leq 3m$

Thus, a different ionic strength results in each case and the W factor would have to be adjusted accordingly.

Weak Electrolytes

The problem highlighted in Case 3 above becomes even more significant in the case of weak electrolytes. This can be seen, in the case of CO_2, where:

$$CO_2\,(aq) + H_2O = HCO_3^- + H^+ \quad (8.24)$$

$$HCO_3^- = CO_3^{2-} + H^+ \quad (8.25)$$

The distribution among $CO_2(aq)$, HCO_3^- and CO_3^{2-} is strictly determined by the simultaneous solution of the constituent equilibrium electroneutrality and material balance equations for the system. The total ionic strength is thus obtained only by solving this set. In order to handle weak electrolytes and still utilize the above framework, it would be necessary to:

1) Obtain good density data for the binary (e.g. H_2O - CO_2)
2) Define the apparent molal volumes for all species (e.g. $CO_2\,(aq)$, CO_3^{2-}, HCO_3^-)
3) Regress the required coefficients based upon simultaneously solving a full electrolyte model with the density equation imbedded.

Temperature Effects

The material presented above is based upon binary density data fits and multi-component density prediction at a single temperature. In order to handle a broad spectrum of temperatures, fits should be done across several isotherms and then, ultimately, the fit coefficients (e.g. A, B, C and D in equation (8.15)) can themselves be fit to functions of temperature.

Illustrative Example

Fortunately, there is an appreciable body of good density data available on a broad spectrum of H_2O-salt binary systems. Furthermore, the availability of high quality density measurements in a number of ternary and quarternary systems means that the mixing rules can be adequately tested. The references given at the end of this chapter include a number of important sources for these data.

The specific system chosen for illustration in this chapter is H_2O-$La(NO_3)_3$ - $Mg(NO_3)_2$.

The following represents published (11) density data on the H_2O-$La(NO_3)_3$ binary at 25°C:

Molality	Density
0.03255	1.00587
0.05918	1.01299
0.07429	1.01699
0.12520	1.03040
0.22173	1.05500
0.45720	1.11417
0.91842	1.22134
1.54544	1.35203
2.04584	1.44516
3.34770	1.60000

For the H_2O-$Mg(NO_3)_2$ binary at the same temperature the data is:

Molality	Density
0.13760	1.01190
0.28093	1.02710
0.43036	1.04260
0.58629	1.05840
0.74915	1.07450
0.91942	1.09100
1.09790	1.10790
1.28426	1.12510
1.48004	1.14270
1.68560	1.16070
1.90170	1.17900
2.12918	1.19770

Using the following W and $\overline{V}^{\circ}$ values:

$La(NO_3)_3$:	W = 6	$\overline{V}^{\circ}$ = 50.7548 (12)
$Mg(NO_3)_2$:	W = 3	$\overline{V}^{\circ}$ = 39.9312 (12)

and fitting the binary systems separately, we get:

$La(NO_3)_3$:	A = 6.31109	B = +0.0455	C = -0.01689
$Mg(NO_3)_2$:	A = -1.65176	B = +2.32808	C = -0.35539

Now, based upon the mixing rules presented earlier, the comparison between predicted and published (4) densities is:

$m_{La(NO_3)_3}$	$m_{Mg(NO_3)_2}$	I	Predicted	Published
3.04439	1.02472	21.341	1.65667	1.65200
2.42719	0.83453	17.067	1.55681	1.55600
1.89845	0.65023	13.341	1.45973	1.46500
1.49539	0.51188	10.508	1.37804	1.38000
1.16426	0.39950	8.184	1.30520	1.31000
0.85553	0.29462	6.017	1.23200	1.23700
0.60830	0.20840	4.275	1.16915	1.17100
0.38660	0.13446	2.723	1.10977	1.11400
0.27376	0.08939	1.911	1.07752	1.07300
1.86865	1.94577	17.049	1.53033	1.52600

1.37632	1.43464	12.562	1.41693	1.41900
1.00138	1.04167	9.133	1.31994	1.32000
0.70032	0.72879	6.388	1.23397	1.23600
0.45142	0.46835	4.114	1.15615	1.15800
0.24080	0.25200	2.201	1.08537	1.08700
0.69526	3.22078	13.834	1.41018	1.41200
0.48941	2.26646	9.736	1.30797	1.31300
0.35659	1.65069	7.092	1.23489	1.24100
0.24462	1.13780	4.881	1.16807	1.17100
0.11228	0.52212	2.240	1.08005	1.08300

Thus the resulting fit is seen to be excellent.

ENTHALPY

In this section, we consider the problem of calculating the species partial molal enthalpies, $\bar{H}_i$ in cal/gm-mole, for an electrolyte solution. The total enthlapy is related to the partial molal enthalpy by:

$$H_T = \sum_i n_i \bar{H}_i \tag{8.26}$$

What is needed is a formulation for $\bar{H}_i$. This can be done by starting with the definition of excess free energy:

$$\bar{G}_i^{ex} = \bar{G}_i - \bar{G}_i^I \tag{8.27}$$

where,

$\bar{G}_i^{ex}$ - excess partial molar Gibbs free energy of species i, cal/gm-mole

$\bar{G}_i$ - actual partial molar Gibbs free energy of species i, cal/gm-mole

$\bar{G}_i^I$ - partial molar Gibbs free energy of species i in an ideal solution, cal/gm-mole. The standard state of species i is chosen such that the relationship between the Gibbs free energy of the ideal solution, in an infinitely dilute solution and at the standard state is $\bar{G}^I = \bar{G}^\infty = \bar{G}^o$

From earlier developments we saw that:

$$\bar{G}_i = \left(\frac{\partial G}{\partial n_i}\right)_{T,P,n_{j \neq i}} \tag{2.14}$$

$$\mu_i = \left(\frac{\partial G}{\partial n_i}\right)_{T,P,n_{j \neq i}} \tag{2.17}$$

$$\mu_i = \mu_i^o + R\,T \ln (\gamma_i m_i) \tag{2.22}$$

Also, $\gamma_i \longrightarrow 1$ as $m_i \longrightarrow 0$ (infinite dilution). Using this fact we can combine (8.27), (2.14), (2.17) and (2.22) and obtain:

$$\bar{G}_i^{ex} = R\,T\,[\ln \gamma_i m_i - \ln m_i] \tag{8.28}$$

or

$$\bar{G}_i^{ex} = R\,T \ln \gamma_i \tag{8.29}$$

In order to introduce $\bar{H}_i$ we refer back to the Gibbs-Helmholtz equation:

$$\left(\frac{\partial \bar{G}/T}{\partial T}\right)_{P,n} = -\frac{\bar{H}}{T^2} \tag{2.7}$$

Equations (8.27), (8.29) and (2.7) when properly combined, become:

$$\left(\frac{\partial \bar{G}_i^{ex}/T}{\partial T}\right)_{P,n} = R\left(\frac{\partial \ln \gamma_i}{\partial T}\right)_{P,n} = \frac{-\bar{H}_i^{ex}}{T^2} = \frac{\bar{H}_i^{\infty} - \bar{H}_i}{T^2} \qquad (8.30)$$

where,

$\bar{H}_i^{ex}$ = excess partial molar enthalpy of species i, cal/gm-mole

$\bar{H}_i$ = actual partial molar enthalpy of species i, cal/gm-mole

$\bar{H}_i^{\infty}$ = infinite dilution partial molar enthalpy of species i, cal/gm-mole

Rearranging (8.30) we finally get:

$$\bar{H}_i = \bar{H}_i^{\infty} - R\,T\left(\frac{\partial \ln \gamma_i}{\partial T}\right)_{P,n} \qquad (8.31)$$

What equations (8.30) and (8.31) really tell us is that the total enthalpy can be determined if we know the component heat of formation at infinite dilution and the temperature functionality of the activity coefficient for each constituent species i. Let us now consider each of these in more detail.

HEAT OF FORMATION AT INFINITE DILUTION

Every enthalpy calculation requires a basis. For purposes of this development, we will take, for aqueous species, the pure component heat of formation at 25°C and infinite dilution as the basis. Electrolyte solutions, as we have seen, contain either molecular or ionic species. Thus, in order to compute $\bar{H}_i^{\infty}(T)$ we will need formulations for both molecular and ionic species.

Molecular Species

For molecular species (e.g. CO_2, NH_3, H_2S, H_2O, etc.) we normally can develop $\bar{H}_i$ by using component vapor-liquid equilibrium data in the following manner. The partial molar heat capacity at zero pressure has been correlated for many gases and the coefficients are given in the literature in terms of the Kobe equation (15):

$$\bar{Cp}_i = a + b\,T + C\,T^2 + d\,T^3 \qquad (8.32)$$

This then leads to the equation:

$$\bar{H}_{V,i} = \bar{H}_{V,i}(T^\circ) + \int_{T^\circ}^{T} \bar{C}p \, dT$$

$$= \bar{H}_{V,i}(T^\circ) + 1.8 \;[\, a\,(T - T^\circ) + \frac{b}{2}\,(T^2 - T^{\circ 2}) + \frac{c}{3}\,(T^3 - T^{\circ 3}) + \frac{d}{4}\,(T^4 - T^{\circ 4})] \qquad (8.33)$$

where:

$\bar{H}_{V,i}$ = partial molar enthalpy of the vapor species i, cal/gm-mole

$\bar{C}p$ = partial molar heat capacity, cal/gm-mole°C

T° = basis temperature, 298.15 K

Further, for vapor-liquid equilibrium we saw in Chapter III that

$$\Delta\bar{H}_{RXN} = \bar{H}^{\infty}_{L,i} - \bar{H}_{V,i} = R\,T^2 \frac{d \ln K}{dT} \qquad (3.25)$$

where:

$\bar{H}^{\infty}_{L,i}$ = partial molar enthalpy at infinite dilution for the aqueous molecular species i, cal/gm-mole

For vapor-liquid equilibria, we saw back in Chapter III that

$$K = \gamma_L\, m_L \;/\; f_V\, p_V \qquad (3.21)$$

For many VLE the equilibrium constant K has been fit as a function of temperature to the form:

$$\ln K = A + B/T + C \ln T + DT \qquad (8.34)$$

Combining equations (3.25) and (8.34) above yields

$$\Delta\bar{H}_{RXN} = R(-\,B + C\,T + D\,T^2) \qquad (8.35)$$

But, $\Delta\bar{H}_{RXN}$ is, as noted above:

$$\Delta\bar{H}_{RXN} = \bar{H}^{\infty}_{L,i} - \bar{H}_{V,i} \qquad (8.36)$$

So, by combining equations (8.33), (8.35) and (3.25)

$$\overline{H}^{\infty}_{L,i} = [R (-B + CT + DT^2)] + \overline{H}_{V,i}(T^o) + 1.8 [a (T - T^o) + \frac{b}{2} (T^2 - T^{o2}) + \frac{c}{3} (T^3 - T^{o3}) + \frac{d}{4} (T^4 - T^{o4})] \qquad (8.37)$$

For many species of interest, B, C, D, $\overline{H}_{V,i}(T^o)$, a, b, c and d are readily available (13, 14, 15, 16).

Equation (8.37) is a practical working relationship for the infinite dilution aqueous molecular component partial molar enthalpy as a function of temperature. There is, however, an underlying assumption made. The relationship derived above assumes equilibrium between liquid and vapor, or, in other words a saturated liquid-vapor situation. In a subcooled liquid the enthalpy predicted by equation (8.37) should be corrected for the enthalpy difference between the pure component at the prevailing pressure and the saturation pressure.

Inasmuch as pressure effects on liquids are often minimal, except at quite elevated pressures (above 100 atmospheres), this factor can usually be ignored.

<u>Ionic Species</u>

In Chapter III we reviewed the work of Criss and Cobble (3-9). Applying a technique known as the correspondence principal, a relationship between the "absolute" entropy at temperature t_2 and the "absolute" entropy at 25°C was developed. The relationship is:

$$\overline{S^o_{t_2}} = a_{t_2} + b_{t_2} \overline{S^o_{25}} \qquad (3.32)$$

where,

$\overline{S}^o_{t_2}$ = absolute molar entropy of the species at temperature t_2, cal/°C-gm mole

a_{t_2}, b_{t_2} = coefficients which are functions of temperature

Since,

$$\int_{t_1}^{t_2} d\overline{S^o} = \int_{t_1}^{t_2} \overline{Cp^o} \frac{dT}{T} \qquad (8.38)$$

we can eventually obtain:

$$\overline{Cp}^\circ(t_2) - \overline{Cp}^\circ(25) = \frac{a_{t_2} - \overline{S}^\circ_{25^\circ}[1.0 - b_{t_2}]}{\ln(T_2/298.15)} \tag{8.39}$$

Equation (8.39) can be rewritten as:

$$\overline{Cp}^\circ(t_2) - \overline{Cp}^\circ(25) = \overline{Cp}^\circ_{MEAN}(t_{2,25}) = \alpha_{t_2} + \beta_{t_2}\,\overline{S}^\circ_{25^\circ} \tag{8.40}$$

with α and β defined as in Chapter III:

$$\alpha_{t_2} = \frac{a_{t_2}}{\ln(T_2/298.15)} \tag{3.36a}$$

$$\beta t_2 = \frac{-(1. - b_{t_2})}{\ln(T_2/298.15)} \tag{3.36b}$$

Criss and Cobble further observed that all ions could be separated into four distinct categories, namely,

1) ordinary cations
2) ordinary anions
3) oxy-anions
4) acid oxy-anions

In Appendix 3.1, α and β for various temperatures are tabulated for all four classes of ions.

Inasmuch as

$$\overline{H}^\infty_i(t_2) = \overline{H}^\infty_i(25) + Cp_{MEAN}(t_{2,25})\,(t_2 - 25) \tag{8.41}$$

the partial molal enthalpy at infinite dilution can be estimated for many ionic species.

Range of Applicability

The use of equation (8.40) involves the assumption that there will be good data available for α and β for all ionic species over broad ranges of temperature. As a matter of fact, data on ionic heat capacities for temperatures above 200°C have only recently become available. Such data has shown (5) that extrapolation above 200°C is often quite inaccurate. Furthermore, there are ionic species, such as occur in amine systems, for which entropy data either does not exist or which do not necessarily prove particularly amenable to the estimation techniques outlined above.

The usual procedure for handling systems which do not respond to the $\overline{Cp}^{\circ}_{MEAN}$ approach is to rely on experimentally determined:

1) heat of solution
2) heat capacity

data. The treatment here is much more conventional. As an example, if carbon dioxide is dissolved in an amine solution we simply have

heat of solution = change in sensible heat of the solution

EXCESS ENTHALPY

Equations (8.37) and (8.41) thus give the means to calculate the partial molal enthalpies at infinite dilution for ionic and molecular species. The calculation of the partial molal enthalpy of species i is complete but for the contribution due to the "excess enthalpy" term of equation (8.31):

$$-R\,T^2\left(\frac{\partial \ln \gamma_i}{\partial T}\right)_{P,n}$$

In practice, this term is often small, but where the functionality between the activity coefficient and temperature is known (see discussion in Chapter IV), it is then a matter of straightforward differentiation. As an additional note of caution it should be observed that having a fairly good function to fit γ_i with respect to temperature is no guarantee that the partial derivative will be nearly as accurate.

EXAMPLE

As an example of the detailed calculation of the enthalpy of an electrolyte solution containing both molecules and ions, consider a solution of 100 moles of

H_2O, 1 mole of CO_2, and 2 moles of NaCl at 50°C at 1 atm pressure. Let's further assume that the calculation of the electrolyte equilibrium yields:

Species	n = gm-moles
$H_2O(aq)$	100.98
H^+	$1.9x10^{-11}$
OH^-	0.02
$CO_2(aq)$	$1.0x10^{-7}$
CO_3^{2-}	0.98
HCO_3^-	0.02
Na^+	2.0
Cl^-	2.0

The objective is to use equation (8.31) with equation (8.26) to calculate H_T, the total enthalpy. Thus,

$$H_T = \sum_i \overline{H}_i^{\infty} \cdot n_i - R\,T^2\left(\frac{\partial \ln \gamma_i}{\partial T}\right)_{P,n} \cdot n_i \qquad (8.42)$$

The techniques to be used to determine $\overline{H}_i^{\infty}$ for the various species can be summarized as follows

Species	Enthalpy Calculation Technique
$H_2O(aq)$	Steam tables
$CO_2(aq)$	Equation (8.37)
H^+	Equation (8.41)
OH^-	Equation (8.41)
CO_3^{2-}	"
HCO_3^-	"
Na^+	"
Cl^-	"

In order that the calculations be consistent, all constituent enthalpies will be adjusted to the heat of formation at 25°C. Taking each species individually we have:

$H_2O(aq)$

The steam tables (13) provide the enthalpy of saturated aqueous water at 50°C. Adjusting these tables to reflect a change of reference state to equal the heat of formation at 25°C, $\overline{H}^{\infty}_{H_2O}(25°C) = -68315.$ cal/gm-mole, we get:

$$\overline{H}^{\infty}_{H_2O}(50°C) = -67866.\ \text{cal/gm mole}$$

$CO_2(aq)$

In order to use equation (8.37) to calculate $\overline{H}^{\infty}_{CO_2}$ (50°C) the following values are required:

Parameter	Value	Source
B	17371.2 cal/mole	14
C	43.0607 cal/K mole	14
D	-0.002191 cal/K^2 mole	14
H_{V,CO_2}(25°C)	- 94501.0 cal/gm-mole	17
a	-94.55 cal/mole	15
b	8.979 cal/mole K^2	15
c	1.28502×10^{-2} cal/mole K^2	15
d	-9.4284×10^{-5} cal/mole K^3	15
R	1.987	

The calculation then yields

$\overline{H}^{\infty}_{CO_2}$(50°C) = -106276. cal/gm-mole

Ions

In order to calculate the ionic $\overline{H}^{\infty}_i$ we will need for each ion the quantities $\overline{S}°_{25°}$, $\overline{H}°_{25°}$ and the a_t and b_t term of the correspondence principle for 50°C. Using references (3) and (17) we get:

ion	$\overline{H}^{o}_{25°C}$ cal/gm-mole	$\overline{S}^{o}_{25°C}$ cal/gm-mole	$a_{50°}$	$b_{50°}$	ion type
H^+	0.0	0.0	2.8	0.968	simple cation
Na^+	-57278.9	14.4	2.8	0.968	simple cation
OH^-	-54970.0	-2.57	-13.6	0.978	simple anion
CO_3^{2-}	-161840.0	-13.6	10.0	1.155	oxy-anion
HCO_3^-	-165390.0	21.8	-9.64	1.271	acid oxy-anion
Cl^-	-39952.0	13.5	-3.6	0.978	simple anion

Plugging into equation (8.41), we obtain:

ion	$\overline{H}_i^\infty(50°)$ cal/gm-mole
H^+	70.
Na^+	-56861.
OH^-	-55122.
CO_3^{2-}	-161983.
HCO_3^-	-164938.
Cl^-	-39712.

The overall enthalpy based upon the infinite dilution term only is thus obtained by applying equation (8.42) ignoring the excess enthalpy term. This yields a value of .720933E+07 cal.

As mentioned earlier, the "excess enthalpy" term is small and difficult to obtain. For purposes of illustration only, we can estimate the $\gamma_i(T)$ for the Na^+ and Cl^- ions based upon binary correlation parameters published by Rastogi and Tassios (14). The Bromley formulation for calculating activity coefficients, simplified for a 1:1 salt, is:

$$\log \gamma_i = \frac{\ln \gamma_i}{2.303} = -\frac{A\sqrt{I}}{1+\sqrt{I}} + \frac{(0.06+0.6B)I}{(1+1.5I)^2} + BI \qquad (8.43)$$

The Debye-Hückel constant and the Bromley "B" are both functions of temperature. Regressing data available in Pitzer and Brewer (15) the Debye-Hückel constant may be expressed in terms of the Celsius temperature:

$$A = .49192 + .71445E\text{-}3\ t + .21063E\text{-}5\ t^2 + .11776E\text{-}7\ t^3 \qquad (8.44)$$

The Bromley "B" is given by Rastogi as a function of the absolute temperature:

$$B = \frac{B_1}{T} + B^* \ln\left(1. - \frac{243}{T}\right) \qquad (8.45)$$

where, for NaCl, B = 42.97 and B* = 0.05127.

Differentiating equation (8.43) to obtain the required $-RT^2\ \partial \ln \gamma_i / \partial T$ we get:

$$-R\,T^2 \frac{\partial \ln \gamma_i}{\partial T} = -2.303\ R\,T^2\left[\frac{-\sqrt{I}}{1+\sqrt{I}}\ \frac{dA}{dT} + \left(I + \frac{0.6I}{(1+1.5I)^2}\right)\frac{dB}{dT}\right] \qquad (8.46)$$

The required derivatives are:

$$\frac{dA}{dT} = .71445E\text{-}3 + .42126E\text{-}5\ t + .35328E\text{-}7\ t^2 \qquad (8.47)$$

$$\frac{dB}{dT} = -\frac{42.97}{T^2} + \frac{.05127 * 243/T^2}{(1 - 243/T)} \qquad (8.48)$$

Evaluating this term for the sodium or chloride ion at 50°C = 323.15K at the approximate I=2., we get

$$-R\,T^2\left(\frac{\partial \ln \gamma_i}{\partial T}\right) = 214.731 \text{ cal/gm-mole} \qquad (8.49)$$

Even if we multiply by four moles of total Na^+ and Cl^- we can see that this term is very small compared to the infinite dilution term of .720933E+07. Nevertheless, since most of the solution is H_2O this value is somewhat comparable to the $\overline{H}_i^\infty$ for the ions and should not be dismissed a priori.

In actual practice, the calculation of $\partial \gamma_i(T)/\partial T$ in a multicomponent system is done on a digital computer and is synthesized from binary ion-ion, ion-molecule and molecule-molecule interactions. Because of the complexity of the function involved it is normally impractical to carry analytical derivatives through such a calculation. Because of this, these derivatives are normally developed by perturbing the temperature and recalculating the $\gamma_i(T)$.

REFERENCES/FURTHER READING

1. Rock, P.A., Chemical Thermodynamics, Macmillan Co., New York (1969)

2. Prausnitz, Molecular Thermodynamics of Fluid Phase Equilibria, Prentice-Hall Inc., Englewood Cliffs, NJ (1969)

3. Criss, C.M. and J.W. Cobble, "The Thermodynamic Properties of High Temperature Aqueous Solutions. IV. Entropies of Ions up to 200° and the Correspondence Principal", J. Am. Chem. Soc., v86, pp5385-5389 (1964)

4. Criss, C.M. and J.W. Cobble, "The Thermodynamic Properties of High Temperature Aqueous Solutions. V. The Calculation of Ionic Heat Capacities up to 200°", J. Am. Chem. Soc., v86, pp5390-5393 (1964)

5. Criss, C.M., "Estimations of Thermodynamic Properties of Aqueous Solutions at High Temperatures", 41st International Water Conference (1981)

6. Cobble, J.W., "Empirical Considerations of Entropy I. The Entropies of the Oxy-Anions and Related Species", J. Chem. Phys., v21, #9, pp1443-1446 (1953)

7. Cobble, J.W., "Empirical Considerations of Entropy II. The Entropies of Inorganic Complex Ions", J. Chem. Phys., v21, #9, pp1446-1450 (1953)

8. Cobble, J.W., "Empirical Considerations of Entropy III. A Structural Approach to the Entropies of Aqueous Organic Solutes and Complex Ions", J. Chem. Phys., v21, #9, pp1451-1456 (1953)

9. Cobble, J.W., "The Thermodynamic Properties of High Temperature Aqueous Solutions. VI. Applications of Entropy Correspondence to Thermodynamics and Kinetics", J. Am. Chem. Soc., v86, pp5394-5401 (1964)

10. Teng, T. and F. Lenzi, "Density Prediction of Multicomponent Aqueous Solutions from Binary Data", Can. J. of Chem. Eng., v53, pp673-676 (1975)

11. International Critical Tables, McGraw-Hill, New York (1928)

12. Millero, F.J., "The Partial Molal Volumes of Electrolytes in Aqueous Solutions", Water and Aqueous Solutions, R.A. Horne, ed. Wiley-Interscience, New York (1972) 519-596

13. Perry, R.H. and C.H. Chilton, eds., Chemical Engineers Handbook, 5th ed., McGraw-Hill Book Co., New York (1973)

14. Wilhelm, E., R. Battino, and R.J. Wilcock, "Low Pressure Solubility of Gases in Liquid Water", Chem Rev., v77, #2, pp219-262 (1977)

15. Kobe, K.A. et al, "Thermochemistry for the Petroleum Industry", Petroleum Refiner (Jan. 1949 - Dec. 1959)

16. Schulze G. and J.M. Prausnitz, "Solubilities of Gases in Water at High Temperatures", I&EC Fundamentals, v20, pp175-177 (1981)

17. Wagman, D.D. et. al., Selected Values of Chemical Thermodynamic Properties, NBS Technical Notes 270-3 (1968), 270-8 (1981)

18. Kell, G.S., "Density, Thermal Expansivity, and Compressibility of Liquid Water from 0° to 150°C: Correlations and Tables for Atmospheric Pressure and Saturation Reviewed and Expressed on 1968 Temperature Scale", J. Chem. Eng. Ref. Data, v20, #1, pp97-105 (1975)

19. Rastogi, A. and D. Tassios, "Estimation of Thermodynamic Properties of Binary Aqueous Electrolyte Solutions in the Range 25-100°C", I & EC Process Des. Dev., v19, #3, pp477-482 (1980)

20. Lewis, G.N. and M. Randall, Thermodynamics, McGraw-Hill, New York, (1961)

IX:

Worked Examples

WORKED EXAMPLES

The ultimate objective of this handbook is to enable the practicing engineer to effectively formulate and solve mathematical models for aqueous electrolyte systems. In this chapter, several electrolyte systems of practical interest are examined and the formulation and solution of the attendant simulation models are reviewed.

Figure 9.1 shows a schematic of the basic electrolyte simulation problem. Given the temperature, pressure and a series of inflow rates, the goal is to predict the distribution between phases (vapor, liquid and solid) and the compositions within each phase.

In order to mathematically predict the state of electrolyte systems, there are three essential steps involved. These are:

1) Formulating a proper set of nonlinear algebraic equations in order to represent the system.

2) Obtaining, via published material and/or nonlinear regression, the required coefficients for the equilibrium constants and species interaction parameters for activity coefficient calculations.

3) Solving, by hand if possible, the equations formulated. Hand solutions unfortunately will apply to only simple strong electrolyte systems. In almost any practical case, the power of the digital computer will have to be brought to bear.

Model Formulation

Formulating a model for an isothermal electrolyte flash requires stating, in equation form, the requisite

1) Material balances
2) Electroneutrality
3) Phase equilibria
4) Reaction equilibria

Figure 9.1

Basic Electrolyte Simulation Problem

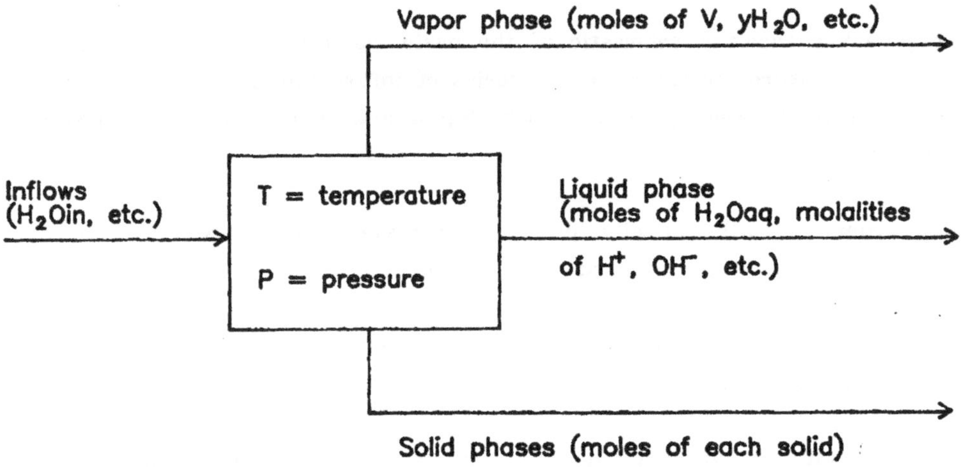

This can be done by systematically carrying out the following sequence:

1) Identify, one phase at a time, the unknown species compositions that will have to be solved for. In carrying out this step it is often very useful to write out the chemical mass action equations which are expected to occur. In Chapter VI of this Handbook we saw that several references (12, 24, 25, 26) are particularly helpful in deciding which reactions and/or species to include. If a vapor phase is present either the total pressure or the total moles of vapor must be specified for the isothermal system to be determined. If the pressure is fixed, then V, the total moles of vapor, and the mole fractions of the vapor species are the unknown variables. If the total vapor moles are fixed then the total pressure and vapor species mole fractions are the unknown variables. For example, the system CO_2-H_2O vapor-liquid equilibria would yield the following unknown species compositions in the liquid phase: (v1) moles $H_2O(aq)$ and molality of (v2) H^+, (v3) OH^-, (v4) CO_3^{2-}, (v5) HCO_3^-, and (v6) $CO_2(aq)$. The vapor phase would contain either the vapor rate and vapor mole fractions, (v7a) V, (v8a) yH_2O and (v9a) yCO_2, or the total vapor pressure and partial vapor pressures, (v7b) P, (v8b) pH_2O and (v9b) pCO_2. Whichever method of classifying the vapor phase is used, the system contains nine variables (unknowns).

2) Write the required equilibrium equations in order to characterize the interphase equilibria (vapor-liquid or solid-liquid equilibria) as well as the intraphase equilibria. The intraphase equilibria are single step dissociation or hydrolysis reactions. The equations should be written in the thermodynamic form:

Liquid phase equilibria:

$$K(P,T) = \prod_{i=1}^{NP} \left(\gamma_i m_i\right)^{\nu_i} \Big/ \prod_{i=1}^{NR} \left(\gamma_i m_i\right)^{\nu_i}$$

For example, the dissociation $HCN(aq) = H^+ + CN^-$ would yield:

$$K(P,T) = \gamma_{H^+} m_{H^+} \gamma_{CN^-} m_{CN^-} \big/ \left(\gamma_{HCN(aq)} m_{HCN(aq)}\right)$$

For phosphoric acid dissociation, $H_3PO_4(aq) = H^+ + H_2PO_4^-$:

$$K(P,T) = \gamma_{H^+} m_{H^+} \gamma_{H_2PO_4^-} m_{H_2PO_4^-} \big/ \left(\gamma_{H_3PO_4(aq)} m_{H_3PO_4(aq)}\right)$$

Vapor-liquid equilibria:

When the vapor phase is described using vapor mole fractions:

$$K(P,T) = \prod_{i=1}^{NP} \gamma_i m_i \; / \; \prod_{i=1}^{NR} f_i y_i P$$

For example, the CO_2 VLE, $CO_2(vap) = CO_2(aq)$, would yield:

$$K(P,T) = \gamma_{CO_2(aq)} m_{CO_2(aq)} / (f_{CO_2(vap)} y_{CO_2(vap)} P) \qquad (9.1a)$$

If the vapor phase is described with vapor partial pressures, the generalized description of the VLE is:

$$K(P,T) = \prod_{i=1}^{NP} \gamma_i m_i \; / \; \prod_{i=1}^{NR} f_i p_i$$

The CO_2 VLE would therefore be:

$$K(P,T) = \gamma_{CO_2(aq)} m_{CO_2(aq)} / (f_{CO_2(vap)} p_{CO_2(vap)}) \qquad (9.1b)$$

where:

- P - total pressure
- T - temperature
- γ_i - activity coefficient
- m_i - species molality
- NP - number of product species for the reaction
- NR - number of reactant species for the reaction
- f_i - fugacity coefficient
- y_i - vapor phase mole fraction
- p_i - vapor partial pressure
- K - equilibrium constants

For the H_2O-CO_2 system there are five equilibrium equations:

- the carbon dioxide vapor-liquid equilibrium, equation (9.1) shown above
- the water vapor-liquid equilibrium, $H_2O(vap) = H_2O$:

$$K_{H_2O(vap)} = a_w / (f_{H_2O} y_{H_2O} P) \qquad (9.2)$$

- the carbon dioxide hydrolysis, $CO_2(aq) + H_2O = H^+ + HCO_3^-$:

$$K_{CO_2(aq)}(P,T) = \gamma_{H^+}\ m_{H^+}\ \gamma_{HCO_3^-}\ m_{HCO_3^-}\ /\ (\gamma_{CO_2(aq)}\ m_{CO_2(aq)}\ a_w) \quad (9.3)$$

- the water dissociation, $H_2O = H^+ + OH^-$:

$$K_{H_2O}(P,T) = \gamma_{H^+}\ m_{H^+}\ \gamma_{OH^-}\ m_{OH^-}\ /\ a_w \quad (9.4)$$

- The HCO_3^- dissociation, $HCO_3^- = H^+ + CO_3^{2-}$

$$K_{HCO_3^-}(P,T) = \gamma_{H^+}\ m_{H^+}\ \gamma_{CO_3^{2-}}\ m_{CO_3^{2-}}\ /\ (\gamma_{HCO_3^-}\ m_{HCO_3^-}) \quad (9.5)$$

3) Write the electroneutrality balance. The general form is:

$$\sum_{i=1}^{NC} z_i\ m_i = \sum_{i=1}^{NA} z_i\ m_i$$

where:

NC - number of cations

NA - number of anions

z_i - absolute value of the ionic species charge

For the carbon dioxide-water system this equation is

$$m_{H^+} = m_{OH^-} + 2\ m_{CO_3^{2-}} + m_{HCO_3^-} \quad (9.6)$$

4) Write a number of material balances equal to the number of unknowns (see step 1) less the number of equations written in steps 2 and 3. For CO_2-H_2O there are nine unknowns; five K equations were written by step 2 and one electroneutrality equation resulted from step 3. This leaves three material balances to be written. The rules to follow are:

a) If a vapor phase is present, always include an equation for the sum of the vapor mole fractions to equal unity, $\Sigma y_i=1.0$, or the sum of the partial pressures to equal the total pressure, $\Sigma p_i=P$.

Thus for CO_2-H_2O the vapor phase balance would be:

$$y_{H_2O(vap)} + y_{CO_2(vap)} = 1.0 \quad (9.7a)$$

if the total pressure, P, is fixed, or:

$$p_{H_2O(vap)} + p_{CO_2(vap)} = P \qquad (9.7b)$$

if the total moles of vapor, V, is fixed.

b) The remaining balances are elemental balances of the form equating the amount of an element in the input species with the amount in the products:

$$\sum_{i=1}^{NIC} N_i * n_i = \sum_{i=1}^{NPC} N_i * n_i$$

where,

NIC - number of inflow components containing the particular element being balanced

NPC - number of product components containing the particular element being balanced

n_i - total moles of species i

N_i - number of atoms of element present in species i

If all elemental balances, including hydrogen and oxygen balances, are written, an overall balance would be redundant and should not be written. Conversely, the oxygen balance can be left out in favor of an overall balance. For the carbon dioxide-water system the following material balances may be written:

a carbon balance

$$CO_2(in) = V * y_{CO_2(vap)} + \frac{H_2O(aq)}{55.51}(m_{CO_3^{2-}} + m_{HCO_3^-} + m_{CO_2(aq)}) \qquad (9.8)$$

and a hydrogen balance

$$2.\ H_2O(in) = V * y_{H_2O(vap)} + 2.\ H_2O(aq) + \frac{H_2O(aq)}{55.51}(m_{H^+} + m_{OH^-} + m_{HCO_3^-}) \qquad (9.9)$$

The factor $[H_2O(aq)/55.51]$ is used to convert the molalities to moles. This factor arises since molality is defined as moles per 1000 grams solvent H_2O. Since H_2O has a molecular weight of 18.02, molality is moles per 55.51 moles H_2O.

5) Write specific definitions (equations) for each equilibrium constant and activity coefficient in the system. These K's and γ's then plug directly into the basic equilibrium equations written earlier. Applicable formulations for these thermodynamic parameters were discussed in Chapters I-VII. In the examples below we will use various formulations for illustrative purposes. One example occuring in the system CO_2-H_2O would be

$$\ln \gamma_{H^+} = f^{\gamma} + (B_{H^+-CO_3^{2-}})(m_{CO_3^{2-}}) + (B_{H^+-HCO_3^-})(m_{HCO_3^-}) + (B_{H^+-CO_2(aq)})(m_{CO_2(aq)})$$

where:

f^{γ} - Debye-Hückel term

B - various interaction coefficients involving H^+ with molecules as well as oppositely charged ions

In order to write a deterministic model it is necessary to have the number of equations equal to the number of unknowns. If after writing the equilibrium K equations, electroneutrality and elemental balances there are still not enough equations to match the number of unknowns, one or both of the following two considerations should be brought to bear:

1) The equilibrium equations should be reviewed to see if any further independent equilibria, involving the species being considered, can be written.

2) The possibility of writing independent elemental balances for species subsets of a particular chemical element should be considered. For example, in a model involving CO(vap), CO(aq), CO_2(vap), CO_2(aq), HCN(vap), HCN(aq) and CN^-, separate carbon balances can be written for <u>C</u>O, <u>C</u>O_2 and <u>C</u>N. This is true so long as there is no equilibrium reaction which involves conversion from one form of carbon to another. Thus, instead of writing:

$$CO(in) + CO_2(in) + HCN(in) = V * (y_{CO(vap)} + y_{CO_2(vap)} + y_{HCN(vap)}) + H_2O(aq)/55.51*(m_{CO(aq)} + m_{CO_2(aq)} + m_{HCN(aq)} + m_{HCO_3^-} + m_{CO_3^{2-}} + m_{CN^-})$$

the correct approach would be to write:

$$CO(in) = V*y_{CO(vap)} + H_2O(aq)/55.51 * m_{CO(aq)}$$

$$CO_2(in) = V*y_{CO_2(vap)} + H_2O(aq)/55.51 * (m_{CO_2(aq)} + m_{HCO_3^-} + m_{CO_3^{2-}})$$

$$HCN(in) = V*y_{HCN(vap)} + H_2O(aq)/55.51 * (m_{HCN(aq)} + m_{CN^-})$$

Obtaining Coefficients

Once a model has been developed, it is necessary to provide specific equilibrium constants (K values) as well as specific coefficients for species interactions required to accurately compute activity coefficients.

K values are usually straightforward. In Chapter III it was shown that these can be determined by:

1) published fits for K(T,P), or,
2) estimates based upon $\Delta G°_{RXN}$, $\Delta H°_{RXN}$ and $\Delta Cp°_{RXN}$

Species interactions are considerably more difficult. The specific coefficients which characterize the interaction of two species as a function of ionic strength and temperature are quite dependent upon both model formulation and regression (fitting) of experimental data.

1) Published values for the interaction coefficients are applicable over limited ranges of ionic strength and temperature.

2) Published interaction coefficients were fit based upon a supposition as to the chemical reactions occurring and the species present. If a significant intermediate is left out in formulating the model, the resulting interactions calculated may have very limited validity, particularly if one tries to utilize them as part of calculations involving additional components. These then are model dependent coefficients and may reproduce the experimental data, but do not apply to other systems.

3) In the absence of reasonable published interaction coefficients, it is necessary to utilize one of several possible approaches:

a) For strong electrolytes, the ion-ion interactions can be fit based upon binary system data estimation techniques outlined by Bromley and/or Meissner.

b) For general electrolyte systems the species interactions can be fit based upon binary and ternary systems data and, as species are added, only the new "crossover" interactions need be fit. For example for H_2O-NH_3-CO_2 the approach would be to:

i) Take data for H_2O-NH_3 and, using a model based upon H_2O, H^+, OH^-, NH_4^+ and NH_3(aq) fit the various significant species interactions. The significant interactions are NH_4^+-OH^-, NH_3(aq)-NH_3(aq) and NH_3(aq)-OH^-. The H^+-OH^- interaction is always ignored because the concentration of either H^+ or OH^- will always be quite small and the resulting interaction imposes very little effect on the model. In this case, H-OH and NH_3(aq)-H^+ interactions are ignored because the hydrogen ion concentration will be very small. These calculations can be done by using a combination of nonlinear regression and numerical equation solving.

ii) Do the same for H_2O-CO_2. The significant interactions here are H^+-HCO_3^-, H^+-CO_3^{2-}, CO_2(aq)-H^+, CO_2(aq)-CO_3^{2-}, CO_2(aq)-HCO_3^-, CO_2(aq)-CO_2(aq).

iii) Use ternary data to handle the "crossover" interactions: $NH_2CO_2^-$-CO_2(aq), $NH_2CO_2^-$-NH_3(aq), H^+-$NH_2CO_2^-$, NH^+-$NH_2CO_2^-$.

This part of the work, while tractable so long as adequate data is available, is tedious, time consuming and demands good numerical tools. It is, currently without question, the pivotal step in the entire process of working out an electrolyte simulation.

Model Solution

Solving the general set of nonlinear equations which result from formulating an electrolyte simulation model will usually require a digital computer along with an appropriate numerical technique for solving the equations. For certain simple problems a solution by hand is feasible. Reasonable assumptions can allow hand solutions to be carried out, but the trial and error evaluation, the checking of assumptions and the tedious calculation of activity coefficients can all be avoided by using general computer equation solving.

Two methods for numerical solution which have been used effectively are free energy minimization and the Newton-Raphson Method. The free energy minimization technique has the virtue of directly deciding questions of phase existence, while the Newton-Raphson Method has the virtue of dealing directly and clearly with the equation set formulated. The examples in this chapter will, whenever the computer is required, be treated by the Newton-Raphson technique.

One of the greatest problems in solving electrolyte systems numerically stems from the fact that solution of the resulting nonlinear algebraic equations is an iterative process which only converges if a starting guess is provided which falls within the "circle of convergence" for the problem. Since the hydrogen molality may vary from $10^{1.0}$ to $10^{-20.0}$ for different problems it is a formidable challenge to provide a guess good to within one or two orders of magnitude.

Our recommended approach is to examine the model and the specific inflows and to try to estimate pH. Given this estimate, the actual K values and an initial estimation of all activity coefficients of 1.0, one can usually estimate the other species concentrations. A general application of this approach can be seen in Butler (6) where the concept of "predominance plots" is discussed. In "predominance plots" various species concentrations are plotted versus pH. Normally, these plots are developed based upon actual K values along with a very simple estimate of activity coefficients. Figure 9.2 illustrates a typical predominance plot for phosphoric acid.

Figure 9.2

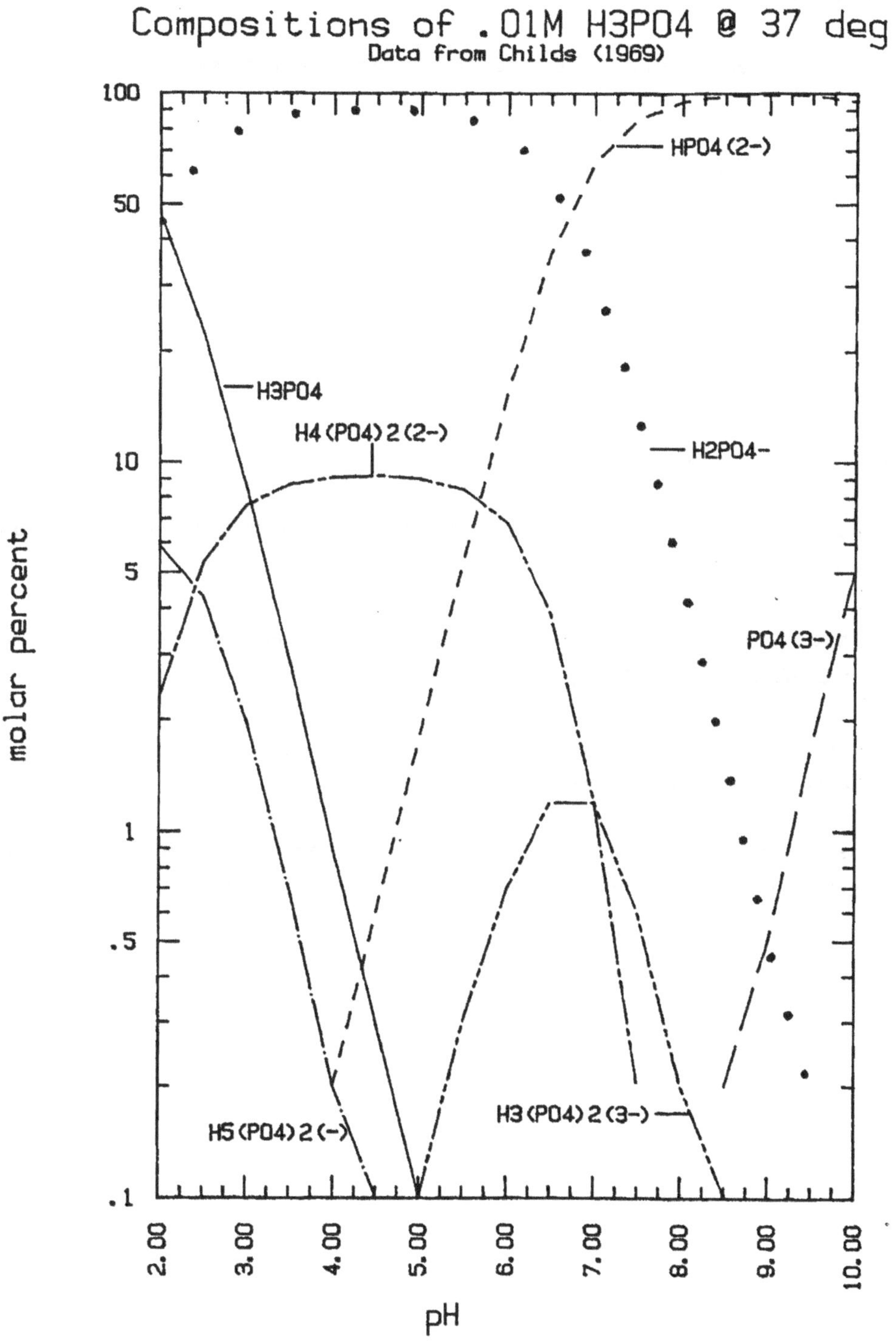
Compositions of .01M H3PO4 @ 37 deg
Data from Childs (1969)
100
50
10
5
1
.5
.1
molar percent
2.00
3.00
4.00
5.00
6.00
7.00
8.00
9.00
10.00
pH
HPO4(2-)
H3PO4
H4(PO4)2(2-)
H2PO4-
PO4(3-)
H5(PO4)2(-)
H3(PO4)2(3-)

Specific Examples

This chapter focuses mainly on specific illustrative problems. The systems considered are:

1) Sodium Chloride Solubility
2) Water - Chlorine
3) Water - Ammonia - Carbon Dioxide
4) Water - Sulfur Dioxide
5) Chrome Hydroxides
6) Gypsum Solubility
7) Phosphoric Acid Titrations

In each case the following steps were taken:

1) a straightforward deterministic model based upon the techniques outlined above was written

2) a specific activity coefficient formulation was chosen and, where applicable, the published interaction parameter values are presented

3) a numerical or hand developed solution for one specific condition is provided for each illustrative system.

4) the results of a series of solutions for various conditions is plotted against actual published data in order to assess the accuracy of the models.

Sodium Chloride Solubility

The binary sodium chloride-water system has been the object of many studies. As a result there is a wealth of published data for a wide range of temperatures. This data includes solubility, density, vapor pressure lowering and heat of solution measurements. Because of this availability of data and the straightforward strong electrolyte behavior of the system, sodium chloride has almost always been included as an example when illustrating activity coefficient modeling techniques. For this application, Meissner's method of activity coefficient calculation will be used.

The equation describing the sodium chloride dissociation/precipitation is:

$$NaCl(c) = Na^+ + Cl^-$$

As this is a binary solution of a 1-1 electrolyte, the following relationships may be noted:

$$m_{Na^+} = m_{Cl^-} = m_\pm$$

$$I = m_{Na^+} = m_{Cl^-} = m_\pm$$

where:

m_i - molality of ion i in solution

$m_\pm$ - mean molality in solution

I - ionic strength

Also,

$$\gamma_\pm = \gamma_{Na^+} = \gamma_{Cl^-}$$

That is, the mean molal activity coefficient, $\gamma_\pm$, equals the individual ionic activity coefficients. This results in the following equilibrium equation for sodium chloride solubility:

$$K_{NaCl} = \gamma_\pm^2 \, m_\pm^2 \qquad \text{(N.1)}$$

The solubility product of NaCl at 25°C may be calculated using values of Gibbs free energies which were presented by the National Bureau of Standards (26) and are tabulated in Appendix B:

$$\Delta G_{RXN} = \Delta G_{Na^+} + \Delta G_{Cl^-} - \Delta G_{NaCl}$$

$$= (-62593. - 31372.) - (-91815.)$$

$$= -2150.$$

$$K_{NaCl} = \exp\left[\frac{-\Delta G_{RXN}}{RT}\right] = \exp\left[\frac{-(-2150)}{(1.987)(298.15)}\right] = \exp 3.629$$

$$= 37.668 \qquad (N.2)$$

Meissner's (14) method for calculating the reduced activity coefficient, Γ, is detailed in Chapter IV. For 1-1 electrolytes in binary solution:

$$\gamma_{\pm} = \Gamma^{o}$$

The activity coefficient of NaCl is calculated using Meissner's method:

$$q = 2.23 \text{ for NaCl} \qquad (N.3a)$$

$$B = .75 - .065\ q \qquad (N.3b)$$

$$C = 1. + .055\ q \exp(-.023\ I^{3}) \qquad (N.3c)$$

$$\Gamma^{*} = 10. \wedge [-.5107 \sqrt{I} \ / \ (1. + C\sqrt{I}\ )\ ] \qquad (N.3d)$$

$$\Gamma^{o} = [1. + B\ (1. + .1\ I)^{q} - B]\ \Gamma^{*} \qquad (N.3e)$$

The saturation molality of sodium chloride can be solved for using equations (N.1) and (N.3) with the solubility product (N.2). As an initial guess for the molality, the activity coefficient can be assumed to equal one, and the guess calculated by rearranging equation (N.1):

$$m_{\pm} = \sqrt{K_{NaCl}} \ / \ \gamma_{\pm} = \sqrt{37.668} \ / \ 1.0$$

$$= 6.13743$$

Using this value of $m_{\pm}$, the activity coefficient of NaCl can be calculated using equations (N.3):

$$B = .60505$$

$$C = 1.000601746$$

$$\Gamma^{*} = .4328318361$$

$$\Gamma^{o} = .932291251$$

The solubility product for this molality is:

$$K = \gamma_{\pm}^{2}\ m_{\pm}^{2}$$

$$= .932291251^{2} * 6.13743^{2}$$

$$= 32.739823$$

which is lower than the K value of 37.668 calculated using Gibbs free energies.

Setting a new guess for the sodium chloride molality of 6.5 results in the following values for the activity coefficient and K value:

$m_{\pm}$ = 6.5

B = .60505

C = 1.000221565

Γ^* = .4297712822

Γ° = .9640990565

K = 39.27083

This K value is larger than the target value of 37.668 so the NaCl molality guess is reduced to 6.4:

$m_{\pm}$ = 6.4

B = .60505

C = 1.00029537

Γ^* = .4305850461

Γ° = .9552083968

K = 37.37284941

Therefore, the saturation molality of sodium chloride using Meissner's method of calculating activity coefficients is slightly greater than 6.4 molal. The actual saturation molality at 25°C is tabulated in Seidell (23) as 6.14618 molal. In order to calculate this molality using the Meissner activity coefficient calculation method, the solubility product should equal 32.8863.

Meissner also presented an equation for adjusting the 25°C q parameter for use at other temperatures:

$$q_t = q_{25}\left(1. - \frac{.0027(t-25)}{|z_+ z_-|}\right)$$

with t indicating degrees Celsius. Using this equation to get new q values, similar hand calculations were done from 10° to 100°C. The deviation between the thermodynamic K value and the solubility product calculated with the saturation molality and activity coefficient ranged from -.002 to .009. Figure 9.3 is a plot comparing the results with experimental data. The poor agreement between the calculated and experimental saturation molalities is the result of two factors. The first is that the 25°C q presented by Meissner for NaCl was given for a maximum concentration of 3 to 4 molal. At 25°C the saturation point of sodium chloride

Figure 9.3

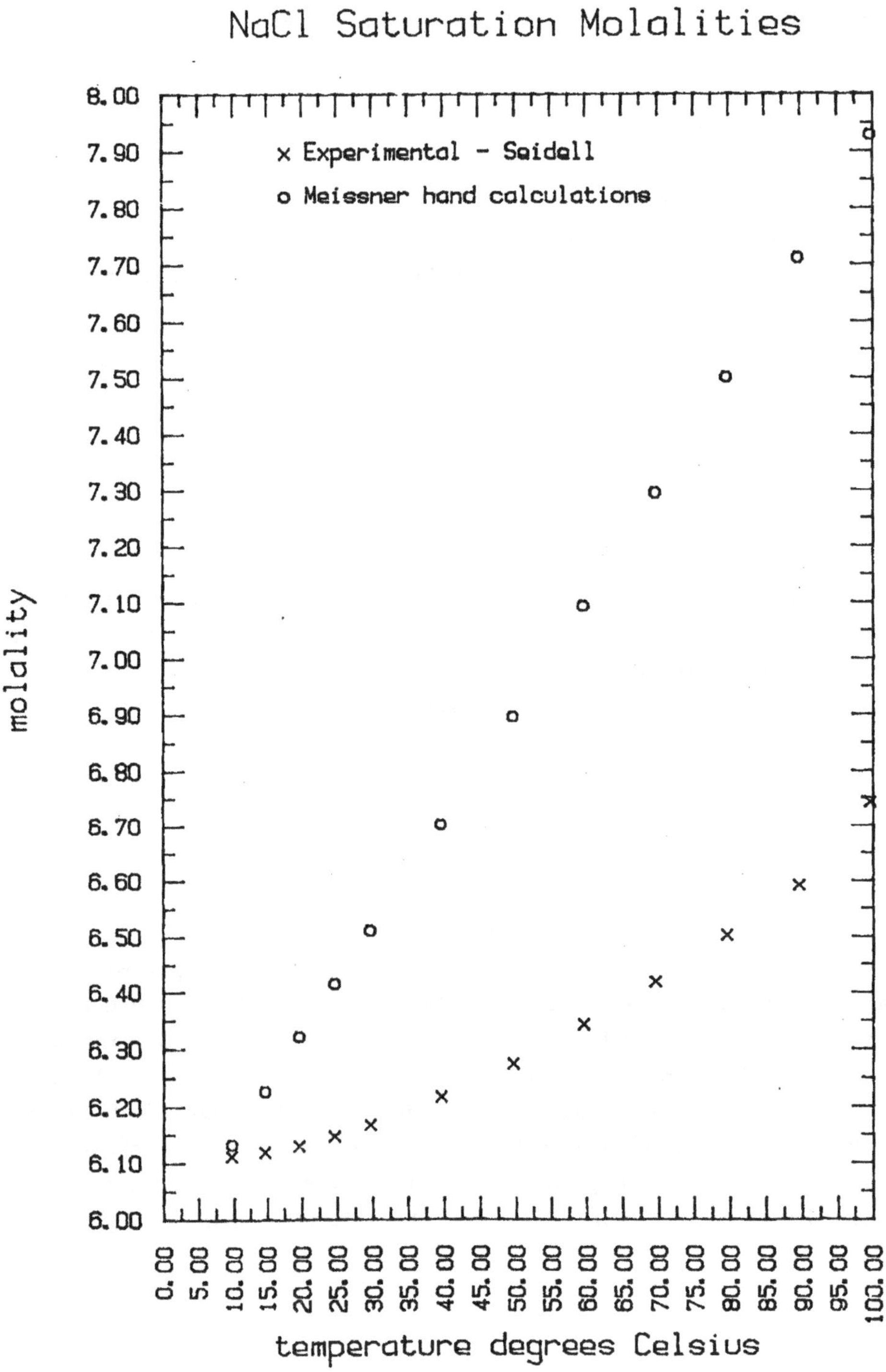
NaCl Saturation Molalities
× Experimental - Seidell
o Meissner hand calculations
molality
8.00
7.90
7.80
7.70
7.60
7.50
7.40
7.30
7.20
7.10
7.00
6.90
6.80
6.70
6.60
6.50
6.40
6.30
6.20
6.10
6.00
0.00
5.00
10.00
15.00
20.00
25.00
30.00
35.00
40.00
45.00
50.00
55.00
60.00
65.00
70.00
75.00
80.00
85.00
90.00
95.00
100.00
temperature degrees Celsius

exceeds the upper limit by more than 2 molal. It should also be noted that the solubility of sodium chloride increases with temperature and consequently moves further beyond the valid concentration range. The second factor contributing to the difference between the calculated and experimental points is that the method of calculating q_t from q_{25} only provides an approximate value of q_t. It should also be expected that for the equation to be most successful, the value of q at 25°C must be very good.

Another way of using Meissner's method is to use the plot he presented of ionic strength versus the reduced activity coefficients for different values of q. By using the plot to get the value of the activity coefficient for a given guess for the molality, the activity coefficient calculations may be avoided. For example, by assuming the activity coefficient to be equal to unity, the starting molality guess for the NaCl saturation point is 6.13743 as seen earlier. From the Meissner plot given in Figure 9.4, the activity coefficient for q = 2.23 and ionic strength = 6.13743 is approximately .9. A new guess for the molality can then be calculated:

$$m_{\pm} = \sqrt{K_{NaCl}} \;/\; .9 = 6.8194$$

This molality has an activity coefficient of approximately .95, resulting in a new molality iteration of 6.46:

$$m_{\pm} = \sqrt{37.668} \;/\; .95 = 6.13743 \;/\; .95 = 6.46$$

The activity coefficient value for 6.46 molal is approximately 0.925, which results in:

$$m_{\pm} = \sqrt{37.668} \;/\; .925 = 6.635$$

This molality gives an activity coefficient of about 0.94, which results in a molality of 6.5.

Figure 9.4

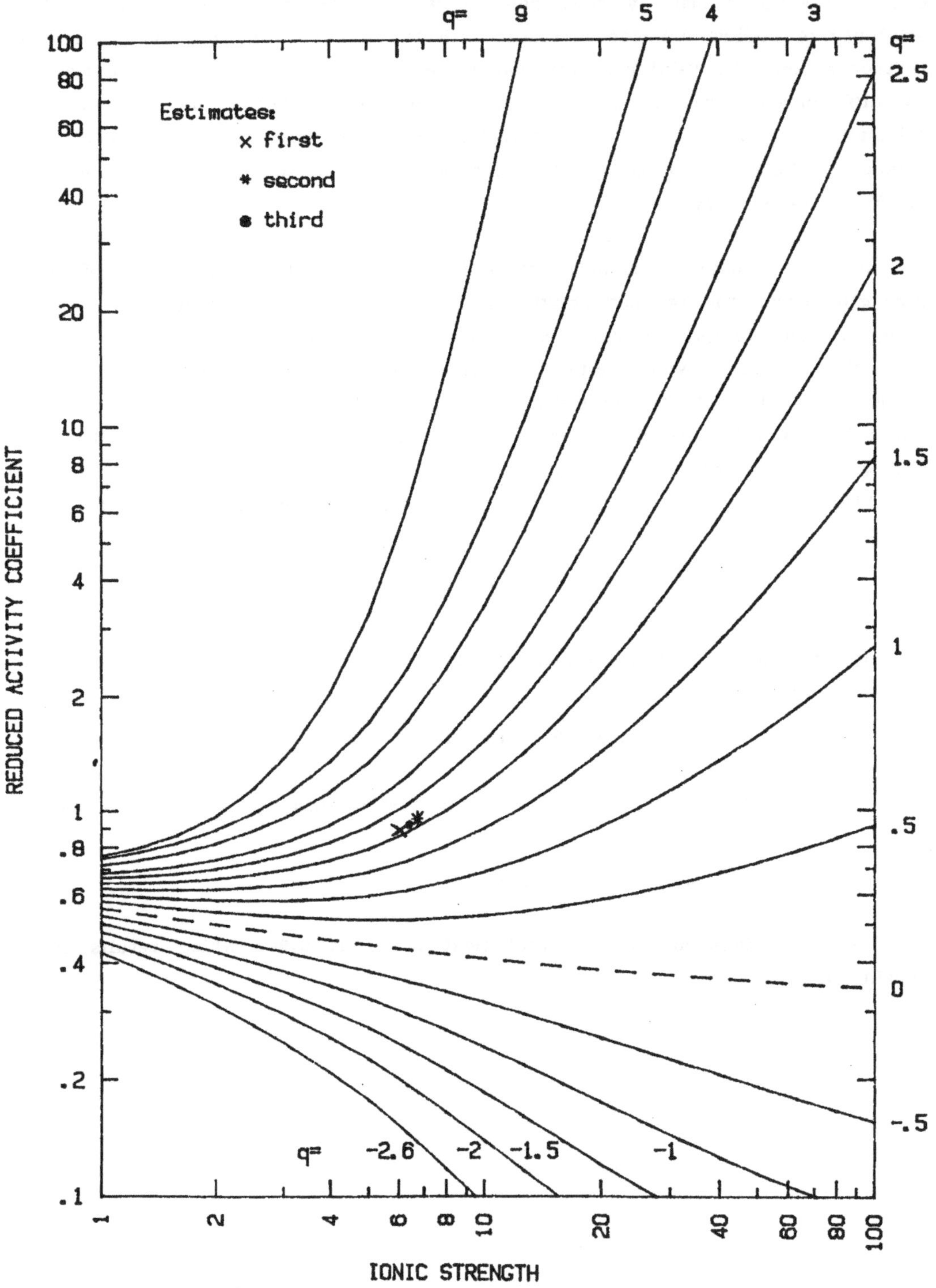
q= 9 5 4 3
q= 2.5
2
1.5
1
.5
0
-.5
q= -2.6 -2 -1.5 -1
Estimates:
× first
* second
● third
100
80
60
40
20
10
8
6
4
2
1
.8
.6
.4
.2
.1
1 2 4 6 8 10 20 40 60 80 100
REDUCED ACTIVITY COEFFICIENT
IONIC STRENGTH

Water - Chlorine

The solubility of chlorine in water at low partial pressures is greater than would be predicted by Henry's law. This deviation is attributed to the chemical reaction of chlorine with water. In order to model this system, the first step is to write out the mass action equations for the interphase and intraphase equilibria. The chlorine hydrolysis reaction is written:

$$Cl_2(aq) + H_2O = H^+ + Cl^- + HClO(aq)$$

The hypochlorous acid, HClO(aq), created by this reaction, has a dissociation equation:

$$HClO(aq) = H^+ + ClO^-$$

In addition, the water dissociation equation:

$$H_2O = H^+ + OH^-$$

and vapor-liquid equilibria:

$$H_2O(vap) = H_2O$$
$$Cl_2(vap) = Cl_2(aq)$$

would be expected to occur as well. Thus, there are five equilibria to consider. By examining these five equations, it is clear that, for a given temperature, pressure and input amounts of water and chlorine, the system would be completely determined in terms of the following nine concentrations (model unknowns):

liquid phase:

moles H_2O, $m_{Cl_2(aq)}$, m_{H^+}, m_{OH^-}, $m_{HClO(aq)}$, m_{ClO^-}, m_{Cl^-}

vapor phase:

H_2O and Cl_2 vapor mole fractions, total moles of vapor

The particular choice of units, moles of H_2O, molalities of liquid phase species and mole fractions of vapor species is a standard which we suggest for convenience. As noted in the introduction to this chapter, V, the total moles of vapor must be added as a tenth unknown. A complete deterministic model for describing this system must contain ten equations in order to determine the ten unknowns. The starting point is to write K equations for each equilibria:

water dissociation:

$$K_{H_2O} = \gamma_{H^+} m_{H^+} \gamma_{OH^-} m_{OH^-} / a_w \tag{C.1}$$

chlorine hydrolysis:

$$K_{Cl_2} = \gamma_{H^+} m_{H^+} \gamma_{Cl^-} m_{Cl^-} \gamma_{HClO(aq)} m_{HClO(aq)} / (\gamma_{Cl_2(aq)} m_{Cl_2(aq)} a_w) \tag{C.2}$$

hypochlorous acid dissociation:

$$K_{HClO(aq)} = \gamma_{H^+} m_{H^+} \gamma_{ClO^-} m_{ClO^-} / (\gamma_{HClO(aq)} m_{HClO(aq)}) \tag{C.3}$$

where:
- K_i - equilibrium constant
- γ_i - activity coefficient of species i
- m_i - molality of species i
- a_w - water activity

The vapor-liquid equilibria may be expressed:

$$K_{H_2O(vap)} = a_w / (y_{H_2O} P f_{H_2O}) \tag{C.4}$$

$$K_{Cl_2(vap)} = \gamma_{Cl_2(aq)} m_{Cl_2(aq)} / (y_{Cl_2} P f_{Cl_2}) \tag{C.5}$$

where:
- y_i - vapor mole fraction of species i
- P - total pressure
- f_i - fugacity coefficient of species i

The remaining five equations are elemental balances, electroneutrality and the vapor balance:

hydrogen balance:

$$2\ H_2O(in) = 2\ H_2O + \frac{H_2O}{55.51} (m_{H^+} + m_{OH^-} + m_{HClO(aq)}) + 2\ y_{H_2O}\ V \tag{C.6}$$

chlorine balance:

$$2\ Cl_2(in) = \frac{H_2O}{55.51}\left(2\ m_{Cl_2(aq)} + m_{Cl^-} + m_{HClO(aq)} + m_{ClO^-}\right) + 2\ y_{Cl_2}V \qquad (C.7)$$

oxygen balance:

$$H_2O(in) = H_2O + \frac{H_2O}{55.51}\left(m_{OH^-} + m_{HClO(aq)} + m_{ClO^-}\right) + y_{H_2O}\ V \qquad (C.8)$$

electroneutrality:

$$m_{H^+} = m_{OH^-} + m_{ClO^-} + m_{Cl^-} \qquad (C.9)$$

vapor phase balance:

$$y_{H_2O} + y_{Cl_2} = 1. \qquad (C.10)$$

where: $H_2O(in)$ - moles of water input
$Cl_2(in)$ - moles of chlorine input
H_2O - moles of water in solution
V - moles of vapor

The species molalities are multiplied by the factor $H_2O/55.51$ in the elemental balances to put them on a moles basis in order to be consistent with the other terms in the equations. This factor, discussed earlier, accounts for the fact that molality is moles per 1000 grams = 55.51 moles of solvent H_2O.

The equilibrium constants for equations (C.1), (C.2), (C.3), (C.4) and (C.5) may be calculated using thermodynamic property data and equation (3.31):

$$\ln K = \left(-\frac{\Delta G^o_{RXN}}{R\,T^o}\right) - \frac{\Delta H^o_{RXN}}{R}\left(\frac{1}{T} - \frac{1}{T^o}\right) - \frac{\Delta Cp^o_{RXN}}{R}\left(\ln\frac{T}{T^o} - \frac{T^o}{T} + 1.\right) \qquad (3.31)$$

where: ΔG^o_{RXN} - Gibbs free energy of reaction, cal/mole at 298.15 K
ΔH^o_{RXN} - enthalpy of the reaction, cal/mole at 298.15 K
ΔCp^o_{RXN} - heat capacity of the reaction, cal/K mole at 298.15 K
R - gas constant, 1.98719 cal/K mole
T - system temperature, Kelvins
T^o - reference temperature, 298.15 K

K - equilibrium constant at temperature T

For the species in the water-chlorine system, the National Bureau of Standards (26) has tabulated the following values for the thermodynamic properties:

	ΔG^o	ΔH^o	S^o	Cp^o
species	cal/mole		cal/K mole	
H_2O	-56687.	-68315.	16.71	17.995
H_2O(vap)	-54634	-57796.	45.104	8.025
H^+	0.	0.	0.	0.
OH^-	-37594.	-54970.	-2.57	-35.5
Cl_2 (aq)	1650.	-5600.	29.	
Cl_2 (vap)	0.	0.	53.288	8.104
HClO(aq)	-19100.	-28900.	34.	
Cl^-	-31372.	-39952.	13.5	-32.6
ClO^-	-8800.	-25600.	10.	

These values are also found in Appendix B.

A number of simplifying assumptions will be made for this system due to a lack of published parameters. The first is the frequently made assumption that at low pressures the vapor phase behaves as an ideal gas. The fugacity coefficients which account for deviations from ideality in the vapor phase are therefore equal to unity and equations (C.4) and (C.5) may be rewritten:

$$K_{H_2O(vap)} = a_w / (y_{H_2O} P) \tag{C.11}$$

$$K_{Cl_2(vap)} = \gamma_{Cl_2(aq)} m_{Cl_2(aq)} / (y_{Cl_2} P) \tag{C.12}$$

A similar assumption is made for the behavior of the molecular species Cl_2(aq) and HClO(aq) in the liquid phase with the result that the activity coefficients for the molecules equal unity. This is not a bad assumption for dilute solutions wherein the ionic strength is less than 1.0.

$$\gamma_{Cl_2(aq)} = 1.0$$

$$\gamma_{HClO(aq)} = 1.0$$

Bromley's (5) method of calculating the activity coefficients of ionic species in multicomponent solutions, detailed in Chapter V, leads to the following equations:

$$\gamma_{H^+} = 10.\wedge\left[-\frac{A\sqrt{I}}{1.+\sqrt{I}} + \dot{B}_{HCl}\, m_{Cl^-} + \dot{B}_{HClO}\, m_{ClO^-}\right]$$

$$\gamma_{OH^-} = 10.\wedge\left[-\frac{A\sqrt{I}}{1.+\sqrt{I}}\right]$$

$$\gamma_{Cl^-} = 10.\wedge\left[-\frac{A\sqrt{I}}{1.+\sqrt{I}} + \dot{B}_{HCl}\, m_{H^+}\right]$$

$$\gamma_{ClO^-} = 10.\wedge\left[-\frac{A\sqrt{I}}{1.+\sqrt{I}} + \dot{B}_{HClO}\, m_{H^+}\right]$$

where: A - Debye-Hückel constant

I - ionic strength = $.5\ \Sigma\, m_i\ z_i^2$

$= .5\ (m_{H^+} + m_{OH^-} + m_{Cl^-} + m_{ClO^-})$

m_i - molality of species i

z_i - charge of species i

$$\dot{B}_{ij} = \frac{(0.06 + .6\, B_{ij})\,|z_i\ z_j|}{\left(1. + \frac{1.5}{z_i\ z_j}\ I\right)^2} + B_{ij} \qquad (5.8)$$

B_{ij} = Bromley interaction parameter for cation i and anion j

The table of Bromley interaction parameters in Appendix 4.2 gives a value of:

B_{HCl} = .1433

but no value for the B_{HClO} interaction, which must be estimated.

In 1972, Bromley (4) suggested that the ion interaction parameter B_{MX} of the extended Debye-Hückel equation:

$$\log \gamma_{\pm} = -\frac{A\,|z_+\ z_-|\,\sqrt{I}}{1.+\sqrt{I}} + 2\left(\frac{\nu_+\ \nu_-}{\nu}\right) B_{MX}\ m + \ldots$$

could be estimated by summing values for the individual ions:

$$B_{\pm} \simeq B_+ + B_- \qquad (4.40)$$

He also showed that the values of the single ion B's could be related to the ionic entropies.

The similar behavior of the Bromley equation B parameters allows for the estimation of the parameter for untabulated species. When Bromley presented his equations in 1973, along with the B values for many species, he presented tabulated values based upon an expanded estimation equation:

$$B_{\pm} = B_{+} + B_{-} + \delta_{+} \delta_{-} \qquad (4.45)$$

However, B_- and δ_- values were not presented for the ClO^- ion. Using Bromley's assumption that $B_{Na^+} = 0.0$, values of B_- for many anions can be found using the values of B_{NaX} tabulated by Bromley in equation (4.40):

$$B_{NaX} = B_{Na^+} + B_{X^-}$$

$$B_{NaX} = B_{X^-}$$

This assumes the δ terms to be zero. These B_- values, plotted versus $z_iS_i°$, the anion's entropy times the anion charge in Figure 9.5 affect the relatively linear behavior seen by Bromley in 1972. The B_- values were then used to get average values for the cation B_+ from the tabulated B_{MX}. The B_+ terms are also plotted in Figure 9.5 along with the lines fit to the B_+, and the anion and oxy-anion B_-'s. The fit resulted in the following values for the ions of interest:

$$B_{H^+} = .08012 \qquad B_{ClO^-} = -.08498$$

The estimated ion interaction term for HClO is therefore:

$$B_{HClO} = B_{H^+} + B_{ClO^-} = -.00486$$

The water activity for a single electrolyte ij in solution is calculated from Bromley's equation for the osmotic coefficient:

$$\phi_{ij} = 1. - 2.303 \left[\frac{A|z_i z_j|}{I} \left(1. + \sqrt{I} - \frac{1.}{1. + \sqrt{I}} - 2. \ln (1. + \sqrt{I}\,)\right) - \frac{(0.06 + .6 B_{ij})|z_i z_j|}{a} \left(\frac{1. + 2aI}{(1. + aI)^2} - \frac{\ln (1. + aI)}{aI}\right) - .5 B_{ij} I \right] \qquad (5.51)$$

Figure 9.5

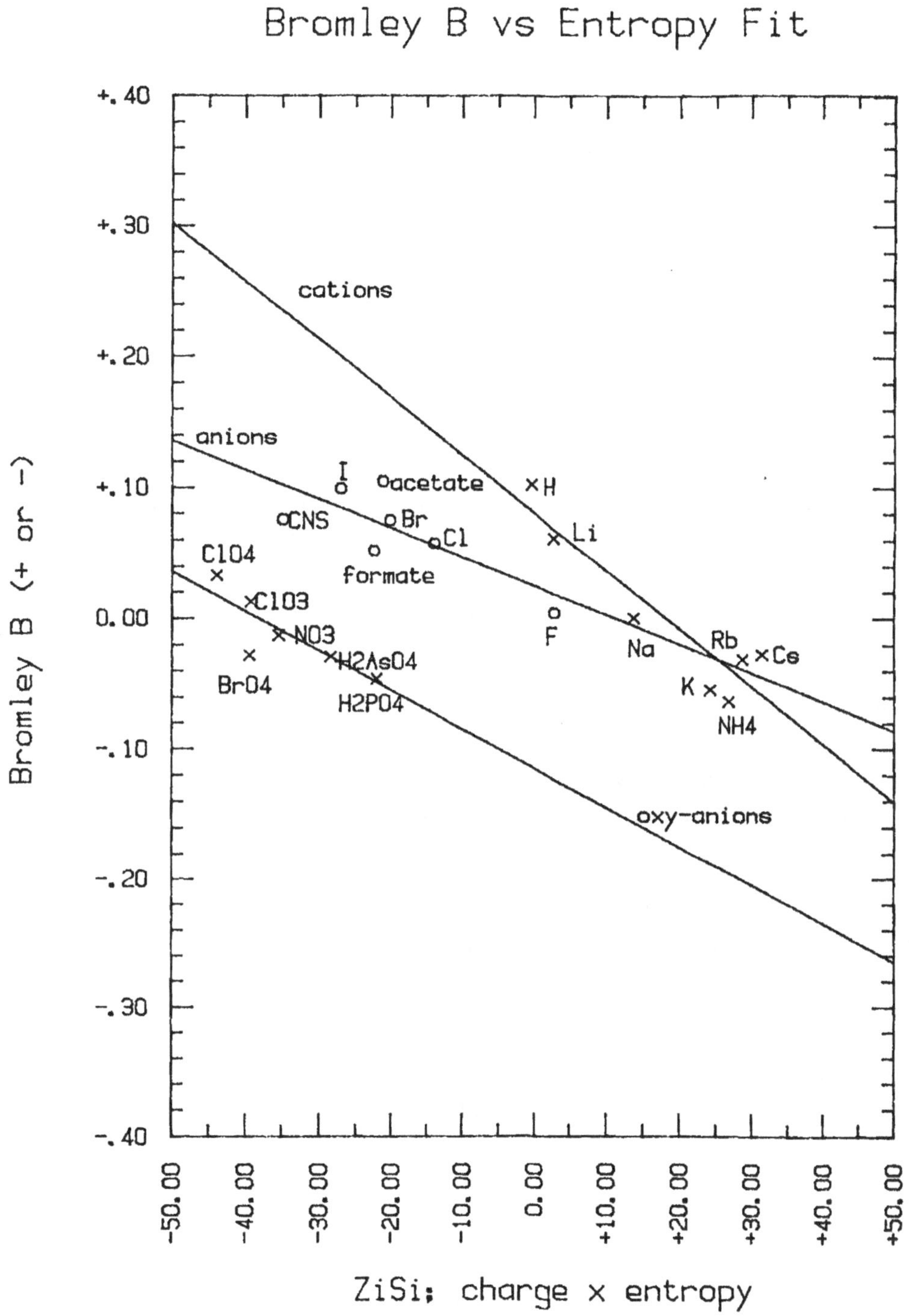
Bromley B vs Entropy Fit
+.40
+.30
+.20
+.10
0.00
-.10
-.20
-.30
-.40
cations
anions
oxy-anions
I
acetate
H
CNS
Br
Cl
Li
ClO4
formate
ClO3
F
Na
NO3
Rb
Cs
H2AsO4
BrO4
H2PO4
K
NH4
-50.00
-40.00
-30.00
-20.00
-10.00
0.00
+10.00
+20.00
+30.00
+40.00
+50.00
Bromley B (+ or -)
ZiSi; charge x entropy

$$\ln (a^\circ_w)_{ij} = -\frac{M_s \nu m}{1000.} \phi_{ij} \tag{2.30}$$

where:

ϕ_{ij} - is the osmotic coefficient for ion pair ij. For the chlorine-water system, the two ion pairs in the solution are H-Cl and H-ClO.

A - the Debye-Hückel constant

I - ionic strength

B_{ij} - Bromley interaction parameter for ion pair ij

z - ionic charge

a $= 1.5 / |z_i z_j|$

M_s - solvent molecular weight, 18.02 for water

m - molality of ion pair ij

ν - number of ions that ion pair ij dissociates into

After calculating the hypothetical binary solution water activities the mixed solution water activity can be calculated using the mixing rule suggested by Meissner (14):

$$\ln (a_w)_{mix} = \sum_i \sum_j W_{ij} \ln (a^\circ_w)_{ij} + r \tag{5.31}$$

For this solution:

r - the residue term = 0.

I_c - cationic strength = $.5\, m_{H^+}$

I_a - anionic strength = $.5\, (m_{OH^-} + m_{ClO^-} + m_{Cl^-})$

I_i - individual ionic strengths = $.5\, m_i z_i^2$

$$W_{ij} = \frac{I_i}{I_c} \frac{I_j}{I_a} \left[\frac{(z_i + z_j)^2}{z_i z_j} \right] \frac{I_c}{I} \frac{I_a}{I}$$

z_i - absolute value of the ionic charge

$$\ln (a_w)_{mix} = 4. \left[\frac{I_{H^+} I_{Cl^-}}{I^2} \right] \ln (a^\circ_w)_{HCl} + 4. \left[\frac{I_{H^+} I_{ClO^-}}{I^2} \right] \ln (a^\circ_w)_{HClO}$$

With these definitions of the system activity coefficients and the equilibrium constants calculated using the tabulated thermodynamic data and equations (3.31), the ten equations, (C.1) - (C.3) and (C.6) - (C.12), may be solved for the ten unknowns with the Newton-Raphson method.

Note that the concentration of $Cl_2(aq)$ is 0.0474 molal. This is the amount of chlorine that is presented by physical solubility only, i.e. Henry's law solubility. The inclusion of the hydrolysis reaction to form H^+, Cl^- and HClO(aq) (.0307 molal) has increased the calculated apparent solubility by about 65% (.0307/.0474). This hydrolysis "sink" for chlorine becomes more important at lower Cl_2 pressures. Perry's Handbook, in the third and fourth editions (17), gives the following equation for calculating chlorine solubility in water:

$$C = H' \; p + (K_e \; H' \; p)^{1/3} \quad \text{lb-moles } Cl_2/\text{cu. ft.}$$

where:
- p - chlorine partial pressure, atm
- H' - tabulated Henry's law coefficient, lb-moles Cl_2/(cu. ft.) (atm)
- K_e - tabulated equilibrium constant, (lb-moles/cu. ft.)^2, representing the hydrolysis equation: $Cl_2 + H_2O = HClO + H^+ + Cl^-$

This equation, from Whitney and Vivian (27), was recommended for use when the Cl_2 pressure was less than 1 atm. This is a good approximation when the dissociation of HClO(aq) is unimportant. Unfortunately the values for the equilibrium constant K_e were incorrectly converted from the Whitney and Vivian units to the Perry's Handbook units. The Henry's law coefficients and the correct values of K_e are:

temp Celsius	H' lb-moles Cl_2 /(cu. ft.) (atm)	K_e (lb-moles/cu. ft.)^2
10	0.00707	4.99E-7
15	.00584	8.84E-7
20	.00469	9.51E-7
25	.00390	13.6E-7

This equation was dropped in the fifth and sixth editions of Perry's Handbook, possibly due the the error in the conversion which could result in an error of 40 to 100 times the expected chlorine solubility.

Figure 9.6

SYSTEM: CHLORINE - WATER

METHOD: L.A. Bromley, AIChE J, v19, #2 (1973)

```
==========================================================================

Temperature:  25.00 degrees Celsius
Pressure:      0.80 atm
H2Oin:        55.51 moles
Cl2in:         0.90 moles

Liquid Phase Equilibrium Constants:
     K(H2O)  =  0.101073E-13 mole/kg
     K(Cl2)  =  0.450617E-03 mole/kg
     K(HClO) =  0.281834E-07 mole/kg

Vapor-Liquid Equilibrium Constants:
     K(H2O)  =  0.319794E+02 mole/kg
     K(Cl2)  =  0.617352E-01 mole/kg

Calculation Results:

     Liquid Phase:

       Ionic strength =  0.307892E-01

                      moles                        activity coefficient

       H2O         0.554459E+02                        0.998940E+00
       Hion        0.307536E-01                        0.855537E+00
       OHion       0.456365E-12                        0.838919E+00
       Cl2aq       0.474049E-01                        0.100000E+01
       HClO(aq)    0.307536E-01                        0.100000E+01
       ClOion      0.390908E-07                        0.841737E+00
       Clion       0.307536E-01                        0.855537E+00

     Vapor phase:

                 mole fraction                             moles

       yH2O        0.390462E-01                        0.333937E-01
       yCl2        0.960954E+00                        0.821841E+00

       Total vapor        0.85524 moles
```

Figure 9.7

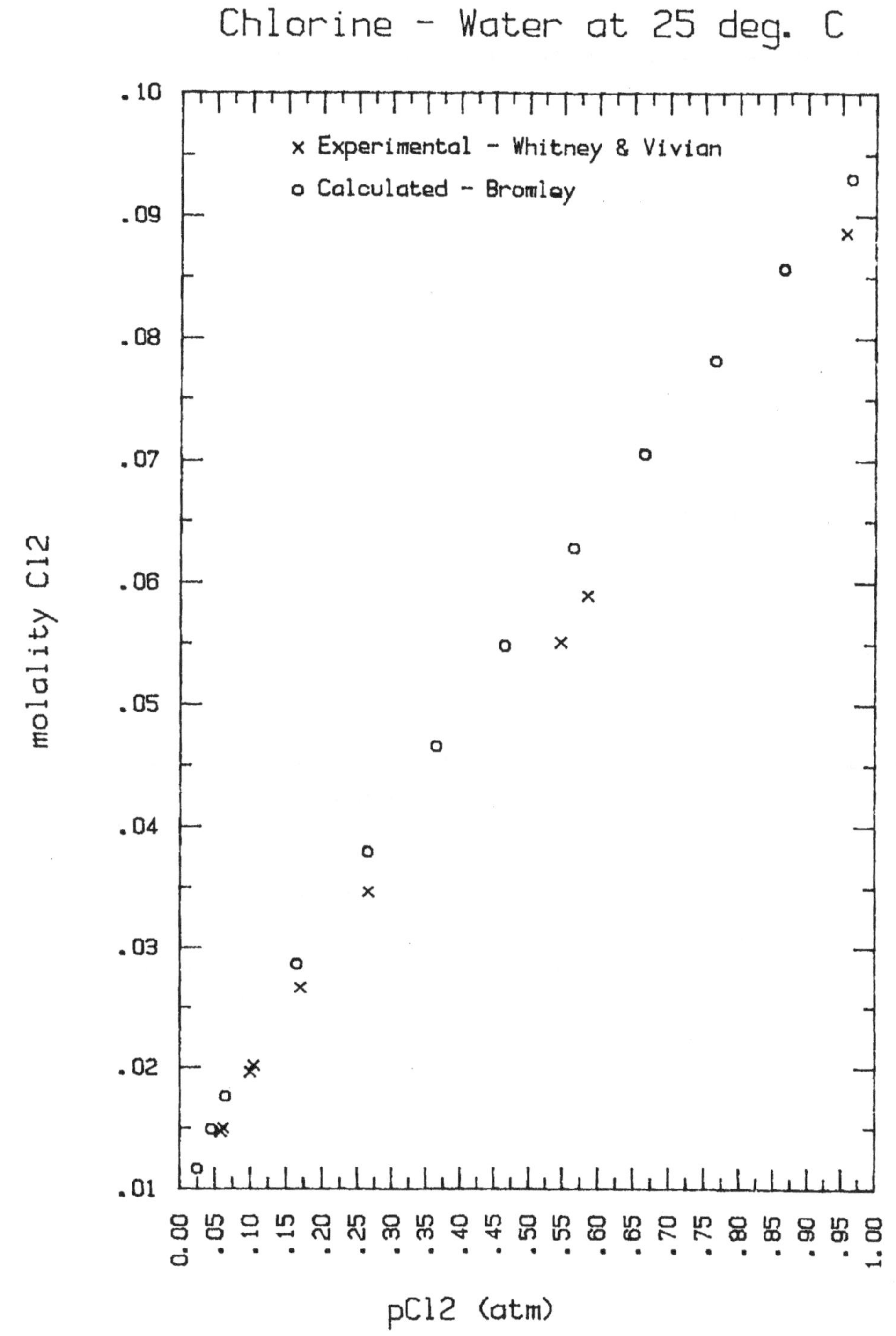
Chlorine - Water at 25 deg. C
x Experimental - Whitney & Vivian
o Calculated - Bromley
molality Cl2
.10
.09
.08
.07
.06
.05
.04
.03
.02
.01
0.00
.05
.10
.15
.20
.25
.30
.35
.40
.45
.50
.55
.60
.65
.70
.75
.80
.85
.90
.95
1.00
pCl2 (atm)

Water - Ammonia - Carbon Dioxide

The binary water-ammonia system contains the following single step equilibria:

vapor-liquid equilibria

$$NH_3(vap) = NH_3(aq) \quad (T.1)$$

$$H_2O(vap) = H_2O(aq) \quad (T.2)$$

water dissociation

$$H_2O = H^+ + OH^- \quad (T.3)$$

ammonia hydrolysis and dissociation

$$NH_3(aq) + H_2O = NH_4^+ + OH^- \quad (T.4)$$

In addition to the water dissociation (T.3) and vapor-liquid equilibrium (T.2), the binary water-carbon dioxide system contains:

vapor-liquid equilibrium

$$CO_2(vap) = CO_2(aq) \quad (T.5)$$

carbon dioxide dissociation

$$CO_2(aq) + H_2O = H^+ + HCO_3^- \quad (T.6)$$

bicarbonate dissociation

$$HCO_3^- = H^+ + CO_3^{2-} \quad (T.7)$$

For the ternary H_2O-NH_3-CO_2 system, all seven of the above equilibria are included. The formation of the carbamate ion must also be described:

$$NH_3 + HCO_3^- = NH_2CO_2^- + H_2O \quad (T.8)$$

The need to include the carbamate equilibria points up the fact that in order to model electrolyte systems one must identify all relevant species. Ignoring the possibility of solids formation (e.g. ammonium bicarbonate, ammonium carbamate, ammonium carbonate, ammonium sesquicarbonate), these eight equilibria define the water-ammonia-carbon dioxide system in terms of the twelve species concentrations:

$H_2O(aq)$, y_{H_2O}, m_{H^+}, m_{OH^-}, $m_{NH_3(aq)}$, y_{NH_3}, $m_{NH_4^+}$, $m_{CO_2(aq)}$, y_{CO_2}, $m_{HCO_3^-}$, $m_{CO_3^{2-}}$, and $m_{NH_2CO_2^-}$

These species concentrations and the total moles of vapor, V, are the system unknowns. Since there are thirteen unknowns in the system and only eight equilibria equations, (T.1) to (T.8), five more equations must be developed (13 - 8 = 5):

material balances:

nitrogen balance

$$NH_3(in) = \frac{H_2O(aq)}{55.51} (m_{NH_3(aq)} + m_{NH_4^+} + m_{NH_2CO_2^-}) + V * y_{NH_3} \quad (T.9)$$

carbon balance

$$CO_2(in) = \frac{H_2O(aq)}{55.51} * (m_{CO_2(aq)} + m_{HCO_3^-} + m_{CO_3^{2-}} + m_{NH_2CO_2^-}) + V * y_{CO_2} \quad (T.10)$$

hydrogen balance

$$2\ H_2O(in) + 3\ NH_3(in) = 2\ H_2O + \frac{H_2O(aq)}{55.51} * (m_{H^+} + m_{OH^-} + 4\ m_{NH_4^+} + 3\ m_{NH_3(aq)} + m_{HCO_3^-} + 2\ m_{NH_2CO_2^-}) + V * (3\ y_{NH_3} + 2\ y_{H_2O}) \quad (T.11)$$

electroneutrality equation:

$$m_{NH_4^+} + m_{H^+} = m_{OH^-} + m_{HCO_3^-} + 2\ m_{CO_3^{2-}} + m_{NH_2CO_2^-} \quad (T.12)$$

vapor mole fraction balance:

$$y_{H_2O} + y_{NH_3} + y_{CO_2} = 1. \quad (T.13)$$

Values for the equilibrium constants for reactions (T.3), (T.4), (T.6), (T.7) and (T.8) were calculated with a temperature fit equation by Edwards et al. (11):

$$\ln K = A/T + B \ln T + C\ T + D$$

The Henry's constants for the vapor-liquid equilibria (T.1), (T.2) and (T.5) were calculated using a similar equation. The parameters for these equations and the activity coefficient interaction parameters used by Edwards et al. are tabulated and plotted in Appendix 9.2.

On the following pages are program listings for the ammonia-carbon dioxide-water system using the Edwards et al. modified Pitzer equation to calculate activity coefficients, and the Nakamura et al. (16) method of calculating fugacity coefficients as outlined in Appendix 9.3. The liquid phase unknowns are molalities, except for water which as always, remains in moles. The vapor phase unknowns are the vapor mole fractions and total moles of vapor. The system model equations in subroutine FUNC reflect these units.

The code is heavily commented with frequent references to equations found within the text of this book. For example, the hydrogen balance just discussed is noted to be equation (T.11).

```
C       *******************************************************************
C
C       PROGRAM SOUR
C       ============
C
C
C       FOR THE AMMONIA-CARBON DIOXIDE-WATER SYSTEM USING THE EDWARDS,
C       MAURER, NEWMAN AND PRAUSNITZ PITZER ACTIVITY COEFFICIENT
C       CALCULATION METHOD AND THE NAKAMURA ET AL. METHOD TO CALCULATE
C       FUGACITY COEFFICIENTS.  THE SYSTEM HAS THIRTEEN UNKNOWNS:
C       MOLES OF WATER, SIX IONIC MOLALITIES, TWO AQUEOUS MOLECULE
C       MOLALITIES, THREE VAPOR MOLE FRACTIONS AND TOTAL MOLES OF VAPOR.
C
C
C       VAX VMS OPERATING SYSTEM
C
C
C       SUBROUTINES:
C       ------------
C
C          ACTCOF - CALCULATES THE ACTIVITY COEFFICIENTS
C
C          BETA   - CREATES THE ARRAYS CONTAINING THE INTERACTION
C                   PARAMETERS NEEDED FOR ACTIVITY COEFFICIENT
C                   CALCULATIONS
C
C          FUNC   - CONTAINS THE SYSTEM MODEL EQUATIONS.
C
C          HENRY  - CALCULATES THE HENRY'S CONSTANTS
C
C          KCAL   - CALCULATES THE EQUILIBRIUM CONSTANTS AND THE
C                   WATER SATURATION PRESSURE
C
C          NAKA   - CALCULATES THE FUGACITY COEFFICIENTS
C
C          PMOLV  - CALCULATES PARTIAL MOLAL VOLUMES
C
C          SETNAK - CREATES THE PARAMETER ARRAYS FOR FUGACITY
C                   COEFFICIENT CALCULATIONS
C
C          VSOLVE - SOLVES FOR THE MOLAR VOLUME FOR FUGACITY
C                   COEFFICIENT CALCULATIONS
C
C
C       SET DATA TYPES:
C       ---------------
C
         IMPLICIT REAL*8 (A-H,O-Z)
C
         REAL*8 MNH3,MNH4,MCO2,MHCO3,MCO3,NH2CO2,MH,MOH
C
         CHARACTER*12 DAT
C
         EXTERNAL FUNC
C
C
```

```
C
C
C
C     COMMON BLOCKS CONTAIN:
C     ----------------------
C
C
C     - THE ACTIVITY COEFFICIENTS AND IONIC STRENGTH
C
        COMMON /ACTIV/AW,AH,AOH,ANH3,ANH4,ACO2,AHCO3,ACO3,ANH2CO,SI
C
C
C     - PARAMETERS FOR CALCULATING THE NAKAMURA FUGACITY COEFFICIENTS
C
        COMMON /CASAP/ANA(3,3),CNA(3),ALNA(3),BENA(3),GANA(3),DENA(3),
     +              ALPHA0(3),ALPHA1(3),BETA0(3),BETA1(3)
C
C
C     - DEBYE-HUCKEL CONSTANT, 5 EQUILIBRIUM CONSTANTS, SATURATION
C       PRESSURE OF WATER AND 3 PARTIAL MOLAR VOLUMES
C
        COMMON /EQCONS/DHCON,EKH2O,EKNH3,EKCO2,EKHCO3,EKCARB,PWS,VMH2O,
     +               VMNH3,VMCO2
C
C
C     - FUGACITY COEFFICIENTS
C
        COMMON /FUGS/FH2O,FNH3,FCO2,FPH2O
C
C
C     - HENRY'S CONSTANTS
C
        COMMON /HENCO/HNH3,HCO2
C
C
C     - CELSIUS TEMPERATURE, PRESSURE IN ATM., AND INPUT MOLES OF
C       EACH SPECIES
C
        COMMON /INPUT/TEMP,PT,TNH3,TH2O,TCO2
C
C
C     - THE UNKNOWNS TO BE SOLVED FOR:  MOLES OF H2O, 8 LIQUID PHASE
C       MOLALITIES, 3 VAPOR PHASE MOLE FRACTIONS AND TOTAL MOLES OF
C       VAPOR
C
        COMMON /NEWRAP/H2O,MH,MOH,MNH3,MNH4,MCO2,MHCO3,MCO3,NH2CO2,
     +               YH2O,YNH3,YCO2,V
C
C
C     - PARAMETERS FOR ACTIVITY COEFFICIENT CALCULATIONS
C
        COMMON /PARAMS/B0(20),B1(20)
C
C
C
C
```

```
C
C
C       WRITE OUTPUT PAGE HEADER
C       ------------------------
C
        CALL DATE(DAT)
        WRITE(6,3) DAT
      3 FORMAT('1'//'        SYSTEM: AMMONIA - CARBON DIOXIDE - WATER'//
     +         5X,'METHOD: Edwards, Maurer, Newman & Prausnitz, AIChE',
     +         ' J (1978)'//5X,70('=')//'        DATE: ',A//)
C
C
C
C
C       SET INPUT DATA
C       --------------
C
C       - TEMPERATURE, DEGREES CELSIUS
C
          TEMP=60.
C
C       - DEBYE-HUCKEL CONSTANT FROM LEWIS & RANDALL, CONVERTED FROM
C         ACTVITY COEFFICIENT LOG 10 BASIS TO THE OSMOTIC COEFFICIENT
C         NATURAL LOG BASED ONE USED BY PITZER'S EQUATIONS
C
          DH=.545
          DHCON=2.303*DH/3.
C
C       - MOLES OF WATER INPUT
C
          TH2O=55.51
C
C       - MOLES OF AMMONIA INPUT
C
          TNH3=9.
C
C       - MOLES OF CARBON DIOXIDE INPUT
C
          TCO2=7.
C
C       - TOTAL SYSTEM PRESSURE, ATMOSPHERES
C
          PT=1.
C
C
C       OUTPUT
C
        WRITE(6,8) TEMP,PT,TH2O,TNH3,TCO2
      8 FORMAT('      Temperature: ',F5.2,' degrees Celsius'/
     +        '      Pressure:    ',F5.2,' atm'/
     +        '      H2Oin:       ',F5.2,' moles'/
     +        '      NH3in:       ',F5.2,' moles'/
     +        '      CO2in:       ',F5.2,' moles'//)
C
C
C
```

```
C
C
C     SET UP TEMPERATURE DEPENDENT CONSTANTS AND PARAMETER ARRAYS
C     -----------------------------------------------------------
C
C
C     - CALCULATE THE TEMPERATURE DEPENDENT EQUILIBRIUM CONSTANTS AND
C       THE SATURATION PRESSURE OF WATER
C
      CALL KCAL(TEMP)
      WRITE(6,20) EKH2O,EKNH3,EKCO2,EKHCO3,EKCARB,PWS
   20 FORMAT('     Equilibrium Constants:'/
     +        10X,'K(H2O)    = ',E13.6,' mole/kg'/
     +        10X,'K(NH3)    = ',E13.6,' mole/kg'/
     +        10X,'K(CO2)    = ',E13.6,' mole/kg'/
     +        10X,'K(HCO3)   = ',E13.6,' mole/kg'/
     +        10X,'K(NH2CO2) = ',E13.6,' mole/kg'//
     +        5X,'Saturation Pressure of Water: ',E13.6,' atm'/)
C
C
C     - CALCULATE THE HENRY'S CONSTANTS FOR THIS TEMPERATURE
C
      CALL HENRY(TEMP)
      WRITE(6,22) HNH3,HCO2
   22 FORMAT('     Henry''s Constants:'/
     +        10X,'H(NH3)  = ',E13.6,' atm kg/mole'/
     +        10X,'H(CO2)  = ',E13.6,' atm kg/mole'/)
C
C
C     - SET UP THE ARRAYS CONTAINING THE BETA0 AND BETA1 ION-ION,
C       ION-MOLECULE AND MOLECULE-MOLECULE SELF INTERACTION PARAMETERS
C       FOR ACTIVITY COEFFICIENT CALCULATIONS
C
      CALL BETA(TEMP)
C
C
C     - SET UP THE ARRAYS CONTAINING THE PARAMETERS FOR THE NAKAMURA
C       FUGACITY COEFFICIENT CALCULATIONS
C
      CALL SETNAK
C
C
C     - CALCULATE THE PARTIAL MOLAR VOLUMES:
C
C       - OF WATER
C
        CALL PMOLV(3,TEMP,VMH2O)
C
C       - OF AMMONIA
C
        CALL PMOLV(1,TEMP,VMNH3)
C
C       - OF CARBON DIOXIDE
C
        CALL PMOLV(2,TEMP,VMCO2)
C
```

```
      WRITE(6,30) VMH2O,VMNH3,VMCO2
 30   FORMAT(5X,'Partial Molar Volumes:'/
     +        10X,'v(H2O)  = ',E13.6,' cm^3/mole'/
     +        10X,'v(NH3)  = ',E13.6,' cm^3/mole'/
     +        10X,'v(CO2)  = ',E13.6,' cm^3/mole'//
     +        5X,'Calculation Results:'/)
C
C
C     SOLVE THE SYSTEM
C     ----------------
C
C     - INITIALIZE THE ACTIVITY COEFFICIENTS
C
      AH2O=1.
      AH=1.
      ANH4=1.
      AOH=1.
      ANH3=1.
      ACO3=1.
      AHCO3=1.
      ACO2=1.
      ANH2CO=1.0
C
C     - SET INITIAL GUESSES FOR THE VAPOR MOLE FRACTIONS.
C
      PH2O=.2*PT
      PNH3=.1*PT
      YH2O=PH2O/PT
      YNH3=PNH3/PT
      YCO2=1.-YNH3-YH2O
C
C     - SET AN INITIAL GUESS FOR THE VAPOR RATE
C
      V=3.
C
C     - SET INTIAL GUESSES FOR THE LIQUID COMPOSITIONS
C
      H2O=TH2O
      MNH4=.95*TCO2
      MHCO3=.5*TCO2
      NH2CO2=.2*TCO2
      MNH3=DABS(.4*(TNH3-TCO2))
      MH=.8E-8
      MOH=EKH2O/MH
      MCO2=MH*MHCO3/EKCO2
      MCO3=(MNH4+MH-MOH-MHCO3-NH2CO2)/2.
C
C     - CALL THE NEWTON-RAPHSON METHOD TO SOLVE FOR THE 13 UNKNOWNS.
C       THE SYSTEM MODEL EQUATIONS ARE IN SUBROUTINE FUNC.  THE
C       SUBROUTINE NEWRAP IS NOT INCLUDED IN THIS LISTING.  THE
C       METHOD IS WELL DOCUMENTED IN TEXTBOOKS AND OTHER EQUATION
C       SOLVERS MAY BE USED
C
      M=13
      CALL NEWRAP(M,FUNC)
C
```

```
C
C
C
C      OUTPUT THE RESULTS
C      ------------------
C
       WRITE(6,110) SI,H2O,AW,MH,AH,MOH,AOH,MNH3,ANH3,MNH4,ANH4,
      +             MCO2,ACO2,MCO3,ACO3,MHCO3,AHCO3,NH2CO2,ANH2CO
  110 FORMAT(7X,'Liquid Phase Molalities and Activity Coefficients:'//
      +       10X,'Ionic strength = ',E13.6,' molal'/
      +       12X,'H2O       ',E13.6,12X,'AW        ',E13.6/
      +       12X,'Hion      ',E13.6,12X,'AH        ',E13.6/
      +       12X,'OHion     ',E13.6,12X,'AOH       ',E13.6/
      +       12X,'NH3aq     ',E13.6,12X,'ANH3      ',E13.6/
      +       12X,'NH4ion    ',E13.6,12X,'ANH4      ',E13.6/
      +       12X,'CO2aq     ',E13.6,12X,'ACO2      ',E13.6/
      +       12X,'CO3ion    ',E13.6,12X,'ACO3      ',E13.6/
      +       12X,'HCO3ion   ',E13.6,12X,'AHCO3     ',E13.6/
      +       12X,'NH2CO2ion ',E13.6,12X,'ANH2CO2   ',E13.6/)
C
C
       WRITE(6,115) YH2O,FH2O,YNH3,FNH3,YCO2,FCO2,V
  115 FORMAT(7X,'Vapor Phase Mole Fractions and Fugacity Coefficients:'/
      +       12X,'yH2O      ',E13.6,12X,'fH2O      ',E13.6/
      +       12X,'yNH3      ',E13.6,12X,'fNH3      ',E13.6/
      +       12X,'yCO2      ',E13.6,12X,'fCO2      ',E13.6//
      +       12X,'Total vapor ',F13.5,' moles'/'1')
C
C
C
       STOP
       END
C
C      *******************************************************************
```

```
C       ******************************************************************
C
        SUBROUTINE ACTCOF
C       =================
C
C
C       THIS SUBROUTINE CALCULATES THE ACTIVITY COEFFICIENTS USING THE
C       ADAPTATION OF PITZER'S EQUATIONS GIVEN BY EDWARDS ET AL.  THE
C       METHOD IS DESCRIBED IN CHAPTER VII.
C
C
C       CALLED BY: FUNC
C
C
C
        IMPLICIT REAL*8 (A-H,O-Z)
C
        REAL*8 MNH3,MNH4,MCO2,MHCO3,MCO3,NH2CO2,MH,MOH
C
C
        COMMON /ACTIV/AW,AH,AOH,ANH3,ANH4,ACO2,AHCO3,ACO3,ANH2CO,SI
C
        COMMON /EQCONS/DHCON,EK(5),PWS,VMOL(3)
C
        COMMON /NEWRAP/H2O,MH,MOH,MNH3,MNH4,MCO2,MHCO3,MCO3,NH2CO2,
       +                Y(3),V
C
        COMMON /PARAMS/B0(20),B1(20)
C
C
        DIMENSION BAW(20),BGAM(20),BPGAM(20)
C
C
C
C       CALCULATE THE IONIC STRENGTH AND ITS SQUARE ROOT
C       ------------------------------------------------
C
        SI=.5*(MH+MOH+MNH4+MHCO3+4.*MCO3+NH2CO2)
        DSI=DSQRT(SI)
C
C
C       SET UP TERMS USED FREQUENTLY IN ACTIVITY COEFFICIENT CALCULATIONS
C       -----------------------------------------------------------------
C
C
C       DEBYE-HUCKEL EXPRESSION
C
C
```

$$f^{\gamma} = -A_{\phi}\left[\frac{\sqrt{I}}{1. + 1.2\sqrt{I}} + \frac{2}{1.2}\ \ln\,(1. + 1.2\sqrt{I})\right]$$

```
C
C
          FAC=-DHCON*(DSI/(1.+1.2*DSI)+(2./1.2)*DLOG(1.+1.2*DSI))
C
C
```

```
      BPSUM=0.0
      DO 20 I=1,20
C
C                   β1
C     B   = β0 +   --   { 1. - (1. + 2. √I ) exp(-2. √I ) }
C      ij          2I
C
      BGAM(I)=B0(I)+B1(I)*(1.-(1.+2.*DSI)*DEXP(-2.*DSI))/(2.*SI)
C
C              β1
C     B'  =   ----    { -1. + (1. + 2. √I + 2. I) exp(-2. √I ) }
C      ij     4 I²
C
      BPGAM(I)=B1(I)*(-1.+(1.+2.*DSI+2.*SI)*DEXP(-2.*DSI))/(4.*SI**2)
C
C     THE FOLLOWING TERM IS DEFINED FOR WATER ACTIVITY CALCULATIONS
C
C
C     B^φ = β0 + β1 exp( -2. √I )
C      ij
C
      BAW(I)=B0(I)+B1(I)*DEXP(-2.*DSI)
   20 CONTINUE
C
C     BPSUM =     Σ          Σ        m  m  B'
C              j≠H2O      k≠H2O       j  k   jk
C
      BPSUM=2.*(MH*(MOH*BPGAM(1)+MHCO3*BPGAM(2)+MCO3*BPGAM(3)+
     +      NH2CO2*BPGAM(4))+MNH4*(MOH*BPGAM(5)+MHCO3*BPGAM(6)+
     +      MCO3*BPGAM(7)+NH2CO2*BPGAM(8))+MNH3*(MH*BPGAM(9)+
     +      MOH*BPGAM(10)+MNH4*BPGAM(11)+MHCO3*BPGAM(12)+
     +      MCO3*BPGAM(13)+MNH3*BPGAM(18))+MCO2*(MH*BPGAM(14)+
     +      MOH*BPGAM(15)+MNH4*BPGAM(16)+NH2CO2*BPGAM(17)+
     +      MCO2*BPGAM(19)+MNH3*BPGAM(20)))
C
C
C
C    THE GENERAL EXPRESSION OF AN IONIC ACTIVITY COEFFICIENT IN A
C    MULTICOMPONENT SOLUTION IS SEEN IN EQUATION (7.23):
C
C
C    ln γ  = z² f^γ + 2   Σ    m  B   + z²   Σ       Σ   m  m  B'
C        i    i         j≠H2O  j   ij    i  j≠H2O  k≠H2O j  k   jk
C
C
C    CALCULATE THE IONIC ACTIVITY COEFFICIENTS FOR:
C
C
C    - H ion
C
C
C     ln γ   = f^γ + 2.    Σ     m  B     + BPSUM
C         H              j≠H2O   j   H -j
C
      AH=DEXP(FAC+2.*(MOH*BGAM(1)+MHCO3*BGAM(2)+MCO3*BGAM(3)+
     +   NH2CO2*BGAM(4)+MNH3*BGAM(9)+MCO2*BGAM(14))+BPSUM)
```

```
C
C      - OH ion
C
C
C        ln γOH- = f^γ + 2.   Σ     mj Bj-OH- + BPSUM
C                           j≠H2O
C
       AOH=DEXP(FAC+2.*(MH*BGAM(1)+MNH4*BGAM(5)+MNH3*BGAM(10)+
     +     MCO2*BGAM(15))+BPSUM)
C
C      - NH4 ion
C
C
C        ln γNH4+  = f^γ + 2.    Σ     mj BNH4+-j + BPSUM
C                              j≠H2O
C
       ANH4=DEXP(FAC+2.*(MOH*BGAM(5)+MHCO3*BGAM(6)+MCO3*BGAM(7)+
     +     NH2CO2*BGAM(8)+MNH3*BGAM(11)+MCO2*BGAM(16))+BPSUM)
C
C      - HCO3 ion
C
C
C        ln γHCO3- = f^γ + 2.    Σ     mj Bj-HCO3- + BPSUM
C                              j≠H2O
C
       AHCO3=DEXP(FAC+2.*(MH*BGAM(2)+MNH4*BGAM(6)+MNH3*BGAM(12))+
     +      BPSUM)
C
C      - CO3 ion
C
C
C        ln γCO3-- = 2² f^γ + 2   Σ    mj Bj-CO3-- + 2² BPSUM
C                              j≠H2O
C
       ACO3=DEXP(4.*FAC+2.*(MH*BGAM(3)+MNH4*BGAM(7)+MNH3*BGAM(13))+
     +     4.*BPSUM)
C
C      - NH2CO2 ion
C
C
C        ln γNH2CO2- = f^γ + 2.   Σ     mj Bj-NH2CO2- + BPSUM
C                               j≠H2O
C
       ANH2CO=DEXP(FAC+2.*(MH*BGAM(4)+MNH4*BGAM(8)+MCO2*BGAM(17))+
C    +      BPSUM)
C
C      CALCULATION OF AQUEOUS MOLECULE ACTIVITY COEFFICIENTS:
C      -------------------------------------------------------
C
C      THE GENERAL EQUATION FOR THE ACTIVITY COEFFICIENT OF AN AQUEOUS
C      MOLECULE IN A MULTICOMPONENT SOLUTION IS EQUATION (7.24):
C
C      ln γm = 2.   Σ    mj Bmj
C                 j≠H2O
C
```

```
C
C     - NH3
C
C       ln γNH3(aq) = 2.    Σ     mj BNH3(aq)-j
C                        j≠H2O
C
C
      ANH3=DEXP(2.*(MH*BGAM(9)+MOH*BGAM(10)+MNH4*BGAM(11)+MHCO3*
     +        BGAM(12)+MCO3*BGAM(13)+MNH3*BGAM(18)+MCO2*BGAM(20)))
C
C     - CO2
C
C       ln γCO2(aq) = 2.    Σ     mj BCO2(aq)-j
C                        j≠H2O
C
C
      ACO2=DEXP(2.*(MH*BGAM(14)+MOH*BGAM(15)+MNH4*BGAM(16)+
     +        NH2CO2*BGAM(17)+MCO2*BGAM(19)+MNH3*BGAM(20)))
C
C
C     CALCULATE THE WATER ACTIVITY USING EQUATION (7.25):
C     ---------------------------------------------------
C
C
C                  [ 2.   Aφ  I^3/2                               ]
C     ln aw = Mw   [ -------------    -    Σ        Σ     mi mj Bφij ] -
C                  [ 1. + 1.2 √I        i≠H2O    j≠H2O              ]
C
C
C             Mw     Σ     mi
C                 i≠H2O
C
C
      WM=.01802
      AW=DEXP(WM*(2.*DHCON*SI**1.5/(1.+1.2*DSI)-2.*(MH*(MOH*BAW(1)+
     +   MHCO3*BAW(2)+MCO3*BAW(3)+NH2CO2*BAW(4))+MNH4*(MOH*BAW(5)+
     +   MHCO3*BAW(6)+MCO3*BAW(7)+NH2CO2*BAW(8))+MNH3*(MH*BAW(9)+
     +   MOH*BAW(10)+MNH4*BAW(11)+MHCO3*BAW(12)+MCO3*BAW(13))+
     +   MCO2*(MH*BAW(14)+MOH*BAW(15)+MNH4*BAW(16)+NH2CO2*BAW(17)))-
     +   2.*(BAW(18)*MNH3**2+BAW(19)*MCO2**2+BAW(20)*MCO2*MNH3))-
     +   WM*(MH+MOH+MHCO3+MCO3+MCO2+MNH3+MNH4+NH2CO2))
C
      RETURN
      END
C
C     *******************************************************************
```

```
C     ******************************************************************
C
      SUBROUTINE BETA(TEMP)
C     =====================
C
C
C     TO SET UP THE ARRAYS CONTAINING THE INTERACTION TERMS FOR THE
C     ACTIVITY COEFFICIENT CALCULATIONS.  VALUES FOR THE SYSTEM, AND
C     OTHER SPECIES, ARE TABULATED IN APPENDIX 9.2, TABLES 3, 4 AND 5.
C
C
C     INPUT:
C       TEMP - TEMPERATURE IN DEGREES CELSIUS
C
C     OUTPUT:
C       B0, B1 - PITZER β0 AND β1 INTERACTION PARAMETER ARRAYS;
C                VALUES FROM EDWARDS ET AL
C
C
      IMPLICIT REAL*8 (A-H,O-Z)
C
C
      COMMON /PARAMS/B0(20),B1(20)
C
C
C
C     CONVERT THE TEMPERATURE TO KELVINS
C     ----------------------------------
C
      TK=TEMP+273.15
C
C
C     SET THE PARAMETERS FOR:
C     =======================
C
C     - THE ION-ION β0 INTERACTIONS
C       ---------------------------
C             (1) H-OH
C             (2) H-HCO3
C             (3) H-CO3
C             (4) H-NH2CO2
C             (5) NH4-OH
C             (6) NH4-HCO3
C             (7) NH4-CO3
C             (8) NH4-NH2CO2
C
        B0(1)=.208
        B0(2)=.071
        B0(3)=.086
        B0(4)=.198
        B0(5)=.06
        B0(6)=-.0435
        B0(7)=-.062
        B0(8)=.0505
C
C
```

```
C       - THE ION-ION β1'S
C         ----------------
C
          DO 5 I=1,8
          B1(I)=.018+3.06*B0(I)
    5     CONTINUE
          B1(7)=0.0
C
C       - THE AMMONIA MOLECULE-ION INTERACTIONS
C         -------------------------------------
C                (9)  NH3-H
C                (10) NH3-OH
C                (11) NH3-NH4
C                (12) NH3-HCO3
C                (13) NH3-CO3
C
          B0(9)=.015
          B0(10)=.227-1.47D-3*TK+2.6D-6*TK**2
          B0(11)=.0117
          B0(12)=-.0816
          B0(13)=.068
C
C       - THE CARBON DIOXIDE MOLECULE-ION INTERACTIONS
C         --------------------------------------------
C                (14) CO2-H
C                (15) CO2-OH
C                (16) CO2-NH4
C                (17) CO2-NH2CO2
C
          B0(14)=.033
          B0(15)=.26-1.62D-3*TK+2.89D-6*TK**2
          B0(16)=.037-2.38D-4*TK+3.83D-7*TK**2
          B0(17)=.017
C
C       - THE MOLECULE-MOLECULE INTERACTIONS
C         ----------------------------------
C                (18) NH3-NH3
C                (19) CO2-CO2
C                (20) NH3-CO2
C
          B0(18)=-.026+12.29/TK
          B0(19)=-.4922+149.2/TK
          B0(20)=.5*(B0(18)+B0(19))
C
C       - FOR THE ION-MOLECULE AND MOLECULE-MOLECULE β1'S
C         -----------------------------------------------
C
          DO 10 I=9,20
          B1(I)=0.0
   10     CONTINUE
          B1(11)=-.02
          B1(12)=.4829
        RETURN
        END
C
C       *****************************************************************
```

```
C      ******************************************************************
C
       SUBROUTINE FUNC(F)
C      ==================
C
C
C      SYSTEM EQUATION MODEL FOR SOLVING FOR THE UNKNOWNS VIA THE
C      NEWTON-RAPHSON METHOD.  THE METHOD STRIVES TO MAKE THE EQUATION,
C      OR FUNCTION, VALUES EQUAL ZERO.
C
C
C      OUTPUT:
C        F - FUNCTION VALUES FOR THE CURRENT VALUES OF THE UNKNOWNS.
C            THESE VALUES ARE PASSED TO THIS SUBROUTINE IN THE
C            NEWRAP COMMON BLOCK.
C
C
C      CALLS: ACTCOF, NAKA
C
C      CALLED BY: NEWRAP
C
C
       IMPLICIT REAL*8 (A-H,O-Z)
C
       REAL*8 MNH3,MNH4,MCO2,MHCO3,MCO3,NH2CO2,MH,MOH,NH3IN
C
C
       COMMON /ACTIV/AW,AH,AOH,ANH3,ANH4,ACO2,AHCO3,ACO3,ANH2CO,SI
C
       COMMON /EQCONS/ALPHA,EKH2O,EKNH3,EKCO2,EKHCO3,EKCARB,PWS,
      +               VMH2O,VMNH3,VMCO2
C
       COMMON /FUGS/FH2O,FNH3,FCO2,FPH2O
C
       COMMON /HENCO/HNH3,HCO2
C
       COMMON /INPUT/T,PT,NH3IN,H2OIN,CO2IN
C
       COMMON /NEWRAP/H2O,MH,MOH,MNH3,MNH4,MCO2,MHCO3,MCO3,NH2CO2,
      +               YH2O,YNH3,YCO2,V
C
C
       DIMENSION F(13),Y(3)
C
C
C      CALCULATE THE ACTIVITY COEFFICIENTS
C      -----------------------------------
C
       CALL ACTCOF
C
C
C
C      CONVERT THE CELSIUS TEMPERATURE TO KELVINS
C
       TK=T+273.15
C
```

```
C
C      SET GAS CONSTANT
C
       R=82.06
C
C
C      CALCULATE THE FUGACITY COEFFICIENTS:
C      -----------------------------------
C
C        - FOR PURE WATER
C
         NVFUG=1
         Y(1)=1.0
         Y(2)=0.0
         Y(3)=0.0
         CALL NAKA(TK,PWS,Y,NVFUG,FPH2O,FDUM,FDUM)
C
C        - FOR THE MIXED SOLUTION
C
         NVFUG=3
         Y(1)=YH2O
         Y(2)=YNH3
         Y(3)=YCO2
         CALL NAKA(TK,PT,Y,NVFUG,FH2O,FNH3,FCO2)
C
C
C
C      FUNCTION CALCULATIONS
C      ---------------------
C
C      CALCULATE THE FUNCTION VALUES FOR THE CURRENT VALUES OF THE
C      UNKNOWNS.  THE EQUATION NUMBERS REFER TO THE GENERAL MODEL
C      EQUATIONS FOR THE SYSTEM GIVEN IN CHAPTER IX.
C
C      THE LIQUID PHASE EQUILIBRIA:
C      ---------------------------
C
C      - THE WATER DISSOCIATION EQUATION, EQUATION (T.3)
C
        F(1)=1.-AH*MH*AOH*MOH/(EKH2O*AW)
C
C
C      - THE AMMONIA DISSOCIATION, EQUATION (T.4)
C
        F(2)=1.-AOH*MOH*ANH4*MNH4/(EKNH3*ANH3*MNH3*AW)
C
C
C      - THE CARBON DIOXIDE HYDROLYSIS, EQUATION (T.6)
C
        F(3)=1.-AH*MH*AHCO3*MHCO3/(EKCO2*ACO2*MCO2*AW)
C
C
C      - THE BICARBONATE DISSOCIATION, EQUATION (T.7)
C
        F(4)=1.-AH*MH*ACO3*MCO3/(EKHCO3*AHCO3*MHCO3)
C
```

```
C       - THE CARBAMATE FORMATION, EQUATION (T.8)
C
          F(5)=1.-ANH2CO*NH2CO2*AW/(EKCARB*ANH3*MNH3*AHCO3*MHCO3)
C
C
C       VAPOR-LIQUID EQUILIBRIA:
C       ------------------------
C
C       - FOR WATER, EQUATION (T.2)
C
          F(6)=YH2O*FH2O*PT-AW*FPH2O*PWS*DEXP(VMH2O*(PT-PWS)/(R*TK))
C
C
C       - FOR AMMONIA, EQUATION (T.1)
C
          F(7)=YNH3*FNH3*PT-MNH3*ANH3*HNH3*DEXP(VMNH3*(PT-PWS)/(R*TK))
C
C
C       - FOR CARBON DIOXIDE, EQUATION (T.5)
C
          F(8)=YCO2*FCO2*PT-MCO2*ACO2*HCO2*DEXP(VMCO2*(PT-PWS)/(R*TK))
C
C
C       ELEMENTAL BALANCES:
C       -------------------
C
C       - NITROGEN BALANCE, EQUATION (T.9)
C
          F(9)=(MNH3+MNH4+NH2CO2)*H2O/55.51+YNH3*V-NH3IN
C
C
C       - CARBON BALANCE, EQUATION (T.10)
C
          F(10)=(MCO2+MHCO3+MCO3+NH2CO2)*H2O/55.51+YCO2*V-CO2IN
C
C
C       - HYDROGEN BALANCE, EQUATION (T.11)
C
          F(11)=2.*H2O+(MH+MOH+4.*MNH4+3.*MNH3+MHCO3+2.*NH2CO2)*H2O/55.51+
     +          3.*YNH3*V+2.*YH2O*V-2.*H2OIN-3.*NH3IN
C
C
C       ELECTRONEUTRALITY, EQUATION (T.12)
C       ----------------------------------
C
        F(12)=MNH4+MH-(MOH+MHCO3+2.*MCO3+NH2CO2)
C
C       VAPOR BALANCE, EQUATION (T.13)
C       ------------------------------
C
        F(13)=YH2O+YNH3+YCO2-1.
C
        RETURN
        END
C
C       ******************************************************************
```

```
C      ******************************************************************
C
       SUBROUTINE HENRY(TEMP)
C      ======================
C
C
C      CALCULATE THE HENRY'S CONSTANTS FOR AMMONIA AND CARBON DIOXIDE
C      USING THE TEMPERATURE FITS GIVEN BY EDWARDS ET AL.:
C
C         ln H  = A/T  +  B ln T  +  C T  +  D
C
C      THE PARAMETERS A, B, C AND D USED HERE, AS WELL AS ONES FOR
C      SPECIES NOT INCLUDED IN THIS SYSTEM, ARE TABULATED IN APPENDIX
C      9.2, TABLE 2
C
C
C      INPUT:
C        TEMP - TEMPERATURE IN DEGREES CELSIUS
C
C
       IMPLICIT REAL*8 (A-H,O-Z)
C
C
       COMMON /HENCO/HNH3,HCO2
C
C
C      CALCULATIONS ARE DONE IN KELVINS
C      --------------------------------
C
       TK=TEMP+273.15
C
C      CALCULATE THE HENRY'S CONSTANT
C      ------------------------------
C
C        - FOR NH3
C
         HNH3=DEXP(-157.552/TK+28.1001*DLOG(TK)-.049227*TK-149.006)
C
C        - FOR CO2
C
         HCO2=DEXP(-6789.04/TK-11.4519*DLOG(TK)-.010454*TK+94.4914)
C
       RETURN
       END
C
C      ******************************************************************
```

```
C       ********************************************************************
C
        SUBROUTINE KCAL(TEMP)
C       ====================
C
C
C       THIS SUBROUTINE CALCULATES THE EQUILIBRIUM CONSTANTS USING THE
C       TEMPERATURE FITS GIVEN BY EDWARDS ET AL.:
C
C             ln K  =  A/T  +  B ln T  +  C T  +  D
C
C       THE PARAMETERS FOR THE SPECIES INVOLVED IN THIS SYSTEM AND FOR
C       SOME NOT IN THIS SYSTEM ARE TABULATED IN APPENDIX 9.2, TABLE 1.
C       THE T OF THE EQUATION IS IN KELVINS.  THE TEMPERATURE PASSED INTO
C       THIS SUBROUTINE IS IN DEGREES CELSIUS.
C
C       THE SATURATION PRESSURE OF WATER IS ALSO CALCULATED IN THIS
C       SUBROUTINE.
C
C
C       THE EQUILIBRIUM CONSTANTS ARE RETURNED IN COMMON EQCONS IN THIS
C       ORDER:
C         EKVAL(1) = K(H2O)
C         EKVAL(2) = K(NH3)
C         EKVAL(3) = K(CO2)
C         EKVAL(4) = K(HCO3ion)
C         KKVAL(5) = K(NH2CO2ion)
C
C
C       CALLED BY: MAIN
C
C
C
        IMPLICIT REAL*8 (A-H,O-Z)
C
C
        DIMENSION A(5),B(5),C(5),D(5)
C
C
        COMMON /EQCONS/DHCON,EKVAL(5),PWS,VMOL(3)
C
C
C
C       DATA FROM APPENDIX 9.2, TABLE 1:
C       --------------------------------
C
        DATA A /-13445.9, -3335.71, -12092.1, -12431.7, 2895.65/
C
        DATA B /-22.4773, 1.4971, -36.7816, -35.4819, 0.0/
C
        DATA C /0.0, -.0370566, 3*0.0/
C
        DATA D /140.932, 2.7608,235.482, 220.067, -8.5994/
C
C
C
```

```
C       CALCULATE USING TEMPERATURE IN KELVINS
C       --------------------------------------
C
        TK=TEMP+273.15
C
C       CALCULATE THE K VALUES:
C       ----------------------
C
        DO 10 I=1,5
        EKVAL(I)=DEXP(A(I)/TK+B(I)*DLOG(TK)+C(I)*TK+D(I))
   10   CONTINUE
C
C
C       CALCULATE THE SATURATION PRESSURE OF WATER AT THIS TEMPERATURE
C       --------------------------------------------------------------
C
        PWS=DEXP(70.434643-7362.6981/TK+.00695208*TK-9.*DLOG(TK))
C
        RETURN
        END
C
C       *********************************************************************
```

```
C     ***************************************************************
C
      SUBROUTINE NAKA(TEMP,PRESS,Y1,NVFUG,FUNA)
C     =========================================
C
C
C     THIS SUBROUTINE CALCULATED THE FUGACITY COEFFICIENTS USING
C     NAKAMURA ET AL'S METHOD AS OUTLINED IN APPENDIX 9.3.
C
C
C     INPUT:
C       NVFUG - NUMBER OF SPECIES TO CALCULATE FUGACITY COEFFICIENTS FOR
C       PRESS - PRESSURE IN ATMOSPHERES
C       TEMP  - TEMPERATURE IN KELVINS
C       Y1    - VAPOR MOLE FRACTIONS OF THE SPECIES
C
C     OUTPUT:
C       FUNA - FUGACITY COEFFICIENTS
C
C     CALLS: VSOLVE
C
      IMPLICIT REAL*8 (A-H,O-Z)
C
C
      COMMON /CASAP/ANA(3,3),CNA(3),ALNA(3),BENA(3),GANA(3),DENA(3),
     +              ALPHA0(3),ALPHA1(3),BETA0(3),BETA1(3)
C
      COMMON /SUM/AM,BM,CM
C
      COMMON /TEMPRES/TEMPA,PRESSA
C
      DIMENSION A(3,3),AIJ1(3,3),ALIJ(3,3),B(3),BIJ(3,3),BIJ0(3,3),
     +          BIJ1(3,3),FUNA(3),Y(3),Y1(3)
C
C
C
C     NORMALIZE THE VAPOR MOLE FRACTIONS
C     ----------------------------------
C
      TOTY=Y1(1)+Y1(2)+Y1(3)
      DO 5 I=1,3
      Y(I)=Y1(I)/TOTY
    5 CONTINUE
C
C
C     INITIALIZATIONS
C     ---------------
C
      AM=0.
      BM=0.
      CM=0.
      TEMPA=TEMP
      PRESSA=PRESS
      V=0.
      ETA=0.
C
```

```
C       CALCULATIONS:
C       -------------
C
C       EQUATION (A9.3c):
C
C       b  = exp {2.30259 (-γ  - δ  T)}
C        i                   i    i
C
        DO 10 I=1,NVFUG
        B(I)=EXP(2.30259*(-GANA(I)-DENA(I)*TEMPA))
     10 CONTINUE
C
        DO 15 I=1,NVFUG
        IP=I
        DO 15 J=IP,NVFUG
        IF(BETA0(I).EQ.0.0) BETA0(I)=BENA(I)
        IF(BETA0(J).EQ.0.0) BETA0(J)=BENA(J)
C
C       EQUATION (A9.3m):
C
C       β°  = .5 (β° + β¹)
C         ij       i    j
C
        BIJ0(I,J)=.5*(BETA0(I)+BETA0(J))
C
        IF(BETA1(I).NE.0.0.AND.BETA1(J).NE.0.0) THEN
C
C         EQUATION (A9.31):
C
C         β¹  =  √ β¹ β¹
C           ij      i  j
C
          BIJ1(I,J)=(BETA1(I)*BETA1(J))**.5
        ELSE
          BIJ1(I,J)=0.0
        END IF
C
C       EQUATION (A9.3j)
C
C       β   = β°  + β¹
C        ij     ij    ij
C
        BIJ(I,J)=BIJ0(I,J)+BIJ1(I,J)
C
        IF(ALPHA1(I).NE.0.0.AND.ALPHA1(J).NE.0.0) THEN
C
C          EQUATION (A9.3k)
C
C          α¹  =  √ α¹ α¹
C            ij      i  j
C
          AIJ1(I,J)=(ALPHA1(I)*ALPHA1(J))**.5
        ELSE
          AIJ1(I,J)=0.0
        END IF
        AIJ0=ANA(I,J)
```

```
C
C
C     EQUATION (A9.3i)
C
C     α   = α°  + α¹
C      ij    ij    ij
C
      ALIJ(I,J)=AIJ0+AIJ1(I,J)
      IF(AIJ0.EQ.0.0.AND.I.EQ.J) ALIJ(I,J)=ALNA(I)
C
C
C     EQUATION (A9.3h)
C
C                        β
C                         ij
C     a   = α     +    ----
C      ij    ij          T
C
      A(I,J)=ALIJ(I,J)+BIJ(I,J)/TEMP
      IF(I.NE.J) A(J,I)=A(I,J)
   15 CONTINUE
C
C
C     EQUATION (A9.3g)
C
C             n      n
C     a   =   Σ      Σ   y  y  a
C      M     i=1    j=1   i  j  ij
C
      DO 17 I=1,NVFUG
      DO 17 J=1,NVFUG
      AM=AM+Y(I)*Y(J)*A(I,J)
   17 CONTINUE
C
C
      DO 20 I=1,NVFUG
C
C
C     EQUATION (A9.3d)
C
C             n
C     b   =   Σ  y  b
C      M     i=1  i  i
C
      BM=BM+Y(I)*B(I)
C
C
C     EQUATION (A9.3e)
C
C             n
C     c   =   Σ  y  c
C      M     i=1  i  i
C
      CM=CM+Y(I)*CNA(I)
   20 CONTINUE
C
C
```

```
C
C     GET THE MOLAR VOLUME v
C
      CALL VSOLVE(V,IER)
      IF(IER.NE.0) THEN
        WRITE(6,28)
   28   FORMAT(' NO CONVERGENCE FOR V AFTER 25 ITERATION STEPS')
        RETURN
      END IF
C
C
C     CALCULATE THE COMPRESSIBILITY FACTOR, EQUATION (A9.3f)
C
C           P v
C     z  = -----
C           R T
C
      Z=PRESS*V/(.082054*TEMP)
C
C
C     CALCULATE THE REDUCED DENSITY:
C
C            b
C             M
C     ξ  =  ----
C           4. v
C
      ETA=BM/(4.*V)
C
C
C     NOW CALCULATE THE FUGACITY COEFFICIENTS.  THE TERMS FACn
C     ARE PORTIONS OF EQUATION (A9.3b):
C
C             4. ξ - 3. ξ²
C     FAC1 = -------------
C              (1. - ξ)²
C
      FAC1=(4.*ETA-3.*ETA**2)/(1.-ETA)**2
C
C
C                   -2.
C     FAC3   =   -------
C                 R T v
C
      FAC3=-2./(.082054*TEMP*V)
C
C
      DO 50 I=1,NVFUG
      FAC4=0.
      FAC5=0.
      FAC7=0.
C
C            [ b  ]   [  4. ξ - 2. ξ²  ]
C            [  i ]   [                ]
C     FAC2 = [ -- ]   [ -------------- ]
C            [ b  ]   [                ]
C            [  M ]   [   (1. - ξ)³    ]
C
      FAC2=B(I)/BM*(4.*ETA-2.*ETA**2)/(1.-ETA)**3
```

```
C
C                 n
C     FAC4  =     Σ   y  a
C               i=1    i  ij
C
      DO 35 J=1,NVFUG
      FAC4=FAC4+Y(J)*A(I,J)
   35 CONTINUE
C
C                 5        (-1.)^j      ( c_M )^j
C     FAC5  =     Σ       ---------     ( --- )        + 1.
C                j=1      (j + 1.)      (  v  )
C
      DO 40 J=1,5
      FAC5=FAC5+((-1)**J/(J+1)*(CM/V)**J)
   40 CONTINUE
      FAC5=FAC5+1.
C
C               a_M  c_i
C     FAC6  =  ----------
C               R T v²
C
      FAC6=AM*CNA(I)/(0.082054*TEMP*V**2)
C
C                 4     (-1.)^j  (j + 1.)      ( c_M )^j
C     FAC7  =     Σ    -------------------     ( --- )       + .5
C                j=1        (j + 2.)           (  v  )
C
      DO 45 J=1,4
      FAC7=FAC7+(-1)**J*(J+1)/(J+2)*(CM/V)**J
   45 CONTINUE
      FAC7=FAC7+.5
C
C
C     EQUATION (A9.3b) FOR FUGACITY COEFFICIENT f OF SPECIES i:
C
C
C     ln f(i) = FAC1 + FAC2 + FAC3 x FAC4 x FAC5 + FAC6 x FAC7 - ln z
C
C
      FUNA(I)=EXP(FAC1+FAC2+FAC3*FAC4*FAC5+FAC6*FAC7-DLOG(Z))
   50 CONTINUE
C
      RETURN
      END
C
C     *****************************************************************
```

```
C       *****************************************************************
C
        SUBROUTINE PMOLV(ICODE,TC,VMOL)
C       ================================
C
C
C       THIS SUBROUTINE CALCULATES PARTIAL MOLAL VOLUME OF SOLUTE IN
C       WATER BY THE METHOD OF BRELVI AND O'CONNELL, AIChE J, v18, #6,
C       pp1239-1243   (1972).  SEE APPENDIX 9.4.
C
C       INPUT PARAMETERS:
C          ICODE - ICODE FOR INDICATING SOLUTE:
C                  1 - NH3
C                  2 - CO2
C                  3 - H2O
C          TC    - TEMPERATURE IN DEGREES CELSIUS
C
C       OUTPUT PARAMETERS:
C          VMOL - PARTIAL MOLAL VOLUME
C
C       CALLED BY: MAIN
C
C
        IMPLICIT REAL*8 (A-H,O-Z)
C
C
C
C       VSTAR - v*, CHARACTERISTIC VOLUME, cm³/g mole
C       ---------------------------------------------
C
        DIMENSION VSTAR(3)
C
        DATA VSTAR/65.2, 80., 0./
C
C
C
C       FOR WATER:
C       ----------
C
        IF(ICODE.EQ.3) THEN
          VMOL=18.02*(1.001508+3.976412D-6*TC**2)
          RETURN
        END IF
C
C
C       SET THE GAS CONSTANT, cm³ atm/mole K
C       ------------------------------------
C
        R=82.06
C
C       CONVERT CELSIUS TO KELVIN TEMPERATURE
C       -------------------------------------
C
        T=TC+273.15
C
C
```

```
C       CALCULATE THE PURE SOLVENT DENSITY, g-mole/cm³
C       -----------------------------------------------
C
        IF(TC.LE.50.) THEN
          DS=1.0-4.8D-6*TC**2
        ELSE
          DS=1.0064-2.5D-4*TC-2.3D-6*TC**2
        END IF
C
C       CALCULATE THE REDUCED DENSITY
C       ----------------------------
C
        RHO=DS*46.4/18.02
        FAC=RHO-1.0
C
C       CALCULATE THE ISOTHERMAL COMPRESSIBILITY, EQUATION (A9.4d)
C       ---------------------------------------------------------
C
        COM=18.02/(DS*R*T*(DEXP(-.42704*FAC+2.089*FAC**2-.42367*FAC**3)
     +      -1.0))
C
C       EQUATION (A9.4b)
C       ----------------
C
        IF(RHO.GE.2.0.AND.RHO.LE.2.785) THEN
          SERIES=-2.4467+2.12074*RHO
C
C       EQUATION (A9.4c)
C       ----------------
C
        ELSE IF(RHO.GE.2.785.AND.RHO.LE.3.2) THEN
          SERIES=3.02214-1.87085*RHO+0.71955*RHO**2
        ELSE
          RETURN
        END IF
C
C       REDUCED VOLUME INTEGRAL OF THE MOLECULAR DIRECT CORRELATION
C       FUNCTION
C       -----------------------------------------------------------
C
        C12=(VSTAR(ICODE)/46.4)**0.62*(-DEXP(SERIES))
C
C       CALCULATE THE MOLAL VOLUME WITH EQUATION (A9.4a)
C       ------------------------------------------------
C
        VMOL=COM*R*T*(1.0-C12)
C
        RETURN
        END
C
C       *****************************************************************
```

```
C     ********************************************************************
C
      SUBROUTINE SETNAK
C     =================
C
C
C     SET VALUES FOR THE ARRAYS IN NAKAMURA FUGACITY COEFFICIENT
C     ROUTINE FOR THE AMMONIA - CARBON DIOXIDE - WATER SYSTEM.
C     ORDER:  1 - WATER
C             2 - AMMONIA
C             3 - CARBON DIOXIDE
C     THE METHOD IS DESCRIBED AND PARAMETERS FOR ADDITIONAL SPECIES
C     ARE TABULATED IN APPENDIX 9.3.
C
C
      IMPLICIT REAL*8 (A-H,O-Z)
C
      COMMON /CASAP/ANA(3,3),CNA(3),ALNA(3),BENA(3),GANA(3),DENA(3),
     +              ALPHA0(3),ALPHA1(3),BETA0(3),BETA1(3)
C
C
C     INTERACTION PARAMETERS FOR POLAR-POLAR AND POLAR-NONPOLAR
C     MIXTURES, TABULATED IN APPENDIX 9.3, TABLES 3 AND 5
C
C       ANA(I,J) = α°ij
C
        ANA(1,1)=0.0
        ANA(2,2)=0.0
        ANA(3,3)=0.0
        ANA(1,2)=1.4
        ANA(2,1)=1.4
        ANA(1,3)=4.36
        ANA(3,1)=4.36
        ANA(2,3)=3.1
        ANA(3,2)=3.1
C
C     PURE COMPONENT PARAMETERS TABULATED IN APPENDIX 9.3, TABLE 1
C
C       CNA(I) = ci
C
        CNA(1)=.01
        CNA(2)=.01
        CNA(3)=0.0
C
C       ALNA(I) = αi
C
        ALNA(1)=3.1307
        ALNA(2)=2.6435
        ALNA(3)=3.1693
C
C       BENA(I) = βi
C
        BENA(1)=1161.7
        BENA(2)=561.63
        BENA(3)=253.17
C
```

```
C
C        GANA(I) = γi
C
         GANA(1)=1.5589
         GANA(2)=1.3884
         GANA(3)=1.234
C
C        DENA(I) = δi
C
         DENA(1)=.593D-4
         DENA(2)=1.47D-4
         DENA(3)=.467D-4
C
C      NONPOLAR AND POLAR CONTRIBUTIONS FROM APPENDIX 9.3, TABLE 2
C
C        ALPHA0(I) = αi°
C
         ALPHA0(1)=1.06
         ALPHA0(2)=1.83
         ALPHA0(3)=0.0
C
C        ALPHA1(I) = αi¹
C
         ALPHA1(1)=2.07
         ALPHA1(2)=.81
         ALPHA1(3)=0.0
C
C        BETA0(I) = βi°
C
         BETA0(1)=8.4
         BETA0(2)=13.3
         BETA0(3)=0.0
C
C        BETA1(I) = βi¹
C
         BETA1(1)=1153.3
         BETA1(2)=548.3
         BETA1(3)=0.0
C
       RETURN
       END
C
C      ****************************************************************
```

```
C      ********************************************************************
C
       SUBROUTINE VSOLVE(VV,IER)
C      =========================
C
C
C      SUBROUTINE TO SOLVE FOR THE MOLAR VOLUME FOR THE NAKAMURA
C      FUGACITY COEFFICIENTS VIA EQUATION (A9.3a):
C
C           R T              a
C      P  = --- BOX  -  ---------
C            v          v (v + c)
C
C
C      WHERE:
C        v    - MOLAR VOLUME, L/MOLE
C        a    - AM SUMMATION TERM PASSED IN COMMON BLOCK
C        c    - CM SUMMATION TERM PASSED IN COMMON BLOCK
C
C                1. + ξ + ξ² - ξ³
C        BOX  =  ----------------
C                    (1. - ξ)³
C
C                 b
C        ξ    =  ---    ; REDUCED DENSITY
C                4 v
C
C        b   - BM SUMMATION TERM PASSED IN COMMON BLOCK
C
C
C      ROUTINE ATTEMPTS TO MAKE LEFT AND RIGHT HAND SIDES OF EQUATION
C      (A9.3a) EQUAL BY SOLVING FOR v.  TEST ON P1 IS CONVERGENCE TEST.
C
C
C      INPUT COMMONS:
C        AM - SUMMATION FROM SUBROUTINE NAKA, SIGNIFIES ATTRACTIVE
C             FORCE STRENGTH IN L² ATM/MOLE²
C        BM - SUMMATION FROM SUBROUTINE NAKA; SIGNIFIES HARDCORE
C             SIZES OF MOLECULES
C        CM - CONSTANT, L/MOLE; FROM SUM IN SUBROUTINE NAKA
C        P  - PRESSURE IN ATM
C        T  - TEMPERATURE IN KELVINS
C
C      OUTPUT PARAMETER:
C        VV - MOLAR VOLUME
C
C      CALLED BY: NAKA
C
C
       IMPLICIT REAL*8 (A-H,O-Z)
C
C
       COMMON /SUM/AM,BM,CM
C
       COMMON /TEMPRES/T,P
C
```

```
C
      IER=0
C
C     GAS CONSTANT, l atm/K mole
C
      R=.082054
C
      IF(V.GT.0.0.AND.V.LT.200.) THEN
        ETA=BM/(4.*V)
        BOX=(1.+ETA+ETA**2-ETA**3)/(1.-ETA)**3
        P1=P-((R*T*BOX/V)-AM/(V*(V+CM)))
        VV=V
        IF(DABS(P1).LT.1.D-5) RETURN
      END IF
C
C     SET INITIAL GUESS FROM RELATION Pv = RT
C
      V=R*T/P
      VLO=BM/4.*1.0000001D0
      VHI=V*100.
      V=VLO+(VHI-VLO)/2.
      VINC=(VHI-VLO)/2.
C
   10 ETA=BM/(4.*V)
      BOX=(1.+ETA+ETA**2-ETA**3)/(1.-ETA)**3
      P1=P-((R*T*BOX/V)-AM/(V*(V+CM)))
      VINC=VINC/2.
      IF(DABS(P1).GT.1.D-6) THEN
        IF(P1.GT.0.) V=V-VINC
        IF(P1.LE.0.) V=V+VINC
        IF(VINC.GT.1.D-15) GO TO 10
      END IF
      VV=V
      IF(DABS(P1).LT.1.D-04) RETURN
C
C     SET NEW GUESS FOR V AND INCREMENT WHEN UNCONVERGED
C
      VLO=1.D-09
      VHI=BM/4.*.99999999
      V=VLO+(VHI-VLO)/2.
      VINC=(VHI-VLO)/2.
C
C     TRY AGAIN WITH NEW GUESS
C
   20 ETA=BM/(4.*V)
      BOX=(1.+ETA+ETA**2-ETA**3)/(1.-ETA)**3
      P1=P-((R*T*BOX/V)-AM/(V*(V+CM)))
      V2=V*1.001
      ETA=BM/(4.*V2)
      BOX=(1.+ETA+ETA**2-ETA**3)/(1.-ETA)**3
      P2=P-((R*T*BOX/V2)-AM/(V2*(V2+CM)))
      DP=(P2-P1)/(V2-V)
      VINC=VINC/2.
      IF(DP.GE.0.) V=V-VINC
      IF(DP.LT.0.) V=V+VINC
      IF(VINC.GT..0001) GO TO 20
```

```
      IF(P1.GT.0.) GO TO 900
      VLO=V
      VHI=BM/4*.9999999
      V=VLO+(VHI-VLO)/2.
      VINC=(VHI-VLO)/2.
   30 ETA=BM/(4.*V)
      BOX=(1.+ETA+ETA**2-ETA**3)/(1.-ETA)**3
      P1=P-((R*T*BOX/V)-AM/(V*(V+CM)))
      VINC=VINC/2.
      IF(DABS(P1).GT.1.D-6) THEN
        IF(P1.GT.0.) V=V-VINC
        IF(P1.LE.0.) V=V+VINC
        IF(VINC.GT.1.D-8) GO TO 30
      END IF
      VV=V
      IF(DABS(P1).LT.1.D-04) RETURN
C
C     UNABLE TO CONVERGE
C
      IER=1
      WRITE(6,70) V,P1,VINC
      WRITE(6,70) T,P,AM,BM,CM
   70 FORMAT(1X,5G16.6)
      RETURN
C
C     SET ERROR FLAG IF UNCONVERGED
C
  900 WRITE(6,966) V,P
  966 FORMAT(' GAS PHASE CONDENSED.  V= ',E12.6,' FOR PRESSURE=',F7.3/)
      WRITE(6,70) T,P,AM,BM,CM
      VV=V
      RETURN
      END
C
C     ******************************************************************
```

Figure 9.8

```
SYSTEM: AMMONIA - CARBON DIOXIDE - WATER

METHOD: Edwards, Maurer, Newman & Prausnitz, AIChE J (1978)

========================================================================

Temperature: 60.00 degrees Celsius
Pressure:     1.00 atm
H2Oin:       55.51 moles
NH3in:        9.00 moles
CO2in:        7.00 moles

Equilibrium Constants:
     K(H2O)     =  0.945751E-13 mole/kg
     K(NH3)     =  0.184361E-04 mole/kg
     K(CO2)     =  0.523043E-06 mole/kg
     K(HCO3)    =  0.724203E-10 mole/kg
     K(NH2CO2)  =  0.109673E+01 mole/kg

Saturation Pressure of Water:  0.196546E+00 atm

Henry's Constants:
     H(NH3)  =  0.701688E-01 atm kg/mole
     H(CO2)  =  0.610014E+02 atm kg/mole

Partial Molar Volumes:
     v(H2O)  =  0.183051E+02 cm^3/mole
     v(NH3)  =  0.308184E+02 cm^3/mole
     v(CO2)  =  0.348117E+02 cm^3/mole


Calculation Results:

  Liquid Phase Molalities and Activity Coefficients:

     Ionic strength =  0.112365E+02 molal
       H2O         0.492487E+02             AW          0.918976E+00
       Hion        0.134369E-07             AH          0.977538E+00
       OHion       0.723010E-05             AOH         0.915178E+00
       NH3aq       0.514194E+00             ANH3        0.891991E+00
       NH4ion      0.925273E+01             ANH4        0.126924E+00
       CO2aq       0.128607E-01             ACO2        0.997687E+00
       CO3ion      0.198375E+01             ACO3        0.130500E-02
       HCO3ion     0.493293E+01             AHCO3       0.951841E-01
       NH2CO2ion   0.352293E+00             ANH2CO2     0.729538E+00

  Vapor Phase Mole Fractions and Fugacity Coefficients:
       yH2O        0.181711E+00             fH2O        0.992995E+00
       yNH3        0.323161E-01             fNH3        0.996795E+00
       yCO2        0.785973E+00             fCO2        0.996863E+00

       Total vapor        0.68644 moles
```

Figure 9.9

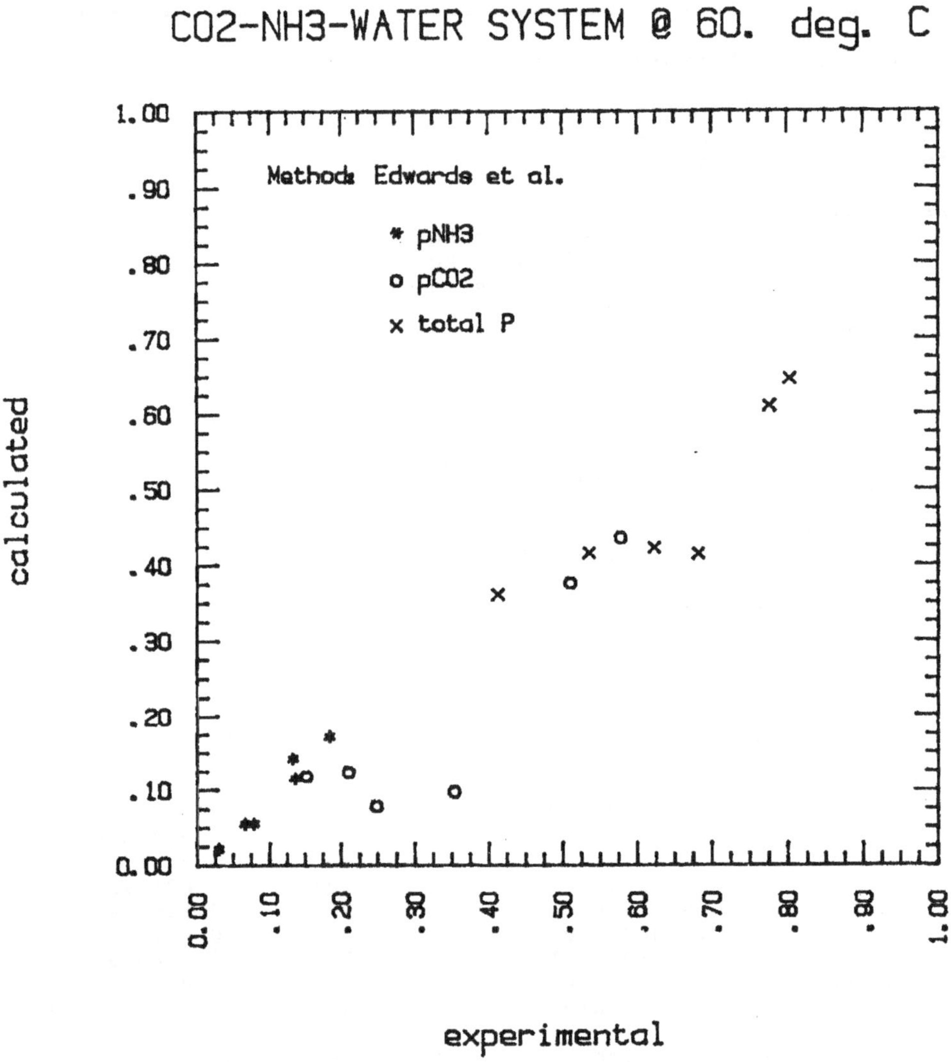
CO2-NH3-WATER SYSTEM @ 60. deg. C
Method: Edwards et al.
* pNH3
o pCO2
x total P
calculated
experimental
1.00
.90
.80
.70
.60
.50
.40
.30
.20
.10
0.00

Water - Sulfur Dioxide

The water-sulfur dioxide binary solution can be characterized by the following equilibria:

vapor-liquid equilibria

$H_2O(vap) = H_2O$

$SO_2(vap) = SO_2(aq)$

water dissociation

$H_2O = H^+ + OH^-$

sulfur dioxide dissociation

$SO_2(aq) + H_2O = H^+ + HSO_3^-$

second dissociation

$HSO_3^- = H^+ + SO_3^{2-}$

Assuming the original amounts of H_2O and SO_2 mixed together are known, the nine unknowns in this system are:

liquid phase:

moles H_2O, m_{H^+}, m_{OH^-}, $m_{SO_2(aq)}$, $m_{HSO_3^-}$, $m_{SO_3^{2-}}$

vapor phase:

H_2O and SO_2 vapor mole fractions, and V, the total vapor rate

The liquid phase reactions are expressed as K equations:

$$K_{H_2O} = \gamma_{H^+} m_{H^+} \gamma_{OH^-} m_{OH^-} / a_w \qquad (S.1)$$

$$K_{SO_2} = \gamma_{H^+} m_{H^+} \gamma_{HSO_3^-} m_{HSO_3^-} / (\gamma_{SO_2(aq)} m_{SO_2(aq)} a_w) \qquad (S.2)$$

$$K_{HSO_3^-} = \gamma_{H^+} m_{H^+} \gamma_{SO_3^{2-}} m_{SO_3^{2-}} / (\gamma_{HSO_3^-} m_{HSO_3^-}) \qquad (S.3)$$

where:

m_i - species molality

γ_i - species activity coefficient

a_w - water activity

K_i - equilibrium constant

For this system, the activity coefficients are calculated using the modified Pitzer equations presented by Beutier and Renon (2).

They used a temperature fit equation for the equilibrium constants:

$$K_i = \exp\,(A_i/T + B_i \ln T + C_i\, T + D_i) \tag{S.4}$$

The appropriate parameters for this equation are tabulated and plotted in Appendix 9.1.

The vapor-liquid equilibria are expressed:

$$y_{H_2O}\ f_{H_2O}\,P = a_w\,P_w^s\ f_w^s\ \exp\left[\frac{\bar{v}_w\,(P - P_w^s)}{RT}\right] \tag{S.5}$$

$$y_{SO_2}\ f_{SO_2}\,P = m_{SO_2}\,\gamma_{SO_2}\,H_{SO_2}\ \exp\left[\frac{\bar{v}_{SO_2}(P - P_w^s)}{RT}\right] \tag{S.6}$$

where:

y_i - vapor mole fraction of i

f_i - fugacity coefficient of vapor i; calculated using the method of Nakamura et al. (16), see Appendix 9.3

P - total pressure, atmospheres

γ_{SO_2} - activity coefficient of SO_2 in solution

m_{SO_2} - molality of SO_2 in solution

a_w - water activity

P_w^s - vapor pressure of pure water, calculated using the correlation in Lange's Handbook (10)

$$P_w^s = [10.\ \wedge\ (7.96681 - \frac{1668.21}{T-45.15})]\ /\ 760.\ \text{atmospheres}$$

f_w^s - fugacity coefficient of pure water; see Appendix 9.3

$\bar{v}_i$ - partial molal volume of i in water at infinite dilution dm^3/mole; from correlation of Brelvi and O'Connell (3), see Appendix 9.4

H_{SO_2} - Henry's constant for SO_2; atm kg/mole

R - gas constant; 82.06 cm^3.atm/(mole.K)

T - temperature, Kelvins

The exponential terms in equations (S.5) and (S.6) are from Krichevsky and Kasarnovsky (13) for pressure correction to the Henry's constant. The constant was presented in a temperature fit equation of the same form as the equilibrium constant:

$$H_i = \exp (A_i/T + B_i \ln T + C_i T + D_i) \qquad (S.7)$$

Parameters for the Henry's constants are tabulated and plotted in Appendix 9.1.

With equations (S.1), (S.2), (S.3), (S.5) and (S.6) defining the phase equilibria, four more equations are required in order to solve for the nine unknowns in the system. These are material balances and the electroneutrality balance:

vapor phase balance:

$$y_{H_2O} + y_{SO_2} = 1.0 \qquad (S.8)$$

sulfur balance:

$$SO_2(in) = \frac{H_2O}{55.51} (m_{SO_2(aq)} + m_{HSO_3^-} + m_{SO_3^{2-}}) + y_{SO_2} V \qquad (S.9)$$

hydrogen balance:

$$2\ H_2O(in) = 2\ H_2O + \frac{H_2O}{55.51} (m_{H^+} + m_{OH^-} + m_{HSO_3^-}) + 2\ y_{H_2O} V \qquad (S.10)$$

electroneutrality:

$$m_{H^+} = m_{OH^-} + m_{HSO_3^-} + 2\ m_{SO_3^{2-}} \qquad (S.11)$$

The general expression for Beutier and Renon's modification of Pitzer's activity coefficient equation is detailed in Chapter VII. The specific activity coefficient equations can be expressed based upon a sequence of definitions as follows:

- ionic strength:

$$I = .5\ (m_{H^+} + m_{OH^-} + m_{HSO_3^-} + 4.\ m_{SO_3^{2-}})$$

- Debye-Hückel term:

$$FAC = -\frac{A}{3}\left[\frac{\sqrt{I}}{1. + 1.2\sqrt{I}} + \frac{2.0}{1.2} \ln (1. + 1.2\sqrt{I})\right]$$

The Debye-Hückel constant, A, is calculated as follows:

$$A = \left(\frac{2.\,\pi N_A d_0}{1000.}\right)^{1/2}\left(\frac{e^2}{D\,k\,T}\right)^{3/2}$$

where:

N_A - Avagadro's number, 6.0232×10^{23} /mole

d_0 - density of water, kg/dm³, at temperature T

e - electron charge, 4.8029×10^{-10} esu

D - dielectric constant of water

$$D = 305.7 \exp[-\exp(-12.741 + .01875\ T) - T/219.]$$

k - Boltzmann's constant, 1.38045×10^{-16} erg/K

T - temperature, Kelvins

- ion-ion interactions using standard Pitzer terms

i = 1 for the H^+-OH^- interaction

i = 2 for the H^+-HSO_3^- interaction

i = 3 for the H^+-SO_3^{2-} interaction

$$B_i = \beta_0(i) + \beta_1(i)\ [1. - (1. + 2.\sqrt{I})\exp(-2.\sqrt{I})]/(2.\ I)$$

$$B'_i = \beta_1(i)\ [1. - (1. + 2.\sqrt{I} + 2.\ I)\exp(-2.\sqrt{I})]$$

$$Bw_i = \beta_0(i) + \beta_1(i)\exp(-2.\sqrt{I})$$

$$Bsum = 2.\ m_{H^+}(m_{OH^-}\ B'_1 + m_{HSO_3^-}\ B'_2 + m_{SO_3^{2-}}\ B'_3)$$

The species specific coefficient values for the water-sulfur dioxide system are shown along with other species interaction values in Appendix 9.1.

- ion-molecule interaction coefficients from dielectric effects:

a. volume of neutral solution excluding ions, $dm^3/kg\ H_2O$

$$V_m = \frac{1.}{d_0} + m_{SO_2}\ \bar{v}_{SO_2}$$

b. volume of ionic solution excluding neutral solutes, $dm^3/kg\ H_2O$

$$V_i = \frac{1.}{d_0} + m_{H^+}\ \bar{v}_{H^+} + m_{OH^-}\ \bar{v}_{OH^-} + m_{HSO_3^-}\ \bar{v}_{HSO_3^-} + m_{SO_3^{2-}}\ \bar{v}_{SO_3^{2-}}$$

c. volume of ionic cavities for ion i, dm /mole

$$v_i^c = \frac{4}{3}\pi N_A r_i^3 \times 10^{-27}$$

d. volume of all ionic cavities, $dm^3/kg\ H_2O$

$$V_c = m_{H^+}\ v^c_{H^+} + m_{OH^-}\ v^c_{OH^-} + m_{HSO_3^-}\ v^c_{HSO_3^-} + m_{SO_3^{2-}}\ v^c_{SO_3^{2-}}$$

e. volume of real solution, $dm^3/kg\ H_2O$

$$V_f = V_i + m_{SO_2}\ \bar{v}_{SO_2}$$

f. dielectric constant of solution without ions

$$D_s = D\left[1. + \frac{\alpha_{SO_2}\, m_{SO_2}}{V_m}\right]$$

g. dimensionless constant of dielectric contribution of ion i in activity coefficient

$$L_i = \frac{e^2\, z_i^2}{2\, r_i\, k\, T\, D} \times 10^8$$

z_i - ionic charge

Values for the following terms, used above, are tabulated in Appendix 9.1:

$\bar{v}_i$ - partial molal volume, dm^3/mole

r_i - ionic cavity radius, Å

α_i - dielectric coefficient of neutral solute i, dm^3/mole

- definition of terms used often to simplify the activity coefficient equations:

$$DIV = \frac{D}{D_s}$$

$$V_{fc} = V_f - V_c$$

$$V_{ic} = V_i - V_c$$

$$\Sigma L = m_{H^+}\, L_{H^+} + m_{OH^-}\, L_{OH^-} + m_{HSO_3^-}\, L_{HSO_3^-} + m_{SO_3^{2-}}\, L_{SO_3^{2-}}$$

$$BRAC = DIV * m_{SO_2}\left[\frac{-\alpha_{SO_2}}{V_m}\;\frac{V_i + .5\, V_c}{V_{ic}} - \frac{1.5\, \bar{v}_{SO_2}\, V_c}{V_{ic}\, V_{fc}}\right]$$

- molecule-molecule interactions:

$\lambda_{SO_2\text{-}SO_2}$ value tabulated in Appendix 9.1 with other self interactions; ternary self interaction term is calculated:

$$\mu_{SO_2\text{-}SO_2\text{-}SO_2} = -\frac{1.}{55.5}\left[\lambda_{SO_2\text{-}SO_2} + \frac{1.}{166.5}\right]$$

Having established the above sequence of definitions, the actual activity coefficients are as follows:

- hydrogen ion activity coefficient:

$$\gamma_{H^+} = \exp\left[FAC + 2.\,(m_{OH^-}\, B_1 + m_{HSO_3^-}\, B_2 + m_{SO_3^{2-}}\, B_3) - \frac{Bsum}{4\, I^2} + BRAC\, L_{H^+} + 1.5\, \Sigma L\left(DIV\, \frac{V_f\, v^c_{H^+} - V_c\, \bar{v}_{H^+}}{V_{fc}^2} - \frac{V_i\, v^c_{H^+} - V_c\, \bar{v}_{H^+}}{V_{ic}^2}\right)\right]$$

- hydroxide ion activity coefficient

$$\gamma_{OH^-} = \exp\left[FAC + 2.\ B_1\ m_{H^+} - \frac{Bsum}{4\ I^2} + BRAC\ L_{OH^-} + 1.5\ \Sigma L \left(DIV\ \frac{V_f\ v^c_{OH^-} - V_c\ \bar{v}_{OH^-}}{V^2_{fc}} - \frac{V_i\ v^c_{OH^-} - V_c\ \bar{v}_{OH^-}}{V^2_{ic}} \right) \right]$$

- bisulfite ion activity coefficient

$$\gamma_{HSO_3^-} = \exp\left[FAC + 2.\ B_2\ m_{H^+} - \frac{Bsum}{4\ I^2} + BRAC\ L_{HSO_3^-} + 1.5\ \Sigma L \left(DIV\ \frac{V_f\ v^c_{HSO_3^-} - V_c\ \bar{v}_{HSO_3^-}}{V^2_{fc}} - \frac{V_i\ v^c_{HSO_3^-} - V_c\ \bar{v}_{HSO_3^-}}{V^2_{ic}} \right) \right]$$

- sulfite ion activity coefficient

$$\gamma_{SO_3^{2-}} = \exp\left[4.\ FAC + 2.\ B_3\ m_{H^+} - \frac{Bsum}{I^2} + BRAC\ L_{SO_3^{2-}} + 1.5\ \Sigma L \left(DIV\ \frac{V_f\ v^c_{SO_3^{2-}} - V_c\ \bar{v}_{SO_3^{2-}}}{V^2_{fc}} - \frac{V_i\ v^c_{SO_3^{2-}} - V_c\ \bar{v}_{SO_3^{2-}}}{V^2_{ic}} \right) \right]$$

- sulfur dioxide activity coefficient

$$\gamma_{SO_2} = \exp\left[2.\ \lambda_{SO_2-SO_2}\ m_{SO_2} + 3.\ \mu_{SO_2-SO_2-SO_2}\ m^2_{SO_2} + \Sigma L\ DIV \left(\frac{-1.5\ \bar{v}_{SO_2}\ V_c}{V^2_{fc}} + \frac{V_f + .5\ V_c}{V_{fc}} \left[\frac{\bar{v}_{SO_2}}{V_m} - \frac{DIV\ (\bar{v}_{SO_2} + \alpha_{SO_2})}{V_m} \right] \right) \right]$$

- water activity

M_w - molecular weight of water, .01802 kg/mole

$$a_w = \exp\left\{ M_w \left(\frac{2.\ A}{3.}\ \frac{I^{1.5}}{1. + 1.2\ \sqrt{I}} - 2.\ m_{H^+}\ [\ Bw_1\ m_{OH^-} + Bw_2\ m_{HSO_3^-} + Bw_3\ m_{SO_3^{2-}}\ ] - \lambda_{SO_2-SO_2}\ m^2_{SO_2} - 2.\ \mu_{SO_2-SO_2-SO_2}\ m^3_{SO_2} - \left(m_{H^+} + m_{OH^-} + m_{HSO_3^-} + m_{SO_3^{2-}} + m_{SO_2} \right) - \Sigma L \left[\frac{-\alpha_{SO_2} m_{SO_2} DIV^2}{d_0\ V^2_m}\ \frac{V_f + .5\ V_c}{V_{fc}} + \frac{1.5\ V_c\ DIV}{d_0\ V^2_{fc}} - \frac{1.5\ V_c}{d_0\ V^2_{ic}} \right] \right) \right\}$$

With these expressions for the activity coefficients, the equilibrium constants and Henry's constant calculated with the fit equations (S.4) and (S.7), and the fugacity coefficients calculated using Nakamura's method as outlined in Appendix 9.3, the nine unknown concentrations may be determined for set input amounts of H_2O and SO_2 with the Newton-Raphson method using equations (S.1), (S.2), (S.3), (S.5), (S.6), (S.8), (S.9), (S.10) and (S.11) as the system model. An example of such a calculation is shown in Figure 9.10.

Figure 9.10

SYSTEM: SULFUR DIOXIDE - WATER

METHOD: Beutier & Renon, IECPDD (1978)

==

```
Temperature:   50.00 degrees Celsius
Pressure:       1.00 atm
H2Oin:         55.51 moles
SO2in:         10.00 moles


Equilibrium Constants:
     K(H2O)  =   0.538103E-13 mole/kg
     K(SO2)  =   0.916327E-02 mole/kg
     K(HSO3) =   0.357133E-07 mole/kg

Saturation pressure of water:  0.121686E+00 atm

Henry's Law Constant:
     H(SO2) =  0.181753E+01 atm kg/mole

Partial Molar Volumes:
     V(H2O) =  0.182263E+02 cm^3/mole
     V(SO2) =  0.424448E+02 cm^3/mole


Calculation Results:

     Liquid Phase:

       Ionic strength =  0.947195E-01

                    moles                      activity coefficient

       H2O       0.541220E+02                      0.988251E+00
       Hion      0.923510E-01                      0.692946E+00
       OHion     0.102733E-11                      0.768934E+00
       SO2aq     0.486806E+00                      0.958079E+00
       SO3ion    0.920291E-07                      0.380459E+00
       HSO3ion   0.923508E-01                      0.696780E+00

     Vapor phase:

                mole fraction                  fugacity coefficient

       yH2O      0.120903E+00                      0.994216E+00
       ySO2      0.879097E+00                      0.990397E+00

       Total vapor      10.71650 moles
```

Figure 9.11

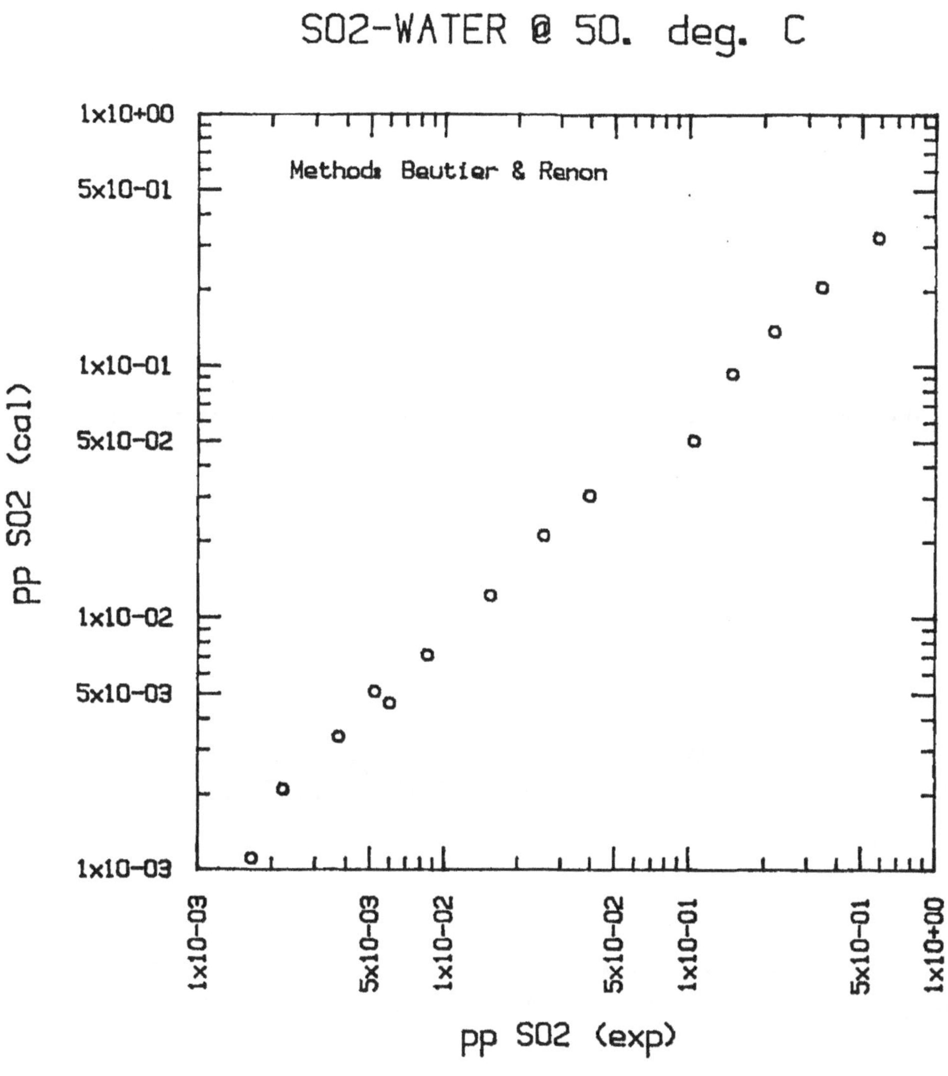
SO2-WATER @ 50. deg. C
Method: Bautier & Renon
1x10+00
5x10-01
1x10-01
5x10-02
1x10-02
5x10-03
1x10-03
pp SO2 (cal)
1x10-03
5x10-03
1x10-02
5x10-02
1x10-01
5x10-01
1x10+00
pp SO2 (exp)

Chrome Hydroxides

The Russian thermodynamic data compilation (12) contains values for the crystal $Cr(OH)_3$ and Cr^{3+}, $CrOH^{2+}$ and $Cr(OH)_2^+$ ions. In their book on cation hydrolysis, Baes and Mesmer (1) presented equilibrium constants for those species and for the ionic $Cr(OH)_4^-$, $Cr_2(OH)_2^{4+}$ and $Cr_3(OH)_4^{5+}$ species and the aqueous molecule $Cr(OH)_3(aq)$. These equilibrium constants can be used to determine values for the Gibbs free energies of the species not included in the Russian data compilation. As a first step, the equilibrium constants presented by Baes and Mesmer should be checked against values calculated using the tabulated ΔG's. For example, for the reaction:

$$Cr^{3+} + 2\ H_2O = Cr(OH)_2^+ + 2\ H^+$$

Baes and Mesmer give a value of log K = -9.7 for the equilibrium constant at 25°C. Using the Russian thermodynamic data, the equilibrium constant is calculated as:

$$\Delta G_{RXN} = \Delta G_{Cr(OH)_2^+} + 2\ \Delta G_{H^+} - (\Delta G_{Cr^{3+}} + 2\Delta G_{H_2O})$$

$$= -\ 153500. + 2\ (0) - (-53312. + 2\ (-56703.)\ )$$

$$= 13218.\ \text{cal/g mole}$$

$$\ln K = -\frac{\Delta G_{RXN}}{R\,T}$$

$$K = \exp\left(-\frac{\Delta G_{RXN}}{R\,T}\right) = \exp\left(-\frac{13218.}{1.98719 * 298.15}\right)$$

$$= 2.046771E{-}10$$

$$\log K = -9.68893$$

This value compares favorably with that presented by Baes and Mesmer as do the equilibrium constants for $CrOH^{2+}$ and $Cr(OH)_3$ (c). It is therefore 'safe' to get the Δ G's from the Baes and Mesmer equilibrium constants and use them with the Russian thermodynamic data. For example, a value of log K = -18. at 25°C is given by Baes and Mesmer for the reaction:

$$Cr^{3+} + 3\ H_2O = Cr(OH)_3\,(aq) + 3\ H^+$$

By converting the equilibrium constant to a natural logarithm base:

$$\log K = -18.$$
$$\ln K = -41.4465317$$

and using this value in the rearranged equilibrium constant equation:

$$\ln K = -\frac{\Delta G_{RXN}}{R\,T}$$

$$\Delta G_{RXN} = -(\ln K)\,R\,T$$

the excess Gibbs free energy of the reaction can be calculated:

$$\Delta G_{RXN} = -(-41.4465317)\,(1.98719)\,(298.15)$$
$$= 24556.27003$$

Since ΔG_{RXN} is defined as the sum of ΔG's of the products minus the sum of the ΔG's of the reactants:

$$\Delta G_{RXN} = (\Delta G_{Cr(OH)_3(aq)} + 3\Delta G_{H^+}) - (\Delta G_{Cr^{3+}} + 3\Delta G_{H_2O})$$

and values are known for water and the Cr^{3+} and H^+ ions, the Gibbs free energy of the $Cr(OH)_3$ aqueous molecule may be calculated:

$$\Delta G_{Cr(OH)_3(aq)} = \Delta G_{RXN} + \Delta G_{Cr^{3+}} + 3\Delta G_{H_2O} - 3\Delta G_{H^+}$$
$$= 24556.27003 + (-53312.) + 3(-56703.) - 3.(0.)$$
$$= -198864.73 \text{ cal/g mole}$$

Values for the Gibbs free energies of the $Cr(OH)_4^-$, $Cr_2(OH)_2^{4+}$ and $Cr_3(OH)_4^{5+}$ ions were calculated in the same manner. The thermodynamic data for the system is found in Table 9.1.

Table 9.1 - Thermodynamic Property Data

	ΔG kcal/g mole	ΔH kcal/g mole	S cal/K mole	Cp cal/K mole
From Russian compilation (12):				
H^+	0.0	0.0	0.0	0.0
OH^-	-37.6	-54.98	-2.595	
H_2O	-56.703	-68.3149	16.75	17.997
Cr^{3+}	-53.312	-56.4	-51.5	
$CrOH^{2+}$	-104.4			
$Cr(OH)_2^+$	-153.5			
$Cr(OH)_3(c)$	-207.3	-242.2		
From Baes and Mesmer (1):				
$Cr(OH)_3(aq)$	-198.865			
$Cr(OH)_4^-$	-242.984			
$Cr_2(OH)_2^{4+}$	-213.127			
$Cr_3(OH)_4^{5+}$	-375.6295			

In addition to the water dissociation equation:

$$H_2O = H^+ + OH^-$$

$$K_{H_2O} = a_{H^+} \, a_{OH^-} \, / \, a_w \qquad \text{(Cr.1)}$$

the reactions and corresponding equilibrium equations in a solution containing chrome hydroxides may be described:

$$Cr^{3+} + OH^- = CrOH^{2+}$$

$$K_{CrOH^{2+}} = a_{Cr^{3+}} \, a_{OH^-} \, / \, a_{CrOH^{2+}} \qquad \text{(Cr.2)}$$

$$CrOH^{2+} + OH^- = Cr(OH)_2^+$$

$$K_{Cr(OH)_2^+} = a_{CrOH^{2+}} \, a_{OH^-} \, / \, a_{Cr(OH)_2^+} \qquad \text{(Cr.3)}$$

$$Cr(OH)_2^+ + OH^- = Cr(OH)_3(aq)$$

$$K_{Cr(OH)_3(aq)} = a_{Cr(OH)_2^+} \; a_{OH^-} \; / \; a_{Cr(OH)_3(aq)} \qquad (Cr.4)$$

$$Cr^{3+} + 3\;OH^- = Cr(OH)_3(c)$$

$$K_{Cr(OH)_3(c)} = a_{Cr^{3+}} \; a_{OH^-} \qquad (Cr.5)$$

$$Cr(OH)_3(aq) + OH^- = Cr(OH)_4^-$$

$$K_{Cr(OH)_4^-} = a_{Cr(OH)_3(aq)} \; a_{OH^-} \; / \; a_{Cr(OH)_4^-} \qquad (Cr.6)$$

$$Cr(OH)_2^+ + Cr^{3+} = Cr_2(OH)_2^{4+}$$

$$K_{Cr_2(OH)_2^{4+}} = a_{Cr(OH)_2^+} \; a_{Cr^{3+}} \; / \; a_{Cr\;(OH)_2^{4+}} \qquad (Cr.7)$$

$$Cr_2(OH)_2^{4+} + Cr(OH)_2^+ = Cr_3(OH)_4^{5+}$$

$$K_{Cr_3(OH)_4^{5+}} = a_{Cr_2(OH)_2^{4+}} \; a_{Cr(OH)_2^+} \; / \; a_{Cr_3(OH)_4^{5+}} \qquad (Cr.8)$$

In the equilibrium equations the term a_i denotes the activity of species i and has been defined as the product of the activity coefficient and molality, $a_i = \gamma_i \, m_i$.

In their 1956 paper, Chamberlin and Day (8) note that, while soluble in both acid and alkaline solutions, chrome hydroxide is virtually insoluble in the 8.5 to 9. pH range. Our desire in this illustrative problem is to study chrome hydroxide solubilities and species distribution over a range of pH. This can be done by adding an acid or a base to the H_2O-$Cr(OH)_3$ system. The addition of an acid, such as HCl, or base, such as NaOH, to the model would allow study of the solubility of chrome hydroxide over a range of pH's and would require the addition of only one equation to the model since the hydrogen and hydroxide ions are already present in the model.

For the binary H_2O-$Cr(OH)_3$ system the conventional unknowns of the model are:

1) H_2O, moles of water
2) $Cr(OH)_3(c)$, moles of solid
3) molalities, m_i, of the species H^+, OH^-, Cr^{3+}, $CrOH^{2+}$, $Cr(OH)_2^+$, $Cr(OH)_3(aq)$, $Cr(OH)_4^-$, $Cr_2(OH)_2^{4+}$ and $Cr_3(OH)_4^{5+}$

In order to assure a final result based upon 55.51 moles of water, the H_2O will be made a known set to 55.51 and H_2O(in) will be made a free (unknown) variable. In performing the actual calculations we are really interested in the distribution of species, at saturation, with changing pH. Thus, $Cr(OH)_3$(in) and NaOH(in) will be made unknowns while fixing m_{OH^-} at a target value and $Cr(OH)_3$(c) at 1.0E-06, or, effectively zero. The water-chrome binary therefore results in eleven unknowns. The inclusion of a small amount of caustic in the solution, NaOH(in), assuming complete dissociation, will add the sodium ion, m_{Na^+}, to the list of unknowns. The addition of hydrochloric acid to the solution, HCl(in), will add the need for the chlorine ion molality, m_{Cl^-}. The total number of unknowns in a chrome hydroxide solution with the acid or the base included is then twelve. Recalling the requirement for a model to have the same number of equations as unknowns, four equations need to be added to the equilibrium equations (Cr.1) to (Cr.8) shown earlier. These equations are:

- hydrogen balance

for the solution with hydrochloric acid added:

$$2H_2O(in) + 3\ Cr(OH)_3(in) + HCl(in) = 2\ H_2O + H_2O\ [m_{H^+} + m_{OH^-} + m_{CrOH^{2+}} + 2\ m_{Cr(OH)_2^+} + 3\ m_{Cr(OH)_3(aq)} + 4\ m_{Cr(OH)_4^-} + 2\ m_{Cr_2(OH)_2^{4+}} + 4\ m_{Cr_3(OH)_4^{5+}}]\ /\ 55.51 + 3\ Cr(OH)_3(c) \qquad (Cr.9.a)$$

for the solution with caustic added:

$$2\ H_2O(in) + 3\ Cr(OH)_3(in) + NaOH(in) = 2\ H_2O + H_2O\ [m_{H^+} + m_{OH^-} + m_{CrOH^{2+}} + 2\ m_{Cr(OH)_2^+} + 3\ m_{Cr(OH)_3(aq)} + 4\ m_{Cr(OH)_4^-} + 2\ m_{Cr_2(OH)_2^{4+}} + 4\ m_{Cr_3(OH)_4^{5+}}]\ /\ 55.51 + 3\ Cr(OH)_3(c) \qquad (Cr.9b)$$

- the chrome balance:

$$Cr(OH)_3(in) = H_2O\ (m_{Cr^{3+}} + m_{CrOH^{2+}} + m_{Cr(OH)_2^+} + m_{Cr(OH)_3(aq)} + m_{Cr(OH)_4^-} + 2\ m_{Cr_2(OH)_2^{4+}} + 3\ m_{Cr_3(OH)_4^{5+}})\ /\ 55.51 + Cr(OH)_3(c) \qquad (Cr.10)$$

- electroneutrality

for the solution with hydrochloric acid added:

$$m_{H^+} + 3\,m_{Cr^{3+}} + 2\,m_{CrOH^{2+}} + m_{Cr(OH)_2^+} + 4\,m_{Cr_2(OH)_2^{4+}} +$$

$$5\,m_{Cr_3(OH)_4^{5+}} = m_{OH^-} + m_{Cl^-} + m_{Cr(OH)_4^-} \qquad \text{(Cr.11.a)}$$

for the solution with caustic added:

$$m_{H^+} + 3\,m_{Cr^{3+}} + 2\,m_{CrOH^{2+}} + m_{Cr(OH)_2^+} + 4\,m_{Cr_2(OH)_2^{4+}} +$$

$$5\,m_{Cr_3(OH)_4^{5+}} + m_{Na^+} = m_{OH^-} + m_{Cr(OH)_4^-} \qquad \text{(Cr.11.b)}$$

- an elemental balance:

for the solution with hydrochloric acid, a chlorine balance:

$$HCl(in) = H_2O\ (\ m_{Cl^-}\ )\ /\ 55.51 \qquad \text{(Cr.12.a)}$$

for the solution with caustic, a sodium balance:

$$NaOH(in) = H_2O\ (\ m_{Na^+}\ )\ /\ 55.51 \qquad \text{(Cr.12.b)}$$

Since a set of interaction parameters for the activity coefficient calculations for chrome hydroxide solutions have not been published, a number of assumptions were made. The first was that the activity coefficient of the aqueous molecule $Cr(OH)_3(aq)$ was unity:

$$\gamma_{Cr(OH)_3(aq)} = 1.$$

This is a fairly safe assumption due to the low solubility of chrome hydroxide; the molality of the aqueous molecule would be expected to be very low and it's activity coefficient would not deviate far from unity.

The second assumption was also made by relying on the low solubility. Recalling from Chapter V Bromley's (5) equations for the activity coefficient of a single ion in a multicomponent solution:

$$\log \gamma_i = -\frac{A z_i^2 \sqrt{I}}{1. + \sqrt{I}} + F_i \tag{5.5}$$

where,

A - is the Debye-Hückel constant

z_i - is the absolute value of the ionic charge

I - is the ionic strength

$F_i = \sum_j \dot{B}_{ij} Z_{ij}^2 m_j$

m_j - is the ionic molality

$Z_{ij} = .5\ (z_i + z_j)$

$$\dot{B}_{ij} = \frac{(0.06 + .6\ B_{ij})\ z_i\ z_j}{\left(1. + \frac{1.5}{z_i\ z_j}\ I\right)^2} + B_{ij} \tag{5.8}$$

B_{ij} is the Bromley interaction parameter. Values are tabulated for the caustic and hydrochloric acid:

$B_{NaOH} = .0747$

$B_{HCl} = .1433$

A value is also tabulated for the chrome ion - chloride ion interaction:

$B_{CrCl_3} = .1026$

which may be used in calculations involving the hydrochloric acid. No parameters are tabulated for the following ion-ion interactions:

- the anion $Cr(OH)_4^-$ with the following cations:
  H^+, Na^+, Cr^{3+}, $CrOH^{2+}$, $Cr(OH)_2^+$, $Cr_2(OH)_2^{4+}$, $Cr_3(OH)_4^{5+}$

- the anion OH^- with the following cations:
  Cr^{3+}, $CrOH^{2+}$, $Cr(OH)_2^+$, $Cr_2(OH)_2^{4+}$, $Cr_3(OH)_4^{5+}$

- the anion Cl^- with the following cations:
  $CrOH^{2+}$, $Cr(OH)_2^+$, $Cr_2(OH)_2^{4+}$, $Cr_3(OH)_4^{5+}$

Because of this, the assumption is made that each of the B_{ij}'s listed above is equal to zero. Equation (5.8) for the $\dot{B}_{ij}$ calculation therefore reduced to:

$$\dot{B}_{ij} = \frac{0.06\ z_i\ z_j}{\left(1. + \frac{1.5}{z_i\ z_j}\ I\right)^2}$$

This reduces the Bromley formulation to one quite similar to that of Davies (9), described in Chapter IV. Davies suggested that the B interaction term of the equation:

$$\log \gamma_{\pm} = -\frac{A\,|z_+\ z_-|\,\sqrt{I}}{1. + \sqrt{I}} + B\,m \qquad (4.39)$$

could be set for any species in solutions of low ionic strength:

$$B = .15\,|z_+\ z_-|$$

By setting B_{ij} to zero, the expression of the interaction effects becomes dependent on the ionic charges and solution concentration.

Using equation (5.5) and the above parameters and assumptions, the activity coefficients for the ionic species were calculated:

$$\log \gamma_{H^+} = FAC + \dot{B}_{H-Cl}\ Z^2_{H-Cl}\ m_{Cl^-} + \dot{B}_{H-Cr(OH)_4}\ Z^2_{H-Cr(OH)_4}\ m_{Cr(OH)_4^-}$$

$$\log \gamma_{Cr^{3+}} = 9\ FAC + \dot{B}_{Cr-OH}\ Z^2_{Cr-OH}\ m_{OH^-} + \dot{B}_{Cr-Cr(OH)_4}\ Z^2_{Cr-Cr(OH)_4}\ m_{Cr(OH)_4^-} + \dot{B}_{Cr-Cl}\ Z^2_{Cr-Cl}\ m_{Cl^-}$$

$$\log \gamma_{CrOH^{2+}} = 4\ FAC + \dot{B}_{CrOH-OH}\ Z^2_{CrOH-OH}\ m_{OH^-} + \dot{B}_{CrOH-Cr(OH)_4}\ Z^2_{CrOH-Cr(OH)_4}\ m_{Cr(OH)_4^-} + \dot{B}_{CrOH-Cl_4}\ Z^2_{CrOH-Cl}\ m_{Cl^-}$$

$$\log \gamma_{Cr(OH)_2^+} = FAC + \dot{B}_{Cr(OH)_2-OH}\ Z^2_{Cr(OH)_2-OH}\ m_{OH^-} + \dot{B}_{Cr(OH)_2-Cr(OH)_4}\ Z^2_{Cr(OH)_2-Cr(OH)_4}\ m_{Cr(OH)_4^-} + \dot{B}_{Cr(OH)_2-Cl}\ Z^2_{Cr(OH)_2-Cl}\ m_{Cl^-}$$

$$\log \gamma_{Cr_2(OH)_2^{4+}} = 16\ FAC + \dot{B}_{Cr_2(OH)_2-OH}\ Z^2_{Cr_2(OH)_2-OH}\ m_{OH^-} + \dot{B}_{Cr_2(OH)_2-Cr(OH)_4}\ Z^2_{Cr_2(OH)_2-Cr(OH)_4}\ m_{Cr(OH)_4^-} + \dot{B}_{Cr_2(OH)_2-Cl}\ Z^2_{Cr_2(OH)_2-Cl}\ m_{Cl^-}$$

$$\log \gamma_{Cr_3(OH)_4^{5+}} = 25\ FAC + \dot{B}_{Cr_3(OH)_4-OH}\ Z^2_{Cr_3(OH)_4-OH}\ m_{OH^-} + \dot{B}_{Cr_3(OH)_4-Cr(OH)_4}\ Z^2_{Cr_3(OH)_4-Cr(OH)_4}\ m_{Cr(OH)_4^-} + \dot{B}_{Cr_3(OH)_4-Cl}\ Z^2_{Cr_3(OH)_4-Cl}\ m_{Cl^-}$$

$$\log \gamma_{Na^+} = FAC + \dot{B}_{Na-OH}\ Z^2_{Na-OH}\ m_{OH^-} + \dot{B}_{Na-Cr(OH)_4}\ Z^2_{Na-Cr(OH)_4}\ m_{Cr(OH)_4^-}$$

$$\log \gamma_{OH^-} = FAC + \dot{B}_{Cr-OH}\ Z^2_{Cr-OH}\ m_{Cr^{3+}} + \dot{B}_{CrOH-OH}\ Z^2_{CrOH-OH}\ m_{CrOH^{2+}} + \dot{B}_{Cr(OH)_2-OH}\ Z^2_{Cr(OH)_2-OH}\ m_{Cr(OH)_2^+} + \dot{B}_{Cr_2(OH)_2-OH}\ Z^2_{Cr_2(OH)_2-OH}\ m_{Cr_2(OH)_2^{4+}} + \dot{B}_{Cr_3(OH)_4-OH}\ Z^2_{Cr_3(OH)_4-OH}\ m_{Cr_3(OH)_4^{5+}} + \dot{B}_{Na-OH}\ Z^2_{Na-OH}\ m_{Na^+}$$

$$
\begin{aligned}
\log \gamma_{Cr(OH)_4^-} = {} & FAC + \dot{B}_{H-Cr(OH)_4} Z^2_{H-Cr(OH)_4} m_{H^+} + \\
& \dot{B}_{Cr-Cr(OH)_4} Z^2_{Cr-Cr(OH)_4} m_{Cr^{3+}} + \\
& \dot{B}_{CrOH-Cr(OH)_4} Z^2_{CrOH-Cr(OH)_4} m_{CrOH^{2+}} + \\
& \dot{B}_{Cr(OH)_2-Cr(OH)_4} Z^2_{Cr(OH)_2-Cr(OH)_4} m_{Cr(OH)_2^+} + \\
& \dot{B}_{Cr_2(OH)_2-Cr(OH)_4} Z^2_{Cr_2(OH)_2-Cr(OH)_4} m_{Cr_2(OH)_2^{4+}} + \\
& \dot{B}_{Cr_3(OH)_4-Cr(OH)_4} Z^2_{Cr\ (OH)_3-Cr(OH)_4} m_{Cr_3(OH)_4^{5+}} + \\
& \dot{B}_{Na-Cr(OH)_4} Z^2_{Na-Cr(OH)_4} m_{Na^+}
\end{aligned}
$$

$$
\begin{aligned}
\log \gamma_{Cl^-} = {} & FAC + \dot{B}_{H-Cl} Z^2_{H-Cl} m_{H^+} + \dot{B}_{Cr-Cl} Z^2_{Cr-Cl} m_{Cr^{3+}} + \\
& \dot{B}_{CrOH-Cl} Z^2_{CrOH-Cl} m_{CrOH^{2+}} + \dot{B}_{Cr(OH)_2-Cl} Z^2_{Cr(OH)_2-Cl} \\
& m_{Cr(OH)_2^+} + \dot{B}_{Cr_2(OH)_2-Cl} Z^2_{Cr_2(OH)_2-Cl} m_{Cr_2(OH)_2^{4+}} + \\
& \dot{B}_{Cr_3(OH)_4-Cl} Z^2_{Cr_3(OH)_4-Cl} m_{Cr_3(OH)_4^{5+}}
\end{aligned}
$$

The FAC term, used for simplification of the equations, is the Debye-Hückel portion:

$$
FAC = - \frac{A\sqrt{I}}{1. + \sqrt{I}}
$$

Figure 9.12

SYSTEM: CHROME HYDROXIDE - WATER

METHOD: L.A. Bromley, AIChE J, v19, #2 (1973)

==

Temperature: 25.00 degrees Celsius

Equilibrium Constants:

K[H2O]	=	0.993810E-14 mole/kg
K[CrOH]	=	0.129765E-09 mole/kg
K[Cr(OH)2]	=	0.371861E-08 mole/kg
K[Cr(OH)3(aq)]	=	0.203317E-05 mole/kg
K[Cr(OH)3(c)]	=	0.643828E-30 mole/kg
K[Cr(OH)4]	=	0.166535E-04 mole/kg
K[Cr2(OH)2]	=	0.234985E-04 mole/kg
K[Cr3(OH)4]	=	0.251811E-06 mole/kg

H2Oin:	0.555000E+02 moles
Cr(OH)3in:	0.562404E-05 moles
HClin:	0.100000E-01 moles
NaOHin:	0.101039E-01 moles

Liquid Phase:

Ionic strength = 0.101040E-01
pH = 0.995781E+01

	moles	activity coefficient
H2O	0.555100E+02	0.999998E+00
Hion	0.121904E-09	0.904017E+00
OHion	0.100000E-03	0.901798E+00
Crion	0.221396E-17	0.396527E+00
CrOHion	0.931969E-12	0.654627E+00
Cr(OH)2ion	0.164516E-07	0.899321E+00
Cr(OH)3(aq)	0.656232E-06	0.100000E+01
Cr(OH)4ion	0.395136E-05	0.899321E+00
Cr2(OH)2ion	0.298029E-20	0.185467E+00
Cr3(OH)4ion	0.448016E-21	0.724906E-01
Cr(OH)3(c)	0.100000E-05	
Clion	0.100000E-01	0.898104E+00
Naion	0.101039E-01	0.898141E+00

Figure 9.13a

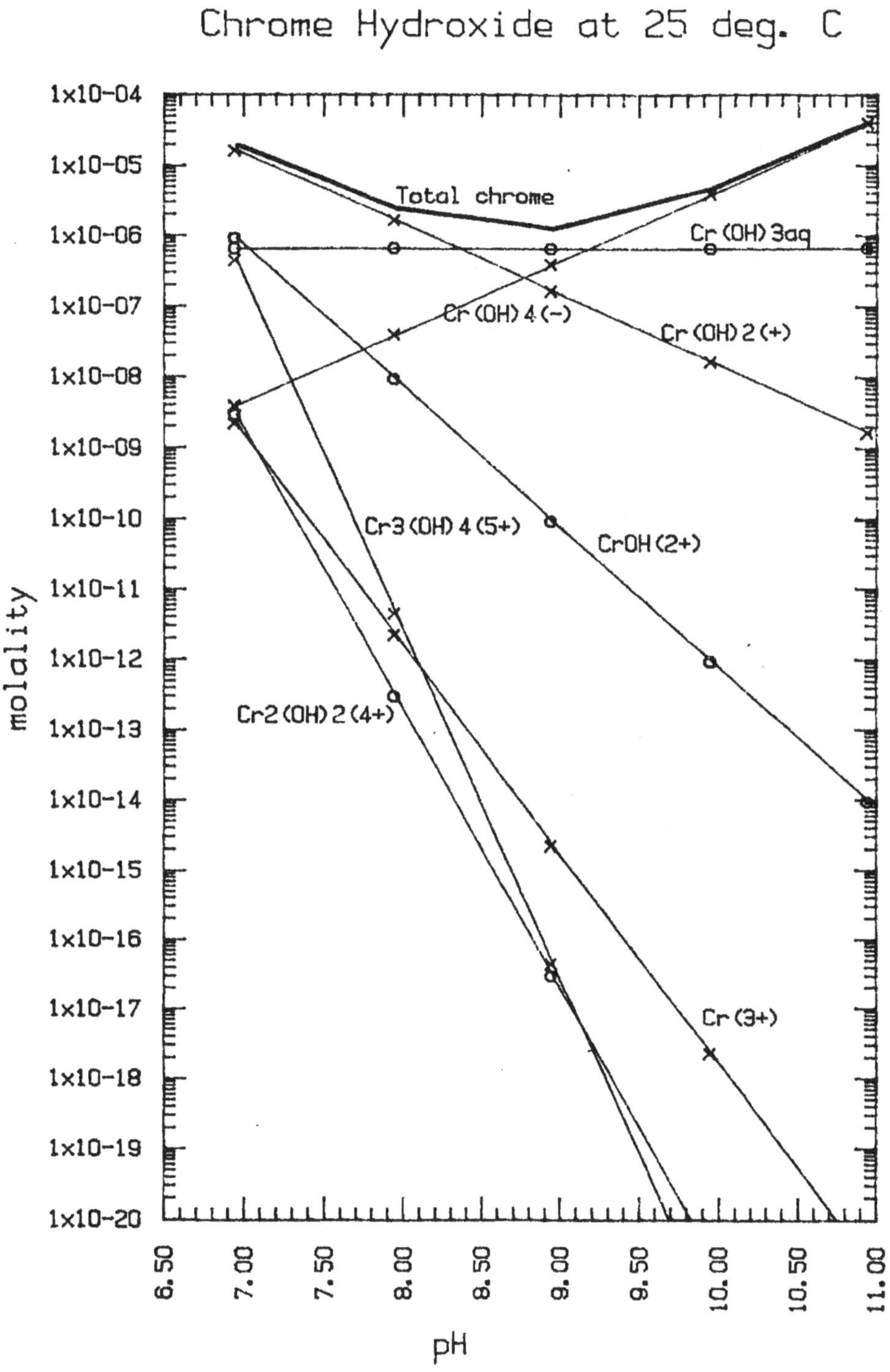
Chrome Hydroxide at 25 deg. C
1x10-04
1x10-05
1x10-06
1x10-07
1x10-08
1x10-09
1x10-10
1x10-11
1x10-12
1x10-13
1x10-14
1x10-15
1x10-16
1x10-17
1x10-18
1x10-19
1x10-20
molality
Total chrome
Cr(OH)3aq
Cr(OH)4(-)
Cr(OH)2(+)
Cr3(OH)4(5+)
CrOH(2+)
Cr2(OH)2(4+)
Cr(3+)
6.50
7.00
7.50
8.00
8.50
9.00
9.50
10.00
10.50
11.00
pH

Figure 9.13b

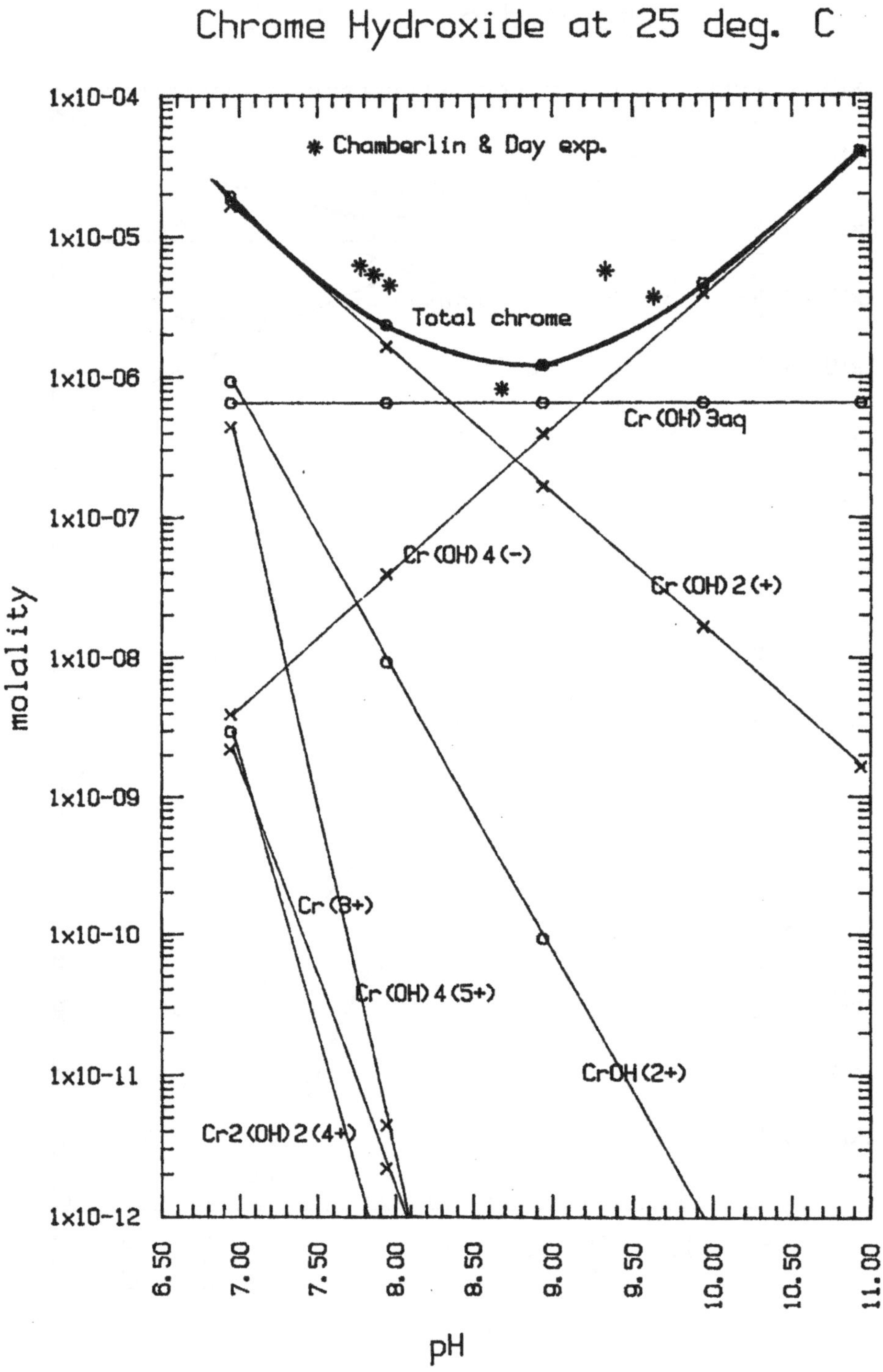
Chrome Hydroxide at 25 deg. C
* Chamberlin & Day exp.
Total chrome
Cr (OH) 3aq
Cr (OH) 4 (-)
Cr (OH) 2 (+)
Cr (3+)
Cr (OH) 4 (5+)
CrOH (2+)
Cr2 (OH) 2 (4+)
molality
pH
1x10-04
1x10-05
1x10-06
1x10-07
1x10-08
1x10-09
1x10-10
1x10-11
1x10-12
6.50
7.00
7.50
8.00
8.50
9.00
9.50
10.00
10.50
11.00

Gypsum Solubility

As discussed in Chapter VI, the gypsum-water system has been a source of disagreement between many investigators. Attempts to mathematically model the system have been hampered by the very small saturation concentration of gypsum, about .0151 molal $CaSO_4$ at 25°C. Interaction parameters based on the binary solution data are not very successful when used in modeling multicomponent solutions as evidenced by the multicomponent solution tests involving gypsum in Chapter V. Additionally, there is disagreement among the investigators as to whether or not the formation of the aqueous molecule $CaSO_4(aq)$ should be considered in the model:

gypsum solubility

$$CaSO_4.2H_2O \; = \; Ca^{2+} + SO_4^{2-} + 2\ H_2O$$

aqueous molecule formation

$$CaSO_4(aq) \; = \; Ca^{2+} + SO_4^{2-}$$

The decision as to whether to include $CaSO_4(aq)$ in the model is all the more crucial in the light of Chapter VII where we discussed just how important the consideration of all species and species interactions could be. In her 1981 doctoral dissertation, P.S.Z. Rogers (22) concluded that the gypsum could be considered to be completely dissociated and that in multicomponent solutions it was more important to consider Pitzer's higher order mixing terms. Pitzer considered these terms to be unimportant in mixtures of doubly and singly charged ions; Rogers success may well have been due simply to the addition of parameters to her regression.

For purposes of illustration we will assume complete dissociation of the gypsum. The system may then be described:

water dissociation

$$H_2O \; = H^+ + OH^-$$

gypsum solubility

$$CaSO_4.2H_2O \; = \; Ca^{2+} + SO_4^{2-} + 2\ H_2O$$

To determine the solubility of gypsum in a sodium chloride solution, the NaCl solubility equation is also necessary:

$$NaCl(c) = Na^+ + Cl^-$$

These three equilibria result in the following equilibrium K equations in a water-gypsum-sodium chloride system model:

$$K_{H_2O} = \gamma_{H^+} m_{H^+} \gamma_{OH^-} m_{OH^-} / a_w \quad (G.1)$$

$$K_{CaSO_4.2H_2O} = \gamma_{Ca^{2+}} m_{Ca^{2+}} \gamma_{SO_4^{2-}} m_{SO_4^{2-}} a_w^2 \quad (G.2)$$

$$K_{NaCl(c)} = \gamma_{Na^+} m_{Na^+} \gamma_{Cl^-} m_{Cl^-} \quad (G.3)$$

Since complete dissociation is being assumed for the gypsum and sodium chloride, the possible formation of complexes such as $NaSO_4^-$ and HSO_4^- is disregarded. Given the initial moles of water, H_2O(in), gypsum as moles of $CaSO_4$, $CaSO_4$(in), and sodium chloride, NaCl(in), in the solution, the model must determine nine unknowns:

liquid phase:

moles of H_2O and the ionic molalities m_{H^+}, m_{OH^-}, $m_{Ca^{2+}}$, $m_{SO_4^{2-}}$, m_{Na^+}, m_{Cl^-}

solid phase:

moles of $CaSO_4.2H_2O$ and NaCl(c)

Of course, depending on the input concentrations the amount of $CaSO_4.2H_2O$ and/or NaCl(c) precipitation may equal zero. It is also possible to find the saturation molality of one in a solution of the other by setting the precipitation amount to a small number and using the input amount as an unknown. For example, to find the saturation point of gypsum in a 4 molal solution of sodium chloride, $CaSO_4.2H_2O$ may be set as $1.x10^{-6}$ and the input amount of gypsum, $CaSO_4$(in), used as an unknown by the model.

Either way, the model still needs six more equations in order for the Newton-Raphson method to be able to solve the system. These would be elemental and electroneutrality balances:

calcium balance

$$CaSO_4(in) = CaSO_4.2H_2O + \frac{H_2O}{55.51}\left(m_{Ca^{2+}}\right) \quad (G.4)$$

hydrogen balance

$$2\ H_2O(in) = 2\ H_2O + \frac{H_2O}{55.51}\left(m_{H^+} + m_{OH^-}\right) + 4\ CaSO_4.2H_2O \quad (G.5)$$

sodium balance

$$NaCl(in) = NaCl(c) + \frac{H_2O}{55.51}\left(m_{Na^+}\right) \quad (G.6)$$

sulfur balance

$$CaSO_4(in) = CaSO_4.2H_2O + \frac{H_2O}{55.51}\left(m_{SO_4^{2-}}\right) \quad (G.7)$$

chlorine balance

$$NaCl(in) = NaCl(c) + \frac{H_2O}{55.51}\left(m_{Cl^-}\right) \quad (G.8)$$

electroneutrality

$$m_{H^+} + m_{Na^+} + 2\, m_{Ca^{2+}} = m_{OH^-} + m_{Cl^-} + 2\, m_{SO_4^{2-}} \quad (G.9)$$

The complete Pitzer equation for the mean activity coefficient of electrolyte CA in a multicomponent solution is:

$$\ln \gamma_{CA} = |z_+ z_-|\, f^{\gamma} + \frac{2\nu_+}{\nu} \sum_a m_a \left[B_{Ca} + (\Sigma mz)\, C_{Ca} + \frac{\nu_-}{\nu_+}\, \Theta_{Aa}\right] +$$

$$\frac{2\nu_-}{\nu} \sum_c m_c \left[B_{cA} + (\Sigma mz)\, C_{cA} + \frac{\nu_+}{\nu_-}\, \Theta_{Cc}\right] +$$

$$\sum_c \sum_a m_c m_a \left[\, |z_+ z_-|\, B'_{ca} + \frac{1}{\nu}\left(2\nu_+ z_+ C_{ca} + \nu_+ \psi_{Cca} + \nu_- \psi_{caA}\right)\right]$$

$$+ .5 \sum_c \sum_{c'} m_c m_{c'} \left[\frac{\nu_-}{\nu}\, \psi_{cc'A} + |z_+ z_-|\, \Theta'_{cc'}\right] +$$

$$.5 \sum_a \sum_{a'} m_a m_{a'} \left[\frac{\nu_+}{\nu}\, \psi_{Caa'} + |z_+ z_-|\, \Theta'_{aa'}\right] \quad (5.32)$$

The terms used in this equation are:

$$f^{\gamma} = -A_{\phi}\left[\frac{\sqrt{I}}{1. + 1.2\sqrt{I}} + \frac{2.}{1.2}\ln(1. + 1.2\sqrt{I})\right]$$

A_{ϕ} = Debye-Hückel constant for osmotic coefficients

$$B_{ij} = \beta_0 + \frac{2\beta_1}{\alpha_1^2 I}\left[1. - (1. + \alpha_1\sqrt{I})\exp(-\alpha_1\sqrt{I})\right] + \frac{2\beta_2}{\alpha_2^2 I}\left[1. - (1. + \alpha_2\sqrt{I})\exp(-\alpha_2\sqrt{I})\right]$$

$$B'_{ij} = \frac{2.\beta_1}{\alpha_1^2 I^2}\left[-1. + (1. + \alpha_1\sqrt{I} + .5\alpha_1^2 I)\exp(-\alpha_1\sqrt{I})\right] + \frac{2.\beta_2}{\alpha_2^2 I^2}\left[-1. + (1. + \alpha_2\sqrt{I} + .5\alpha_2^2 I)\exp(-\alpha_2\sqrt{I})\right]$$

α_1 = 1.4 for bivalent metal sulfates, otherwise = 2.0

α_2 = 12. for bivalent metal sulfates, otherwise = 0.0

$$C_{ij} = \frac{C^{\phi}_{ij}}{2.\ \sqrt{z_+ z_-}} \quad ; \quad C^{\phi} \text{ values tabulated in Appendix 9.5}$$

β - values tabulated in Appendix 9.5

$$\Theta_{ij} = \theta_{ij} + {}^{E}\theta_{ij}(I) \tag{5.33a}$$

$$\Theta'_{ij} = {}^{E}\theta'_{ij}(I) \tag{5.33b}$$

It was the E terms, which are dependent only on the ion charges and total ionic strength, that Pitzer felt could be ignored and Rogers said should not. The calculation of these E terms is described in Appendix 5.1. The Θ_{ij} and Ψ_{ijk} terms are tabulated in Appendix 9.5.

The inclusion of the electrostatic effects of unsymmetrical mixing, the E terms, changes the general form of the ionic activity coefficients given as equations (5.34a) and (5.34b). The ionic activity coefficients of cation C and anion A in a multicomponent solution are now written:

$$\ln \gamma_C = z_+^2\ f^{\gamma} + \sum_a m_a \left[2\ B_{Ca} + (2\ \sum_c m_c\ z_c)\ C_{Ca}\right] +$$

$$\sum_c m_c\ (2\ \Theta_{Cc} + \sum_a m_a\ \psi_{Cca}) + \sum_c \sum_a m_c\ m_a\ (z_+^2\ B'_{ca} + |z_+|\ C_{ca}) +$$

$$.5 \sum_a \sum_{a'} m_a\ m_{a'}\ (\psi_{Caa'} + z_+^2\ \Theta'_{aa'}) + .5\ z_+^2 \sum_c \sum_{c'} m_c\ m_{c'}\ \Theta'_{cc'}$$

$$\ln \gamma_A = z_-^2\ f^{\gamma} + \sum_c m_c \left[2\ B_{cA} + (2 \sum_a m_a\ z_a)\ C_{cA}\right] + \sum_a m_a\ (2\ \Theta_{Aa} + \sum_c m_c\ \psi_{Aac}) +$$

$$\sum_c \sum_a m_c\ m_a\ (z_-^2\ B'_{ca} + |z_-|\ C_{ca}) + .5 \sum_c \sum_{c'} m_c\ m_{c'}\ (\psi_{cc'A} + z_-^2\ \Theta'_{cc'}) +$$

$$.5\ z_-^2 \sum_a \sum_{a'} m_a\ m_{a'}\ \Theta'_{aa'}$$

The water activity is calculated from the osmotic coefficient calculated by the following equation:

$$\phi - 1. = (\sum_i m_i)^{-1} \Big[2\ I\ f^{\phi} + 2 \sum_c \sum_a m_c\, m_a \Big(B^{\phi}_{ca} + \frac{(\sum mz)}{\sqrt{z_c\, z_a}}\, C^{\phi}_{ca}\Big) +$$

$$\sum_c \sum_{c'} m_c\, m_{c'} \Big(\Theta_{cc'} + I\, \Theta'_{cc'} + \sum_a m_a\, \psi_{cc'a} \Big) +$$

$$\sum_a \sum_{a'} m_a\, m_{a'} \Big(\Theta_{aa'} + I\, \Theta'_{aa'} + \sum_c m_c\, \psi_{caa'} \Big) \Big] \qquad (5.54)$$

where:

$$f^{\phi} = - A_{\phi}\ \frac{\sqrt{I}}{1. + 1.2\sqrt{I}}$$

$$B^{\phi}_{ca} = \beta_0 + \beta_1 \exp(-\alpha_1 \sqrt{I}\,) + \beta_2 \exp(-\alpha_2 \sqrt{I}\,)$$

$$(\sum mz) = \sum_a m_a \left| z_a \right| = \sum_c m_c \left| z_c \right|$$

Using the parameters tabulated in Appendix 9.5 and the electrostatic effect parameters calculated with Pitzer's method as outlined in Appendix 5.1, the ionic activity coefficients for the gypsum - sodium chloride - water system are calculated as follows:

$$Z = 2.\ (m_{H^+} + 2.\ m_{Ca^{2+}} + m_{Na^+})$$

$$Bsum = m_{Ca^{2+}} \left(m_{Cl^-}\ B'_{Ca^{2+}-Cl^-} + m_{SO_4^{2-}}\ B'_{Ca^{2+}-SO_4^{2-}} \right) +$$

$$m_{Na^+} \left(m_{Cl^-}\ B'_{Na^+-Cl^-} + m_{SO_4^{2-}}\ B'_{Na^+-SO_4^{2-}} \right)$$

$$Csum = m_{Ca^{2+}} \left(m_{Cl^-}\ C^{\phi}_{Ca^{2+}-Cl^-} / 2\sqrt{2} + m_{SO_4^{2-}}\ C^{\phi}_{Ca^{2+}-SO_4^{2-}} / 4. \right) +$$

$$m_{Na^+} \left(m_{Cl^-}\ C^{\phi}_{Na^+-Cl^-} / 2 + m_{SO_4^{2-}}\ C^{\phi}_{Na^+-SO_4^{2-}} / 2\sqrt{2} \right)$$

$$\Theta_{i-i'} = \theta_{i-i'} + {}^{E}\theta_{i-i'}(I)$$

$$\Theta'_{i-i'} = {}^{E}\theta'_{i-i'}(I)$$

for the hydrogen ion:

$$\ln \gamma_{H^+} = f^{\gamma} + Bsum + Csum$$

for the hydroxide ion:

$$\ln \gamma_{OH^-} = f^{\gamma} + Bsum + Csum$$

for the calcium ion:

$$\ln \gamma_{Ca^{2+}} = 4\ f^{\gamma} + \left(m_{Cl^-}\right) \text{BRAC1} + \left(m_{SO_4^{2-}}\right) \text{BRAC3} + \left(m_{Na^+}\right) \text{BRAC5} + 4\ \text{Bsum} + 2\ \text{Csum} + \text{BRAC7} + 4\ \text{BRAC8} + 4\ \text{BRAC9}$$

for the sodium ion:

$$\ln \gamma_{Na^+} = f^{\gamma} + \left(m_{Cl^-}\right) \text{BRAC2} + \left(m_{SO_4^{2-}}\right) \text{BRAC4} + \left(m_{Ca^{2+}}\right) \text{BRAC5} + \text{Bsum} + \text{Csum} + \text{BRAC7} + \text{BRAC8} + \text{BRAC9}$$

for the sulfate ion:

$$\ln \gamma_{SO_4^{2-}} = 4\ f^{\gamma} + \left(m_{Ca^{2+}}\right) \text{BRAC3} + \left(m_{Na^+}\right) \text{BRAC4} + \left(m_{Cl^-}\right) \text{BRAC6} + 4\ \text{Bsum} + 2\ \text{Csum} + \text{BRAC10} + 4\ \text{BRAC11} + 4\ \text{BRAC12}$$

for the chloride ion:

$$\ln \gamma_{Cl^-} = f^{\gamma} + \left(m_{Ca^{2+}}\right) \text{BRAC1} + \left(m_{Na^+}\right) \text{BRAC2} + \left(m_{SO_4^{2-}}\right) \text{BRAC6} + \text{Bsum} + \text{Csum} + \text{BRAC10} + \text{BRAC11} + \text{BRAC12}$$

The following terms were defined to simplify the equations:

$$\text{BRAC1} = 2\ B_{Ca^{2+}-Cl^-} + Z\ \frac{C^{\phi}_{Ca^{2+}-Cl^-}}{2.\sqrt{2.}}$$

$$\text{BRAC2} = 2\ B_{Na^+-Cl^-} + Z\ \frac{C^{\phi}_{Na^+-Cl^-}}{2.}$$

$$\text{BRAC3} = 2\ B_{Ca^{2+}-SO_4^{2-}} + Z\ \frac{C^{\phi}_{Ca^{2+}-SO_4^{2-}}}{4.}$$

$$\text{BRAC4} = 2\ B_{Na^+-SO_4^{2-}} + Z\ \frac{C^{\phi}_{Na^+-SO_4^{2-}}}{2.\sqrt{2.}}$$

$$\text{BRAC5} = 2\ \theta_{Ca^{2+}-Na^+} + \left(m_{Cl^-}\right) \psi_{Ca^{2+}-Na^+-Cl^-} + \left(m_{SO_4^{2-}}\right) \psi_{Ca^{2+}-Na^+-SO_4^{2-}}$$

$$\text{BRAC6} = 2\ \theta_{SO_4^{2-}-Cl^-} + \left(m_{Ca^{2+}}\right) \psi_{Ca^{2+}-SO_4^{2-}-Cl^-} + \left(m_{Na^+}\right) \psi_{Na^+-SO_4^{2-}-Cl^-}$$

$$\text{BRAC7} = m_{Cl^-} \left(m_{SO_4^{2-}}\right) \psi_{Ca^{2+}-Cl^--SO_4^{2-}}$$

$$\text{BRAC8} = m_{Cl^-} \left(m_{SO_4^{2-}}\right) \theta'_{Cl^--SO_4^{2-}}$$

$$\text{BRAC9} = m_{Ca^{2+}} \left(m_{Na^+} \right) \, \Theta'_{Ca^{2+}-Na^+}$$

$$\text{BRAC10} = m_{Ca^{2+}} \left(m_{Na^+} \right) \, \psi_{Ca^{2+}-Na^+-SO_4^{2-}}$$

$$\text{BRAC11} = m_{Ca^{2+}} \left(m_{Na^+} \right) \, \Theta'_{Ca^{2+}-Na^+}$$

$$\text{BRAC12} = m_{SO_4^{2-}} \left(m_{Cl^-} \right) \, \Theta'_{SO_4^{2-}-Cl^-}$$

The water activity of the solution is calculated from the osmotic coefficient

$$\ln \, a_w = - \, \frac{M_s \, \sum_i m_i}{1000.} \, \phi$$

Using Pitzer's equation, the osmotic coefficient is:

$$\phi = 1. + \left(2 \, I \, f^{\phi} + \left[\, 2 \, m_{Na^+} \, m_{Cl^-} \left(B^{\phi}_{Na^+-Cl^-} + (\Sigma mz) \, C^{\phi}_{Na^+-Cl^-} \right) + \right. \right.$$

$$m_{Na^+} \, m_{SO_4^{2-}} \left(B^{\phi}_{Na^+-SO_4^{2-}} + \frac{(\Sigma \, mz)}{\sqrt{2.}} \, C^{\phi}_{Na^+-SO_4^{2-}} \right) +$$

$$m_{Ca^{2+}} \, m_{Cl^-} \left(B^{\phi}_{Ca^{2+}-Cl^-} + \frac{(\Sigma \, mz)}{\sqrt{2.}} \, C^{\phi}_{Ca^{2+}-SO_4^{2-}} \right) +$$

$$\left. m_{Ca^{2+}} \, m_{SO_4^{2-}} \left(B^{\phi}_{Ca^{2+}-SO_4^{2-}} + \frac{(\Sigma mz)}{2.} \, C^{\phi}_{Ca^{2+}-SO_4^{2-}} \right) \right] +$$

$$2 \left[m_{Na^+} \, m_{Ca^{2+}} \left(\Theta_{Na^+-Ca^{2+}} + I \, \Theta'_{Na^+-Ca^{2+}} + m_{Cl^-} \, \psi_{Na^+-Ca^{2+}-Cl^-} + \right. \right.$$

$$\left. m_{SO_4^{2-}} \, \psi_{Na^+-Ca^{2+}-SO_4^{2-}} \right) \right] + 2 \left[m_{Cl^-} \, m_{SO_4^{2-}} \left(\Theta_{Cl^--SO_4^{2-}} + \right. \right.$$

$$\left. \left. \left. I \, \Theta'_{Cl^--SO_4^{2-}} + m_{Na^+} \, \psi_{Na^+-Cl^--SO_4^{2-}} + m_{Ca^{2+}} \, \psi_{Ca^{2+}-Cl^--SO_4^{2-}} \right) \right] \right) \Bigg/$$

$$\left(m_{H^+} + m_{OH^-} + m_{Na^+} + m_{Cl^-} + m_{Ca^{2+}} + m_{SO_4^{2-}} \right)$$

The system model, equations (G.1) to (G.9), may be solved for the nine species concentrations using the above equations for the osmotic and activity coefficients with a nonlinear equation solving method such as the Newton-Raphson. The results of such a calculation for the solubility of gypsum in a 4 molal sodium chloride solution may be seen in Figure 9.14. Figure 9.15 is a plot of the experimental data versus the solubilities calculated using the Rogers/Pitzer method. The results using Pitzer's method, detailed in Chapter V, are also included in Figure 9.15. It can be seen that Rogers' inclusion of the electrostatic effects terms has resulted in an improvement of the fit.

Pitzer (19) has noted that there are systems in which the association equilibria are so strong that they must not be ignored. He felt that it was much simpler however to incorporate the effects into the interaction coefficients in multicomponent solutions. In their 1985 paper, Rafal, Clark, Rastogi and Scrivner (20) developed a model including $CaSO_4(aq)$ and based upon extremely close fits to published data, inferred that the inclusion of the $CaSO_4(aq)$ was quite significant. As with the Rogers/Pitzer method, the gypsum-sodium chloride-water system activity coefficients were calculated using binary solution parameters for NaCl, $CaCl_2$, etc. The parameters for the sodium sulfate binary differed in that the intermediate $NaSO_4^-$ ion was considered with the result that the ion-ion interactions Na^+-SO_4^{2-} and Na^+-$NaSO_4^-$ were used. A major difference between the Rogers and Rafal et al. models is in the number of parameters added in order to account for interactions in the multicomponent solution. For the 25°C NaCl-$CaSO_4$.$2H_2O$-H_2O solution, Rogers added the following terms:

$$\theta_{Na-Ca} \qquad \theta_{SO_4-Cl}$$

$$\psi_{Na-Ca-Cl} \qquad \psi_{SO_4-Cl-Ca}$$

$$\psi_{Na-Ca-SO_4} \qquad \psi_{SO_4-Cl-Na}$$

$$^E\theta_{Na-Ca} \qquad ^E\theta_{SO_4-Cl}$$

$$^E\theta'_{Na-Ca} \qquad ^E\theta'_{SO_4-Cl}$$

Of these ten additional terms, Pitzer chose to ignore the E terms which must be recalculated for each new ionic strength.

By including the $CaSO_4(aq)$ and $NaSO_4^-$ complexes in their model, Rafal et al. found it necessary to add only the β_0 and β_1 terms for the interaction of the $CaSO_4(aq)$ molecule with the sodium or chlorine ion:

$$\beta_0\,(Na\text{-}CaSO_4) \;=\; \beta_0\,(Cl\text{-}CaSO_4)$$

$$\beta_1 \, (Na\text{-}CaSO_4) = \beta_1 \, (Cl\text{-}CaSO_4)$$

In fact by fitting the β_0 and β_1 terms to a simple expression in temperature:

$$\beta = a + bt + ct^2$$

a total of only six parameters were needed in order to model the system at any temperature between 0 and 80°C. Figure 9.16 gives a plot of their results versus experimental data at 40°C.

Figure 9.14

```
SYSTEM: GYPSUM - SODIUM CHLORIDE - WATER

METHOD: Rogers Thesis using Pitzer Equations

=====================================================================

Temperature: 25.00 degrees Celsius

H2Oin:        55.51 moles
NaClin:        1.50 moles
CaSO4in:       1.00 moles

Equilibrium Constants:
  K(H2O)          =  0.101073E-13 moles/kg
  K(CaSO4.2H2O) =  0.261500E-04 moles/kg


Calculation Results:

Liquid Phase:
  Ionic strength =  0.170102E+01

  moles of water and water activity
    H2O        0.536071E+02     AH2O      0.948308E+00

  ionic molalities and activity coefficients
    H          0.236171E-06     AH        0.414539E+00
    OH         0.236171E-06     AOH       0.414539E+00
    Ca         0.502557E-01     ACa       0.206296E+00
    SO4        0.502557E-01     ASO4      0.558101E-01
    Na         0.150000E+01     ANa       0.636897E+00
    Cl         0.150000E+01     ACl       0.658859E+00

Solid Phase:
    CaSO4.2H2O:  0.951467E+00 moles
```

Figure 9.15

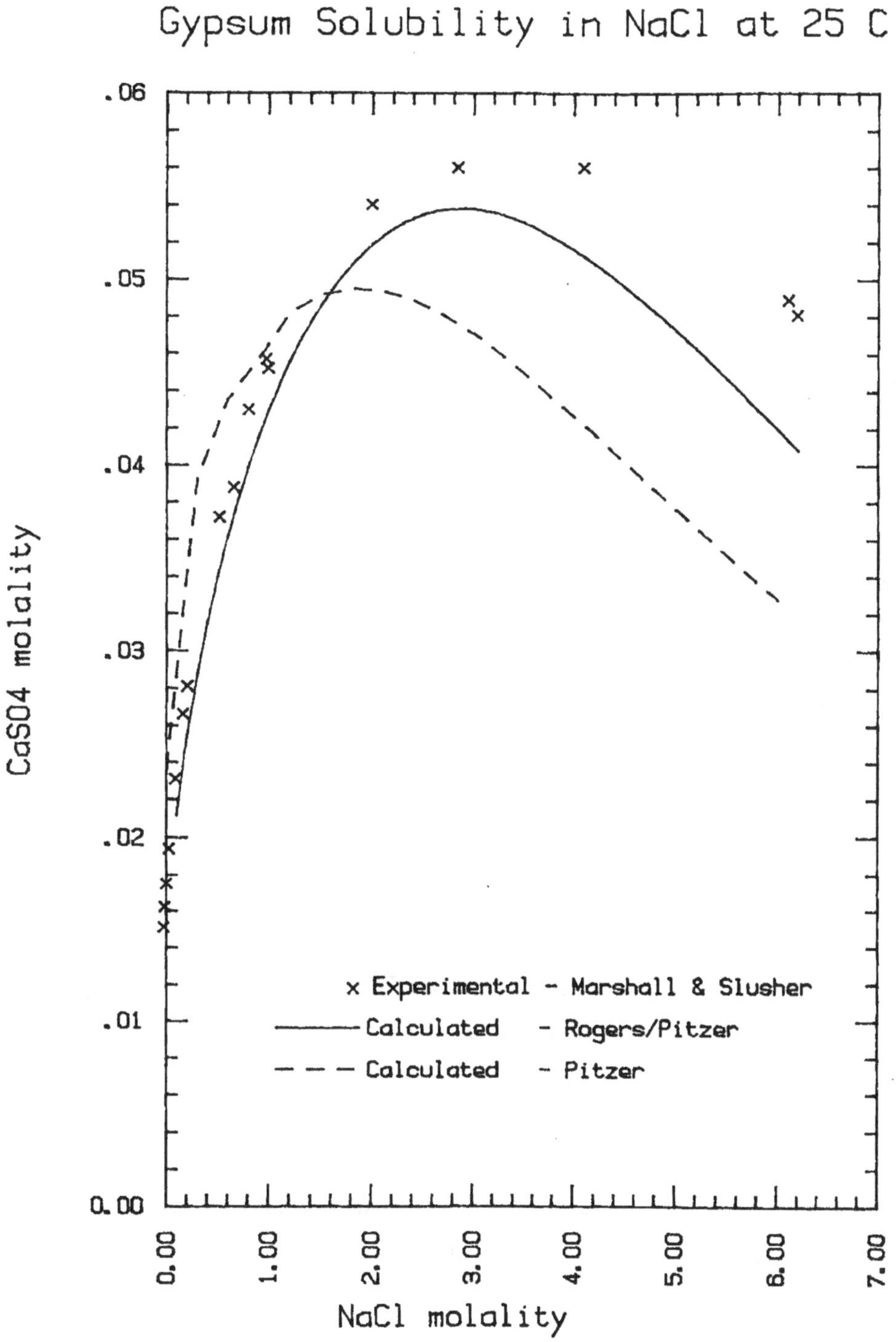
Gypsum Solubility in NaCl at 25 C
.06
.05
.04
.03
.02
.01
0.00
CaSO4 molality
0.00
1.00
2.00
3.00
4.00
5.00
6.00
7.00
NaCl molality
× Experimental - Marshall & Slusher
Calculated - Rogers/Pitzer
Calculated - Pitzer

Figure 9.16

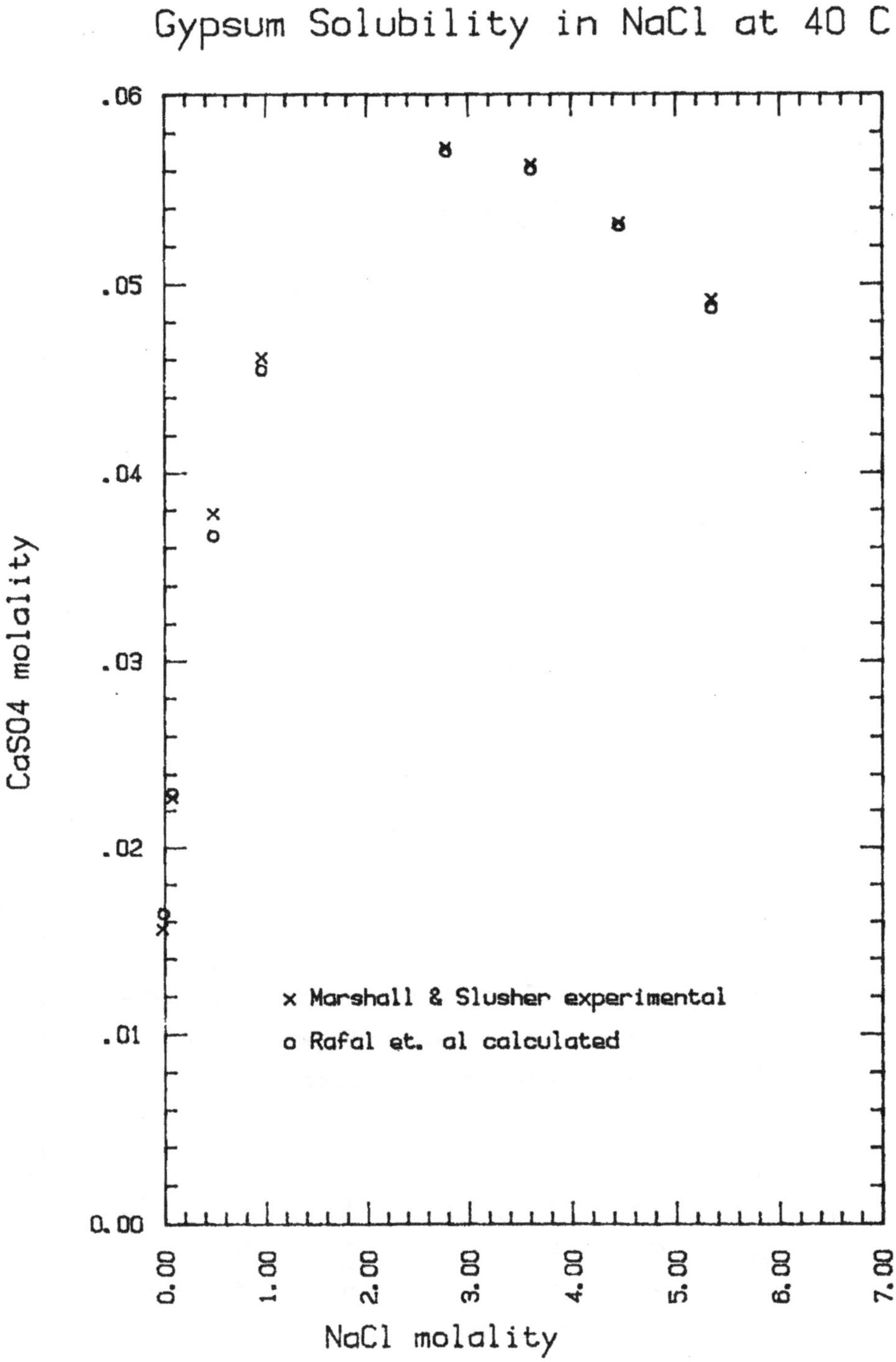
Gypsum Solubility in NaCl at 40 C
.06
.05
.04
.03
.02
.01
0.00
CaSO4 molality
0.00
1.00
2.00
3.00
4.00
5.00
6.00
7.00
NaCl molality
× Marshall & Slusher experimental
o Rafal et. al calculated

Water - Phosphoric Acid

As discussed in Chapter VI, the equilibria in a water-phosphoric acid solution have been studied by numerous authors resulting in nine suggested reactions to explain the system behavior:

$H_3PO_4(aq) = H^+ + H_2PO_4^-$
$H_2PO_4^- = H^+ + HPO_4^{2-}$
$HPO_4^{2-} = H^+ + PO_4^{3-}$
$H_2PO_4^- + H_3PO_4(aq) = H_5(PO_4)_2^-$
$H_2PO_4^- + H_2PO_4^- = H_4(PO_4)_2^{2-}$
$H_2PO_4^- + HPO_4^{2-} = H_3(PO_4)_2^{3-}$
$H_3PO_4(aq) + H_3PO_4(aq) = H_6(PO_4)_2(aq)$
$H_3P_2O_7^- + H_2O = H_3PO_4(aq) + H_2PO_4^-$
$H_2P_2O_7^{2-} + H_2O = 2\ H_2PO_4^-$

Although there has been much study of this system and a large body of good experimental data has been published, the inclusion of all nine equilibria in a mathematical model would be unwarranted since it is unlikely that all species are actually present at the same time. There is as yet insufficient knowledge of the species present in the solution. The investigator must choose a subset of the equilibria.

In 1976, Pitzer and Silvester (18) published the result of their study of modeling the phosphoric acid-water system using the Pitzer equations for activity and osmotic coefficients. As discussed in Chapter IV, Pitzer found the following equation for the activity coefficient of ion i by taking the appropriate derivative of his definition of the excess Gibbs free energy:

$$\ln \gamma_i = \frac{1}{RT}\frac{\partial G^{ex}}{\partial n_i} = \frac{z_i^2}{2.} f' + 2. \sum_j \lambda_{ij} m_j + \frac{z_i^2}{2.} \sum_j \sum_k \lambda'_{jk} m_j m_k + 3. \sum_j \sum_k \mu_{ijk} m_j m_k \qquad (4.58)$$

where:

$f' = \frac{\partial f(I)}{\partial I}$, f(I) describes the long-range electrostatic effects as a function of ionic strength and temperature

λ_{ij} - term for describing the short-range effects between species i and j as a function of ionic strength

$\lambda'_{ij} = \frac{\partial \lambda_{ij}}{\partial I}$, equals zero if i and j are not both ions

μ_{ijk} - triple body interaction term

An expression for the osmotic coefficient was also determined:

$$\phi - 1. = -\frac{\partial G^{ex} / \partial n_w}{RT \sum_i m_i}$$

$$= \left(\sum_i m_i\right)^{-1} \left[(If' - f) + \sum_i \sum_j m_i \; m_j \, (\lambda_{ij} + I \lambda'_{ij}) + 2. \sum_i \sum_j \sum_k m_i \, m_j \, m_k \, \mu_{ijk} \right] \quad (P.1)$$

For the long-range electrostatic effects, the following correlations were chosen:

$$\frac{1}{2} f' = f^{\gamma} = -A_{\phi} \left[\frac{\sqrt{I}}{1. + 1.2\sqrt{I}} + \frac{2.0}{1.2} \ln (1. + 1.2\sqrt{I}) \right] \quad (4.62)$$

$$\frac{1}{2} (f' - \frac{f}{I}) = f^{\phi} = -A_{\phi} \left[\frac{\sqrt{I}}{1. + 1.2\sqrt{I}} \right] \quad (P.2)$$

The virial coefficients for a binary strong electrolyte solution were lumped together and assigned an empirical equation to show the ionic strength dependence:

$$B^{\gamma}_{ij} = 2. \; \lambda_{ij} + I \lambda'_{ij} + \frac{\nu_i}{2 \nu_j} (2 \lambda_{ii} + I \lambda'_{ii}) + \frac{\nu_j}{2 \nu_i} (2 \lambda_{jj} + I \lambda'_{jj})$$

$$= 2. \beta_0 + \frac{2\beta_1}{4I} \left[1. - (1. + 2. \sqrt{I} - 2. I) \exp (-2. \sqrt{I}) \right] \quad (4.63)$$

$$C^{\gamma}_{ij} = \frac{9}{2\sqrt{\nu_i \nu_j}} (\nu_i \mu_{iij} + \nu_j \mu_{ijj})$$

$$= \frac{3}{2} C^{\phi}_{ij} \quad (P.3)$$

$$B^{\phi}_{ij} = \lambda_{ij} + I \lambda'_{ij} + \frac{\nu_i}{2 \nu_j} (\lambda_{ii} + I \lambda'_{ii}) + \frac{\nu_j}{2 \nu_i} (\lambda_{jj} + I \lambda'_{jj})$$

$$= \beta_0 + \beta_1 \exp (-2. \sqrt{I}) \quad (P.4)$$

The β_0, β_1 and C^{ϕ} terms are the familiar Pitzer interaction parameters tabulated in Appendix 4.4.

Pitzer and Silvester used only the first dissociation to model the phosphoric acid-water system:

$$H_3PO_4(aq) = H^+ + H_2PO_4^-$$

$$K = \exp\left(-\frac{\Delta G^\circ}{RT}\right) = \frac{\gamma_{H^+}\ m_{H^+}\ \gamma_{H_2PO_4^-}\ m_{H_2PO_4^-}}{\gamma_{H_3PO_4(aq)}\ m_{H_3PO_4(aq)}}$$

Excluding the water dissociation, the phosphoric acid solution would contain the following species: H^+, $H_2PO_4^-$ and H_3PO_4 (aq). Unlike the treatment given similar solutions containing ions and molecules presented by other authors, Pitzer and Silvester chose to disregard ion-ion interaction terms and modeled the system using only ion-molecule and molecule self interaction terms. Their justification for this was that they felt that by including the association equilibrium in the model the interaction term became redundant. They hypothesized that using a small value for $B_{H^+-H_2PO_4^-}$ in the calculation would be reflected as a small change to the dissociation constant:

$$\Delta\left(\frac{1}{K}\right) = 2\ B_{H^+-H_2PO_4^-}$$

and were able to prove this to themselves.

For the ion-molecule interactions and molecule self interactions, Pitzer and Silvester were able to determine the λ and μ terms. The following equations resulted for the activity and osmotic coefficients for the binary phosphoric acid solution:

- H^+ activity coefficient

$$\ln \gamma_{H^+} = f^\gamma + 2\ \lambda_{H^+-H_3PO_4}\ m_{H_3PO_4} \qquad (P.5)$$

- $H_2PO_4^-$ activity coefficient

$$\ln \gamma_{H_2PO_4^-} = f^\gamma + 2\ \lambda_{H_2PO_4^--H_3PO_4}\ m_{H_3PO_4} \qquad (P.6)$$

- H_3PO_4(aq) activity coefficient

$$\ln \gamma_{H_3PO_4(aq)} = 2\ \lambda_{H^+-H_3PO_4}\ m_{H^+} + 2\ \lambda_{H_2PO_4^--H_3PO_4}\ m_{H_2PO_4^-} + 2\ \lambda_{H_3PO_4-H_3PO_4}\ m_{H_3PO_4} + 3\ \mu_{H_3PO_4-H_3PO_4-H_3PO_4}\ m^2_{H_3PO_4} \qquad (P.7)$$

- the osmotic coefficient

$$\phi - 1. = (m_{H^+} + m_{H_2PO_4^-} + m_{H_3PO_4})^{-1} \; [2\ I\ f^{\phi} + 2\ m_{H_3PO_4}\left(m_{H^+}\ \lambda_{H^+-H_3PO_4} + m_{H_2PO_4^-}\ \lambda_{H_2PO_4^--H_3PO_4} \right) + \lambda_{H_3PO_4-H_3PO_4}\ m^2_{H_3PO_4} + 2\ \mu_{H_3PO_4-H_3PO_4-H_3PO_4}\ m^3_{H_3PO_4}] \quad (P.8)$$

The following values were presented for the parameters of phosphoric acid solutions at 25°C:

$\lambda_{H^+-H_3PO_4} = 0.290$ $\lambda_{H_2PO_4^--H_3PO_4} = -0.400$

$\lambda_{H_3PO_4-H_3PO_4} = 0.05031$ $\mu_{H_3PO_4-H_3PO_4-H_3PO_4} = 0.01095$

$K_{H_3PO_4} = 7.1425\text{x}10^{-3}$ $pK = 2.14615$

The parameters were determined by fitting osmotic coefficients from solvent vapor pressure measurements with a least-squares method. The maximum concentration was about 6 molal and the standard deviation of fit was .005. They were surprised to find a minimum in the degree of dissociation at about 2 molal and an increase in dissociation as the concentration increased. This behavior was attributed to the strong negative value of the $H_2PO_4^--H_3PO_4$ interaction. Additional parameters were presented based on a study of the $KCl-KH_2PO_4-H_2O$ system:

$\lambda_{K^+-H_3PO_4} = -0.070$ $\lambda_{Cl^--H_3PO_4} = 0.0$

$\theta_{Cl^--H_2PO_4^-} = 0.10$ $\psi_{K^+-Cl^--H_2PO_4^-} = -0.0105$

where the parameters used for the binary HCl, KCl and KH_2PO_4 solutions were those tabulated in Appendix 4.4. The mixture parameters were defined:

$$\theta_{Cl^--H_2PO_4^-} = \lambda_{Cl^--H_2PO_4^-} - .5\ \lambda_{H_2PO_4^--H_2PO_4^-} - .5\ \lambda_{Cl^--Cl^-}$$

$$\psi_{K^+-Cl^--H_2PO_4^-} = 6\ \mu_{K^+-Cl^--H_2PO_4^-} - 3\ \mu_{K^+-H_2PO_4^--H_2PO_4^-} - 3\ \mu_{K^+-Cl^--Cl^-}$$

In their conclusions, Pitzer and Silvester note that the strong $H_2PO_4^--H_3PO_4$ interaction term gives credence to the suggested association equilibria:

$H_2PO_4^- + H_3PO_4(aq) = H_5(PO_4)_2^-$

They also suggest, based upon the interaction parameters of NaH_2PO_4 and KH_2PO_4 salts and the $Cl^--H_2PO_4^-$ interaction parameter found in their study, that the tendency of $H_2PO_4^-$ to dimerize

$H_2PO_4^- + H_2PO_4^- = H_4(PO_4)_2^{2-}$

is about half as great as the $H_5(PO_4)_2^-$ formation. However, they felt these reactions to be weak enough to be ignored in the concentration range of their study which was maximum six molal.

To model a phosphoric acid solution using Pitzer and Silvester's method there would be five unknowns in the liquid phase:

$$H_2O,\ m_{H^+},\ m_{OH^-},\ m_{H_3PO_4(aq)} \text{ and } m_{H_2PO_4^-}$$

The equilibria equations are:

- water dissociation

$$K_{H_2O} = \gamma_{H^+}\ m_{H^+}\ \gamma_{OH^-}\ m_{OH^-} \ /\ a_w \qquad (P.9)$$

- phosphoric acid dissociation

$$K_{H_3PO_4} = \gamma_{H^+}\ m_{H^+}\ \gamma_{H_2PO_4^-}\ m_{H_2PO_4^-} \ /\ (\gamma_{H_3PO_4(aq)}\ m_{H_3PO_4}) \qquad (P.10)$$

The activity coefficient equations for the system are (P.5), (P.6) and (P.7) with the addition of:

$$\ln \gamma_{OH^-} = f^{\gamma}$$

The water activity would be calculated using the osmotic coefficient from equation (P.8) in the following:

$$\ln a_w = - \frac{\phi\ M_s\ \Sigma m_i}{1000.}$$

A deterministic model requires three more equations in order to solve for the five unknowns. These equations would be:

- hydrogen balance

$$2\ H_2O(in) + 3\ H_3PO_4(in) = 2\ H_2O + \frac{H_2O}{55.51}\ (m_{H^+} + m_{OH^-} + 2\ m_{H_2PO_4^-} + 3\ m_{H_3PO_4})$$

- phosphorous balance

$$H_3PO_4(in) = \frac{H_2O}{55.51}\ (m_{H_2PO_4^-} + m_{H_3PO_4})$$

- electroneutrality

$$m_{H^+} = m_{OH^-} + m_{H_2PO_4^-}$$

The results using Pitzer and Silvester's model can be seen in Figures 9.17 and 9.18. While they successfully fit the experimental osmotic coefficient it can be seen in Figure 9.18 that as the concentration of phosphoric acid increases, the model begins to deviate from the measured pH values. This supports the idea that it is necessary to account for the complexing behavior that occurs in many solutions.

Figure 9.17

```
SYSTEM: H3PO4-H2O

METHOD: Pitzer, J. Sol. Chem., v5, #4, p269  (1976)

======================================================================

Temperature: 25. degrees Celsius

H2Oin:         55.51 moles
H3PO4in:        2.50 moles

Equilibrium Constants:
  K(H2O)    =  0.101073E-13 moles/kg
  K(H3PO4)  =  0.714250E-02 moles/kg

Calculation Results:

  Ionic strength =   0.295530E+00

  moles of water and water activity:
    H2O          0.555100E+02      AH2O          0.955328E+00

  molalities and activity coefficients:
    Hion         0.295530E+00      AHion         0.227654E+01
    OHion        0.226411E-13      AOHion        0.633855E+00
    H3PO4        0.220447E+01      AH3PO4        0.137212E+01
    H2PO4        0.295530E+00      AH2PO4        0.108662E+00

  pH =  0.172123E+00
```

Figure 9.18

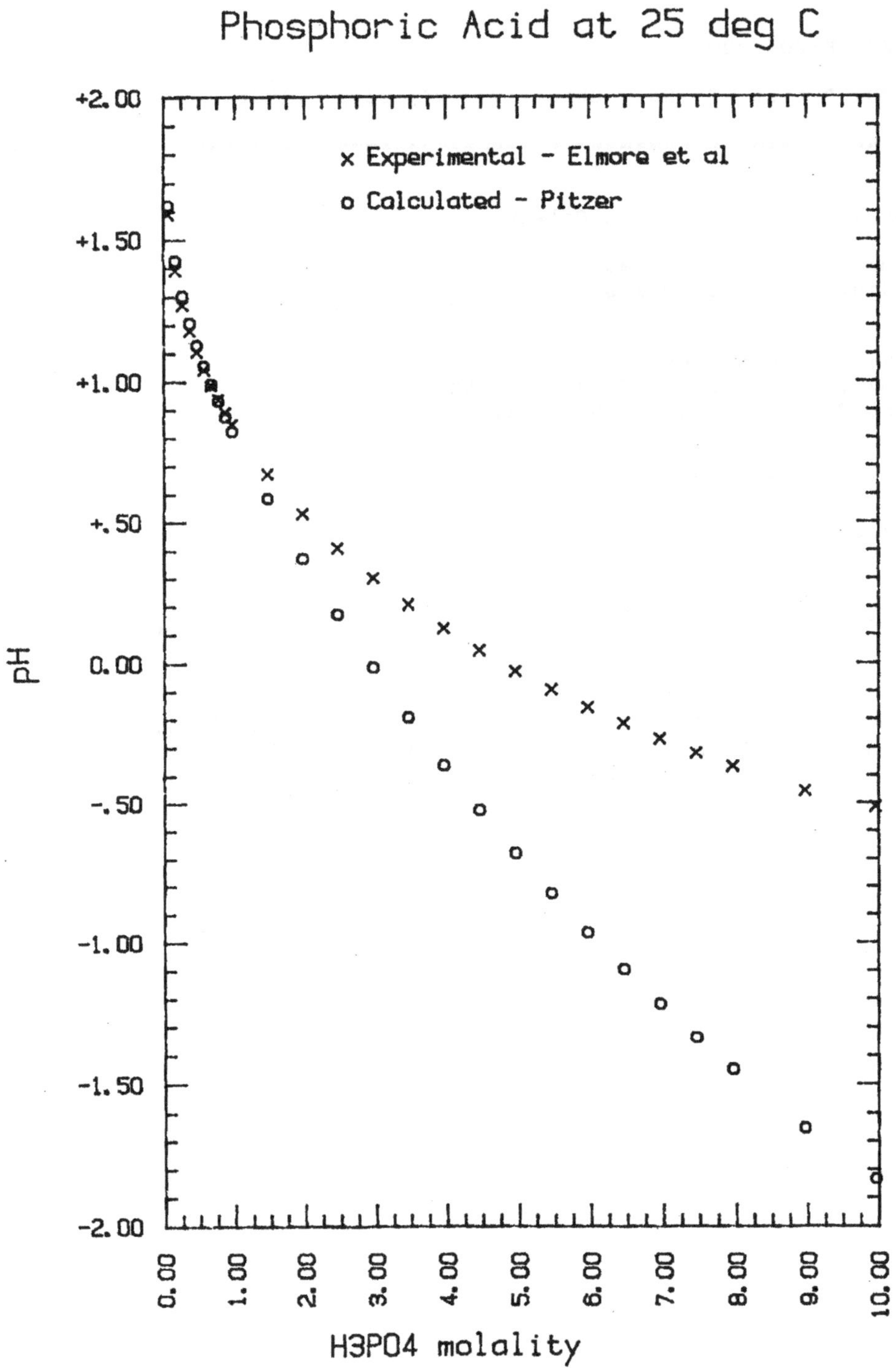
Phosphoric Acid at 25 deg C
× Experimental - Elmore et al
o Calculated - Pitzer
pH
+2.00
+1.50
+1.00
+.50
0.00
-.50
-1.00
-1.50
-2.00
0.00
1.00
2.00
3.00
4.00
5.00
6.00
7.00
8.00
9.00
10.00
H3PO4 molality

APPENDIX 9.1

Parameters for Beutier and Renon's Method

From: H. Renon, "Representation of NH_3-H_2S-H_2O, NH_3-SO_2-H_2O, and NH_3-CO_2-H_2O Vapor-Liquid Equilibria", Thermodynamics of Aqueous Systems with Industrial Applications, S.A. Newman ed., ACS Symposium Series 133, Washington D.C., pp173-186 (1980)

Table 1: Temperature fit parameters for equilibrium constants

	A	B	C	D
K_{CO_2}	-12092.1	-36.7816	0.0	235.482
$K_{HCO_3^-}$	-12431.7	-35.4819	0.0	220.067
K_{H_2O}	-13455.9	-22.4773	0.0	140.932
K_{HS^-}	-2048.99	15.65	0.0	-114.45
K_{H_2S}	-12995.40	-33.5471	0.0	218.5989
$K_{HSO_3^-}$	1333.4	0.0	0.0	-21.274
K_{NH_3}	-3335.71	1.4971	-.0370566	2.76080
$K_{NH_2CO_2^-}$	2900.	0.0	0.0	-8.6
K_{SO_2}	-3768.	-20.	0.0	122.53

Table 2: Temperature fit parameters for Henry's constants

	A	B	C	D
H_{CO_2}	-6789.04	-11.4519	-.010454	94.4914
H_{H_2S}	-13236.8	-55.0551	.0595651	342.595
H_{NH_3}	-157.552	28.1001	-.049227	-149.006
H_{SO_2}	-5160.4	-7.61	0.0	60.538

Table 3: Pitzer ion-ion interaction parameters

	β_0	β_1	C^{ϕ}
H^+ - CO_3^{2-}	0.0	0.0	0.0
- HCO_3^-	.126	.294	0.0
- HS^-	.18	.32	0.0
- HSO_3^-	-.06	-.54	0.0
- $NH_2CO_2^-$	.085	.255	0.0
- OH^-	.04	.12	0.0
- S^{2-}	0.0	0.0	0.0
- SO_3^{2-}	.12	1.08	0.0
NH_4^+ - CO_3^{2-}	.0413	.6587	0.0
- HCO_3^-	-.054	.594	-.08
- HS^-	.05456	.19344	.002
- HSO_3^-	0.0	.45	.0014
- $NH_2CO_2^-$	0.0	.5	-.006
- OH^-	.115	.345	0.0
- S^{2-}	.0413	.6587	.0255
- SO_3^{2-}	.0413	.6587	.0045

Ternary ion interactions:

$\mu_{NH_4^+ - S^{2-} - S^{2-}} = .004$

$\mu_{NH_4^+ - SO_3^{2-} - SO_3^{2-}} = .0007$

Table 4: Temperature fit molecule self interaction parameters

$\lambda_{mm} = A + B/T$; T in Kelvins

	A	B
$\lambda_{CO_2-CO_2}$	-.4922	149.20
$\lambda_{H_2S-H_2S}$	-.2106	61.56
$\lambda_{NH_3-NH_3}$	-.026	12.29
$\lambda_{SO_2-SO_2}$	-.05	0.0

Table 5: Dielectric effect parameters

species	$\bar{\bar{v}}$	α	r
CO_2	.035	-.035	
H_2S	.0349	-.032	
NH_3	.0245	-.019	
SO_2	.036	-.030	
CO_3^{2-}	.0065		4.0
H^+	-.0047		3.8
HCO_3^-	.0288		2.7
HS^-	.018		2.3
HSO_3^-	.035		2.7
NH_4^+	.0134		2.5
$NH_2CO_2^-$	.0459		2.7
OH^-	.0005		3.5
S^{2-}	-.0037		3.3
SO_3^{2-}	.0197		2.8

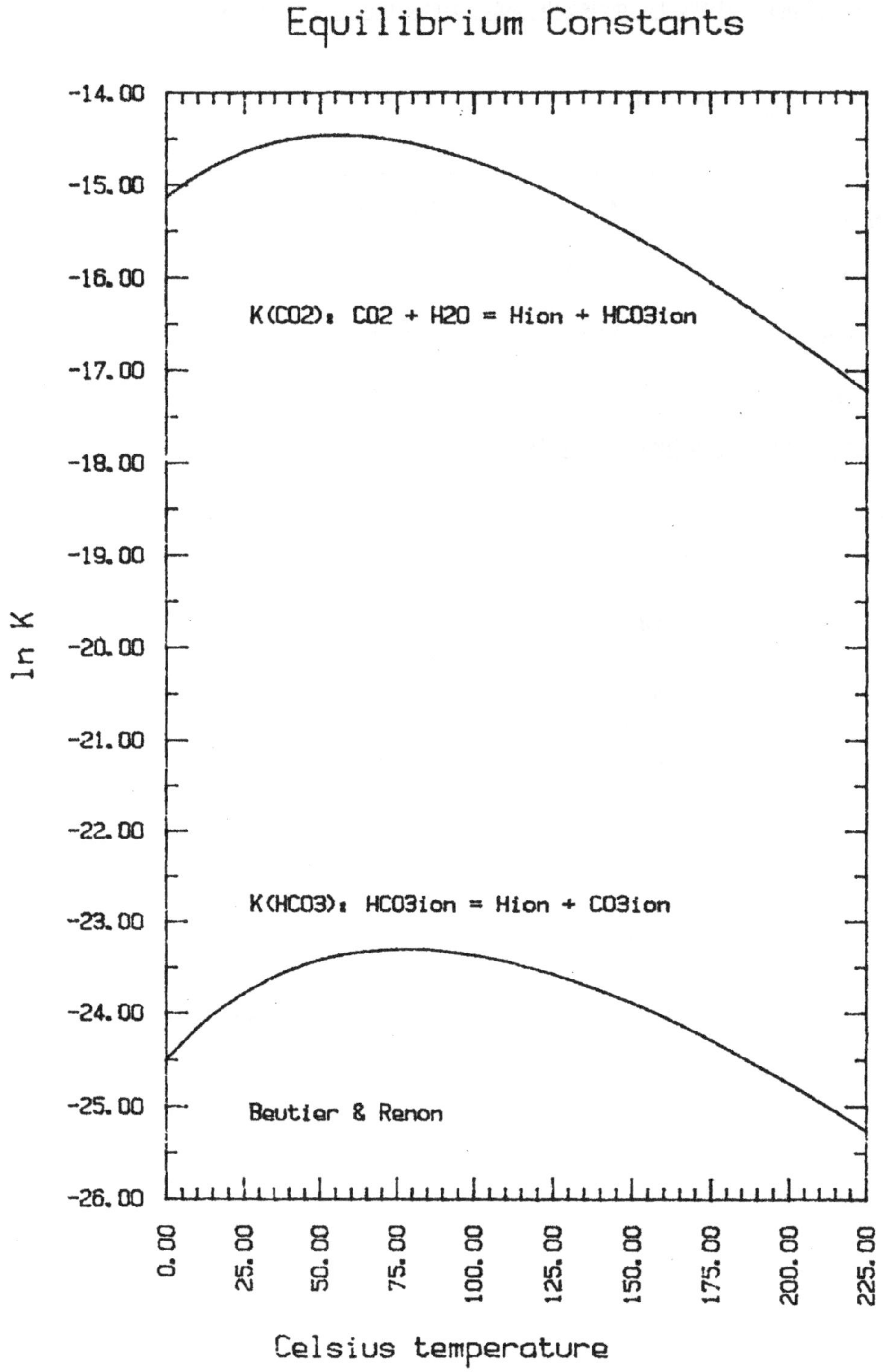
Equilibrium Constants
-14.00
-15.00
-16.00
-17.00
-18.00
-19.00
-20.00
-21.00
-22.00
-23.00
-24.00
-25.00
-26.00
ln K
K(CO2): CO2 + H2O = Hion + HCO3ion
K(HCO3): HCO3ion = Hion + CO3ion
Beutier & Renon
0.00
25.00
50.00
75.00
100.00
125.00
150.00
175.00
200.00
225.00
Celsius temperature

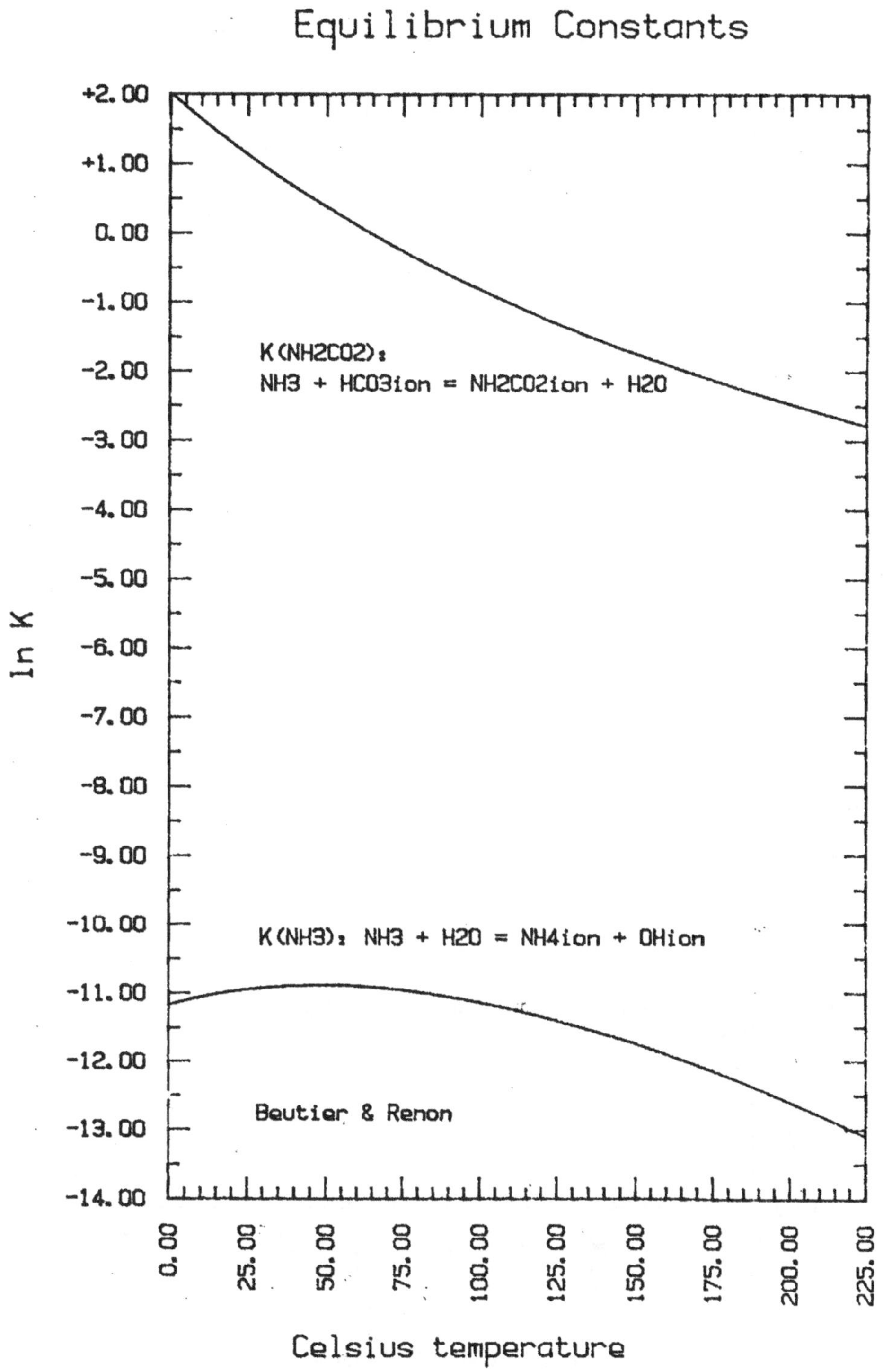
Equilibrium Constants
+2.00
+1.00
0.00
-1.00
-2.00
-3.00
-4.00
-5.00
-6.00
-7.00
-8.00
-9.00
-10.00
-11.00
-12.00
-13.00
-14.00
ln K
K(NH2CO2):
NH3 + HCO3ion = NH2CO2ion + H2O
K(NH3): NH3 + H2O = NH4ion + OHion
Beutier & Renon
0.00
25.00
50.00
75.00
100.00
125.00
150.00
175.00
200.00
225.00
Celsius temperature

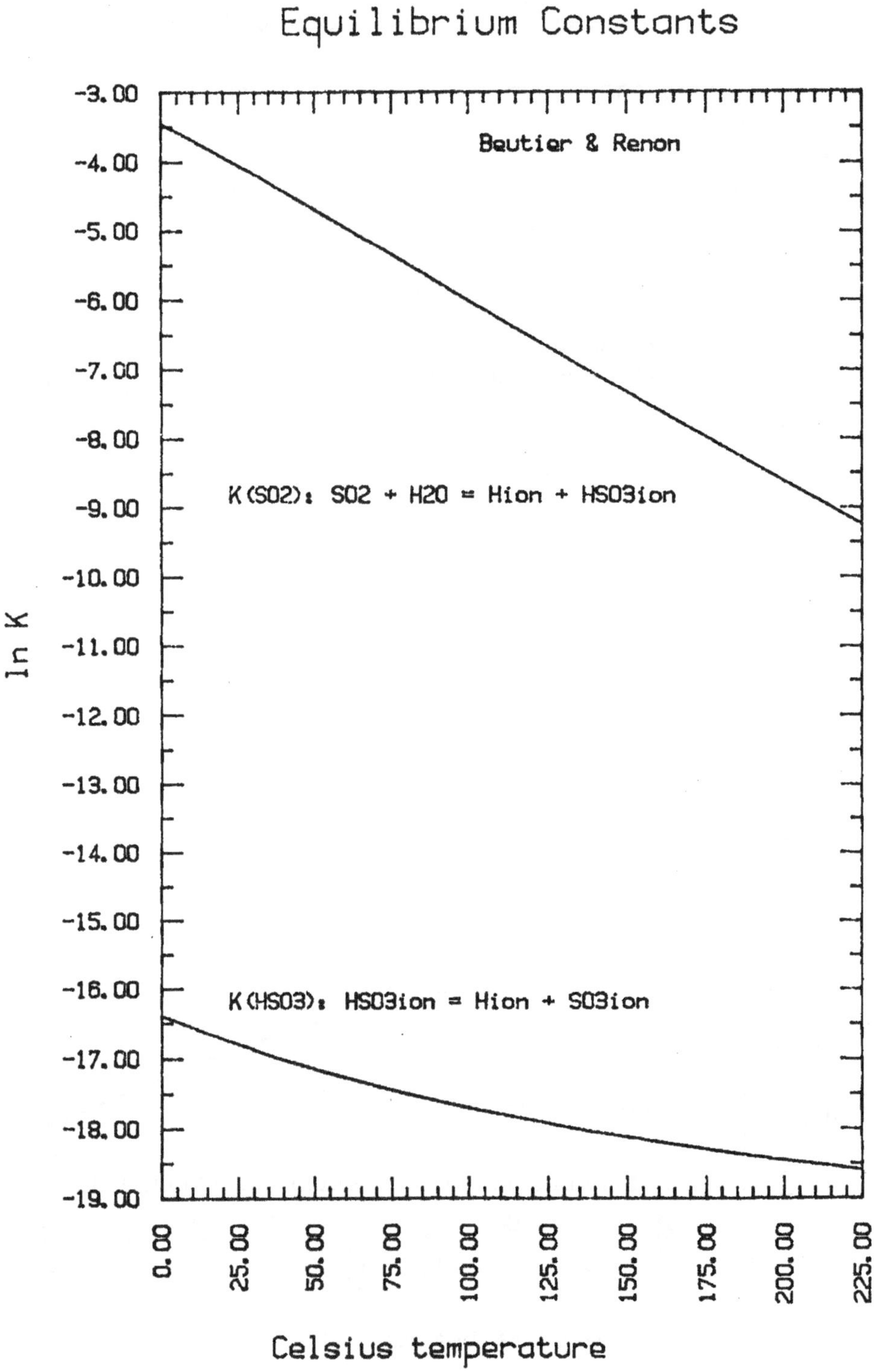
Equilibrium Constants
Beutier & Renon
K(SO2): SO2 + H2O = Hion + HSO3ion
K(HSO3): HSO3ion = Hion + SO3ion
ln K
-3.00
-4.00
-5.00
-6.00
-7.00
-8.00
-9.00
-10.00
-11.00
-12.00
-13.00
-14.00
-15.00
-16.00
-17.00
-18.00
-19.00
0.00
25.00
50.00
75.00
100.00
125.00
150.00
175.00
200.00
225.00
Celsius temperature

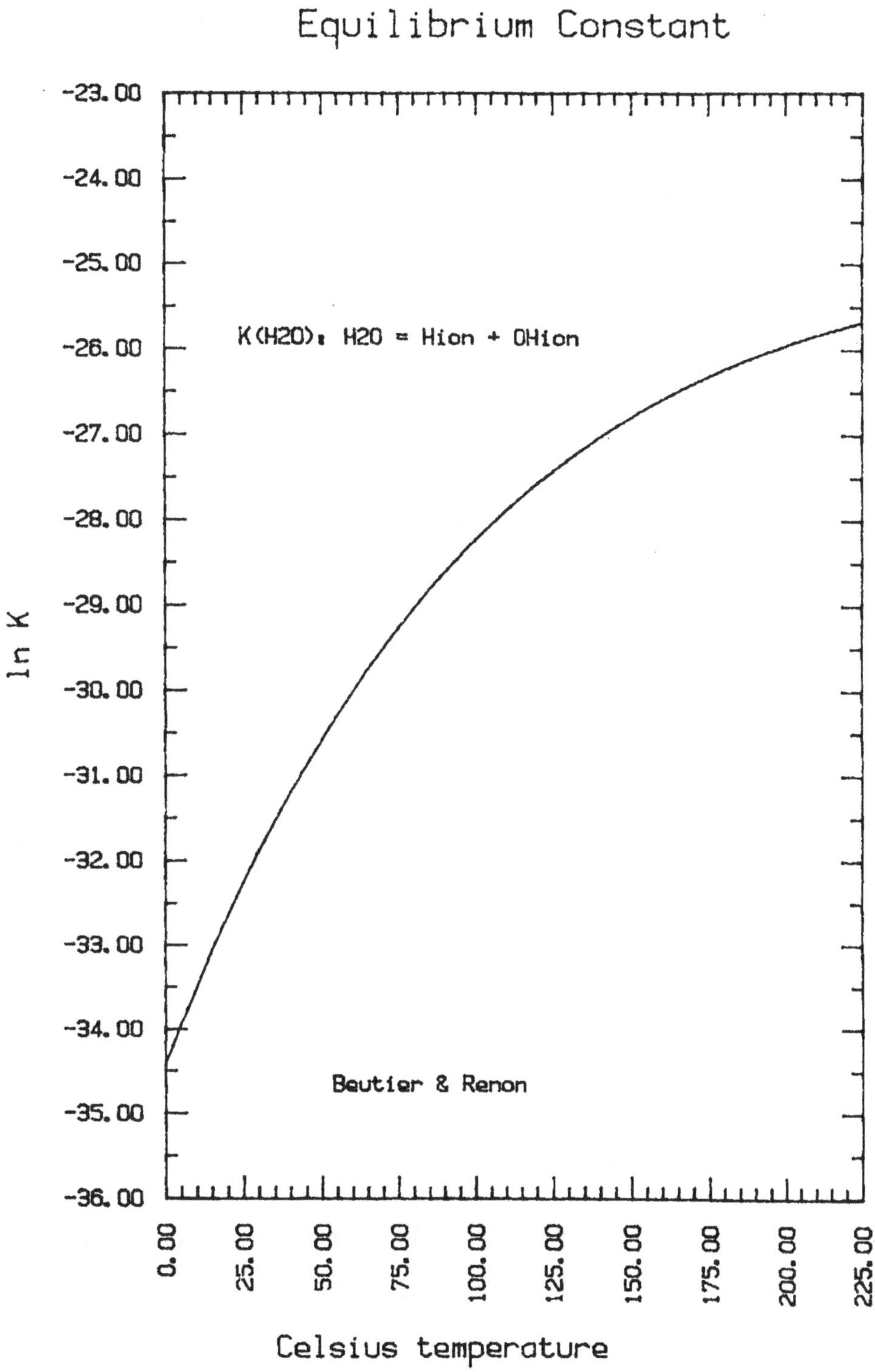
Equilibrium Constant
-23.00
-24.00
-25.00
-26.00
-27.00
-28.00
-29.00
-30.00
-31.00
-32.00
-33.00
-34.00
-35.00
-36.00
ln K
K(H2O): H2O = Hion + OHion
Beutier & Renon
0.00
25.00
50.00
75.00
100.00
125.00
150.00
175.00
200.00
225.00
Celsius temperature

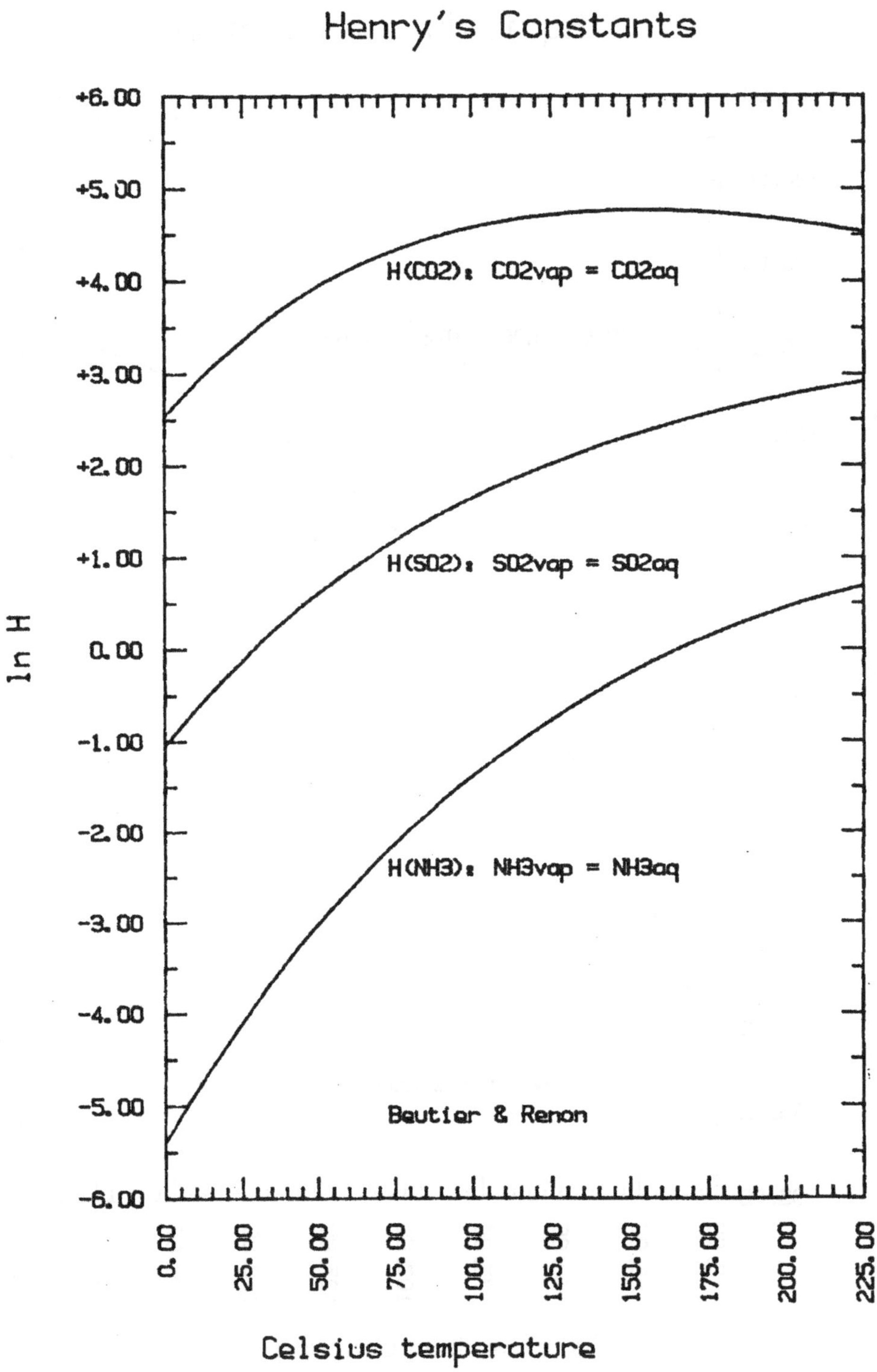
Henry's Constants
+6.00
+5.00
+4.00
+3.00
+2.00
+1.00
0.00
-1.00
-2.00
-3.00
-4.00
-5.00
-6.00
ln H
H(CO2): CO2vap = CO2aq
H(SO2): SO2vap = SO2aq
H(NH3): NH3vap = NH3aq
Beutier & Renon
0.00
25.00
50.00
75.00
100.00
125.00
150.00
175.00
200.00
225.00
Celsius temperature

APPENDIX 9.2

Parameters for Edwards, Maurer, Newman and Prausnitz' Method

From: G. Maurer, "On the Solubility of Volatile Weak Electrolytes in Aqueous Solutions", Thermodynamics of Aqueous Systems with Industrial Applications, S.A. Newman ed., ACS Symposium Series 133, Washington D.C., pp139-172 (1980)

Table 1: Temperature fit parameters for equilibrium constants

	A	B	C	D	T range
K_{CO_2}	-12092.1	-36.7816	0.0	235.482	0-225
$K_{HCO_3^-}$	-12431.7	-35.4819	0.0	220.067	0-225
K_{H_2O}	-13445.9	-22.4773	0.0	140.932	0-225
K_{HS^-}	-7211.2	0.0	0.0	-7.489	20-100
K_{H_2S}	-12995.40	-33.5471	0.0	218.5989	0-150
$K_{HSO_3^-}$	1333.4	0.0	0.0	-21.274	0-50
K_{NH_3}	-3335.7	1.4971	-0.0370566	2.76	0-225
$K_{NH_2CO_2^-}$	2895.65	0.0	0.0	-8.5994	20-60
K_{SO_2}	637.396	0.0	-0.0151337	-1.96211	0-50

Table 2: Temperature fit parameters for Henry's constants

	A	B	C	D	T range
H_{CO_2}	-6789.04	-11.4519	-.010454	94.4914	0-250
H_{H_2S}	-13236.8	-55.0551	0.0595651	342.595	0-150
H_{NH_3}	-157.552	28.1001	-.049227	-149.006	0-150
H_{SO_2}	-5578.8	-8.76152	0.0	68.418	0-100

Table 3: Pitzer ion-ion interaction parameters

	β_0	β_1
H^+ - CO_3^{2-}	0.086	*
- HCO_3^-	0.071	*
- HS^-	0.194	*
- HSO_3^-	0.085	*
- $NH_2CO_2^-$	0.198	*
- OH^-	0.208	*
- S^{2-}	0.127	*
- SO_3^{2-}	0.103	*
NH_4^+ - CO_3^{2-}	-0.062	0.0
- HCO_3^-	-0.0435	*
- HS^-	0.0638	*
- HSO_3^-	-0.0466	.0876
- $NH_2CO_2^-$	0.0505	*
- OH^-	0.06	*
- S^{2-}	-0.021	*
- SO_3^{2-}	-0.045	*

* $\beta_1 = 0.018 + 3.06\ \beta_0$

Table 4: Temperature fit molecule self interaction parameters

$\beta_0 = E + F/T$; T in Kelvins

	E	F
CO_2	-0.4922	149.2
H_2S	-0.2106	61.56
NH_3	-0.0260	12.29
SO_2	0.0275	0.0

Table 5: Molecule-ion interaction parameters

		β_0
CO_2	- H^+	0.033
	- HSO_3^-	-0.03
	- NH_4^+	$0.037 - 2.38E{-4}\ T + 3.83E{-7}\ T^2$
	- $NH_2CO_2^-$	0.017
	- OH^-	$0.26 - 1.62E{-3}\ T + 2.89E{-6}\ T^2$
	- S^{2-}	0.053
	- SO_3^{2-}	0.068
H_2S	- CO_3^{2-}	0.077
	- H^+	0.017
	- HCO_3^-	-0.037
	- HSO^-	-0.045
	- NH_4^+	$0.120 - 2.46E{-4}\ T + 3.99E{-7}\ T^2$
	- $NH_2CO_2^-$	-0.032
	- OH^-	$0.26 - 1.72E{-3}\ T + 3.07E{-6}\ T^2$
	- SO_3^{2-}	0.051
NH_3	- CO_3^{2-}	0.068
	- H^+	0.015
	- HCO_3^-	-0.0816
	- HS^-	-0.0449
	- HSO_3^-	-0.038
	- NH_4^+	0.0117
	- OH^-	$0.227 - 1.47E{-3}\ T + 2.6E{-6}\ T^2$
	- S^{2-}	0.032
	- SO_3^{2-}	-0.044

		β_1
NH_3	- HCO_3^-	0.4829
	- HS^-	0.406
	- NH_4^+	-0.20

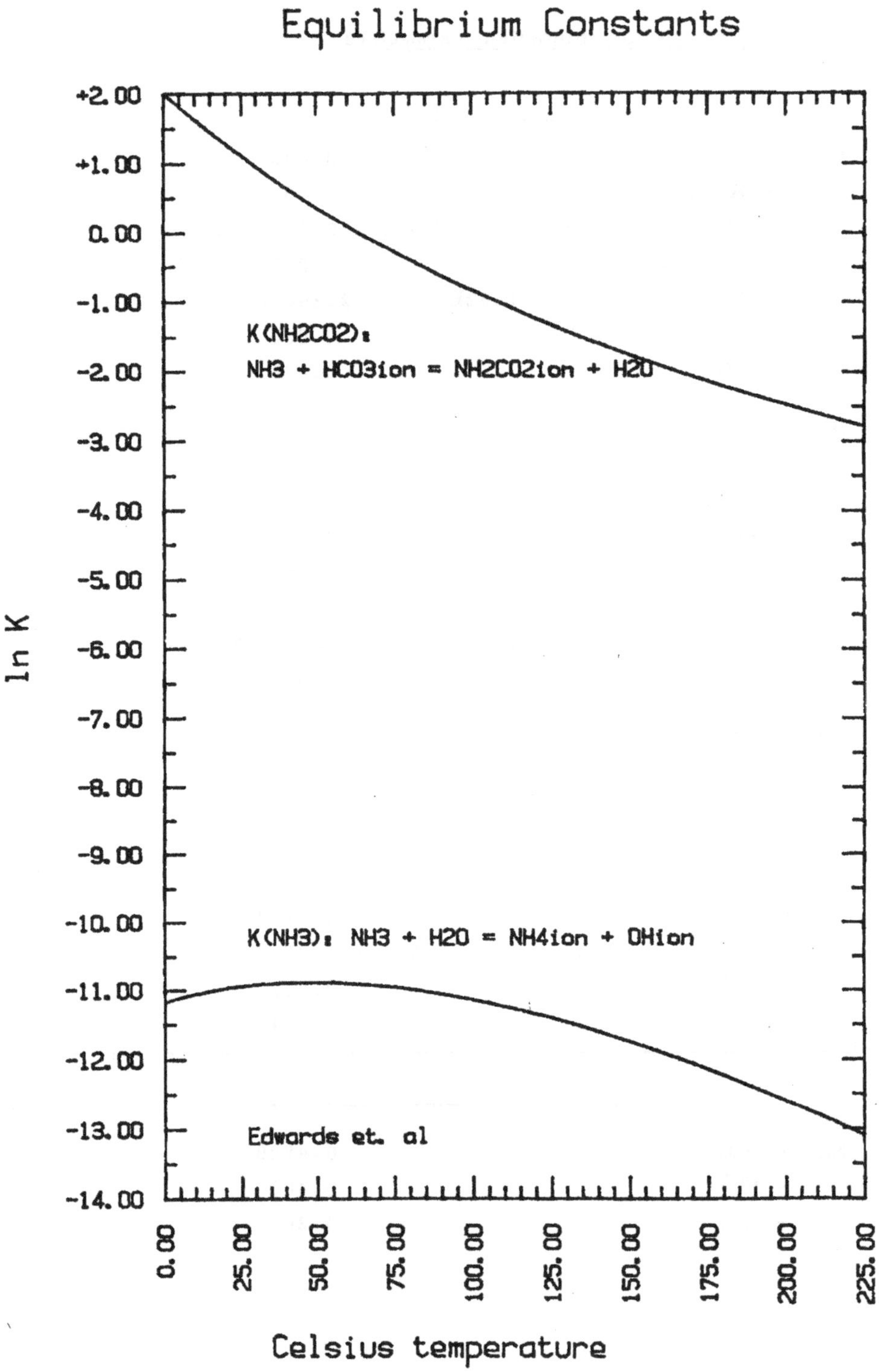
Equilibrium Constants
+2.00
+1.00
0.00
-1.00
-2.00
-3.00
-4.00
-5.00
-6.00
-7.00
-8.00
-9.00
-10.00
-11.00
-12.00
-13.00
-14.00
ln K
K(NH2CO2):
NH3 + HCO3ion = NH2CO2ion + H2O
K(NH3): NH3 + H2O = NH4ion + OHion
Edwards et. al
0.00
25.00
50.00
75.00
100.00
125.00
150.00
175.00
200.00
225.00
Celsius temperature

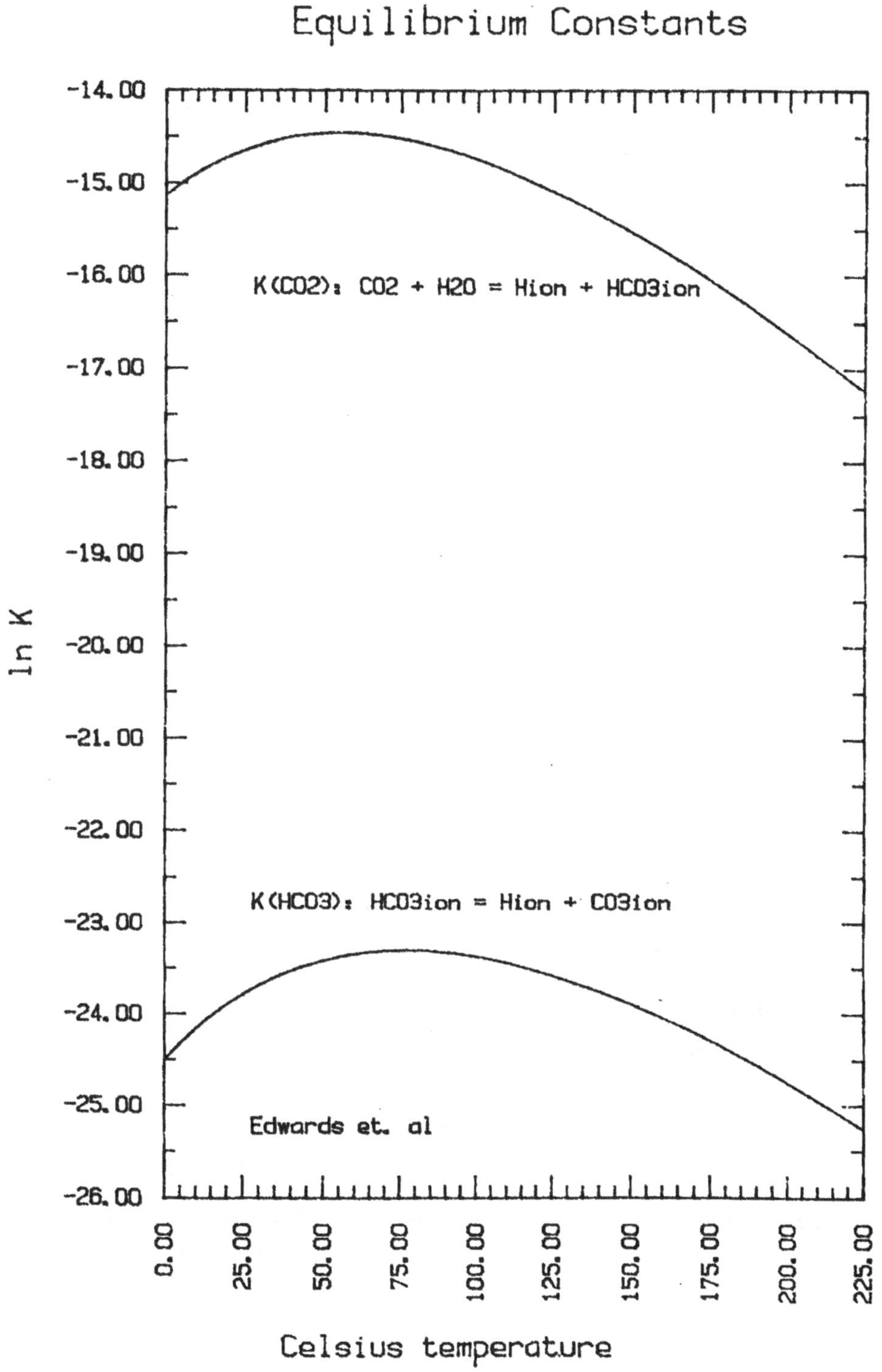
Equilibrium Constants
-14.00
-15.00
-16.00
-17.00
-18.00
-19.00
-20.00
-21.00
-22.00
-23.00
-24.00
-25.00
-26.00
ln K
K(CO2): CO2 + H2O = Hion + HCO3ion
K(HCO3): HCO3ion = Hion + CO3ion
Edwards et. al
0.00
25.00
50.00
75.00
100.00
125.00
150.00
175.00
200.00
225.00
Celsius temperature

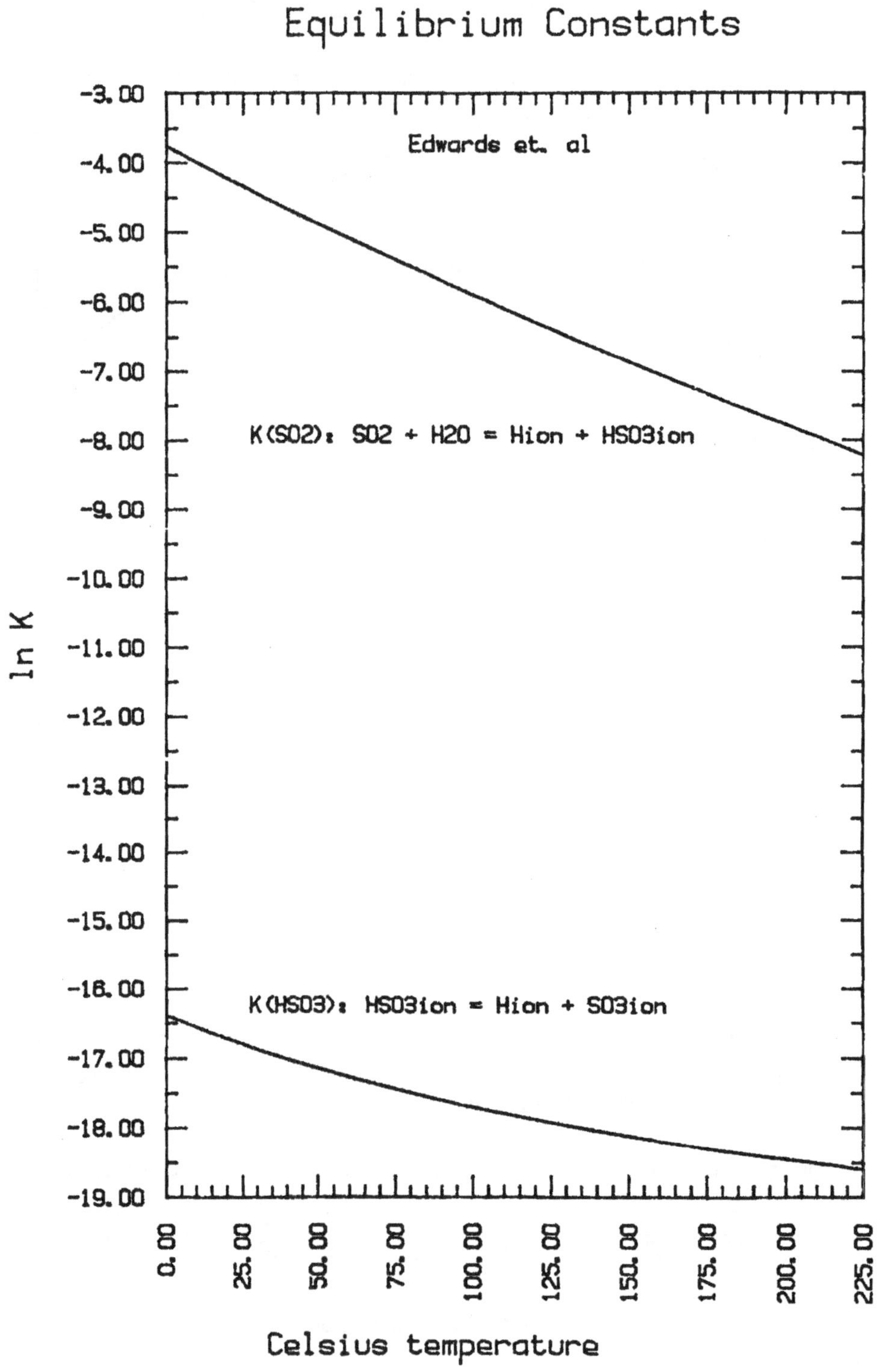
Equilibrium Constants
Edwards et. al
K(SO2): SO2 + H2O = Hion + HSO3ion
K(HSO3): HSO3ion = Hion + SO3ion
ln K
-3.00
-4.00
-5.00
-6.00
-7.00
-8.00
-9.00
-10.00
-11.00
-12.00
-13.00
-14.00
-15.00
-16.00
-17.00
-18.00
-19.00
0.00
25.00
50.00
75.00
100.00
125.00
150.00
175.00
200.00
225.00
Celsius temperature

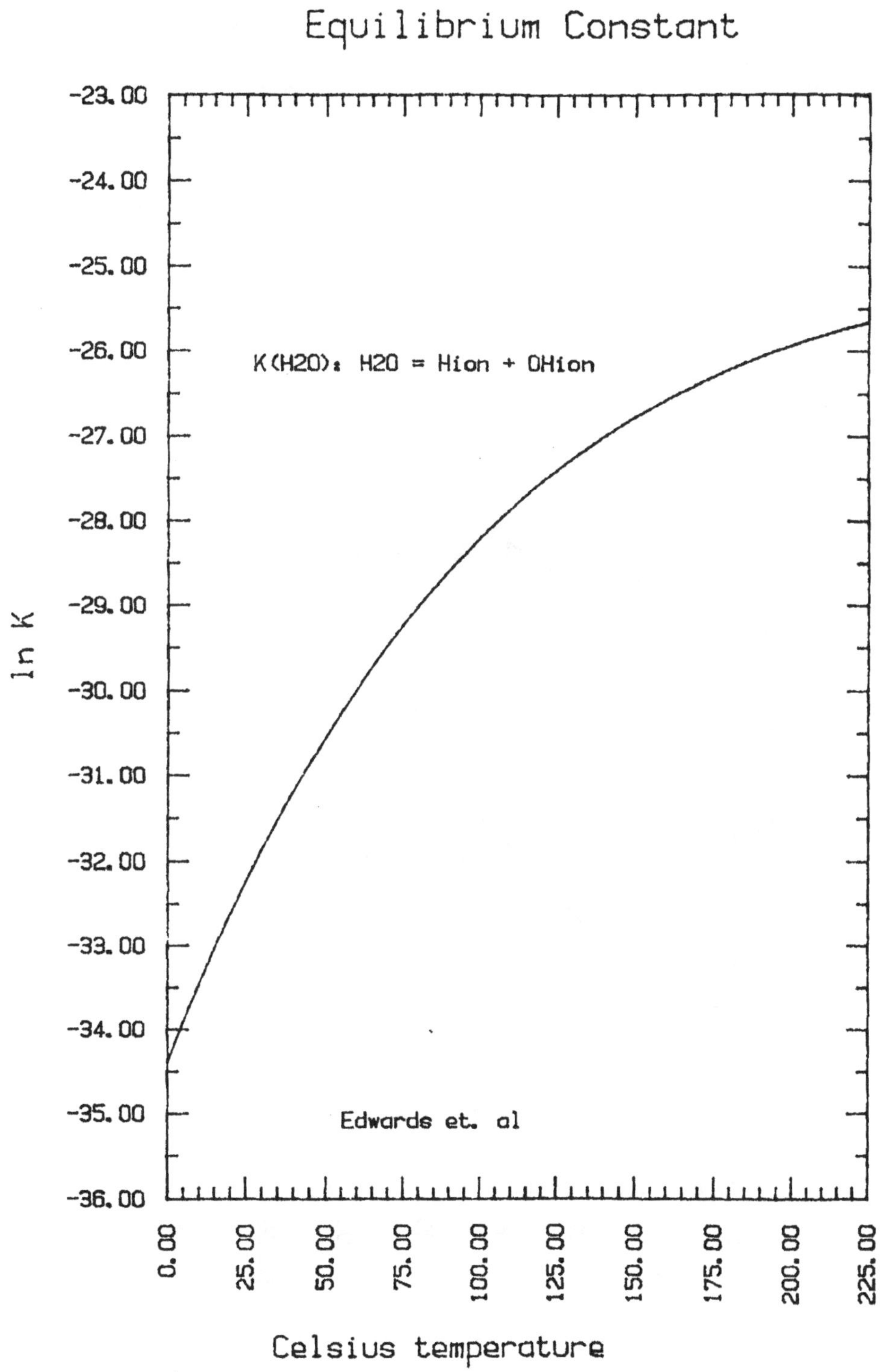
Equilibrium Constant
-23.00
-24.00
-25.00
-26.00
-27.00
-28.00
-29.00
-30.00
-31.00
-32.00
-33.00
-34.00
-35.00
-36.00
ln K
K(H2O): H2O = Hion + OHion
Edwards et. al
0.00
25.00
50.00
75.00
100.00
125.00
150.00
175.00
200.00
225.00
Celsius temperature

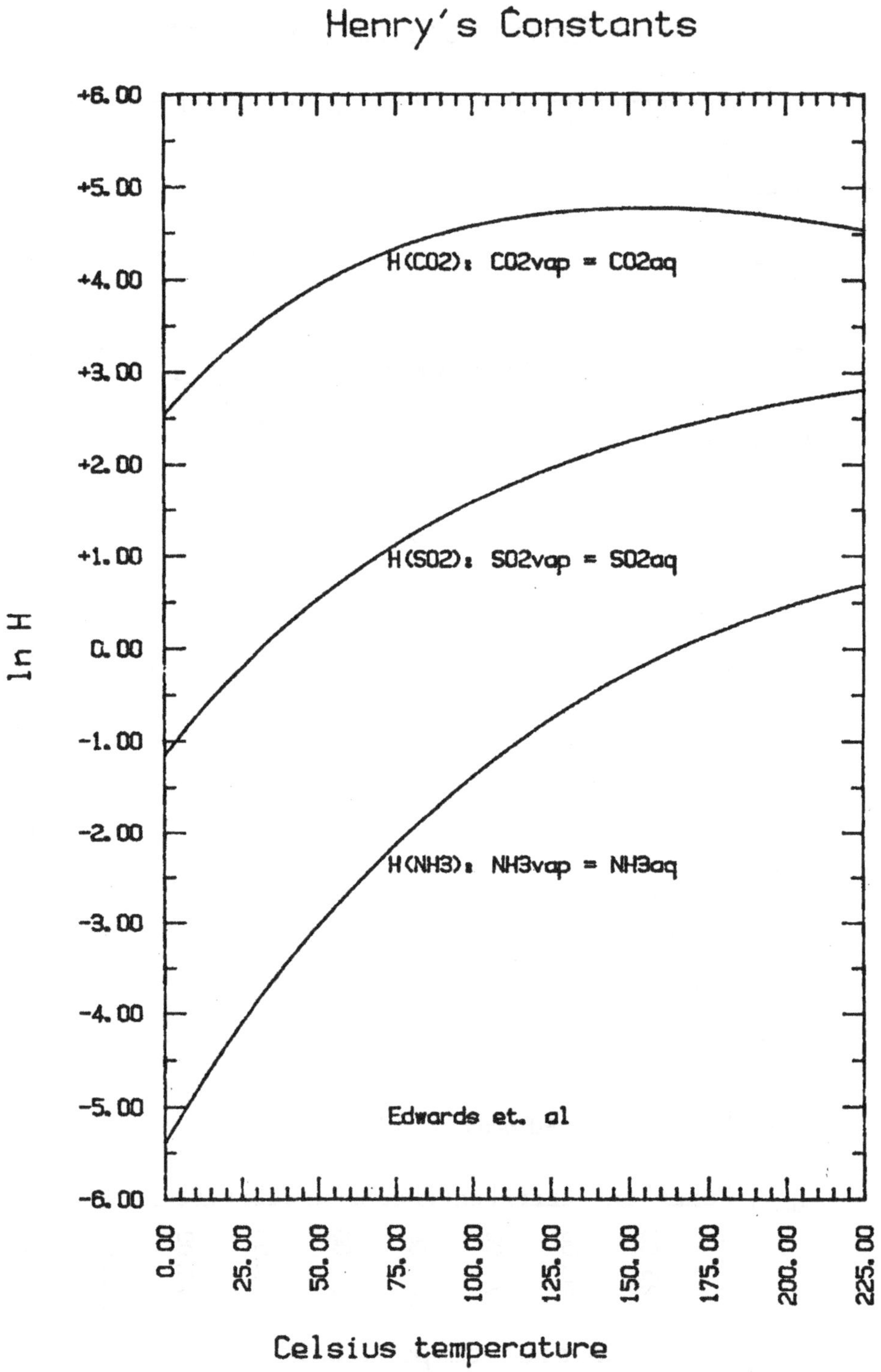
Henry's Constants
+6.00
+5.00
+4.00
+3.00
+2.00
+1.00
0.00
-1.00
-2.00
-3.00
-4.00
-5.00
-6.00
ln H
H(CO2): CO2vap = CO2aq
H(SO2): SO2vap = SO2aq
H(NH3): NH3vap = NH3aq
Edwards et. al
0.00
25.00
50.00
75.00
100.00
125.00
150.00
175.00
200.00
225.00
Celsius temperature

APPENDIX 9.3

Fugacity Coefficient Calculation

From: R. Nakamura, G.J.F. Breedveld, J.M. Prausnitz, "Thermodynamic Properties of Gas Mixtures Containing Common Polar and Nonpolar Components", Ind. Eng. Chem. Process Des. Dev., v15, #4, pp557-564 (1976)

Nakamura et al. proposed the following perturbed-hard-sphere equation of state for gas mixtures:

$$P = \frac{RT}{v}\left[\frac{1. + \varepsilon + \varepsilon^2 - \varepsilon^3}{(1. - \varepsilon)^3}\right] - \frac{a}{v\,(v + c)} \qquad \text{(A9.3a)}$$

where:
- P - pressure, atmospheres
- R - gas constant, .082056 l. atm./K mole
- T - temperature, Kelvins
- v - molar volume, l/mole
- ε - reduced density, b/4v
- b - parameter signifying the hard-core size of the molecule, l/mole
- a - parameter signifying the attractive force strength, l atm/mole
- c - constant, l/mole

In addition to equations for the reduced enthalpy difference and entropy difference, the following equation was presented for calculating the fugacity coefficient of a species k in the gas mixture:

$$\ln f_k = \left[\frac{4.\,\varepsilon - 3.\varepsilon^2}{(1. - \varepsilon)^2}\right] + \frac{b_k}{b_M}\left[\frac{4.\,\varepsilon - 2.\varepsilon^2}{(1. - \varepsilon)^3}\right] -$$

$$\frac{2.}{RTv}\left[\sum_{j=1}^{n} y_j\, a_{kj}\right]\left[\sum_{m=1}^{5} \frac{(-1.)^m}{(m + 1.)}\left(\frac{c_M}{v}\right)^m + 1.\right] +$$

$$\frac{a_m\, c_k}{RT\, v}\left[\sum_{m=1}^{4} \frac{(-1.)^m\,(m+1.)}{(m + 2.)}\left(\frac{c_M}{v}\right)^m + \frac{1.}{2.}\right] - \ln z \qquad \text{(A9.3b)}$$

With the P, R, T, v and ε terms defined earlier, the following definitions and calculations apply:

n - number of species

b_k - $\exp\,[2.30259\,(-\,\gamma - \delta T)]$ (A9.3c)

γ - constant tabulated below

δ - constant tabulated below

$$b_M = \sum_{i=1}^{n} y_i\, b_i \qquad \text{(A9.3d)}$$

y_i - vapor mole fraction of i

c_k - tabulated below

$$c_M = \sum_{i=1}^{n} y_i c_i \tag{A9.3e}$$

$$z = \frac{Pv}{RT}\text{, compressibility factor} \tag{A9.3f}$$

$$a_M = \sum_{i=1}^{n} \sum_{j=1}^{n} y_i y_j a_{ij} \tag{A9.3g}$$

$$a_{ij} = \alpha_{ij} + \frac{\beta_{ij}}{T} \tag{A9.3h}$$

$$\alpha_{ij} = \alpha^0_{ij} + \alpha^1_{ij} \tag{A9.3i}$$

$$\beta_{ij} = \beta^0_{ij} + \beta^1_{ij} \tag{A9.3j}$$

$$\alpha^1_{ij} = \sqrt{\alpha^1_i \; \alpha^1_j} \tag{A9.3k}$$

$$\beta^1_{ij} = \sqrt{\beta^1_i \; \beta^1_j} \tag{A9.3l}$$

$$\beta^0_{ij} = \frac{1}{2}\left(\beta^0_i + \beta^0_j\right) \tag{A9.3m}$$

$\alpha^1_i = \beta^1_i = 0.0$ for nonpolar gases

Values of α and β are tabulated below.

Table 1: Pure component parameters

gas	c	α	β	γ	$\delta \times 10^4$
Ar	0.0	1.3931	21.17	1.3169	2.201
CH_4	0.0	2.4328	17.09	1.2146	1.089
C_2H_4	0.0	4.7617	129.99	1.0774	0.696
C_2H_6	0.0	4.9758	326.02	1.0208	1.307
C_3H_6	0.0	7.2010	762.25	0.9294	0.792
C_3H_8	0.0	7.6434	1023.4	0.8829	0.923
CO	0.0	1.3276	34.82	1.2709	0.808

gas	c	α	β	γ	$\delta \times 10^4$
CO_2	0.0	3.1693	253.17	1.2340	0.467
H_2	0.0	0.2107	5.57	1.5853	1.335
H_2O	0.01	3.1307	1161.7	1.5589	0.593
H_2S	0.0	3.6194	454.32	1.1823	1.699
N_2	0.0	1.5324	8.56	1.2458	1.199
NH_3	0.01	2.6435	561.63	1.3884	1.470
SO_2	0.017	2.8730	1815.4	1.1043	2.721

Units: c - l/mole
α - atm l^2/mole^2
β - atm l^2/mole^2 K
γ - dimensionless
δ - K^{-1}

Table 2: Nonpolar and polar contribution to parameters α and β for four polar gases

gas	α_i^0	α_i^1	β_i^0	β_i^1
H_2O	1.06	2.07	8.4	1153.3
H_2S	2.52	1.10	16.6	437.7
NH_3	1.83	0.81	13.3	548.3
SO_2	2.86	0.01	21.9	1793.5

Table 3: Interaction parameter α_{12}^0 for polar-nonpolar mixtures Numbers in parentheses are estimates.

	α_{12}^0			
	polar			
nonpolar	H_2O	NH_3	H_2S	SO_2
Ar	1.86	(1.9)	(2.1)	(2.0)
CH_4	2.89	2.76	3.09	(2.6)
C_2H_4	4.57	(3.4)	(3.9)	(3.8)
C_2H_6	3.54	(3.7)	3.9	(3.9)
C_3H_6	6.00	(4.4)	(4.6)	(4.6)
C_3H_8	3.75	(4.5)	4.6	(4.7)

Table 3 continued

nonpolar	H_2O	NH_3	H_2S	SO_2
CO	(2.2)	(2.0)	(3.1)	(2.1)
CO_2	4.36	(3.1)	2.8	(3.8)
H_2	(0.8)	(0.80)	(0.9)	(0.8)
N_2	2.19	(1.18)	2.02	(2.0)

Table 4: Parameter α_{12} for binary mixtures of nonpolar gases Numbers in parentheses are estimates.

mixture	α_{12}
Ar-CH_4	(1.90)
Ar-C_2H_4	(2.25)
Ar-C_2H_6	(2.35)
Ar-C_3H_6	(2.92)
Ar-C_3H_8	(2.95)
Ar-CO	(1.42)
Ar-CO_2	1.72
Ar-N_2	(1.55)
CH_4-C_2H_4	3.08
CH_4-C_2H_6	3.43
CH_4-C_3H_6	(3.75)
CH_4-C_3H_8	3.84
C_2H_4-C_2H_6	(4.75)
C_2H_4-C_3H_6	(5.70)
C_2H_4-C_3H_8	(5.80)
C_2H_6-C_3H_6	5.86
C_2H_6-C_3H_8	(6.05)
C_3H_6-C_3H_8	7.41
CO-CH_4	(1.85)
CO-C_2H_4	(2.20)
CO-C_2H_6	(2.40)
CO-C_3H_6	(2.80)
CO-C_3H_8	(2.75)
CO-CO_2	(1.85)

mixture	α_{12}
CO_2-CH_4	2.45
CO_2-C_2H_4	3.85
CO_2-C_2H_6	3.62
CO_2-C_3H_6	(4.15)
CO_2-C_3H_8	4.28
H_2-Ar	0.529
H_2-CH_4	0.732
H_2-C_2H_4	(0.94)
H_2-C_2H_6	0.972
H_2-C_3H_6	(1.15)
H_2-C_3H_8	(1.18)
H_2-CO	0.617
H_2-CO_2	0.731
H_2-N_2	0.685
N_2-CH_4	1.60
N_2-C_2H_4	2.46
N_2-C_2H_6	2.47
N_2-C_3H_6	(3.00)
N_2-C_3H_8	3.05
N_2-CO	(1.42)
N_2-CO_2	2.00

Table 5: Interaction parameter α_{12}^{0} for polar-polar mixtures
Numbers in parentheses are estimates.

mixture	α_{12}^{0}
H_2O-NH_3	(1.4)
H_2O-H_2S	2.2
H_2O-SO_2	(1.7)
NH_3-H_2S	(2.1)
NH_3-SO_2	(2.3)
H_2S-SO_2	(2.7)

APPENDIX 9.4

Brelvi and O'Connell Correlation for Partial Molal Volumes

From: S.W. Brelvi & J.P. O'Connell, "Corresponding States Correlations for Liquid Compressibility and Partial Molal Volumes of Gases at Infinite Dilution in Liquids", AIChE J, v18, #6, pp1239-1243 (1972)

The partial molal volume of solute 1 at infinite dilution in solvent 2 can be calculated using the equation presented by Brelvi and O'Connell:

$$\frac{\bar{v}_1^o}{K_2^o\, R\, T} = 1. - C_{12}^o \tag{A9.4a}$$

where: $\bar{v}^o$ - partial molar volume at infinite dilution, cm³/g mole

R - gas constant, 82.06 atm cm³/g mole K

T - temperature, Kelvins

C_{12}^o - reduced volume integral of the molecular direct correlation function c_{ij} at infinite dilution

K_2^o - isothermal compressibility at infinite dilution, atm^{-1}

The following terms were also defined and used to calculate the C_{12}^o and K_2^o terms:

v_i^* - characteristic volume of i, cm³/g mole. For nonpolar species the critical volume, v_c, may be used; experimental compressibility data may be used to determine v* for polar compounds. Some values are tabulated below.

ρ - pure solvent density, g mole/cm³

$\tilde{\rho}$ - reduced density = ρv_2^*

The reduced volume integral may be calculated using one of the following equations:

when $2. \le \tilde{\rho} \le 2.785$

$$\ln - C_{12}^o \left[\frac{v_2^*}{v_1^*}\right]^{.62} = -2.4467 + 2.12074\ \tilde{\rho} \tag{A9.4b}$$

when $2.785 \le \tilde{\rho} \le 3.2$

$$\ln - C_{12}^o \left[\frac{v_2^*}{v_1^*}\right]^{.62} = 3.02214 - 1.87085\ \tilde{\rho} + .71955\ \tilde{\rho}^2 \tag{A9.4c}$$

The isothermal compressibility may be calculated from the following equation:

$$\ln\left[1. + \frac{1.}{\rho K R T}\right] = -.42704\,(\tilde{\rho} - 1.) + 2.089\,(\tilde{\rho} - 1.)^2 - .42367\,(\tilde{\rho} - 1.)^3 \qquad \text{(A9.4d)}$$

Table 1: Characteristic volumes*

species	v*, cm^3/g mole
H_2O	46.4
H_2S	90.0
NH_3	65.2
SO_2	115.0
CO_2	80.0

* These values, along with characteristic volumes for an additional 84 compounds, are tabulated in:

J. Prausnitz, T. Anderson, E. Grens, C. Eckert, R. Hsieh, J. O'Connell, Computer Calculations for Multicomponent Vapor-Liquid and Liquid-Liquid Equilibria, Prentice-Hall Inc., Englewood Cliffs, NJ (1980) Appendix C-1, pp145-149

APPENDIX 9.5

Gypsum Solubility Study Parameters at 25°C

From: Rogers, P.S.Z., "Thermodynamics of Geothermal Fluids", Doctoral Dissertation, University of California, Berkeley (1981)

Table 1: Binary solution parameters for the Pitzer equations

salt	β_0	β_1	β_2	C^{ϕ}
$CaCl_2$	.31590	1.6140		-.00034
$CaSO_4$	.2	3.1973	-54.24	0.0
NaCl	.07650	.2664		.00127
Na_2SO_4	.019575	1.113		.00570

Table 2: Mixed electrolyte solution parameters for the Pitzer equations

i	j	k	θ_{ij}	ψ_{ijk}
Na^+	Ca^{2+}	Cl^-	.07	-.007
		SO_4^{2-}		-.067
SO_4^{2-}	Cl^-	Ca^{2+}	.03	-.027
		Na^+		0.0

Table 3: Gypsum solubility product at 25°C

$$K_{CaSO_4 \cdot 2H_2O} = 2.615 \times 10^{-5}$$

REFERENCES

1. Baes, C.F., Jr.; R. E. Mesmer, The Hydrolysis of Cations, J. Wiley and Sons, New York (1976)

2. (a) Beutier, D.; H. Renon, "Representation of NH_3-H_2S-H_2O, NH_3-CO_2-H_2O, and NH_3-SO_2-H_2O Vapor-Liquid Equilibria", Ind. Eng. Chem. Process Des. Dev., v17, #3, pp220-230 (1978)

   (b) Beutier, D.; H. Renon, "Corrections", Ind. Eng. Chem. Process Des. Dev., v19, #4, p722 (1980)

3. Brelvi, S.W.; J.P. O'Connell, "Corresponding States Correlations for Liquid Compressibility and Partial Molal Volumes of Gases at Infinite Dilution in Liquids", AIChE J, v18, #6, pp1239-1243 (1972)

4. Bromley, L.A., "Approximate individual ion values of β (or B) in extended Debye-Hückel theory for uni-univalent aqueous solutions at 298.15K", J. Chem. Thermo., v4, pp669-673 (1972)

5. Bromley, L.A., "Thermodynamic Properties of Strong Electrolytes in Aqueous Solutions", AIChE J, v19, #2, pp313-320 (1973)

6. Butler, J.N., Ionic Equilibrium - A Mathematical Approach, Addison-Wesley, Reading, Mass. (1964)

7. Campbell, H.J., Jr; N.C. Scrivner, K. Batzar, R.F. White, "Evaluation of Chromium Removal from a Highly Variable Wastewater Stream", Proceedings of the 32nd Purdue Industrial Waste Conference, Purdue University, Ann Arbor Science (1977)

8. Chamberlin, N.S.; R.V. Day, "Technology of Chrome Reduction with Sulfur Dioxide", Proceedings of the Eleventh Purdue Industrial Waste Water Conference, v91, Lafayette, Ind., Purdue Univ., pp129-156 (1957)

9. Davies, C.W., Ion Association, Butterworths Scientific Publications, London (1962)

10. Dean, J.A., ed., Lange's Handbook of Chemistry, 11th edition, McGraw-Hill Book Co., New York (1973)

11. Edwards, T.J.; G. Maurer, J. Newman, J.M. Prausnitz, "Vapor-liquid Equilibria in Multicomponent Aqueous Solutions of Volatile Weak Electrolytes", AIChE J, v24, #6, pp966-976 (1978)

12. (a) Glushko, V.P., dir.; Thermal Constants of Compounds - Handbook in Ten Issues, VINITI Academy of Sciences of the U.S.S.R., Moscow

    Volume I (1965) - O, H, D, T, F, Cl, Br, I, At, ^{3}He, He, Ne, Ar, Kr, Xe, Rn

    (b) Volume II (1966) - S, Se, Te, Po

(c) Volume III (1968) - N, P, As, Sb, Bi

(d) Volume IV-1 (1970) - C, Si, Ge, Sn, Pb
Volume IV-2 (1971)

(e) Volume V (1971) - B, Al, Ga, In, Tl

(f) Volume VI-1 (1972) - Zn, Cd, Hg, Cu, Ag, Au, Fe, Co, Ni, Ru, Rh Pd, Os, Ir, Pt
Volume VI-2 (1973)

(g) Volume VII-1 (1974) - Mn, Tc, Re, Cr, Mo, W, V, Nb, Ta, Ti, Zr, Hf
Volume VII-2 (1974)

(h) Volume VIII-1 (1978) - Sc, Y, La, Ce, Pr, Nd, Pm, Sm, Eu, Gd, Tb, Dy, Ho, Er, Tm, Yb, Lu, Ac, Th, Pa, U, Np, Pu, Am, Cm, Bk, Cf, Es, Fm, Md, No
Volume VIII-2 (1978)

(i) Volume IX (1979) - Be, Mg, Ca, Sr, Ba, Ra

(j) Volume X-1 (1981) - Li, Na, K, Rb, Cs, Fv
Volume X-2 (1981)

13. Krichevsky, I.R.; J.S. Kasarnovsky, "Thermodynamical Calculations of Solubilities of Nitrogen and Hydrogen in Water at High Pressures", JACS, v57, pp2168-2171 (1935)

14. Meissner, H.P.; C.L. Kusik, "Electrolyte Activity Coefficients in Inorganic Processing", AIChE Symposium Series 173, v74, pp14-20 (1978)

15. Maurer, G., "On the Solubility of Volatile Weak Electrolytes in Aqueous Solutions", Thermodynamics of Aqueous Systems with Industrial Applications, S.A. Newman, ed., ACS Symposium Series 133, Washington D.C., pp139-172 (1980)

16. Nakamura, R.; G.J.F. Breedveld, J.M. Prausnitz, "Thermodynamic Properties of Gas Mixtures Containing Common Polar and Nonpolar Components", Ind. Eng. Chem. Process Des. Dev., v15, #4, pp557-564 (1976)

17. (a) Perry, J.H., ed., Chemical Engineers' Handbook, 3rd edition, McGraw-Hill Book Co., NY (1950), pp674-675

(b) Perry, R.H.; C.H. Chilton, S.D. Kirkpatrick, eds., Chemical Engineers' Handbook, 4th edition, McGraw-Hill Book Co., NY (1963), pp14-3,4,5,

18. Pitzer, K.S.; L.F. Silvester, "Thermodynamics of Electrolytes. VI. Weak Electrolytes Including H_3PO_4", J. Sol. Chem., v5, #4, pp269-278 (1976)

19. Pitzer, K.S., "Theory: Ion Interaction Approach", Activity Coefficients in Electrolyte Solutions, vol. I, R.M. Pytkowicz, ed., CRC Press, Boca Raton, FLA, p162 (1979)

20. Rafal, M.; D.M. Clark, A. Rastogi, N.C. Scrivner, "Proper Classification and Modeling of Aqueous Electrolytes", AIChE Meeting (November 1984)

21. Renon, H; "Representation of NH_3-H_2S-H_2O, NH_3-SO_2-H_2O, and NH_3-CO_2-H_2O Vapor-Liquid Equilibria", Thermodynamics of Aqueous Systems with Industrial Applications, S.A. Newman, ed., ACS Symposium Series 133, Washington, D.C., pp173-186 (1980)

22. Rogers, P.S.Z., "Thermodynamics of Geothermal Fluids", Doctoral dissertation, Univ. California, Berkeley (1981)

23. Seidell, A.; W.F. Linke, eds., Solubilities - Inorganic and Metal-Organic Compounds, vII, ACS, Washington, D.C. (1965)

24. Sillen, L.G.; Stability Constants of Metal-Ion Complexes, Section I: Inorganic Ligands, The Chemical Society, London, Special Publication No. 17 (1964)

25. Smith, R.M.; A.E. Martell, Critical Stability Constants, Vol. 4: Inorganic Complexes, Plenum Press, New York (1976)

26. (a) Wagman, D.D.; W.H. Evans; V.B. Parker; I. Halow; S.M. Bailey; R.H. Schumm; Selected Values of Chemical Thermodynamic Properties - Tables for the First Thirty-Four Elements in the Standard Order of Arrangement, National Bureau of Standards Technical Note 270-3, 1968

(b) Wagman, D.D.; W.H. Evans; V.B. Parker; I. Halow; S.M. Bailey; R.H. Schumm; Selected Values of Chemical Thermodynamic Properties - Tables for Elements 35 through 53 in the Standard Order of Arrangement, NBS Tech Note 270-4, 1969

(c) Wagman, D.D.; W.H. Evans; V.B. Parker; I. Halow; S.M. Bailey; R.H. Schumm; K.L. Churney; Selected Values of Chemical Thermodynamic Properties - Tables for Elements 54 through 61 in the Standard Order of Arrangement, NBS Tech Note 270-5, 1971

(d) Parker, V.B.; D.D. Wagman and W.H. Evans; Selected Values of Chemical Thermodynamic Properties - Tables for the Alkaline Earth Elements (Elements 92 through 97 in the Standard Order of Arrangement), NBS Tech Note 270-6, 1971

(e) Schumm, R.H.; D.D. Wagman; S. Bailey; W.H. Evans; V.B. Parker, Selected Values of Chemical Thermodynamic Properties - Tables for the Lanthanide (Rare Earth) Elements (Elements 62 through 76 in the Standard Order of Arrangement), NBS Tech Note 270-7, 1973

(f) Wagman, D.D.; W.H. Evans; V.B. Parker; R.H. Schumm; R.L. Nuttall; Selected Values of Chemical Thermodynamic Properties - Compounds of Uranium, Protactinium, Thorium, Actinium, and the Alkali Metals, NBS Tech Note 270-8, 1981

(g) Wagman, D.D.; W.H. Evans, V.B. Parker, R.H. Schumm, I. Halow, S.M. Bailey, K.L. Churney, R.L. Nutall, The NBS tables of chemical thermodynamic properties - Selected values for inorganic and C_1 and C_2 organic substances in SI units, J. Phys. Chem. Ref. Data, v11, suppl. 2 (1982)

27. Whitney, R.P.; J.E. Vivian, "Solubility of Chlorine in Water", I&EC, v33, #6, pp741-744 (1941)

X: Appendices

APPENDIX A

COMPUTER PROGRAMS FOR SOLVING EQUILIBRIA PROBLEMS

With the parallel evolution of electrolyte formulations and the digital computer there have been several significant developments in terms of general purpose computer programs which calculate electrolyte equilibria. In this Appendix we will give a brief synopsis of six such programs. It should be emphasized that this list is a sampling and is not intended to be comprehensive. Given the rapid growth in electrolyte applications, new programs will undoubtedly be released after this Handbook goes to press.

The particular programs to be reviewed in this section are:

1) WATEQ
2) REDEQL.EPA
3) MINEQL
4) ENIVEL
5) ASPEN Plus
6) ECES

In each case we will organize the description to cover the following characteristics:

1) general description
2) source language
3) hardware limitations (if known)
4) basis for activity coefficients
5) species supported
6) conditions supported
7) availability

WATEQ

The WATEQ program was developed by the U.S. Geological Survey in order to predict both trace and major element speciation and mineral equilibria in natural waters. The program, along with several evolutionary versions, WATEQF and WATEQ2, is limited to low temperature (25°C) and moderate ionic strength (0-3 molal). The package contains an extensive databank which, in the WATEQ2 extension, includes Ag, As, Cd, Cs, Cu, Mn, Ni, Pb, Rb and Zn. The limitation of 3.0 molal ionic strength arises primarily due to the activity coefficient formulation which is based upon an extended Debye-Hückel which incorporates a term linear in ionic strength.

Source language: FORTRAN (WATEQF, WATEQ2), PL/1 (WATEQ)

Hardware: IBM (WATEQ), IBM + others (WATEQ2, WATEQF)

Activity Coefficients: Debye-Hückel with two parameter extension term linear in ionic strength.

Conditions supported: Temperature limited to 25°C. Ionic strength limited to 3.0 molal.

Species supported: Species limited to an available public databank which is quite extensive.

Availability: Available from the U.S. Geological Survey,

REDEQL

This program was developed by the Environmental Protection Agency. It is based upon the Davies equation and is intended for water quality use. The program includes an extensive databank for metal-ligand complexes. Based upon the Newton-Raphson technique it allows prediction of solid formation. The program is limited in that all equilibrium constant values are based upon 25°C and the activity coefficients on Davies equation which, as we have seen, is limited to moderate ionic strength of less than 1.0 molal.

Source Language: FORTRAN

Hardware: Several

Activity coefficients: Davies extension to Debye-Hückel

Species supported: Wide range of metal-ligand complexes

Conditions supported: 25°C and up to 1.0 molal ionic strength

Availability: Available from the EPA,

MINEQL

MINEQL is a second generation version of REDEQL. The primary difference over and above REDEQL seems to be with respect to clarity and flexibility, otherwise, the description given above for REDEQL should suffice.

Source Language: FORTRAN

Hardware: Several

Activity coefficients: Davies extension to Debye-Hückel

Species supported: Wide range of metal-ligand complexes

Conditions supported: 25°C and up to 1.0 molal ionic strength

Availability: Available through the Department of Civil Engineering, Massachusetts Institute of Technology, Cambridge, Massachusetts

ENIVEL

The ENIVEL Program is a general purpose vapor-liquid-solid aqueous electrolyte simulation program in which the model is specified as a set of chemical equations in standard form. All necessary equations for equilibrium, electroneutrality and material balance are automatically generated and solved. The program can also perform nonaqueous thermodynamic calculations.

Source Language: FORTRAN

Hardware: Several

Activity coefficients: Based upon the Pitzer formulation

Species supported: Extensive public databank

Conditions supported: Broad range based upon Pitzer fits

Availabiltiy: Proprietary program available from Jaycor, Inc.,

ASPEN Plus

ASPEN Plus is a general purpose simulator which contains, among many other facilities, an aqueous electrolyte vapor-liquid-solid simulation capability. The package has a number of special features, including:

1) A facility for the prediction of multistage (tower) systems

2) A facility for flowsheet simulations based upon the ASPEN Plus Simulator

3) A nonlinear regression option to allow users to develop their own interaction coefficients

Source Language: FORTRAN

Hardware: IBM/MVS or CMS, VAX/VMS, CDC/NOS

Activity coefficients: Based upon either Chen's formulation or Pitzer's formulation

Species supported: Limited public databank with respect to species interactions. Relies more on regression facilities for fitting specific systems.

Conditions supported: Limited primarily by data regressed since formulations are applicable over wide ranges of conditions.

Availability: Proprietary package available, as a consequence of licensing ASPEN Plus, from Aspen Technology Inc., 251 Vassar Street, Cambridge, Massachusetts, 02139

ECES

The Equilibrium Compositions of Electrolyte Solutions (ECES) Program is an extensive general purpose vapor-liquid-solid prediction program for aqueous electrolyte systems. The package has a number of special features, including:

1) Model specification is done by user writing standard mass action chemical equations along with chemical species to be considered.

2) A facility for the prediction of multistage (tower) systems

3) A facility for easy integration with most Flowsheet Simulator systems

4) An extensive public databank containing the information necessary to calculate equilibrium constants and activity coefficients

5) A nonlinear regression option to allow users to generate their own Bromley and/or Pitzer and/or NRTL interaction parameter values in order to develop private databanks.

6) The ability to "switch the roles" of variables in order to solve design variations (eg. fixed pH, fixed enthalpy, etc.) of the conventional vapor-liquid-solid simulation.

7) The ability to routinely obtain the sensitivity of the calculated unknowns with respect to the input variables.

Source Language: FORTRAN

Hardware: VAX/VMS, IBM/MVS or CMS, CDC/NOS, UNIVAC/EXEC, PRIME

Activity coefficients: Ion-ion interactions by extended Bromley or Pitzer β_0 , β_1 formulations. Ion-molecule by modified Setschénow or Pitzer formulation. Molecule-molecule by Pitzer or, in the case of water-molecule by NRTL.

Species supported: Public databank of almost 300 species. Private databank concept together with nonlinear regression allows for open ended extensions.

Conditions supported: Depends on range of fits as well as extrapolation capability, but generally useful for:

- Temperature 0°C - 250°C
- Pressure 0 - 200 atmospheres
- Ionic strength 0 - 30.0

Availability: Proprietary program available from OLI Systems Inc., 52 South Street, Morristown, NJ 07960

This completes the survey of six prominent examples of computer programs for solving electrolyte equilibrium problems. There are, as noted earlier, a number of others which may have specific or general merit. A partial list of such programs along with appropriate references are:

1. EQ3/EQ6 (A.2)
2) SOLMNEQ (A.3)
3) GEOCHEM (A.4)
4) PATHI (A.5)

REFERENCES

GENERAL

A1. Nordstrom, D.K.; L.N. Plummer; T.M.L. Wigley; T.J. Wolery; J.W. Ball; E.A. Jenne; R.L. Bassett; D.A. Crerar; T.M. Florence; B. Fritz; M. Hoffman; G.R. Holdren Jr.; G.M. Lafon; S.V. Mattigod; R.E. McDuff; F. Morel; M.M. Reddy; G. Sposito; and J. Thrailkill, Comparison of Computerized Chemical Models for Equilibrium Calculations in Aqueous Systems", Chemical Modeling in Aqueous Systems, E.A. Jenne ed., ACS Symposium Series 93, pp 857-892 (1979)

A2. Wolery, T.J., "Some chemical aspects of hydrothermal processes at mid-oceanic ridges - a theoretical study. I. Basalt - sea water reaction and chemical cycling between the oceanic crust and the oceans. II. Calculation of chemical equilibrium between aqueous solutions and minerals", Ph.D. Thesis, Northwestern Univ., Evanston, IL, (1978)

A3. Kharaka, Y.K. and I. Barnes, "SOLMNEQ: Solution-mineral equilibrium computations", NTIS Tech. Rept. PB213-899, Springfield, VA (1973)

A4. Mattigod, S.V. and G. Sposito, "Chemical modeling of trace metal equilibria in contaminated soil solutions using the computer program GEOCHEM", E.A. Jenne, ed., Chemical Modeling in Aqueous Systems, ACS Sum. Series 93, pp (1979)

A5. Helgeson, H.C., T.H. Brown, A. Nigrini, and T.A. Jones, "Calculation of mass transfer in geochemical processes involving aqueous solutions", Geochim. Cosmochim. Acta, 34, pp 569-592 (1970)

WATEQ2

A6. Ball, J.W., E.A. Jenne and D.K. Nordstrom, "WATEQ2 - A Computerized Chemical Model for Trace and Major Element Speciation and Mineral Equilibria of Natural Waters", Chemical Modeling in Aqueous Systems, E.A. Jenne, ed., ACS Symposium Series 93, pp 815-835, (1979)

A7. Truesdell, A.H., and B.F. Jones, "WATEQ, a computer program for calculating chemical equilibria of natural waters", NTIS-PB2-20464, (1973)

A8. Truesdell, A.H., and B.F. Jones, "WATEQ, a computer program for calculating chemical equilibria of natural waters", U.S. Geol. Survey J. Res. 2(2), pp 233-274, (1974)

A9. Plummer, L.N., B.F. Jones, and A.H. Truesdell, "WATEQF - A Fortran IV version of WATEQ, a computer program for calculating chemical equilibrium of natural waters", U.S. Geol. Survey Water Resour. Invest. 76-13, (1976)

A10. Ball, J.W., E.A. Jenne, and D.K. Nordstrom, "Additional and revised thermochemical data and computer code for WATEQ2-- A computerized chemical model for trace and major element speciation and mineral equilibria of natural waters", U.S. Geol. Survey Water Resour. Invest. 78-116

REDEQL.EPA

A11. Morel, F. and J.J. Morgan, "A numerical method for computing equilibria in aqueous chemical systems", Env. Sci. Tech. v6, pp 58-67 (1972)

A12. Ingle, S.E., J.A. Keniston and D.W. Schults, "REDEQL.EPA Aqueous Chemical Equilibrium Computer Program", EPA Report EPA-600/3-80-049 (1980)

ENIVEL

A13. Levine, H.B. "High-Speed, High-Accuracy Calculation of Multicomponent, Multiphase, Nonideal Chemical and Physical Equilibria by the Method of Univariant Descent", Report J510-81-027, JAYCOR, Inc., San Diego, CA (1981)

MINEQL

A14. Westall, J.C., J.L. Zachary, and F.M.M. Morel, "MINEQL - A Computer Program for the Calculation of Chemical Equilibrium Composition of Aqueous Systems", Technical Note No. 18, Ralph M. Parsons Laboratory, Dept. of Civil Eng., Mass. Inst. of Tech. (1976)

ASPEN Plus

A15. Chen, C-C, H.I. Britt and J.F. Boston, "Process Simulation of Electrolyte Systems", Proceedings of the 1984 Summer Computer Simulation Conference, Vol. 1, Soc. for Comp. Simulation, pp 552-557 (1984)

ECES

A16. Zemaitis, J.F., Jr., "Predicting Vapor-Liquid-Solid Equilibria in Multicomponent Aqueous Solutions of Electrolytes", Thermodynamics of Aqueous Systems with Industrial Applications, S.A. Newman, Ed. ACS Symposium Series 133, pp 227 (1980)

A17. Sanders, S.J., "Case Studies in Modeling Aqueous Electrolyte Solutions", Computers and Chemical Engineering, Accepted for Publication

A18. Sanders, S.J., "Modeling Organics in Aqueous Sulfuric Acid Solutions", I&EC Process Design and Development, Accepted for Publication

A19. Rafal, M., "Multistage Simulation Involving Electrolytes", Proceedings of the 1984 Summer Computer Simulation Conference, Vol. 1, Soc. for Comp. Simulation, pp 624-629 (1984)

APPENDIX B

SELECTED THERMODYNAMIC DATA

Conventions:

ΔH_f^o - enthalpy of formation at the standard state

ΔG_f^o - Gibbs energy of formation at the standard state

S^o - entropy at the standard state

Cp^o - heat capacity at the standard state

State:

g - gaseous

liq - liquid

c - crystalline solid

amorp - glassy or amorphous

aq - in water solution. The "aq" entry denotes the hypothetical ideal solution of unit molality.

aqu - an aqueous solution as above. However, here the concentration is not unit molality but is assumed to be dilute.

The aqueous species conform to the following conventions:

molecules: An "undissociated" or "unionized" notation of a neutral molecule indicates that the values given are for the undissociated molecule as opposed to the sum of the ionic values. Those denoted as ionized or having no further notation are the sum of the values for the ions. For example, aqueous H_3PO_4 has sets of values:

H_3PO_4 ionized values which are equal to the sum of the H^+ ion and PO_4^{3-} ion values

$$\Delta H_{H_3PO_4} = 3\,\Delta H_{H^+} + \Delta H_{PO_4^{3-}}$$
$$-305.3 = 3*0. + (-305.3)$$

H_3PO_4 undissociated values which are not summed values of the ions:

$\Delta H_{H_3PO_4} \neq 3\,\Delta H_{H^+} + \Delta H_{PO_4^{3-}}$
$-307.92 \neq 3*0 + (-305.3)$
$\neq 2\,\Delta H_{H^+} + \Delta H_{HPO_4^{2-}}$
$\neq 2*0 + (-308.83)$
$\neq \Delta H_{H^+} + \Delta H_{H_2PO_4^-}$
$\neq 0 + (-309.82)$

ions: The values given for the aqueous ions are for the undissociated ion and are not derived from the sum of the constituent species. For example, the zinc and chloride ions have the following values for the energy of formation:

$\Delta G_{Zn^{2+}} = -35.14$
$\Delta G_{Cl^-} = -31.372$

The $ZnCl^+$ aqueous ion value is not the sum of the values:

$\Delta G_{ZnCl^+} \neq \Delta G_{Zn^{2+}} + \Delta G_{Cl^-}$
$-65.8 \neq -35.14 - 31.372$
$-65.8 \neq -66.512$

The tables are arranged according to the Standard Order of Arrangement of the periodic table as given by the National Bureau of Standards.

Substance		298.15K; 25°C				
Formula & Description	State	Formula Weight	ΔH°_{f} kcal/mole	ΔG°_{f} kcal/mole	S° cal/deg mole	Cp° cal/deg mole
1 - OXYGEN						
O_2	g	31.9988	0	0	49.003	7.016
	aq		-2.8	3.0	26.5	
2 - HYDROGEN						
H^+	g	1.0080	367.161			
	aq		0	0	0	0
H_2	g	2.0159	0	0	31.208	6.889
	aq		-1.0	4.2	13.8	
OH^-	g	17.0074	-33.67			
	aq		-54.970	-37.594	-2.57	-35.5
H_2O	liq	18.0153	-68.315	-56.687	16.71	17.995
	g		-57.796	-54.634	45.104	8.025
9 - FLOURINE						
F^-	g	18.9984	-64.7			
	aq		-79.50	-66.64	-3.3	-25.5
F_2	g	37.9968	0	0	48.44	7.48
HF	liq	20.0064	-71.65		18.02+x	12.35
	g		-64.8	-65.3	41.508	6.963
undissoc.	aq		-76.50	-70.95	21.2	
ionized	aq		-79.50	-66.64	-3.3	-25.5
HF_2^-	aq	39.0048	-155.34	-138.18	22.1	
10 - CHLORINE						
Cl^-	g	35.453	-58.8			
	aq		-39.952	-31.372	13.5	-32.6
Cl_2	g	70.906	0	0	53.288	8.104
	aq		-5.6	1.65	29.	
Cl_3^-	aq	106.359		-28.8		

+x undetermined residual enthalpy

Substance			298.15K; 25°C			
Formula & Description	State	Formula Weight	ΔH_f° kcal/mole	ΔG_f°	S° cal/deg mole	Cp°
ClO^-	aq	51.4524	-25.6	-8.8	10.	
ClO_2	g	67.4518	24.5	28.8	61.36	10.03
	aq		17.9	28.7	39.4	
ClO_2^-	aq		-15.9	4.1	24.2	
ClO_3	g	83.4512	37.			
ClO_3^-	aq		-24.85	-1.92	38.8	
ClO_4^-	aq	99.4506	-30.91	-2.06	43.5	
HCl	g	36.4610	-22.062	-22.777	44.646	6.96
	aq		-39.952	-31.372	13.5	-32.6
$HClO_2$	g	52.4604			56.54	8.88
undissoc.	aq		-28.9	-19.1	34.	
$HClO_2$ undissoc.	aq	68.4598	-12.4	1.4	45.0	
$HClO_3$	aq	84.4592	-24.85	-1.92	38.8	
$HClO_4$	liq	100.4586	-9.70			
	aq		-30.91	-2.06	43.5	
$HClO_4.H_2O$	c	118.4739	-91.35			
$HClO_4.2H_2O$	liq	136.4892	-162.04			
14 - SULFUR						
S^{2-}	aq	32.064	7.9	20.5	-3.5	
SO_2	liq	64.0628	-76.6			
	g		-70.944	-71.748	59.30	9.53
undissoc.	aq		-77.194	-71.871	38.7	

Substance				298.15K; 25°C			
Formula & Description		State	Formula Weight	ΔH_f° kcal/mole	ΔG_f° kcal/mole	S° cal/deg mole	Cp° cal/deg mole
SO_3^{2-}		aq	80.0622	-151.9	-116.3	-7.	
SO_4^{2-}		aq	96.0616	-217.32	-177.97	4.8	-70.
HS^-		aq	33.0720	-4.2	2.88	15.0	
H_2S		g	34.0799	-4.93	-8.02	49.16	8.18
	undissoc.	aq		-9.5	-6.66	29.	
HSO_3^-		aq	81.0702	-149.67	-126.15	33.4	
HSO_4^-		aq	97.0696	-212.08	-180.69	31.5	-20.
H_2SO_3	undissoc.	aq	82.0781	-145.51	-128.56	55.5	
H_2SO_4		c	98.0775				
		liq		-194.548	-164.938	37.501	33.20
		aq		-217.32	-177.97	4.8	-70.
$H_2SO_4.1H_2O$		liq	116.0929	-269.508	-227.182	50.56	51.35
$H_2SO_4.2H_2O$		liq	134.1082	-341.085	-286.770	66.06	62.34
$H_2SO_4.3H_2O$		liq	152.1236	-411.186	-345.178	82.55	76.23
$H_2SO_4.4H_2O$		liq	170.1389	-480.688	-403.001	99.09	91.35
$H_2SO_4.6.5H_2O$		liq	215.1772	-653.264	-546.403	140.51	136.30
18 - NITROGEN							
N_2		g	28.0134	0	0	45.77	6.961
NO_3^-	nitrate	aq	62.0049	-49.56	-26.61	35.0	-20.7
NH_3		g	17.0306	-11.02	-3.94	45.97	8.38
	undissoc.	aq		-19.19	-6.35	26.6	
NH_4^+		aq	18.0386	-31.67	-18.97	27.1	19.1

Substance			298.15K; 25°C			
Formula & Description	State	Formula Weight	ΔH°_f kcal/mole	ΔG°_f kcal/mole	S° cal/deg mole	Cp° cal/deg mole
HNO_3	liq	63.0129	-41.61	-19.31	37.19	26.26
	g		-32.28	-17.87	63.64	12.75
	aq		-49.56	-26.61	35.0	-20.7
$HNO_3.H_2O$	liq	81.0282	-113.16	-78.61	51.84	43.61
$HNO_3.3H_2O$	liq	117.0589	-252.40	-193.91	82.93	77.71
NH_4OH	liq	35.0460	-86.33	-60.74	39.57	37.02
undissoc.	aq		-87.505	-63.04	43.3	
ionized	aq		-86.64	-56.56	24.5	-16.4
19 - PHOSPHORUS						
PO_4^{3-}	aq	94.9714	-305.3	-243.5	-53.	
$P_2O_7^{4-}$	aq	173.9434	-542.8	-458.7	-28.	
HPO_4^{2-}	aq	95.9794	-308.83	-260.34	-8.0	
$H_2PO_4^-$	aq	96.9873	-309.82	-270.17	21.6	
H_3PO_4	c	97.9953	-305.7	-267.5	26.41	25.35
	liq		-302.8			
ionized	aq		-305.3	-243.5	-53.	
undissoc.	aq		-307.92	-273.10	37.8	
$H_3PO_4.0.5\ H_2O$	c	107.0030	-342.1	-296.9	30.87	30.12
$H_3PO_4.H_2O$	c	116.0106	-374.96			
$HP_2O_7^{3-}$	aq	174.9514	-543.7	-471.4	11.	
$H_3P_2O^{7-}$	aq	175.9593	-544.6	-480.5	39.	
$H_3P_2O_7^-$	aq	176.9673	-544.1	-483.6	51.	

Substance				298.15K; 25°C			
Formula & Description		State	Formula Weight	ΔH_f° kcal/mole	ΔG_f° kcal/mole	S° cal/deg mole	Cp° cal/deg mole
20 - CARBON							
CO_2		g	44.0100	-94.051	-94.254	51.06	8.87
	undissoc.	aq		-98.90	-92.26	28.1	
CO_3^{2-}		aq	60.0094	-161.84	-126.17	-13.6	
HCO_3^-		aq	61.0174	-165.39	-140.26	21.8	
H_2CO_3	undissoc.	aq	62.0253	-167.22	-148.94	44.8	
NH_4HCO_3		c	79.0559	-203.0	-159.2	28.9	
NH_2COONH_4	ammonium carbamate	c	78.0712	-154.17	-107.09	31.9	
29 - ALUMINUM							
Al^{3+}		g	26.9815	1310.70			
		aq		-127.	-116.	-76.9	
AlF^{2+}		aq	45.9799		-192.		
AlF_2^+		aq	64.9783	-266.			
AlF_3		c	83.9767	-359.5	-340.6	15.88	17.95
		g		-287.9	-284.0	66.2	14.97
	un-ionized	aq		-363.	-338.	-6.	
	ionized	aq		-366.	-316.	-86.8	
33 - ZINC							
Zn^{2+}		g	65.37	665.09			
		aq		-36.78	-35.14	-26.8	11.
$ZnCl^+$		aq	100.823		-65.8		
$ZnCl_2$		c	136.276	-99.20	-88.296	26.64	17.05
		g		-63.6			
		aq		-116.68	-97.88	0.2	-54.
	undissoc.	aq			-96.5		

Substance			298.15K; 25°C			
Formula & Description	State	Formula Weight	ΔH_f° kcal/mole	ΔG_f°	S° cal/deg mole	Cp°
$ZnCl_3^-$	aq	171.729		-129.2		
$ZnCl_4^{2-}$	aq	207.182		-159.2		
$ZnClO_4^+$	aq	164.821		-39.0		
$Zn(ClO_4)_2$	aq	264.271	-98.60	-39.26	60.2	
$Zn(ClO_4)_2 \cdot 6H_2O$	c	372.363	-509.89	-371.8	130.4	
36 - COPPER						
Cu^+	g	63.54	260.513			
	aq		17.13	11.95	9.7	
Cu^{2+}	g		729.93			
CuCl	c	98.993	-32.8	-28.65	20.6	11.6
$CuCl^+$	aq			-16.3		
$CuCl_2$	c	134.446	-52.6	-42.0	25.83	17.18
undissoc.	aq			-47.3		
$CuCl_2 \cdot 2H_2O$	c	170.477	-196.3	-156.8	40.	
$CuCl_2^-$	aq	134.446		-57.4		
$CuCl_3^{2-}$	aq	169.899		-90.		
$Cu(ClO_4)_2$	aq	262.441	-46.34	11.54	63.2	
$Cu(ClO_4)_2 \cdot 6H_2O$	c	370.533	-460.9			

Substance			298.15K; 25°C			
Formula & Description	State	Formula Weight	ΔH_f° kcal/mole	ΔG_f° kcal/mole	S° cal/deg mole	Cp° cal/deg mole
39 - NICKEL						
Ni^{2+}	g	58.71	700.32			
	aq		-12.9	-10.9	-30.8	
$NiOH^+$	aq	75.717	-68.8	-54.4	-17.	
$Ni(OH)_2$	c	92.725	-126.6	-106.9	21.	
	aq		-122.8	-86.1	-35.9	
$NiCl_2$	c	129.616	-72.976	-61.918	23.34	17.13
$NiCl_2.2H_2O$	c	165.647	-220.4	-181.7	42.	
$NiCl_2.4H_2O$	c	201.677	-362.5	-295.2	58.	
$NiCl_2.6H_2O$	c	237.708	-502.67	-409.54	82.3	
$Ni(ClO_4)_2$	aq	257.611	-74.7	-15.0	56.2	
$Ni(ClO_4)_2.6H_2O$	c	365.703	-486.6			
$Ni(NH_3)_2^{2+}$	aq	92.771	-58.9	-30.6	20.4	
$Ni(NH_3)_4^{2+}$	aq	126.832	-104.9		61.8	
$Ni(NH_3)_6^{2+}$	aq	160.894	-150.6	-61.2	94.3	
40 - COBALT						
Co^{2+}	g	58.9332	679.17			
	aq		-13.9	-13.0	-27.	
$Co(OH)_2$ blue, precipitated	c	92.9479		-107.6		
pink, precipitated	c		-129.0	-108.6	19.	
pink, precip., aged	c			-109.5		
	aq		-123.8	-88.2	-32.	
undissoc.	aq		-100.8			

Substance			298.15K; 25°C			
Formula & Description	State	Formula Weight	ΔH_f^o kcal/mole	ΔG_f^o	S° cal/deg mole	Cp°
$CoCl_2$	c	129.8392	-74.7	-64.5	26.09	18.76
	aq		-93.8	-75.7	0.	
$CoCl_2.H_2O$	c	147.8545	-147.			
$CoCl_2.2H_2O$	c	165.8699	-220.6	-182.8	45.	
$CoCl_2.6H_2O$	c	237.9312	-505.6	-412.4	82.	
$Co(ClO_4)_2$	aq	257.8344	-75.7	-17.1	60.	
$Co(ClO_4)_2.6H_2O$	c	365.9264	-487.2			
41 - IRON						
Fe^{2+}	g	55.847	657.8			
	aq		-21.3	-18.85	-32.9	
Fe^{3+}	g		1365.9			
	aq		-11.6	-1.1	-75.5	
$FeOH^+$	aq	72.8544	-77.6	-66.3	-7.	
$FeOH^{2+}$	aq		-69.5	-54.83	-34.	
$Fe(OH)_2$ precipitated	c	89.8617	-136.0	-116.3	21.	
$Fe(OH)_2^+$	aq			-104.7		
$Fe(OH)_3$ precipitated	c	106.8691	-196.7	-166.5	25.5	
undissoc.	aq			-157.6		
$Fe(OH)_3^-$	aq			-147.0		
$Fe(OH)_4^{2-}$	aq	123.8765		-184.0		
$Fe_2(OH)_2^{4+}$	aq	145.7087	-146.3	-111.68	-85.	
$FeCl^{2+}$	aq	91.300	-43.1	-34.4	-27.	
$FeCl_2$	c	126.753	-81.69	-72.26	28.19	18.32
	g		-35.5			
	aq		-101.2	-81.59	-5.9	

Substance		298.15K; 25°C				
Formula & Description	State	Formula Weight	ΔH_f^o kcal/mole	ΔG_f^o kcal/mole	S° cal/deg mole	Cp° cal/deg mole
$FeCl_2.2H_2O$	c	162.7837	-227.8			
$FeCl_2.4H_2O$	c	198.8143	-370.3			
$FeCl_2^+$	aq			-66.7		
$FeCl_3$	c	162.206	-95.48	-79.84	34.0	23.10
	g		-60.7			
	aq		-131.5	-95.2	-35.0	
undissoc.	aq			-96.7		
$FeSO_4$	c	151.9086	-221.9	-196.2	25.7	24.04
	aq		-238.6	-196.82	-28.1	
$FeSO_4^+$	aq		-222.7	-184.7	-31.	
$FeSO_4.H_2O$	c	169.9239	-297.25			
$FeSO_4.4H_2O$	c	223.9700	-508.9			
$FeSO_4.7H_2O$	c	278.0160	-720.50	-599.97	97.8	94.28
$Fe(SO_4)_2^-$	aq	247.9702		-364.4		
$Fe_2(SO_4)_3$	c	399.8788	-617.0			
	aq		-675.2	-536.1	-136.6	

48 - MANGANESE

Formula & Description	State	Formula Weight	ΔH_f^o	ΔG_f^o	S°	Cp°
Mn^{2+}	g	54.9380	602.1			
	aq		-52.76	-54.5	-17.6	12.
$MnOH^+$	aq	71.9454	-107.7	-96.8	-4.	
$Mn(OH)_2$ precipitated	amorp	88.9527	-166.2	-147.0	23.7	
$Mn(OH)_3^-$	aq	105.9601		-177.9		
$MnCl^+$	aqu	90.3910		-86.7		

Substance			298.15K; 25°C			
Formula & Description	State	Formula Weight	ΔH_f^o kcal/mole	ΔG_f^o kcal/mole	S° cal/deg mole	Cp° cal/deg mole
$MnCl_2$	c	125.8440	-115.03	-105.29	28.26	17.43
	g		-63.0			
	aq		-132.66	-117.3	9.3	-53.
undissoc.	aq			-117.6		
$MnCl_2.H_2O$	c	143.8593	-188.8	-166.4	41.6	
$MnCl_2.2H_2O$	c	161.8747	-261.0	-225.2	52.3	
$MnCl_2.4H_2O$	c	197.9054	-403.3	-340.3	72.5	
$MnCl_3^-$	aq	161.2970		-148.2		
51 - CHROMIUM						
Cr^{2+}	g	51.996	634.2			
	aq		-34.3			
Cr^{3+}	g		1350.			
$Cr(OH)_3$ precipitated	c	103.0181	-254.3			
$[Cr(H_2O)_4(OH)_2]^+$	aqu	158.0721	-455.8			
$[Cr(H_2O)_5OH]^{2+}$	aqu	159.0801	-463.6			
$[Cr(H_2O)_4(OH)_2]OH$	c	175.0795	-514.4			
$[Cr(H_2O)_6]^{3+}$	aqu	160.0880	-477.8			
$[Cr(H_2O)_5OH](OH)_2$	c	193.0948	-586.7			
76 - LANTHANUM						
La^{3+}	g	138.91	933.2			
	aq		-169.0	-163.4	-52.0	-3.
$La(NO_3)_3$	c	324.925	-299.8			
$La(NO_3)_3.3H_2O$	c	378.971	-520.0			
$La(NO_3)_3.4H_2O$	c	396.986	-592.3			
$La(NO_3)_3.6H_2O$	c	433.017	-732.23			

Substance		298.15K; 25°C				
Formula & Description	State	Formula Weight	ΔH_f^o kcal/mole	ΔG_f^o kcal/mole	S^o cal/deg mole	Cp^o cal/deg mole

93 - MAGNESIUM

Formula & Description	State	Formula Weight	ΔH_f^o	ΔG_f^o	S^o	Cp^o
Mg^{2+}	g	24.312	561.299			
	aq		-111.58	-108.7	-33.0	
$MgOH^+$	aq	41.3194		-149.8		
$Mg(OH)_2$	c	58.3267	-220.97	-199.23	15.10	18.41
precipitated	amorp		-220.0			
	g		-134.			
	aq		-221.52	-183.9	-38.1	
$MgCl_2$	c	95.218	-153.28	-141.45	21.42	17.06
$MgCl_2.H_2O$	c	113.2333	-231.03	-205.98	32.8	27.55
$MgCl_2.2H_2O$	c	131.2487	-305.86	-267.24	43.0	38.05
$MgCl_2.4H_2O$	c	167.2793	-453.87	-388.03	63.1	57.70
$MgCl_2.6H_2O$	c	203.3100	-597.28	-505.49	87.5	75.30
$Mg(ClO_4)_2$	c	223.2132	-135.97			
	aq		-173.40	-112.8	54.0	
$Mg(ClO_4)_2.2H_2O$	c	259.2439	-291.3			
$Mg(ClO_4)_2.4H_2O$	c	295.2746	-439.1			
$Mg(ClO_4)_2.6H_2O$	c	331.3052	-584.5	-445.3	124.5	
$MgSO_4$	c	120.3736	-307.1	-279.8	21.9	23.06
undissoc.	aq		-324.1	-289.74	-1.7	
	aq		-328.90	-286.7	-28.2	
$MgSO_4.H_2O$	c	138.3889	-382.9	-341.5	30.2	
	amorp		-376.4	-335.8	33.0	
$MgSO_4.2H_2O$	c	156.4043	-453.2			
$MgSO_4.4H_2O$	c	192.4350	-596.7			
$MgSO_4.6H_2O$	c	228.4656	-737.8	-629.1	83.2	83.20
$MgSO_4.7H_2O$	c	246.4810	-809.92	-686.4	89.	

Substance		298.15K; 25°C				
Formula & Description	State	Formula Weight	ΔH_f^o kcal/mole	ΔG_f^o kcal/mole	S° cal/deg mole	Cp° cal/deg mole
$Mg(NO_3)_2$	c	148.3218	-188.97	-140.9	39.2	33.92
	aq		-210.70	-161.9	37.0	
$Mg(NO_3)_2.2H_2O$	c	184.3525	-336.8			
$Mg(NO_3)_2.6H_2O$	c	256.4138	-624.59	-497.3	108.	
94 - CALCIUM						
Ca^{2+}	g	40.08	460.29			
	aq		-129.74	-132.30	-12.7	
$CaOH^+$	aq	57.087		-171.7		
$Ca(OH)_2$	c	74.095	-235.68	-214.76	19.93	20.91
	g		-130.			
	aq		-239.68	-207.49	-17.8	
$CaCl_2$	c	110.986	-190.2	-178.8	25.0	17.35
	g		-112.7	-114.54	69.35	14.18
	aq		-209.64	-195.04	14.3	
$CaCl_2.H_2O$	c	129.001	-265.1			
$CaCl_2.2H_2O$	c	147.017	-335.3			
$CaCl_2.4H_2O$	c	183.047	-480.3			
$CaCl_2.6H_2O$	c	219.078	-623.3			
$Ca(ClO_4)_2$	c	238.981	-176.09			
	aq		-191.56	-136.42	74.3	
$Ca(ClO_4)_2.4H_2O$	c	311.043	-465.8	-352.97	103.6	
$CaSO_4$ insol., anhydrite	c	136.142	-342.76	-315.93	25.5	23.82
sol., α	c		-340.64	-313.93	25.9	23.95
sol., β	c		-339.58	-312.87	25.9	23.67
	aq		-347.06	-310.27	-7.9	
$CaSO_4.1/2H_2O$ macro; α	c	145.149	-376.85	-343.41	31.2	28.54

Substance				298.15K; 25°C			
Formula & Description		State	Formula Weight	ΔH_f^o	ΔG_f^o	S°	Cp°
				kcal/mole		cal/deg mole	
$CaSO_4.1/2H_2O$	micro; β	c		-376.35	-343.18	32.1	29.69
$CaSO_4.2H_2O$	selenite	c	172.172	-483.42	-429.60	46.4	44.46

99 - SODIUM

Formula & Description		State	Formula Weight	ΔH_f^o	ΔG_f^o	S°	Cp°
Na^+		aq	22.9898	-57.39	-62.593	14.1	11.1
NaOH		c	39.9972	-101.723	-90.709	15.405	14.23
		g		-49.5	-50.2	54.57	11.56
		aq		-112.360	-100.187	11.5	-24.4
$NaOH.H_2O$		c	58.0126	-175.560	-150.435	23.780	21.55
$NaOH.2H_2O$		liq	76.0280	-243.565	-208.705	46.840	57.22
$NaOH.3.5H_2O$		liq	103.0511	-348.900	-295.545	68.377	84.71
$NaOH.4H_2O$		liq	112.0588	-383.64	-324.30	76.17	
$NaOH.5H_2O$		liq	130.0742	-452.75	-381.60	92.27	
$NaOH.7H_2O$		liq	166.1050	-590.11	-495.74	125.79	
NaCl		c	58.4428	-98.268	-91.815	17.24	12.07
		g		-42.22	-47.00	54.90	8.55
		aq		-97.34	-93.965	27.6	-21.5
$NaSO_4^-$		aq	206.2996	-109.7	-109.5	61.8	
Na_2SO_4	(v, orthorhombic)	c	142.0412	-331.52	-303.59	35.75	30.64
	(III, metastable)	c2				37.030	30.90
$Na_2SO_4.10H_2O$		c	322.1952	-1034.24	-871.75	141.5	

Substance			298.15K; 25°C			
Formula & Description	State	Formula Weight	ΔH_f^o	ΔG_f^o	S°	Cp°
			kcal/mole		cal/deg mole	
100 - POTASSIUM						
K^+	aq	39.1020	-60.32	-67.70	24.5	5.2
KOH	c	56.1094	-101.521	-90.61	18.85	15.51
	g		-55.2	-55.6	56.92	11.76
	aq		-115.29	-105.29	21.9	-30.3
$KOH.H_2O$	c	74.1248	-179.0	-154.2	28.	
$KOH.2H_2O$	c	92.1402	-251.2	-212.1	36.	
KCl	c	74.5550	-104.385	-97.79	19.74	12.26
	g		-51.18	-55.69	57.12	8.72
	aq		-100.27	-99.07	38.0	-27.4

1. (a) Wagman, D.D.; W.H. Evans; V.B. Parker; I. Halow; S.M. Bailey; R.H. Schumm; Selected Values of Chemical Thermodynamic Properties - Tables for the First Thirty-Four Elements in the Standard Order of Arrangement, National Bureau of Standards Technical Note 270-3, 1968

(b) Wagman, D.D.; W.H. Evans; V.B. Parker; I. Halow; S.M. Bailey; R.H. Schumm; Selected Values of Chemical Thermodynamic Properties - Tables for Elements 35 through 53 in the Standard Order of Arrangement, NBS Tech Note 270-4, 1969

(c) Wagman, D.D.; W.H. Evans; V.B. Parker; I. Halow; S.M. Bailey; R.H. Schumm; K.L. Churney; Selected Values of Chemical Thermodynamic Properties - Tables for Elements 54 through 61 in the Standard Order of Arrangement, NBS Tech Note 270-5, 1971

(d) Parker, V.B.; D.D. Wagman and W.H. Evans; Selected Values of Chemical Thermodynamic Properties - Tables for the Alkaline Earth Elements (Elements 92 through 97 in the Standard Order of Arrangement), NBS Tech Note 270-6, 1971

(e) Schumm, R.H.; D.D. Wagman; S. Bailey; W.H. Evans; V.B. Parker, Selected Values of Chemical Thermodynamic Properties - Tables for the Lanthanide (Rare Earth) Elements (Elements 62 through 76 in the Standard Order of Arrangement), NBS Tech Note 270-7, 1973

(f) Wagman, D.D.; W.H. Evans; V.B. Parker; R.H. Schumm; R.L. Nuttall; Selected Values of Chemical Thermodynamic Properties - Compounds of Uranium, Protactinium, Thorium, Actinium, and the Alkali Metals, NBS Tech Note 270-8, 1981

APPENDIX C

NBS Special Publication 685

Compiled Thermodynamic Data Sources for Aqueous and Biochemical Systems:
An Annotated Bibliography (1930-1983)

Robert N. Goldberg
Center for Chemical Physics
National Measurement Laboratory
National Bureau of Standards
Gaithersburg, MD 20899

Sponsored by:
Design Institute for
Physical Property Data
Project 811
American Institute of
Chemical Engineers
New York, NY

December 1984

U.S. Department of Commerce
Malcolm Baldrige, Secretary
National Bureau of Standards
Ernest Ambler, Director

Library of Congress
Catalog Card Number: 84-601131
National Bureau of Standards
Special Publication 685
Natl. Bur. Stand. (U.S.).
Spec. Publ. 685.
106 pages (Dec. 1984)
CODEN: XNBSAV

U.S. Government Printing Office
Washington: 1984

For sale by the Superintendent
of Documents,
U.S. Government Printing Office,
Washington, DC 20402

ABSTRACT

This is a selected and annotated bibliography of sources of compiled and evaluated chemical thermodynamic data relevant to biochemical and aqueous systems. The principal thermodynamic properties considered herein are Gibbs energy and equilibrium data, enthalpies of formation and reaction, heat capacities and entropies, and the corresponding partial molar and excess properties. Derived quantities used in calculating the above are also included. Transport and mechanical data have also been identified to a lesser degree. Included in the annotations to the data sources are brief descriptions of the types of properties tabulated, the classes of materials dealt with, and the degree of completeness of the compilations.

Keywords: Aqueous systems; bibliography; biochemical systems; enthalpy data; entropy data; equilibrium data; excess properties; Gibbs energy data; heat capacity data; partial molar properties; review articles; thermochemistry; thermodynamics.

INTRODUCTION

"There is a growing need for reliable thermodynamic data for both scientific and practical purposes. However, the existence of a desired piece of data in the primary literature does not guarantee its recovery by an interested user. Indeed the recovery process may be a decidedly non-trivial as well as a time consuming matter. At one time this problem was, in large part, managed by periodic review articles, monographs, and the International Critical Tables. The National Standard Reference Data System, CODATA, and other coordinating organizations are helping to provide current compilations. However, the preparation of new reviews is sufficiently complex that there are at any give time many subject areas that do not have current compilations available. In this circumstance earlier reviews are an important resource for the technologist or scientist. Unfortunately, the proliferation of such works and the enormity of the available literature have made even the recovery of review and compilation articles a matter of some difficulty. It is to help solve this problem for one small part of our science that we have prepared this annotated bibliography of thermodynamic data sources relevant to biochemical and aqueous systems. These two areas are mutually complementary. In environmental problems such as water quality control, and in the utilization of aqueous systems with respect to energy resources and the processing of material resources, the aqueous systems of interest very often include the same substances and require the same information as in the study of biochemical systems. In addition to information on the aqueous solutions, the properties of the pure substances are often required in order to establish reference points for interrelating various systems. Thus, this bibliography is a selective listing of sources of thermodynamic information for pure substances and aqueous solutions, selected for their particular relevance to these kinds of problems."

The above quotation is from NBS Special Publication 454 and it contains the principle upon which both that publication and this one, the successor to it, are based. The identification and collection of sources of compiled thermodynamic data was begun by the late George Armstrong at the National Bureau of Standards at least ten years ago. The author joined him in this task several years ago and with him co-authored NBS Special Publication 454 on this subject in 1976. Many of the descriptions of older works in this Bibliography are taken directly from Special Publication 454. The intervening years have seen a remarkable growth in the subject matter of this Bibliography and a revision of it was suggested by the Design Institute for Physical Property Data of the American Institute of Chemical Engineers which also helped in providing the necessary funding for the project. A "review of reviews" which the reader might also wish to consult is the chapter by E. F. G. Herrington in "Chemical Thermodynamics: Volume I" (see item [90]).

The scope of the present work remains essentially that of Special Publication 454. The general aim is to assist the reader in locating those publications which contain thermochemical data which can best serve his needs. Equilibrium data is taken in its most general sense and includes equilibrium constants, enthalpies, entropies, heat capacities, volumes, and partial molar and excess property data. To a much lesser extent, transport and other properties have been included. Unfortunately, much of the data on biochemical systems is scattered throughout much of the literature and there is a need for definitive reviews and compilations in several areas such as enzyme-catalyzed reactions and on the denaturation of enzymes and nucleic acids. This need is all the more pressing due to the growing importance of bio-engineering and its industrial importance. Because of this, various chapters and short reviews in books on biochemistry have been included in this bibliography.

The entries are listed in alphabetical order by first author or by the editor. There are a few cases in which the work of a research group, or the volumes of a multi-volume work are kept together and entered under the name of the principal researcher, the editor, or the institution. All authors have been cited in the author index and it should be consulted if an article is not found where expected.

Each annotation contains a brief summary of the contents of the book or article cited. Specifically, the nature and state of the substances dealt with, the extent of coverage, and whether the given property values have been compiled, selected, or critically evaluated is stated. Also, whenever a meaningful statement on temperature range can be made, it has been included. For some large compilations the ranges are too different for individual systems to make any statement practical, and for them, the authors have usually attempted to cover the full range of the data. It is hoped that the contents of the annotation is sufficient to permit the reader to select those references which most directly apply to his own needs.

The present bibliography identifies 162 different sources for the reader to consult. Unfortunately, many of the books cited are out-of-print, not in English, or in report form (many of the U.S. government sponsored reports are available from the National Technical Information Service, Springfield, Virginia 22161). The reader is urged to exercise care and judgment when combining formation properties from different sources when calculating equilibrium constants and enthalpies of reaction. There are several reasons for this precaution: (1) different reference states may be used for the elements, (2) there are different conventions regarding standard states, (3) equilibrium "constants" may be concentration dependent, and (4) different thermodynamic pathways may have been used to calculate tabulated formation properties. The safest approach here is to calculate equilibrium constants (or enthalpy and entropy changes) from the formation properties given in one reference and then to combine them with similar data calculated or found in other references.

The alphabetical subject index that follows the main section of this bibliography gives an indication of thermodynamic properties, physical or chemical processes, classes of substances, and, in a few cases, individual substances for which information is to be found herein. It is not practical to give an exhaustive index to the contents of the individual references. Hence the absence of a piece of information in the index does not necessarily mean its absence in the references. This is particularly true with respect to particular substances, which the reader should assume are not listed except by chance of title or abstract.

It is not possible to claim completeness for this bibliography and somewhat arbitrary decisions have been made as to whether or not to include a specific item. The author would appreciate comments from interested readers concerning data sources omitted from this bibliography.

The author thanks Drs. William Evans, David Garvin, and Vivian Parker for their comments on this bibliography and Beverly S. Geisbert for her excellent clerical work on the manuscript.

BIBLIOGRAPHY

[1] Alberty, R. A.
Standard Gibbs Free Energy, Enthalpy, and Entropy Changes as a Function of pH, and pMg for Several Reactions Involving Adenosine Phosphates
Journal of Biological Chemistry 244, 3290 (1969).

The standard Gibbs energy, enthalpy, and entropy changes for the hydrolysis of adenosine-5'-triphosphate to adenosine-5'-diphosphate are computed as a function of pH and magnesium ion concentration at 25 °C. A critical evaluation of the relevant literature data is included. Also see item [120]

* * * * * * * * * *

[2] Armstrong, G. T., Domalski, E. S., Furukawa, G. T., Reilly, M. L., Wilhoit, R. C. and others
A Survey of Thermodynamic Properties of the Compounds of the Elements CHNOPS - A Series of Eighteen Reports
(National Bureau of Standards Reports No. 8521, 8595, 8641, 8906, 8992, 9043, 9089, 9374, 9449, 9501, 9553, 9607, 9883, 9968, 10070, and 10291, published during the years 1964 to 1970, U. S. Department of Commerce, Washington, D.C.)

This series of reports is a survey of the thermodynamic properties of selected compounds of biological importance containing the elements carbon, hydrogen, nitrogen, oxygen, phosphorus, and sulfur. Included in these reports are heat capacity data; heats and Gibbs energies of formation; vapor pressure data; tables of thermodynamic functions; Gibbs energies, entropies and enthalpies of solution and dilution; and thermodynamic properties of mixed solvent systems. Except for the tables of thermal functions, the data refer to 25 °C or nearby temperatures. References to sources of data in the literature are included.

Some of the data contained in these reports formed the basis of Domalski's enthalpy of combustion tables (see item [29]) and of Wilhoit's tables (see item [15]). The remainder of the material in these reports has not appeared in press elsewhere. The NBS Chemical Thermodynamics Data Center has a complete set of these reports available for examination.

* * * * * * * * * *

[3] Ashcroft, S. J. and Mortimer, C. T.
Thermochemistry of Transition Metal Complexes
(Academic Press, London and New York, 1970).

This book surveys the literature to 1968 on energy changes for processes involving transition metal complexes including both organic and inorganic ligands. A critical review of the thermochemical data for over 1500 systems of complexes is given. Comparable data from various sources are shown in juxtaposition. Values of ΔH, ΔG, and ΔS for various stages of complex formation are usually listed for processes in aqueous solution at or near 25 °C. $\Delta_f H°$, $\Delta_f G°$, $\Delta_f S°$ for crystalline complex substances are given where available. In many instances correlations of the data for various metals with a single ligand, and for various related complexes of a given metal, are given graphically or by means of bond-energy estimates based on the data.

* * * * * * * * * *

[4] Baes, C. F., Jr. and Mesmer, R. E.
The Hydrolysis of Cations
(John Wiley and Sons, New York, 1976)

This book on the chemistry of hydrolysis of inorganic cations contains a substantial amount of equilibrium data pertinent to hydrolysis reactions. For each of the elements which produces a cation or cations in aqueous solution, the available equilibrium data for the hydrolysis reaction(s) at or about 298 K has been critically assessed in order to obtain "best" values for equilibrium constants and quotients applicable to a given medium. When available, ΔH and ΔS data for the hydrolysis reactions are also presented. The data, with references and comments, is arranged under the element of interest.

* * * * * * * * * *

[5] Barner, H. E. and Scheurman, R. V.
Handbook of Thermochemical Data for Compounds and Aqueous Species
(John Wiley and Sons, New York, 1978)

This book presents thermodynamic properties for a wide variety of ions and complexes in aqueous solution over a range of temperature (typically 25 to 300 °C). Most of the properties were calculated from data at 25 °C, much of which was taken from the National Bureau of Standards Technical Note 270 (see item [149]). Phase transition data were taken from the book by I. Barin and O. Knacke, "Thermochemical Properties of Inorganic Substances" (Springer-Verlag, Berlin, 1973). The extension of the 25 °C data to higher temperatures was done using estimation schemes such as those developed by C. M. Criss and J. W. Cobble (J. Am. Chem. Soc. 86, 5390 (1964)) and H. C. Helgeson (J. Phys. Chem. 71, 3121 (1967)).

* * * * * * * * * *

[6] Bates, R. G.
Determination of pH, Theory and Practice (Second Edition)
(John Wiley and Sons, New York, 1973)

This book, the primary topic of which is the establishment of an operational pH scale, contains several tables of interest. Tabulated are the ion product of water from 0 to 60 °C, the vapor pressure, density, and dielectric constant of water from 0 to 100 °C, dielectric constants of pure liquids, and pH values of several aqueous buffer systems.

* * * * * * * * * *

[7] Battino, R.
Volume Changes on Mixing for Binary Mixtures of Liquids
Chemical Reviews 71, 5 (1971).

This critical and extensive review deals with both the volume changes associated with the mixing of binary mixtures of liquids and partial molar volumes at infinite dilution of various solvent systems, aqueous and non-aqueous. The temperature range cited is that at which the experimental measurements have been performed. Included is a detailed discussion of the experimental methods used for measurements and associated theoretical developments. The coverage of the available literature appears to be very thorough. There are 427 references to the primary literature.

* * * * * * * * * *

[8] Battino, R., and Clever, H. L.
The Solubility of Gases in Liquids
Chemical Reviews 66, 395 (1966)

This thorough and detailed review contains a discussion of experimental methods for the measurement of gas solubilities in all types of liquids, including water. Tabulated are the solubilities of oxygen, nitrogen, and argon in water at one atmosphere pressure and from 0 to 50 °C. Also to be found herein is an extensive table listing sources of gas solubility data in the primary literature. Although the tables do not give selected values, the authors have given their assessment of the reliability of the data to be found in the listed sources by means of a coding scheme. There are 686 references to the primary literature. Also see item [154].

* * * * * * * * * *

[9] Beezer, A. E. (editor)
Biological Microcalorimetry
(Academic Press, London and New York, 1980)

This book contains fourteen chapters on various topics related to calorimetry as applied to biochemical and biological systems. The coverage of topics is very broad and includes bacteria, cells, microorganisms, drugs, membranes, and enzymes. While there are few tables of thermodynamic data in this book, many of the chapters contain useful collections of references to the literature pertinent to the topic of the individual chapter.

* * * * * * * * * *

[10] Benson, S. W.
Thermochemical Kinetics:
Methods for the Estimation of Thermochemical Data and Rate Parameters
(Second Edition)
(John Wiley and Sons, New York and London 1976)

This monograph gives tables of necessary data and descriptions of methods of their use for calculating ΔH_f, C_p, and S at 25 °C for gas phase molecules and radicals and for extrapolating them to higher temperatures. The procedures can be applied to hydrocarbons, oxygen-containing compounds, nitrogen-containing compounds, polycyclic structures, haloalkanes, organo-sulfur compounds, organo-metallic compounds, polycyclic substances, and deal principally with organic compounds.

* * * * * * * * * *

[11] Benson, S. W., Cruickshank, F. R., Golden, D. M., Haugen, G. R., O'Neal, H. E., Rogers, A. S., Shaw, R. and Walsh, R.
Additivity Rules for the Estimation of Thermochemical Properties
Chemical Reviews 69, 279 (1969)

This lengthy technical article gives procedures for calculating the properties $\Delta_f G°$, $S°$, $C_p°$, for organic compounds in the gas phase. Parameters for calculating C_p are given for the temperature range from 300 to 1500 K. The availability of C_p as a function of temperature allows calculation of $\Delta_f G°$ and $S°$ at the same temperatures. The necessary constants for making the calculations are given for individual chemical groupings in some 38 tables. Many classes of functional groups and molecular conformations are included. Examples are given comparing calculated and observed values. Agreements of 1 $kcal \cdot mol^{-1}$ (4.184 $kJ \cdot mol^{-1}$) or better in $\Delta_f G°$ and 1 $cal \cdot mol^{-1} \cdot K^{-1}$ or better in $C_p°$ and $S°$ are generally found. Also see item [10] which supersedes this article.

* * * * * * * * * *

[12] Bichowsky, F. R. and Rossini, F. D.
The Thermochemistry of the Chemical Substances
(Reinhold Publishing Corporation, New York, 1936)

Although outdated, this book still provides useful references to the older thermochemical literature. Tabulated are $\Delta_f H°$ and enthalpy of transition values for the elements and their compounds, with the data for carbon-containing compounds being terminated at two carbon atoms. It should be noted that the data pertain to a temperature of 18 °C and to diamond, rather than graphite, as the standard state for carbon; the yellow form is the reference state for phosphorus. The yellow form is thermochemically identical to the white form which is the reference state used in the NBS Thermochemical Tables [149]. The data upon which this book was based were used in preparing NBS Circular 500, see item [131].

* * * * * * * * * *

[13] Bondi, A.
Physical Properties of Molecular Crystals, Liquids, and Glasses
(John Wiley and Sons, New York and London, 1968)

This monograph is designed for use by chemical engineers in estimating physical properties needed in design calculations, as well as by physical chemists and synthetic chemists who need to understand the relationship between structure and physical properties. Correlations of several kinds are described in the text. These are then restated as methods for estimation of the properties.

Substances considered include non-polar and polar gases, non-polar and polar liquids, associated liquids, crystalline solids, glasses, polymers, and polymer melts, as well as others. Procedures given include many variants, depending upon the properties given as initial information. Among the properties for which procedures are given are: density, heat capacity, enthalpy, entropy, enthalpy and entropy of fusion, enthalpy of vaporization, vapor pressure, cubical thermal expansion coefficient, bulk modulus, Young's modulus compressibility, thermal conductivity, rotational diffusion constant, relaxation times, mass diffusion, viscosity, and others.

* * * * * * * * * *

[14] Boublik, T., Fried, V. and Hala, E.
The Vapour Pressures of Pure Substances
(Elsevier, New York, 1973)

Data are presented in the form of constants for the Antoine equation for the temperature dependence of the vapor pressures of 806 substances in the normal and low pressure region. Almost all of the substances contain carbon. Experimental data from selected original sources are given, together with smoothed values obtained from the Antoine equations at the same temperature, and the absolute and percentage deviations. Standard deviations are calculated. A standard boiling point is calculated for each substance.

* * * * * * * * * * *

[15] Brown, H. D. (editor)
Biochemical Microcalorimetry
(Academic Press, New York and London, 1969)

This book contains seventeen chapters on subjects concerning heat measurements and their relationship to biology and biochemistry. Also see the 1980 volume of this same series, item [63]. Several of the articles contain tabulations of data:

Author(s)	Title of Chapter
R. C. Wilhoit	Thermodynamic Properties of Biochemical Substances

This chapter (no. II) includes several tables: (1) Selected values of thermodynamic properties ($\Delta_f G°$, $\Delta_f H°$, $S°$, and $C_p°$) for about 120 important biochemical species or compounds; (2) Enthalpies and Gibbs energies of formation of adenosine phosphoric acid species relative to H_2ADP^{1-} at 25 °C; (3) $\Delta H°$ and $\Delta G°$ for six important biochemical processes; and (4) Partial molar properties (L_2, L_1, C_1) of aqueous glucose, glycerol, glycine, and urea at 25 °C.

Author(s)	Title of Chapter
S. Ono and K. Takahashi	Chemical Structure and Reactions of Carbohydrates

This chapter (no. IV) has a table of enthalpies of isomerization of eleven carbohydrates and a table given enthalpies of hydrolysis of aqueous α-1,4 and α-1,6 glucosidic linkages in several glucosides at 25 °C.

Author(s)	Title of Chapter
T. Ackermann	Physical States of Biomolecules: Calorimetric Study of Helix-Random Coil Transitions in Solution

This chapter (no. VI) contains a tabulation of calorimetrically determined enthalpy values accompanying conformational changes of macromolecules in solution (26 references) and their transition temperatures.

H. D. Brown Calorimetry of Enzyme-Catalyzed Reactions

This chapter (no. VII) contains a summary of calorimetric enthalpy values for enzyme catalyzed systems (21 references) at or near 25 °C.

* * * * * * * * * *

[16] Chapman, T. W. and Newman, J.
A Compilation of Selected Thermodynamic and Transport Properties of Binary Electrolytes in Aqueous Solution
(U. S. Atomic Energy Commission Report UCRL-17767 (1968))

Data from the literature on the properties of sixty-one common binary inorganic electrolytes at various temperatures are tabulated with appropriate references. The properties include the density, viscosity, transference number, diffusion coefficient, and the activity coefficient.

* * * * * * * * * *

[17] Charlot, G.
Selected Constants - Oxidation-Reduction Potentials of Inorganic Substances in Aqueous Solution
(Butterworths, London, 1971)

This reference work, prepared under the auspices of the IUPAC, contains selected values of electrochemical potentials relative to the assigned zero value of the standard hydrogen electrode at or near 25 °C. Entries are given for about 350 inorganic systems. The literature coverage is through 1967.

* * * * * * * * * *

[18] Christensen, J. J., Eatough, D. J. and Izatt, R. M.
Handbook of Metal Ligand Heats and Related Thermodynamic Quantities
(Second edition)
(Marcel Dekker, New York, 1975)

This handbook gives tabulated values of thermodynamic functions in aqueous solution. Enthalpies are given for equilibria involving metal ions and ligands, together with the related thermodynamic quantities log K, ΔS, and ΔC_p, where available. The body of the book consists of a table (414 pp.) in which are summarized the published literature values up to 1974, classified according to ligand. In addition, the appropriate reaction, the method, and conditions of measurement of ΔH are given. The temperatures are also specified and are in the vicinity of 25 °C. Both inorganic and organic ligands, and complexes of about seventy metallic elements are given. A seven page guide to the use of the table and indexes is given. The table is indexed by author, by ligand formula, and by metal. An index of synonyms and a chronological list of references are also given.

* * * * * * * * * *

[19] Christensen, J. J., Eatough, D. J. and Izatt, R. M.
The Synthesis and Ion Binding of Synthetic Multidentate Macrocyclic Compounds
Chemical Reviews 74, 351 (1974)

This review article contains, in addition to a discussion on the synthesis, kinetic, and structural parameters, a referenced compilation of log K, ΔH, ΔS, and ΔC_p data at or near 25 °C for the interaction of inorganic cations with synthetic multidentate macrocyclic compounds. Each entry in the table of thermodynamic data includes the ionic strength, solvent system, and method of measurement used in obtaining the data.

* * * * * * * * * *

[20] Christensen, J. J., Hansen, L. D. and Izatt, R. M.
Handbook of Proton Ionization Heats and Related Thermodynamic Quantities
(John Wiley and Sons, New York, 1976)

This book is a compilation of enthalpies, entropies, and pK values for the proton ionization of (mostly) organic compounds in water. For each of the approximately 600 compounds covered in this handbook is given the ΔH, ΔS, and pK for the proton ionization, the method and conditions of measurement, and some brief remarks. The temperatures are also specified and are at or near 25 °C. Literature references are also given for each compound.

* * * * * * * * * *

[21] Clark, W. M.
Oxidation-Reduction Potentials of Organic Systems
(The Williams and Wilkins Company, Baltimore, 1960)

This 584 page monograph contains a comprehensive discussion on the determination of the electrochemical potentials of organic systems with emphasis both on theory and experimental practice. Included are approximately 100 tables of critically evaluated oxidation and reduction potentials for organic and biochemical systems through about 1960. Included are the quinones, phenols, anilines, porphyrins, nicotinamide-adenine dinucleotide, and nicotinamide-adenine dinucleotide phosphate systems, and several others. The author has, in most cases, specified the conditions to which the data refer.

* * * * * * * * * *

[22] Clarke, E. C. W. and Glew, D. N.
Journal of the Chemical Society Faraday Transactions I 76, 1911 (1980)

Debye-Hückel limiting slopes for the osmotic coefficient in water are presented as a function of temperature (0 to 150 °C). The recommended values were derived from literature measurements of the static dielectric constant and density of water.

* * * * * * * * * *

[23] CODATA Recommended Key Values for Thermodynamics 1977
(CODATA Bulletin No. 28 (1978) and Tentative Set of Key Values for Thermodynamics: Parts VI, VII, and VIII, CODATA Special Reports: No. 4 (March 1977), No. 7 (April 1978), and No. 8 (April 1980))
(CODATA Secretariat, Paris)

These are reports by the CODATA (Committee on Data for Science and Technology of the International Council of Scientific Unions) Task Group on Key Values for Thermodynamics in which are presented recommended values for the quantities $\Delta_f G°$ (298.15 K), S° (298.15 K), $\Delta_f H°$ (298.15 K), and H° (298.15 K) - H° (0 K) for ≈180 of the thermochemically more important elements and compounds, including some aqueous species, mostly electrolytes. These bulletins supersede earlier CODATA reports of this group. The recommended values are based on a completely new evaluation of all pertinent data available at the time of publication. Consequently, they should not be used indiscriminately with data from earlier self-consistent tables, such as Wagman et al. or Medvedev et al., items [93] and [149]. The earlier publications in this series include CODATA Bulletins 5, 10, 17, and 22 published, respectively, in 1971, 1973, 1976, and 1977. They are useful for the references which they contain to the source literature from which the recommended key values were obtained.

* * * * * * * * * *

[24] Coetzee, J. F. and Ritchie, C. D. (editors) (in two volumes)
Solute-Solvent Interactions
(Marcel-Dekker, New York and London, 1969 and 1976).

Relevant chapters are:

Author(s)	Title of Chapter
J. W. Larson and L. G. Hepler	Heats and Entropies of Ionization (Vol 1, Chapter 1)

This is a detailed review and evaluation of the enthalpies, Gibbs energies, entropies, and heat capacity changes accompanying ionization of organic acids. Included are eleven tables of data on various types of acids, including the carboxylic acids, phenols, anilinium ions, ammonium ions, the amino acids, barbituric acids, and several inorganic acids. The authors also discuss the interpretation of the data in terms of molecular considerations. The tabulated data refer to 25 °C and standard state conditions. There are 224 references.

E. M. Arnett and D. R. McKelvey	Solvent Isotope Effect on Thermodynamics of Non-reacting Solutes (Vol. 1, Chapter 6)

This is a general and extensive review dealing with differences in thermodynamic properties between light and heavy water systems. The properties dealt with include Gibbs energies and enthalpies of transfer and solubilities; systems for which data are tabulated include the more common inorganic electrolytes and ions, alcohols, amides, amino acids, and several nonelectrolytes.

P. M. Laughton and R. E. Robertson	Solvent Isotope Effects for Equilibria and Reactions (Vol. 1, Chapter 7)

Included in this review are tabulated values giving differences in pK values for weak acids in light and heavy water.

C. V. Krishnan and H. L. Friedman Enthalpies of Transfer for Solutes in Polar Solvents (Vol. 2, Chapter 3)

The chapter reviews the existing experimental data for standard enthalpies and entropies of solution of pure substances in various solvents and gives derived entropies of transfer of molecules, ions, and groups from one solvent to another and from the gas phase to water. The data are summarized in 22 tables and cover both inorganic and organic substances, electrolytes and non-electrolytes. There are 146 references to the literature.

* * * * * * * * * *

[25] Cohn, E. J. and Edsall, J. T. (including chapters by J. G. Kirkwood, H. Mueller, J. L. Oncley, and G. Scatchard)
Proteins, Amino Acids, and Peptides: As Ions and Dipolar Ions
(Hafner Publishing Co., New York and London, 1965)

This book contains a general discussion of the physical chemistry of proteins, amino acids, and peptides. Most of the chapters contain references to the source literature where one can find specific pieces of thermodynamic data. Tables in this book of particular interest to this bibliography are: (1) activity coefficients of amino acids and peptides, (2) solubilities of amino acids in water and other solvents, (3) the influence of CH_2 groups upon the interactions between dipolar ions, (4) enthalpies of ionization of amino acids and peptides, and (5) thermodynamic functions ($\Delta G°$, $\Delta H°$, $\Delta S°$, and $\Delta C_p°$) for the ionization of amino acids. Also see entry [24].

* * * * * * * * * *

[26] Conway, B. E.
Electrochemical Data
(Elsevier, New York, 1952)

This monograph is a comprehensive (359 pp.) collection of data on various aspects of pure and applied electrochemistry, relating to organic and inorganic substances, both solid and in solution. Data are presented in tabular form, in ten chapters, introduced with references to data sources and explanations of approach. Physical properties include densities and vapor pressures of various aqueous solutions, dielectric constants, dipole moments, and other properties. Relative partial molar enthalpies and activity and osmotic coefficients, conductance values, ionic mobilities, transference numbers and diffusion coefficients are tabulated. There is a chapter on dissociation constants, solubilities, and buffer solutions. The chapter on properties of electric double layers contains tables of electrokinetic potentials and properties of various interfaces including ones of such biological interest as the mammalian red blood cell. The biologist would also find relevant the section on transport and general properties of colloids and macromolecular electrolytes, including extensive tables of mobilities for such compounds as

hemoglobins, serum albumins, and red blood cells, among others. Electrode chemistry tables include a compilation of data on reversible electrode processes:
liquid junction potentials, half-cell potentials, and electrochemical equivalents for certain elements. A chapter on electrode kinetics gives a critical selection of the available determinations of the parameters of a number of electrode reactions.

* * * * * * * * * * *

[27] Covington, A. K.
Electrolyte Solutions Bulletin
(University of Newcastle-Upon-Tyne, 1971 to present)

This is a current awareness bulletin that provides titles and references of "recent papers covering all aspects of the physical chemistry and structure of electrolyte solutions, the methods used in these studies including spectroscopy, equilibrium (but not kinetic processes), electrode systems, pH, ion selective and reference electrodes."

* * * * * * * * * * *

[28] Cox, J. D. and Pilcher, G.
Thermochemistry of Organic and Organometallic Compounds
(Academic Press, London and New York, 1970)

This monograph is a critical compilation of thermochemical data for the title field published since 1930. The enthalpies of formation at 298.15 K of some 3000 substances are listed, with estimates of error. Where enthalpies of vaporization are known or can be reliably estimated these are listed and in these cases the enthalpies of formation of both gaseous and condensed phases are given. Extensive introductory material presents experimental procedures for reduction of experimental data of the type found in the book. Applications of thermochemical data are given, and there is a section on methods of estimating enthalpies of formation of organic compounds. Also see items [29] and [116].

* * * * * * * * * * *

[29] Domalski, E. S.
Selected Values of Heats of Combustion and Heats of Formation of Organic Compounds Containing the Elements C, H, N, O, P, and S
Journal of Physical and Chemical Reference Data 1, 221 (1972)

Selected values of the enthalpies of combustion and enthalpies of formation of 719 organic compounds at 298.15 K are reported. The selected values are augmented by commentary and original source references. The Wiswesser Line Notation is also given for each compound. The methods used in updating older work are described.

* * * * * * * * * * *

[30] Domalski, E. S., Evans, W. H. and Hearing, E. D.
Heat Capacity, Entropies, and Some Phase Transition Properties of Organic Compounds in the Liquid and Solid Phases
Journal of Physical and Chemical Reference Data, Volume 13, Supplement No. 1, 1984. (286 pp.)

This extensive review gives heat capacities and entropies for approximately 1400 organic compounds in the liquid and solid phases at or near 25 °C. Values for enthalpies and entropies of phase transitions are also included. The literature coverage is from 1881 through most of 1982 and there are detailed references given to the source literature.

* * * * * * * * * *

[31] Domalski, E. S., W. H. Evans and T. L. Jobe, Jr., (principal contributors)
Thermodynamic Data for Waste Incineration
(The American Society of Mechanical Engineers, New York, 1979)

This 160 page monograph presents thermodynamic data for materials which are chemical mixtures, polymers, composite materials, solid wastes, and substances not easily identifiable by a single stoichiometric formula. The thermodynamic properties given are heats of reaction and transitions, heat capacities and heat contents, and vapor and sublimation pressures. Included is a material name and property index which covers approximately 600 materials, many of which are of biological interest, and including data on animals.

* * * * * * * * * *

[32] Dorsey, N. E.
Properties of Ordinary Water-Substance
(Reinhold Publishing Corporation, New York, 1940)

This classic book is an exhaustive and critical compilation of the physical properties of water as reported in the literature through the year 1938. Essentially every physical property of pure water is covered. Although there is little emphasis on aqueous solutions, the solubilities and diffusion constants of selected gases in water are treated. Also see [134].

* * * * * * * * * *

[33] Dymond, J. H. and Smith, E. B.
The Virial Coefficients of Gases
(Clarendon Press, Oxford, 1980)

This book is a critical compilation of the virial coefficients of about 300 gases at various temperatures. Included are references to the original sources as well as, in some cases, standard deviations and estimated inaccuracies for the data. This revised edition also includes data on gaseous mixtures.

* * * * * * * * * *

[34] Fasman, G. D. (editor)
Handbook of Biochemistry and Molecular Biology
(CRC Press, Cleveland, Ohio, 1976)

This compilation is a revision and an expansion of the 1968 edition (see item [140]). Many of the earlier contributions have been revised. The three sections on the solubilities of fatty acids by K. S. Markley are in the volume "Lipids, Carbohydrates, and Steroids" on pages 496 and 497. All of the other articles cited in item [140] appear in the volume "Physical and Chemical Reference Data" between pages 107 and 366. Items of interest which did not appear in the earlier edition are as follows:

Author(s) or Source	Item
E.D. Mooz	Data on the Naturally Occurring Amino Acids (Proteins, p. 111-174)

These tables include pK values for the amino acids at or near 25 °C.

D. B. Dunn and R. H Hall	Purines, Pyrimidines, Nucleosides, and Nucleotides: Physical Constants and Spectral Properties (Nucleic Acids, p. 65-215)

These tables include pK values for these materials at 25 °C.

Interunion Commission on Biothermodynamics	Recommendations for Measurement and Presentation of Biochemical Equilibrium Data. (Physical and Chemical Reference Data, p. 93-103)

This important report gives a set of recommendations pertinent to the measurement and treatment of biochemical equilibrium data.

N. Langerman	Enthalpy, Entropy, and Free Energy Values for Biochemical Redox Reactions. (Physical and Chemical Reference Data, p. 121)

This table summarizes ΔG, ΔH, and ΔS values for 14 biochemical redox reactions at 25 °C.

* * * * * * * * * *

[35] Florkin, M. and Mason, H. (editors)
Comparative Biochemistry, Volume II. Free Energy and Biological Function
(Academic Press, New York and London, 1960)

This book contains several chapters dealing with the use of Gibbs energy data in biochemistry. The following chapters contain references to and/or tabulations of thermochemical data:

Authors	Title of Chapter
M. R. Atkinson and R. K. Morton,	Free Energy and Biosynthesis of Phosphates (Vol. 1, Chapter 1)

This chapter contains a tabulation of Gibbs energies and equilibrium constants for various metabolic processes involving phosphates. The temperature, pH, magnesium ion concentration, and appropriate literature references are given.

L. F. Leloir, C. E. Cardini and E. Cabib — Utilization of Free Energy for the Biosynthesis of Saccharides (Volume 1, Chapter 2)

Included in the discussion are some references to equilibrium data relevant to the biosynthesis of saccharides.

P. P. Cohen and G. W. Brown, Jr. — Ammonia Metabolism and Urea Biosynthesis (Volume 1, Chapter 4)

In their discussion of ammonia metabolism, the authors have used other compilations of thermodynamic data to compute Gibbs energy changes for these processes.

* * * * * * * * * *

[36] Fox, D., Labes, M. M. and Weissberger, A. (editors)
Physics and Chemistry of the Organic Solid State
(John Wiley and Sons, Interscience Publishers, New York 1963)

E. F. Westrum, Jr. and J. P. McCullough — Chapter 1. Thermodynamics of Crystals

In addition to discussion of thermodynamic properties of organic substances and their measurement, this chapter (178 pp.) gives tables of entropy of fusion, and vapor pressures and a table of thermodynamic data sources for about 800 organic compounds (798 references).

* * * * * * * * * *

[37] Franks, F. (editor)
Water-A Comprehensive Treatise (in five volumes)
(Plenum Press, New York and London, 1972 to 1975)

This treatise consists of forty-six chapters dealing with water and aqueous solutions. Although concerned with all aspects of water, several chapters contain extremely useful summaries of thermodynamic data. Among these are:

Author(s)	Title of Chapter
F. Franks	The Properties of Ice (Vol. 1, Chapter 4)

Tabulated are the ionic equilibrium constant and transport properties of ice at -10 °C.

G. S. Kell — Thermodynamic and Transport Properties of Fluid Water (Vol. 1, Chapter 10)

Included is the density of water as a function of temperature, the vapor pressure, specific heat, partial molar volume, critical properties, and viscosity of water from 0 to 500 °C and to 8 kbar. Also see items [43] and [134].

K. Tödheide — Water at High Temperatures and Pressures (Vol. 1, Chapter 13)

This chapter is an extensive tabulation of the high temperature properties of water. Included are the specific volume, fugacity, Gibbs energy and enthalpy of formation, entropy, viscosity, thermal conductivity, dielectric constant, and ion product of water. The temperature and pressure ranges are 0 to 1000 °C and up to 250 kbar. Also see items [43] and [134].

F. Franks and D. S. Reid — Thermodynamic Properties (Vol. 2, Chapter 5)

Given are ΔG°, ΔH°, ΔS°, and ΔC_p° for the solution of hydrocarbons, alcohols, and rare gases in water at 25 °C.

H. L. Friedman and C. V. Krishnan — Thermodynamics of Ionic Hydration (Vol. 3, Chapter 1)

Tabulated are single-ion entropies of about 110 diatomic and polyatomic ions in water; Gibbs energies, enthalpies, and entropies of hydration of monatomic ions at 25 °C; partial molar volumes of about 120 common ions at 25 °C; ionic partial molar heat capacities of ions; Gibbs energies of transfer of inorganic electrolytes from H_2O to D_2O; and calorimetrically determined enthalpies of solution of salts in H_2O and D_2O.

D. Eagland — Nucleic Acids, Peptides, and Proteins (Vol. 4, Chapter 5)

Thermodynamic parameters for coil → helix and homopolymer → coil helix transitions of amino acids in aqueous solution are tabulated.

H. L. Anderson and R. H. Wood — Thermodynamics of Aqueous Mixed Electrolytes (Vol. 3, Chapter 2)

Included are data on the enthalpies and excess Gibbs energies of mixing of about 24 mixed electrolyte systems in water at 25 °C.

* * * * * * * * * * *

[38] Fredenslund, A., Gmehling, J. and Rasmussen, P.
Vapor-liquid Equilibria Using UNIFAC
(Elsevier Scientific Publishing Company, Amsterdam, 1977)

This book describes the UNIFAC (an acronym formed from Universal Quasi-Chemical Functional Group Activity Coefficient) procedure for estimating activity coefficients of non-electrolytes in fluid mixtures and the use of these activity coefficients for the prediction of vapor-liquid equilibria. The estimation scheme is based upon a group-contribution method. The methodology is clearly described with examples of its use and with computer programs listed in the Appendix to the volume. There is also a chapter which describes a procedure for estimating the second virial coefficients of gases. The coverage of materials and functional groups includes alcohols, alkanes, ethers, organic acids, ketones, aldehydes, amines, hydrocarbons, and nitrogen compounds.

* * * * * * * * * * *

[39] Freeman, R. D. (editor)
Bulletin of Chemical Thermodynamics
(Thermochemistry, Inc., Oklahoma State University, Stillwater, OK 74078)

Each annual issues (volume 25 appeared in 1982) of the Bulletin is a current awareness index to, and a comprehensive bibliography of, articles that pertain to chemical thermodynamics (broadly interpreted) and that were published in the previous calender year. The index is arranged by chemical substance and identifies the thermodynamic property reported for each substance; numerical data are not given. Four subdivisions of the index cover inorganic substances, organic substances, organic mixtures, and biochemical and macromolecular systems. Also included in each Bulletin is a section of Reports, on "work completed but not yet published", from a large number of laboratories located worldwide, a bibliography of recently published books related to thermodynamics, and a section of miscellaneous items of interest to the thermodynamic community. The early volumes of the "Annual Review of Physical Chemistry", published since 1950 by Annual Reviews, Inc., contained bibliographic information for general thermochemistry and the thermodynamics of electrolyte solutions; these articles were, in part, the forerunners of the Bulletin. From 1965 to 1976 the Bulletin had the title "Bulletin of Thermodynamics and Thermochemistry" and was published by the University of Michigan under the editorship of E.F. Westrum, Jr.

* * * * * * * * * * *

[40] Freir, R. K.
Aqueous Solutions: Data for Inorganic Compounds (in two volumes)
(Walter de Gruyter, Berlin and New York, 1976 and 1977)

This very well organized and easy-to-use handbook contains a wealth of thermodynamic data on aqueous solutions. There are 1300 solutions for which data is given. The solutes are both inorganic and organic, and include data on biochemical substances when available. Properties given are densities, solubilities, equilibrium constants, Gibbs energies of formation, electrochemical potentials, conductivities, pH values, species-composition diagrams, and vapor pressures. The most serious shortcoming of this handbook is the absence of specific references to the literature sources of the data. The bulk of the text is in German.

* * * * * * * * * *

[41] Goldberg, R. N., Staples, B. R., Nuttall, R. L. and Arbuckle, R. A Bibliography of Sources of Experimental Data Leading to Activity or Osmotic Coefficients for Polyvalent Electrolytes in Aqueous Solution (NBS Special Publication 485, U. S. Government Printing Office, Washington, D.C., 1977)

This is a bibliography of sources of experimental data that can be used to calculate either activity or osmotic coefficients of polyvalent electrolyes in water at the temperatures (0 to 100 °C) for which the data exist. The compounds are arranged according to the standard thermochemical order of arrangement. There are approximately 400 references to the source literature.

* * * * * * * * * *

[42] Gurvich, L. V., Veits, I. V., Medvedev, V. A., Khachkuruzov, G.A., Yungman, V.S., Bergman, G.A. et al. "Termodinamicheskie Svoistva Individual'nykh Veschestv" (Thermodynamic Properties of Individual Substances) V.P. Glushko, general editor Volume I, parts 1 and 2 (1978); Volume II, parts 1 and 2 (1979); Volume III, parts 1 and 2 (1981); Volume IV, in press. (Izdatel'stvo "Nauka", Moscow, 1979).

This extensive work is a critical evaluation and collection of the thermodynamic properties of the elements and their compounds. The properties tabulated are C_p°, $(G^\circ-H^\circ(0))/T$, S°, and $H^\circ-H^\circ(0)$ and $\log_{10}K$ where K for gases is the equilibrium constant for the reaction of forming the given compound from its atoms; for solids K refers either to vaporization or to atomization. The temperature range for which the data are given is 100 K to the highest temperature for which data exist, the temperature intervals being in steps of 100 K. There are extensive discussions of the sources from which the data have been taken, the computational procedures, and also, references to the source literature. Also see item [144]. The contents of the four volumes are:

Volume	Contents
I	O, H(D,T), F, Cl, Br, I, He, Ne, Ar, Kr, Xe, Rn, S, N, and P
II	C, Si, Ge, Sn, and Pb
III	B, H, Ga, In, Tl, Be, Mg, Ca, and Ba
IV	Cr, Mo, W, V, Nb, Ta, Ti, Zr, Hf, Sc, Y, La, Th, U, Pu, Li, Na, K, Rb, and Cs.

* * * * * * * * * *

[43] Haar, L., Gallagher, J. S. and Kell, G. S.
A Thermodynamic Surface for Water: The Formulation and Computer Programs
(NBSIR 81-2253, National Bureau of Standards, Washington, D.C., 1981)

Haar, L., Gallagher, J. S. and Kell, G. S.
The Anatomy of the Thermodynamic Surface of Water: The Formulation and Comparisons with Data
(The American Society of Mechancial Engineers, New York, 1982)

Haar, L., Gallagher, J. S., and Kell, G. S.
NBS-NRC Steam Tables
(Hemisphere Press, Washington, DC, 1984)

The 1981 article summarizes the development of a thermodynamic surface for water with which all thermodynamic properties for the fluid states can be calculated from the freezing line to 1000 K and up to 1 GPa in pressure. The discussion is very brief, but gives references to earlier work and indicates that a more detailed publication is forthcoming. Given are coefficients of the Helmholtz function which define the surface. Plots of heat capacity, enthalpy, and speed of sound are included.

The 1982 report contains a computer program which can be used to calculate twelve thermodynamic properties of water and steam over the range $0°\ C \le t \le 1000\ °C$ and $0 \le p \le 1000$ MPa. The properties include the Helmoltz function, the pressure, entropy, isochoric and isobaric heat capacity, the 2nd virial coefficient, and the speed of sound. Also see item [134].

The book in press contains tables, with documentation, for all of the equilibrium properties of water and steam for which reliable data exist. Also included in the book are tables of transport and mechanical property data such as the surface tension, the viscosity, and the thermal conductivity. The temperature range which the data tables extend is $0\ °C \le t \le 2500\ °C$ and the pressure range is $0 \le p \le 3000$ MPa.

* * * * * * * * * *

[44] Hala, E., Wichterle, I., Polak, J. and Boublik, T.
Vapour-Liquid Equilibrium Data at Normal Pressures
(Pergamon Press, Oxford, 1968)

This book contains correlated vapor-liquid-equilibrium data on mixtures (mostly binary but with some multicomponent systems). The correlations are done using any one of several approaches: The Antoine equation for a pure component, van Laar and Margules equations for mixtures. In all cases the source(s) of the experimental data is given. The coverage of systems (400) is very broad and it includes aqueous, nonaqueous, and organic-inorganic mixtures.

* * * * * * * * * *

[45] Hamer, W. J. (editor)
The Structure of Electrolytic Solutions
(John Wiley and Sons, New York, 1959)

This book contains several chapters relevant to this bibliography. They are:

Author(s)	Title of Chapter
C. W. Davies	Incomplete Dissociation in Aqueous Salt Solutions (Chapter 3)

Tabulated are values of $-\log_{10}K$ at 25 °C for the pairing in aqueous solution of eighteen common inorganic cations with thirty-eight of the more common inorganic and organic anions.

E. Lange	Heats of Dilution of Dilute Solutions of Strong and Weak Electrolytes (Chapter 9)

This chapter contains a discussion of the theoretical interpretation and calculations of heats of dilution of electrolytes of various charge types with some information on non-electrolytes and weak electrolytes. Data from the literature are presented in graphical form. Included are data for eighteen inorganic electrolytes and seven inorganic non-electrolytes at 25 °C.

H. S. Harned	Diffusion and Activity Coefficients of Strong and Weak Electrolytes (Chapter 10)

Tabulated are the activity coefficients, obtained from diffusion data, for twenty common aqueous electrolyte systems at 25 °C.

F. H. Spedding and G. Atkinson	Properties of Rare Earth Salts in Electrolytic Solutions (Chapter 22)

Tabulated are the equivalent conductances, transference numbers, activity coefficients, densities and partial molar volumes, apparent molar compressibilities, heats of solution and dilution for the rare earth salts in aqueous solution at 25 °C.

* * * * * * * * * * *

[46] Hamer, W. J., and DeWane, H. J.
Electrolytic Conductance and the Conductances of the Halogen Acids in Water
(NSRDS-NBS 33, U. S. Government Printing Office, Washington, D.C., 1970)

This monograph contains a detailed evaluation of equivalent conductance data for hydrofluoric, hydrochloric, hydrobromic, and hydroiodic acids in water at various concentrations from -20 to +65 °C.

* * * * * * * * * * *

[47] Hamer, W. J. and Wu, Y. C.
The Activity Coefficients of Hydrofluoric Acid in Water from 0 to 35 °C
Journal of Research of the National Bureau of Standards 74A, 761 (1970)

This very detailed review on hydrofluoric acid contains critically evaluated data for the activity coefficient of HF as a function of molality and temperature (0 to 35 °C), equilibrium constants for the ionic association reactions characteristic of HF, calculated pH values, and calculated concentrations of the pertinent ions.

* * * * * * * * * *

[48] Hamer, W. J. and Wu, Y. C.
Osmotic Coefficients and Mean Activity Coefficients of Uni-Univalent Electrolytes in Water at 25 °C
Journal of Physical and Chemical Reference Data 1, 1047 (1972)

This evaluation gives values for the osmotic coefficients and mean activity coefficients of seventy-nine uni-univalent electrolytes in aqueous solution at 25 °C, with values expressed on the molality scale. The data from the literature were fitted, by statistical procedures, to equations which express the quantities as functions of electrolyte concentration. Literature references are given to fifty-one additional uni-univalent electrolytes. Also see item [159].

* * * * * * * * * *

[49] Harned, H. S. and Owen, B. B.
The Physical Chemistry of Electrolytic Solutions (3rd Edition)
(Reinhold Publishing Corporation, New York, 1958)

This book (about 800 pp.) is a treatise on the physical chemistry of electrolytic solutions with coverage of both equilibrium and non-equilibrium properties. The book includes tables of values of the equivalent conductance, dissociation constants, transference numbers, diffusion coefficients, relative apparent molar heat contents, activity coefficient, pH values, densities, and activity coefficients for many of the more common inorganic and organic electrolyte solutions.

* * * * * * * * * *

[50] Harned, H. S. and Robinson, R. A.
Multicomponent Electrolyte Solutions
(Pergamon Press, Oxford, 1968)

This monograph deals with the theoretical and experimental aspects of multicomponent and largely inorganic electrolyte solutions, with emphasis upon the measurement and interpretation of activity coefficients, heats of mixing, and volume changes accompanying mixing. There is a useful bibliography of activity coefficient data for mixed electrolyte systems. We note, for the reader's information, the following monographs published in the same series of books (The International Encyclopedia of Physical Chemistry and Chemical Physics) and which deal principally with theory or methods of measurement pertinent to electrolyte solutions:

Author	Title of Volume
E. A. Guggenheim and R. H. Stokes	Equilibrium Properties of Aqueous Solutions of Single Strong Electrolytes
R. H. Stokes and R. Mills	Viscosity of Electrolytes and Related Properties
E. J. King	Acid-Base Equilibria

* * * * * * * * * * *

[51] Hawkins, D. T.
Physical and Chemical Properties of Water:
A Bibliography: 1957-1974
(Plenum Press, New York, 1976)

This bibliography consists of 3600 references to the literature that deal with physical properties of pure water or dilute aqueous solutions. The papers are arranged by category of properties. Categories include thermodynamic properties, transport data, densities, acoustical, electrical, magnetic, and radiation properties. The bibliography covers the years 1957 to 1974. There is an author and keyword index. A bibliographic listing covering the years 1969 to 1974 appeared in the Journal of Solution Chemistry, 4, 621 (1976).

* * * * * * * * * * *

[52] Helgeson, H. C., Kirkham, D. H. and Flowers, G. C.
Theoretical Prediction of the Thermodynamic Behavior of Aqueous Electrolytes at High Pressures and Temperatures: IV. Calculation of Activity Coefficients, Osmotic Coefficients, and Apparent Molal and Standard Relative Partial Molal Properties to 600 °C and 5 kb
American Journal of Science 281, 1249 (1981)

This 268 page article is concerned with the prediction of the thermodynamic properties of aqueous electrolyte solutions at high temperatures and pressures. There is an extensive discussion of the fundamental thermodynamics of solutions and a discussion of theoretical concepts and models which have been used to describe electrolyte solutions. There is a very extensive bibliography (600 citations) which contains valuable references to specific systems of interest. Some specific tables of interest to this bibliography contain Debye-Hückel parameters at 25 °C, standard state partial molar entropies and heat capacities at 25 °C, and parameters for calculating activity coefficients, osmotic coefficients, relative apparent and partial molar enthalpies, heat capacities, and volumes at 25 °C.

* * * * * * * * * * *

[53] Hepler, L. G. and Hopkins, H. P., Jr.
Thermodynamics of Ionization of Inorganic Acids and Bases in Aqueous Solution.
Reviews in Inorganic Chemistry 1, 303 (1979)

The review contains a general discussion of the thermodynamics of aqueous acid-base chemistry. The systems which are discussed are pure water, CO_2 + H_2O, NH_3 + H_2O, hydrofluoric acid, phosphoric acid, hydrogen sulfide, SO_2 + H_2O, sulfuric acid, H_2CrO_4, $H_2Cr_2O_7$, iodic acid, and aqueous metal cations. This article cites values $\Delta G°$, $\Delta H°$, and $\Delta C_p^°$ for the ionizations of these various acids.

* * * * * * * * * *

[54] Hepler, L. G. and others
Thermochemistry of the Transition Metal Elements and Their Compounds

This series of papers contain critical reviews and selections of thermochemical data for the transition metal elements and their compounds. Tabulated are selected values of $\Delta_f G°$, $\Delta_f H°$, S°, and electrode potentials at 25 °C. Each article contains a discussion of the data upon which the selections have been made and the references to the source literature. The papers in the series are:

Authors	Title of Paper
R. N. Goldberg and L. G. Helper	Thermochemistry and Oxidation Potentials of the Platinum Group Metals and Their Compounds

Chemical Review 68, 229 (1968)

L. M. Gedansky and L. G. Hepler	Thermochemistry of Silver and Its Compounds

Engelhard Industries Technical Bulletin IX, No. 4, 117 (1969)

L. M. Gedansky and L. G. Hepler	Thermochemistry of Gold and Its Compounds

Englehard Industries Technical Bulletin X, No. 1, 5 (1969)

L. M. Gedansky, E. M. Woolley and L. G. Hepler	Thermochemistry of Compounds and Aqueous Ions of Copper

Journal of Chemical Thermodynamics 2, 561 (1970)

J. O. Hill, I. G. Worsley and L. G. Hepler	Thermochemistry and Oxidation Potentials of Vanadium, Niobium, and Tantalum

Chemical Reviews 71, 127 (1971)

L. G. Hepler and G. Olofsson — Mercury: Thermodynamic Properties, Chemical Equilibria, and Standard Potentials

Chemical Reviews 75, 585 (1975)

J. G. Travers, I. Dellien and L. G. Hepler — Scandium: Thermodynamic Properties, Chemical Equilibria, and Standard Potentials

Thermochimica Acta 15, 89 (1976)

L. G. Hepler and P. P. Singh — Lanthanum: Thermodynamic Properties, Chemical Equilibria, and Standard Potentials

Thermochemica Acta 16, 95 (1976)

I. Dellien, F. M. Hall and L. G. Hepler — Chromium, Molybdenum and Tungsten: Thermodynamic Properties, Chemical Equilibria, and Standard Potentials

Chemical Reviews 76, 283 (1976).

T. A. Zordan and L. G. Hepler — Thermochemistry and Oxidation Potentials of Manganese and Its Compounds

Chemical Reviews 68, 737 (1968)

* * * * * * * * * *

[55] Hirata, M., Ohe, S. and Nagakama, K.
Computer Aided Data Books of Vapor-Liquid Equilibria
(Kodansha Limited and Elsevier Scientific, Tokyo and New York, 1975)

This large volume (933 pages) contains tables, plots, and parameters pertinent to vapor-liquid-equilibria (VLE). VLE data for ≈1000 binary systems have been collected and treated by computer to obtain the tables and system graphs presented. References to the primary literature are included along with a detailed compound index. The coverage of compounds is very broad and includes a wide variety of organic mixtures and 30 organic-water mixtures.

* * * * * * * * * *

[56] Horsley, L. H.
Azeotropic Data (in three volumes)
Advance in Chemistry Series No. 6, 35 and 116
(American Chemical Society, Washington, D.C., 1952, 1962 and 1973)

This series is concerned with the properties of azeotropic mixtures, specifically boiling temperatures, pressures, vapor-liquid equilibrium data, and compositions. Each of the volumes contains references to the source literature for each of the system catalogued. In terms of coverage of mixtures, the series is comprehensive for both binary and ternary mixtures and covers both inorganic and organic materials. The most recent volume contains a section dealing with the prediction and calculation of azeotropic mixtures.

* * * * * * * * * *

[57] Horvath, A. L.
Reference Literature to the Critical Properties of Aqueous Electrolyte Solutions
Journal of Chemical Information and Computer Sciences 15, 245 (1975)

This bibliography gives references to the critical properties of aqueous electrolyte solutions. The bibliography covers ≈ 75 solutions and has 85 references.

* * * * * * * * * *

[58] Horvath, A. L.
Reference Literature to Solubility Data between Halogenated Hydrocarbons and Water
Journal of Chemical Documentation 12, 163 (1972)

This is a bibliography giving 103 references to solubility data for halogenated hydrocarbons (C_1 to C_6) in water. References are given for approximately 100 compounds.

* * * * * * * * * *

[59] Ingraham, L. L. and Pardee, A. B.
Free Energy and Entropy in Metabolism
in Metabolic Pathways, Volume I
(D. M. Greenberg, editor)
(Academic Press, New York, 1967)

This chapter contains a general discussion of the thermodynamics of metabolic processes, with the (unevaluated) data itself being presented in the course of the discussion. The emphasis is almost entirely upon Gibbs energy changes measured under physiological or near physiological conditions. There are 143 references to the primary literature.

* * * * * * * * * *

[60] Izatt, R. M., Christensen, J. J. and Rytting, J. H.
Sites and Thermodynamic Quantities Associated with Proton and Metal Ion Interaction with Ribonucleic Acid, Deoxyribonucleic Acid, and Their Constituent Bases, Nucleosides, and Nucleotides
Chemical Reviews 71, 439 (1971)

This review contains twenty-three pages of tables of thermodynamic data (log K, ΔH, ΔS, and ΔC_p) pertinent to the interaction of protons and metal ions with the nucleic acids and their molecular components together with the methods and experimental conditions (pH, temperature, ionic strength) used in their determination. There are 229 references.

* * * * * * * * * *

[61] Janz, G. J.
Thermodynamic Properties of Organic Compounds
Estimation Methods, Principles, and Practice (revised edition)
(Academic Press, New York and London, 1967)

This well-established monograph discusses computation of thermodynamic properties such as heat capacities, entropies, enthalpies and Gibbs energies by statistical-mechanical methods, by methods of structural similarity, by methods of group contributions, by methods of group equations, and by methods of generalized vibrational assignments. The chemical properties: enthalpy of formation, and enthalpy of combustion are treated in terms of bond energies and group increments. Some 78 tables are given of increments, group contributions, and bond contributions as specifically needed for estimation of particular properties.

* * * * * * * * * *

[62] Janz, G. J. and Tomkins, R. P. T.
Non-Aqueous Electrolytes Handbook (in two volumes)
(Academic Press, New York, 1972)

These volumes contain extensive tabulations of physical data pertinent to non-aqueous solvents, both single solvent and mixed solvent systems. The properties that are tabulated include melting point, boiling point, dielectric constant, viscosity, specific conductance, density, transference number, solubility, enthalpy of solution and dilution, E° values for electrochemical cells, vapor pressure, polarographic data, ligand exchange rate, and spectroscopic data. The vast majority of the approximately 300 solvent systems dealt with are organic. There is a substance-property index, and sources of data are referenced.

* * * * * * * * * *

[63] Jones, M. N. (editor)
Biochemical Thermodynamics
(Elsevier Scientific Publishing Co., Amsterdam, 1979)

This book is divided into eleven chapters, each on a topic pertinent to biochemical thermodynamics and written by an expert on that topic. Each chapter contains useful references to papers in the literature which contain tabular summaries of thermodynamic data pertinent to the theme of that chapter. The chapters and authors are:

Author(s)	Title of Chapter
M. N. Jones	The Scope of Thermodynamics in Biochemistry
F. Franks	Aqueous Solution Interactions of Low Molecular Weight Species - The Applicability of Model Studies in Biochemical Thermodynamics
W. Pfeil and P. L. Privalov	Conformational Changes in Proteins
H. J. Hinz	Conformation Changes in Nucleic Acids
D. S. Reid	Thermodynamics of Aqueous Polysacharide Solutions
M. N. Jones	The Thermal Behavior of Lipid Systems and Biological Membranes
S. J. Gill	Ligand Binding of Gases of Hemoglobin
M. Monti and I. Wadsö	Calorimetric Studies in Blood Cells
G. C. Krescheck	Thermochemical Studies on Bacterial and Mammalian Cells
A. G. Lowe	Energetics of Muscular Contraction
B. Crabtree and D. J. Taylor	Thermodynamics and Metabolism

* * * * * * * * * *

[64] Jordan, T. E.
Vapor Pressure of Organic Compounds
(John Wiley and Sons, Interscience Publishers, New York, 1954)

This is a comprehensive compilation (266 pages) of vapor pressure data for organic compounds. Included are tables on the hydrocarbons, alcohols, aldehydes, esters, ketones, acids, phenols, and metal organic compounds. Data for each compound are shown in graphical form, i.e. vapor pressure as a function of temperature. References to the data sources in the literature are given.

* * * * * * * * * *

[65] Joshi, R. M. and Zwolinski, B. J.
Heats of Polymerization and Their Structural and Mechanistic Implications
in Vinyl Polymerization, Volume 1, Part I, edited by G. E. Ham
(Marcel Dekker, New York, 1967)

The authors discuss experimental methods used to measure and derive enthalpies of polymerization. A listing of experimental data on enthalpies of polymerization is provided for 81 organic polymerization reactions. Other tables give enthalpies of formation, enthalpies of vaporization, entropies, Gibbs energies, equilibrium constants, rate constants and activation energies at 25 °C for a variety of polymerization processes. A discussion of the structural influence upon the enthalpy of polymerization is also given. At the end of the chapter, 164 references are cited.

* * * * * * * * * *

[66] Kaimakov, E. A. and Varshavskaya, N. L.
Measurement of Transport Numbers in Aqueous Solutions of Electrolytes
Russian Chemical Reviews (Uspekhi Khimii) 35, 89 (1966)

This review article summarizes the various methods available for the measurement of transference numbers. The authors have included a table which summarizes available (thru 1966) transport number data for aqueous electrolyte solutions, including references to the source literature.

* * * * * * * * * *

[67] Karapet'yants, M. Kh. and Karapet'yants, M. L.
Thermodynamic Constants of Inorganic and Organic Compounds
(Ann Arbor Humphrey Science Publishers, Ann Arbor and London, 1970)
Translated from the Russian:
"Osnovnye Termodinamicheskie Konstanty Neorganicheskikh i Organicheskikh Veshchestv"
(Izdatel'stvo "Khimiya", Moscow, 1968)

This book is a compilation of $\Delta_f G°$, ΔH_f^o, $S°$, and C_p values at 298.15 K for about 4000 substances in the condensed and gaseous phases, and in aqueous solution. Covered are not only the inorganic elements and their compounds, but also data for the organic compounds through 34 carbon atoms. The authors point out that their tabulated values do not always form a self-consistent system of thermodynamic data. There are 2733 references.

* * * * * * * * * *

[68] Kaufmann, D. W. (editor)
Sodium Chloride
(Reinhold Publishing Corporation, New York, 1960)

This monograph, published under the auspices of the American Chemical Society, has in its Appendix a useful compilation of the physical properties of aqueous sodium chloride solutions. Included are essentially all of the measured equilibrium and transport properties of this system at various temperatures and pressures. The data are well referenced.

* * * * * * * * * *

[69] Kazavchinskii, Ya. Z., Kessel'man, P. M., Kirillin, V. A., Riukin, S., Sheindlin, A. E., Shpil'rain, E. E., Sychev, V. V. and Timrot, D. L. (edited by V. A. Kirillin)
Heavy Water-Thermophysical Properties
(U. S. Department of Commerce, National Technical Information Service, Springfield, VA 1971)
Translated from the Russian:
"Tyazhelaya voda. Teplofizicheskie Svoistva"
(Gosudarstvennoe energeticheskoe izdatel'stvo, Moskva-Leningrad, 1963)

This treatise is an exhaustive compilation of physical data on heavy water (deuterium oxide). Some of the more relevant properties that are covered include densities, critical constants, vapor pressures, enthalpies of transition, viscosity, and thermal conductivity, equation of state, and tables of thermodynamic properties as functions of temperature and pressure.

* * * * * * * * * *

[70] Keenan, H. J., Keyes, G. F., Hill, P. G. and Moore, J. G.
Steam Tables - Thermodynamic Properties of Water, Including Vapor, Liquid, and Solid Phases (International Edition, Metric Units)
(John Wiley and Sons, New York, 1969)

This book presents the results of a reassessment and correlation of the thermodynamic data for water. It supersedes the Keenan and Keyes Tables of 1936. Values are tabulated for the specific volume, internal energy, and enthalpy, as functions of temperature and pressure. Also given are data for vapor-liquid and vapor-solid equilibrium, superheated vapor, and the compressed liquid. Mollier and temperature-entropy charts are included along with charts of heat capacity of liquid and vapor, Prandtl number, and isentropic expansion coefficient. The data and tables are discussed in an appendix of 25 pages and a list of 37 references is given. Also see items [43] and [134] for other correlations.

* * * * * * * * * *

[71] Kertes, A. S. (editor-in-chief)
Solubility Data Series
(Pergamon Press, New York, 1979 to 1981)

This eighteen volume series, prepared under the auspices of the IUPAC, is a part of a continuing project concerned with the preparation of a comprehensive, critical compilation of data on solubilities in all physical systems including gases, solids, and liquids. The volumes which have been issued to date, with their titles and editors, are:

Editor(s)	Title	Volume Numbers
H. L. Clever	Helium and Neon - Gas Solubilities	1
H. L. Clever	Krypton, Xenon, and Radon - Gas Solubilities	2
M. Salomon	Silver Azide, Cyanide, Cyanamides, Cyanate Selenocyanate and Thiocyanate	3
H. L. Clever	Argon	4
C. L. Young	Hydrogen and Deuterium	5/6
R. Battino	Oxygen and Ozone	7
C. L. Young	Oxides of Nitrogen	8
W. Hayduk	Ethane	9
R. Battino	Nitrogen and Air	10
B. Scrostai and C. A. Vincent	Alkali Metal and Alkaline-Earth Metal and Ammonium Halides. Amide Solvents	11
O. Popovych	Tetraphenylborates	18

The following volumes are planned for future publication:

Z. Galus and C. Guminski	Metals in Mercury	12
C. L. Young	Oxides of Nitrogen, Sulfur and Chorine	13
R. Battino	Nitrogen	14
H. L. Clever and W. Gerrard	Hydrogen Halides in Non-Aqueous Solvents	15
A. L. Horvath	Halogenated Benzenes	16
E. Wilhelm and C. L. Young	Hydrogen, Deterium, Fluorine and Chlorine	17

In general, for each system covered in this series, the preparer, who is identified, gives the following information: the components in the system, the intensive variables, the experimental values of the solubility, the experimental methodology and source and purity of materials, a discussion of the evaluation procedure, estimates of error, and literature citations.

* * * * * * * * * *

[72] Kirgintsev, A. N., Trushnikova, L. N., and Lavrenteva, V. F.
Solubilities of Inorganic Substances in Water
(in Russian)
(Izdatelstvo Khimiya, Leningrad, 1972)

This book consists of aqueous phase diagrams of pure salts in water. The phase diagrams cover the composition and temperature ranges for which the data exist. The diagrams are particularly useful in assessing the regions of stability of the various inorganic hydrates and in obtaining solubilities. The authors have included references to the source literature from which the data have been obtained. The coverage of inorganic salts spans the periodic chart. There are data for $\approx$1200 inorganic salts which are indexed in the Appendix to the book.

* * * * * * * * * *

[73] Kortum, G., Vogel, W. and Andrussow, K.
Dissociation Constants of Organic Acids in Aqueous Solution
(Butterworths, London, 1961)

This book is a compilation of 1056 dissociation constants of organic acids in aqueous solution, presented in tabular form. Introductory and explanatory remarks are in both German and English. Remarks in the Table are in German. Part I is a critical discussion of techniques for measurements of dissociation constants by conductance, electrometric, catalytic and optical methods. Each method is classified and assigned a code in Part II, which deals with use of the tables, and methods of calculation. The Tables themselves are arranged by acid class including: aliphatic and alicyclic carboxylic acids, aromatic carboxylic acids, phenolic acids, and other acids and special classes. The Tables contain the name, chemical formula, and thermodynamic dissociation constant K of each acid, the temperature (°C) of measurement, the range of concentration over which the measurements were made, code for the method of measurement, calculation procedure and any corrections made, a critical evaluation of the quality of the measurement, and the source reference. All data were drawn from the literature, covering the period between 1927 and 1956, and are referenced in a classified reference list. A compound index is provided.

* * * * * * * * * *

[74] Kragten, J.
Atlas of Metal-Ligand Equilibria in Aqueous Solution
(Ellis Horwood, Chichester, England, 1978)

This book contains ≈750 pages of predominance, or species-composition diagrams, for aqueous solutions involving 45 metals and 25 common ligands. Most of the thermodynamic data was taken from Sillen and Martell and from Smith and Martell (see items [89] and [137]).

* * * * * * * * * *

[75] Krebs, H. A. and Kornberg, H. L., with appendix by K. Burton
Energy Transformations in Living Matter
(Springer-Verlag, Berlin, 1957)

The main part of this monograph surveys the various biochemical pathways by which living systems utilize energy. In the Appendix are (1) Tables of Gibbs energies of formation of ninety-eight compounds of biological importance, (2) Gibbs energies and electrochemical potentials of important biological oxidation-reduction reactions, and (3) Gibbs energy changes accompanying the processes of glycolysis and alcoholic fermentation, the tricarboxylic acid cycle, and hydrolysis. The source of data is given for each entry.

* * * * * * * * * *

[76] Landolt-Börnstein
Numerical Data and Functional Relationships in Science and Technology

Several volumes in the Landolt-Börnstein series contain data relevant to this bibliography. They are:

Author(s)	Title of Volume	Location
G. Beggerow	Heats of Mixing and Solution	New Series, Group IV, Volume 2

This very extensive volume (695 pages) contains data on heats of mixing, solution, and dilution for both organic and inorganic substances. The coverage of substances spans the periodic chart and includes both binary and multi-component mixtures. The data on each system include the components of the mixture, the temperature, the composition, the enthalpies, and the appropriate literature citations.

(K.-H. Hellwege, editor, Springer-Verlag, Berlin, 1976)

J. D'Ans, H. Surawski and C. Synowietz	Densities of Binary, Aqueous Systems and Heat Capacities of Liquid Systems	New Series, Group IV, Volume 1

This 123 page section contains extensive tables of the specific heats of binary, and a few ternary, aqueous systems at the temperatures and molalities for which data exist. The authors include plots of the data and references to the source literature.

(K.-H. Hellwege, editor, Springer-Verlag, Berlin, 1977)

W. Auer, H. D. Baer, K. Bratzler, F. Burhorn, H. Kientz, O. Kubachevski, Fr. Losch, A. Neckel, H. Nelkanski, Kl. Schäffer, and R. Wienecke	Kalorische Zustandsgrössen	6th Edition, Volume 2, Part 4

This volume contains thermal properties: molar heat capacity, entropy, enthalpy, enthalpy of formation, Gibbs energy of formation, in the standard state, and enthalpies of phase changes for many organic and inorganic substances in SI units. The dependence of thermal functions and heat capacity upon temperature is given for many substances. Some other thermodynamic quantities are given. Extensive tables are given of group contributions to enthalpies of formation and the Gibbs energies of formation of organic substances (gases) in $kcal \cdot mol^{-1}$ and $kJ \cdot mol^{-1}$. Many of the heat capacity data are presented in diagrams. This series also includes tables of freezing point depressions, conductivities, transference numbers, and densities for aqueous systems.

(K. Schäfer and E. Lax, editors, Springer-Verlag, Berlin, 1961)

J. Weishaupt	Thermodynamic Equilibria of Boiling Mixtures	New Series, Group IV, Volume 3

This volume deals with the thermodynamic vapor-liquid equilibrium of binary and multicomponent mixtures. The coverage of systems extends across the periodic chart. The mixtures are adequately indexed and the reference to the primary literature given. The information given on a typical system includes tables of equilibrium temperatures as a function of the mole fractions of the components at a given pressure with accompanying figures.

(H. Hausen, editor, Springer-Verlag, Berlin, 1976)

* * * * * * * * * *

[77] J. A. Larkin (editor)
International Data Series B. Thermodynamic Properties of Aqueous Organic Systems
(Engineering Sciences Data Unit, London, 1978 and 1979)

This part of the International Data Series deals with the thermodynamic properties of approximately 100 aqueous organic systems. Tabulated are excess properties (G, H, C, and V), vapor-liquid equilibrium data, and vapor pressures as a function of composition. Each table is prepared by an individual contributor and includes, in a very neat tabular form, the tables of property values, the correlating equations and their coefficients, the methods of measurement, a brief discussion, and the appropriate references to the literature.

* * * * * * * * * *

[78] Latimer, W. M.
The Oxidation States of the Elements and Their Potentials in Aqueous Solution (second edition)
(Prentice-Hall, Englewood Cliffs, New Jersey, 1959)

This book contains extensive tables of $\Delta_f G°$, $\Delta_f H°$, $S°$ values for the elements and their compounds as well as electrode potential diagrams calculated from the tabulated $\Delta_f G°$ values, from other measurements, and estimates when appropriate. Many of the tabulated data were taken from National Bureau of Standards Circular 500 (see item [131]). The appendices include activity coefficient data for 77 strong electrolytes and a discussion of methods whereby entropies may be estimated.

* * * * * * * * * *

[79] Lewis, G. N. and Randall, M.
(revised by K. S. Pitzer and L. Brewer)
Thermodynamics
(McGraw-Hill, New York, 1961)

This standard textbook on chemical thermodynamics contains an Appendix (no. 4) of selected data for aqueous electrolyte solutions. Compiled are activity coefficients, Debye-Hückel parameters, relative partial molar enthalpies, and relative partial molar heat capacities for about 70 of the most common electrolytes in aqueous solution at 25 °C. More recent Debye-Hückel parameters are to be found in the pages of Pitzer, Peiper, and Busey and of Bradley and Pitzer (see item [121]) and the paper of Clarke and Glew, item [22].

* * * * * * * * * *

[80] Leyendekkers, J. V.
Thermodynamics of Seawater
(Marcel Dekker, New York and Basel, 1976)

This book deals with the thermodynamic properties of seawater as a multicomponent electrolyte solution. Each chapter covers a thermodynamic property of seawater and its constituents. There are chapters on the fundamentals of thermodynamics, the entropy, the volume, the expansibility, and the compressibility. Included in each chapter is a discussion of the methods for calculating or estimating the desired property with tables of values of experimental and calculated properties. There are extensive references to the primary literature. The author has indicated that a second volume was to be published, but it has not yet appeared in print. Also see [126].

* * * * * * * * * * *

[81] Linke, W. F.
Solubilities: Inorganic and Metal-Organic Compounds--A Compilation of Solubility Data from the Periodical Literature. Volume I: A-Ir, Volume II: K-Z
(Volume I: D. Van Nostrand Co., Princeton, New Jersey, 1958)
(Volume II: American Chemical Society, Washington, D.C., 1965)

These two volumes (total of 3401 pages) are comprehensive compilations of mostly unevaluated solubility data for inorganic and metal-inorganic compounds. Both aqueous and non-aqueous solvent systems are included. The temperatures and compositions given cover the ranges for which experimental data exist. References are given to the data sources. These two volumes had their origins in the solubility compilations begun by A. Seidell in 1907.

* * * * * * * * * * *

[82] Lobo, V. M. M.
Electrolyte Solutions: Literature Data on Thermodynamic and Transport Properties
(Coimbra Editora, Coimbra, Portugal, 1975)

This monograph is a compilation of physical property data on aqueous salt solutions. The properties tabulated are the density, the conductance, transference number, viscosity, diffusion coefficient and the activity coefficient. The monograph contains data on approximately 70 binary systems at temperatures and compositions for which experimental data exist and includes an index and references to the sources of the data.

* * * * * * * * * * *

[83] Long, C. (editor)
Biochemists Handbook
(D. Van Nostrand Co., Princeton, New Jersey, 1961)

This reference book contains several tables of thermochemical data:

Author(s)	Title of Section
S. P. Datta and A. K. Grzybowski	pH and Acid-Base Equilibria (pages 19 to 58)

This section contains a discussion of pH scales and the electrometric measurement of pH, tables of assigned pH values for various buffered solutions, and tables (about 600 entries) of thermodynamic acid dissociation constants of weak organic acids (some as a function of temperature).

K. Burton — Free Energy Data and Oxidation-Reduction Potentials (pages 93 to 95)

This section contains some revisions and additions to Burton's earlier tables (see, in this bibliography, the book by Krebs and Kornberg, item [75]).

* * * * * * * * * *

[84] Long, F. A. and McDevit, W. F.
Activity Coefficients of Non-electrolyte Solutes in Aqueous Salt Solutions
Chemical Reviews 51, 119 (1952)

This review article is concerned with the effect of salts on the activity coefficients of non-electrolytes. There is an extensive discussion of both the theoretical and experimental procedures used in the study of this phenomena. There is an extensive appendix to the article which lists salt effects on non-polar electrolytes ($\approx$100 systems). The article has 180 references.

* * * * * * * * * *

[85] MacKay, D. and Shiu, W. Y.
A Critical Review of Henry's Law Constants for Chemicals of Environmental Interest
Journal of Physical and Chemical Reference Data 10, 1175 (1981)

This is a critical review and tabulation of Henry's Law constants, vapor pressures, and solubilities in water of 150 organic substances of environmental importance. The compounds include gaseous, liquid, and solid alkanes, cycloalkanes, alkenes, alkynes, monoaromatics, polynuclear aromatics, halogenated alkanes, alkenes and aromatics, and selected pesticides. Nearly all of the data refer to 25 °C.

* * * * * * * * * *

[86] Maczynski, A., Maczynska, Z., Rogalski, M., Skrzecz, A., and Dunajska, K.
Verified Vapor-Liquid Equilibrium Data (in four volumes)
(PWN - Polish Scientific Publishers, Warsaw, 1976 to 1979)

This four volumes sèries contains vapor-liquid equilibrium data (i.e., compositions of vapor and liquid phases as functions of temperature and pressure) covering the règions for which experimental measurements exist. The systems are almost entirely organic. The data have been recalculated into SI units and have been smoothed in many cases, often using the Redlich-Kister equation. There are detailed references to the source literature and there is an alphabetic index of the systems covered. The four volumes are:

Author(s)	Title
A. Maczynski	Binary Hydrocarbon Systems
A. Maczynski, Z. Maczynska, and M. Rogalski	Binary Systems of Hydrocarbons and Related Non-Oxygen Compounds
A. Maczynski, Z. Maczynska, and A. Skrzecz	Binary Systems of Organic Compounds Containing Halogen, Nitrogen and Sulfur
A. Maczynski, Z. Maczynska, T. Treszczanowicz, and K. Dunajska	Binary Systems of Hydrocarbons and Oxygen Compounds Without Alcohols and Acids

* * * * * * * * * *

[87] Marcus, Y., Kertes, A. and Yanir, E.
Equilibrium Constants of Liquid-Liquid Distribution Reactions (in three volumes)
(Butterworths, London, 1974 and 1977)

This three volume series contains equilibrium data on liquid-liquid distribution reactions. The equilibria of concern include: distribution, dissociation and aggregation of the extractant, reactions of the extractant with diluents and with other solvents, extraction of the water, equilibria of the extraction metal ions, and extraction of metal ions with the extractant as the ligand. Volumes 1 and 2 cover organophosphorous extractions and alkylammonium salt extractants, respectively. Volume 3 deals with distribution reactions of carboxylic and sulfonic acid extractants, and the distribution of inorganic acids, salts, and complexes between aqueous solutions and both inert solvents and solvents which have oxygen donor atoms. Each reaction entry includes the equilibrium constant, the temperature for which the data exists, the conditions, and a reference(s) to the source literature.

* * * * * * * * * *

[88] Marshall, W. L. and Franck, E. U.
Ion Product of Water Substance, 0-1000 °C, 1-10,000 bars
New International Formulation and its Background
Journal of Physical and Chemical Reference Data 10, 295 (1981)

The ion-product of water is represented as a function of temperature (0 to 1000 °C) and pressure (1 to 10,000 bars) by an equation with adjustable parameters which have been determined by least-squares procedures using data from the literature. The paper also contains the background for the international formulation for the ion-product of water as issued by the International Association for the Properties of Steam in May 1980. Also see [108].

* * * * * * * * * *

[89] Martell, A. E. and Smith, R. M.
Critical Stability Constants. Volume 1: Amino Acids
(Plenum Press, New York, 1974)

Smith, R. M. and Martell, A. E.
Critical Stability Constants. Volume 2: Amines
(Plenum Press, New York, 1975)

Martell, A. E. and Smith, R. M.
Critical Stability Constants. Volume 3: Other Organic Ligands
(Plenum Press, New York, 1977)

Smith, R. M. and Martell, A. E.
Critical Stability Constants. Volume 4: Inorganic Complexes
(Plenum Press, New York, 1976)

and

Martell, A. E. and Smith, R. M.
Critical Stability Constants. Volume 5: First Supplement
(Plenum Press, New York, 1982)

Volume 1 contains selected values of log K, ΔH, and ΔS at 25 °C for the interaction of inorganic metal and hydrogen ions with several classes of organic ligands. The organic ligands dealt with are the aminocarboxylic acids, iminodiacetic acid and its derivatives, peptides, aniline carobyxlic acids, pyridine carboxylic acids, peptides, and several other miscellaneous ligands. The data are critically selected from the literature rather than being simply compiled. Each selected datum contains a reference to the primary literature. This 469 page volume contains a ligand formula and name index.

Volume 2 is similar in arrangement to Volume 1. The organic ligands dealt with are the aliphatic, secondary, and tertiary amines, azoles, azines, and the amino phosphorous acids.

Volume 3 contains the organic ligands which are not contained in Volumes 1 and 2. These ligands include carboxylic and phosphorous acids, phenols, alcohols, amides, amines, halides, and many others.

Volume 4 is similar in arrangement to the earlier volumes. The ligands dealt with are the hydroxide ion, some of the transition metal ligands (vanadium, chromium, molybdenum, tungsten, and others) and ligands of the Groups III through VII elements.

Volume 5 is a supplement to the first four volumes and serves to make the coverage of the literature more current. Also see item [137].

* * * * * * * * * *

[90] McGlashan, M. L. (editor)
Chemical Thermodynamics (in two volumes)
(The Chemical Society, London, 1971 and 1978)

This two volume series contains reviews of the current literature pertinent to several specialized areas of chemical thermodynamics. Each chapter is written by a specialist in that area and includes many useful references to the primary literature. The chapters and their authors are:

Volume 1	Title of Chapter
M. L. McGlashan	The Scope of Chemical Thermodynamics
E.F.G. Herrington	Thermodynamic Quantities, Thermodynamic Data, and their Uses
A. J. Head	Combustion and Reaction Calorimetry
J. F. Martin	The Heat Capacities of Organic Compounds
J.D. Cox and I. W. Lawrenson	The p, V, T Behavior of Single Gases
J. F. Counsell	Modern Vapourflow Calorimetry
D. Ambrose	Vapour Pressures
S. G. Frankiss and J. H. S. Green	Statistical Methods for Calculating Thermodynamic Functions
O. Kubaschewski, P. J. Spencer and W. A. Dench	Metallurgical Thermochemistry at High Temperatures
Volume 2	
K. N. Marsh	The Measurment of Thermodynamic Excess Functions of Binary Liquid Mixtures
T. M. Letcher	Activity Coefficients at Infinite Dilution from Gas-Liquid Chromatography
C. L. Young	Experimental Methods for Studying Phase Behavior of Mixtures at High Temperatures and Pressures
G. M. Schneider	High-Pressure Phase Diagrams and Critical Properties of Fluid Mixtures
F. L. Swinton	Mixtures Containing a Fluorocarbon
A. G. Williamson	Specific Interactions in Nonelectrolyte Mixtures
C. M. Knobler	Volumetric Properties of Gaseous Mixtures
R. L. Scott	Critical Exponents for Binary Fluid Mixtures
C. P. Hicks	A Bibliography of Thermodynamic Quantities for Binary Fluid Mixtures

* * * * * * * * * *

[91] McMeekin, T. L.
The Solubility of Biological Compounds
in "Solutions and Solubilities," Part I. (M. R. J. Dack, editor)
(John Wiley and Sons, New York, 1975)

This chapter contains a discussion on systematic studies on the solubilities of proteins and lipids. The author has attempted to relate solubility to structure and has included several tables and plots giving solubilities of amino acids, proteins, lipids and fatty acids. There are 62 references to the literature.

* * * * * * * * * *

[92] McMillen, D. F. and Golden, D. M.
Hydrocarbon Bond Dissociation Energies
Annual Reviews of Physical Chemistry 33, 493 (1982)

This review article contains "best" values of bond-dissociation energies at 298.15 K of hydrocarbons and their nitrogen, oxygen, sulfur, halogen, and silicon derivatives. There is also some limited data on inorganic molecules. The tables include data on $\Delta_f H^\circ$ of the related radicals. The authors provide references to a dozen earlier reviews on this subject and there are 242 references to the source literature. Also see item [162].

* * * * * * * * * *

[93] Medvedev, V. A., Bergman, G. A., Gurvich, L. V., Yungman, V. S., Vorob'ev, A. F., Kolesov, V. P. and others (V. P. Glushko, general editor)
Thermal Constants of Substances (Volumes 1 to 10)
(in Russian)
(Viniti, Moscow, 1965 to 1982)

This extensive series represents many years of effort by numerous Russian thermodynamicists engaged in the critical evaluation of thermodynamic data. Included in the tables are carefully selected values of $\Delta_f G^\circ$, $\Delta_f H^\circ$, S°, $H_T - H_0$, and C_p at 298.15 K for the elements and their compounds including many aqueous species. Also given are dissociation energies of gases and enthalpies of phase changes. There are extensive references to the primary literature from which the tabulated values were obtained. Carbon and its compounds are covered up to two carbon atoms. The coverage and arrangements of substances is similar to that used in the NBS Tables of Chemical Thermodynamic Properties (see item [149]).

* * * * * * * * * *

[94] Meites, L. (editor)
Handbook of Analytical Chemistry
(McGraw-Hill Book Co., New York, 1963)

This treatise on analytical chemistry contains several useful tabulations of thermochemical data. Some of these tabulations are taken directly from other sources cited in this bibliography and are not included below:

Author(s)	Title of Section
V. E. Bowers and R. G. Bates	Equilibrium Constants of Proton-transfer Reactions in Water (Section 1, Table 1-10)

Tabulated are pK values at 15, 25, and 35 °C for about 200 organic and inorganic acids; no literature references are given.

L. Meites	Formal Equilibrium Constants of Proton-transfer Reactions at Finite Ionic Strength (Section 1, Table 1-11)

This section contains a tabulation of pK values at 25 °C for a selected series of about 150 acids, and bases. These data were taken from Bjerrum, Schwarzenbach, and Sillen, "Stability Constants of Metal Complexes," Part I, The Chemical Society, London, 1957 which was later revised by Sillen and Martell (see item [137]). Also given are tables of acid dissociation data pertinent to ethanol-water and methanol-water mixtures.

D. A. Aikens and C. N. Reilley	Formation Constants of Metal Complexes (Section 1, Table 1-17)

Tabulated are log K values for the binding of the more common inorganic metal ions to 55 ligands. ΔH and ΔS values are also given for four ligands. Literature references are given.

* * * * * * * * * * *

[95] Merrill, A. L. and Watt, B. K.
Energy Value of Foods: Basis and Derivation
Agriculture Handbook No. 74
(U.S. Government Printing Office, Washington, D.C., 1955)

Watt, B. K. and Merrill, A. L.
Composition of Foods: Raw, Processed, Prepared
Agriculture Handbook No. 8 (revised)
(U. S. Government Printing Office, Washington, D.C., 1963)

This pair of monographs provides numerous composition and energy values for food and foodstuff ingredients. It should be noted, of course, that the Calorie used in food energy values is one kilocalorie (4.184 kJ) as used in thermochemistry. In Agriculture Handbook No. 74, Part I gives a discussion of the sources of food energy in terms of organic compound class, and of the experimental determination of enthalpies of combustion. Parts II, III, and IV apply the data to physiological processes. An appendix gives composition and enthalpy of combustion of foods. Tables 1 to 5 and table 24 give enthalpies of combustion of specific food items or component substances. Care should be used in taking values from the numerous tables, as correction factors have sometimes been applied to adjust for physiological processes. These adjustments are indicated by footnotes. In Agriculture Handbook No. 8, Appendix A is of particular interest as it gives notes on energy values and nutrients, including (adjusted) enthalpies of combustion of many foods and food ingredients.

* * * * * * * * * *

[96] Milazzo, G. and Caroli, S.
Tables of Standard Electrode Potentials
(John Wiley and Sons, New York, 1978)

This book contains extensive tables of standard electrode potentials covering the periodic chart. For each electrode reaction is given the standard potential, the temperature and the pressure, the solvent, and a literature reference. Occasionally the temperature coefficient of the electrode potential is given together with an estimate of uncertainty. Much of the tabulated data is taken from secondary sources (such as item [149]).

* * * * * * * * * *

[97] Miller, D. G.
Application of Irreversible Thermodynamics to Electrolyte Solutions.
I. Determination of Ionic Transport Coefficients ℓ_{ij} for Isothermal Vector Transport Processes in Binary Electrolyte Systems.
Journal of Physical Chemistry 70, 2639 (1966)
II. Ionic Coefficients ℓ_{ij} for Isothermal Vector Transport Processes in Ternary Systems.

These papers derive equations relating fundamental isothermal transport coefficients (ℓ_{ij}'s) to experimentally measurable quantities for electrolytes in a neutral solvent. ℓ_{ij}'s for the most common aqueous ionic solutions are calculated from critically reviewed data.

* * * * * * * * * *

[98] Millero, F. J.
The Partial Molal Volumes of Electrolytes in Aqueous Solutions.
Compilation of the Partial Molal Volumes of Electrolytes at Infinite Dilution, $\bar{V}^{\circ}$ and the Apparent Molal Volume Concentration Depéndence Constants, S_v^* and b_v, at Various Temperatures, in
Water and Aqueous Solutions, R. A. Horne (editor)
(John Wiley and Sons, Interscience Publishers, New York, 1972)

This chapter (no. 13 in this book) is concerned with the measurement and interpretation of partial molar volumes and their concentration and temperature dependence. Included are tables of the partial molar volumes of the common inorganic and organic electrolytes (about 200 systems) as well as values of the partial molar volumes of the more common inorganic and organic ions (about 100 species). The data refer to temperatures from 0 to 200 °C. Also see items [99] and [123].

* * * * * * * * * *

[99] Millero, F. J.
The Molal Volumes of Electrolytes
Chemical Reviews 71, 147 (1971)

This is a detailed and thorough review article dealing with molar volumes of electrolytes in water. Included is a history and discussion of theoretical developments associated with molar volumes. Tabulated are the partial molar volumes of the common (about 50) inorganic and organic ions in water at temperatures ranging from zero to 200 °C. Also is given partial molar volume data for non-aqueous systems. There are 366 references to the literature. Also see item [98] and [123].

* * * * * * * * * *

[100] Mishchenko, K. P. and Poltoratzkii, G. M.
Aspects of the Thermodynamics and Structure of Aqueous and Non-Aqueous Electrolyte Solutions
(in Russian)
(Izdatelstvo Khimia, Leningrad, 1968)

This 350 page monograph contains extensive discussions and correlations (theoretical and empirical) of existing experimental data on enthalpies, Gibbs energies, and entropies, of solution and ionization of inorganic acids, bases, and salts in water and selected organic solvents. Heat capacities, enthalpies, Gibbs energies, and entropies of the substances and their ions in solution are also discussed. Extensive use is made of diagrams relating observed properties to periodic groupings of the elements. The monograph contains numerous small tables of properties of limited groups of substances. A summary compilation of selected values of thermodynamic properties occupies 43 pages, giving $\Delta_f G°$, $\Delta_f H°$, $S°$ and $C_p°$ of pure and dissolved inorganic substances in their standard state at 25 °C, selected enthalpies of solution at 25 °C, enthalpies of dilution of common acids, bases, and salts at 25 °C, and heat contents and partial molar heat capacities for selected salts vs. concentration and temperature in water, methanol, ethanol, and a few other organic solvents.

* * * * * * * * * *

[101] Morss, L. R. Thermochemical Properties of Yttrium, Lanthanum, and the Lanthanide Elements and Ions
Chemical Reviews <u>76</u>, 827 (1976)

This review presents tables of values of $\Delta_f H°$, and $S°$ at 25 °C for the aqueous ions of yttrium, lanthanum, and the lanthanide elements based upon both experimental measurement data and a group correlation scheme.

* * * * * * * * * *

[102] Nancollas, George H.
Interactions in Electrolyte Solutions
(Elsevier, Amsterdam, 1966)

This monograph contains a general discussion of the methodology, kinetics, mechanisms, structural features, and general thermodynamic trends of ionic association reactions in solution. In the Appendix there are several tables of thermodynamic functions for ion-association reactions with references to the source literature.

* * * * * * * * * *

[103] Naumov, G. B., Rhyzhenko, B. N. and Khodakovsky, I. L.
Handbook of Thermodynamic Quantities
(in Russian)
(Atomizadat, Moscow, 1971)

This reference book contains a compilation of thermodynamic data for about 2000 chemical compounds and aqueous ions (mostly inorganic). The thermodynamic properties tabulated are $\Delta_f G°$, $\Delta_f H°$, $S°$, and $C_p°$ at 298.15 K, electrode potentials, enthalpies and entropies for phase transitions, $\Delta_f G°$ of inorganic aqueous ions from 25 to 350 °C, partial molar heat capacities from 10 to 130 °C, and the partial molar volumes of aqueous electrolytes at high temperatures and pressures. There are 1550 references given to the primary literature and to the literature evaluations of others.

* * * * * * * * * *

[104] National Bureau of Standards Electrolyte Data Center
A series of papers, published in the Journal of Physical and Chemical Reference Data from 1977 to 1981.

This series of papers contains evaluations leading to recommended values of activity and osmotic coefficients and excess Gibbs energies of aqueous electrolyte solutions at 298.15 K. The data are presented both in tabular form as a function of molality and as coefficients of several correlating equations. The correlations are detailed and include a careful statistical analysis of the data and references to the source literature. The data cover the entire composition range for which data exist and include most of the electrolytes of charge type 12 and 21 (also see items [48], [121], [124], and [128]). The papers in the series are:

Author(s)	Title of Paper
B. R. Staples and R. L. Nuttall	The Activity and Osmotic Coefficients of Aqueous Calcium Chloride at 298.15 K

Journal Physical and Chemical Reference Data 6, 385 (1977)

R. N. Goldberg and R. L. Nuttall	Evaluated Activity and Osmotic Coefficients for Aqueous Solutions: The Alkaline Earth Metal Halides

Journal of Physical and Chemical Reference Data 7, 263 (1978)

R. N. Goldberg, B. R. Staples and R. L. Nuttall	Evaluated Activity and Osmotic Coefficients for Aqueous Solutions: Iron Chloride and the Bi-univalent Compounds of Nickel and Cobalt

Journal of Physical and Chemical Reference Data 8, 923 (1979)

R. N. Goldberg — Evaluated Activity and Osmotic Coefficients for Aqueous Solutions: Bi-univalent Compounds of Lead, Copper, Manganese, and Uranium

Journal of Physical and Chemical Reference Data 8, 1005 (1979)

R. N. Goldberg — Evaluated Activity and Osmotic Coefficients for Aqueous Solutions: Bi-univalent Compounds of Zinc, Cadmium, and Ethylene Bis (Trimethylammounium) Chloride and Iodide

Journal of Physical and Chemical Reference Data 10, 1 (1981).

R. N. Goldberg — Evaluated Activity and Osmotic Coefficients for Aqueous Solutions: Thirty-Six Uni-Bivalent Electrolytes

Journal of Physical and Chemical Reference Data 10, 761 (1981)

B. R. Staples — Activity and Osmotic Coefficients of Aqueous Metal Nitrites

Journal of Physical and Chemical Reference Data 10, 765 (1981)

B. R. Staples — Activity and Osmotic Coefficients of Aqueous Sulfuric Acid

Journal of Physical and Chemical Reference Data 10, 779 (1981)

* * * * * * * * * *

[105] Nývlt, J. Solid-Liquid Phase Equilibria (Elsevier, Amsterdam, 1977)

This 246 page monograph contains tables of solubilities as a function of temperature of approximately 350 compounds to form binary and ternary mixtures. The substances are mostly inorganic and include only 22 organic compounds. Also included are enthalpies of solution of the substances into pure water. The author does not give references to the primary literature.

* * * * * * * * * *

[106] Oetting, F.L., Medvedev, V., Rand, M.H. and Westrum, E.F., Jr. (editors)
The Chemical Thermodynamics of Actinide Elements and Compounds
(International Atomic Energy Agency, Vienna, 1976 to present)

This series of books contains thermodynamic data (heat capacities, formation properties, thermal functions, vapor pressures, and many other properties) for the actinide elements and their compounds. In addition to extensive tables of data there is also a discussion of the sources from which the data have been taken and the calculations performed. Parts 2 and 3 contain data on the aqueous ions of the actinide elements. To date, the parts which have been published are:

Author(s)	Title	Part No.
F.L. Oetting, M.H. Rand and R. J. Ackermann	The Actinide Elements	1
J. Fuger and F. L. Oetting	The Actinide Aqueous Ions	2
E. H. P. Cordfunke and P. A. G. O'Hare	Miscellaneous Actinide Compounds	3
P. Chioti, V. V. Akhachinskij, I. Ansara, and M. H. Rand	The Actinide Binary Alloys	5

Parts 4 (Chalcogenides), 6 (Carbides), 7 (Pnictides), and 8(Halides) are scheduled for publication in 1983.

* * * * * * * * * *

[107] Ohe, S. (editor)
Computer Aided Data Book of Vapor Pressure
(Data Book Publishing Co., Tokyo, 1976)

This lengthy book (2000 pages) contains extensive plots of vapor pressures for various pure substances, both organic and inorganic, as a function of temperature as well as the constants of the Antoine equation. Included is a substance index and references to the literature.

* * * * * * * * * *

[108] Olofsson, G. and Hepler, L. G.
Thermodynamics of Ionization of Water Over Wide Ranges of Temperature and Pressure
Journal of Solution Chemistry 4, 127 (1975)

This review ties together equilibrium and thermal data to obtain "best" values for ΔG°, ΔH°, ΔC_p°, ΔV°, and $\Delta\kappa^\circ$ (the isothermal compressibility change) for the ionization of water over the temperature range 0 to 300 °C and the pressure range 1 to 8000 atmospheres. Also see item [88].

* * * * * * * * * *

[109] Parker, V. B., Staples, B. R., Jobe, T. L., Jr. and Neumann, D. B. A Report on Some Thermodynamic Data for Desulfurization Processes (NBSIR 81-2345, National Bureau of Standards, Washington, D.C., 1981)

This report contains values of thermochemical properties and processes pertinent to coal gas desulfurization. Substances covered include solutions formed from the aqueous ions: OH^-, SO_3^{2-}, HSO_3^-, SO_4^{2-}, CO_3^{2-}, HCO_3^-, H^+, Mn^{2+}, Fe^{2+}, Mg^{2+}, Ca^{2+}, Na^+, and K^+ and include solid, liquid, aqueous and gaseous compounds or species formed from these ions. Properties given are $\Delta_fG°$, $\Delta_fH°$, $S°$, $C_p°$ (H_T-H_0), L_ϕ, $\gamma_\pm$, and ϕ at 298.15 K. Predicted values of $\Delta G°$, $\Delta H°$, $\Delta S°$, and $\Delta C_p°$ for important desulfurization processes were calculated from the formation properties. These data are consistent with the data in item [149].

* * * * * * * * * * *

[110] Parker, V. B.
The Thermochemical Properties of the Uranium-Halogen Containing Compounds (NBSIR 80-2029, National Bureau of Standards, Washington, D.C., 1980)

This report contains a detailed evaluation of the thermochemistry of 142 uranium-halogen containing compounds. The properties given are $\Delta_fH°$, $\Delta_fG°$, $S°$, $C_p°$, and (H_T-H_0), all at 298.15 K, and $\Delta_fH°$ at 0 K. The analysis of much of the data involves the consideration of the aqueous chemistry of uranium containing compounds (also see item [149]). The recommended values are consistent with the CODATA scale (item [23]).

* * * * * * * * * * *

[111] Parker, V. B.
Thermal Properties of Aqueous Uni-univalent Electrolytes (NSRDS NBS 2, U.S. Government Printing Office, Washington, D.C., 1965)

This monograph is a review of the heat-capacity, enthalpy-of-solution, and enthalpy-of-dilution data on simple 1-1 electrolytes, organic and inorganic, in aqueous solutions. From the critical analysis of this data, tables of selected "best" values of apparent heat capacities, and enthalpies of dilution are given, as well as selected values of the enthalpies of solution to the infinitely dilute solution. Also included is a review of data on the enthalpies of neutralization of monobasic acids which has led to a selected "best" value for the enthalpy of ionization of water. Data on each property are introduced with a discussion of methods employed in reducing the data to a standard form and are listed by compound, in the order: acids, ammonium and amine salts, silver salts, and salts of the alkali metals. For each compound are listed the various investigations, with the temperature and range of concentrations measured. Graphs of molar heat capacity and molar enthalpy as functions of concentration are also included, for aqueous solutions of many of the compounds discussed. In addition, there is an abbreviated listing, by compound, of review and compilation papers on the thermal properties of the aqueous uni-univalent electrolytes. These and other references are also listed alphabetically in a separate reference section with 652 entries. The chosen "best" values at 25 °C for each parameter and compound are arranged in a series of 21 tables.

* * * * * * * * * * *

[112] Parker, V.B., Wagman, D. D. and Garvin, D.
Selected Thermochemical Data Compatible with the CODATA Recommendations
(NBSIR 75-968, National Bureau of Standards, Washington, D.C., 1976)

Selected thermochemical properties, $\Delta_f G°$, $\Delta_f H°$, $S°$, (all at 298.15 K), $\Delta_f H°$ (0 K), and H (298.15 K) - H (0 K), are given for 384 substances (almost entirely inorganic) including many of the more commonly encountered aqueous species. The selected values are intended to be compatible with the current CODATA recommendations on key values for thermodynamics (see item [23]).

* * * * * * * * * *

[113] Parsons, R.
Handbook of Electrochemical Constants
(Academic Press, New York, 1959)

This handbook contains extensive tables of data for the more common inorganic and organic aqueous electrolyte solutions. Properties covered include dielectric constants, activity coefficients, relative partial molar enthalpies, equilibrium constants, solubility products, conductivities, electrochemical potentials, Gibbs energies and enthalpies of formation, entropies, heat capacities, viscosities, and diffusion coefficients, Unfortunately, only a few of the tables contain references to the sources of the data.

* * * * * * * * * *

[114] Pauling, L.
The Nature of the Chemical Bond (third edition)
(Cornell University Press, Ithaca, New York, 1960)

This well established monograph provides general information about the nature of chemical bonding in (principally) inorganic compounds which is fundamentally very important for the estimation of enthalpies of formation, but not always easily applied.

* * * * * * * * * *

[115] Pedley, J. B. (editor)
Computer Analysis of Thermochemical Data (CATCH Tables)
(University of Sussex, Brighton, England, 1972 to 1974)

These thermochemical tables consist of enthalpies of formation at 298.15 K calculated from thermochemical data networks. Included are appropriate references to the literature and estimated errors in the enthalpies of formation. An interesting and important feature of this scheme is that the tables can be readily updated by computer. These tables contain a substantial amount of data for aqueous species. The following tables have been published:

Author(s)	Element(s)	Year
J. D. Cox	Halogen Compounds (Fluorine, Chlorine, Bromine, Iodine)	1972

G. Pilcher	Nitrogen	1972
A. J. Head	Phosphorus	1972
J. B. Pedley and B. S. Iseard	Silicon	1972
D. S. Barnes	Chromium, Molybdenum and Tungsten	1974

* * * * * * * * * *

[116] Pedley, J. B. and Rylance, J.
Sussex - N. P. L. Computer Analysed Thermochemical Data: Organic and Organometallic Compounds
(Sussex University, Brighton, England, 1977)

This book contains enthalpy data on ≈ 4000 compounds and 5000 chemical processes involving organic and organometallic compounds. Much of the processing of the data was done by computer (see item [150]). Properties which are presented include enthalpies of combustion, formation, sublimation, vaporization, and reaction. It is also a revision of the values published by Cox and Pilcher (item [28]). There are ≈ 450 references to the sources from which the data were taken.

* * * * * * * * * *

[117] Perrin, D. D.
Dissociation Constants of Inorganic Acids and Bases in Aqueous Solution
(Butterworths, London, 1969)

This short (163 pp.) monograph is a compilation of dissociation constants of 217 inorganic acids and bases. The classes of compounds include not only conventional acids and bases, but also hydrated metal ions and free radicals, such as hydroxyl, the only criterion being gain or loss of a proton or hydroxyl ion. The data are organized into a single table listing the compounds, preceded by a brief introduction to the use of table, and a section on methods of measurement and calculation. The methods are classified as conductometric, electrometric, optical, or other. Elements and compounds are listed in decreasing extent of protonation. pK values are, wherever possible, obtained by extrapolation to zero ionic strength. The table also gives the temperature of each measurement, remarks as to ionic strength, concentration, and any other factors relating to pK, coded references to method of measurement, the procedure used in evaluating the constants and any corrections taken into consideration, and the literature references. There are approximately 1100 references listed alphabetically by author.

* * * * * * * * * *

[118] Perrin, D. D.
Dissociation Constants of Organic Bases in Aqueous Solution
(Butterworths, London, 1965)

This book contains values of dissociation constants for organic bases in aqueous solution. The bases are arranged under the headings aliphatic, alicyclic, aromatic, heterocyclic, natural products, dyes and indicators, substances lacking a basic nitrogen atom, and miscellaneous. Accompanying the data entry for each base is the temperature, method of measurement, formula, assessment of the measurement, and the appropriate reference. There are 3790 data entries in this book. Also see item [73].

* * * * * * * * * *

[119] Phillips, R.
Adenosine and the Adenine Nucleotides. Ionization, Metal Complex Formation, and Conformation in Solution
Chemical Reviews 66, 501 (1966)

This article is a detailed review of the thermodynamics, kinetics, and structural characteristics of adenosine and the adenine nucleotides in solution. Both log K and enthalpy data are tabulated for protonation and metal-ion binding reactions to adenosine and the adenine nucleotides. The conditions are given under which the tabulated data are applicable, namely, ionic strength, temperature, supporting electrolyte, pH, method of measurement, as well as references to the original data source.

* * * * * * * * * *

[120] Phillips, R. C., George, P., and Rutman, R. J.
Thermodynamic Data for the Hydrolysis of Adenosine Triphosphate as a Function of pH, Mg^{2+} Ion Concentration, and Ionic Strength
Journal of Biological Chemistry 244, 3330 (1969)

This article deals with the computation of the Gibbs energy change for the hydrolysis of adenosine-5'-triphosphate to adenosine-5'-diphosphate as a function of magnesium ion concentration, pH, and ionic strength at 25 °C. A critical evaluation of the existing data pertinent to this computation is included. References to 24 papers are given. Also see item [1].

* * * * * * * * * *

[121] Pitzer, K. S. and others
Thermodynamics of Electrolytes
(A series of papers published in several journals)

This series of papers contains an extensive array of correlated data on aqueous electrolyte solutions, much of it having been calculated using the system of equations given in paper I in this series. The contents of these papers have been summarized by Pitzer in a chapter in the book edited by Pytkowicz (see item [123]). The data include activity and osmotic coefficients, relative apparent molar enthalpies and heat capacities, excess Gibbs energies, entropies, heat capacities, volumes, and some equilibrium constants and enthalpies. Systems of interest include both binary solutions and multicomponent mixtures. While most of the data pertain to 25 °C, the papers on sodium chloride, calcium chloride, and sodium carbonate cover the data at the temperatures for which experiments have been performed. Also see items [48], [104], and [124].

The papers in this series which contain data relevant to this bibliography are:

Pitzer, K.S.
Thermodynamic Properties of Aqueous Solutions of Bivalent Sulphates
Journal of the Chemical Society, Faraday Transactions II 68, 101 (1972).

Pitzer, K.S. and Mayorga, G.
Thermodynamics of Electrolytes. II. Activity and Osmotic Coefficients for Strong Electrolytes with One or Both Ions Univalent
Journal of Physical Chemistry 77, 2300 (1973).

Pitzer, K.S. and Mayorga, G.
Thermodynamics of Electrolytes. III. Activity and Osmotic Coefficients for 22 Electrolytes
Journal of Solution Chemistry 3, 539 (1974):

Pitzer, K.S. and Kim, J.J.
Thermodynamics of Electrolytes IV. Activity and Osmotic Coefficients for Mixed Electrolytes
Journal of the American Chemical Society 96, 5701 (1974).

Pitzer, K.S. and Silvester, L.F.
Thermodynamics of Electrolytes. VI. Weak Electrolytes Including H_3PO_4
Journal of Solution Chemistry 5, 269 (1976).

Pitzer, K.S., Roy, R.N. and Silvester, L.F.
Thermodynamics of Electrolytes VII. Sulfuric Acid
Journal of the American Chemical Society 99, 4930 (1977).

Silvester, L.F. and Pitzer, K.S.
Thermodynamics of Electrolytes. 8. High-Temperature Properties, Inclduing Enthalpy and Heat Capacity, with Application to Sodium Chloride
Journal of Physical Chemistry 81, 1822 (1977).

Pitzer, K.S., Peterson, J.R. and Silvester, L.F.
Thermodynamics of Electrolytes. IX. Rare Earth Chlorides, Nitrates and Perchlorates.
Journal of Solution Chemistry 7, 45 (1978).

Silvester, L.F. and Pitzer, K.S.
Thermodynamics of Electrolytes. X. Enthalpy and the Effect of Temperature on the Activity Coefficients.
Journal of Solution Chemistry 7, 327 (1978).

Pitzer, K.S. and Silvester, L.F.
Thermodynamics of Electrolytes. II. Properties of 3:2, 4:2, and Other High-Valence Types
Journal of Physical Chemistry 82, 1239 (1978).

Bradley, D.J. and Pitzer, K.S.
Thermodynamics of Electrolytes 12. Dielectric Properties of Water and Debye-Hückel Parameters to 350 °C and 1 kbar
Journal of Physical Chemistry 83, 1599 (1979).

Pitzer, K.S. and Peiper, J.C.
Activity Coefficients of Aqueous $NaHCO_3$.
Journal of Physical Chemistry 84, 2396 (1980).

Peiper, J.C. and Pitzer, K.S.
Thermodynamics of Aqueous Carbonate Solutions Including Mixtures of Sodium Carbonate, Bicarbonate, and Chloride
Journal of Chemical Thermodynamics 14, 613 (1982).

Rogers, P.S.Z. and Pitzer, K.S.
Volumetric Properties of Aqueous Sodium Chloride Solutions
Journal of Physical and Chemical Reference Data 11, 15 (1982).

Pitzer, K.S.
Theory: Ion Interaction Approach in "Activity Coefficients in Electrolyte Solutions," R.M. Pytokowicz (editor)
(CRC press, Boca Raton, Florida, 1979).

See item [123]

Roy, R.N., Gibbons, J.J., Peiper, J.C. and Pitzer, K.S.
Thermodynamics of the Unsymmetrical Mixed Electrolyte $HCl-LaCl_3$
Journal of Physical Chemistry, 87, 2365 (1983).

---

deLima, M.C.P and Pitzer, K.S.
Thermodynamics of Saturated Aqueous Solutions Including Mixtures of NaCl, KCl, and CsCl
Journal of Solution Chemistry, 12, 171 (1983).

deLima, M.C.P. and Pitzer, K.S.
Thermodynamics of Saturated Electrolyte Mixtures of NaCl with Na_2SO_4 and with $MgCl_2$
Journal of Solution Chemistry, 12, 187 (1983).

Phutela, R.C. and Pitzer, K.S.
Thermodynamics of Aqueous Calcium Chloride
Journal of Solution Chemistry, 12, 201 (1983).

Pitzer, K.S., Peiper, J.C. and Busey, R.H.
Thermodynamic Properties of Aqueous Sodium Chloride Solutions
Journal of Physical and Chemical Reference Data, in review

* * * * * * * * * *

[122] Pourbaix, M. (and others)
Atlas of Electrochemical Equilibrium in Aqueous Solution
(Pergamon Press, Oxford, 1966)

This book contains tabulations of Gibbs energy of formation data for many of the principal compounds of the inorganic elements. Electrochemical potentials and their dependence on pH, are calculated for many important couples. Much of the Gibbs energy data is taken from the evaluations and compilations of others. Of particular utility are the species-composition or predominance diagrams which appear in this book.

* * * * * * * * * *

[123] Pytkowicz, R. M. (editor) Activity Coefficients in Electrolyte Solutions (in two volumes) (CRC Press, Boca Raton, Florida, 1979)

This two volume series has several chapters dealing with various aspects of electrolyte solutions. The chapters which contain tables of thermodynamic data and/or a particularly useful collection of references to such literature are:

Author	Title of Chapter
K. S. Pitzer	Theory: Ion Interaction Approach

This chapter (no. 7 in Volume I) contains a summary of much of the research done by K. S. Pitzer and his collaborators on the development of models for electrolyte solutions and on the representation of thermodynamic properties of electrolyte solutions in terms of these models. The tables in this chapter include: Debye-Hückel parameters for the osmotic coefficient, enthalpy, and heat capacity as a function of temperature; parameters for the activity and osmotic coefficients of approximately 270 aqueous strong electrolytes at 25 °C; parameters for the relative apparent molar and excess enthalpy of ≈ 90 strong electrolytes at 25 °C; a table of parameters for the activity and osmotic coefficients of ≈75 binary mixtures with and without common ions and with up to three solutes present; and parameters for the thermodynamic properties of aqueous NaCl and H_2SO_4 as a function of temperature. The author has included references to his earlier papers which also contain valuable data on electrolyte solutions (also see item [121]).

F. J. Millero — Effects of Pressure and Temperature on Activity Coefficients

This very extensive (99 pages) chapter (no. 2 in Volume II) contains a general discussion of the effects of temperature and pressure on activity coefficients for both binary and mixed electrolyte solutions. Properties of interest are the partial molar volume, expansibility, compressibility, heat capacity, and enthalpy. There is also an excellent discussion of methods of estimating partial molar properties in mixed electrolyte solutions. There are 226 references to the literature. Tables of data are presented for: Debye-Hückel limiting law slopes for the apparent molar volume, enthalpy, heat capacity, expansibility, and compressibility as a function of temperature; parameters for the partial molar volumes of 30 aqueous electrolyes at 25 °C; parameters for the partial molar expansibility of ten electrolytes at 25 °C; parameters for the partial molar compressibilities of 33 electrolytes at 25 °C; values of the activity coefficients of aqueous NaCl solutions at 25 °C as a function of pressure (up to 1000 bars); parameters for the partial molar enthalpies of 59 electrolytes at 25 °C; parameters for the partial molar heat capacities of 140 electrolytes at 25 °C; and tables giving compositions and the partial molar properties of average seawater.

M. Whitfield — Activity Coefficients in Natural Waters

This long (147 pages and 266 references) and very detailed chapter (no. 3 in Volume II) contains a very extended discussion of the thermodynamics of mixtures of aqueous electrolyte solutions with emphasis upon the properties of natural waters. There is a very extensive collection of tables in this article (116 in total) and the contents of this many tables is not easily summarized in our survey. Included are the following: association constants for the formation of ion pairs and weak acids; activity coefficients for both the ions and major components of sea water; formation constants for the binding of heavy metals to several anions found in sea water; gas solubilities including carbon dioxide in sea water; solubilities of polychlorinated biphenyls in sea water;

Setchenow coefficients for hydrocarbons and for volatile solutes in sea water; the osmotic coefficient and density of sea water as a function of temperature and salinity. Thermodynamic solubility products of minerals in brines; the activity coefficient of carbon dioxide in sea water; speciation calculations on copper, zinc, cadmium, and lead in sea water; excess Gibbs energies of mixing of electrolyte solutions at 25 °C; and pairwise and triplet interaction terms for electrolyte solutions in terms of various models.

* * * * * * * * * * *

[124] Rard, J. A., Habenschuss, A. and Spedding, F. H.
A Review of the Osmotic Coefficients of Aqueous H_2SO_4 at 25 °C
Journal of Chemical and Engineering Data 21, 374 (1976)
and
A Review of the Osmotic Coefficients of Aqueous $CaCl_2$ at 25 °C
Journal of Chemical and Engineering Data 22, 180 (1977)

These two articles are critical evaluations of experimental data leading to values of the osmotic coefficients at 25 °C of the two reference electrolytes calcium chloride and sulfuric acid. Also see items [104] and [121].

* * * * * * * * * * *

[125] Reid, R. C., Prausnitz, J. M., and Sherwood, T. K.
The Properties of Gases and Liquids, Their Estimation and Correlation (third edition)
(McGraw-Hill, New York, 1977)

This lengthy monograph discusses various methods available for calculation of estimating properties of materials, and then provides recommendations for action with respect to each kind of property. Included in the book are procedures for making estimates of critical constants, normal boiling temperatures, Lennard-Jones potential parameters, compressibility factors and equations of state, liquid molar volumes and densities, and vapor pressures. Estimates of enthalpies of vaporization, of ideal-gas heat capacities, and of enthalpies and Gibbs energies of formation are treated. For real fluids variations of enthalpy, entropy, internal energy, and heat capacity with pressure are treated. Some methods are given for estimating the properties of fluid mixtures. Surface tension and the transport properties-viscosity, diffusion coefficient and thermal conductivity-are discussed. Numerous tables present comparisons of observed and calculated properties. The Appendix contains tables of thermodynamic properties of many organic compounds.

* * * * * * * * * * *

[126] Riley, J. P. and Skirrow, G. (editors)
Chemical Oceanography (in six volumes)
(Academic Press, New York and London, 1975 and 1976)

This six volume series contains 34 chapters on various topics related to chemical oceanography. Contained in these chapters is a wealth of chemical and physical information on the properties of sea water and its constituents measurement and interpretation of chemical potentials; the theory of diffusion (emphasizing conductance and viscosity in concentrated solutions), and methods of measurement of diffusion coefficients. The final third of the text deals primarily with characteristics of specific electrolyte solutions, including weak and mixed electrolytes, and strong acids, and includes an extensive (98 pp.) appendix with approximately 75 tables of osmotic and activity coefficients, standard cell potentials (E°) in various organic solvents, ionic radii, and ionization constants of organic acids in aqueous solution, and other information. The narrative is supplemented with graphs, tables, equations, and references.

* * * * * * * * * * *

[129] Rosenblatt, David H. (editor)
Research and Development of Methods for Estimating Physicochemical Properties of Organic Compounds of Environmental Concern (in two volumes)
(Arthur D. Little, Cambridge, Massachusetts, 1981)

These two volumes contain twenty-six chapters, each one dealing with a method for estimating a physico-chemical property of environmentally important organic compounds. For the most part, each chapter provides: "(1) a general discussion of the property and its importance in environmental considerations, (2) an overview of available estimation methods, (3) a description plus step--by-step instructions for each selected method, (4) worked-out examples for each method, (5) a listing of sources of available data on the property, (6) a list of symbols used, and (7) the cited references." The chapters in the book are:

Author(s)	Title of Chapter
W.J. Lyman	Octanol/Water Partition Coefficient
W.J. Lyman	Solubility in Water
W.J. Lyman	Solubility in Various Solvents
W.J. Lyman	Adsorption Coefficient for Soils and Sediments
S.E. Bysshe	Bio-concentration Factor in Aquatic Organisms
J.C. Harris and M.J. Hayes	Acid Dissociation Constant
J.C. Harris	Rate of Hydrolysis
J.C. Harris	Rate of Aqueous Photolysis
K.M. Scow	Rate of Biodegradation
W.J. Lyman	Atmospheric Residence Time
C. F. Grain	Activity Coefficient
C.E. Rechsteiner, Jr.	Boiling Point

as well as useful references to the primary literature. Several topics of thermodynamic interest are the solubilities of gases and salts, chemical speciation of the major and minor constituents of sea water, densities, activity and osmotic coefficients, expansibilities, isothermal compressibilities, velocity of sound, heat capacities, and transport and optical properties of sea water. The Appendix contains a useful series of tables of many of these properties. Also see [80].

* * * * * * * * * * *

[127] Geological Survey Bulletins

Robie, R. A., Hemingway, B. S. and Fisher, J. R.
Thermodynamic Properties of Minerals and Related Substances at 298.15 K and 1 Bar (10^5 Pascals) Pressure and at Higher Temperatures
Geological Survey Bulletin 1452
(U.S. Government Printing Office, Washington, D.C., 1978)

and

Hemingway, B. S., Haas, J. L., Jr., and Robinson, G. R., Jr.
Thermodynamic Properties of Selected Minerals in the System Al_2O_3-CaO-SiO_2-H_2O at 298.15 K and 1 Bar (10^5 Pascals)
Pressure and at Higher Temperature
Geological Survey Bulletin 1544
(U.S. Government Printing Office, Washington, D.C., 1982)

The 1978 publication is a 456 page monograph containing selected values for the entropy, molar volume, and for the enthalpy and Gibbs energy of formation for the elements, 133 oxides, and 212 other minerals and related substances at 298.15 K. Thermal functions are also given for those substances for which heat-capacity or heat-content data are available. The thermal functions are tabulated at 100 K intervals for temperatures up to 1800 K. The monograph includes detailed references to the source literature and a compound index. Also see items [42] and [144].

The 1982 publication is a supplement to Geological Survey Bulletin 1452. It contains a survey of thermodynamic data for minerals in the system Al_2O_3-CaO-SiO_2-H_2O and tabulated values for C_p°, $(H^\circ-H^\circ$ (298.15 K))/T, $(G^\circ-H^\circ$ (298.15 K))/T, $\Delta_f H^\circ$, and $\Delta_f G^\circ$ from 298.15 to 1800 K. The calculations are well documented both as to computational procedures and the sources of data.

* * * * * * * * * * *

[128] Robinson, R. A. and Stokes, R. H.
Electrolyte Solutions. The Measurement and Interpretation of Conductance,
Chemical Potential and Diffusion in Solution of Simple Electrolytes
(Second edition)
(Butterworths, London, 1965)

In this, the revised second edition of a monograph first published in 1955, the first part presents a fundamental discussion of aqueous organic and inorganic electrolyte solutions. Included is a discussion of ionizing solvents (i.e. water), electrolytic conductivities and transport numbers, the

C.E. Rechsteiner, Jr.	Heat of Vaporization
C.F. Grain	Vapor Pressure
R.G. Thomas	Volatilization from Water
R.G. Thomas	Volatilization from Soil
W.A. Tucker and L.H. Nelken	Diffusion Coefficients in Air and Water
J.H. Hagopian	Flash Points of Pure Substances
L.H. Nelken	Densities of Vapors, Liquids and Solids
C.F. Grain	Surface Tension
C.F. Grain	Interfacial Tension with Water
C.F. Grain	Liquid Viscosity
J.D. Birkett	Heat Capacity
J.D. Birkett	Thermal Conductivity
L.H. Nelken and J.D. Birkett	Dipole Moment
L.H. Nelken	Index of Refraction

* * * * * * * * * *

[130] Rossini, F. D., Pitzer, K. S., Taylor, W. J., Ebert, J. P. Kilpatrick, J. E., Beckett, C. W., Williams, M. D. and Werner, H. C.
Selected Values of Properties of Hydrocarbons
National Bureau of Standards Circular C461
(U. S. Government Printing Office, Washington, D.C. 1947)

and

Rossini, F. D., Pitzer, K. S., Arnett, R. L., Brown, R. M. and Pimentel, G. C.
Selected Values of Physical and Thermodynamic Properties of Hydrocarbons and Related Compounds
(Carnegie Press, Pittsburgh, Pennsylvania, 1953)

This monograph resulted from the work of American Petroleum Institute (API) Research Project 44. In the 1953 revision, values are given for 40-odd physical and thermodynamic properties of several hundred hydrocarbons in metric and U. S. Customary units. The data in most instances represent selected values from careful studies, many of which were done in connection with the same API Research Project. Experimental data are supplemented by theoretical calculations or empirical correlations. References to the source data and a bibliography are given.

* * * * * * * * * *

[131] Rossini, F. D., Wagman, D. D., Evans, W. H., Levine, S., and Jaffe, I.
Selected Values of Chemical Thermodynamic Properties
National Bureau of Standards Circular 500
(U. S. Government Printing Office, Washington, D.C., 1952)

This was for many years the most comprehensive authoritative compilation of thermochemical data at 298.15 K for inorganic substances. All inorganic substances and organic substances containing two carbon atoms or fewer per molecule are included if thermodynamic data exist for calculating one of the properties tabulated. Properties tabulated in Part I are $\Delta_f G°$ (0), $\Delta_f H°$ (0), $\Delta_f H°$ (298.15 K), $\Delta_f G°$ (298.15 K), log K_f, S° (298.15 K), and $C_p°$ (298.15 K). Properties tabulated in Part II are temperature, pressure, enthalpy change, entropy change and heat capacity change for transition, fusion, and vaporization processes. The data from original sources were critically evaluated and functions tabulated, maintaining internal consistency by the relationship: $\Delta_f G°$ (298.15 K) = $\Delta_f H°$ (298.15 K) - $T\Delta_f S°$ (298.15K). The sources of data for each data item are listed and a bibliography is included. Part I has been superseded by the NBS Tables of Chemical Thermodynamic Properties (see [149]). Much of the data in Part II cannot be found readily elsewhere.

* * * * * * * * * * *

[132] Saline Water Conversion Engineering Data Book - 1975
(National Technical Information Service, Springfield, Virginia, 1975)

This very large (707 pages) book was produced by the M. W. Kellogg Company for the Office of Saline Water of the U. S. Department of the Interior. The book contains a large amount of miscellaneous engineering information pertinent to examples of engineering and cost calculations, materials of construction, and process flow diagrams for distillation, freezing, reverse osmosis, and electrodialysis. Included in the tables of physical property data are heat capacities, enthalpies, vapor pressures, solubilities, equilibrium constants, critical properties, and mechanical and transport properties. Materials of interest include sea water and aqueous solutions formed from sodium chloride, magnesium chloride, calcium sulfate, calcium carbonate, magnesium sulfate, and hydrocarbons. Much of the data is presented in graphical form with references to the source from which it was taken.

* * * * * * * * * * *

[133] Sanderson, R. T.
Chemical Bonds and Bond Energy
(Academic Press, New York and London, 1971)

This book contains a general discussion of the calculation of bond energies and gives details of calculations performed on 850 different kinds of bonds in more than 500 compounds. Although the bond energies refer to the gaseous state, they are still useful in applications to aqueous systems. The Appendix contains a table of bond energies and enthalpies of formation. The chemical bonds are both inorganic and organic in nature. Also see items [11] and [92].

* * * * * * * * * * *

[134] Schmidt, E.
Properties of Water and Steam in SI Units
(Springer-Verlag, Berlin, 1982)

This book contains tables of the properties of water and steam from 0 to 800 °C and from 0 to 1000 bar which have been calculated using a set of equations accepted by the members of the Sixth International Conference on the Properties of Steam in 1967. Properties which are tabulated include the pressure, specific volume, density, specific enthalpy, specific heat of evaporation, specific entropy, specific isobaric heat capacity, dynamic viscosity, thermal conductivity, the Prandtl number, the ion-product of water, the dielectric constant, the isentropic exponent, the surface tension and Laplace coefficient. Also see items [43] and [70].

* * * * * * * * * *

[135] Serjeant, E. P. and Dempsey, B.
Ionization Constants of Organic Acids in Aqueous Solution
(Pergamon Press, New York, 1979)

This critical compilation of pK values is a supplement to an earlier monograph by Kortum, Vogel, and Andrussow (see item [73]). This volume extends the literature coverage to the end of 1970 and summarizes data for some 4500 acids. In conjunction with the Kortum compilation, the total number of acids covered is $\approx$ 5500. Each entry includes the name of the substance, its molecular formula, pK value; the temperature, method of measurement, an assessment of reliability, and references to the literature.

* * * * * * * * * *

[136] Silcock, H. (editor) Solubilities of Inorganic and Organic Compounds (in three volumes) (Pergamon Press, Oxford, 1979)
Translated from the Russian: "Spravochnik po rastvorimosti"
(Izdatel'stvo Nauka, Moscow, 1969)

This three volume series is an extension of the work of Stephen and Stephen (see item [142]). These volumes cover ternary and multicomponent systems of inorganic substances.

* * * * * * * * * *

[137] Sillen, L. G. and Martell, A. E.
Stability Constants of Metal Ion Complexes, Section I: Inorganic Ligands, Section II: Organic Ligands (Second edition)
(Special Publication No. 17, The Chemical Society, London, 1964
and Supplement No. 1, Special Publication No. 25, 1971)

With the publication of this second edition of Stability Constants, the Chemical Society combined two previously separate volumes into one, two-part volume (745 pp.), the first covering constants of inorganic ligands, the second, organic ligands. Both sections include all data published up to the end of 1960 and some from 1961 to 1963; the scope of the inorganic section has been extended to cover redox equilibria and the extraction of inorganic ligands into nonaqueous solvents. The data are organized into separate tables, each table summarizing the data for the association of one particular ligand with all the metallic ions which have been studied in conjunction with it. Method of measurement, composition, and temperature of the media to which the data refer, are given for each ligand-metal pair. Acid dissociation constants of the ligands are recorded by including the hydrogen ion as one of the cations with which the ligands associate. Redox equilibria are represented by including the electron as a ligand, and hydrolysis of metallic ions is described by regarding the hydroxyl ion as one of the ligands.

The arrangement of material is now more uniform than in the two parts of the first edition, but there remain minor differences of presentation between the inorganic and the organic section. In the inorganic table, 80 ligands are ordered according to group in the periodic system; metal ions are arranged within each inorganic ligand table, in the same order. In the organic section, the ligands (1028) are in order of their empirical formulae, and the metallic ions in the alphabetical order of their international symbols. Methods of measurement are given, with 42 separate methods alphabetized and coded in the introduction; the medium is usually aqueous. Equilibrium constants are given in both tables. The organic section includes consecutive or stepwise constants, K, whenever possible, and cumulative or gross constants, β, if they are the only quantities determined, or if the sequence of stepwise constants is incomplete. The inorganic table includes equilibrium constants for consecutive and cumulative reactions, solubility constants, acid constants and base constants. Certain special constants are also given; e.g. K_p - equilibria involving a gas. Both tables give enthalpy and entropy changes, and symbols relating to the references which are listed at the end of each table. Each of the metals, inorganic ligands and organic ligands are indexed alphabetically, with appropriate table number, at the end of the book.

The first edition was authored by J. Bjerrum, G.Schwarzenbach, and L. G. Sillen and was published in two parts by the Chemical Society of London: (1) "Organic Ligands," Special Publication No. 6 (1957) and "Inorganic Ligands," Special Publication No. 7 (1958).

The supplement to the second edition of "Stability Constants of Metal-Ion Complexes" is a review of pertinent literature published between the completion of the 1964 Tables and the end of 1968, including also some data published before the completion of the 1964 Tables, but omitted from them. In the organic part, a change of policy has been initiated; rather than being omitted, results seeming incomplete or of dubious validity are now included in the tables, with critical comments. The section "How to Use the Tables" has been brought up-to-date; a few methods have been added. Limits of error are now sometimes given. A new feature in the organic section is a Functional Group Index, covering also the 1964 Tables, and a table of 34 macromolecular organic ligands, including albumin, DNA, RNA, and insulin. Also see item [89] for the continuation of these compilations by Martell and Smith.

* * * * * * * * * *

[138] Skinner, H. A. (editor)
MTP International Review of Science - Volume 10
Thermochemistry and Thermodynamics
(Butterworths, London, 1972)

Although this book contains few tables of thermodynamic data, it does contain several chapters of interest. These chapters are:

Author(s)	Chapter
I. Wadsö	Biochemical Thermochemistry
R. F. Jameson	Thermodynamics of Metal-Complex Formation
G. Pilcher	Thermochemistry of Chemical Compounds
K. P. Mishchenko	Thermodynamics of Electrolyte Solutions
B. J. Zwolinski and J. Chao	Critically Evaluated Tables of Thermodynamic Data

The last article provides a very useful guide to much of the evaluated thermodynamic data that are not covered in this bibliography.

* * * * * * * * * *

[139] Smith-Magowan, D. and Goldberg, R. N. A Bibliography of Sources of Experimental Data Leading to Thermal Properties of Binary Aqueous Electrolyte Solutions (NBS Special Publication 537, U.S. Government Printing Office, Washington, D.C., 1979)

This is a bibliography of sources of experimental data that can be used to calculate either relative apparent molar enthalpies or apparent molar heat capacities of aqueous electrolyte solutions. The compounds are arranged according to the standard thermochemical order of arrangement. There are approximately 300 references to the source literature.

* * * * * * * * * *

[140] Sober, H. A. (editor)
Handbook of Biochemistry-Selected Data for Molecular Biology
(The Chemical Rubber Co., Cleveland, Ohio, 1968)

This compendium contains the following twenty-three tables of thermochemical data. This Handbook appeared in revised form in 1976 (see item [34]):

Author(s) or Source	Table
From "The Chemistry of the Amino Acids and Proteins," C.L.A. Schmidt (editor) Charles C. Thomas Co., Springfield, IL	Coefficients of Solubility Equations of Certain Amino Acids in Water (Section B, page B-3)

Solubility data for 34 amino acids are fitted to equations giving the solubility as a function of temperature.

J. O. Hutchens

Heat Capacities, Absolute Entropies and Entropies of Formation of Amino Acids and Related Compounds
(Section B, page B-5)

C°_p, S°, and $\Delta_f S^\circ$ at 25 °C are tabulated for 28 amino acids, 3 peptides, 4 proteins, and 3 related substances.

J. O. Hutchens and E. P. K. Hade, Jr.

Solubilites of Amino Acids in Water at Various Temperatures
(Section B, page B-10)

The solubilities of 18 amino acids are tabulated at four different temperatures from 1 to 40 °C.

J. O. Hutchens

Heats of Combustion, Enthalpy and Free Energy of Formation of Amino Acids and Related Compounds
(Section B, page B-7)

Enthalpies of combustion and formation and Gibbs energies of formation at 25 °C are tabulated for 45 amino acids and related compounds.

J. O. Hutchens

Heats of Solution of Amino Acids in Aqueous Solution at 25 °C
(Section B, page B-11)

The heats of solution for 37 amino acids at 25 °C are tabulated.

J. O. Hutchens

Free Energies of Solution and Standard Free Energies of Formation of Amino Acids in Aqueous Solution at 25 °C
(Section B, page B-13)

Gibbs energies of solution and formation at 25 °C are given for 18 amino acids.

J. O. Hutchens

Activities of Amino Acids and Peptides at 25 °C
(Section B, page B-14)

Molar activity coefficients at 25 °C are tabulated as a function of molality for 14 amino acids.

From K. S. Markley, "Fatty Acids. Part I," second edition, Interscience, New York (1960)

Solubility of Fatty Acids in Water (Section E, page E-13)

The solubilities of 13 fatty acids in water at five temperatures from 0 to 60 °C are tabulated.

From K. S. Markley, "Fatty Acids. Part I," second edition, Interscience, New York (1960)

Approximate Solubilities of Water in Saturated Fatty Acids at Various Temperatures (Section E, page E-13)

The approximate solubilities of water in thirteen saturated fatty acids at various temperatures is given.

From K. S. Markley, "Fatty Acids. Part I," second edition, Interscience, New York (1960)

Solubility of Simple Saturated Triglycerides (Section E, page E-14)

Tabulated are solubilities of five saturated triglycerides in various non-aqueous solvents and at a variety of temperatures.

From K. S. Markley, "Fatty Acids. Part I," second edition, Interscience, New York (1960)

Solubilities of Mixed Triacid Triglycerides at 25 °C (Section E, page E-15)

Solubilities of four mixed triacid triglycerides in four non-aqueous solvents are tabulated.

P. A. Loach

Oxidation-Reduction Potentials, Absorbance Bands and Molar Absorbance of Compounds in Biochemical Studies (Section J, page J-27)

Oxidation-reduction potentials at ambient temperatures are tabulated for an assortment of 253 couples frequently encountered in biochemical studies with appropriate references.

R. W. Henderson and T. C. Morton

Oxidation-Reduction Potentials of Hemoproteins and Metalloporphyrins (Section J, page J-35)

Oxidation-reduction potentials are tabulated for 241 hemoproteins and metalloporphyrins at various temperatures.

R. M. Izatt and J. J. Christensen

Heats of Proton Ionization and Related Thermodynamic Quantities (Section J, page J-49)

This section is an extensive tabulation of enthalpy and entropy changes and pK values for organic and biochemical systems with 323 references to the literature.

G. C. Krescheck

Calorimetric ΔH Values Accompanying Conformational Changes of Macromolecules in Solution (Section J, page J-140)

Tabulated are ΔH values accompanying the conformation changes of 25 macromolecular systems.

W. P. Jencks

Free Energies of Hydrolysis and Decarboxylation (Section J, page J-144)

A discussion is given of standard states appropriate to biochemical thermodynamics. Tabulated are Gibbs energies of hydrolysis of esters of acetic acid and related compounds, of thiol esters, amides, phosphates, and of glycolysis (and of decarboxylation).

W. P. Jencks and J. Regenstein

Ionization Constants of Acids and Bases (Section J, page J-151)

Given is an extensive tabulation of pK values for ionization of several hundred acids and bases with 116 references to the literature.

R. G. Bates

Measurement of pH (Section J, page J-190) and Buffer Solutions (Section J, page J-195)

pH values have been assigned to several important buffer systems.

N. F. Good, G. D. Winget, W. Winter, T. N. Connolly, S. Izawa, and R. M. M. Singh, 5, 472 (1966)

Properties of Some New Buffers for Biochemistry Biological Research (Section J, page J-195)

Tabulated are pK values with temperature coefficients and metal-buffer binding constants for several buffers useful for biological research.

* * * * * * * * * *

[141] Stary, J. and Freiser, H.
Equilibrium Constants of Liquid-Liquid Distribution Reactions.
Part IV: Chelating Extractants
(Pergamon Press, Oxford and New York, 1978)

This 228 page volume summarizes equilibrium constants for liquid-liquid distribution equilibrium constants up to the end of 1972. Each table includes: the extractant and its extractable metal complexes, and the distribution equilibrium constants and extraction constants. The aqueous phase metal complex formation constants are not included and the reader is referred to items and in this bibliography for sources of this type of data. References to the primary literature are included.

* * * * * * * * * *

[142] Stephen, H. and Stephen, T. (editors)
Solubilities of Inorganic and Organic Compounds (in five volumes)
(Pergamon Press, Oxford, 1963 and 1964)

This series, which has been translated from the Russian, consists of five volumes (about 5500 pages) and is a selection from the literature of data on the solubilities of elements, inorganic compounds, and organic compounds in binary, ternary, and multi-component systems. References are given to sources of data in the literature. The data are unevaluated and refer to temperatures for which experimental data exist. Also see the three volume extension edited by Silcock [136].

* * * * * * * * * *

[143] Stull, D. R.
Vapor Pressure of Pure Substances. Organic Compounds
Industrial and Engineering Chemistry 39, 517 (1947)
and
Vapor Pressure of Pure Substances. Inorganic Compounds
Industrial and Engineering Chemistry 39, 540 (1947)

These articles contain evaluated vapor pressure data on over 1200 organic and 300 inorganic compounds. Given for each compound are those temperatures at which the compounds has a given vapor pressure.

* * * * * * * * * *

[144] Stull, D.R. and Prophet, H. (project directors) JANAF Thermochemical Tables (second edition) NSRDS-NBS 37 (U.S. Government Printing Office, Washington, D.C., 1971)

This extensive volume contains values of C_p°, S°, $(G^\circ-H^\circ$ (298.15 K))/T, $H^\circ-H^\circ$ (298.15 K), $\Delta_f H^\circ$, $\Delta_f G^\circ$, and $\log_{10}K$, where K is the equilibrium constant for the reaction of forming the given compound from its elements. The temperature range for which the data are given is 0 K to the highest temperature for which the data exist, the temperature intervals being in steps of 100 K. For each compound given there is a concise and detailed discussion of the data upon which the accompanying table is based and references to the source literature. The volume contains data for ≈1200 elements and compounds in the crystalline, liquid and gaseous states. It should be noted that the reference state for phosphorus is the red form. While most of the substances are inorganic, there is also some coverage of organic materials up to five carbon atoms. Three supplements to the JANAF tables have appeared in the Journal of Physical and Chemical Reference Data for a total of 1322 tabulations involving 35 elements and their compounds. Also see [42]. The supplements are:

Author(s)	Reference
M.W. Chase,Jr., J.L. Curnutt, A.T. Hu H. Prophet, A.N. Syverud and L.C. Walker	Journal of Physical and Chemical Reference Data 3, 311 (1974).
M.W.Chase, Jr.,J.L. Curnutt, H. Prophet R.A. McDonald and A.N. Syverud	Ibid 4, 1 (1975)
M.W. Chase, Jr., J.L. Curnutt, R.A. McDonald and A.N. Syverud	Ibid 7, 793 (1978)
M.W. Chase, Jr., J.L. Curnutt, J.R. Downey, Jr., R.A. McDonald, A.N. Syverud, and E.A. Valenzuela	Ibid 11, 695 (1982)

* * * * * * * * * *

[145] Stull, D. R., Westrum, E. F. and Sinke, G. C. The Chemical Thermodynamics of Organic Compounds (John Wiley and Sons, New York, 1969)

This monograph is divided into three parts. The first part gives the theoretical basis and principles of thermodynamics and thermochemistry, some experimental and computational methods used, and some applications to industrial problems. The second part gives thermal and thermochemical properties in the ideal gas state from 298.15 to 1000 K. In this section, the sources of data are listed and discussed and standardized tables are presented for 918 organic compounds. Values of C_p°, S°,$-(G - H^\circ$(298.15 K))/T, $H^\circ-H^\circ$(298.15 K), $\Delta_f H^\circ$, $\Delta_f G^\circ$, and log K_p are given enthalpy of formation, entropy, and consistent values of $\Delta_f G^\circ$ and log K_p of organic compounds at 298.15 K. In excess of 4000 compounds are listed, including a few inorganic compounds. In one chapter there is a brief discussion of methods of estimating thermodynamic quantities.

* * * * * * * * * *

[146] Tatevskii, V. M., Benderskii, V. A. and Yarovoi, S. S.
Rules and Methods for Calculating the Physico-Chemical Properties of Paraffinic Hydrocarbons
(Pergamon Press, New York and Oxford, 1961)
Translated from the Russian:
"Zakonomernosti i metody rascheta fizikokhimicheskikh svoisty parafionvykh uglevodorodov"
(Gostoptekhizdat, Moscow, 1960)

This monograph is a summary and a consolidation of the results of some years of work by Tatevskii and others, extending and elaborating some procedures introduced by Rossini and others. Properties calculated include: molar volume, molar refraction, vapor pressure, enthalpy of formation from atoms or elements, Gibbs energy of formation, and enthalpy of combustion. Three different methods are used. Tables of constants and illustrations of the accuracy of the methods are given.

* * * * * * * * * *

[147] Timmermans, J.
Physico-Chemical Constants of Pure Organic Compounds (in two volumes)
(Elsevier, Amsterdam, 1950 and 1965)

These two volumes contain tables of physical data for pure organic compounds. The arrangement of the data is by compound. Properties tabulated include vapor pressure, boiling point, triple point, viscosity, specific heat, critical constants, density, compressibility, refractive index, enthalpy of vaporization, and dielectric constant.

* * * * * * * * * *

[148] Timmermans, J.
The Physico-Chemical Constants of Binary Systems in Concentrated Solutions (in four volumes)
(John Wiley and Sons, Interscience Publishers, New York, 1959)

These volumes contain extensive tabulations of physical data relevant to concentrated solutions of binary systems, both organic and inorganic. The properties that are tabulated include dielectric constant, viscosity, equivalent conductivity, surface tension, diffusion and thermal diffusion coefficients, vapor pressure, specific heat, electrochemical data, enthalpy of combustion, enthalpy of dilution and solution, transition enthalpies, and other properties. These books contain extensive tabulations of data pertinent to water and electrolyte solutions. The data are well organized and there is a general compound index as well as references to the original data sources. The literature coverage is through the year 1957.

* * * * * * * * * *

[149] Wagman, D. D., Evans, W. H., Parker, V. B., Schumm, R. H., Halow, I., Bailey, S. M., Churney, K. L., and Nuttall, R. L.
The NBS Tables of Chemical Thermodynamic Properties
Journal of Physical and Chemical Reference Data 11, Supplement No. 2 (1982)

This 392 page volume is the most recent comprehensive compilation in English of critically evaluated thermochemical data at 298.15 K for inorganic substances and for organic substances containing one or two carbon atoms. The values are in joules and refer to a standard state of 0.1 MPa. The coverage is approximately 14,000 substances. The properties tabulated are $\Delta_f H°$ (0), $\Delta_f G°$ (298.15 K), $\Delta_f H°$ (298.15 K), $H°$ (298.15 K) - $H°$ (0), $S°$ (298.15 K), and $C_p°$ (298.15 K). The data from original sources have been critically evaluated and "best" values selected for the properties tabulated, maintaining internal consistency by the relationships $\Delta_f G°$ (298.15 K) = $\Delta_f H°$ (298.15 K) - $T\Delta_f S°$ (298.15 K), and $\Delta_f H°$ (298.15 K) - $\Delta_f H°$ (0) = Σ ($H°$ (298.15 K) - $H°$ (0)). This volume contains an excellent discussion dealing with the construction, use, and history of thermochemical tables. This volume supersedes NBS Circular 500 (item [131]). Also see item [93].

The NBS Tables of Chemical Thermodynamic Properties is a new edition of NBS Technical Note 270, "Selected Valued of Chemical Thermodynamic Properties." The values in the Technical Notes were in calories and refer to a standard state of one atmosphere. The 1982 NBS Tables also contain some changes, additions, and corrections to the Technical Note. Although superseded by the 1982 NBS Tables, the NBS Technical Note 270 is still relevant to thermochemists since many papers in the literature from the years 1965 to 1982 made use of the values tabulated in the Technical Note. The Technical Note was issued in eight parts from 1965 to 1981 and was published by the U.S. Government Printing Office, Washington, D.C. The parts of the Technical Note, the years of publication and their contents are:

Technical Note	Authors	Year of Publication
270-1	Wagman, D. D., Evans, W. H., Parker, V. B., Halow, I., Bailey, S. M., and Schumm, R. H.	1965

Tables for the elements O, H, He, Ne, Ar, Kr, Xe, Rn, F, Cl, Br, I, At, S, Se, Te, Po, N, P, As, Sb, Bi, and C.

270-2	Wagman, D. D., Evans, W. H., Halow, I., Parker, V. B., Bailey, S. M., and Schumm, R. H.	1966

Tables for the elements Si, Ge, Sn, Pb, B, Al, Ga, In, and Tl.

270-3	Wagman, D. D., Evans, W. H., Parker, V. B., Halow, I., Bailey, S. M., and Schumm, R. H.	1968

This part combined parts 1 and 2 into one volume and superseded them. Tables for Zn and Cd were also added.

270-4 Wagman, D. D., Evans, W. H., Parker, V. B. Halow, I., Bailey, S. M., and Schumm, R. H. 1969

Tables for the elements Hg, Cu, Ag, Au, Ni, Co, Fe, Pd, Rh, Ru, Pt, Ir, Os, Mn, Te, Re, Cr, Mo, and W.

270-5 Wagman, D. D., Evans, W. H., Parker, V. B. Halow, I., Bailey, S. M., Schumm, R. H., and Churney, K. L. 1971

Tables for the elements V, Nb, Ta, Ti, Zr, Hf, Sc, and Y.

270-6 Parker, V. B., Wagman, D. D., and Evans, W. H. 1971

Tables for the elèments Be, Mg, Ca, Sr, Ba, and Ra.

270-7 Schumm, R. H., Wagman, D. D., Bailey, S. M., Evans, W. H., and Parker, V. B. 1973

Tables for the elements Lu, Yb, Tm, Er, Ho, Dy, Tb, Gd, Eu, Sm, Pm, Nd, Pr, Ce, and La.

270-8 Wagman, D. D., Evans, W. H., Parker, V. B., Schumm, R. H., and Nuttall, R. L. 1981

Tables for the elements U, Pa, Th, Ac, Li, Na, K, Rb, Cs, and Fr. Additions and corrections to the series are given in an appendix.

* * * * * * * * * *

[150] Wagman, D. D., Schumm, R. H. and Parker, V. B.
A Computer-Assisted Evaluation of the Thermochemical Data of the Compounds of Thorium
(NBSIR 77-1300, National Bureau of Standards, Washington, D.C., 1977)

This report presents a computer-assisted evaluation of the thermochemical properties of compounds of thorium. Values of $\Delta_f H°$, $\Delta_f G°$, S° are tabulated at 298.15 K for the compounds and aqueous species of thorium. There is also a "reaction catalog" which gives values of ΔG°, ΔH°, and S° for individual processes or substances and upon which the tabulated formation properties are based. This report also contains thermal functions and includes references to the primary literature.

* * * * * * * * * *

[151] Washburn, E. W. (editor-in-chief)
International Critical Tables of Numerical Data, Physics, Chemistry, and Technology
(McGraw-Hill Book Company, New York, 1930)

At the time of publication this seven volume series was the most comprehensive source of thermochemical and other data in existence. Essentially all equilibrium and transport properties and classes of materials were covered. The Critical Tables, although now superseded by modern compilations and evaluations of data, still remain useful for references to the older literature. The tables are arranged according to property with groups of tables being arranged according to discipline. One volume of this series is an index of the materials whose properties are dealt with.

* * * * * * * * * *

[152] Wichterle, I., Linek, J. and Hala, E.
Vapor-Liquid Equilibrium Data Bibliography (and two Supplements)
(Elsevier, Amsterdam, 1973, 1976, and 1979)

These three books contain references to the source literature on vapor-liquid equilibrium data for inorganic and organic mixtures. The first volume covers the literature through 1972, the second through 1975, and the third through 1978. The compounds are arranged alphabetically except for the carbon containing compounds. The preparation of the two Supplements has been computerized to make updating more efficient. The series contains 6700 references to the source literature.

* * * * * * * * * *

[153] Wilhelm, E. Battino, R. and Wilcock, R. J.
Chemical Reviews <u>77</u>, 219 (1977)

This paper is a critical review of the low-pressure solubility of 57 gases in liquid water and in heavy water. The solubility of each gas is given in terms of the coefficients of an equation which can be used to calculate the solubility at a given temperature. Also given are ΔG°, ΔH°, ΔS°, and ΔC_p° values for the solution process at various temperatures and at one atmosphere partial pressure of the gas. The authors give extensive details concerning their correlations and include references to the literature.

* * * * * * * * * *

[154] Wilhelm, E. and Battino, R.
Thermodynamic Functions of the Solubilities of Gases in Liquids at 25 °C
Chemical Reviews <u>73</u>, 1 (1973)

This review article contains selected values for the solubility, entropy, and enthalpy changes in solution for 16 gases in 39 non-aqueous solvents. Also given are the coefficients of polynomial expressions for the Gibbs energy of solution for a gas in a given solvent as a function of temperature. See also item [8].

* * * * * * * * * *

[155] Wilhoit, R. C. and Zwolinski, B. J.
Handbook on Vapor Pressures and Heats of Vaporization of Hydrocarbons and Related Compounds (API 44 - TRC)
(Thermodynamics Research Center, Texas A and M Research Foundation, College Station, Texas, 1971)

This handbook gives data for 680 hydrocarbons, 95 carbon-sulfur compounds, and water from 0 to 150 °C. It is indexed by compound name, and by boiling point.

* * * * * * * * * *

[156] Wilhoit, R. C. and Zwolinski, B. J.
Physical and Thermodynamic Properties of Aliphatic Alcohols
Journal of Physical and Chemical Reference Data, Volume 2, Supplement No. 1, 1973

This review contains critically evaluated values of the vapor pressure, heat capacity, enthalpies of transition, entropies, thermodynamic functions for the real and ideal gases, densities, refractive indexes, and critical properties for 722 alcohols in the carbon range C_1 to C_{50}. This comprehensive review is 420 pages long and lists 2036 references.

* * * * * * * * * *

[157] Wisniak, J.
Phase Diagrams, A Literature Source Book (Parts A and B)
(Elsevier, New York, 1981)

These two volumes contain extensive references (circa 18,000) to phase diagrams in the literature. The systems covered include organic and inorganic materials and aqueous systems. The coverage extends over the entire periodic chart. The systems are arranged in order of increasing carbon and hydrogen with the remaining elements in alphabetical order.

* * * * * * * * * *

[158] Wisniak, J. and Tamir, A.
Mixing and Excess Thermodynamic Properties: A Literature Source Book
(Elsevier, Amsterdam, 1978)

This book contains approximately 6000 references to excess mixing property data for mixtures of all types, including electrolyte, non-electrolyte, and metallurgical systems. The appropriate references to the literature are given for each mixture. The coverage of materials encompasses the entire periodic chart and also includes ternary, quaternary, and multi-component mixtures.

* * * * * * * * * *

[159] Wu, Y. C. and Hamer, W. J.
Electrochemical Data. Part XIV. Osmotic Coefficients and Mean Activity Coefficient of a Series of Uni-Bivalent and Bi-Univalent Electrolytes in Aqueous Solutions at 25 °C. Part XVI. Osmotic Coefficients and Mean Activity Coefficients of a Number of Uni-Trivalent and Tri-Univalent Electrolytes in Aqueous Solution at 25 °C
(NBSIR No. 10052 and 10088, National Bureau of Standards, 1969)

These reports give values for the osmotic and mean ionic activity coefficients of uni-bivalent, bi-valent, uni-trivalent, and tri-univalent electrolytes in aqueous solution at 25 °C. In each case, the values tabulated are those calculated by fitting the literature data to the equation for the excess Gibbs energy, and represents a good fit to the experimental data. Literature references are included. Also see item [48].

* * * * * * * * * * *

[160] Yatsimirskii (or Jazimirski), K. B.
Thermochemie von Komplexverbidungen
(Akademie-Verlag, Berlin, 1956)
translated by Georg Crull from the Russian
"Thermochimia Complexnik Coedinenie"
(Akad. Nauk. CCCP, Moscow, 1951)

This monograph contains correlations of thermochemical data of complex compounds in terms of gas, crystal, and solution models emphasizing ionic radii. More than 1400 substances are considered. In summary tables, 52 metal cations, 33 neutral ligands, and 25 anions have enthalpies of formation listed, usually for gaseous and aqueous state. The compounds formed from these ions and ligands are listed usually with the enthalpy of formation of the crystal, and for many of them enthalpies of solution at infinite dilution are given. The chapters discuss and give tables and correlations of hydration enthalpies of cations, anions and some amine salts, ion entropies in solution, enthalpies of formation and solution, energies of gas and crystal ions, binding energies of H_2O, NH_3 and other molecules to central metal ions in gaseous complex ions and similar binding energies.

* * * * * * * * * * *

[161] Yatsimirskii, K. B., and Vasil'ev, V. P.
Instability Constants of Complex Compounds
(Consultants Bureau, New York, 1966 and Pergamon Press, New York, 1960)

This volume contains an extensive tabulation of equilibrium data (K and -log K) and enthalpies of formation for processes involving the more common ions and ligands in aqueous solution. Specified are the temperature, ionic strength, and method of measurement with reference(s) to the appropriate literature. The tables contain instability constants for 138 predominantly inorganic complexes.

* * * * * * * * * * *

[162] Kerr, J. A.
Strengths of Chemical Bonds
in CRC Handbook of Chemistry and Physics, 64th Edition, 1983-1984
(CRC Press, Inc., Boca Raton, Florida, 1983)

This 20 page compilation (pages F-176 to F-195 in the "Handbook of Chemistry and Physics") contains values of bond strength in diatomic and polyatomic molecules, enthalpies of formation of gaseous atoms from the elements in their standard states, and enthalpies of formation of free radicals. The data refer to 298.15 K. The coverage of bond types is extensive and there are references to the source literature. Also see item [92].

* * * * * * * * * * *

AUTHOR INDEX

SUBJECT INDEX

This index gives an indication of thermodynamic properties, physical or chemical processes, classes of substances, and, in a few cases, individual substances for which information is to be found herein. It is not practical to give an exhaustive index to the contents of the individual references. Hence the absence of a piece of information in the index does not necessarily mean its absence in the references. This is particularly true with respect to particular substances, which the reader should assume are not listed except by chance of title or abstract.

NBS-114A (REV. 2-8C)

U.S. DEPT. OF COMM. BIBLIOGRAPHIC DATA SHEET (See instructions)	1. PUBLICATION OR REPORT NO.	2. Performing Organ. Report No.	3. Publication Date
	NBS/SP-685		December 1984

4. TITLE AND SUBTITLE

Compiled Thermodynamic Data Sources for Aqueous and Biochemical Systems: An Annotated Bibliography (1930-1983)

5. AUTHOR(S)

Robert N. Goldberg

6. PERFORMING ORGANIZATION *(If joint or other than NBS, see instructions)*

NATIONAL BUREAU OF STANDARDS
DEPARTMENT OF COMMERCE
GAITHERSBURG, MD 20899

7. Contract/Grant No.

8. Type of Report & Period Covered

Final

9. SPONSORING ORGANIZATION NAME AND COMPLETE ADDRESS *(Street, City, State, ZIP)*

Design Institute for Physical Property Data
Project 811
American Institute of Chemical Engineers
New York, NY

10. SUPPLEMENTARY NOTES

Library of Congress Catalog Card Number: 84-601131

☐ Document describes a computer program; SF-185, FIPS Software Summary, is attached.

11. ABSTRACT *(A 200-word or less factual summary of most significant information. If document includes a significant bibliography or literature survey, mention it here)*

This is a selected and annotated bibliography of sources of compiled and evaluated chemical thermodynamic data relevant to biochemical and aqueous systems. The principal thermodynamic properties considered herein are Gibbs energy and equilibrium data, enthalpies of formation and reaction, heat capacities and entropies, and the corresponding partial molar and excess properties. Derived quantities used in calculating the above are also included. Transport and mechanical data have also been identified to a lesser degree. Included in the annotations to the data sources are brief descriptions of the types of properties tabulated, the classes of materials dealt with, and the degree of completeness of the compilations.

12. KEY WORDS *(Six to twelve entries; alphabetical order; capitalize only proper names; and separate key words by semicolons)*

aqueous systems; bibliography; biochemical systems; enthalpy data; entropy data; equilibrium data; excess properties; Gibbs energy data; heat capacity data; partial molar properties; review articles; thermochemistry; thermodynamics.

13. AVAILABILITY

☒ Unlimited
☐ For Official Distribution. Do Not Release to NTIS
☒ Order From Superintendent of Documents, U.S. Government Printing Office, Washington, D.C. 20402.
☐ Order From National Technical Information Service (NTIS), Springfield, VA. 22161

14. NO. OF PRINTED PAGES

106

15. Price

USCOMM-DC 6043-P80

Index

INDEX

Printed and bound in the UK by
CPI Antony Rowe, Eastbourne

Printed and bound by CPI Group (UK) Ltd, Croydon, CR0 4YY
16/04/2025
14658420-0005